# X-RAY AND INNER-SHELL PROCESSES

# Previous Proceedings in the Series of International Conferences of X-Ray and Inner-Shell Processes

|      | Year | Held in                    | Publisher                          | ISBN          |
|------|------|----------------------------|------------------------------------|---------------|
| 17th | 1996 | Hamburg, Germany           | AIP Conference Proceedings 389     | 1-56396-563-1 |
| 16th | 1993 | Debrecen, Hungary          | Nucl. Instr. & Methods **87**, 1994 |               |
| 15th | 1990 | Knoxville, Tennessee, USA  | AIP Conference Proceedings 215     | 0-88318-790-6 |

# Other Related Titles from AIP Conference Proceedings

**507** X-Ray Microscopy: VI International Conference on X-Ray Microscopy
Edited by Werner Meyer-Ilse, Tony Warwick, and David Atwood, March 2000, 1-56396-926-2

**500** The Physics of Electronic and Atomic Collisions: XXI International Conference
Edited by Yukikazu Itikawa, Kazuhiko Okuno, Hiroshi Tanaka, Akira Yagishita, and Michio Matsuzawa, February 2000, 1-56396-777-4

**477** Atomic Physics 16: Sixteenth International Conference on Atomic Physics
Edited by William E. Baylis and Gordon W. F. Drake, May 1999, 1-56396-752-9

**467** Spectral Line Shapes: Volume 10, 14$^{th}$ ICSLS
Edited by Roger M. Herman, March 1999, 1-56396-754-5

**454** Resonance Ionization Spectroscopy: Ninth International Symposium
Edited by J. C. Vickerman, I. Lyon, N. P. Lockyer, and J. E. Parks, December 1998, 1-56396-810-X

**434** Atomic and Molecular Data and Their Applications: ICAMDATA - First International Conference
Edited by Peter J. Mohr and Wolfgang L. Wiese, June 1998, 1-56396-751-0

**417** Synchrotron Radiation Instrumentation: Tenth US National Conference
Edited by Ernest Fontes, December 1997, 1-56396-742-1

To learn more about these titles, or the AIP Conference Proceedings Series, please visit the webpage
http://www.aip.org/catalog/aboutconf.html

# X-RAY AND INNER-SHELL PROCESSES

18th International Conference

*Chicago, Illinois    August 1999*

■ **Invited Presentations**

*EDITORS*
R. W. Dunford
D. S. Gemmell
E. P. Kanter
B. Krässig
S. H. Southworth
L. Young
*Argonne National Laboratory*

**AMERICAN INSTITUTE OF PHYSICS**

Melville, New York
AIP CONFERENCE PROCEEDINGS ■ 506

**Editors:**

R. W. Dunford
D. S. Gemmell
E. P. Kanter
B. Krässig
S. H. Southworth
L. Young

Argonne National Laboratory
Physics Division, Bldg. 203
9700 S. Cass Avenue
Argonne, IL 60439-4803
USA

E-mail: dunford@anlphy.phy.anl.gov
gemmell@anl.gov
kanter@anl.gov
kraessig@anl.gov
southworth@anl.gov
young@anl.gov

The articles on pp. 444–466, 479–485, 585–589, and 677–683 were authored by U. S. Government employees and are not covered by the below mentioned copyright.

Authorization to photocopy items for internal or personal use, beyond the free copying permitted under the 1978 U.S. Copyright Law (see statement below), is granted by the American Institute of Physics for users registered with the Copyright Clearance Center (CCC) Transactional Reporting Service, provided that the base fee of $17.00 per copy is paid directly to CCC, 222 Rosewood Drive, Danvers, MA 01923. For those organizations that have been granted a photocopy license by CCC, a separate system of payment has been arranged. The fee code for users of the Transactional Reporting Service is: 1-56396-713-8/00/$17.00.

© 2000 American Institute of Physics

Individual readers of this volume and nonprofit libraries, acting for them, are permitted to make fair use of the material in it, such as copying an article for use in teaching or research. Permission is granted to quote from this volume in scientific work with the customary acknowledgment of the source. To reprint a figure, table, or other excerpt requires the consent of one of the original authors and notification to AIP. Republication or systematic or multiple reproduction of any material in this volume is permitted only under license from AIP. Address inquiries to Office of Rights and Permissions, Suite 1NO1, 2 Huntington Quadrangle, Melville, N.Y. 11747-4502; phone: 516-576-2268; fax: 516-576-2450; e-mail: rights@aip.org.

L.C. Catalog Card No. 99-069958
ISBN 1-56396-713-8
ISSN 0094-243X
Printed in the United States of America

## CONTENTS

Preface ........................................................... xi
Committees and Sponsors ........................................... xiii
Excerpts from the X99 Business Meeting ............................ xv

### I. HISTORICAL REVIEWS

Opening Remarks ................................................... 3
    B. Crasemann
X-Ray and Inner-Shell Processes: Their Impact on Our Understanding
of Atomic Physics and Atoms Interacting with Solids ............... 5
    J.-P. Briand
Exploiting Compton Scattering: Studies of Spin Density Distributions ........ 18
    M. J. Cooper
Atomic Auger Spectroscopy: Historical Perspective and Recent Highlights .... 33
    W. Mehlhorn

### II. ATOMS

Some Frontiers of X-Ray/Atom Interactions ......................... 59
    R. H. Pratt
Current Status of (e,2e) Measurements of Energy-Momentum Densities ...... 81
    E. Weigold
Photo- and Charged-Particle-Ionization of He and $D_2$ Studied
with COLTRIMS .................................................... 101
    M. A. Abdallah, M. Achler, H. Braeuning, A. Braeuning-Deminian,
    C. L. Cocke, A. Czasch, R. Doerner, A. Landers, V. Mergel,
    T. Osipov, M. Prior, H. Schmidt-Boecking, M. Singh, T. Weber,
    W. Wolff, and H. E. Wolf
Three-Electron Photo-Processes in Lithium ......................... 116
    Y. Azuma
Unique Features of Photoelectron/Auger-Electron
Coincidence Experiments .......................................... 132
    V. Schmidt
Angular Distributions and Correlations in Auger Cascades
of Atomic Argon Following 2p→4s Excitation ....................... 148
    K. Ueda, Y. Shimizu, H. Chiba, Y. Sato, M. Kitajima, H. Tanaka,
    S.-M. Huttula, H. Aksela, I. P. Sazhina, and N. M. Kabachnik
K-Shell Ionization and Double-Ionization of Au Atoms
with 1.33 MeV Photons ............................................ 153
    A. Belkacem, D. Dauvergne, B. Feinberg, D. Ionescu, J. Maddi,
    and A. H. Sorensen

## III. MOLECULES

**Dynamics of Competing Ultra-Fast Fragmentation and Resonant Auger Processes** .................................................. 161
    O. Björneholm

**Dynamical Effects and Selective Fragmentation after Inner Shell Excitation.** .................................................. 177
    M. Simon, C. Miron, R. Guillemin, K. Le Guen, D. Ceolin,
    E. Shigemasa, N. Leclercq, and P. Morin

**High-Resolution Molecular Inner-Shell Electron Spectroscopies** .............. 188
    J. D. Bozek, N. Berrah, E. Kukk, T. D. Thomas, T. X. Carroll,
    L. J. Saethre, J. A. Sheehy, and P. W. Langhoff

**Photoelectron Emission from Oriented Molecules.** ........................ 205
    U. Becker

**Oscillating Partial Cross Sections in $C_{60}$: Evidence for a Beating Frequency** ........................................ 217
    A. Rüdel, R. Hentges, and U. Becker

**Beyond the Dipole Approximation: Angular-Distribution Effects in the 1s Photoemission from Small Molecules** ........................ 222
    O. A. Hemmers, H. Wang, D. W. Lindle, P. Focke, I. A. Sellin,
    J. D. Mills, J. A. Sheehy, and P. W. Langhoff

## IV. SOLIDS AND SURFACES

**Atomic Effects Seen in Solid Phases** ....................................... 231
    B. Sonntag

**Multi-Atom Resonant Photoemission.** ....................................... 251
    C. S. Fadley, E. Arenholz, A. W. Kay, J. Garcia de Abajo,
    B. S. Mun, S.-H. Yang, Z. Hussain, and M. Van Hove

**X-Ray Resonant Raman Scattering** ......................................... 273
    P. Carra

**A Band-Structure-Based Approach to Modeling X-Ray Absorption, Fluorescence, and Resonant Inelastic Scattering** .......................... 283
    E. L. Shirley, J. A. Carlisle, S. R. Blankenship, R. N. Smith,
    L. J. Terminello, J. J. Jia, T. A. Callcott, and D. L. Ederer

**Resonant Inelastic Scattering at the 3d and 4d Resonances of $LaAlO_3$** ........ 304
    A. Moewes

**Resonant Inelastic X-Ray Scattering from Transition Metal Oxides** .......... 312
    J. P. Hill

**Electronic States of Metals and Alloys Investigated by High-Resolution Bloch-$k$ Selective X-Ray Raman Scattering** ............................ 327
    A. Kaprolat, H. Enkisch, M. H. Krisch, and W. Schülke

**X-Ray Magnetic Circular Dichroism of Model Heisenberg Ferromagnets** ..... 336
    A. Rogalev and J. Goulon

**Diffracted Anomalous Fine Structure (DAFS) to Study Surfaces and Interfaces** ............................................................ 351
    M. Benfatto, R. Felici, and F. Comin

**Charge Stripes Formation by X-Ray Illumination in High $T_c$ Superconductors** ................................................. 358
    G. Bianconi, D. Di Castro, N. L. Saini, A. Bianconi, M. Colapietro, and A. Pifferi

**Charge Transfer at Surfaces on Femtosecond Timescales: New Information from Electron Spectroscopies** ........................... 372
    D. Menzel and W. Wurth

**Probing the Nature of Hydrogen Bonds with X-Rays** ..................... 385
    P. M. Platzman

## V. HIGHLY CHARGED IONS

**Inverse Photoionization Studied via Radiative Electron Capture into Highly Charged Ions** ............................................. 389
    T. Stöhlker, O. Brinzanescu, A. Krämer, T. Ludziejewski, X. Ma, P. Swiat, and A. Warczak

**Atoms in Extreme Virtual Photon Fields of Fast, Highly Charged Ions** ....... 403
    J. Ullrich, B. Bapat, A. Dorn, S. Keller, H. Kollmus, R. Mann, R. Moshammer, R. E. Olson, W. Schmitt, and M. Schulz

**X-Rays and Inner-Shell Processes with Heavy Ions Channeled in Thin Crystals** ..................................................... 418
    D. Dauvergne, M. Chevallier, C. Cohen, N. Cue, J. Dural, R. Kirsch, A. L'Hoir, D. Lelièvre, P. H. Mokler, J.-C. Poizat, H.-T. Prinz, J.-M. Ramillon, J. Remillieux, P. Roussel-Chomaz, J.-P. Rozet, F. Sanuy, D. Schmaus, C. Stephan, M. Toulemonde, D. Vernhet, and A. Warczak

**Two- and Three-Body Effects in Fast Ion-Atom Collisions: Analogies Between Photon and Charged Particle Impact** .................. 427
    N. Stolterfoht, B. Sulik, J. A. Tanis, J.-Y. Chesnel, L. Gulyás, F. Frémont, D. Lecler, D. Hennecart, X. Husson, J. P. Grandin, M. Grether, C. Koncz, and B. Skogvall

**High-Resolution Measurements of the K-Shell Spectral Lines of Hydrogenlike and Heliumlike Xenon** ................................. 444
    K. Widmann, P. Beiersdorfer, G. V. Brown, J. R. Crespo López-Urrutia, A. L. Osterheld, K. J. Reed, J. H. Scofield, and S. B. Utter

**First Experimental Results and New Theoretical Calculations on Photoionization Processes in Multiply-Charged $Xe^{3+}$ to $Xe^{7+}$ Ions** ........... 467
    J.-M. Bizau, C. Blancard, R. Marmoret, D. Hitz, J.-M. Esteva, D. Cubaynes, C. Couillaud, P. Ludwig, C. Rémond, A. Compant La Fontaine, J. Delaunay, J. Bruneau, J. Lachkar, and F. J. Wuilleumier

**Charge Exchange Induced X-Ray Transitions of Hollow Ions in Laser Field Ionized Plasmas** ........................................ 472
    F. B. Rosmej, D. H. H. Hoffmann, A. Y. Faenov, T. A. Pikuz, A. I. Magunov, I. Y. Skobelev, T. Auguste, P. D'Oliveira, S. Hulin, and P. Monot

## VI. NUCLEAR EFFECTS

**Inelastic Scattering of Synchrotron Radiation from Electrons and Nuclei for Lattice Dynamics Studies** .................................... 479
    E. E. Alp, W. Sturhahn, H. Sinn, T. Toellner, M. Hu, J. Sutter, and A. Alatas

**Nuclear Excitation by Electronic Transition Between Atomic Shells** .......... 486
    E. V. Tkalya

**Non-Resonant Excitation of Nuclear Levels by Photons.** .................... 496
    E. G. Drukarev

## VII. FUNDAMENTAL PHYSICS

**Testing Cosmological Variability of Fundamental Constants** ................ 503
    D. A. Varshalovich, A. Y. Potekhin, and A. V. Ivanchik

**Energies and QED Effects in Few-Electron Atoms and Ions.** ................ 512
    G. W. F. Drake and Z.-C. Yan

**Atomic Collisions with Ultra Slow Antiprotons.** ........................... 533
    Y. Yamazaki

## VIII. IMAGING AND MEDICAL APPLICATIONS

**Atom-Resolving X-Ray Holography.** ...................................... 549
    B. Adams, T. Hiort, G. Materlik, Y. Nishino, and D. V. Novikov

**Live X-Ray Refraction Imaging Using Vertically and Horizontally Wide X-Rays** ............................................................ 565
    J. Matsui, Y. Kagoshima, Y. Tsusaka, K. Yokoyama, K. Takai, S. Takeda, and K. Yamasaki

**Frontiers of X-Ray Spectromicroscopy in Biology and Medicine: Gadolinium in Brain Cancer** ............................................. 577
    G. De Stasio, B. Gilbert, P. Perfetti, G. Margaritondo, D. Mercanti, M. T. Ciotti, P. Casalbore, L. M. Larocca, A. Rinelli, and R. Pallini

**High-Resolution X-Ray Imaging for Microbiology at the Advanced Photon Source** ........................................................... 585
    B. Lai, K. M. Kemner, J. Maser, M. A. Schneegurt, Z. Cai, P. P. Ilinski, C. F. Kulpa, D. G. Legnini, K. H. Nealson, S. T. Pratt, W. Rodrigues, M. L. Tischler, and W. Yun

**The Possibility of Using X-Ray Diffraction with Hair to Screen for Pathologic Conditions such as Breast Cancer** ........................ 590
    V. James and D. Cookson

## IX. NEW SOURCES AND TECHNIQUES

**Prospects for an X-Ray FEL Light Source and Some Possible Scientific Applications** ................................................. 597
    J. Arthur

**X-Ray Lasers Driven by Optical Lasers** .................................. 613
    Y. Kato, A. Nagashima, K. Nagashima, M. Kado, T. Kawachi,
    N. Hasegawa, M. Tanaka, A. Sasaki, and K. Moribayashi

**X-Rays in Curved Spaces** ............................................... 621
    J. A. Golovchenko and C. Liu

**Thermal Calorimeters for High Resolution X-Ray Spectroscopy** ............ 638
    M. Galeazzi, D. McCammon, and W. T. Sanders

**Coherent X-Ray Generation via Ultrafast Coster-Kronig Decay in Solid Targets Excited by Table-Top Lasers** ............................. 651
    C. Tóth, S. H. Son, D. Kim, and C. P. J. Barty

**Time-Resolved X-Ray Photoabsorption and Diffraction on Timescales from ns to fs** .......................................................... 664
    P. A. Heimann, T. Missalla, A. Lindenberg, I. Kang, S. Johnson,
    Z. Chang, H. C. Kapteyn, R. W. Lee, R. W. Falcone, R. W. Schoenlein,
    T. E. Glover, A. A. Zholents, M. S. Zolotorev, and H. A. Padmore

**A Novel Type of X-Ray Interferometer** ................................. 669
    O. Kettig, H. Backe, N. Clawiter, S. Dambach, T. Doerk, N. Elbai,
    H. Euteneuer, F. Hagenbuck, P. Holl, H. Jacobs, K.-H. Kaiser,
    J. Kemmer, T. Kerschner, G. Kube, H. Koch, W. Lauth, H. Mannweiler,
    H. Matthäy, H. Schöpe, D. Schroff, M. Schüttrumpf, R. Stötter,
    L. Strüder, T. Walcher, A. Wilms, C. v. Zanthier, and M. Zemter

## X. CONFERENCE SUMMARY

**Closing Remarks for X99** .............................................. 677
    R. D. Deslattes

**Scientific Program** ................................................... 685

**Author Index** ......................................................... 691

## PREFACE

The 18th International Conference on X-ray and Inner-Shell Processes was held at the Drake Hotel in downtown Chicago from August 23 to August 27, 1999. This conference, known colloquially as "X99", is the most recent in a long series of international "X" conferences, which trace their roots to two originally separate series on "X-ray Physics" and "Inner-Shell Ionization". These combined in 1978 into the present triennial series. The development over recent years of tunable high-brilliance hard x-ray beams from dedicated synchrotron sources has given added impetus to the science that is at the focus of this conference series. Argonne National Laboratory, home of the recently constructed and commissioned Advanced Photon Source (APS), hosted X99. The APS is one of three third-generation high-energy synchrotron x-ray sources, the other two being the ESRF located near Grenoble, France and SPring-8 near Himeji, Japan. The x-ray beams now available at these facilities constitute exquisitely fine tools. The precision and power of these sources and also of the modern particle accelerators used in collision studies permit the exploration of x-ray and inner-shell processes with unprecedented detail and resolution. At X99 we heard about many of the results that imaginative applications of these new tools are now producing.

Interest in this conference series continues at a high level. There were 320 attendees with representation from 35 countries. Especially gratifying was the interest displayed by young scientists. There were 55 students in attendance. A total of 350 contributed abstracts were received, of which 300 were presented in four poster sessions. The program contained 11 Plenary Lectures (intended to be in the nature of broad reviews), 28 Progress Reports (intended to have a narrower focus on topics of current active interest), and 16 talks on "Hot Topics". Including the Conference Summary talk, there were thus 56 oral presentations. This book contains the written versions of these presentations.

Abstracts of all the oral presentations and of the contributed papers were distributed as a Book of Abstracts to all participants at the start of the conference. Additional copies may be obtained by contacting the conference office.

The conference organizers gratefully acknowledge the suggestions and advice of the International Scientific Committee in preparing the program.

D.S. Gemmell<br>
Conference Chairman

## The Editors

R.W. Dunford  
D.S. Gemmell  
E.P. Kanter  

B. Krässig  
S.H. Southworth  
L. Young  

Argonne National Laboratory  
Argonne, Illinois 60439

### International Scientific Committee

S.O. Aksela (Finland)  
M. Ya. Amusia (Russia)  
Y. Awaya (Japan)  
P. Beiersdorfer (USA)  
N. Berrah (USA)  
J.E. Burgdörfer (USA)  
C. Cohen (France)  
B. Crasemann (USA)  
D.S. Gemmell (USA)  
N.M. Kabachnik (Russia)  
F.P. Larkins (Australia)  
D.W. Lindle (USA)  
A. Marcelli (Italy)  

G. Materlik (Germany)  
P.H. Mokler (Germany)  
T. Mukoyama (Japan)  
J.E. Nordgren (Sweden)  
R.C.C. Perera (USA)  
V. Schmidt (Germany)  
M. Simon (France)  
B.F. Sonntag (Germany), chairman  
T. Surić (Croatia)  
R. Szargan (Germany)  
A. Warczak (Poland)  
F. Wuilleumier (France)  

### Local Organizing Committee

R.W. Dunford  
D.S. Gemmell, chairman  
E.P. Kanter  
B. Krässig  
P.A. Montano  

G.K. Shenoy, co-chairman  
S.H. Southworth  
G.B. Stephenson  
S.B. Strasser  
L. Young  

Conference secretary: Barbara Weller

### Sponsors

We are grateful for the generous financial support provided by Argonne National Laboratory, the National Science Foundation, The International Union of Pure and Applied Physics, and Roper Scientific.

EXCERPTS FROM THE X99 BUSINESS MEETING

The International Scientific Committee selected Rome, Italy as the site for the next conference in the series, The 19th International Conference on X-ray and Inner-Shell Processes ("X02"). The meeting will be held in August, 2002 and Dr. Antonio Bianconi will be the Conference Chairman.

The Committee elected 12 new members to replace those retiring. The composition of the new Committee (for X02) is as follows:

    S.O. Aksela (Finland)
\* Y. Azuma (Japan)
\* U. E. Becker (Germany)
    P. Beiersdorfer (USA)
    N. Berrah (USA)
\* A. Bianconi (Italy)
    J. Burgdörfer (USA)
    C. Cohen (France)
    D.S. Gemmell (USA), Chairman
\* A. N. Grum-Grzhimailo (Russia)
    D.W. Lindle (USA)
    A. Marcelli (Italy)
    G. Materlik (Germany)

\* N.O.T. Martenson (Sweden)
    P.H. Mokler (Germany)
    T. Mukoyama (Japan)
\* M. Pajek (Poland)
\* S. A. Sheinerman (Russia)
    M. Simon (France), Treasurer
\* S. H. Southworth (USA)
    T. Surić (Croatia)
\* K. Ueda (Japan)
\* J. Ullrich (Germany)
\* E. Weigold (Australia)
\* Y. Yamazaki (Japan)

(Asterisks denote new committee members.)

# I. HISTORICAL REVIEWS

# Opening Remarks

B. Crasemann

*Physics Department, University of Oregon, Eugene, Oregon 97403, USA*

It is a great pleasure to see so many old friends gathered for this 18th International Conference on X-ray and Inner-Shell Processes --- and especially, so many young friends: a manifestation of the extraordinary vigor with which our field is making major advances in the understanding of atomic structure and dynamics, and of photon-atom interactions. Indeed, hardly anyone could have foreseen the extent of these quite astonishing developments at the time of the forerunner of this conference series, held in 1972 in Atlanta at the instance of a small group of dedicated investigators, including the late Dick Fink, Steve Manson (who is here) and P. Venugopala Rao.

As we all know, this renaissance of atomic inner-shell physics is primarily due, on the experimental side, to the development of synchrotron-radiation sources, which in its impact can be compared with the introduction of the Bragg spectrometer. Complemented by collision experiments (as Nico Stolterfoht likes to remind me), synchrotron radiation has opened a vast field of exploration. On the theoretical side, great advances in computational physics have made it possible to perform calculations of unprecedented sophistication. The continuous, symbiotic interaction among theory and experiment greatly accelerated progress. The very key to this progress, however, is the hard work and dedication of a large number of very gifted scientists, and that is why this Conference promises to be so fascinating as many of our colleagues report on their latest work. As Mike Knotek once remarked when he chaired the Users' Organization at Stanford some twenty years ago, "The progress of this Laboratory rests upon the sweat and toil of the experimenters" ("and theorists" was implied).

It seems most appropriate that the keynote address at this Conference should be given by a colleague who has been involved with these developments almost since their beginning, and has contributed more than his share of sweat and toil, plus admirable ingenuity and leadership.

Jean-Pierre Briand, Head of the *Laboratoire de Physique Atomique et Nucléaire* in the *Université Pierre et Marie Curie* in Paris, has led a range of landmark studies in atomic physics and related fields. At the 1972 Conference alluded to earlier, he reported on the first observation of x-ray *hypersatellites,* resulting from double inner-shell ionization. This was part of his group's pioneering studies of multi-ionization processes that accompany nuclear decay. Realizing early the potential of synchrotron radiation, he performed experiments on sources at LURE, in Berkeley, and elsewhere. He was among the first to study so-called "hollow" atoms, and measured their decay above and below solid surfaces. In recognition of his accomplishments, he was awarded the Holweck Medal for 1997.

Jean Pierre's co-workers, friends and colleagues are particularly aware that his scientific skills are complemented by unusual generosity and human warmth. In his Laboratory in Paris he has welcomed many of us, for short or longer periods of pleasant collaboration.

Let me just mention one incident which may illustrate some of Jean-Pierre's personal qualities. In May of 1985, during days of almost unbelievable heat, he journeyed to the annual meeting of what was then called DEAP (now DAMOP) in Norman, Oklahoma, where he was to receive the certificate of his election to Fellow of the American Physical Society. Characteristically, he carried some special gifts for his friends: jars of the most exquisite French goat cheese, as pungent as it comes, put up in olive oil. He had placed the glass jars well-cushioned among his clothes in a suitcase. On arrival, however, just before the ceremony, Jean-Pierre discovered that some jars had broken, be it because of the heat or some other physical cause, and the penetratingly aromatic *chèvre* and oil had thoroughly soaked his white shirts and dark suit. Unabashed, Jean-Pierre quickly borrowed the necessary clothing from others and appeared on the stage, with impeccable aplomb. This time, hot as it may be outside, I have noticed no smell of goat cheese.

It gives me great pleasure to introduce our opening speaker, Professor Jean-Pierre Briand, who will talk on "X-ray and inner-shell processes: their impact on our understanding of atomic physics and of atoms interacting with solids."

# X-Ray and Inner-Shell Processes: Their Impact on our Understanding of Atomic Physics and Atoms Interacting with Solids

Jean-Pierre Briand

*ERIS. Equipe de Recherche Ions-Surface, Université Pierre et Marie Curie,
4 Place Jussieu, 75252 Paris Cedex 05, France*

**Abstract.** Atomic physics and the basic concepts of quantum theory have been probed in the last fifty years by using the techniques of optics and lasers in the visible range. The new powerful accelerators, storage rings, and various large scale devices, such as ion beams, synchrotron radiation, plasma confinement machines, powerful lasers, etc. developed by the nuclear physicists and high technology engineers have allowed, in the past three decades, new, unexpected and more general insights of atomic structure and more accurate checks of quantum mechanics. It is now possible to prepare any kind of atom or ion, having any number of electrons in any quantum states (atomic manipulations), and to trap or set them in defined places on surfaces. The study of these atomic species having electrons in any deep, or highly excited levels requires the use of electromagnetic transitions in a much wider range of wavelengths than in optics, and, because most of the time they are in auto-ionizing states, of Auger spectrometry. It is the purpose of this talk to review some of the most salient discoveries in the field since this time, to present some of the most recent and exciting results obtained in the last decade, and future prospects.

## THE X-RAY GOLDEN TIME

In the seventies, like many colleagues in the audience, I moved from nuclear physics to atomic physics. It was in 1970 at the time of the first Inner-shell meeting in Atlanta. Atlanta actually is the starting point of a new era for X rays and the birthplace of a new community. Soon I realized that I moved into a field which was considered by "true" atomic physicists as reserved for second row scientists. They named it chemical physics... Times have changed, and I am now convinced that I moved at the right time into the right field.

I would like in this talk to review some results and list some problems this community has encountered during this X ray golden era and, unfortunately, will only concentrate on a few limited aspects of the impact of inner-shell physics on atomic physics. Atomic physics means for most physicists the hydrogen atom, Schrödinger equation, and optics. The observation of the excitation of the valence electron is a very restrictive tool for studying atomic physics, and I would like in this talk to review some significant results showing that what we name *inner-shell physics* actually provides a more generalized view of the atom, and has led to many, sometimes unexpected, discoveries.

## THE LIVING ATOM

When I started working in the field, it was often taught in the universities that it was not possible to remove more than one electron from an atom with a single photon. We were then only able to study atomic structure through the ionization or excitation of a single electron, either in valence shells (optics) or inner-shells (X ray). In fact, this wrong statement was not of much importance.

In the mid-sixties, M. O. Krause and T. A. Carlson (1), with the very first prototype of high resolution-high transmission electrostatic electron spectrometers, directly observed the simultaneous removal of two electrons in a single event, opening a new era in the field. These two electrons share the available energy in a very original way: a manifestation of what we now name *electron correlation*, giving the first physical illustration of the instantaneous interaction of atomic electrons in a stationary state. The study of multi-ionization processes became very popular and still is.

The only type of Hartree-Fock calculations then available was based on the frozen-orbital approximation, i.e., the assumption that all orbitals, except that considered, stayed unchanged after any ionization process, which is obviously not true. Double (multiple) ionization or excitation was first explained by using the shake model, invented in nuclear physics by Migdal and Feinberg (2,3) to explain the spontaneous ionization of atoms accompanying a nuclear decay. This "extra" ionization was then considered as being due to the relaxation of the whole atomic cloud during the change of the central charge: the first attempt to move away from the frozen-orbital model. This model was later extended to any change of the inner central charge of an atom, e.g., the removal of a K electron. The shake model is based on the sudden approximation, i.e., the idea that the "first" ionizing event is so fast that the atomic cloud realizes only "later" that it has to re-adjust. This relaxation cannot be perfect, in principle, because the overlap of two different wavefunctions is, by definition, never equal to one. This leads to "accidents", such as the ejection of an electron which does not want to stay in the same quantum state of the new atom or ion as before. Actually, this stepwise model is not acceptable, because the two events always take place at the same time. It is one of the very basic assumptions of quantum mechanics, and this model only constitutes the zero-order approximation of the description of double ionization processes, as first noted, long ago in a famous meeting in Debrecen, by B. Crasemann (4). Some time later, it was demonstrated that it was not necessary that the removal of the 'first' electron be sudden. This type of calculation, which fairly fits with most experimental results, became very popular and still is. This model means that the change of the atomic cloud follows the central-field approximation. More refined calculations were then introduced, such as the Many Body Perturbation Theory (MBPT) developed by H. Kelly (5). The relative movement of, e.g., two electrons was also studied, for instance, by the Fano school using more appropriate variables to describe a two-electron system. This was the beginning of the study of what one can name the "living" atom, where a certain number of internal relative movements of the electrons inside a stationary state becomes observable, as in the COLTRIMS experiments introduced by H. Schmidt-Böcking (6). The most interesting concept of the shake model is that the "extra" ionization (or excitation) comes from internal atomic forces and is more or less independent of the primary excitation. It thus really provides insight on the relative movements of electrons inside the atom and reflects the internal life of a stationary state. This model also leads to the conclusion that it is impossible to produce selectively a single ionization state and then to observe pure X-ray diagram lines. However, one must be very careful not to violate the uncertainty principle and the theory of measurement, which is now the main problem atomic physicists are facing. There were, in the past few years, many controversies among atomic physicists about what "correlation effect" means. These controversies actually come from the different origins of the parties: those coming from atomic spectroscopy and those from collision physics. The expression "correlation effect" comes from spectroscopists. It describes, by definition, the difference between an exact calculation, based on the central field approximation, and experiment. These correlation effects actually provide a limited amount of information about the time and spatial correlation among the relative movements of the electrons inside an atom, in a stationary state, which, by definition, cannot be observed without being disturbed.

The collision physicists describe the change of a stationary state into another during a given external interaction. Quantum mechanics states that the probability for such a change is given by the time-dependent perturbation theory, e.g., in first order, the Fermi Golden Rule No. 2. Owing to basic principles of quantum mechanics, we cannot describe what happens during the time of this interaction. In principle, the form of the Hamitonian which is introduced in these calculations comes from the classical description of the process involved in the interaction, e.g., a simple Coulomb interaction between an incoming electron and a given complex atom. This description sometimes contains simplifications which ignore some of the ingredients of the true interaction, e.g., magnetic effects. The quality of a theoretical prediction of a given collision then depends, in a correlated way, on the quality of the approximations in calculating a many-body problem and of the physical assumptions contained in the model Hamiltonian. Sometimes, the writing of the supposed events involved in the collision process as, for instance, a pre- or post-collisional interaction, may lead to some confusion, or dangerous extrapolations. We are sometimes allowed to take these changes into account, but not in all cases, because the way we introduce them through many approximations may, in some situations, violate, in an invisible way, the basic principles of quantum mechanics.

At the same time, electron spectroscopy, and then Auger spectroscopy, only used before by nuclear physicists, also began to be a very popular field in our community.

The detection of low-energy electrons, which has some meaning only in atomic physics at very high resolution, became possible after the pioneering work of Carlson and Krause and the beautiful techniques developed by K. Siegbahn for the ESCA method (7). The study of Auger electrons, first observed in 1926, encountered incredible difficulties due to the use of cloud chambers. Later on, nuclear physicists developed magnetic electron spectrometers to study β decay, but the resolution was not good enough for most of the atomic physics purposes. The starting point was the availability of spherical analyzers and the needs in microelectronics of Auger spectroscopy for surface analyses. It is today a market of billions of dollars. In a more general way, nuclear physics allowed the development of new techniques for studying charged-particle spectroscopy, either directly or coincidence experiments.

The Auger process was at that time "rediscovered", in fact more generalized, by Madden and Codling (8) in doubly-excited two-electron atoms such as helium using the first synchrotron light. In many-electron systems, Auger spectroscopy has provided, in the past few decades, a fantastic tribute to our understanding of electron correlation, as discussed by one of the main contributors in the field, W. Mehlhorn (9). The Auger process is nothing but the simple coulombic repulsion of atomic electrons. In innermost shells, where the orbitals are purely quantized, it is a quantum-mechanical process; in outermost shells it becomes, as we will discuss later, a classical problem of diffusion between a few electrons located in a small region of space. Recently, many experiments have also been carried out following the pioneering work of F. Wuilleumier (10) by crossing or merging laser, ion and X-ray beams to study these interactions, as for instance, atomic or molecular photo-ionization in selected charge states and levels of excitation.

One of the great discoveries back then was the first observation by T. Åberg (11) of the radiative Auger effect (RAE), i.e., the simultaneous, spontaneous emission by an auto-ionizing system of one electron and one photon sharing the available energy of the state. The emission of a photon is a quantum electrodynamics problem, whereas the ejection of an electron is a purely coulombic diffusion problem. This process was first explained by T. Åberg (12) as a shake process "following" the sudden change of the innermost shell charges during the emission of the photon. There are not today many studies of this radiative Auger effect, but one can expect that this very fundamental process, which provides deep and relevant insights on the inner life of an atom, will soon be re-examined.

The last thirty years have been devoted to the study of these electron-correlation processes. In other words, to the study of a quantum mechanical many-body problem, in a case where the interaction force is very well known. This constitutes a powerful technique to check the methods used to describe more complex systems, like the atomic nucleus, whose fundamental interaction is not really known. This inner life of the atom is, however, explored through measurements of transition probabilities at a precision level of a few %. Study of the relativistic many-body problem needs, however, a much higher precision, a point we will discuss in the next section.

# X-RAY SPECTROSCOPY AND METROLOGY

One of the advantages of the techniques of optics is the possibility to make direct comparisons of the wavelengths of the transitions with the unit of length, the meter. The precision of these measurements is exceptional: the Rydberg constant is now measured with an accuracy of $10^{-12}$, which allows essential tests of the most fundamental theories.

At the level of precision of $10^{-10}$, the measurement of, e.g., the energy of the Lyman α line of hydrogen, does not allow a check on the differences between the Schrödinger eigenvalues, corrected for relativistic effects through perturbation techniques, and the "exact" Dirac eigenvalues. There was then no real numerical quantitative check of the Dirac equation, which was mainly used for its capacity to predict qualitative effects like the existence of the spin of the electron.

With relativistic and QED corrections scaling like $Z^4$, studies of the energies of the innermost shells would have been a way to make accurate checks of the Dirac equation. Unfortunately, it is very difficult to directly compare X-ray wavelengths with the meter. In 1967, R. Deslattes (13) carried out a very nice experiment with an interferometer, which allowed direct comparison of X-ray and visible wavelengths. Since then, this experiment serves as a reference. Unfortunately, these comparisons are only at the level of precision of $10^{-7}$ in the best cases, i.e., much lower than in optics. The most important difficulty in X-ray measurements is that, owing to the irreducible multiple-ionization processes, X-ray lines, instead of being purely Lorentzian, as in Doppler-free spectroscopy, consist of a complex array of unresolved satellite lines. One thus compares a given point of a wide, non-mathematically defined X-ray line, which is difficult to reproduce, with a given standard. The second difficulty is that one cannot directly compare the "exact" theoretical prediction of the energy of a given transition in a multi-electron atom and a two-body atom (hydrogen-like), even if the present capacities of Dirac-Fock (14) calculations for atoms having many electrons are now exceptional.

# FEW HOLES VERSUS FEW ELECTRONS

The study of innermost shell transitions of two-body systems, such as hydrogen-like ions, is, a priori, the best way to check the most fundamental theories. These were carried out (and still are) by using the powerful ion accelerators of nuclear physics, which allow preparation of any ion of any charge. The most commonly used technique in the past two decades has been the so-called beam-foil technique by passing very energetic low-charge-state ions through thin foils (the higher the energy, the higher the charge state). The only difficulty, now overcome, was the very high velocity of these ions (more than 80% of the velocity of light for uranium) and the corresponding huge Doppler aberrations. It is now possible to prepare, at lower velocities, these highly-stripped ions inside ion sources such as EBIS or ECR, in tokomaks, or in laser-produced high-temperature plasmas. The same year, for instance, the Lyman α line of hydrogen-like iron was observed, for the first time, inside a tokomak, a solar flare in an embarked experiment, and in beam-foil spectroscopy. The study of hydrogenlike ions is now complete for most elements up to uranium, at a level of precision down to $10^{-5}$. These experiments have led to many exciting results in fundamental physics, like the first measurements of Lamb shifts in high fields or the first quantitative checks of the Dirac equation, namely, of the unbalanced action-reaction forces in relativity (15).

The next step will be to directly compare the X rays emitted by hydrogen-like ions with the meter at a higher level of precision. This could be achieved now with the use of ion sources, and we proposed some time ago to use the Lyman α X-ray line of hydrogen-like Ar ions emitted inside an EBIS trap as a new X-ray standard. As a matter of fact, this line, which is a pure Lorentzian, has a natural width that is more than 100 times smaller than any X ray emitted by neutral atoms.

However, one of the problems is our limited knowledge of the atomic nucleus, because the nuclear corrections, for example, of the (1s) Lamb shift of uranium (e.g., size, structure, or even magnetic effects) are already much higher than the true QED effects. Moreover, with the QED calculations still being done through perturbative techniques, the highest-order diagrams are still just at the limit of the present experimental precision.

Studies of the excited states of helium-like ions, the transition energies of which can be directly compared with those of the corresponding hydrogen-like ions, which allows getting rid of most of, e.g., nuclear size and structure problems, or, at a lower level of precision, of QED corrections, still remains a very puzzling and exciting problem. These ions actually constitute the only experimental prototype of a three-body relativistic and quantum system (there is no other fully relativistic three-body system experimentally available). One can then observe, for example, the relativistic electron-electron correlation, few-body QED effects, etc., and this field of research is still very active. In thirty years, X-ray spectroscopy has changed sources from X-ray tubes used for the study of singly-ionized neutral atoms to more sophisticated instruments used to study more elementary objects, leading to much more general and fundamental atomic physics research.

Beyond the study, at the highest level of precision, of the energy of the states of elementary quantum systems, the use of few-electron ions, instead of few-hole atoms, to study most of the basic processes of atomic physics has also led to very exciting results. There is a very famous theorem quoted in the Condon and Shortley text book which states that, by time reversal, a two-electron ion generates the same sequence of states as a two-hole atom having all its shells closed. This is of little interest in precise energy measurements, as discussed above, owing to the need to approximate the screening effects through, e.g., the central-field approximation in many-electron atoms, but may be very interesting for studying the cross sections or transition rates of some direct elementary processes which are known at a much lower level of precision. In this respect, the use of these time-reversal properties (the inverse balance or micro-reversibility theorem) has led in the last decade to very exciting results. Dielectronic recombination (DR), for instance, simultaneous capture of a free electron accompanied by an electron excitation, is nothing but the inverse process of Auger decay, and radiative electron capture (REC) is the inverse process of photoionization. These questions will be addressed in the next sections.

## ATOMIC MANIPULATIONS

One of the virtues of the visible light provided by lasers is its capacity to easily push atoms through radiation pressure. The first demonstration of this effect was given long ago by J.-L. Piquet. This property is now extensively used to trap and cool atoms. One can now reduce the thermal kinetic energy of atoms to extremely low temperatures by using a few lasers in different directions and/or to confine them in a given region of space. One can thus manipulate the atoms as a whole. By using the tip of an STM it is also possible to set or move atoms at a given place on a surface.

The virtues of the physics of inner-shells or accelerator-based physics is to allow manipulation of the "interior" of atoms. As discussed above, we are able to remove the inner-shell electrons without changing the occupation states of all the others. Until 1967, we were able to observe the removal of just one electron, and this was the only way, at that time, to produce X rays. Since 1967, it is possible to observe doubly- and multiply-ionized atoms. At the same time, the first ion accelerators became available, and it became possible to remove in energetic collisions many electrons from a single atom at the same time. These violent collisions are not very selective, but at the highest energies (hundreds of GeV) it became possible, in 1987, to fully ionize uranium atoms, and then to produce any atom having any number of electrons (first "intra-atomic" manipulations).

The most interesting techniques of intra-atomic manipulations appeared when it became possible to dress, in a selective way, fully stripped ions prepared at low velocities in ion sources such as ECR or EBIS. The first experiments in this field were done in studying the interactions of highly-charged ions with isolated atoms or molecules. In this case, the valence electron of the target is captured in a well-defined highly-excited state of the ion in order to insure energy conservation, through a simple over barrier resonant transfer. It was then possible, by choosing carefully the charge of the ion and the binding energy of the valence electron of the target, to selectively populate a given state of an ion. Inside EBIT sources, it is now possible, in tuning properly the energy of the electron beam at resonance, to excite selectively, through dielectronic recombination processes, a given ion among many others present in the trap, in a given excited state.

When all quantum numbers tend to infinity, a quantum-mechanical object tends to be classical. In optics, it is possible by combining multi-photon absorption with RF excitation to prepare excited species whose active electron is in a state where all quantum numbers are maximum, i.e., planetary atoms. In capture processes (the dressing technique) the angular momentum state in which the electron is captured is not yet totally understood, but one can say that a large part of the interaction leads to the population of the highest angular momenta $l$ but, always, for obvious kinematic reasons, the lowest projections $m$ of these momenta, contrary to what happens during the absorption of photons. With the "dressing" technique it now becomes possible to prepare any ion in any given excited state having large $n$ and $l$ quantum numbers, but with the lowest states of the projection of the angular momentum $m$. This leads to an amazing and unexpected view of the atom, because, in this case, one prepares atoms in which the electron is either confined on the north or the south poles of a sphere of large radius (16). We are here very far from the usual vision of atoms with electrons rotating around a nucleus. This was actually imagined by White in his old text book of atomic physics, but if one reads it carefully, one may discover how much the author was puzzled when realizing this exotic property.

One of the most interesting properties of the dressing technique is the possibility of building any exotic state of the ion, e.g., with two very well-localized electrons like in the above example. In this case, the repulsion of these two electrons dramatically depends on the place where they are located, and the expulsion of one of them, an Auger effect, may be treated as a classical system. In consequence, one can study with near-classical objects the correlation effects between the two electrons, and follow their evolution by decreasing their quantum numbers towards "more quantum objects". It is also now possible to prepare even more exotic objects by sending highly-charged ions onto surfaces: the study of hollow atoms.

## HOLLOW ATOMS

Highly-charged ions interacting with isolated atoms or molecules capture, most of the time, only one electron, because the second one to be removed is usually two-times more bound than the first one, and therefore its capture becomes very unlikely. Single capture then dominates in the interactions of highly-charged ions with isolated atoms.

Above a metal surface, for instance, all (conduction) electrons are equivalent, and there is no reason not to capture as many electrons as there are positive charges inside the nucleus. To insure energy conservation, these electrons have still to be captured in Rydberg states of the ion, which leads to the formation of "Rydberg" hollow atoms. Most of the time, these Rydberg hollow atoms do not live long enough to be easily observed before the hollow atom hits the surface where it is peeled off. Below the surface, this re-ionized ion will be re-neutralized inside the solid. The capture process then takes place at closer distances and in lower-energy excited states of the ion, which leads to the formation of "below the surface" hollow atoms. These hollow atoms have been clearly identified (17,18). It has been observed, for instance, at the SuperEBIT source of Livermore, Xe hollow atoms with more than 45 excited electrons in the O, P, ... shells, while the K, L and M shells remain empty. In this case, the whole decay, i.e., all the steps of these exotic species, may be observed from the shell where capture occurs down to the ground state. Capture above the surface is more difficult to identify, and, to date, only the remnant of these Rydberg hollow atoms have been observed.

These hollow atoms provide, no doubt, a very exciting inner-shell process to study. They have also been observed in different experimental situations, such as inside an ECR source, laser-produced plasmas, or through the multiple (up to three) excitation of Li atoms by synchrotron radiation. The main relevant questions are now: how do theses species form and how do they decay?

The way they are formed above a surface seems now quite clear: it must be a resonant transfer, above the Coulomb barrier, of conduction or valence electrons of the solid (the most tightly bound) into excited states of the ion having an energy close to that of the work function of the surface. This mechanism is actually an unknown and puzzling process.

Below the surface, where so many electrons are available, there were controversies on the actual mechanism of capture, but this process is now quite well understood. Hollow atoms below the surface are, in principle, only formed at low velocity (even though hollow atoms at MeV energies have been observed). The ion velocity, to allow the use of the Born-Oppenheimer approximation, must be lower than the Fermi velocity of the electrons of the solid to be captured, and the ions are part of the solid, in a way similar to what happens during the formation of quasi molecules in charge-exchange processes. One observes, for instance, the radiative transitions of conduction electrons of the solid into the holes of the traveling ions. This is nothing but a REC process, except that the relative velocity of the collision is that of the electron to be captured and not that of the ion, as previously observed. The formation of hollow atoms below surfaces has been studied from $C^{5+}$ up to $U^{89+}$, and it has been found that capture processes always populate the highest level of the ion, inside the solid (the top "individual" stationary level of the ion inside the target, the higher states merging into the continuum or the bands of the solid). It has also been found that these electrons, in most cases, come from the conduction or valence band of the solid. The only very selective process which may explain the formation of hollow atoms below the surface is then Auger transitions of valence or conduction electrons of the solid into the top stationary hole state of the ion (the Auger emission, contrary to what happens for radiative transitions, mainly populates the closest available level in order to minimize the energy of the emitted electron). This process is, thus, nothing but the Auger neutralization mechanism, first introduced by Hagstrum (19) in 1952 to explain the interaction, at surfaces, of singly-charged ions.

Today, the main question is "How do these hollow atoms decay?" The main experimental difficulty in answering this question is that they quickly decay and are believed to remain inside the capture area, and, then, at the same time, lose their electrons by Auger emission and recapture new ones. Different attempts have been made to get rid of these difficulties. Hollow atoms "below the surface" have been prepared by passing relatively fast ions through very thin C foils, which may then decay, at least partly, outside the foil, i.e., outside the capture area. "Above the surface", or Rydberg, hollow atoms have also been prepared by sending relatively fast bare ions near $C_{60}$ molecules; in this situation, most of the ions cross the capture area, which is much larger than the buckyball radius, and leave it to decay outside. More recently, it has been demonstrated that the capture process, above a surface, proceeds in separate steps of capture and spontaneous free decay. In the first step, a certain number of electrons are captured to form some hollow ions, followed by a very fast decrease of the range of the capture area, coming from the quick changes of charges (the ion and the image) and a slow removal of the holes created above insulators hindering any further capture process. The ions then also capture and decay in separate steps. The spontaneous decay is here faster than the re-feeding. Below the surface, one observes the opposite situation, the ions being much faster re-fed than re-ionized by, e.g., Auger transitions (they stay quasi neutral). It is actually nothing but a problem of tapping and plugging a hole in a sink. When so many events take place in such a short time, the formation of quantum stationary states at each step of the capture and decay processes lengthens the lifetimes of the states (each process re-starts a stationary state and then lengthens all following processes); this is what people name the *Zeno effect*. Such a lengthening, sometimes huge, of the transitions has already been shown in many cases when all the steps of the decay are observed (18).

Another amazing property of hollow atoms is their stepwise decay. Owing to the time ordering of these steps (e.g., the step-by-step filling of the eight holes of the L shell), hollow atoms may be used as a clock of very short period (approximately a few $10^{-16}$ s). Because of the continuous change in energy of the transitions at each step of the decay of the hollow ion, it has even been possible, by changing the velocity of the ions, to create along their trajectory some milestones to study their interaction above a surface along distances of the order of atomic size, or even lower. The study of these effects, as well as these extreme cases of quantum systems, is no doubt an exciting field of research of inner-shell physics.

## IONS MOVING INSIDE A SOLID

A very specific property of X rays is their capacity to be transmitted well inside a solid. This is the magic property discovered a hundred years ago by Roentgen. This property allows deep analysis of solids by photography or X-ray fluorescence, i.e., the study of the atoms of the solid. The fast highly-charged ions passing through a solid may emit in flight some characteristic X rays, which can escape the solid, and may give us information on their behavior and therefore on their interactions with the solid. This property was first used to study the time evolution of the stripping processes of energetic, e.g., singly ionized, atoms passing through a foil. It can also be used to study the dressing processes of fully stripped energetic ions in a solid. Contrary to what happens with very slow ions traveling inside matter, as discussed above, and which interact at the same time with many atoms (formation of hollow atoms), fast ions, the interaction cross sections of which are smaller by orders of magnitude, only interact with a single atom of the solid at a time, and may travel long distances inside the target without experiencing any collision. Under these conditions, there is enough time to study the spontaneous decay of an ion between two collisions. This property was extensively used to study the spectroscopy of the few-electron ions discussed in a previous section. It can also be used to study each collision process separately.

Most of the time, fast ions may deeply penetrate the atoms of a solid at any impact parameter and interact with the target inner-shells, where they can experience, e.g., Fano-Lichten charge-exchange processes. These processes have been extensively studied in the past decades. This is what happens in non-organized matter. Inside crystals, the trajectory of the ions may be changed in some specific conditions and these ions may be channeled. An ion entering a crystal along one crystallographic direction or plane experiences a grazing incidence collision (the normal velocity of the ion does not allow the penetration of the jellium) and may be specularly reflected, not deeply penetrating the atoms of the solid. In fact, the first studies of the interaction of highly charged ions with surfaces (at surface) were done at the time of the discovery of channeling (very clean surfaces). When an ion is channeled, it does not penetrate the interior of the atoms, and one only observes the interaction taking place at the periphery of these atoms, as reviewed in detail by D. Gemmell (20). This superficial interaction has been studied looking at the X rays emitted in flight by the reflected ions (21), and found to be only due to radiative electron capture (REC): the ion which does not penetrate the atom enough may only capture the *free* electrons through a REC process and emits a very characteristic X ray at an energy of, e.g., $B_K + E_e$, where $E_e$ is the energy of the electron to be captured in the reference frame of the ion). When the ion is not channeled, it only radiates the characteristic X rays of the ion following any Fano-Lichten type collision process.

## HOW TO BUILD AN ATOM WITH A NUCLEUS AND A FREE ELECTRON?

Linking an electron to a nucleus to form an atom seems, at first view, an easy task. It is, in fact, one of the most puzzling open questions today that may only be solved by using the giant machines of nuclear physics. Capturing an electron requires that the system formed with the nucleus and the electron expels some extra energy to the external medium to ensure energy conservation. In atomic physics, one way to allow the system to give the outer world some energy is to create a photon. It is one of the mysteries of quantum electrodynamics. The X-ray community has, in this field, provided an essential contribution: the observation by H. Schnopper (22) of REC, discussed above. This process, theoretically predicted long ago, has never been directly observed before, because the only experimental device that could allow its observation was a plasma, where all the electrons have all the energies and led to the emission of a continuous REC spectrum, impossible to separate from the, always present, bremsstrahlung continua. In order to observe a discrete line at the right energy $h\nu = E_e + B_K$ (where $E_e$ is the energy of the electron to be captured from an atom at rest, and $B_K$ is the binding energy of this, e.g., K electron), Schnopper had the ingenious idea to observe this process in the collision of a moving ion, delivered by an accelerator, onto an electron, quasi at rest, inside a target. In such an experiment, the relative velocity of the collision is that of the ion, which is very large, plus that of the very low (conduction) electron to be captured. The main interest today is to study this process at the lowest possible relative velocities, the cross section for this reaction tending to infinity when the relative velocity of the two partners tends to zero. This is what is named the *infra-red divergence*. It is, no doubt, one of the most exciting experiments on the way in storage rings merging electron and ion beams having the same velocity (in electron coolers for instance). By time reversal, it is nothing but the same process as the photo-ionization at threshold, but in all experiments carried out with synchrotron radiation, one uses multi-electron atom targets, which cannot be exactly calculated and have large natural widths. To study this infra-red divergence, one needs to precisely compare experimental energies and cross sections with exact calculations in a true two-body problem. The main trouble to date is the quality of the beams to be merged (their transverse velocity), which broadens the width of the emitted photons. Physicists and engineers have made tremendous efforts in the past few years to increase the luminosity and the mono-chromatization of these beams. The present status in this adventure leads to very puzzling results, where the cross section of this REC process, as recently demonstrated, saturates at the lowest energies (23). I am convinced that this problem of knowing how an electron and a nucleus make an atom, at the lowest temperatures, is today one of the most exciting challenges in the field.

## CONCLUSIONS

Atomic physics started with the study of hydrogen atoms, looking at visible transitions, which were compared with predictions of the Schrödinger equation. However, the techniques of optics, even though the availability of lasers has led to some tremendous improvements in the field, are still insufficient to explore the whole field of atomic structure and, e.g., electron-ion interactions. The study of few-electron ions, which allows in a different range of wavelengths, especially in the X-ray range, the exploration of very exotic processes in various shells and at very accurate levels of precision, constituted a revolution in the field, which now makes inner-shell and accelerator-based physics the most general tool to study atomic physics.

## REFERENCES

1. M. O. Krause, T. A. Carlson and R. D.Dismuke, Phys. Rev. **170**, 37 (1968).
2. A. Migdal, J. Phys. USSR **4**, 449 (1941).
3. E. Feinberg, J.Phys. USSR **4**, 423 (1941).
4. B. Crasemann and P. Stephas, Proceedings of the Conference on the Electron Capture and Higher Order Processes in Nuclear Decay, Debrecen , Hungary , July 15-18, 1968, p 349.
5. H. P. Kelly, Photoionization and Other Probes of Many-Electron Interactions, F. J. Willeumier, Ed., NATO Advanced Institute Series , Plenum Press, 1976.
6. R. Moshammer, J. Ulrich, K. Kollmus, W. Schmidt, M. Unverzagt, H. Schmidt-Böcking, and R. E.Olson, Proceedings of the X ray and Inner shell Processes, 17th International Conference, Hamburg, Germany, Sept 9-13, 1996, AIP Conference Proceedings 386, p 153.

7. K. Siegbahn, C. Nordling, G. Johansson, J. Heidman, P. F. Heden, K. Hamrin, U. Gelius, T. Bergmark, L. O. Werme, R. Manne, Y. Baer, ESCA Applied to Free Molecules, North Holland Pub (1971).
8. R. P. Madden and K. Codling, Phys. Rev. Lett **10**, 516 ( 1963).
9. W. Mehlhorn, Photoionization and Other Probes of Many-Electron Interactions, F. J. Willeumier, Ed., NATO Advanced Institute Series , Plenum Press, 1976.
10. F. Willeumier, Photoionization of Atomic Ions, Inst. Phys. Conf. Series n°122, 1992, eds. P. J. O. Teubner and E. Weigold, (Institute of Physics: Bristol) pp. 203-212.
11. T. Åberg, J. Utrianen, Phys. Rev. Lett. **22**, 1346 (1969).
12. T. Åberg, Phys. Rev. A **75**, 1737 (1971).
13. R. D. Deslattes, Rev. Sci. Instr. **38**, 815 (1967).
14. J. P. Desclaux, Comput. Commun. **9**, 31 (1975).
15. J. P. Briand, Proceeding of the Nobel Symposium 85, Saltsjobaden, June 29- July 3 ,1992, p157.
16. J. P.Briand, Comments At. Mol. Phys. **33**, 1 (1996).
17. J. P.Briand, L. deBilly, P. Charles, E. Essabaa, P. Briand, R. Geller, J. P. Desclaux, S. Bliman, and C. Ristori, Phys. Rev. Lett. **65**, 159 (1990).
18. J. P. Briand, Comments At. Mol. Phys. **33**, 9 (1996).
19. H. D.Hagstum, Phys. Rev. **96**, 325 (1954).
20. D. S.Gemmell, Review of Modern Physics **46** (1974).
21. S. Andriamonje, M. Chevallier, C. Cohen, J. Dural, M. J. Gaillard, R. Genre, M. Hage-Ali, R. Kirssh, A. Lhoir, B. Mazui, J. Mory, J. Moulin, J. C. Poizat, J. Remilleux, Dschmauss and M. Toulemonde, Phys. Rev. Lett. **59**, (1987).
22. H. Schnopper, H. D. Betz, J. P. Delvaille, K. Kalota, A. R. Sahval, K. W. Jones and H. E. Wagner, Phys. Rev. Lett. **29**, 898 (1972).
23. G. Gwinner, Electron - Ion Recombination at Very Low Energies, Invited Talk at the XVIth ISIAC Meeting, July 19-30 ( 1999), Kyoto, Japan.

# Exploiting Compton Scattering: Studies of Spin Density Distributions

Malcolm J. Cooper

*Department of Physics, University of Warwick, Coventry CV4 7AL, UK*

**Abstract.** Studies of electron density distributions through the Compton effect exploit the Doppler shift that occurs because the target electron, bound in a potential, must be moving. This phenomenon was first exploited by DuMond and co-workers within a decade of Compton's discovery, then largely forgotten for thirty years. After a patchy renaissance it has acquired a new potency with the availability of synchrotron radiation. In particular the ability to produce high fluxes of circularly polarised x-rays has led to studies of the spin density in ferromagnets. Magnetic Compton scattering experiments provide the only direct probe of the bulk ground state spin density distribution in ferro/ferri-magnetic materials. They are often sufficiently characteristic of the element that site-specific contributions can be extracted and conduction electron polarisation studied; such work is the focus of this paper.

## INTRODUCTION

More than 75 years have elapsed since the publication of Compton's paper demonstrating and analysing the effect that bears his name (1,2) and another 10 since Gray gave the first correct description of the phenomenon (3). The importance of the Compton's work, which has been dealt with by my colleague M. Blume at this meeting is perhaps best summarised by the title of Roger Stuewer's book "The Compton Effect -Turning Point in Physics" (4). Compton's desire to explain his measurements in terms of what we would now call classical physics led him to many curious inventions, e.g a ring-shaped electron (5) to which we shall return later. His attitude mirrored the reluctance of perplexed physicists in general to the new physics. The aversion of most physicists, Compton included, throughout the first three decades of this century, to accept the ideas behind the quantum description of the effect serves to remind us that scientists are no more objective and dispassionate than others when faced with uncomfortable, challenging and revolutionary ideas. Eventually Compton's own experimental evidence that x-radiation was indeed inelastically scattered with the wavelength shift predicted by the quantum theory was enough to convince most, but not all, of the inadequacy of the classical approach.

In retrospect Compton pointed to his work under Rutherford at the Cavendish

Laboratory, as the strongest evidence for the failure of classical electrodynamics. At Cambridge he convinced himself that the (Compton) scattering cross-section fell below the classical (Thomson) value: four decades later Compton described this as "an observation comparable in importance with the Michelson Morley experiment"(6). It undoubtably motivated his continued interest when he returned to the USA.

The exploitation of the Compton effect really begins with DuMond, who established the relationship between the broadening of the modified line and electron velocities and performed what is perhaps the most important Compton scattering experiment ever, a study of beryllium (7,8) which provided the first direct evidence that Fermi-Dirac statistics were appropriate to the description of the electron momentum distribution of the conduction electrons. Together with Kirkpatrick and others he produced not only a focussing spectrometer optimised for Compton experiments (9) but also tailor-made x-ray tubes. The spectrometer was typical of DuMond's ingenuity: it comprised 50 calcite analysing crystals oriented to simulate a large, single curved crystal, which was then thought to be impossible to achieve, and it worked superbly.

The lack of activity in the next thirty years is not simply explained by the pursuit of other scientific priorities developed during the Second World War: there is a real problem in that Compton scattering relates to the momentum representation of electron behaviour. This may be just as complete as the quantum mechanical position space representation, but it is an unfamiliar concept, which we find difficult to interpret, and, even worse, it appears to require an integral rather than differential Schrodinger's equation to be solved! At the same time the rest of the x-ray world, studying electron density through x-ray diffraction, were making some progress in showing that electron density distributions were different from those in free atoms, even if at times their results were more to do with extinction in the sample than electron density. To that body of physicists and crystallographers Compton scattering was no more than a decided nuisance.

The renaissance of Compton work in the period 1965 - 1985 was faltering because the compromise between resolution and statistical accuracy had to be struck at too low a level. For example, as DuMond had shown, focussing x-ray spectrometers were capable of useful resolution of the electron momentum, but intensities were desperately low. Gamma rays experiments, on the other hand, could produce statistical accuracy generally only limited by the patience of the observer or occasionally by the half life of the source. Generally speaking the photon energies were higher ($^{241}$Am -60 keV, $^{123m}$Te 159 keV and $^{198}$Au - 412 keV, were popular) than characterisitc x-ray emissions and this lessened worries over the validity of the impulse approximation necessary to relate the scattering cross-section to the electron's ground state momentum density. Unfortunately the resolution was three or four times worse: see (10) for a review of the work in that period. The problems associated with low flux and polychromatic tube sources were greatly diminished, if not vanquished, by the advent of synchrotron radiation. Furthermore, new possibilities, arising from the polarisation properties of the source, emerged and it is that aspect of Compton scattering with which this article is primarily concerned.

# The Basic Theory

The nomenclature adopted is indicated in Figure 1 which shows a schematic Compton scattering event. The Compton profile (ie. the spectral broadening of the Compton-shifted line) can be usefully considered as a Doppler broadening due to the motion of the electrons along the direction of the x-ray scattering vector. Thus if that direction is chosen as the z- axis of a set of Cartesian coordinates, we can define the Compton profile, $J(p_z)$, and its spin-dependent variant, $J_{mag}(p_z)$, as:

$$J(p_z) = \iint n(p) \, dp_x dp_y \quad : \quad J_{mag}(p_z) = \iint [n\uparrow(p) - n\downarrow(p)] \, dp_x \, dp_y \quad (1)$$

where $n(p)$ is the probability distribution of the electron momenta, i.e. it is the electron momentum density distribution. The relationship between electron momentum $p_z$ and the experimental parameters is given by the equation (2),

$$\frac{p_z}{mc} = \frac{(E_s - E_i) + (E_i E_s / mc^2)(1 - \cos\phi)}{(E_i^2 + E_s^2 - 2 E_i E_s \cos\phi)^{\frac{1}{2}}} \quad (2)$$

and $p_z$ is normally quoted in atomic units of momentum ($c = 137; e = \hbar = m = 1$) where 1 a.u.=$1.99 \times 10^{-24}$ kg m s$^{-1}$. For $p_z = 0$ this reduces to the expression for the Compton shift:

$$\frac{E_s}{E_i} = \frac{1}{1 + \frac{E_i}{mc^2}(1 - \cos\phi)} \quad (3)$$

which is equivalent to the much more familiar equation for the wavelength shift $\Delta\lambda$

$$\Delta\lambda = \frac{2h}{mc} \sin^2\frac{\phi}{2} \quad (4)$$

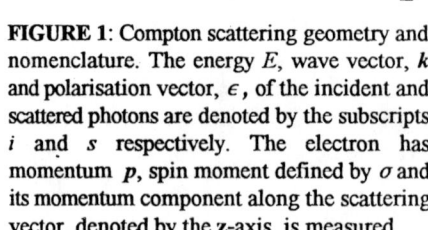

**FIGURE 1**: Compton scattering geometry and nomenclature. The energy $E$, wave vector, $k$ and polarisation vector, $\epsilon$, of the incident and scattered photons are denoted by the subscripts $i$ and $s$ respectively. The electron has momentum $p$, spin moment defined by $\sigma$ and its momentum component along the scattering vector, denoted by the z-axis, is measured.

# Interpreting Compton Data

The fundamental problems associated with studying electron density through Compton scattering are three-fold. In the first place the quantity probed is an alien one, the momentum space density. The second is that it is the projection of that quantity along a line, not the three dimensional density, that can be measured, and the third is that all the electrons contribute, including the core electrons that have little or no sensitivity to their condensed matter environment. In fact momentum space is the natural starting point for the description of "free" electrons in metals, each $k$-state being a plane wave with unique momentum $p=\hbar k$. As mentioned earlier the first success of Compton scattering was the validation of Fermi Dirac statistics for conduction electron states in metallic beryllium. In the presence of any real potential this direct relationship is lost and there is merely a probability than an electron with wave vector $k$ has momentum $p=\hbar(k+G)$ which is given by $|V_{G,k}|^2$ where $V_{G,k}$ is a Fourier coefficient of the potential and $|V_{0,k}|^2$ is no longer unity. This gives rise to the so-called umklapp features in the electron momentum density and the Compton profile.

The calculation of Compton profiles from electronic structure models is complicated because of the inherent 1-D projection. This means, for example, that in order to determine the Compton profile, $J(p_z)$ at any momentum it is necessary to know $n(p)$ at all momenta, which makes computation labourious. It is also a difficulty that is largely avoided in positron annihilation studies of electron momenta because the positively charged positron, excluded from the ion cores, samples little of the high momentum component of the electron wavefunction and direct two dimensional measurements are practicable with area detectors. The compensating advantage of the "clean" photon probe is the certainty of knowing that all target electrons are to be equally weighted.

Energy considerations provide unexpected help in the interpretation of Compton data. The second moment of $J(p_z)$ gives the kinetic energy of the electron distribution and this equates numerically, with a change of sign, to the total energy by the virial theorem. The $p^2$ weighting to the energy implies that the truly small energy differences between free atom and solid (i.e. the cohesive energies) all derive from the behaviour of slowly moving electrons. In other words the core electrons can be modelled accurately by free atom wavefunctions and these in turn can be calculated to more or less arbitrary accuracy. This conclusion is not as negative as it seems: the fact that the high momentum tails to the profiles are predictable means that the "correct" behaviour at high momentum can be taken as a sign that corrections for background, multiple scattering, resolution etc., which would disproportionally affect the profile tails, have been applied properly. Also, as we shall see later in compounds and alloys, fitting the spin-resolved profiles in the tails allows the relative spin moment on each atomic site to be extracted, this specificity is surprising in an incoherent measurement.

The difficulty of understanding electron momentum density distributions is illustrated by the case of a simple diatomic molecule. It was first attempted by Coulson and coworkers in the nineteen forties (11). Remembering that the momentum and position space wavefunctions form a Fourier pair it is easy to see that a simple bonding orbital for a diatomic molecule written as:

$$\Psi(r) = \psi(r) + \psi(r-R) \qquad (6)$$

becomes:

$$X^2(p) = \chi^2(p)\,[(1 + \cos(p \cdot r))] \qquad (7)$$

where $\chi(p)$ is the Fourier transform of $\psi(r)$ and normalisation factors have been neglected. This result shows that the "bond" is represented by cosine fringes which are perpendicular to the bond's direction. The expression is exactly analogous to that for the Fraunhofer diffraction pattern of a pair of long slits, where again the diffracted amplitudes and the transmission function form a Fourier pair.

Although equation (7) contains the signature of the covalently bonded density the interpretation of $n(p)$ will be difficult if the material does not exhibit a unique bonding direction. The interpretation of the Compton profile is further complicated by the projection of the distribution onto a line (the scattering vector). There have been a number of attempt to relate the momentum and position space pictures of electron density in order to permit the interpretation of momentum space data in terms of the familiar position space concepts of bonding. They centre around a "reciprocal form factor", $B(r)$, where

$$B(r) = \int n(p)\,\exp(-i p \cdot r)\,dp \text{ where } B(0,0,z) = \int J(p_z)\,\exp(-i p_z z)\,dp_z \qquad (8)$$

so-called because its relationship to the real space wavefunction mirrors that of the conventional form factor $f(K)$, vis:

$$B(r) = \int \psi^*(r')\,\psi(r+r')\,dr' : \qquad f(K) = \int \chi^*(p)\,\chi(K+p)\,dp' \qquad (9)$$

In equation (9) both form factors have been written as autocorrelation functions, although $f(K)$ is more familiar as the Fourier transform of the charge density $\rho(r)$.

Despite the difficulties of interpretation there have been a number of successful studies of bonding in polar materials beginning with urea (12) LiH (13) and most recently ice (14): the last result is reproduced in Figure 2 and has been discussed fully by P. M. Platzman at this conference. The oscillations in the difference profile formed by taking measurements parallel and perpendicular to the hydrogen bond direction. The authors interpret the period of the oscillations as providing the first direct evidence for hydrogen bonding in ice.

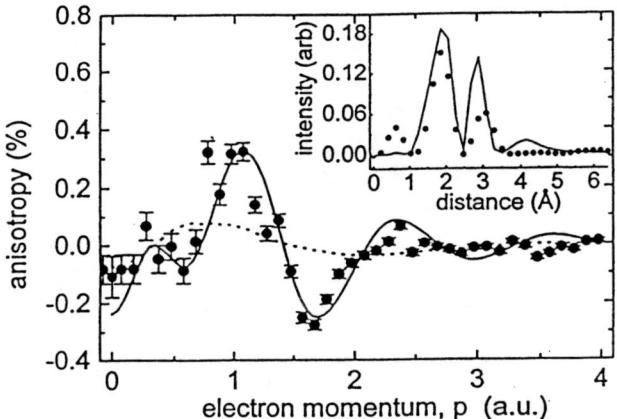

**FIGURE 2**: The difference between Compton profiles of ice, $I_h$, measured with the scattering vector parallel and perpendicular to the hexagonal basal planes (the hydrogen bonds lie close the latter). The solid line shows the prediction for hydrogen bonding. The inset is the power spectrum which clearly shows two dominant distances which relate to the hydrogen and 0-0 bond lengths ( 1.75 Å and 2.75 Å respectively). Reproduced from (14).

# THE CROSS SECTION FOR MAGNETIC COMPTON SCATTERING

For a full account the reader is referred to a comprehensive review of magnetic Compton scattering by Sakai (15), only a summary is given here. The Compton profile contains contributions from all the electrons in the target whereas $[n\uparrow(p) - n\downarrow(p)]$ relates to just the spin-dependent momentum distribution. The Magnetic Compton profile, $J_{mag}(p_z)$, is accessible in experiments in which the incoming radiation is circularly polarised. The relevant cross-section (16), developed from the expressions of Lipps and Tolhoek (17) is:

$$\frac{d^2\sigma}{\Omega\, dE_s} = \left(\frac{e^2}{mc^2}\right)^2 \cdot \frac{E_s}{E_i} \cdot \left(\frac{m}{2\hbar K}\right) [1+\cos^2\phi + P_l \sin^2\phi + \frac{(k_i-k_s)}{mc}(1-\cos\phi)][J(p_z)]$$
$$+ [(\cos\phi-1)\, P_c\, \hat{\sigma} \cdot \frac{(k_i\cos\phi + k_s)}{mc}][J_{mag}(p_z)] \quad (10)$$

Inspection of equation (10) shows that the final "magnetic term" can be isolated by changing its sign. This in turn can be achieved either by reversing the hand of polarisation of the incident photon beam (i.e. flipping $P_c$), or by changing the direction of the unit spin vector $\hat{\sigma}$, which is done by reversing the direction of the spin vector with an external magnetic field.

This equation, which relates the cross section to the ground state of the scatterer, is derived within the impulse approximation. In a simple-minded way the potential is assumed to be unchanged during the instantaneous (i.e. "impulsive") interaction and the energy transfer is wholly kinetic and large - so large in fact that the final electron state can safely be taken as a plane wave. There have been a number of interesting studies of failures of the impulse approximation using other final state wave functions (for example continuum state solutions of the potential) occasioned by the x-ray energies available from both tube and second generation synchrotron sources. In general the impulse approximation works exceedingly well. In the studies reported below, with energy transfers of the order of 100 keV there is little expectation of its failure.

High photon energies on the other hand do pose some problems with regard to the extraction of the Compton profile from the measured scattering cross-section. The relativistic cross-section has been calculated for a free moving electron semi-relativistically (18) by Grotch et al, then relativistically by Blatt et al (19). Their latter result contains terms in $p/mc$ which are not present in the semi-relativistic approach. They show that whereas the scattering from unpolarised electrons can be predicted by combining the cross-section for a stationary electron with the scattering factor $S(K, E)$ as has been done some time ago by Ribberfors (20), for the moving electron, the same generalisation does not (quite) hold for an electron with spin. The invocation of the impulse approximation allows their result to be applied to bound electrons since in the limit of $E_i - E_s >> E_B$ the electron can be treated as freely moving. As Blatt et al point out their result is analogous to that for the charge scattering cross-section in which the ground state Compton profile can be disentangled from other momentum dependent factors in the cross-section by iterative methods as was done for charge scattering by Ribberfors. To a first approximation the momentum dependence of the cross-section is linear across the energy range of interest and simply folding the (magnetic) Compton profile about $p_z=0$ effects the correction. Studies of the magnetic Compton profile of Fe at a number of energies up to ½MeV (16) confirm that the ground state momentum density can be extracted without undue difficulty.

## What is measured?

The very first measurements of magnetic Compton scattering were probably those published in 1916 by Forman (21), i.e. while Compton was wrestling with the problem of the forward peaking of the scattered radiation evident with radioisotopes, which we now know to be high energy (~ 2-3 MeV) gamma-ray sources. Forman used x-rays and found that the transmission through an iron foil was approximately ½% less when the foil was magnetised in the direction of the beam. The difficulties associated with such an experiment at that time cannot be exaggerated. Incredibly it appears to be a valid result. Much latter, in 1954, Gunst and Page (22) made a similar measurement with gamma rays which was related to the Klein-Nishina cross section which was published 13 years after Forman's work. At the time the importance of Forman's work was that it spurred Compton on in his development of the ring model for the electron (7). He had previously adopted a model of the electron as a thin spherical shell, in order to explain the excess forward total scattering. It allowed him to decrease the backscattering through interference effects but his electron, at $\sim 10^{-10}$ cm radius, was a thousand times too large and it could not explain any magnetic effect. The ring electron model got round both difficulties: the magnetic effect was considered to arise because the circulating electron would align with the ring perpendicular to the field direction and this would allow "the incident x-rays can get a better hold on the electrons in this position" (23).

The exploitation of magnetic Compton scattering effects began not with the study of the spin distribution but the polarisation state of the radiation (24, 25). In fact its use to study the scatterer rather than the radiation was suggested by Platzman and Tzoar in 1970 (26) and taken up by Sakai and Ono (27) in the pre-synchrotron era by using the same kind of circularly polarised gamma ray sources as had been studied themselves twenty years earlier. This all happened at roughly the same time that deBergevin and Brunel (28) were demonstrating the magnetic diffraction effects also predicted by (26), also without the benefit of polarised synchrotron sources.

The difference in approach between magnetic Compton scattering, which has to be dealt with relativistically and magnetic diffraction, which can be approached semi-classically, is well illustrated by the question of whether orbital magnetisation contributes to the Compton cross section as it does in diffraction. After some confusing results (29), convincing experimental evidence provided the answer "no" (30, 31) in situations where the impulse approximation is valid, i.e. $L$ is not a good quantum number. This accords with common-sense intuition that orbital momentum can hardly be defined in an instantaneous interaction. A quantitative theoretical argument was put forward by Carra and collaborators (32) and a recent model calculation of hydrogen atom in the *2p* state (33) indicates just how orbital effects can creep into the cross-section as the impulse approximation fails at low energy transfers but no measurements in this low momentum transfer region have yet been attempted. Figure 3 shows typical magnetic Compton data for the material $HoFe_2$ which was studied to establish the unique sensitivity of spin polarised Compton scattering to spin density. These results illustrate the opposed spin moments on Ho and Fe. The left hand diagram shows the temperature dependence of the spin moments (31). At low temperatures the total spin magnetisation, which is equal to $\int J_{mag}(p_z) \, dp_z$, is negative. This reflects the dominance of the spin moment on Ho, but at higher temperatures the Ho moment "softens", the Fe moment is virtually unchanged, and the spin magnetisation is reversed. NB the total magnetisation remains negative because of large orbital moment on Ho which is parallel to its spin moment. The right hand diagram shows the decomposition of the total spin moment into Fe *3d* and Ho *4f* components results from the characteristic difference in their momentum distributions. The *4f* electrons are more tightly bound to the Ho nucleus than the Fe 3d electrons and hence have a characteristically broader momentum distribution. As argued earlier the high momentum tails of the spin resolved profile can be fitted by free atom profiles (34), thanks to the virial theorem, allowing the two contributions to be differentiated (e.g. $CeFe_2$ and $UFe_2$, refs 35 and 36, respectively). This approach has been used in a number of investigations to determine site-specific moments, indeed knowledge of the total magnetisation, together with the spin-resolved data, can allow estimation of the size of the orbital moment. The fit is only valid at higher momenta ( say ($p_z$ >2.0 a.u.) and the composite curve does not fit the data at low momenta because of the presence of a diffuse contribution to the spin density resulting from the polarisation of the conduction electrons. This requires an accompanying band model for its analysis. The diffuse moment cannot for example be taken as the difference between the data and the two curves. Although the atomic Ho profile is likely to describe the *4f* electrons at all momenta the Fe *3d* free atom profile will not describe the behaviour at low momentum as is evident from a number of studies, examples of which will be given below.

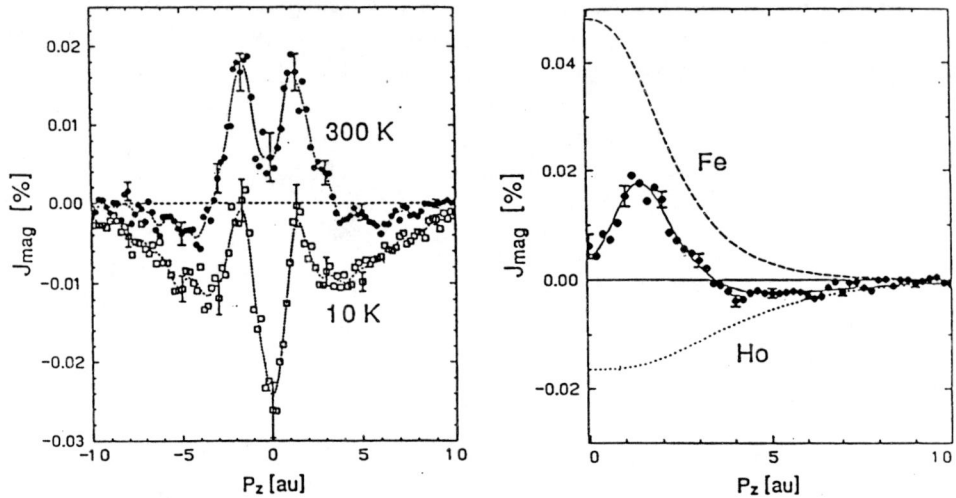

**FIGURE 3:** The spin resolved Compton profile of HoFe$_2$. The left hand diagram shows the full magnetic Compton profile at two temperatures (31). The right hand diagram shows the room temperature data averaged left/right (Compton profiles are symmetric about $p_z = 0$ and fitted for $p_z > 2$ a.u. with free atom Compton profiles taken from (34). The ~50% difference in width of the Ho *4f* and Fe *3d* profiles allows their contributions to be separated. The *4f* function will describe the Ho spin moment at all momenta. Taken from (30).

# RECENT STUDIES

## 1: Testing Electronic Structure Models In Transition Metals

In many cases the analysis of Compton profiles can only proceed hand in hand with electronic structure calculations because contributions from valence and conduction electrons are engulfed by the large contributions from core electrons. The situation is more favourable with spin-resolved data since the inner shell paired electron contributions are automatically removed, but the separation of, say, the conduction electron spin polarised contribution from the 3d density in a transition ferromagnet still requires both inputs. The recent study of nickel (37) is a good example of the necessary marriage of experiment and theory. The measurement was performed at room temperature in a 1Tesla field which was sufficient to saturate the disc shaped sample notwithstanding any demagnetisation effects. Interestingly it shows clearly that there are features of the magnetic Compton profile that defy description by either the Local Spin Density Approximation (LSDA) or Generalised Gradient Approximation (GGA) based calculations as is evident from Figure 4. Not least of these discrepancies is the fact that the Linearised Muffin Tin Orbitals (LMTO) and Full Potential Linearised Augmented Plane Wave (FLAPW) models overestimate seriously the total moment: the curves lie consistently above the data at low and intermediate momenta. This is not a discrepancy that can simply be removed by "scaling": changes in the exchange splitting for example would also change the spin-resolved momentum density distribution. On the other hand there is a satisfactory degree of agreement between experiment and theory at high momenta, which indicates that the data processing and the modelling are correctly based, convergence to free atom values indicating both that the theory has the correct cohesive energy and that the experimental data have been correctly processed. Interestingly some of the "umklapp" bumps which, in the models, are related to features at low momentum in the first Brillouin zone appear in the data (indicated by arrows in Figure 4) at the correct reciprocal lattice vectors despite the fact that the low momentum peaks are conspicuously absent.

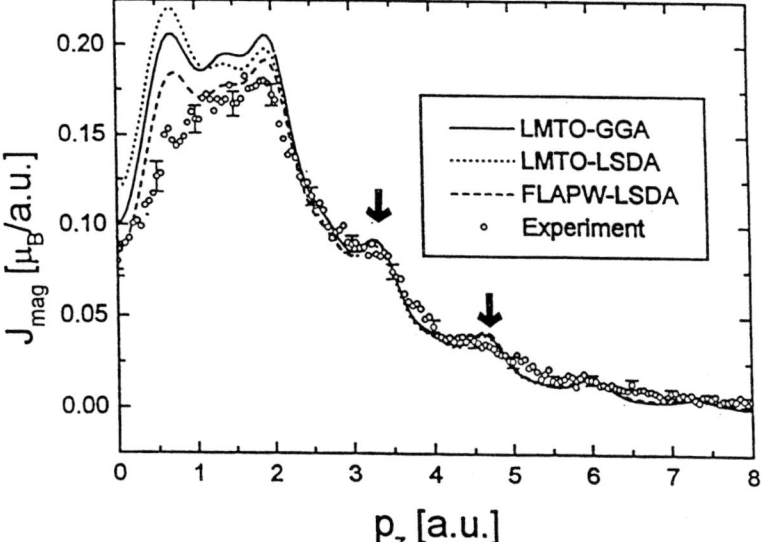

**FIGURE 4**: The magnetic Compton profile of single crystal nickel for the [110] direction. The diagram shows the experimental results (open circles), obtained at a momentum resolution of 0.43 a.u., together with a number of electronic structure calculations as indicated in the box. The area under the data curve represents the spin moment per atom and the models all predict a larger spin moment than the measured $0.56\mu_B$. This difference is particularly evident in the overestimation of the spin-resolved profile at low momentum. The arrows indicate the "umklapp" features at high momentum related to the predicted structure in the first Brillouin zone. Results taken from (37).

## 2: Gadolinium and Gd-Y alloys

Gadolinium has a relatively simple magnetic structure compared to many other rare earths. However the fact that it is hexagonal had deterred electronic structure calculations of the magnetic Compton profile. Recently Duffy et al (38) carried out LMTO calculations (in both GGA and SLDA approximations) to compare with FLAPW calculations by Kubo and Asano (39) and their own measurements. In contrast to Ni (37) where the FLAPW describes the data better than the LMTO models, both models describe the data adequately as is evident from Figure 5 with the LMTO model providing a better value for the observed moment. Indeed the FLAPW model appears to overestimate the induced s-p conduction electron moment. In this instance the GGA does not produce any significant improvement over the LSDA.

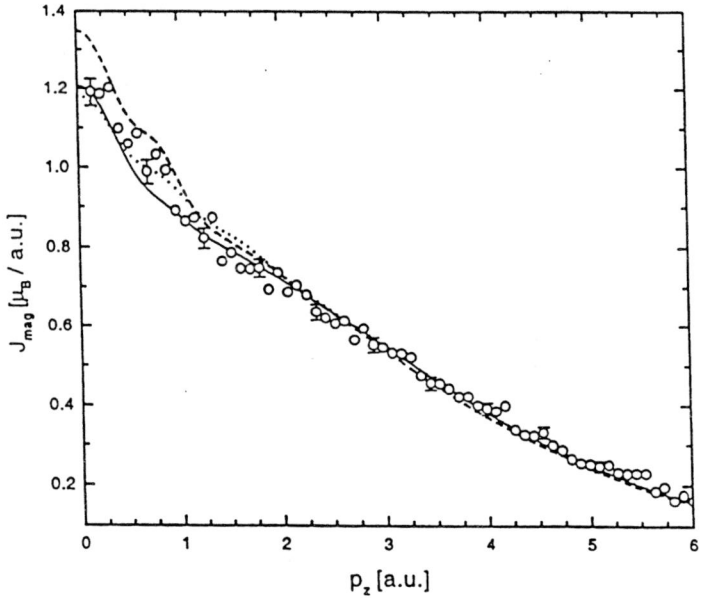

**FIGURE 5**: The magnetic Compton profile of Gd. The area under the data curve is normalised to the known moment of 7.62 $\mu_B$. The results are compared with LMTO electronic structure calculations in the SLDA (solid line) and GGA (dotted line) approximations. The FLAPW prediction (39) is also shown (dashed line). Taken from (38).

The success of the LMTO method for Gd was subsequently exploited by using it in a study of the Gd-Y alloy $Gd_{62.4}Y_{37.6}$ (40) by using a 16 atom (10-Gd, 6-Y) supercell which accurately mimics the known composition. Various random site occupancies of the supercell produced essentially unchanged results for the magnetic Compton profile of the alloy and the result with Gd atoms on all sites did not differ significantly from the

LMTO calculation for the real unit cell. The interest in these alloys derives from the fact that the total magnetic moment is greater than would be expected if yttrium were simply diluting the moment per unit cell (41). The alternative source of the extra moment could be orbital and the suggestion had been made (42) that the change in the crystal field upon alloying would permit spin-orbit coupling to induce such a moment. In fact the LMTO supercell calculation predicts a slightly reduced moment associated with the Gd *5d-like* and *6d-like* conduction electrons and an extra contribution to the delocalised spin density from the yttrium *p-like* and *d-like* electrons. Overall an extra induced spin moment of $0.14\mu_B$, which represents an increase of just 2%, is predicted. The data confirm that a spin moment ($0.16 \pm 0.03\ \mu_B$) of the correct magnitude does exist and this is reproduced in Figure 6 as demonstrating the sensitivity of the experimental technique.

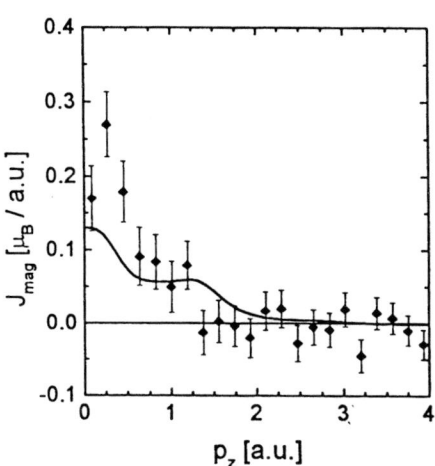

**FIGURE 6**: The difference between the magnetic Compton profile of Gd and the alloy $Gd_{62.4}Y_{37.6}$. There would be a zero result if all the moment were due to Gd. The extra moment of $0.16\mu_B$ is associated with the induced polarisation of the yttrium conduction electrons. Taken from (40).

## 3: The Samarium Moment in $SmMn_2Ge_2$: an example of direct interpretation

This material is an example of a set of layered compounds of the type *RE-Ge-Mn₂-Ge-RE* which, broadly speaking, exhibit ferromagnetism for *Mn-Mn* interplanar spacings greater than 0.287 nm, and antiferromagnetism otherwise. This samarium compound, which actually has that critical spacing, exhibits reentrant ferromagnetism. The material is difficult to study with neutrons because of its high capture cross-section. Although the existence of a moment of order 3 $\mu_B$ on Mn is established from neutron work (43) the size and direction of any moment on the samarium site was not well-established because the neutron measurement is sensitive to $M_L+M_S$ and these two moments are of similar size, furthermore Hund's rules predict that they are coupled antiparallel. The Compton result shows immediately that there is a spin moment on the Sm site. Our data (44) are shown in Figure 7. The positive direction is that of the external field and the manganese spin moment. The fact that the profile is negative above 2 a.u. means that there is a samarium spin moment which is antiparallel to that on the manganese site.

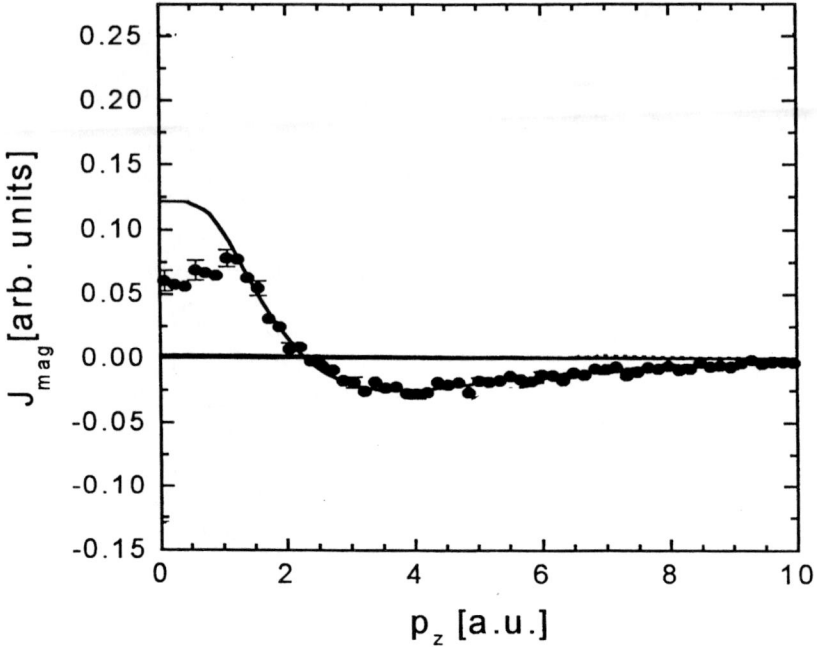

**FIGURE 7**: The magnetic Compton profile of Sm Mn$_2$Ge$_2$. The existence of a negative part to the curve proves immediately that there is a spin moment on the samarium site.
The analysis of these data in terms of the size of the samarium spin and orbital moments is described in (44).

Again the characteristic greater width of the *4f* momentum distribution is sufficient to allow the Sm 4f contribution to be isolated. Inspection of the data shows that the total spin moment in the material is approximately zero because the area under the data curve is zero, i.e. the spin moments on Mn and Sm are equal in size. Given that there is unlikely to be a significant orbital moment on Mn this implies that the orbital moment on the Sm site is approximately equal to the observed bulk magnetisation. The analysis of these data is detailed in (44).

## ACKNOWLEDGEMENTS

I am very grateful to many colleagues who have contributed to the research described in this paper, especially Jon Duffy (Warwick University and Joanne McCarthy (ESRF) who have been involved in the majority of the Compton scattering studies, carried out by the Warwick-based group, that are discussed in this article. I am also grateful to other staff at ESRF and to long term collaborators David Timms (Portsmouth University) and

Gienek Zukowski (Warsaw University). The UK Engineering and Physical Sciences Research Council provided financial support and ESRF is thanked for the provision of beamtime. Last, but not least, I thank the organisers of X99 for inviting me to present a paper on Compton scattering at a location so closely linked with the work of Arthur Holly Compton.

## REFERENCES

1. Compton, A. H., *Phys Rev.* **21**, 483-502 (1923).
2. Compton, A. H., *Phys Rev.* **22**, 409-413 (1923).
3. Gray, J.A., *Phil. Mag.* **26** 613 (1913).
4. Stuewer, R. H., Compton Effect- Turning Point in Physics New York: Science History Publications 1975.
5. Compton A. H., *Phys. Rev.* **14**, 23 (1919).
6. Unpublished letter, dated 28th April 1958 from Compton to F Miller, quoted in The history of the Compton Effect by Stuewer R., H., dissertation for the Degree of PhD, University of Wisconsin 1968.
7. DuMond, J. W. M., *Phys Rev.* **33**, 643 (1929).
8. DuMond J. W. M., *Rev Mod Phys.* **5**, 33 (1930).
9. DuMond J. W. M. and Kirkpatrick P., *Rev Sci Instr.*, **1**, 88 (1930).
10. Cooper M. J., *Rep. Prog. Phys.* **48**, 415 (1985).
11. Coulson C. A., *Proc. Camb. Phil. Soc.* **37** 55 (1941), see also *Proc. Camb. Phil. Soc.* **37**, 67, 74, 97 and 406 (1941).
12. Reed W. A.,Snyder L. C., Guggenheim H. J., Weber T. A. and Wasserman Z. R., *J Chem Phys.* **69**, 288-196 (1978).
13. Pattison P and Weyrich W., *J. Phys. Chem. Solids*, **40**, 213-22 (1979).
14. Isaacs E. D., Shukla A., Platzman P.M., Hamann D. R., Barbiellini B and Tulk C. A., *Phys. Rev. Letts.*, **82**, 600 (1998).
15. Sakai N., *J Appl Cryst.* **29**, 81 (1996).
16. McCarthy J. E., Cooper M. J., Honkimäki V., Tschetscher T., Suortti P., Gardelis S., Hämäläinen K. and Timms D. N., *Nucl. Instr. & Meths* **A401**, 463 (1997).
17. Lipps F. W. and. Tolhoek H. A., *Physica* **20**, 85 & 395 (1954).
18. Grotch H., Kazes E., Bhatt G. and Owen D. A. *Phys. Rev.* **A 27**, 243 (1983).
19. Bhatt G., Grotch H., Kazes E. and Owen D., *Phys. Rev.* **A 28**, 2195 (1983).
20. Ribberfors R., *Phys. Rev.* **B 12**, 2067 & 3136 (1975).
21. Forman A. H., Phys. Rev. **7**, 119 (1916).
22. Gunst S. B. and Page L. A., *Phys. Rev.* **92**, 970 (1954).
23. Compton A. H., J. Wash. Acad. Sci. **8**, 1 (1918)
24. Wheatley J. C., Huiskamp W. D.,. Diddens A. N, Steenland M. J. and Tolhoek H. A., *Physica* **55**, 395 (1954).
25. Goldhaber M., Grodzins L. and Sunyar A. W. , *Phys. Rev.* **109**, 1015 (1958).
26. Platzman P. M and Tzoar N. *Phys. Rev.* **B2**, 3556 (1970).
27. Sakai N. and Ono K., *Phys. Rev. Lett.* **37**, 351 (1976).
28. DeBergevin F. and Brunel, M. *Acta Cryst.* **A37**, 314 & 324 (1981).
29. Collins S.P., Cooper M. J., Lovesey S.W. and Laundy D., *J. Phys Cond. Matter* **2**, 6439 (1990).
30. Cooper M. J., Zukowski E, Collins S. P., Timms D. N., Itoh F. and Sakurai Y. *J. Phys.*

30. *Cond. Matter* **4,** 399 (1992).
31. Cooper M. J., Zukowski E., Timms D. N., Armstrong R., Itoh F., Tanaka Y., Ito M., Kawata H, and Bateson R., *Phys. Rev. Letts.* **71,** 1095 (1993).
32. Carra P., Fabrizio M., Santoro G., and Thole B.T. *Phys. Rev.* **B53,** R5994 (1996).
33. Kekchrakos D. Trohidou K. N., and Taddei S., *Phys. Rev.* **B56,** 10812 (1997).
34. Biggs F., Mendelsohn L. B. and Mann J. B. *At. Nucl. Data Tables* **16,** 201 (1975).
35. Cooper M J, Lawson P K, Dixon M A G, Zukowski E, Timms D N, Itoh F, Sakurai H, Kawata H, Tanaka Y & Ito M, *Phys. Rev.* **B54,** 4068 (1996).
36. Lawson P. K., Cooper M. J. Dixon M. A. G., Timms D. N., Zukowski E., Itoh F. and Sakurai H. *Phys. Rev.* **B56,** 3239 (1997).
37. Dixon M. A. G., Duffy J. A., Gardelis S., McCarthy J. E., Cooper M. J. , Dugdale S. B., Jarlborg T. and Timms D. N., *J Phys. Cond. Matter* **10,** 2759 (1998)
38. Duffy J. A., McCarthy J. E., Dugdale, S. B., Honkimaki V., Cooper M. J., Alam M. A., Jarlborg T. and Palmer S. B. *J. Phys. Cond. Matter* **10,** 10391 (1998).
39. Kubo Y. and Asano S., *J. Magn. Magn. Mat.* **115,** 117 (1992).
40. Duffy J. A., Dugdale S. B., McCarthy J. E., Cooper M. J., Palmer S. B. and Jarlborg T. *submitted to Phys. Rev. Letts.* (1999).
41. Chappert C. and Renard J. P., *Europhys Letts.* **15,** 553 (1991).
42. Foldeaki M., Chahine R. and Bose T. K. *Phys. Rev.* **B52,** 3471 (1995).
43. Tomka G. J., Ritter C., Reidi P. and Kapusta Cz., *Phys. Rev.* **B58,** 6330 (1998).
44. McCarthy J. E., Duffy J.A., Detlefs C., Cooper M. J. and Canfield P., *submitted to Phys. Rev. Letts.* (1999).

# Atomic Auger Spectroscopy: Historical Perspective And Recent Highlights

## W. Mehlhorn

*Fakultät für Physik, Universität Freiburg, D-79104 Freiburg, Germany*

**Abstract:** The non-radiating decay of an inner-shell ionized atom by the emission of an electron was discovered by Pierre Auger in cloud-chamber experiments in the years 1923 to 1926. The first spectroscopic investigation of Auger electrons was performed by Robinson and Cassie in 1926, marking the birth date of Auger spectroscopy. The following seven decades of Auger spectroscopy will be divided into three periods. In the first period (1926-1960) Auger spectroscopy was mainly connected with β-ray spectroscopy where inner-shell ionization of atoms in the solid state was caused either by γ-conversion or by electron capture. The second period (beginning in 1960) is characterized by the external excitation of gas-phase or free metallic atoms, opening Auger spectroscopy to electron energies in the range of few eV to few keV. The third period (beginning in 1977/78) is characterized by the use of synchrotron radiation with its outstanding properties of tunability, polarization and narrow-band high intensity for the excitation and ionization of inner-shell electrons. Finally, two recent highlights of Auger spectroscopy, the interference between photo- and Auger electron with equal energies and an "almost" complete experiment for Auger decay, will be presented.

## THE DISCOVERY

Nicht jeder, der nach Indien fährt,
entdeckt Amerika.[1]
Erich Kästner (1899-1974) (*)

In 1922, Pierre Victor Auger, at age 23, together with his friend Francis Perrin constructed the first cloud chamber in France and studied α-particle tracks for a report in connection with their teacher's examination at the Ecole Normale Supérieure. But Auger wanted to do more research and started his doctorial work in Prof. Jean Perrin's laboratory at the Sorbonne. In 1975 Auger recalls what he intended to investigate in his work:
"My intention was to try to visualize the whole story of an atomic photoexcitation: first, the production of a photoelectron, then the consequent

---

[1] Not everybody, who travels to India,
discovers America.

> emission of a radiation quantum, and the absorption of this quantum, with production of another photoelectron all in the same cloud chamber" (1).

In the following years 1923 to 1926, he did observe the short tracks of secondary electrons in a series of beautiful cloud-chamber investigations on the photoeffect of monochromatic X-radiation in nobel gas atoms (2-4). But the tracks were not outside the primary X-ray beam, as Auger had expected it, they all started from the same points as the primary photoelectrons. At age of 90, Auger recalled the days of his discovery:

> "I ... noticed, in the cloud-chamber snapshots, that photoelectron trajectories created along a beam of X-rays contained a tiny group of small droplets at the start of each trajectory ... I replaced the gas in the chamber by wet hydrogen in order to lengthen the electron tracks, and observed in my photographs that the little group of drops lengthened into a new, very short trajectory, quite well visible, and which originated at the same point as that of the photoelectron ... By adding to the hydrogen atmosphere heavy atoms of such elements as krypton or xenon, I saw the additional tracks lengthening, indicating that the new electron had an energy that was characteristic of the atom absorbing the X-rays, and increased with atomic number" (5).

Auger interpreted these secondary electrons in his thesis, dated 1926 (4), as being due to

> "a non-radiating or adiabatic transition, involving the fall of one electron from L to K and the simultaneous departure of an L electron" (1).

In fact, the possibility of such non-radiating transitions had been predicted theoretically by Rosseland in 1923 (6).

After the discovery of radiationless (Auger) transitions it became clear that Auger electrons had already been seen earlier. M. de Broglie (7) and H.R. Robinson (8) had observed such lines in the magnetic spectra of photoelectrons by absorption of homogeneous X-radiation. But these additional lines, called "fluorescent electron lines", were believed to be due to the external absorption of fluorescence X-radiation following inner-shell photoionization. Only Lise Meitner pointed out in 1923 that (loosely translated)

> "also in the de Broglie experiment (7) the emission of an L electron is from the same Cu atom in which the fluorescent K-radiation was excited by the primary X-rays. According to Rosseland's view it is quite obvious that the full process occurs in one and the same atom" (9).

The first spectroscopic investigation with emphasis on the "fluorescent electron lines", now already known as being due to non-radiating transitions, was by Robinson and Cassie in 1926 (10); this marks the birth date of Auger electron spectroscopy.

The following historical review deals with the development of Auger spectroscopy (AS) of free atoms. Three periods will be distinguished. These began in 1926 with the first spectroscopic investigation of Auger electrons, in 1960 with the Auger spectroscopy of free atoms and in 1977/78 due to the use of synchrotron radiation in AS. The review closes with the presentation of two recent highlights: the interference between photo- and Auger electron as a consequence of the one-step process of photoionization – Auger decay and the performance of an "almost" complete experiment on Auger decay.

## THE FIRST PERIOD

The first spectroscopic investigation of Auger electrons was made by Robinson and Cassie in 1926 (10). The inner-shell ionization of solid-state atoms was produced by Mo K$\alpha$ radiation, the energy analysis and the detection of electrons was done by a magnetic spectrograph with Helmholtz coils for the magnetic field (with momentum resolution of $1 \times 10^{-3}$) and by photographic plates, respectively. The electron lines appeared as heads of bands with long tails on the low-energy side due to energy losses of the electrons in the solid-state target. Many electron lines could be identified as Auger transitions with an initial K vacancy and the two final vacancies being either in LL' or LM', where the prime indicates a level in an atom already ionized in the L shell. The energy resolution of the Robinson spectrograph was good enough to show clearly that the energy of a K-LL Auger electron is correctly given by

$$E(Auger) = E(K) - E(L) - E(L') \qquad (1)$$

instead of

$$E = \{(E(K) - E(L)\} - E(L) = E(K\alpha) - E(L) \qquad (2)$$

as in the case of an external conversion of K$\alpha$ radiation. The electron lines for the two different cases are shown in Fig. 1. The lines marked 'F' due to 'internal absorption' (lower trace in Fig. 1) are Auger electrons and are shifted to smaller energies compared with the lines due to Cu L shell absorption of Cu K$\alpha$ radiation impinging externally on the Cu target (upper trace in Fig. 1).

Experimentally, developments in Auger spectroscopy in the following three and a half decades were mainly connected with developments in $\beta$-ray spectroscopy. The main developments were
- the reduction of the energy losses of the electrons in the solid-state target by using ultra thin sources,
- the design and construction of high-resolution and high-transmission magnetic spectrometers,
- the detection of electrons by counters instead of photographic plates.

As long as Auger spectroscopy was a byline of $\beta$-ray spectroscopy the goal of ultra thin sources could be fulfilled rather easily: the radioactive material was

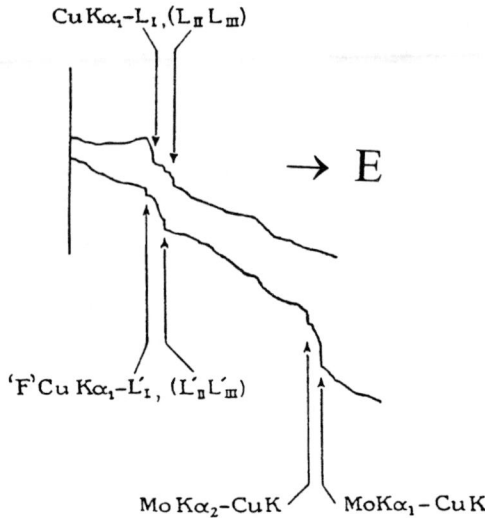

**FIGURE 1.** Electron lines due to 'internal absorption' of Cu K$\alpha_1$ radiation ('F' Cu K$\alpha_1$L$_i$, (L$_{II}$,L$_{III}$)) (lower trace) and due to external absorption of Cu K$\alpha_1$ radiation (upper trace) (from Ref. (10)).

deposited electrochemically or by mass spectrometry on support strips. But the excitation of Auger electrons by external X-radiation in solid-state targets yielded always electron lines with more or less strong low-energy tails, even with very thin evaporated sources. One example is shown in Fig. 2 for Mo K-LL Auger electrons excited by W K$\alpha$ radiation in a 100 Å thick target, energy analysed by an iron-free, double-focusing magnetic spectrometer (momentum resolution of $5 \times 10^{-4}$) and detected by a Geiger-Müller counter (11).

The performance of magnetic spectrometers was greatly improved by introducing the principle of double focusing (12) and using an iron-free coil arrangement. Several iron-free $\pi\sqrt{2}$-magnetic spectrometers were constructed with the radii of the central orbit between 30 cm (13) and 100 cm (14) and with momentum resolution up to $1 \times 10^{-4}$ (14). In Fig. 3 the K-LL Auger spectrum of iodine (15) is shown, taken with the Chalk River instrument (14): The source is prepared by mass separation of radioactive $^{125}$Xe (which decays by electron capture to $^{125}$I) and by depositing it on an aluminium strip with 1.5 keV energy. Although the great improvement of the shape of Auger lines of Fig. 3 compared with those of Fig. 2 is obvious, solid-state line-broadening could never be completely avoided.

On the theoretical side it was G. Wentzel (16) who formulated in 1927 the quantum mechanical theory of radiationless transitions in the nonrelativistic approximation. According to Wentzel the probability of an Auger transition $(n\ell)^{-1} \rightarrow (n_1\ell_1, n_2\ell_2)^{-1} \varepsilon_A \ell_A$ is given by

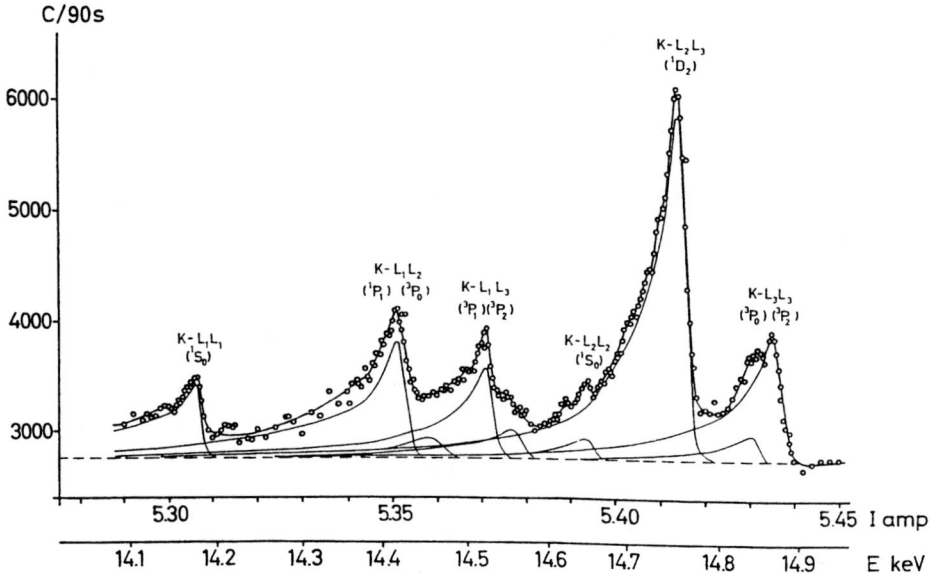

**FIGURE 2.** K-LL Auger spectrum of Mo excited by W Kα radiation (plotted in counts C versus kinetic energy E), thickness of Mo target is 100 Å (from Ref. (11)).

$$P = \frac{2\pi}{\hbar} \left| \langle \Psi_f(n\ell, \varepsilon_A \ell_A; \vec{r}_1, \vec{r}_2) | \frac{e^2}{|\vec{r}_1 - \vec{r}_2|} | \Psi_i(n_1 \ell_1, n_2 \ell_2; \vec{r}_1, \vec{r}_2) \rangle \right|^2 \times \rho(\varepsilon_A) \quad (3)$$

where $\Psi_i$ and $\Psi_f$ are two-electron product wavefunctions, $\varepsilon_A \ell_A$ stands for the Auger electron and $\rho(\varepsilon_A)$ is the density of states per unit of energy. Formula (3) is now commonly referred to as Fermi's Golden Rule although Fermi never claimed this rule for himself (for details see Crasemann (17))

In the first period all calculations of Auger transitions were done with independent-particle wavefunctions using various approximations (mostly hydrogenic and screened hydrogenic wavefunctions) and coupling schemes (jj or LS coupling) (18). An important step was done by Rubenstein in 1955 (19) who calculated Auger rates in LS coupling by using Hartree SCF wavefunctions and employing an electronic computer. The interest in Auger spectroscopy was renewed after Asaad and Burhop (20) in 1958 introduced intermediate coupling in the Auger theory thereby predicting a 9-line K-LL Auger spectrum (instead of a 6-line spectrum for jj coupling or a 5-line spectrum for LS coupling). The 9-line K-LL spectrum was subsequently measured by the research groups in Uppsala (11) (see Fig. 2), Chalk River (15) (Fig. 3) and Stockholm-Brookhaven (21).

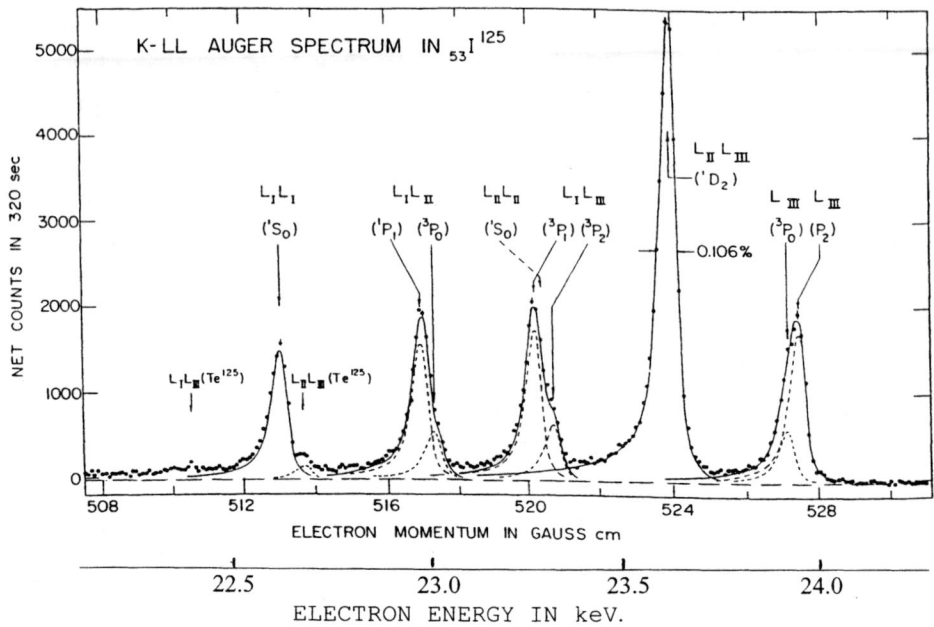

**FIGURE 3.** K-LL Auger spectrum of iodine, target preparation by mass separation of $^{125}$Xe decaying into $^{125}$I. (Reprinted from Graham, R.L., Bergström, I. and Brown, F., Satellites in the K-LL Auger spectra of $_{53}$I$^{125}$ and $_{52}$Te$^{125}$, *Nuclear Physics*, 1962, **39**, 107-123, with kind permission of Elsevier Science – NL, Sara Burgerhartstraat 25, 1055 kV Amsterdam, The Netherlands, and Atomic Energy of Canada Limited).

The main goal of AS in the first period was the determination of intensities and energies of Auger transitions. Due to the lack of low-energy electron detectors and due to severe solid-state line-broadening effects only medium and high-energy Auger spectra could be studied.

## THE SECOND PERIOD

The second period can be considered as a renaissance of the Auger effect. The beginning of this period is connected with the introduction of a novel technique to AS: the external excitation of gaseous targets, the use of electrostatic energy analysers and of detectors for low-energy electrons (22). Honestly, my original intention was not to investigate Auger electrons in the Ph.D. work (22). Indeed, my India, to use an analogy from Erich Kästner, was, in the 1957 planning of my Ph.D., the measurement of photoabsorption coefficients of gases in the soft X-ray region between 10 to 200 Å. In my earlier Diplomarbeit I had measured the intensity of continuous soft X-radiation of a tungsten anode excited by 10 keV electrons. And it was this X-radiation which I intended to use on various gases. The intensities of incident and transmitted X-radiation and its wavelengths ought to be measured via the intensities and energies of photoelectrons of the converter atoms He (a method very similar to PAX suggested later by M.O. Krause (23)). Because He has a small photo cross section I started with Ar in order to get an intense photoelectron signal. Indeed, I got lots of photoelectrons in my cylindrical mirror analyzer but also the Ar $L_{2,3}$-MM Auger electrons. This was then my America.

The use of gas-phase targets and the detection of low-energy electrons were revolutionary, because now AS became possible for electrons in the range of 10 eV to several keV, and this without any solid-state line-broadening effects. This led to many important investigations and new applications of AS:
- AS of K-shell of low-Z elements, e.g. Ne KLL (24,25) (Fig. 4);
- AS of outer shells, e.g. of Ar L (26-28), Kr M (29-31) and Xe N (31) shells (Fig. 5);
- satellite Auger transitions due to multiple ionization, excitation and ionization-excitation (24,25) (Fig. 6);
- satellite Auger transitions due to many-electron effects (electron correlation (30-32) and relaxation (33))
- double Auger transitions, e.g. Ne K-LLL (34) and Ar L-MMM (35);
- AS of molecules, e.g. KLL of $N_2$ (36-38), $O_2$ and CO (37,38).

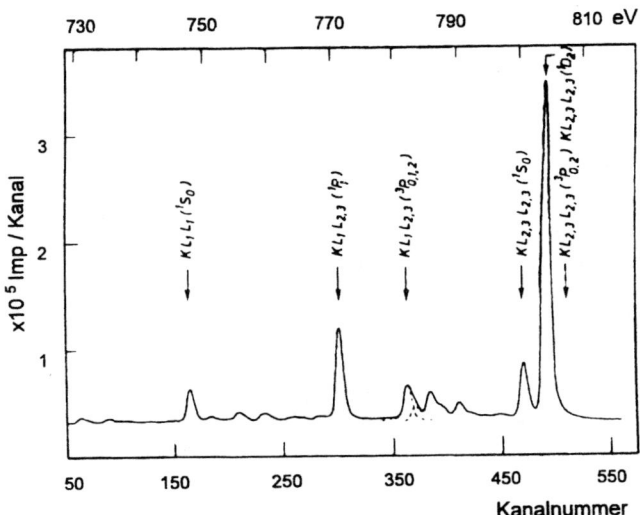

**FIGURE 4.** K-LL Auger spectrum of neon (plotted in counts/channel versus channel number or kinetic energy (upper axis)), excited by 5.7 keV electrons and measured with 0.16 % energy resolution (from Ref. 24 and reprinted from W. Mehlhorn, 70 years of Auger spectroscopy, a historical perspective, *Journal of Electron Spectroscopy and Related Phenomena*, **90**, 1-15 (1998), with permission from Elsevier Science).

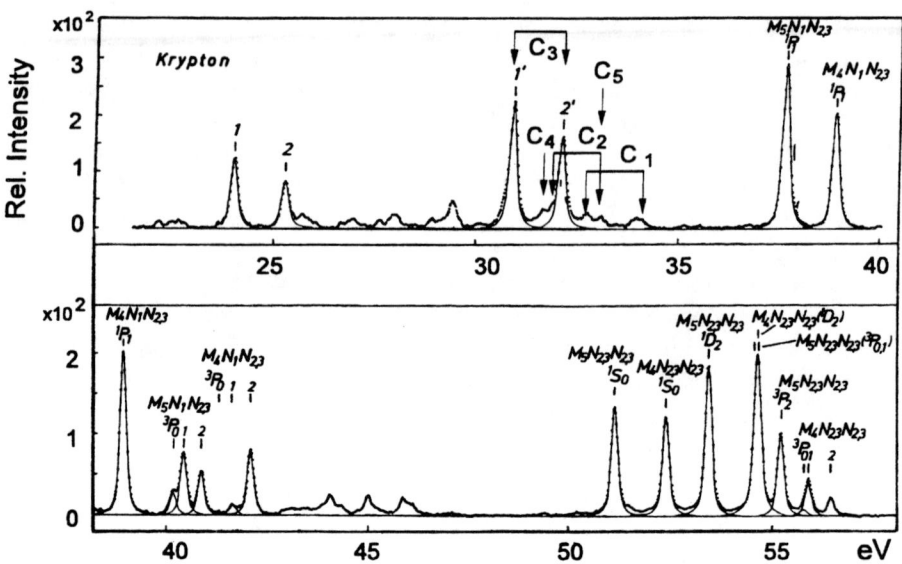

**FIGURE 5.** $M_{4,5}NN$ Auger spectrum of krypton excited by 2.0 keV electron impact (x-axis is kinetic energy). The lines designated by $C_1$ through $C_5$ are satellite lines due to FISCI between the parent configuration $4s4p^{5\ 1,3}P$ and the satellite configuration $4s^24p^3(nd,ns)^{1,3}P$ (from Ref. (30), with permission from Springer-Verlag GmbH & Co. KG).

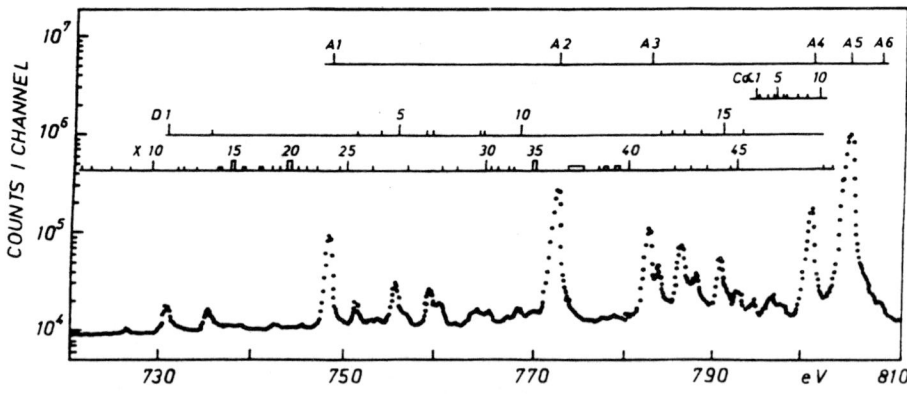

**FIGURE 6.** K Auger spectrum of neon (in counts/channel versus kinetic energy) with diagram lines (A lines) and satellite lines due to KL double ionization (D lines) or K ionization plus $2p \rightarrow 3p$ excitation (3p spectator transitions = $C\alpha$ lines). The satellite lines due $1s \rightarrow nl$ excitations (B-lines) have their positions on the high-energy side of line A5 (from Ref. (25)).

The importance of such investigations was the study not only of the structure and the dynamics of atoms via the diagram Auger transitions, but especially also of electron correlation via the satellite and double Auger transitions.

During this period the experimental technique of AS was continuously improved by increasing the energy resolution of electrostatic analysers to $3 \times 10^{-4}$, using position-sensitive detectors (39) and introducing the metal-vapour target (33,40). The metal-vapour targets opened not only the study of the structure and dynamics of many more elements (41), but also the investigation of solid-state effects on line positions, intensities and widths by comparison of free atomic with solid-state Auger spectra (42,43).

From the middle of the sixties another branch of AS of free atoms developed strongly: the study of ion-atom collisions by using satellite Auger transitions as fingerprint for multiple inner-shell ionization in the target atom (44,45) as well as in the projectile atom (46). The first electron-spectroscopic investigations were made by Rudd using protons or $He^+$ projectiles on He or heavier elements (47,48). He was also the first to observe the doubly-excited autoionizing states of He via electron

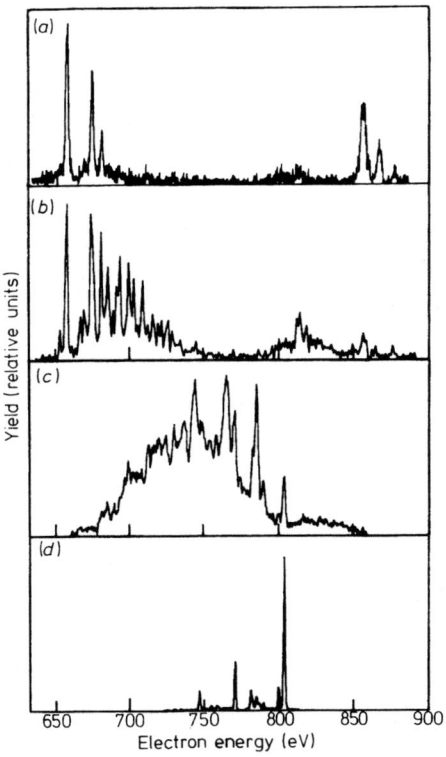

**FIGURE 7.** Comparison of Ne K Auger spectra produced by various projectiles: (a) 200 MeV $Xe^{31+}$, (b) 45 MeV $Cl^{12+}$, (c) 30 MeV $O^{5+}$, (d) 4.2 MeV $H^+$ (from Ref. (45), with permission from IOP Publishing Limited).

spectroscopy (47), but only in later experiments were the asymmetric Fano lineshapes observed (49). The collision of heavy ions with target atoms, e.g. 30 MeV $O^{5+}$ + Ne, led to a distribution of $KL^n$ vacancies in Ne with n ranging from 0 to 6 resulting in a very complex Auger spectrum (Fig. 7c). Useful Auger spectra of target atoms in terms of reasonably resolved lines were obtained either for energetic light projectiles (e.g. 4.2 MeV $H^+$ + Ne, see Fig. 7d) or for energetic highly-ionized heavy projectiles (e.g. 200 MeV $Xe^{31+}$ + Ne, see Fig. 7a). The former collision leads mainly to singly K-shell vacancies whereas the latter collision produces $KL^n$ vacancies with $n \geq 6$.

In case of AS of projectile ions the Doppler effect of the moving emitter causes a kinematic broadening of Auger lines. Performing AS at extreme forward angles of emission relative to the projectile beam reduces this broadening (50) and brings it to a minimum at zero-degree emission. Mainly single inner-shell ionization of the projectile ion can be achieved in collisions with He atoms. This method (called needle ionization) yields, together with the zero-degree AS, high resolution Auger spectra for the different charge states of the incident projectile (51) (Fig. 8). This subject matter has been reviewed (52).

Finer details of the Auger decay are the anisotropic angular distribution and the spin-polarization of Auger electrons. These properties had been predicted in 1968/74 (53,54) and in 1980/81 (55,56) and they were confirmed in 1974 (57) and in 1985/90 (58,59) and 1993 (60), respectively. The necessary condition for an anisotropic Auger emission is the alignment of the initial Auger state with total angular momentum $J > \frac{1}{2}$. An alignment is generally obtained in the ionization by a directed beam of particles (53) or photons (61). Spin-polarization of Auger electrons occur if the initial Auger state is either aligned or oriented (55,56). For example, treating the process ionization-Auger decay as a two-step process, the angular distribution of Auger electrons from the decay of Auger state with total angular momentum $J \leq 3/2$ is given by (54,62)

$$I(\theta) = I_0 (1 + A_{20} \alpha_2 P_2(\cos\theta)). \qquad (4)$$

Here $A_{20}$ is the alignment parameter of the initial Auger state, given by the cross sections of magnetic substates $|M|$; $\alpha_2$ is the Auger decay parameter, given by the amplitudes and relative phases of Auger partial waves $\ell_A$; and $P_2(\cos\theta)$ is the second Legendre polynomial with angle $\theta$ relative to the quantization axis. A measurement of the angular distribution (4) could then give information either on the ionization process via the alignment parameter $A_{20}$ or on the dynamics of the decay process via the decay parameter $\alpha_2$.

On the theoretical side the renaissance of the Auger effect was caused by
- the availability of increasingly powerful computers which led to systematic calculations of Auger transition probabilities in the independent-particle model using various approximations and coupling schemes of the two-electron wavefunctions of eq. (3) (screened hydrogenic wf. (63); Hartree-Fock-Slater wf. (64-66)); a review is given by Chen (67);

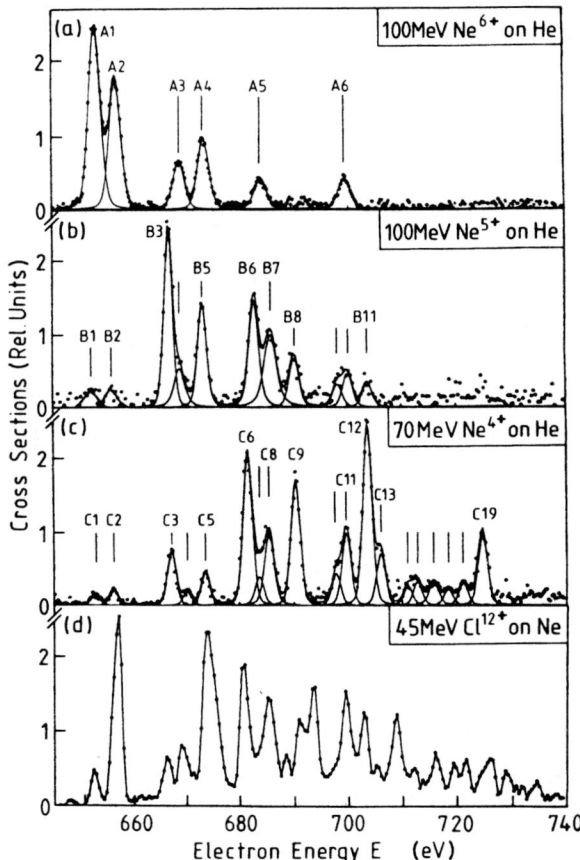

**FIGURE 8.** (a)-(c): K Auger spectra of Ne projectile ions produced in collisions of Ne$^{q+}$ on He for zero-degree observation angle relative to the projectile beam. (d): K Auger spectrum of Ne target atoms produced by 45 MeV Cl$^+$ + Ne collisons (from Ref. (51)).

- introduction of many-electron concepts into the theory of Auger transitions, e.g. configuration interaction (CI), many-body perturbation theory (MPBT) or random-phase approximation (RPA), in order to bring the theoretical results into better agreement with the experiment. The first step in this direction was done by W.N. Asaad (68) who introduced in 1965 the CI between the final Auger states 2s$^{-2}$ $^1S_0$ and 2p$^{-2}$ $^1S_0$ and removed thereby the main discrepancy between theoretical and experimental K-LL Auger intensities. Full agreement of theoretical and experimental K-LL intensities was only obtained by a MBPT calculation (69) where CI between the final continuum channels (FCSCI) was also included.

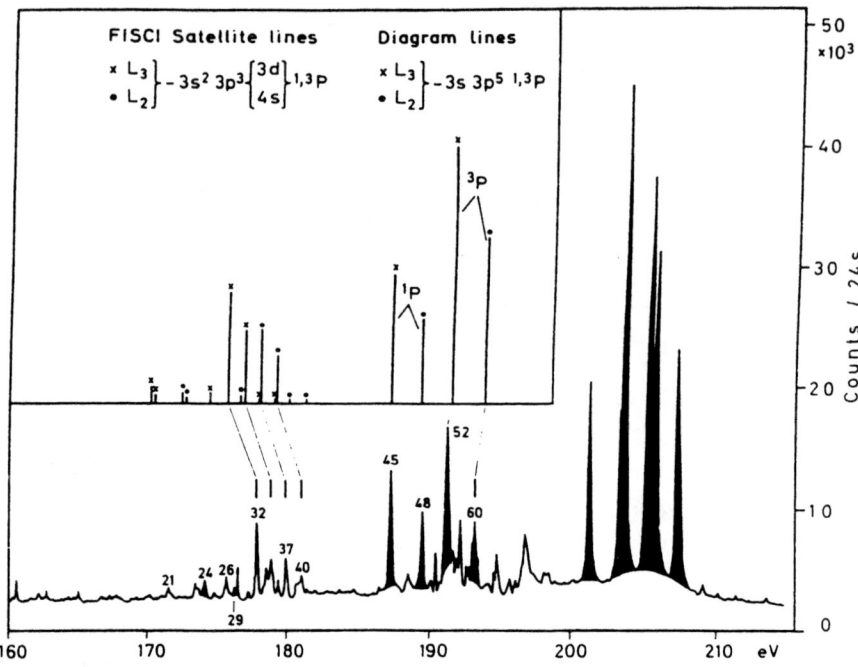

**FIGURE 9.** $L_{2,3}$-MM spectrum of Ar (plotted in counts versus kinetic energy), excited by 3 keV electrons (28)). The diagram lines are the black solid peaks. The FISCI satellite lines and their parent diagram lines $L_{2,3}$-3s3p$^5$($^{1,3}$P) are indicated by the bar spectrum, their intensities and energies are relative to the experimental $L_{2,3}$-3s3p$^5$ $^1$P diagram lines 45 and 48 (71). (From Ref. (32)).

The introduction of CI into the theory of the Auger transitions did not only improve the intensities of diagram transitions but explained also the main features of the satellite Auger spectrum. Specifically, strong CI exists between the final ionic diagram states sp$^5$($^{1,3}$P) and excited satellite states s$^2$p$^3$(s',d)$^{1,3}$P (70). Due to this final ionic state CI (FISCI) an energy shift of diagram lines results and intensity is transferred from the diagram lines to the FISCI satellite lines that have s$^2$p$^3$(s',d)$^{1,3}$P as final states (71). One example is the $L_{2,3}$-3s3p$^5$ spectrum of Ar (Fig. 9). In a multi-configuration final-state calculation the positions and intensities of diagram and FISCI satellite lines (bar spectrum in Fig. 9) were obtained (71) explaining correctly the main features of satellite lines between 171 and 182 eV. Another example is the $M_{4,5}$-$N_1N_{2,3}$ spectrum of Kr (Fig. 5).

Until then, all calculations of Auger transitions have been performed in the two-step model where the decay process is treated independently from the ionization process by using Wentzel's *ansatz* (3). Åberg (72) and Åberg and Howat (73) introduced a treatment of the Auger transition in the broader context of resonance scattering theory. In the Åberg approach the ionization followed by the Auger decay is treated as a coherent one-step process (like the excitation process-autoionization decay in the Fano theory (74)). Accordingly, the transition amplitude T for the process ionization (for simplicity we consider photoionization) – Auger decay has a direct part and a resonance part and is given (in atomic units) by (73)

$$T = \langle A^{++}\varepsilon\varepsilon_A, 0 | H_{int} | A, \omega \rangle$$
$$+ \int_0^\infty \frac{\langle A^{++}\varepsilon\varepsilon_A, 0 | H_{el} - E | A^+\tau, 0 \rangle \langle A^+\tau, 0 | H_{int} | A, \omega \rangle}{\varepsilon + \varepsilon_A - \varepsilon_A^0 - \tau + i\Gamma/2} d\tau. \quad (5)$$

The wavefunction $A^+\tau$ stands for the joint (N-1)-electron Auger state $A^+$ and the outgoing photoelectron with yet unspecified energy $\tau$. The operator $H_{int}$ stands for the interaction of the photon with energy $\omega$ with the atom. The operator $H_{el}$ is the full N-electron Hamiltonian and leads to the final double continuum state $A^{++}\varepsilon\varepsilon_A$, with $E_{A^{++}}$ the energy of the doubly-charged ion and two electrons with energy $\varepsilon$ (photoelectron) and $\varepsilon_A$ (Auger electron). The energy $\varepsilon_A^0 = E_{A^+} - E_{A^{++}}$ is the nominal energy of the Auger electron, the total width of the Auger state $A^+$ is denoted by $\Gamma$.

The direct amplitude $\langle A^{++}\varepsilon\varepsilon_A, 0|H_{int}|i,\omega\rangle$ in eq. (5) describes double photo-ionization. Since $H_{int}$ is a one-electron operator it vanishes if $|i\rangle$ and $|A^{++}\varepsilon\varepsilon_A\rangle$ are constructed in the independent-electron approximation from orthogonal orbitals. It can thus be neglected in comparison with the resonance amplitude. The denominator of the resonance amplitude can be written with the excess energy $E_{exc} = \omega - E_{A^+}$ as $E_{exc} - \tau + i\Gamma/2$ leading to peaking of the amplitude for $\tau \approx E_{exc}$.

According to Åberg (75) there are three regions of excess energy $E_{exc}$:
- $E_{exc} < 0$, i.e. $\omega$ is tuned resonantly to a Rydberg state below the inner-shell ionization threshold. This leads to resonant Auger transitions and to the resonant Raman Auger effect (RRAE).
- $E_{exc} > 0$, but $< \varepsilon_A$. This is the region of post collision interaction (PCI), where the outgoing photoelectron interacts with the Auger electron yielding line shifts and asymmetric line shapes of the Auger electron and a recapture of the photoelectron leading to singly-charged ions.
- $E_{exc} > \varepsilon_A$. Here the energy $\tau$ of the photoelectron is always larger than the energy $\varepsilon_A$ of the Auger electron, i.e. the photoelectron always moves in the field of an effective charge +1, and, as a consequence, there is no PCI effect. This region is called two-step region. The differential cross section of an Auger line is then given by the product of photoionization cross section $\sigma(\tau)$ and a Lorentzian Auger line shape

$$\frac{d\sigma}{d\varepsilon_A} = \sigma(\tau) \frac{\Gamma/2\pi}{(\varepsilon_A - \varepsilon_A^0)^2 + \Gamma^2/4}. \quad (6)$$

Here $\Gamma$ is the level width due to the decay of Auger state $A^+$ into the continuum

$A^{++}\varepsilon\varepsilon_A$, as given by eq. (3).

From the above classification of regions of excess energy it follows that the Auger decay following near-threshold excitation or ionization is correctly described by the single-step approach of eq. (5). Experimentally, such processes could be studied successfully only after tunable and narrow-band synchrotron radiation with large enough intensity became available.

## THE THIRD PERIOD

The third period is characterized by the availability of synchrotron radiation (SR) and its use in Auger spectroscopy. Due to the outstanding properties of SR (wide range of tunability, polarization characteristics, high-intensity narrow-band radiation from undulators) many details of the dynamics of atoms and even new physics could be studied for the first time. Although the first photoabsorption experiments using SR date back to the mid sixties (Madden and Codling (76)) the third period of AS started only in 1977/78, when SR was tuned closely above the Xe 4d threshold for the first study of the post-collision interaction (PCI) following photoionization (77) or was tuned to the excitation of Xe $4d_{3/2,5/2} \to 6p$ or Kr $3d_{3/2,5/2} \to 5p$ for the first study of resonant Auger transitions (78).

Since the first investigation (77) of PCI via the shift of the Auger line to higher energies many more investigations followed. The first calculation of the PCI lineshape and shift of Auger electrons following photoionization was done by Niehaus (79) using a semi-classical theory. Our experimental and theoretical understanding of PCI after photoionization is well demonstrated with Fig. 10. Here, the "angle-averaged" PCI energy shift and shape for the Xe $N_5O_{2,3}O_{2,3}$ ($^1S_0$) Auger line, measured in non-coincidence experiments (80,81), are compared to theoretical results predicted by a semiclassical theory (82) and by a quantum theory (83) based on Åberg's one-step model (eq. (5)). Both theories took into account the time it takes for the Auger electron to pass the previously emitted slow photoelectron (84) which gives zero "angle-averaged" PCI effect if $E_{exc} > \varepsilon_A$. In this case the Auger electron never overtakes the photoelectron and therefore no energy will be exchanged ("no passing" effect). On the other hand, the theoretical predictions without the "no passing" effect (79,85,86) (dashed curves in Fig. 10) do not agree with the experiment. In contrast to the angle-averaged PCI, the angle-dependent PCI effects do not vanish for $E_{exc} > \varepsilon_A^0$. These have been predicted by theory (87,88) and confirmed by coincidence experiment (89). Also the third PCI effect, the recapture of the photoelectron, has been studied for the case Ar 2p ionization (90-92). The quantum-mechanical calculation (93) on the basis of eq. (5) yielded results in excellent agreement with the experiments.

After the first pioneering experiment on resonant Auger transitions of Xe($4d_{3/2,5/2}6p$) and Kr($3d_{3/2,5/2}5p$) by Eberhardt et al. in 1978 (78), little was done until 1986 when the interest in studying these resonantly excited Auger transitions increased rapidly (for references up to 1990 see (94)). This was due to the availability of high-intensity

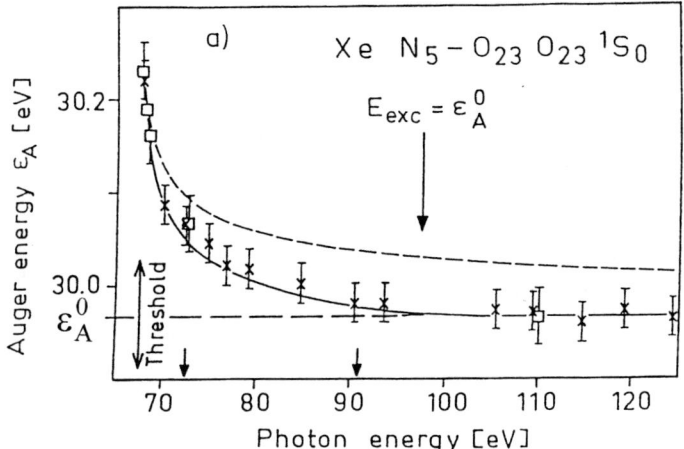

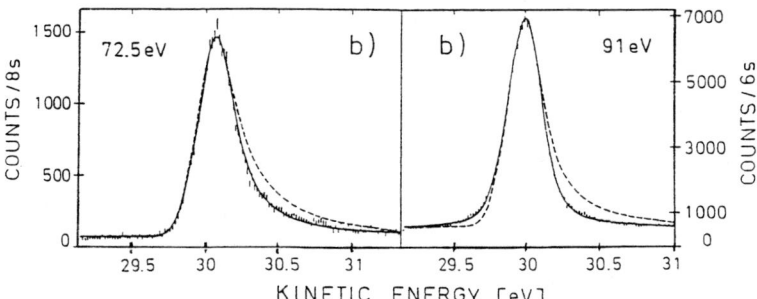

**FIGURE 10.** (a) Angle-averaged PCI shift of the $N_5$-$O_{2,3}O_{2,3}(^1S_0)$ Auger lines of Xe. Experiment: x = (80), ☐ = (81). Theory: ___ = with "no passing" effect (82,83), --- = without "no passing" effect (79,85,86). (b) Experimental and theoretical PCI-distorted Auger line-shapes corresponding to the photon energies marked in (a) by arrows. For explanation of theoretical curves see (a). (Reprinted (abstracted) with permission from Mehlhorn, W., Recent developments in radiationless transitions. In *Proc. Fiftheenth Intern. Conf. on X-Ray and Inner-Shell Processes*, AIP Conf. Proceedings 215, 1990, p. 465. Copyright 1990 American Institute of Physics).

synchrotron radiation from wigglers or even undulators with sufficiently small bandpass. Also the angular distribution of the resonant Auger electrons was studied. The first experiment, performed in 1988/89 by Carlson et al. (95), was followed by experiments by the groups of V. Schmidt (96) and U. Becker (97,98).

A real breakthrough in the investigation of resonant Auger transitions and their angular distribution was the use of the resonant Raman Auger effect (RRAE). The

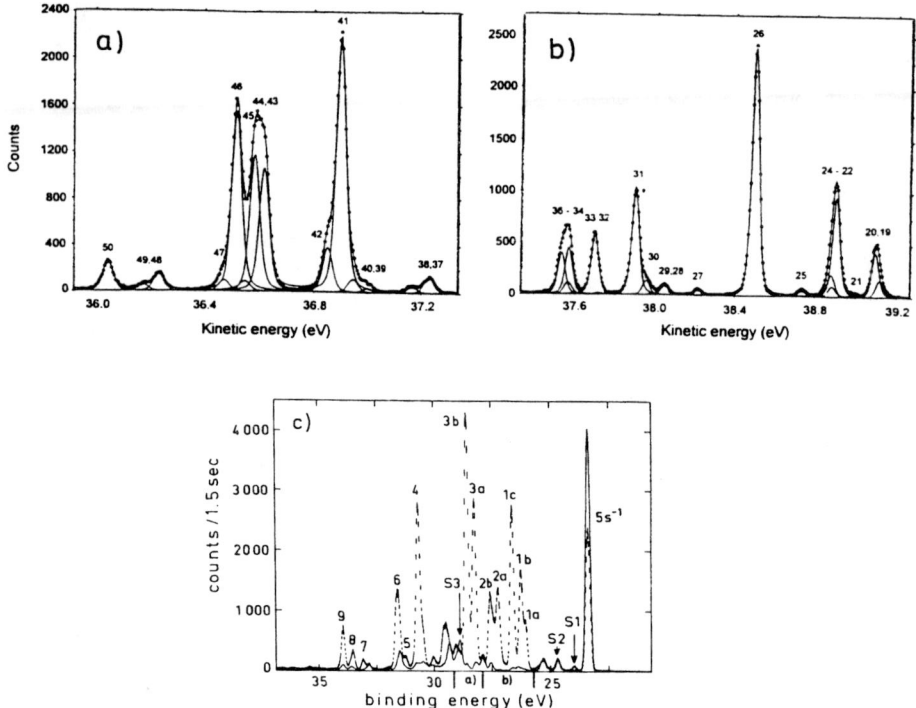

**FIGURE 11.** Comparison of the resonant Auger spectra for Xe $4d_{5/2} \to 6p$ excitation obtained (a,b) with 8 meV bandwidth of incident radiation (106) and (c) with 130 meV bandwidth of incident radiation (96). The energy ranges of (a) and (b) are indicated below the energy axis of (c).
(a) and (b) from Ref. (106), (c) from Ref. (96).

characteristics of RRAE are the linear dispersion of the resonant electron lines with incident photon energy and, more important, the narrowing of the resonant Auger lines to subnatural widths if the bandwidth of the exciting radiation is smaller than the natural width of excited state. Both characteristics are predicted by the Åberg one-step model of eq. (5) and were suggested by first experiments on $2p \to 5d$ excitation in Xe (99,100). Using the very narrow high-intensity radiation from the MAX source in Lund the RRAE was beautifully demonstrated in an investigation of the Kr $M_{4,5}$ and Xe $N_{4,5}$ resonant Auger spectra (101). Many more investigations of resonant Auger electrons (102-107) and of their angular distributions (108-111) utilizing RRAE followed. In Fig. 11 the power of the RRAE is demonstrated on the resonant Auger spectrum for Xe $4d_{5/2} \to 6p$ excitation using radiation with bandwidth of 8 meV (Fig. 11a,b) (106) and of 130 meV (Fig. 11c) (96), the Xe 4d width is 120 meV.

A complete mapping of Auger electron intensities has been measured for excitation/ionization in the vicinity of the Xe $4d_{5/2}$ threshold (112). The incident radiation was changed in small steps from below the ionization threshold at 67.55 eV

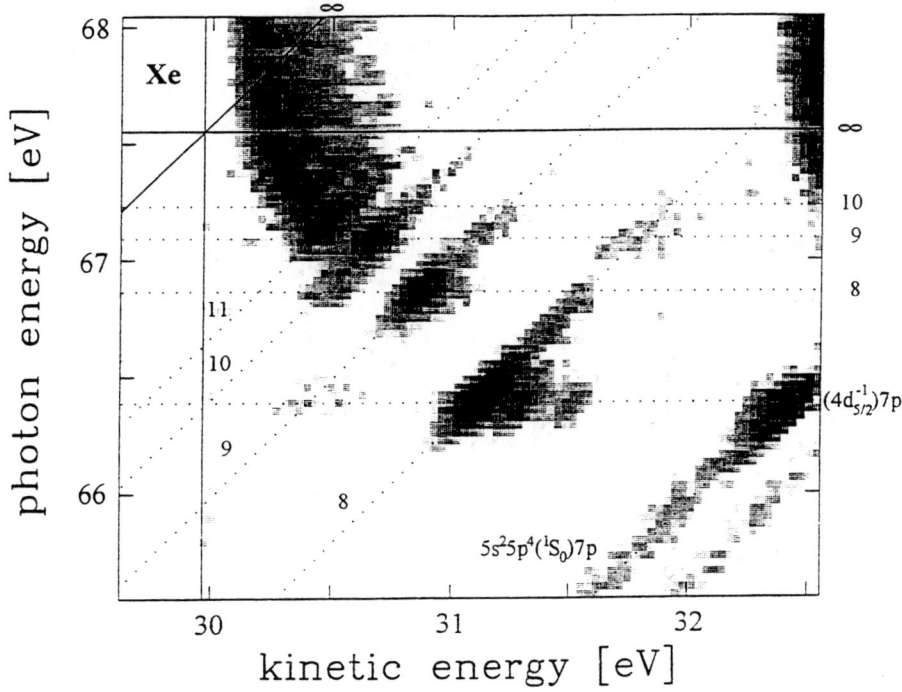

**FIGURE 12.** Grey-scale plot of the Auger electron yield as function of the kinetic energy of the Auger electron in the vicinity of the Xe $N_5$-$O_{2,3}O_{2,3}$ $^1S_0$ Auger line and of the photon energy in the vicinity of the Xe $4d_{5/2}$ ionization threshold. The electron intensity is shown increasing from light to dark. The electrons were detected at $\theta = 55°$ relative to the poton polarization direction (from Ref. (112), with permission from IOP Publishing Limited.).

to above the threshold; its bandpass was 140 meV at 67 eV photon energy and therefore slightly larger than the $4d_{5/2}$ width of 120 meV. Fig. 12 shows the two-dimensional grey-scale plot of Auger electron intensities (with increasing intensity from light to dark) as function of the kinetic energies of Auger electrons in the vicinity of the $N_5$-$O_{2,3}O_{2,3}$ $^1S_0$ Auger line for increasing energy of incident photons (from bottom to top). The region of resonant Auger transitions is below the $4d_{5/2}$ threshold. Here mainly the resonant Auger lines $4d_{5/2}$ np $^1P \to 5s^24p^4(^1S_0)$mp can be seen. The importance of np $\to$ mp shake-up spectator transitions and the linear dispersion of Auger energy with incident photon energy (diagonal intensity distributions), a typical feature of RRAE, can be clearly seen. Above the $4d_{5/2}$ threshold the PCI line shift of the $N_5$-$O_{2,3}O_{2,3}$ $^1S_0$ Auger line is clearly visible (the vertical line at 29.97 eV is the nominal energy $\varepsilon_A^0$ without PCI). Also, there is a continuous evolution from the region of the resonant Auger transitions to the region of PCI distorted Auger lines; a fact pointed out by V. Schmidt as early as 1982 (113). Unquestionably, this rather complex behaviour of electron intensity distribution can be understood quantitatively only within the one-step model by Åberg.

More recently, the evolution of the post-collision-interaction profile of the Xe $N_5$-$O_{2,3}O_{2,3}(^1S_0)$ Auger line from the resonant Auger shake-up/shake-down lines was measured utilizing the RRAE (114), the bandpass of incident radiation was about 6 meV. The RRAE allowed to resolve not only the np → mp spectator shake-up transitions up to m = 16, but also new intense fine structures. These fine structures were suggested to be due to shake transitions np → ms and np → (m-1)d, besides np → mp, caused via final continuum state CI (FCSCI).

# RECENT HIGHLIGHTS OF AUGER SPECTROSCOPY

## Interference effects between photo- and Auger electron

In the photoionization process with subsequent Auger decay one can clearly distinguish between photo- and Auger electron if the energies of both electrons are different. But if the incident SR is tuned such that both electrons have equal energies they are indistinguishable and one expects interference. This interference effect has been predicted by Vegh et al. (115) and Vegh and Macek (116). It is due to the exchange amplitude of the antisymmetric description of the two electrons $e_a$ and $e_b$ ejected with momenta $\vec{k}_a$ and $\vec{k}_b$ and can properly be understood only in the one-step model of Åberg. Vegh and Macek (116) calculated the interference effect in the angular correlation between photoelectrons and Xe $N_5$-$O_{2,3}O_{2,3}$ $^1S_0$ Auger electrons of the same energy $\varepsilon_A^0 = 29.97$ eV ($\varepsilon_A^0$ is the nominal energy of Xe $N_5$-$O_{2,3}O_{2,3}$ $^1S_0$ Auger electrons). The incident photons have thus an energy $h\upsilon = E_B(Xe\ N_5) + 29.97$ eV = 67.55 eV + 29.97 eV = 97.52 eV. Vegh and Macek predict for the observation at the relative angle $\theta_{ab} = 180°$ between the two electrons $e_a$ and $e_b$ a vanishing coincidence signal for $\varepsilon_a = \varepsilon_b = \varepsilon_A^0$. This interference effect occurs only if $\varepsilon_a - \varepsilon_b < \Gamma(Xe\ 4d_{5/2}) = 120$ meV.

The first observation of a reduction of the coincidence signal between the two electrons $e_a$ and $e_b$ at relative angle $\theta_{ab} = 180°$ was made by Schwarzkopf and Schmidt (117). Because the bandpass $\Delta(h\upsilon) = 225$ meV of the incident photons and the energy resolutions of the two spectrometers were considerably larger than $\Gamma(Xe-N_5)$ the observed effect was smaller than predicted. More recently, Viefhaus et al. (118) also measured the reduction of coincidence signal between the same pair of photoelectron-Auger electron at relative angles $\theta_{ab} = 180°$. Due to a small bandpass of incident photons, $\Delta(h\upsilon) < 20$ meV, and high energy resolutions of the time-of-flight electron analyzers the overall coincidence resolution was 80 meV and thus smaller than $\Gamma(Xe\ N_5) = 120$ meV. The obtained results are shown in Fig. 13. From the two-dimensional coincidence map of Fig. 13a the coincidence intensity of the final $O_{2,3}O_{2,3}$ $^1S_0$ state versus the kinetic energy of electron $e_a$ can be evaluated (Fig. 13c). Here a clear vanishing of the coincidence signal has been obtained, the dashed line through the

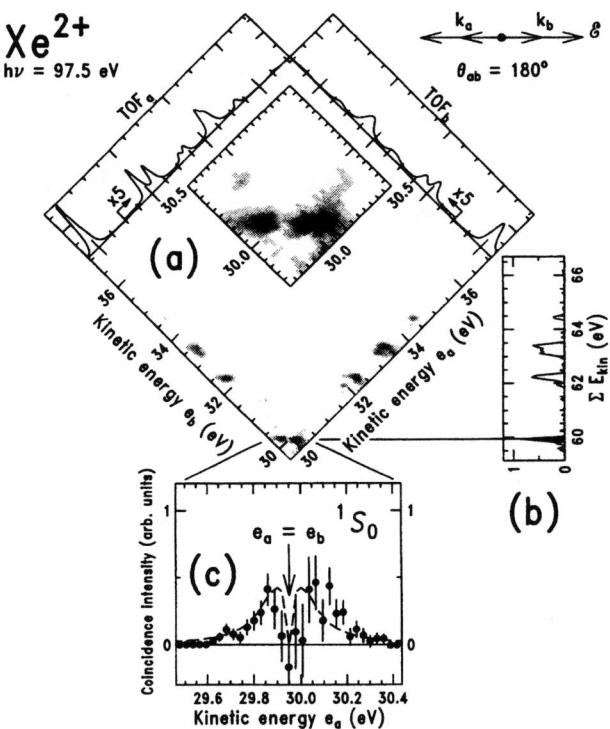

**FIGURE 13.** Coincidence spectra obtained for electrons $e_a$ and $e_b$ of the process $\gamma$ (h$\upsilon$ = 97.5 eV) + Xe $\rightarrow$ Xe$^{2+}$(5p$^{-2}$ $^1S_0$) + $e_a$ + $e_b$. (a) Two-dimensional coincidence map together with an enlarged inset of the interference region and the two non-coincident spectra of the corresponding electron analyzers, (b) integrated coincidence counts versus the sum of the kinetic energies of the two electrons and (c) the coincidence intensity of the $^1S_0$ final state (dark shaded area in (b)) versus the kinetic energy of electron $e_a$. The broken curve is the scaled theoretical result of Sheinerman and Schmidt (119) (from Ref. (118)).

experimental values is the expected coincidence signal if the PCI effect between electrons $e_a$ and $e_b$ is taken into account (119). For more details see V. Schmidt (120).

## An "almost" complete experiment for Auger decay

A complete experiment refers to the experimental determination of the quantum-mechanical amplitudes characterizing the process. The first "almost" complete experiments for Auger decay were performed only recently (121-123). In these investigations the ratio of the absolute values of the amplitudes and the relative phases have been determined; hence the term "almost" is used. Here I will briefly report on our investigation (123) of the Auger decay

$$\text{Na}^+(2s2p^64p\ ^3P) \to \text{Na}^{++}(2s^22p^5\ ^2P_{3/2},^2P_{1/2}) + e_A(\varepsilon s,\varepsilon d) \qquad (6)$$

following 2s-ionization of laser-excited $\text{Na}^*(3p\ ^2P_{3/2})$ atoms by electron impact. In LS coupling the Auger decay (6) is described by two amplitudes, $V_s = |V_s|\exp(i\Delta_s)$ and $V_d = |V_d|\exp(i\Delta_d)$. We measured two independent quantities, the ratio R of integral intensities of transitions $^3P \to ^2P_{1/2}$ and $^3P \to ^2P_{3/2}$ and the decay parameter $\alpha_2$ of transition $^3P \to ^2P_{3/2}$. These quantities are expressed by two theoretical quantities, the ratio $r = |V_s|/|V_d|$ of moduli $|V_s|$ and $|V_d|$ and the relative phase $\Delta_{sd} = \Delta_s - \Delta_d$ (123):

$$R = \frac{W(^3P \to ^2P_{1/2})}{W(^3P \to ^2P_{3/2})} = \frac{4+r^2}{5+8r^2}, \qquad (7)$$

$$\alpha_2(^3P \to ^2P_{3/2}) = \frac{8\sqrt{2}\,r\cos\Delta_{sd} - 1}{5+8r^2}. \qquad (8)$$

From the experimental values $R_{exp} = 1 : 6.6(3)$ and $(\alpha_2)_{exp} = -0.30(6)$ the values $r_{exp}^2 = 15.3\ ^{+5.2}_{-3.4}$ and $(\Delta_{sd})_{exp} = 147°\ ^{+33°}_{-21°}$ or $213°\ ^{+21°}_{-33°}$ have been evaluated. The theoretical values, obtained with wavefunctions which include both the correlation and the core polarization(124), are $r_{th}^2 = 16.0$ and $(\Delta_{sd})_{th} = 143°$ and agree well with the experimental results.

## OUTLOOK

I had the privilege to do Auger research from the very beginning of the second period and due to the novel experimental technique the field of AS was wide open at that time: almost everything which was investigated led to new physics. At that time Auger intensities were measured. Nowadays we will measure the individual Auger amplitudes by performing complete experiments. A first step was done by performing an "almost" complete experiment for the Auger decay $\text{Na}^+(2s2p^64p\ ^3P) \to \text{Na}^{++}(2s^22p^5\ ^2P) + e_A$ following 2s ionization of laser excited $\text{Na}^*(3p\ ^2P_{3/2})$ by electron impact. Other possibilities for an "almost" complete experiment for Auger decay have been suggested and carried out: the combined determination of angular distribution and spin-polarization of resonant Auger electrons (125), the measurement of angular correlation in the cascade autoionization-photoemission (121,122) or in the cascade of two Auger transitions (126). I am very optimistic that with the further development of incident beams (e.g. circularly polarized photons, spin-polarized electrons) and with the preparation of target atoms (e.g. laser-excited atoms (127-130)) new interesting physics is still waiting to be discovered.

In the present historical review I dealt only with atomic Auger spectroscopy but the tree of Auger spectroscopy has many branches (131). The full range and importance of Auger spectroscopy in atomic, molecular, solid state and surface physics was demonstrated in a Symposium on the Auger Effect held in 1989 in honor of Pierre Auger and on the occasion of his 90$^{th}$ birthday. I believe that Pierre Auger, who left the field very soon after his discovery (5,132) and who died in 1993 at the age of 94, could have been very satisfied with the expanding range of scientific advances generated by his original work.

## ACKNOWLEDGEMENTS

I am indebted to B. Crasemann and to N.M. Kabachnik for sending me a copy of reference (5) and a preprint of reference (126), respectively. I thank L.J. Dubé for reading the manuscript. This work is supported in part by the Deutsche Forschungsgemeinschaft.

## REFERENCES

(*) from: Erich Kästner, *Kurz und bündig*, © Atrium Verlag, Zürich und Thomas Kästner.
1. Auger, P., *Surface Science* **48**, 1 (1975).
2. Auger, P., *Compt. Rend.* (Paris) **177**, 169 (1923).
3. Auger, P., *Journ. Physique Radium* **6**, 205 (1925).
4. Auger, P., *Ann. Phys.* (Paris) **6**, 183 (1926).
5. Auger, P., "La Recherche et la Creation du Nouveau", address at the Symposium on the Auger Effect, l'Université Pierre et Marie Curie, Paris, 1989 (unpublished).
6. Rosseland, S., *Z. Physik* **14**, 173 (1923).
7. de Broglie, M., *J. Phys. (France)* **2**, 265 (1921).
8. Robinson, H.R., *Proc. Roy. Soc. London* **A104**, 455 (1923).
9. Meitner, L., *Z. Physik* **17**, 54 (1923).
10. Robinson, H.R., and Cassie, A.M., *Proc. Roy. Soc. London* **A113**, 282 (1926).
11. Hörnfeldt, O., Fahlman, A., and Nordling, C., *Ark. Fysik* **23**, 155 (1962).
12. Svartholm, N., and Siegbahn, K., *Arkiv f. Nat. Astr. Fys.* **33A**, 21 (1946).
13. Siegbahn, K., and Edvarson, K., *Nucl. Phys.* **1**, 137 (1956).
14. Graham, R.L., Ewan, G.T., and Geiger, J.S., *Nucl. Instr. Meth.* **9**, 245 (1960).
15. Graham, R.L., Bergström, I., and Brown, F., *Nucl. Phys.* **39**, 107 (1962).
16. Wentzel, G., *Z. Physik* **43**, 524 (1927).
17. Crasemann, B., A Century of X-Rays in Atomic Physics, in *Proceedings of 17$^{th}$ International Conference on X-Ray and Inner-Shell Processes*, (ed. R.L. Johnson, H. Schmidt-Böcking and B.F. Sonntag), Woodbury, N.Y., AIP Press, 1997, p. 3.
18. For a review see Crasemann, B., Auger and Radiative Transition Probabilities, in *Proc. Intern. Conf. on Inner-Shell Ionization Phenomena and Future Applications* (United States Atomic energy Commission, CONF-720404, Vol. 1, 1972) p. 9.
19. Rubenstein, R.A., Ph.D. Thesis, University of Illinois, 1955 (unpublished).
20. Asaad, W.N., and Burhop, E.H.S., *Proc. Phys. Soc. London* **71**, 369 (1958).
21. Erman, P., Bergström, I., Chu, Y.Y., and Emery, G.T., *Nucl. Phys.* **62**, 401 (1965).
22. Mehlhorn, W., *Z. Physik* **160**, 247 (1960).
23. Krause, M.O., Photoelectron Spectroscopy: A New Approach to X-Ray Analysis, in *Advances in*

*X-ray analysis*, Vol. 16 (Plenum , New York, 1973) p. 74.
24. Körber, H., and Mehlhorn, W., *Z. Physik* **191**, 217 (1966).
25. Krause, M.O., Carlson, T.A., and Moddeman, W.E., *J. Phys. (Paris)* **32**, C4-139 (1971).
26. Mehlhorn, W., *Z. Physik* **208**,1 (1968).
27. Mehlhorn, W., *Z. Physik* **217**, 294 (1968).
28. Werme, L.O., Bergmark, T., and Siegbahn, K., *Physica Scripta* **8**, 149 (1973).
29. Mehlhorn, W., *Z. Physik* **187**, 21 (1965).
30. Mehlhorn, W., Schmitz, W., and Stalherm, D., *Z. Physik* **252**, 399 (1972).
31. Werme, L.O., Bergmark, T., and Siegbahn, K., *Physica Scripta* **6**, 141 (1972).
32. Mehlhorn, W., Auger-electron spectrometry of core levels of atoms, in B. Crasemann (Ed.), *Atomic Inner-Shell Physics*, Plenum Press, 1985, p.119 .
33. Hillig, H., Cleff, B., Mehlhorn, W., and Schmitz, W., *Z. Physik* **268**, 225 (1974).
34. Carlson, T.A., and Krause, M.O., *Phys. Rev. Lett.* **14**, 390 (1965).
35. Carlson, T.A., and Krause, M.O., *Phys. Rev. Lett.* **17**, 1079 (1966).
36. Stalherm, D., Cleff, B., Hillig, H., and Mehlhorn, W., *Z. Naturforsch.* **24a**, 1728 (1969).
37. Siegbahn, K. et al., *ESCA Applied to Free Molecules*, North Holland, Amsterdam, 1971.
38. Moddeman, W.E., Carlson, T.A., Krause, M.O., Pullen, B.P., Bull, W.E., and Schweitzer, G.K., *J. Chem. Phys.* **55**, 2317 (1971).
39. Gelius, U., Basilier, E., Svensson, S., Bergmark, T., and Siegbahn, K., *J. Electr. Spectr. Relat. Phenom.* **2**, 405 (1973).
40. Aksela, S., Väyrynen, J., and Aksela, H., *Phys. Rev. Lett.* **33**, 999 (1974).
41. Mehlhorn, W., Breuckmann, B., and Hausamann, D., *Physica Scripta* **16**, 177 (1977).
42. Väyrynen, J., Aksela, S., and Aksela, H., *Phys. Scripta* **16**, 452 (1977).
43. Kumpula, R., Väyrynen, J., Rantala, T., and Aksela, S., *J. Phys. (Paris)* **C12**, L809 (1979).
44. Matthews, D.L., Johnson, B.M., Mackey, J.J., and Moore, C.F., *Phys. Rev. Lett.* **31**, 1331 (1973).
45. Stolterfoht, N., Schneider, D., Mann, R., and Folkmann, F., *J. Phys.* **B10**, L281 (1977).
46. Sellin, J.A., in: S. Bashkin (Ed.), *Topics in Current Physics*, **Vol, I**: Beam Foil Spectroscopy, Springer, Heidelberg, 1976, p. 265 .
47. Rudd, M.E., *Phys. Rev. Letters* **13**, 503 (1964); **15**, 580 (1965).
48. Volz, D.J., and Rudd, M.E., *Phys. Rev.* **A2**, 1395 (1970).
49. Bordenave-Montesquieu, A., Gleizes, A., Rodiere, M., and Benoit-Cattin, P., *J. Phys.* **B6**, 1997 (1973).
50. Rødbro, M., Bruch, R., and Bisgaard, P.,*J. Phys.* **B10**, 1275 (1977).
51. Itoh, A., Schneider, D., Schneider, T., Zouros, T.J.M., Nolte, G., Schiewitz, G., Zeitz, W., and Stolterfoht, N., *Phys.Rev.* **A31**, 684 (1985).
52. Stolterfoht, N., *Phys. Rep.* **146**, 315 (1987).
53. Mehlhorn, W., *Phys. Lett.*, **26A**, 166 (1968).
54. Cleff., B., and Mehlhorn, W., *J. Phys.* **B7**, 593 (1974).
55. Klar, H., *J. Phys.* **B13**, 4741 (1980).
56. Kabachnik, N.M:, *J. Phys.* **B14**, L337 (1981).
57. Cleff., B., and Mehlhorn, W., *J. Phys.* **B7**, 605 (1974).
58. Hahn, U., Semke, H., Merz, H., and Kessler, H., *J. Phys.* **B18**, L417 (1985).
59. Merz, H., and Semke, J., Spin-polarized Auger Electrons, in *X-Ray and inner-shell processes*, T.A. Carlson, M.O. Krause and S.T. Manson (Eds.) AIP Conf. Proc., 215, AIP, New York, p. 719.
60. Kuntze, R., Salzmann, M., Böwering, N., and Heinzmann, U., *Phys. Rev. Lett.* **70**, 3716 (1993).
61. Flügge, S., Mehlhorn, W., and Schmidt, V., *Phys. Rev. Lett.* **29**, 7 (1972).
62. Berezhko, E.G., and Kabachnik, N.M., *J. Phys.* **B10**, 2467 (1977).
63. Callan, E.J., *Phys. Rev.* **124**, 793 (1961).
64. McGuire, E.J., *Phys. Rev.* **185**, 1 (1969); **A2**, 273 (1970); **A3**, 587 (1971); **A5**, 1043 (1972); **A9**, 1840 (1974).
65. Walters, D.L., and Bhalla, C.P., *Atomic Data* **3**, 301 (1971); *Phys. Rev.* **A3**, 519 and 1919 (1971).
66. Chen, M.H., and Crasemann, B., *Phys. Rev.* **A10**, 2232 (1974).

67. Chen, M.H., Relativistic calculations of atomic transition probabilities, in: B. Crasemann (Ed.) *Atomic Inner-Shell Physics*, Plenum Press, 1985, p. 31.
68. Asaad, W.N., *Nucl. Phys.* **66**, 494 (1965).
69. Kelly, H.P., *Phys. Rev.* **A11**, 556 (1975).
70. Mehlhorn, W., Correlation effects in Auger electron spectroscopy, in: F. Wuilleumier (Ed.), *Photoionization and other Probes of Many-Electron Interactions*, NATO ASI-Series, Plenum Press, 1976, p. 309.
71. Dyall, K.G., and Larkins, F.P., *J. Phys.* **B15**, 2793 (1982).
72. Åberg T., Theory of atomic decay following inner-shell ionization, in: F. Wuilleumier (Ed.), *Photoionization and Other Probes of Many-Electron Interactions*, NATO ASI-Series, Plenum Press, 1976, p. 273.
73. Åberg, T., and Howat, G., Theory of the Auger effect, in: W. Mehlhorn (Ed.), *Handbuch der Physik*, Vol. XXXI, Springer, 1982, p. 469.
74. Fano, U., *Phys. Rev.* **124**, 1866 (1961).
75. Åberg, T., *Phys. Scripta* **T41**, 71 (1992).
76. Madden, R.P. and Codling, K., *Phys. Rev. Lett.* **10**, 516 (1963); *Astrophys. J.*, **141**, 364 (1965).
77. Schmidt, V., Sandner, N., Mehlhorn, W., Wuilleumier, F., and Adam, M.Y., *Phys. Rev. Lett.* **38**, 63 (1977).
78. Eberhardt, W., Kalkoffen, G., and Kunz, C., *Phys. Rev. Lett.* **41**, 156 (1978).
79. Niehaus, A., *J. Phys.* **B10**, 1845 (1977).
80. Borst, M., and Schmidt, V., *Phys. Rev.* **A33**, 4456 (1986).
81. Schmidt, V., Krummacher, S., Wuilleumier, F., and Dhez, P., *Phys. Rev.* **A24**, 1803 (1981).
82. Russek, A., and Mehlhorn, W., *J. Phys.* **B19**, 911 (1986).
83. Armen, G.B., Tulkki, J., Åberg, T., and Crasemann, B., *Phys. Rev.* **A36**, 5606 (1987).
84. Ogurtsov, G.N., *J. Phys.* **B16**, L745 (1983)
85. Niehaus, A., and Zwakhals, C.J., *J. Phys.* **B16**, L135 (1983).
86. Helenelund, K., Hedman, S., Asplund, L., Gelius, U., and Siegbahn, K., *Phys. Scripta* **27**, 245 (1983).
87. Kuchiev, M.Yu., and Sheinerman, S.A., *Zh. Eksp. Teor. Fiz.* **90**, 1680 (1986).
88. Armen, G.B., *Phys. Rev.* **A37**, 995 (1988).
89. Kämmerling, B., Krässig, B., and Schmidt, V., *J. Phys.* **B26**, 261 (1993).
90. van der Wiel, M., Wight, G. R., and Tol, R.R., *J. Phys.* **B9**, L5, (1976).
91. Eberhardt, W., Bernstorff, S., Jochims, H.W., Whitfield, S.B., and Crasemann, B., *Phys. Rev.* **A38**, 3808 (1988).
92. Samson, J.A.R., Stolte, W.C., He, Z.X., Cutler, J.N., and Hansen, D., *Phys. Rev.* **A54**, 2099 (1996).
93. Tulkki, J., Åberg, T., Whitfield, S.B., and Crasemann, B., *Phys. Rev.* **A41**, 181 (1990).
94. Mehlhorn, W., Recent Developments in Radiationless Transitions, in *Proc. Fifteenth Intern. Conf. on X-Ray and Inner-Shell Processes*, (ed. T.A. Carlson, M.O. Krause and S.T. Manson),AIP Conf. Proceedings 215, 1990, p. 465.
95. Carlson, T.A., Mullins, D.R., Beall, C.E., Yates, B.W:, Taylor, J.W., Lindle, D.W., and Grimm, F.A., *Phys. Rev. Lett.* **60**, 1382 (1988); *Phys. Rev.* **A39**, 1170 (1989).
96. Kämmerling, B., Krässig, B., and Schmidt, V., *J. Phys.* **B231**, 4487 (1990).
97. Hergenhahn, U., Kabachnik, N.M., and Lohmann, B., *J. Phys.* **B24**, 4759 (1991).
98. Hergenhahn, U., Lohmann, B., Kabachnik, N.M., and Becker, U., *J. Phys.* **B26**, L117 (1993).
99. Brown, G.S., Chen, M.H., Crasemann, B., and Ice, G.E., *Phys. Rev. Lett.* **45**, 1937 (1980).
100. Armen, G.B., Åberg, T., Levin, J.C., Crasemann, B., Chen, M.H., Ice, G.E., and Brown, G.S., *Phys. Rev. Lett.* **54**, 1142 (1985).
101. Kivimäki, A.,. Naves de Brito, A., Aksela, S., Aksela, H., Sairanen, O.-P., Ausmees, A., Osborne, S.J., Dantas, L.B., and Svensson, S., *Phys. Rev. Lett.* **71**, 4307 (1993).
102. Aksela, H., and Mursu, J., *Phys. Rev.* **A54**, 2882 (1996).
103. Mursu, J., Aksela, H., Sairanen, O.-P., Kivimäki, A., Nömmiste, E., Ausmees, A., Svensson, S.,

and Aksela, S., *J. Phys.* **B29**, 4387 (1996).
104. Aksela, H., Jauhiainen, J., Kukk, E., Nömmiste, E., Aksela, S., and Tulkki, J., *Phys. Rev.* **A53**, 290 (1996).
105. Jauhiainen, J., Aksela, H., Sairanen, O.-P., Nömmiste, E., and Aksela, S., *J. Phys.* **B29**, 3385 (1996).
106. Aksela, H., Sairanen, O.-P., Aksela, S., Kivimäki, A., Naves de Brito, A., Nömmiste, E., Tulkki, J., Ausmees, A., Osborne, S.J., and Svensson, S., *Phys. Rev.* **A51**, 1291 (1995).
107. Sairanen, O.-P., Aksela, H., Aksela, S., Mursu, J., Kivimäki, A., Naves de Brito, A., Nömmiste. E., Osborne, S.J., Ausmees, A., and Svensson, S., *J. Phys.* **B28**, 4509 (1995).
108. Langer, B., Berrah, N., Farhat, A., Hemmers, O., and Bozek, J.D., *Phys. Rev.* **A53**, 1946 (1996).
109. Caldwell, C.D., and Hallman, S., *Phys. Rev.* **A53**, 3344 (1996).
110. Aksela, H., Jauhiainen, J., Nömmiste, E., Aksela, S., Sundin, S., Ausmees, A., and Svensson, S., *Phys. Rev.* **A54**, 605 (1996).
111. Aksela, H., Jauhiainen, J., Nömmiste, E., Sairanen, O.-P., Karvonen, J., Kukk, E., and Aksela, S., *Phys. Rev.* **A54**, 2874 (1996).
112. Cubric, D., Wills, A.A., Sokell, E., Comer, J., and MacDonald, M.A., *J. Phys.* **B26**, 4425 (1993).
113. Schmidt, V., Post-collision interaction in inner-shell ionization, in: B. Crasemann (Ed.), *Proc. Int. Conf. X-Ray and Atomic Inner-Shell Physics*, API Conf. Proc. 94, 1982, p. 544.
114. Aksela, H., Kivilompolo, M., Nömmiste, E., and Aksela, S., *Phys. Rev. Lett.* **79**, 4970 (1997).
115. Vegh, L., Becker, R., and Macek, J.H., Contributed paper D11, in: *15$^{th}$ Intern. Conf. X-Ray and Inner-Shell Processes*, Knoxville, 1990, (Abstracts).
116. Vegh, L., and Macek, J.H., *Phys. Rev.* **A50**, 4031 (1994).
117. Schwarzkopf, O., and Schmidt, V., *J. Phys.* **B29**, 3023 (1996).
118. Viefhaus, J., Snell, G., Hentges, R., Wiedenhöft, M., Heiser, F., Geßner, O., and Becker, U., *Phys. Rev. Lett.* **80**, 1618 (1998).
119. Sheinerman, S.A., and Schmidt, V., *J. Phys.* **B30**, 1677 (1997).
120. Schmidt, V., Proceedings of this Conference.
121. West, J.B., Ross, K.J., and Beyer, H.J., *J. Phys.* **B31**, L647 (1998).
122. Ueda, K., West, J.B., Ross, K.J., Beyer, H.J., and Kabachnik, N.M., *J. Phys.* B31, 4801 (1998).
123. Grum-Grzhimailo, A.N., Dorn, A., and Mehlhorn, W., *Comm. At. Mol. Phys. Comm. Mod. Phys.* **D1**, 29 (1999).
124. Zatsarinny, O.I., *J. Phys.* **B28**, 4759 (1995); and private communication (1998).
125. Snell, G., Drescher, M., Müller, N., Heinzmann, U., Hergenhahn, U., and Becker, U., *J. Phys.* **B32**, 2361 (1999).
126. Ueda, K., Shimizu, Y., Chiba, H., Sato, Y., Kitajima, M., Tanaka, H., and Kabachnik, N.M., *Phys. Rev. Lett.*, submitted.
127. Baier,S., Schulze, M., Staiger, H., Zimmermann, P., Lorenz, C., Pahler, M., Rüder, J., Sonntag, B., Costello, J.T., and Kiernan, L., *J. Phys.* **B27**, 1341 (1994).
128. Balashov, V.V., Grum-Grzhimailo, A.N., and Kabachnik, N.M., *J. Phys.* **B30**,1269 (1997).
129. Dorn, A., Zatsarinny, O.I., and Mehlhorn, W., *J.Phys.* **B30**, 2975 (1997).
130. Grum-Grzhimailo, A.N., and Mehlhorn, W., *J.Phys.* **B30**, L9 (1997).
131. Mehlhorn, W., *J. Electr. Spectrosc. Relat. Phenom.* **93**, 1 (1998).
132. Briand, J.-P., *Comments At. Mol. Phys.* **33**, 1 (1996).

# II. ATOMS

# Some Frontiers of X-Ray/Atom Interactions

## R. H. Pratt

*Department of Physics and Astronomy*
*University of Pittsburgh, Pittsburgh PA 15260 USA*

**Abstract.** Recent experiments have examined (1) the high energy behavior of single photoionization (including persistence of non-IPA features, i.e. beyond independent particle approximation), particularly in neon, and (2) the persistence of higher multipole effects even at low energies, (3) the ratio of double-to-single photoionization in helium (both in photoabsorption and Compton scattering regimes), and (4) the angular distributions of elastic and inelastic x-ray scattering from neon, exhibiting the simultaneous importance of non-local, correlation and dynamic effects. There are also now extensive experiments in ion-atom scattering which can be used to study direct and inverse photon-atom processes at high energy, including bound electron pair production and inverse photoeffect (direct radiative capture). Recent theoretical developments include the use of the theory of asymptotic Fourier transforms to discuss the high energy behavior of single and double photoionization (including the quasi-free mechanism for double ionization), the identification of effects beyond impulse approximation which should be considered in Compton scattering and in ion-atom scattering, estimation of the importance of higher multipole effects which involve atomic effects (including the first calculations of higher multipole correlation effects). There are some insights regarding the observed cancellations among relativistic, retardation, and higher multipole effects. In Rayleigh scattering a new formulation has been developed, no longer assuming a summation (or average) over filled magnetic substates, permitting discussion of scattering from excited states and ions, including dichroism in scattering. Parameterization of the angular distribution of anomalous scattering is needed to permit a compact description of scattering cross sections. The Delbrück scattering amplitude still resists an adequate treatment, but a full treatment of the forward scattering amplitude has now been completed with the inclusion of bound pair production contributions.

## PRELIMINARIES

Our purpose here is to discuss some frontiers in x-ray atom interactions, focusing on new experiments and their theoretical understanding, with no pretense at completeness. The field is very active, reflecting the new experimental possibilities and the fundamental and applied interests of the subject. The latter include, on the one hand, the problems of entanglement (as seen in correlations) and light-light scattering (as seen in Delbrück scattering), and the other hand practical applications of photoabsorption, anomalous scattering, transparency, and Compton profiles in the study of biological systems and materials.

Experiments in the field are now being performed using the modern synchrotrons. COLTRIMS systems (cold target recoil ion momentum spectroscopy [1]) are making much more detailed observations possible. However higher energy data still relies on older experiments with nuclear sources. The highest energy direct measurements of photoionization (at 6756 keV) were performed using the high flux neutron reactor at Grenoble [2]. There are indirect measurements of the interactions using related processes, such as bound pair production and annihilation, inelastic electron atom scattering (generalized oscillator strength), and relativistic ion-atom collisions (see presentation by Stöhlker [3] in this volume, also Eichler [4])).

We begin by reviewing the dominant photon atom processes of photoabsorption (photoexcitation, photoionization and pair production) and photon scattering (Rayleigh and Compton), with photoionization dominating at lower energies, pair production at higher energies, and Compton scattering at intermediate energies. In summary, we have:

photoabsorption

excitation
$$\gamma + A \to A^*$$

ionization
$$\gamma + A \to A^+ + e$$

pair production
$$\gamma + A \to A + (e^+ + e^-)$$

photon scattering

coherent scattering (Rayleigh)
$$\gamma + A \to \gamma + A$$

inelastic scattering (Compton)
$$\gamma + A \to \gamma + A^+ + e$$

We illustrate in Fig. 1 the total cross sections for these processes, as a function of photon energy, showing the cases $Z = 10, 92$.

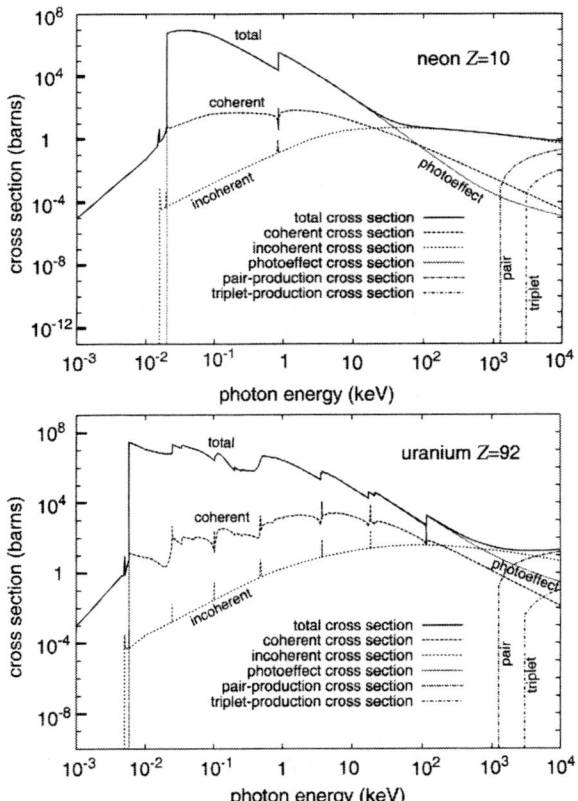

**FIGURE 1.** Cross sections for various photon atom processes (integrated over angles and energies of outgoing particles) for $Z = 10$ (Top) and $Z = 92$ (Bottom). The total photon atom cross section, the sum of these, is also shown. Data are based on results tabulated in [5], but we have schematically included additional resonance structure and also schematically extended the results to lower energies to show the behavior below outer-shell atomic thresholds. (From Ref. [6].)

We can similarly identify the dominant processes for more particular situations. Thus for photoionization the classification of processes is the same: Photoabsorption dominates at lower energies; ionization in Compton scattering dominates at intermediate energies. Very recently it was pointed out that at high energy the dominant ionization process is the rare pair production mode in which the atom is ionized [7]; see Fig. 2, examining the particular case of $K$-shell ionization. Likewise, for multiple ionization (at least of two electron systems) the situation is again similar. Thus it was pointed out a few years ago [8] that while, at low energies, double ionization of two electron systems proceeds through photoabsorption, well above 5 keV double ionization of helium proceeds primarily through Compton scattering, not photoabsorption. See Fig. 3 for estimates of the magnitudes [9].

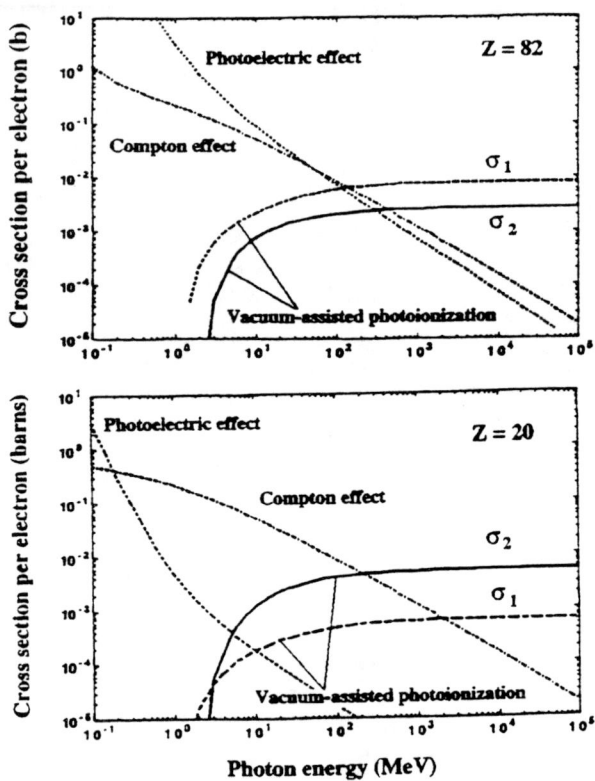

**FIGURE 2.** Top: Cross section for creation of a $K$-shell vacancy for lead ($Z = 82$) for different photoionization processes as a function of the photon energy. The dashed-dotted line corresponds to the Compton effect and the dotted line to the photoelectric effect. Vacuum-assisted photoionization is represented by the dashed line (mechanism 1, pair production in the nuclear field with subsequent $e^{\pm} - e^-$ encounter) and by the solid line (mechanism 2, pair production in the electron field), respectively. The cross sections are per $K$-shell electron. Bottom: Corresponding results for calcium ($Z = 20$). (From Ref. [7].)

After this preliminary discussion we will now, in the following sections, discuss (2) high energy photoionization (focusing on persistent correlation effects at high energy), (3) low energy photoionization (focusing on persistent higher multipole effects at low energies), (4) photon scattering (considering the range of effects now observable in new experiments), and (5) pair production. We will end (6) by giving a summary of our discussions.

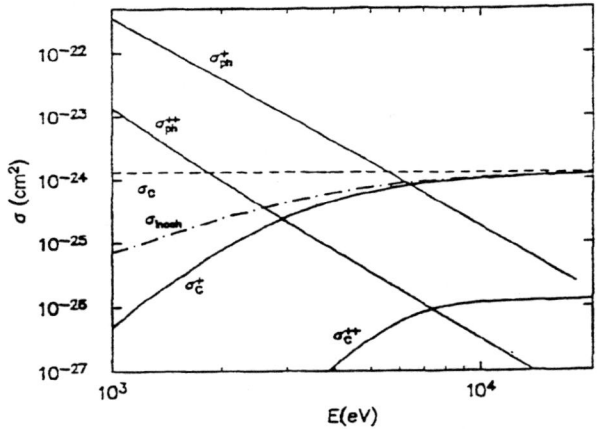

**FIGURE 3.** Cross sections for ionization of He by photon impact. $\sigma_{ph}^+$ and $\sigma_{ph}^{++}$: calculation for single and double ionization by photoabsorption. $\sigma_C^+$ and $\sigma_C^{++}$: single and double ionization by Compton scattering. $\sigma_C$ and $\sigma_{incoh}$ signify Compton scattering by two free electrons and incoherent Compton scattering by helium, respectively. (From Ref. [9].)

# HIGH ENERGY PHOTOIONIZATION

We may summarize present knowledge of the asymptotic behaviors of total photoionization cross sections for ejection of electrons of angular momentum $l$ from an atom due to absorption of a photon of energy $k$:

non-relativistic
$$\sigma \sim 1/k^{(7/2+l)},$$

with correlations
$$\sigma \sim 1/k^{7/2} \ (l=0), \quad \sigma \sim 1/k^{9/2} \ (l>0),$$

relativistic
$$\sigma \sim 1/k.$$

We will begin by discussing the behaviors in the absence of correlations. These follow from the fact that the smallest allowed momentum transfers give the dominant cross sections, and these minimum momentum transfers are large enough at high energy that the cross sections are determined at small distances: at the electron Compton wave length (relativistic case) or smaller (non- relativistic case). In summary, the matrix element $M$ is of the form

$$M \sim \int e^{-i\boldsymbol{p}\cdot\boldsymbol{r}} e^{i\boldsymbol{k}\cdot\boldsymbol{r}}\, r\, r^l\, d^3r \sim 1/\Delta^{4+l},$$

and $e^{-i\boldsymbol{p}\cdot\boldsymbol{r}} e^{i\boldsymbol{k}\cdot\boldsymbol{r}} = e^{i\boldsymbol{\Delta}\cdot\boldsymbol{r}}$, where $\Delta$ is the momentum transfer to the nucleus, one power of $r$ represents the dipole interaction, the other powers the small distance behavior of the bound state wave function. The dominant minimum momentum transfer $\Delta_{\min}$ corresponds to a small distance on an atomic scale:

$$r \sim 1/\Delta_{\min} \text{ (non-relativistic dipole)}, \quad r \sim \lambda_e \text{ (relativistic multipole)},$$

where $\lambda_e$ is the Compton wavelength. At such distances wave functions are nuclear point Coulombic in shape, independent of screening and of energy, except in their normalization (strength at the origin). As a result cross sections $\sigma$, angular distributions, $d\sigma/d\Omega$, and polarization correlations $C_{ij}$ are largely independent of screening and bound state principal quantum number $n$, except for bound state normalization constants, which do not enter $(d\sigma/d\Omega)/\sigma$ and $C_{ij}$.

There has been considerable recent attention to the high energy double ionization of helium, the simplest system in which to study electron-electron correlations, and an example of the persistence of correlations at high energy. In particular it

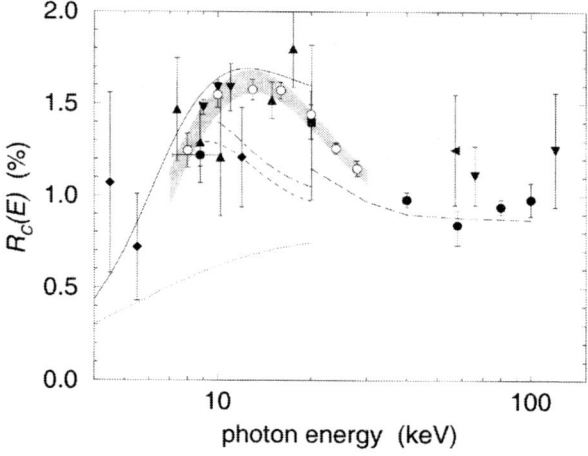

**FIGURE 4.** Experimental and theoretical results for $R_C(E)$. The shaded band is a smooth curve fitted to the experimental data 3r and broadened by the experimental uncertainty. Other experiments: Spielberger et al., solid circles; Levin et al., triangles up; Morgan and Bartlett, diamonds; Samson et al., square; Becker et al., triangles down; Wehlitz et al., triangle left. Theory: Bergstrom et al., MBPT, solid curve; Andersson and Burgdörfer, 3C final state, short-dashed curve, and Byron-Joachain type CI final state, dot-dashed curve; Spielberger et al., 3C final state, long-dashed curve; Surić et al., IA, dotted curve. See [12]. (From Ref. [11].)

has been shown that the high energy limit of the ratio of the cross sections for double and single ionization is a constant, which can be understood in terms of the ejection of one fast electron with the shake-off of the second electron. However the constant is different (smaller by a factor of two) in the Compton regime than in the photoabsorption regime. The constants may be obtained as [10]

$$R_{PE} = 1 - \frac{\sum_B \left| \int \Phi_B^*(\mathbf{r}_1) \Psi_i(\mathbf{r}_1, 0) d^3 r_1 \right|^2}{\int |\Psi_i(\mathbf{r}_1, 0)|^2 d^3 r_1},$$

$$R_C = 1 - \sum_B \int d^3 r \left| \int \Phi_B^*(\mathbf{r}_1) \Psi_i(\mathbf{r}, \mathbf{r}_1) d^3 r_1 \right|^2,$$

in terms of the ground state wave function of the helium atom and the hydrogenic bound state wave functions of once ionized helium. The difference of the ratios for photoabsorption and Compton scattering can be understood by realizing that high energy photoabsorption takes place at small distances (as discussed above), while high energy Compton scattering is basically off the electrons as though free, and so takes place anywhere in the atomic volume. Thus different regions of the initial state wave function contribute to the shake-off for the two processes. Experiment confirms the two shake-off constants. The convergence toward the Compton result is illustrated in Fig. 4, from [11].

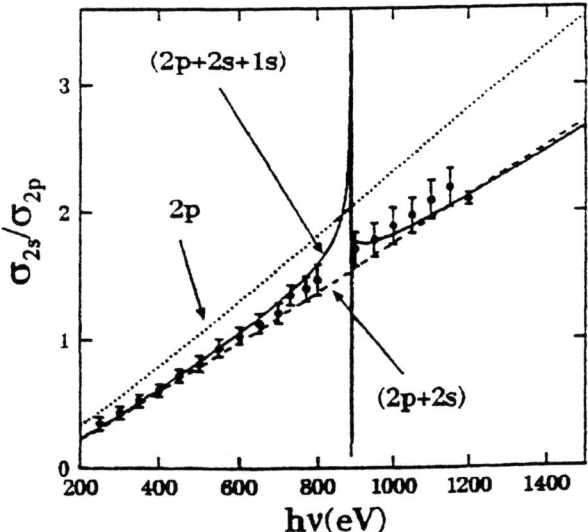

**FIGURE 5.** Ratio of the $2s$ to $2p$ cross section for Ne. The calculations employed the RRPA formalism with the single excitation channels arising from $2p$, $2s$, and $1s$ coupled (solid curve); $2p$ and $2s$ coupled (dashed curve); and $2p$ and $2s$ uncoupled to each other (dotted curve). (From Ref. [13].)

Another new development has been the observation of persistent (high energy limit) correlation effects (beyond initial state normalization) in 2p single photoionization of neon [13], as illustrated in Fig. 5. We have calculated these effects and find that this correction is of order $1/Z$ for 2p ionization, so that its importance diminishes with $Z$. However we also find that for higher $l$ correlations change the energy dependence, so that all cross sections except for $s$-states fall as the (9/2) power of photon energy, although this form is reached more slowly with increasing $Z$. These behaviors arise because it only takes one inverse power of energy to transfer the vacancy following $s$-state ionization to any other angular momentum. This additional power, combined with the (7/2) power of $s$-state ionization, leads to the general (9/2) power of photoionization. Some of these issues have also been discussed by Manson and co-workers [13,14]. (In the relativistic domain all cross sections will fall linearly.) It should be noted that, while the neon cross sections approach their asymptotic behaviors very slowly, ratios of these cross sections have become asymptotic at the energies of the experiment. This reflects the fact that the leading energy correction to the high energy behavior is the same for all subshells, dominantly having the character of the Stobbe factor [15].

Recently a simple unified viewpoint has been presented for understanding high energy photoionization processes and their fall-off with inverse powers of photon energy [16]. The key points are:

- For dominant kinematics photoionization matrix elements $M$ are often asymptotic Fourier transforms (AFT).
- An AFT is determined by the strength of the e-N and e-e singularities of the Hamiltonian, which lead to singularities of wave functions and electron-photon interaction, which determine singularities of the integrand of $M$, and so the inverse power fall-off of $M$ with energy.

These results follow from the Fourier transform theorem, which states:

"A function $f(x, y, z)$, infinitely differentiable at all points $(x, y, z)$, which decreases for large $r = (x^2 + y^2 + z^2)^{1/2}$ faster than any power of $r$, has a Fourier transform which decreases faster than any power of $p = (p_x^2 + p_y^2 + p_z^2)^{1/2}$ for large $p$."

Thus fall-off as a power of $p$ is connected with singularities of the function $f$.

Regarding singularities, note that, not only is $1/r$ singular at $r = 0$, but so (in three dimensions) is $r$, which along the $x$-axis is $|x|$, while $r^2$ is not singular. Consider as an example expansion of a function in powers of $r$: For $n > -2$

$$\lim_{\epsilon \to 0^+} \int e^{-i\mathbf{p}\cdot\mathbf{r}} r^n e^{-\epsilon r} d^3r = \begin{cases} 0, & \text{for } n \text{ even,} \\ (-)\, 4\pi(n+1)!/(ip)^{n+3} & \text{for } n \text{ odd.} \end{cases}$$

Increasing powers of $r$ result in increasing powers of $1/p$.

All the photoionization processes involve high energy continuum electrons in their final state, for which wave functions, except at the smallest distances, have (modified) plane wave character. The high energy matrix elements are asymptotic Fourier transforms, and their fall-off with powers of photon energy is determined by the strength of the singularities of the integrand. The singularities of the integrand are the singularities of the wave functions and the electron-photon interaction, and these correspond to the singularities of the Hamiltonian. Those singularities are the singularities of the interaction, i.e. the singularities of the electron-nucleus and electron-electron Coulomb potential when either electron is at the nucleus or when the two electrons coincide. One calculates high energy matrix elements by expanding the integrands about the singularities; in the FT higher order terms of the expansions lead to higher inverse powers in photon energy.

Our interest (for single ionization) is in integrals of the type

$$\int e^{-i\boldsymbol{p}\cdot\boldsymbol{r}}\, F\, I\, \Psi_{\text{bound}}\, d^3r,$$

where $I$ is the interaction, $e^{-i\boldsymbol{p}\cdot\boldsymbol{r}}F$ is the continuum state, and the integrand is singular at $r = 0$. If the leading small $r$ term in the expansion of $I$ is singular (acceleration form), the leading (non-singular) terms in the expansion of $\Psi_{\text{bound}}$ and $F$ (including in a many-electron system correlation terms, as for Ne 2p, where it explains the persistent correlation behavior discussed earlier) are sufficient to obtain the leading term in $1/p$. If the leading small $r$ term in $I$ is not singular (velocity and length forms) further terms in the wave function expansions will be needed. (In these forms it is not adequate to take a continuum state as a plane wave, when the initial state is a $p$-state.) Results obtained in this way will be independent of the form of electron-photon interaction (length, velocity, acceleration), but since the singularities of the interaction forms are different, the terms needed in the expansions of wave functions about the singularities in order to obtain the full lowest-order result will differ (explaining inconsistencies in some earlier treatments).

In double ionization

$$M \sim \int e^{-i\boldsymbol{p}_1\cdot\boldsymbol{r}_1 - i\boldsymbol{p}_2\cdot\boldsymbol{r}_2}\, (F\, I\, \Psi_{\text{bound}})\, d^3r_1\, d^3r_2.$$

$(F\, I\, \Psi_{\text{bound}})$ is singular when $\boldsymbol{r}_1 = 0$ or $\boldsymbol{r}_2 = 0$ or $\boldsymbol{r}_1 = \boldsymbol{r}_2$. We want this to reduce to a *single* asymptotic Fourier transform (otherwise there will be more powers of $1/p$).

We can achieve this for the e-N singularity at $\boldsymbol{r}_1$ $(\boldsymbol{r}_2) = 0$ *if* we take the kinematics (SHAKE-OFF) for which $\boldsymbol{p}_1$ $(\boldsymbol{p}_2) \approx 0$, i.e.

| $\boldsymbol{p}_1\cdot\boldsymbol{r}_1$ | $+$ | $\boldsymbol{p}_2\cdot\boldsymbol{r}_2$ | or | $\boldsymbol{p}_1\cdot\boldsymbol{r}_1$ | $+$ | $\boldsymbol{p}_2\cdot\boldsymbol{r}_2$ |
|---|---|---|---|---|---|---|
| singular | | 0 | | 0 | | singular |

We can achieve this for the e-e singularity at $r_1 = r_2$ *if* we take the kinematics (QUASI-FREE) for which $(p_1 + p_2) \approx 0$, i.e.

$$p_1 \cdot r_1 + p_2 \cdot r_2 = \tfrac{1}{2}(p_1 + p_2) \cdot (r_1 + r_2) + \tfrac{1}{2}(p_1 - p_2) \cdot (r_1 - r_2)$$

$$\phantom{p_1 \cdot r_1 + p_2 \cdot r_2 = \tfrac{1}{2}}\ 0 \phantom{(p_1 + p_2) \cdot (r_1 + r_2) + \tfrac{1}{2}} \text{singular}$$

The quasi-free term [17] results from the possibility that a photon can be absorbed by two free electrons, though not by one, and not in dipole approximation. This modification of the double/single ratio in photoabsorption, becoming significant in helium above 10 keV, has not yet been observed.

## LOW ENERGY PHOTOIONIZATION

In low energy photoionization one may ask about the importance of relativistic, retardation, and higher multipole effects, which are usually neglected. In fact higher multipole effects are often persistent, even at threshold, and such effects have been seen in several recent experiments, as we shall discuss below.

An opposite phenomena had been observed earlier with regard to relativistic and retardation effects, which were found, while separately large, to substantially cancel in total cross sections, particularly for *s*-states [18]. As a result non-relativistic non-retarded dipole cross sections were fairly good well into relativistic regimes (over 100 keV). Non-relativistic retarded multipole cross sections are much worse. Similar features were found in other processes, such as bremsstrahlung. The cancellation does not occur in angular distributions, so that non-relativistic retarded multipole effects are important below 100 keV. Some explanation of these features has been offered [19], studying the singularities in complex energy of a total cross section. The positions of the singularities are as follows:

| singularities of $\sigma$ in $\kappa$ | retarded multipole | non-retarded dipole |
|---|---|---|
| relativistic kinematics | 0 | $0,\ -2E_b$ spurious |
| non-relativistic kinematics | $0,\ 2(1 \pm P_b)$ spurious | 0 |

Thus the nonrelativistic non-retarded dipole and relativistic retarded multipole cross sections have the same singularities, while the alternatives have additional (spurious) singularities, which lead (in the non-relativistic retarded multipole case, with spurious complex singularities near 511 keV) to very poor predictions. The argument on singularities cannot be made at the level of the matrix element or

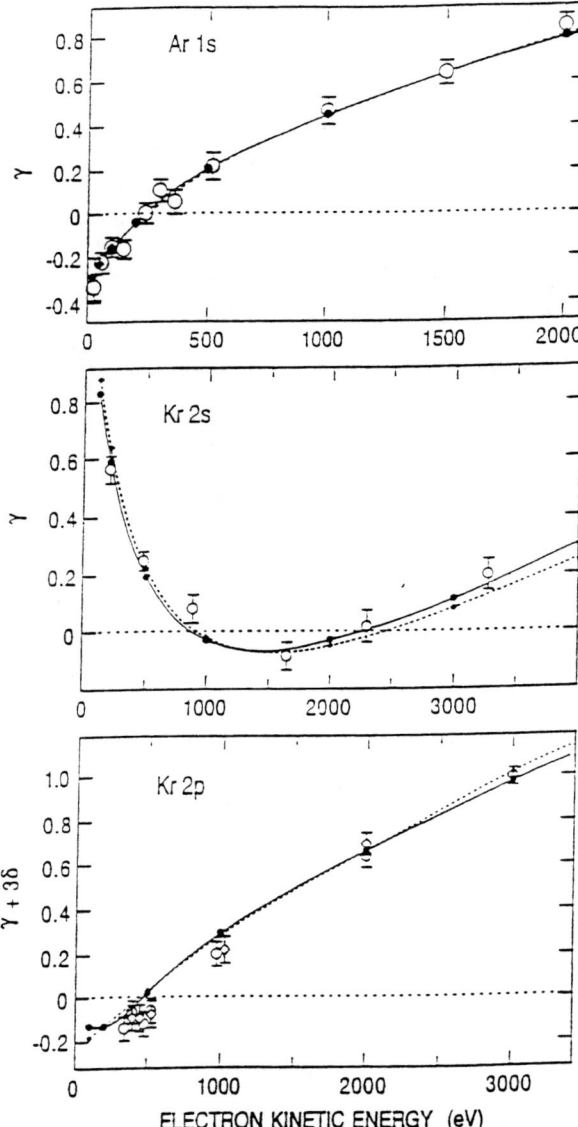

**FIGURE 6.** Energy dependence of the nondipole angular distribution parameter $\gamma$ for Ar $1s$ (top), Kr $2s$ (middle), and of the combined quantity $\gamma + 3\delta$ for Kr $2p_j$ (bottom). Open circles/diamonds, experimental results; in the case of Kr $2p_j$ the circles and diamonds refer to the $j = 1/2$ and $j = 3/2$ fine-structure components. Dashed and solid lines, theoretical predictions from [20] and [21], respectively. (From Ref. [22].)

differential cross section, consistent with the fact the cancellation is not present in the differential cross section.

It has often been supposed that low energy photoionization can be understood within dipole approximation, and the standard discussion of the observable parameters of "complete" experiments is based on that assumption, likewise the ability to experimentally determine such parameters from measurements at "magic angles". The impression is encouraged by the fact that in a simple calculation in the point Coulomb potential, quadrupole effects vanish at threshold with the velocity ($v/c$) of the outgoing electron. In fact there are persistent higher multipole effects, primarily quadrupole effects. These have been calculated [20,21], and one finds that

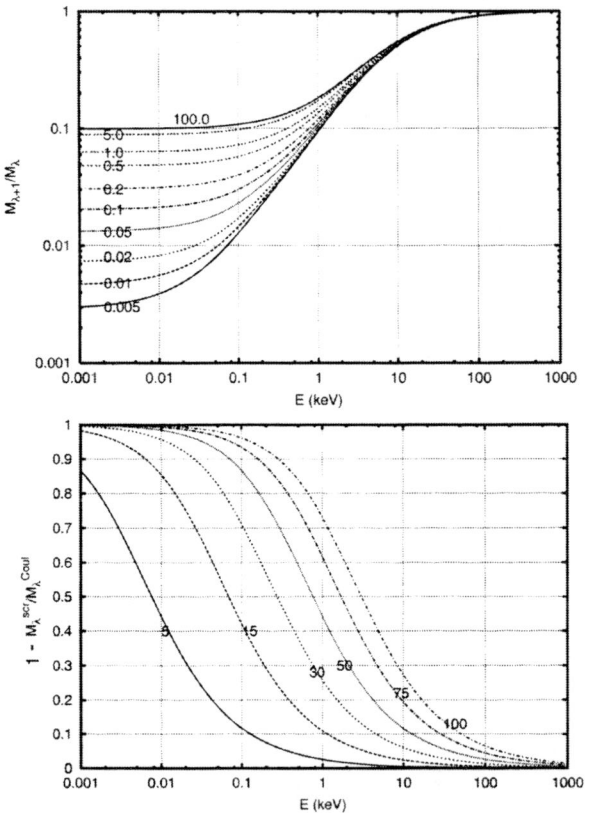

**FIGURE 7.** Top: Approximate electron energy dependence $E$ of the ratio $R$ of two successive reduced multipole matrix elements, $R = M_{\lambda+1}/M_\lambda$, independent of $\lambda$, for various choices of binding energy (in keV). Bottom: Approximate electron energy dependence $E$ of $1-Q$, for the ratio $Q$ of screened and Coulomb reduced matrix elements of given multipole, $Q = M_\lambda^{\text{screened}}/M_\lambda^{\text{Coulomb}}$, independent of $\lambda$, for various choices of nuclear charge $Z$. (From Ref. [23]).

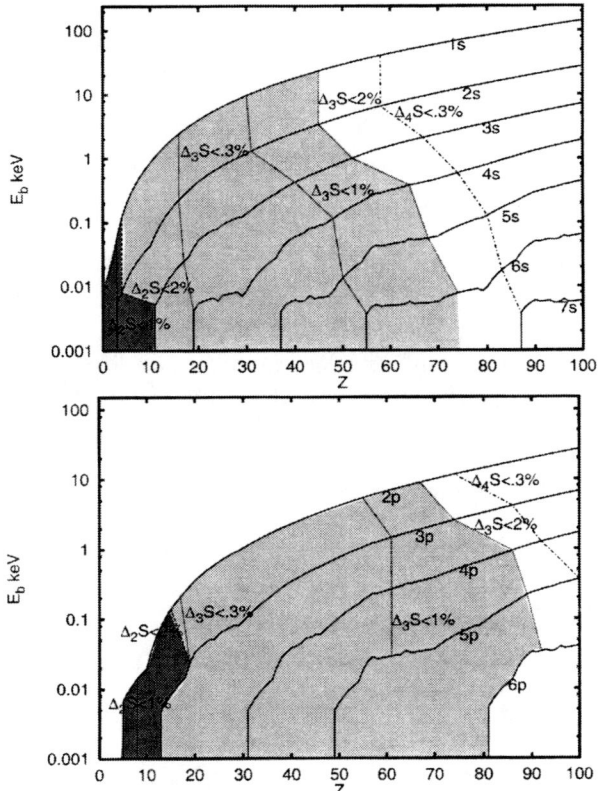

**FIGURE 8.** Maximum screening contribution $\Delta_j S$. Domain in $(Z, n, l)$ in which $\Delta_j S(E, \theta)$ for low multipoles is sufficient to achieve a given level of accuracy. The solid curves correspond to binding energies for $s_{1/2}$ (top) and $p_{1/2}$ (bottom) states for $Z$ ranging from 1 to 100. The dashed lines correspond to the boundary at which quadrupole screening contributions can reach 1% (leftmost line) and 2% (boundary for dark shaded area, within which dipole screening is sufficient for 2% accuracy). The dotted lines correspond to the 0.3%, 1%, and 2% limits for octupole screening contributions (the latter is the boundary for the lightly shaded area, within which screening corrections through quadrupole are sufficient for 2% accuracy). In the white area octupole screening contributions to $S(E, \theta)$ can exceed 2%. The dot-dashed line is the 0.3% boundary for $2^4$-pole screening contributions. The $p_{3/2}$ results are almost the same as the $p_{1/2}$ values. (From Ref. [24].)

the quadrupole contribution is (near the 1s threshold)

$$\kappa \sim 2Z\alpha \cos\Delta, \quad \Delta = \delta_2 - \delta_1.$$

(The quadrupole parameter $\kappa$ of Bechler and Pratt is related to the parameter $\gamma$

of Cooper.) In the Coulomb case

$$\Delta \sim \frac{\pi}{2} - \frac{2}{Z\alpha}\frac{v}{c}, \quad \kappa \sim 4\frac{v}{c},$$

while in the screened case

$$\cos\Delta \text{ is finite}(\pm), \quad \kappa \sim \pm 2Z\alpha.$$

Measurements of these non-dipole parameters are shown in Fig. 6, from [22].

It is clear that with increasing energy increasing numbers of multipoles will contribute (see Fig. 7: Top). On the other hand, with increasing energy multipole matrix elements become increasingly nuclear point Coulomb in character, and at any given level of precision give less information about atomic properties (see Fig. 7: Bottom).

It has been found [24] that measurements of cross sections accurate to 1% are only generally sensitive to atomic information in dipole, quadrupole, and octupole matrix elements, for all energies, elements, and shells; see Fig. 8. This result (only obtained within independent particle approximation) suggests that it should not be necessary to include correlations in the calculation of high multipoles. The first calculations of correlations in quadrupole matrix elements have recently been performed, shown in Figs. 9 and 10, from [25,26], indicating that in these cases

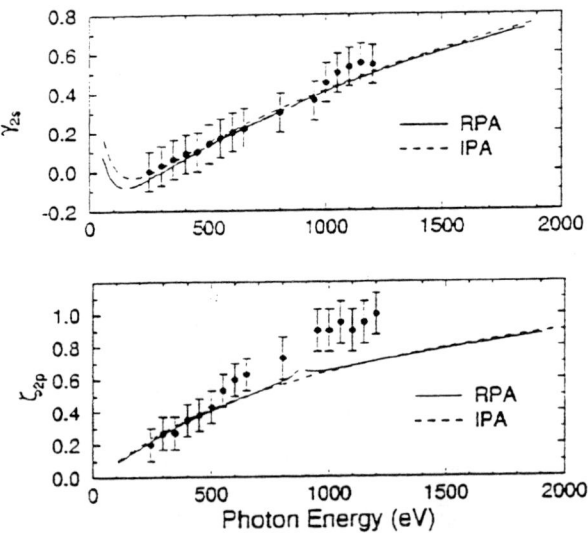

**FIGURE 9.** Top: Asymmetry parameter $\gamma_{2s}$ for the $2s$ subshell of Ne is compared with IPA calculations from [21] and with experimental data from [27]. Bottom: Asymmetry parameter $\gamma_{2p} + 3\delta_{2p}$ for the $2p$ subshell of Ne is compared with IPA calculations from [21] and with experimental data from [27]. (From Ref. [25].)

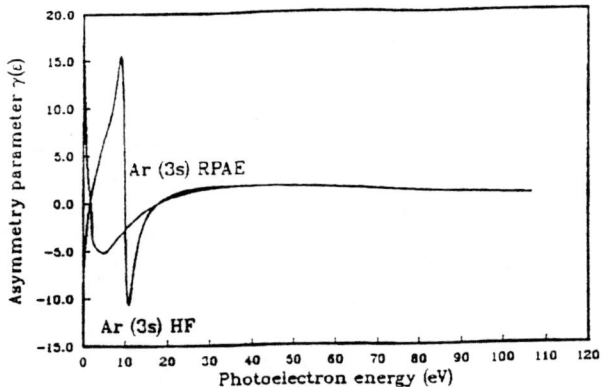

**FIGURE 10.** Nondipole asymmetry parameter $\gamma$ for photoionization out of the $3s$ subshell of Ar as a function of photoelectron energy, calculated in the HF approximation and the RPAE. (From Ref. [26].)

quadrupole correlation effects were unimportant except in the immediate vicinity of threshold.

Another characteristic of the photoionization matrix elements, considered as a function of energy, is the presence of zeroes, i.e. energies at which a matrix element vanishes. These lead to features in cross sections and distributions, including the well-known low-energy Cooper minima [28]. While those minima, due to zeroes in dipole matrix elements, only occur in a screened potential [29], there are minima in quadrupole matrix elements at x-ray energies, also present in a Coulomb potential, leading to observable consequences (changes from forward to backward peaking) in angular distributions [30]. Minima associated with relativistic spin-orbit splitting also occur in super-high-$Z$ elements [31], but are not known to have observable consequences. We have recently found yet another class of zeroes, occurring at higher energies [32]. In dipole matrix elements they occur at a photon energy of $mc^2/(l+1)$, independent of $Z$, principal quantum number $n$, and choice of potential. Even at these higher energies it appears they have some observable consequences, at least in polarization correlations.

## PHOTON SCATTERING

It has recently become possible to perform precision experiments on the scattering of photons from free atoms [33]. The initial experiments were performed for Ne and He in the energy range 11–22 keV, utilizing the Advanced Photon Source. Explanation of the results has required development of more sophisticated theory for these processes, which are of second order in the interaction of electrons and atoms with the radiation field. As in the first order photoionization processes discussed in the previous sections, it becomes necessary to combine (1) a full relativistic retarded multipole account of interactions with electrons in independent particle approximation (IPA) and (2) an account of electron-electron non-local exchange and correlations. In this way agreement between theory and experiment could be achieved, as shown in Fig. 11.

Simple descriptions of Rayleigh and Compton scattering are based on $A^2$ approximation, based on the $A^2$ term of the non-relativistic Hamiltonian. This leads to form factors (FF) for Rayleigh scattering and Compton profiles based on impulse approximation (IA). FF and IA calculations are available within IPA, but there are also treatments which include non-local exchange and correlations. In more recent years full dynamic $S$-matrix calculations have become available, both for Rayleigh scattering [35] and for Compton scattering [36]. These go beyond FF and IA approaches and $A^2$ approximation, utilizing a full relativistic retarded multipole description, but within independent particle approximation (IPA).

To explain the new experiments it was necessary to include both dynamic effects from the full $S$-matrix calculations and non-local and exchange effects, estimated in FF and IA calculations. It was assumed that these effects were small and additive, and they were treated perturbatively. In Rayleigh scattering from a light atom non-local exchange effects are significant, while dynamic effects (anomalous scattering) are also important at lower energies; see Fig. 12, based on [37]. Correlation effects in both Rayleigh and Compton scattering can also be seen at the level of precision

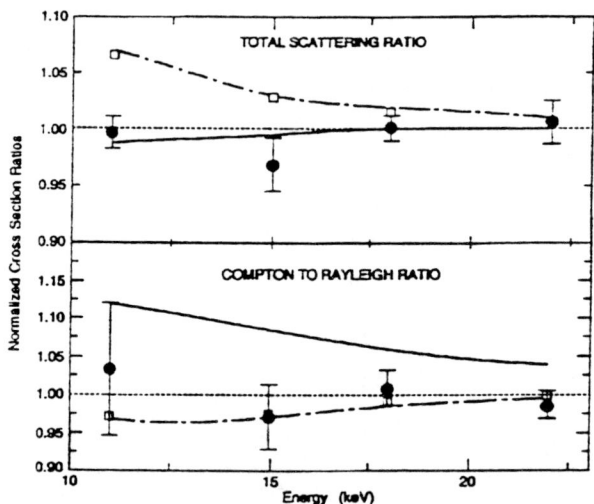

**FIGURE 11.** Top: Ratio of total scattering cross section in neon to helium normalized to the best prediction. Bottom: Compton to Rayleigh cross section ratio in neon normalized to the best prediction for that ratio. Experiment (filled circles); Hubbell [34] (solid line); the most sophisticated theory for which comprehensive tabulations are currently available (MFF + ASF for Rayleigh, Hubbell [34] for Compton) (dot-dashed line). Squares show $S$-matrix Rayleigh and agreement with MFF + ASF results. (From Ref. [33].)

of the experiment. At lower energies the perturbative treatment of these effects would begin to fail, and a full calculation, at least of the lowest multipoles, would be needed, as in photoionization.

There are other new theoretical results in photon-atom scattering. $S$-matrix calculations of Rayleigh scattering from atoms with partially-filled subshells have been improved, to correctly average over cross sections rather than matrix elements [38]. This shows that some transparency features are spurious; however the characteristic near-zero below an edge is a real feature which may have applications. Delbrück scattering still resists a full dynamic treatment. But recently the calculation of forward Delbrück scattering via dispersion relations has been improved, finding a substantial contribution from the bound pair production amplitude which had not previously been considered. Discrepancies between theory and experiment for scattering still remain at energies above 1 MeV where the Delbrück amplitude is becoming important [39].

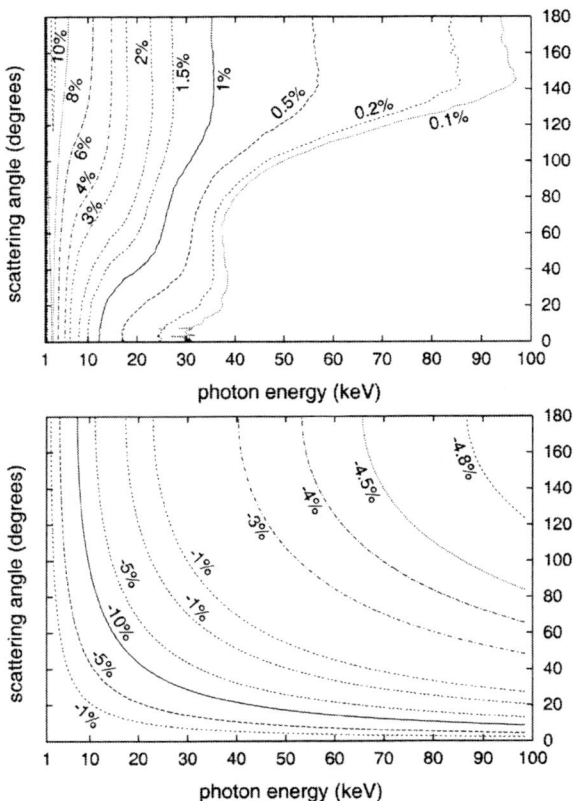

**FIGURE 12.** Top: Dynamic correction for Rayleigh scattering from neon. Bottom: Non-local correction for Rayleigh scattering from neon. (From Ref. [37].)

# PAIR PRODUCTION

Among issues in pair production, we have already mentioned the recent realization [7] that ionization in pair production is the dominant mode for ionization of atoms by photons at high energies. There has long been an anomaly in low energy pair production in Ge, with a resonance-like discrepancy, in comparison to theoretical predictions from relativistic numerical calculations, some 60 keV above threshold. The most recent experimental results [40], shown in Fig. 13, are consistent with the earlier experiments. However these authors also noted that the theoretical predictions did not consider the final state interaction between electron and positron. Using a very simple theory for such effects, they stated that agreement between theory and experiment was improved. However, since all three experiments continue to show a discrepancy in the same energy region, further work would be welcome.

In addition to ordinary pair production, bound pair production on ions (the electron produced is captured by the ion) is of considerable recent interest, important for the next generation of heavy ion colliders as a limiting factor in beam lifetime [42]. The process is formally related to relativistic atomic photoeffect and has been calculated with similar methods [43]. We have already noted the connection of the

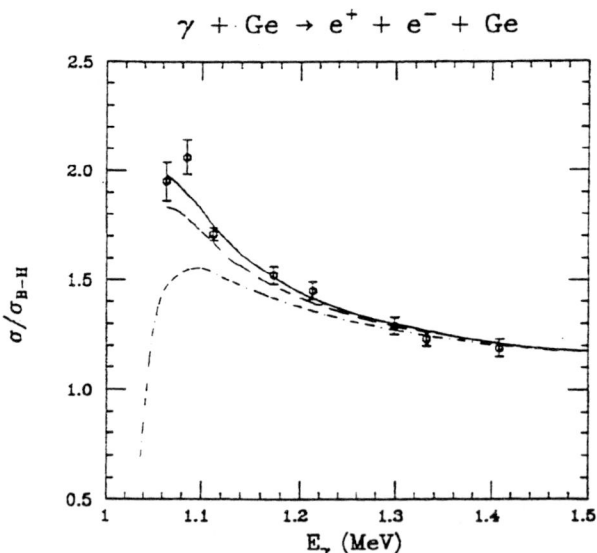

**FIGURE 13.** The pair production cross section in Ge in units of the Bethe-Heitler cross section. The error bars are relative; absolute cross sections were normalized to the prediction [41] at 1836 keV. The solid curve shows a calculation including the $e^+ - e^-$ final-state interaction; the dashed and dash-dotted curves show the standard theory, which neglects this interaction, for a screened and bare point nucleus, respectively. (From Ref. [40].)

process to Delbrück scattering. The inverse process of bound pair annihilation has also been studied.

## SUMMARY

The dominant photon-atom processes are photoexcitation and photoionization, elastic and inelastic photon-atom scattering (the latter also dominating in higher energy double ionization of helium), and pair production (which also provides the dominant high energy ionization mode).

In high energy photoionization one now sees persistence of certain correlations effects: (1) modifications of high energy behavior for $l > 0$ single ionization, as seen in Ne $2p$ experiments, and (2) double ionization, as in helium, through shakeoff mechanisms (observed) and quasi-free mechanisms. High energy photoionization behaviors can be understood to simply follow from the e-N and e-e Coulomb singularities of the interaction Hamiltonian, using the theory of asymptotic Fourier transforms.

In low energy photoionization, higher multipole effects generally persist to threshold. There are quadrupole effects of magnitude $Z\alpha$ and some octupole effects. While higher multipole contributions grow with increasing energy, they also become increasingly nuclear Coulombic in character. At the 1% level screening effects are only needed in dipole, quadrupole and some octupole terms, for any element, any shell, and any energy. There is a cancellation between relativistic and retardation effects on total (but not differential) cross sections, particularly for $s$-states, with the result that nonrelativistic non-retarded dipole results are fairly good for total cross sections up to 100 keV. Zeroes as a function of energy are an important feature of multipole matrix elements, in the dipole case for example responsible for the well-known Cooper minima in cross sections. Recent a new class of zeroes has been found, in the dipole case occurring at a photon energy of $mc^2/(l+1)$, independent of $Z$, principal quantum number $n$, and choice of potential.

New precision x-ray scattering experiments at the Advanced Photon Source have required improving existing theories of photon atom scattering. As in photoionization, one must combine a full relativistic retarded multipole account of interactions with electrons in independent particle approximation (IPA) with an account of electron-electron non-local exchange and correlations. Dynamic, non-local, and correlation effects are all observable. New results have also been obtained for forward Delbrück scattering, using dispersion relations to calculate the contribution from bound pair production.

## ACKNOWLEDGMENTS

This work was supported in part by NSF Grant 9970293.

# REFERENCES

1. Ullrich, J., Moshammer, R., Dörner, R., Jagutzki, O., Mergel, V., Schmidt-Böcking, H., and Spielberger, L., *J. Phys. B* **30**, 2917-2974 (1997).
2. Blakeway, S. J., Gelletly, W., Faust, H. R., and Schreckenbach, K., *J. Phys. B* **16**, 3751-3765 (1983).
3. Stöhlker, Th., (this volume).
4. Eichler, J., for publication in Proceedings of 21st International Conference on the Physics of Electronic and Atomic Collisions, Sendai, Japan, 1999.
5. D. E. Cullen, M. H. Chen, J. H. Hubbell, S. T. Perkins, E. F. Plechaty, J. A. Rathkopf and J. H. Scofield, Lawrence Livermore National Laboratory Report No. UCRL-50400, 1989 (unpublished).
6. Carney, J. P. J., (private communication).
7. Ionescu, D. C., Sørensen, A. H., and Belkacem, A., *Phys. Rev. A* **59**, 3527-3537 (1999).
8. Samson, J. A. R., Greene, C., and Bartlett, R., *Phys. Rev. Lett.* **71**, 201 (1993).
9. Andersson, L. R., and Burgdörfer, J., *Phys. Rev. Lett.* **71**, 50-53 (1993).
10. Surić, T., Logan, B. A., Pisk, K., and Pratt, R. H., *Phys. Rev. Lett.* **73**, 790-793 (1994).
11. Krässig, B., Dunford, R. W., Gemmell, D. S., Hasegawa, S., Kanter, E. P., Schmidt-böcking, H., Schmitt, W., Southworth, S. H., Weber, Th., and Young, L., *Phys. Rev. Lett.* **83**, 53-56 (1999).
12. Here are the references to results presented in Fig. 4. Other experiments: Solid circles - Spielberger, L. *et al.*, *Phys. Rev. Lett.* **74**, 4615-4618 (1995); Spielberger, L. *et al.*, **76** 4685-4688 (1996); Spielberger, L. *et al.*, *Phys. Rev. A* **59**, 371-379 (1999). Triangles up - Levin, J. C. *et al.*, *Phys. Rev. Lett.* **76**, 1220-1223 (1996). Diamonds - Morgan, D. V. *et al.*, *Phys. Rev. A* **59**, 4075-4078 (1999). Square - Samson, J. A. R. *et al.*, *J. Electron Spectrosc. Relat. Phenom.* **78**, 19 (1996). Triangles down - Becker, U. *et al.*, *Aust. J. Phys.* (to be published). Triangle left - Wehlitz, R. *et al.*, *Phys. Rev. A* **53**, R3720-R3722 (1996). Theory: Solid curve - P. M. Bergstrom, Jr. *et al.*, *Phys. Rev. A* **51**, 3044-3052 (1995). Short-dashed curve - Andersson, L. R. *et al.*, *Phys. Rev. A* **50**, R2810-R2813 (1994). Long-dashed curve - Spielberger, L. *et al.*, *Phys. Rev. A* **59**, 371-379 (1999). Dotted-curve - Surić, T. *et al.*, *Phys. Rev. Lett.* **73**, 790-793 (1994).
13. Dias, E. W. B., Chakraborty, H. S., Deshmukh, P. C., Manson, S. T., Hemmers, O., Glans, P., Hansen, D. L., Wang, H., Whitfield, S. B, Lindle, D. W., Wehlitz, R., Levin, J. C., Sellin, I. A., and Perera, R. C. C., *Phys. Rev. Lett.* **78**, 4553-4556 (1997).
14. Amusia, M. Ya, and Manson, S. T., (private communication, also presentation at X-99).
15. Stobbe, M., *Ann. d. Phys.* **7**, 661 (1930); See for example Heitler, W., *The Quantum Theory of Radiation*, 3rd edition, Oxford, 1954, pp 207-208.
16. Surić, T., Drukarev, E. G., and Pratt, R. H., (to be published), also *Viet. J. Phys.* (to be published); *Bull. Amer. Phys. Soc* **44**, 131 (1999).
17. Amusia, M. Ya, Drukarev, E. G., Gorshkov, V. G., and Kazachkov, M. O., *J. Phys.*

*B* **8**, 1248-1266 (1975); Drukarev, E. G., *Phys. Rev. A* **51**, R2684-R2686 (1995); **52**, 3910-3922 (1995).
18. Oh, S. D., McEnnan, J., and Pratt, R. H., Phys. Rev. A14, 1428 (1976); Ron, A., Goldberg, I. B., Stein, J., Manson, S. T., Pratt, R. H., and Yin, R. Y., *Phys. Rev. A* **50**, 1312-1320 (1994).
19. Pratt, R. H., and Kim, Y. S., *Rom. J. Phys.* **38**, 353-370 (1993).
20. Bechler, A., and Pratt, R. H., *Phys. Rev. A* **39**, 1774-1779 (1989); **42**, 6400-6413 (1990).
21. Cooper, J. W., *Phys. Rev. A* **42**, 6942-6945 (1990); **47**, 1841-1851 (1993).
22. Krässig, B., Jung, M., Gemmell, D. S., Kanter, E. P., LeBrun, T., Southworth, S. H., and Young, L., AIP Proc. **389**, 659-670 (Hamburg, 1996), also Jung, M., Krässig, B., Gemmell, D. S., Kanter, E. P., LeBrun, T., Southworth, S. H., and Young, L., *Phys. Rev. A* **54**, 2127-2136 (1996).
23. LaJohn, L. A, (private communication).
24. LaJohn, L. A., and Pratt, R. H., *Phys. Rev. A* **58**, 4989-4992 (1998).
25. Johnson, W. R., Derevianko, A., Cheng, K. T., Dolmatov, V. K., and Manson, S. T., *Phys. Rev. A* **59**, 3609-3613 (1999).
26. Amusia, M. Ya, Baltenkov, A. S., Felfli, Z., Msezane, A. Z., *Phys. Rev. A* **59**, R2544-R2547 (1999).
27. Hemmers, O., Fisher, G., Glans, P., Hansen, D. L., Wang, H., Whitfield, S. B, Wehlitz, R., Levin, J. C., Sellin, I. A., Perera, R. C. C., Dias, E. W. B., Chakraborty, H. S., Deshmukh, P. C., Manson, S. T., and Lindle, D. W., *J. Phys. B* **30**, L727-L733 (1997).
28. Cooper, J. W., *Phys. Rev.* **128**, 681 (1962).
29. Oh, S. D., and Pratt, R. H., *Phys. Rev. A* **34**, 2486-2487 (1986); **37**, 1524-1526 (1988); **45**, 1583-1586 (1992).
30. Wang, M. S., Kim, Y. S., Pratt, R. H., and Ron, A., *Phys. Rev A* **25**, 857-861 (1982).
31. Yin, R. Y., and Pratt, R. H., *Phys. Rev. A* **35**, 1154-1158 (1987).
32. LaJohn, L. A., and Pratt, R. H., *Bull. Am. Phys. Soc.* **44**, 1534 (1999).
33. Jung, M., Dunford, R. W., Gemmell, D. S., Kanter, E. P., Krässig, B., LeBrun, T. W., Southworth, S. H., Young, L., Carney, J. P. J., LaJohn, L. A., Pratt, R. H., and Bergstrom, Jr., P. M., *Phys. Rev. Lett.* **81**, 1596-1599 (1998).
34. Hubbell, J. H., Veigele, Wm. J., Briggs, E. A., Brown, R. T., Cromer, D. T., and Howerton, R. J., *J. Phys. Chem. Ref. Data* **4**, 471-538 (1975).
35. Kissel, L., Pratt, R. H., and Roy, S. C., *Phys. Rev. A* **22**, 1970-2004 (1980); Kane, P. P., Kissel, L., Pratt, R. H., and Roy, S. C., *Phys. Rep.* **140**, 75-159 (1986); S. Roy et al., Rad. Phys. Chem. 41, 725 (1992); Pratt, R. H., Kissel, L., and Bergstrom, Jr., P. M., in *Resonant Anomalous X-ray Scattering - Theory and Applications*, edited by Fischer, K., Materlik, G., and Sparks, C., North Holland: Elsevier, 1994, pp 9-33.
36. Bergstrom, Jr., P. M., Surić, T., Pisk, K., and Pratt, R. H., *Phys. Rev. A* **48**, 1134-1162 (1993); Bergstrom, Jr., P. M., and Pratt, R. H., *Radiat. Phys. Chem.* **50**, 3-29 (1997).
37. Carney, J. P. J., and Pratt, R. H., *Bull. Am. Phys. Soc.* **44**, 1534 (1999).
38. Carney, J. P. J., Pratt, R. H., Manakov, N. L., and Meremianin, A. V., (to be published).

39. Carney, J. P. J., and Pratt, R. H., *Phys. Rev. A* **60**, 3020-3024 (1999).
40. De Braeckeleer, L., Adelberger, E. G., and Garćia, A., *Phys. Rev. A* **46**, R5324-R5326 (1992).
41. Tseng, H. K., and Pratt, R. H., *Phys. Rev. A* **21**, 454-457 (1980); **24**, 1127-1128 (1981).
42. Palathingal, J. C., Asoka-Kumar, P., Lynn, K. G., Posada, Y., and Wu, X. Y., *Phys. Rev. Lett.* **67**, 3491-3494 (1991); Palathingal, J. C., Asoka-Kumar, P., Lynn, K. G., and Wu, X. Y., *Phys. Rev. A* **51**, 2122-2130 (1995).
43. Bergstrom, Jr., P. M., Kissel, L., and Pratt, R. H., *Phys. Rev. A* **53**, 2865-2868 (1996); Agger, C. K., and Sørensen, A. H., *Phys. Rev. A* **55**, 402-413 (1997).

# Current Status of (e,2e) Measurements of Energy-Momentum Densities

## Erich Weigold

*Research School of Physical Sciences and Engineering, Institute of Advanced Studies, Australian National University, Canberra ACT 0200, Australia*

**Abstract.** Electron Momentum Spectroscopy (EMS), or (e,2e) spectroscopy, is based on measuring the relative differential cross sections for kinematically complete high energy, high momentum transfer, electron impact ionisation events. Under these EMS conditions the ejected electron is knocked cleanly out of the target. Knowing the energies and momenta of the incident and two emitted electrons one can infer the separation (or binding) energies and momenta of the ejected electrons within target or sample. The momentum profile of the differential cross section is directly related to the momentum probability distribution of the Dyson orbital, i.e. square of the (quasi) particle amplitude obtained by overlapping the N electron target with the N-1 electron ion state. Transitions to different ion states can give detailed information on initial as well as final state correlation effects. For condensed matter targets it amounts to a measurement of the quasi-particle energy-momentum density of occupied bands. The principles of the measurement and its application to atoms, molecules, and amorphous, polycrystalline, and single crystal materials are discussed.

## I INTRODUCTION

Electron momentum spectroscopy (EMS) is a powerful technique for investigating the dynamic electronic structure of atoms, molecules, and condensed matter. It is unique in its ability to measure both the momentum densities and energies of the occupied electronic states of the target samples [1,2]. It is particularly sensitive to the valence electrons which are more localised in momentum space, although they are quite diffuse in coordinate space. Its sensitivity to the low momentum region (i.e. essentially the outer position region) of the wavefunction has proved particularly fruitful in quantum chemistry as a tool for developing accurate wavefunctions capable of describing a broad range of chemical and physical properties of molecules [3,4]. In the case of atoms and molecules it provides very sensitive quantitative tests for electron-electron correlation effects, both in the initial (usually ground) state of the target sample or in the ion. This is particularly true for transitions to ion states which are forbidden in the frozen-core independent particle model. Initial state correlations can, however, have significant influence on the momentum densities even for the "main" transitions, particularly in the low-momentum region, which must be largely due to long-range correlations [1–4]. In the case of solids it allows one to make a direct measurement of the complete spectral momentum density $A(\mathbf{q}, \epsilon)$. The peak in $A(\mathbf{q}, \epsilon)$ as a function of the electron momentum $\mathbf{q}$ and energy $\epsilon$ gives the band dispersion, whereas the magnitude of $A(\mathbf{q}, \epsilon)$ gives the momentum density or probability distribution. The width in energy of $A(\mathbf{q}, \epsilon)$ gives the quasiparticle lifetime. Since EMS measures real momenta it is applicable not only to single crystals, but applies equally well to amorphous or polycrystalline samples.

This paper is organised as follows. The theoretical background is outlined in section II. This is followed by the discussion of a range of atomic examples, demonstrating how EMS gives information on independent particle orbitals, quasiparticle or Dyson orbitals, correlation effects, relativistic effects, and momentum densities of excited states and oriented states. Section IV outlines some molecular examples. The application of EMS to solids pointing out experimental difficulties as well as its contribution to the understanding of the structure of solids, is discussed in section V. In conclusion some experimental developments as well as an extended range of applications is summarised.

## II  THEORETICAL BACKGROUND

The two outgoing electrons in an (e,2e) collision are energy and angle analyzed before coincidence detection in an (e,2e) spectrometer [1,2]. We will denote the incident electron by 0, and one of the outgoing electrons by $f$ (the faster or "scattered" electron in asymmetric kinematics), and the other by $s$ (the slower or "ejected" electron), although of course they are indistinguishable. The kinematics chosen for EMS corresponds to Bethe-ridge kinematics, that is for free electron-electron collisions. Often noncoplanar symmetric kinematics is used in which $E_f = E_s$, $\theta_f = \theta_s \sim 45°$, and $\phi = \phi_f - \phi_s + \pi$ is varied to vary the recoil momentum.

For each pair of detected electrons the separation energy $\epsilon$ and recoil momentum $\mathbf{p}$ are recorded. In a clean knockout the recoil momentum $\mathbf{p}$ is equal and opposite to the momentum $\mathbf{q}$ of the bound electron when it is struck. From energy and momentum conservation

$$\epsilon_i = E_o - E_f - E_s \tag{1}$$

$$\mathbf{p} = \mathbf{k_o} - \mathbf{k_f} - \mathbf{k_s}, \tag{2}$$

$$\text{and } \mathbf{q} = -\mathbf{p} = \mathbf{k_f} + \mathbf{k_s} - \mathbf{k_o}. \tag{3}$$

To ensure clean knock-out, the energies of the electrons must be high and the momentum transferred by the incident electron,

$$\mathbf{K} = \mathbf{k_o} - \mathbf{k_f}, \tag{4}$$

must be high, and $\mathbf{K} = \mathbf{k}_s$. The free electrons can then be described as plane waves and the differential cross section is given by [1,2]

$$\sigma(\mathbf{q}, \epsilon_i) = \frac{d^5\sigma}{d\Omega_f d\Omega_s dE_f} = C f_{ee} \sum_{av} |\langle i | \mathbf{a_q} | t \rangle|^2 \delta(\epsilon_i - E_o + E_f + E_s), \quad (5)$$

where $C$ is a constant depending on the momenta of the electrons, $f_{ee}$ is the electron-electron collision factor, which is also essentially constant under EMS conditions [1,2], $|i\rangle$ and $|t\rangle$ are the electronic final ionic and initial (usually ground) target state, and the operator $\mathbf{a_q}$ annihilates an electron of momentum $\mathbf{q}$ in the initial many-body target state $|t\rangle$. The sum over final-state and average over initial state degeneracies is indicated by $\sum_{av}$. For a non-oriented atomic or molecular target, this average over initial (rotational) degeneracies is equivalent to spherical averaging of the differential cross section, i.e. integrating over directions $\hat{\mathbf{q}}$. For a molecule the average over initial vibrational states can be accurately approximated by taking the initial and final electronic states at the initial-state equilibrium positions [5].

For valence orbitals of atoms and molecules the plane wave impulse approximation (5) is generally valid for energies above about 1keV, $K \gtrsim 5au$ and $q \lesssim 1.5au$ [1,2]. The energy independence of the structure information provides a test for the validity of (5).

We can see from (5) that the EMS cross section provides a direct measurement of the spectral density function

$$A(\mathbf{q}, \epsilon) = \sum_{av} |\langle i | \mathbf{a_q} | t \rangle|^2 \delta(\epsilon_i - E_o + E_f + E_s). \quad (6)$$

For atoms and molecules it is usual to make the weak-coupling expansion for the one-electron target-ion overlap amplitude [1,2],

$$\langle i | \mathbf{a_q} | t \rangle = \sum_\alpha \langle i | \alpha \rangle \langle \alpha | \mathbf{a_q} | t \rangle$$
$$= \langle i | \alpha \rangle \langle \alpha | \mathbf{a_q} | t \rangle, \quad (7)$$

where $|\alpha\rangle$ is assumed to be a unique one-hole state formed by annihilating an electron from the orbital $|\alpha\rangle$ in the target state. One can define the experimental orbital to be the quasiparticle amplitude or Dyson orbital

$$\psi_\alpha(\mathbf{q}) \equiv \langle \alpha | \mathbf{a_q} | t \rangle. \quad (8)$$

The differential cross section is then proportional to the absolute square of the Dyson orbital multiplied by the spectroscopic factor (or pole strength)

$$S_i^{(\alpha)} = |\langle i | \alpha \rangle|^2. \quad (9)$$

This is the probability that the final ion-state $|i\rangle$ contains the one-hole state $|\alpha\rangle$. Thus the spectroscopic factors give the intensities of transitions (main and satellite lines) for a given manifold of ion states belonging to orbital $|\alpha\rangle$.

The spectroscopic factor satisfies the manifold sum rule

$$\sum_i S_i^{(\alpha)} = \sum_i \langle \alpha | i \rangle \langle i | \alpha \rangle = 1. \quad (10)$$

# III  ATOMS

For the simplest target, atomic hydrogen, $\psi_\alpha(\mathbf{q}) = \psi_{1s}(\mathbf{q})$, and $S_i^{(1s)} = 1$ since there are no electron-electron correlations and only one final state with $\epsilon_i = 13.6\text{eV}$. Figure 1 shows the measurements of $|\psi_{1s}(q)|^2$ obtained by Lohmann and Weigold [6] normalised to the absolute square of the Schrödinger momentum space wave function (solid curve). Atomic units of momentum are used. The noncoplanar symmetric measurements are independent of energy, which is a check on the validity of the plane-wave theory.

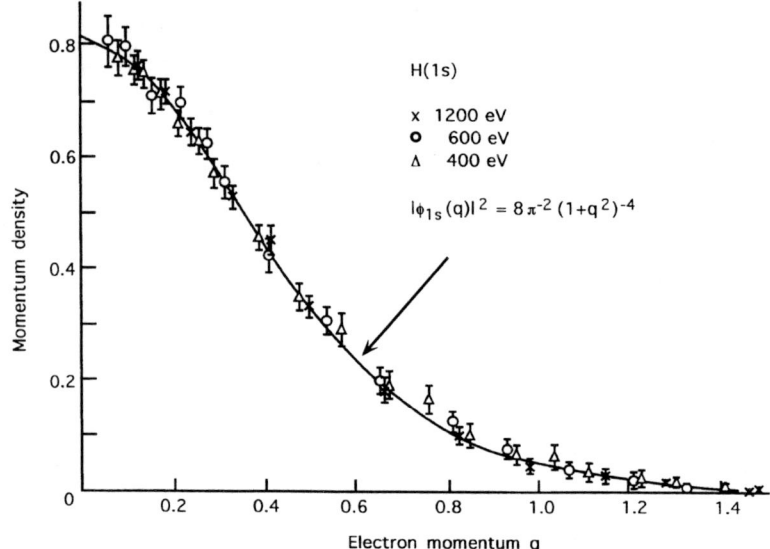

**FIGURE 1.** The momentum profile for atomic hydrogen measured at the indicated energies compared with the square of the exact Schrödinger momentum space wavefunction (solid line).

Electron correlations in many-electron targets show up in two ways in the EMS measurements. If the initial target state is well described by the independent particle Hartree-Fock (or Dirac-Fock) model, then the Dyson orbital (8) is essentially the Hartree-Fock orbital. Correlations in the ion then lead to multiple transitions (a manifold) for ionisation from this orbital, the intensities of the lines being proportional to $S_i^{(\alpha)}$.

Ionisation from the valence $ns$ orbital of the nobel gases provide examples. Table 1 gives the spectroscopic factors for the argon 3s transitions determined from EMS compared with satellite intensities from photo-electron spectroscopy and calculated spectroscopic factors from a variety of many-body calculations. The better calculations tend towards the EMS measurements. The disagreement between EMS and photo-electron spectroscopy can be attributed [1,7] to the fact that photo-electron spectroscopy does not measure spectroscopic factors as defined by (9). The EMS

# IV MOLECULES

The simple linear molecule ethyne, $C_2H_2$, is an example where the target Hartree-Fock approximation works very well, although strong correlations are observed in the ion for ionisaton from the inner valence $2\sigma_g$ orbital [19]. In the binding-energy spectrum shown in figure 7, there is some structure at high energy in addition to the peaks corresponding to the valence $1\pi_u$, $3\sigma_g$, $2\sigma_u$ and $2\sigma_g$ orbitals. The momentum profile of this structure shows that it belongs to the $2\sigma_g$ orbital. The momentum profiles for all transitions are very well described by self-consistent field (SCF) orbital momentum densities.

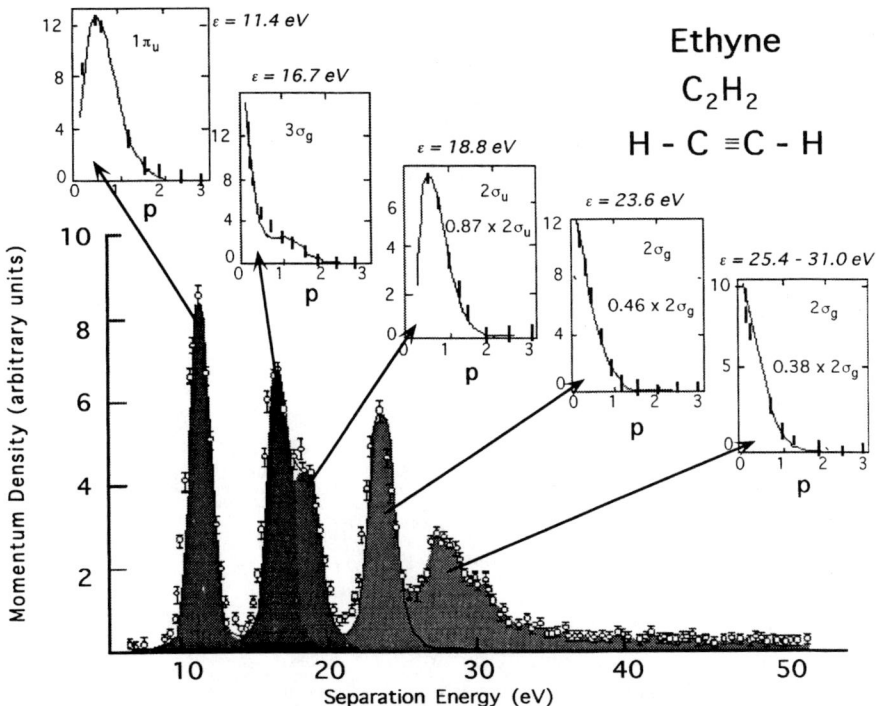

**FIGURE 7.** The binding-energy spectrum for valence electrons of ethyne and the corresponding measured and calculated SCF independent-particle momentum densities.

Correlations in the initial state can lead to orbital densities significantly different from calculated Hartree-Fock orbital densities. Such is the case for the outermost orbitals of water, where electron correlations significantly enhance the density at low momentum relative to the Hartree-Fock orbital. Full CI overlap calculations of the Dyson orbital, Green's function many-body techniques, or Kohn-Sham orbitals derived from density-functional-theory (DFT), which includes correlation and exchange effects, give a much improved description of the experimental orbital

[1,2,4,20,21]. The simple second row hydride HF shows a similar discrepancy, particularly in the momentum distribution of its highest occupied molecular orbital. This can be seen in figure 8, where the 400 and 1200eV data of Brion et al. [22] and the 1500eV data of Braidwood et al. [23] are compared with simple SCF orbitals [24], SCF limit orbitals [25], CI Dyson orbitals [25], DFT Kohn-Sham orbitals [4] and many-body perturbation theory Dyson orbitals(GOA) [22]. The best SCF wave functions are unable to explain the observed large low-momentum density in the experimental orbital. Full CI overlap or DFT Kohn-Sham orbitals are required to describe the results, which show the importance of long-range electron-electron correlations in these molecules. The recent trend in molecular applications of EMS [3,4,25–27] has been to use EMS momentum profiles to fine-tune the quantum chemical many-body calculations, and then use these optimised wave functions to predict important chemical and physical properties of the molecules. Nicholson et al. [28] have used EMS momentum profiles to derive Dyson orbitals for molecules direct from the data using SCF basis orbitals.

## V  SOLIDS

In solids, as in atoms and molecules, the quasiparticle spectral density $A(\mathbf{q}, \epsilon_i)$ for band $i$ contains much more information than just the band peak position as a function of $\mathbf{q}$ and $\epsilon$, i.e. the band dispersion. The energy width in $A(\mathbf{q}, \epsilon_i)$ gives the quasiparticle lifetime at momentum $\mathbf{q}$ and energy $\epsilon_i$. The magnitude of $A(\mathbf{q}, \epsilon)$ is the probability of the quasiparticle having momentum $\mathbf{q}$ and energy $\epsilon$. Correlations can give rise to additional structures (satellites or intrinsic plasmons) [29], which are observable by EMS.

Since the momentum $\mathbf{q}$ measured in EMS is the real electron momentum, EMS of solids does not require the target sample to be a single crystal, as for instance in angle-resolved photo-electron spectroscopy. Thus it can equally well be used to measure $A(q, \epsilon)$ for amorphous and polycrystalline samples.

Although the first EMS measurement was on a solid target [30], technical problems severely limited its application to solids until the recent development of the Flinders University high resolution multiparameter spectrometer [31]. The severest limitation was the poor energy and momentum resolution and the low coincident count rates in the earlier spectrometers.

In order to measure the complete momentum range from $q = 0$ it is necessary to employ transmission kinematics. This requires the preparation of thin (10–20nm) self-supporting membranes of the target sample of around 0.3mm diameter. The thinner the specimen is the smaller are the multiple-scattering effects. The procedures for preparation of a range of these target samples is discussed by Fang et al. [32]. The composition of the specimens has to be well-known, and the exit surface has to be clean for "thicker" samples, when the entrance surface in effect does not contribute to coincidences.

The target sample preparation and characterisation facilities used at Flinders

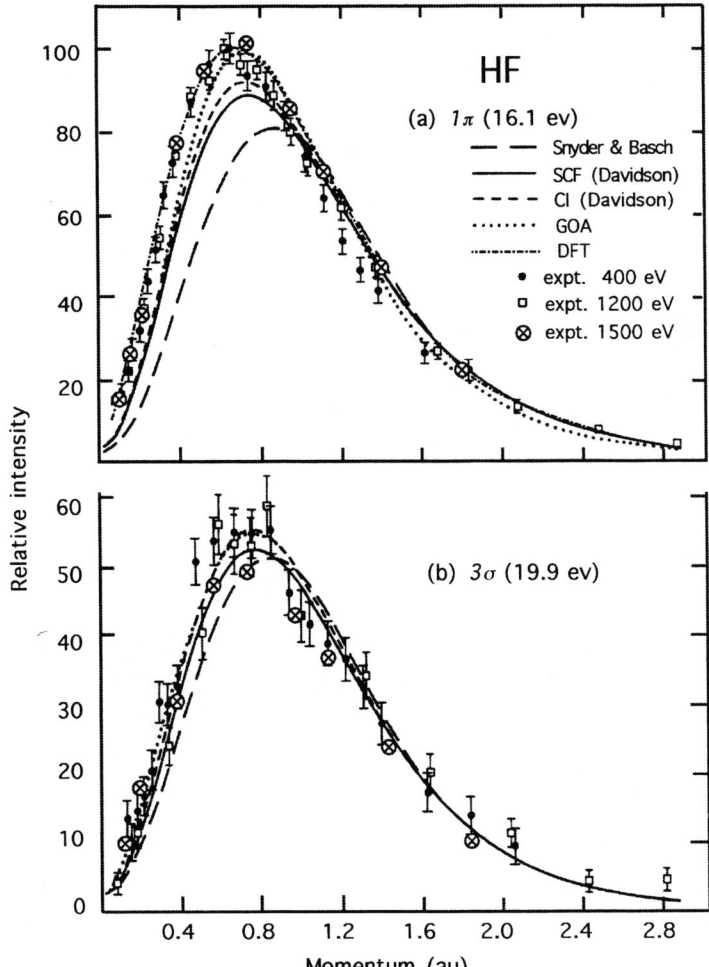

**FIGURE 8.** The momentum distributions for the outermost $1\pi(\epsilon = 6.1\text{eV})$ and $3\sigma(\epsilon = 19.9\text{eV})$ orbitals of HF at total energies of 400eV (□), 1200eV (●) and 1500eV (×) compared with calculated densities, using Snyder and Basch SCF orbitals [24] (long dashes), SCF limit orbitals [25] (solid curve), CI Dyson orbitals [25] (short dashes), density functional theory (DFT) orbitals [4] (dash-dots), and generalised overlap amplitude many-body perturbation theory Dyson orbitals (GOA) [23] (dots).

University are shown schematically in figure 9. Samples can be thinned by cleaving, chemical and/or electrochemical etching. Further thinning can be achieved by reactive-ion (plasma) etching or ion-beam etching. The thickness of the specimens can be measured by a laser interferometer. Other targets can be prepared by evaporation (amorphous carbon) or evaporation onto a thin free-standing film.

Surface characterisation by Auger spectroscopy and annealing can take place in the facility, which is maintained under ultra-high vacuum conditions. The samples can be transferred from one chamber to another and to the spectrometer under ultra-high vacuum conditions.

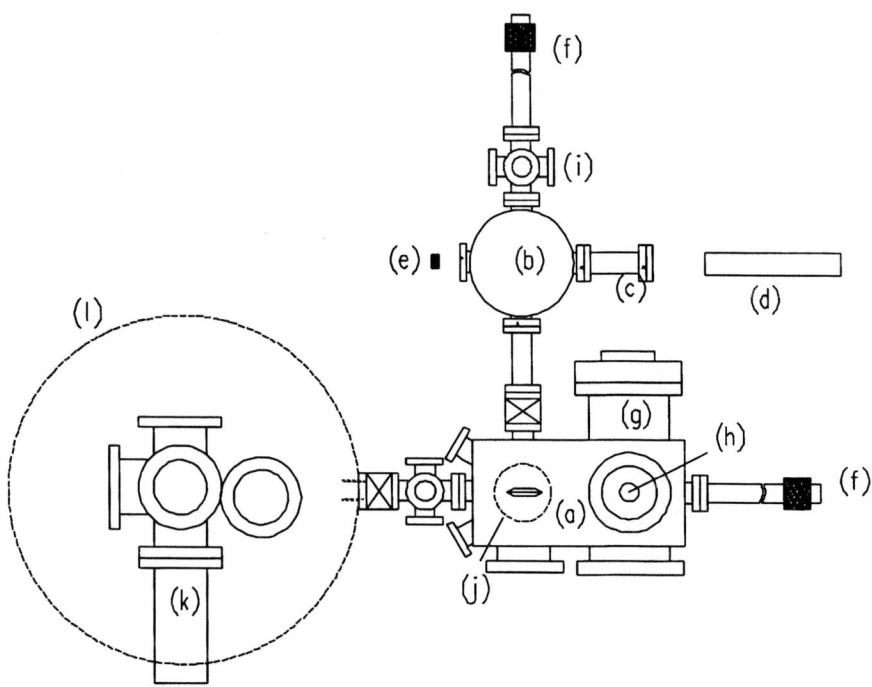

**FIGURE 9.** Plan of the target preparation facilities consisting of ultra-high vacuum preparation chamber (a), (reactive) ion etching chamber (b), ion etching gun (c), laser (d), photon detector (e), transfer arms (f), Auger system for surface analysis (g), sample manipulator and annealing facility (h), load lock and optical microscope for viewing sample (i), evaporator (j), transmission diffractometer (k), and vacuum tank for main spectrometer (l).

A transmission electron-diffraction facility, mounted on top of the spectrometer, is used to further characterise the sample. The sample is mounted on a manipulator with movement in all directions and rotation about the vertical (y) direction (perpendicular to the mean scattering plane). The diffractometer and manipulator permit one to set up a single-crystal specimen so that spectral densities can be measured along chosen crystallographic directions.

The noncoplanar asymmetric kinematics of the Flinders spectrometer is shown schematically in figure 10. The horizontal laboratory $x - z$ plane is defined by $\mathbf{k}_o$ (z-direction) and the mean direction of $\mathbf{k}_f$, and the $y$ direction is the vertical. The azimuthal range ($-18° < \phi_f < 18°$, $180° - 7° < \phi_s < 180° + 7°$) is sam-

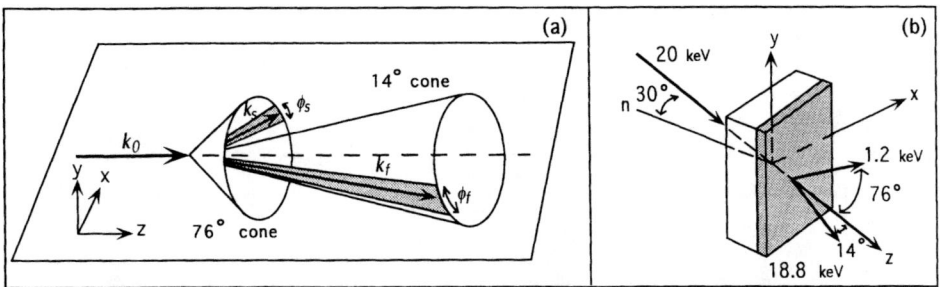

**FIGURE 10.** Schematic representation of the kinematics of the Flinders high-energy EMS spectrometer, showing in (a) the range of angles viewed simultaneously and in (b) the incoming and outgoing energies and sample orientation. The shaded area shows the effective depth (~2nm) from which EMS data are obtained, limited by the small mean free path of the slow (1.2keV) electrons.

pled simultaneously to give cross-section measurements over a range of $q_y$, with $q_x$ and $q_z$ fixed and essentially zero for polar angles $\theta_f = 14°$ and $\theta_s = 76°$ with $E_f = 18.8 \pm 0.01$keV, $E_s = 1.2 \pm 0.02$keV and $E_o = 20$keV $+\epsilon$. $\theta_s$ can be varied over a small range about 76° so that one can sample events as a function of $q_y$ with $q_x$ and $q_z$ not being zero. Hemispherical and toroidal electrostatic deflector analyzers provide energy analysis of respectively the fast and slow emitted electrons. Two-dimensional electron detectors consisting of a set of microchannel plates followed by a resistive anode, mounted at the exit plane of each analyzer, give the azimuthal angle and energy of each detected electron. From this the momentum **q** and separation energy $\epsilon$ corresponding to each coincident event can be recorded. Typical true coincidence count rates with this multiparameter spectrometer are 2–4Hz with an energy resolution of 0.9eV [31,33]. This compares with the earlier measurements [30] with energy resolution of around 100eV and coincidence count rates of the order of $10^{-3}$ Hz. The arrival position on the two-dimensional detector allows one also to infer the trajectory of the detected electron through the analyzer, allowing transit time corrections to be made, significantly improving the coincidence timing resolution. The whole experiment is under complete computer control. The data-reduction techniques are outlined in references [1,34].

It is interesting to see how EMS differentiates the electronic structure of three different forms of amorphous carbon, diamond-like, graphitic and fullerene. The measured energy-momentum densities are shown in figure 11.

Both the graphitic and diamond samples show a parabolic sigma band, the bandwidth for diamond being somewhat larger. All of the electrons in the tetrahedral structure contribute to the sigma band. However in graphite some electrons occupy pi orbitals, which change sign as they cross the plane of the graphite sheet. They therefore have no density at zero momentum. The centre panel in figure 11 shows this pi band, cradled inside the sigma band with much smaller average energy. The pi band and the smaller sigma bandwidth, distinguishes graphite from diamond.

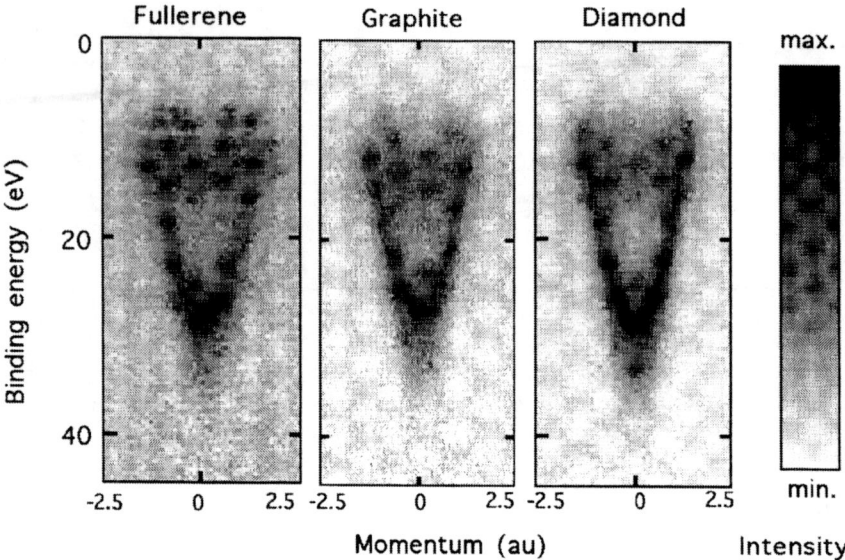

**FIGURE 11.** EMS energy-momentum densities for randomly-oriented fullerene and amorphous graphite and diamond. Plasmon density has been deconvoluted from the diagrams.

The energy momentum density of the fullerene film agrees quite well with that calculated for the molecule [35]. This is because the molecules in the film are bonded mainly by weak van der Waals forces with only minor overlap of orbitals between adjacent molecules. Lower-orbital energies are well spaced, so that they appear as discrete dispersionless momentum densities. The single pi band of graphite is split into two for fullerene. This is because in graphite the pi orbitals are symmetric about the plane of the graphite sheet, while for fullerene the lobes inside and outside the shell are different.

Figure 12 shows the measured spectral-momentum density [36] for a single crystal of graphite along three different crystal directions, matched with the corresponding calculated linear-muffin-tin-orbital (LMTO) densities broadened by the experimental energy resolution, 0.9eV. In the first two panels momentum is measured through the $\Gamma(\mathbf{q} = 0)$ point in the $\Gamma - M$ and $\Gamma - K$ directions. In the third panel the momentum was offset from zero by moving the slow-electron detector angle $\theta_s$. The displacement is 0.55au in the $\Gamma - M$ direction and 0.41 in $\Gamma - A$ direction normal the the basal plane. The line in the Brillouin zone that is scanned in each measurement is shown schematically above each corresponding panel in fig. 12.

In every case the density for the sigma band has an energy maximum at the $\Gamma$ point of the second zone in the scanning direction, confirming the LMTO calculation. As expected for $q_z = 0$ no pi-band density is observed in the first two panels. Pi-band density is observed in the third panel, in agreement with the LMTO calculation.

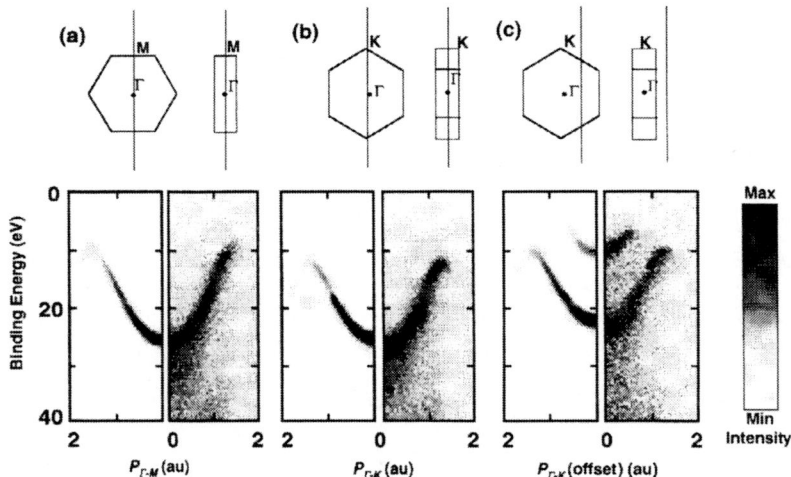

**FIGURE 12.** EMS of crystalline graphite. Three experiments are represented by grey-scale panels. The right side of each panel shows the experimental energy-momentum density. The left side shows the LMTO calculation, convoluted with experimental energy resolution. Above each panel is a schematic diagram showing the path through the Brillouin zone that is scanned.

The first two panels show the dependence of the sigma band near the zone edge on the crystal direction, the band in the $\Gamma - M$ direction going considerably higher in energy. The width of the band, $22 \pm 0.5$eV, exceeds the bandwidth obtained in earlier photo-electron spectroscopy measurements. This width has recently been confirmed through detailed angle-resolved photo-electron spectroscopy measurements by Heske et al. [37].

There are two main difficulties in accurately interpreting all of the information contained in the data. One is experimental and the other is theoretical. The experimental one is due to the presence of multiple scattering by the free electrons. Both elastic and inelastic scattering (mainly plasmon excitation) contribute to a smooth background on which the clean events are superimposed. Assumptions on the shape of the background make large differences to the area attributed to a peak. As the transport of keV electrons through solids is quite well understood it is possible to simulate these multiple scattering effects, using Monte Carlo procedures [1,38]. In this way it is possible to compare the experiment with theory that includes these multiple scattering effects.

The theory problem arises because in a solid all electrons interact strongly with each other due to the long-range nature of the Coulomb field. This leads to screening which reduces the effective range of the electron-electron interaction. The final result is that, even for simple metals, which appear at first sight as free-electron metals (as far as dispersion is concerned), the full many-body calculations of the spectral-momentum density (the spectral function) give considerable intensity away from the single-particle branch [29].

Let us illustrate these effects for the case of aluminum [39]. In figure 13 we show the energy spectrum near zero momentum. In the upper panel are the spectral functions at zero momentum given by various theories. The first is the LMTO model with lifetime broadening included, since this independent particle model has the quasiparticle lifetime as being infinite. The other three are many-body calculations, which give lifetime broadening and intrinsic plasmon (satellite) structure. The lower panel shows these theories corrected for multiple scattering effects compared with the measurements. Quite clearly the cumulant expansion model with

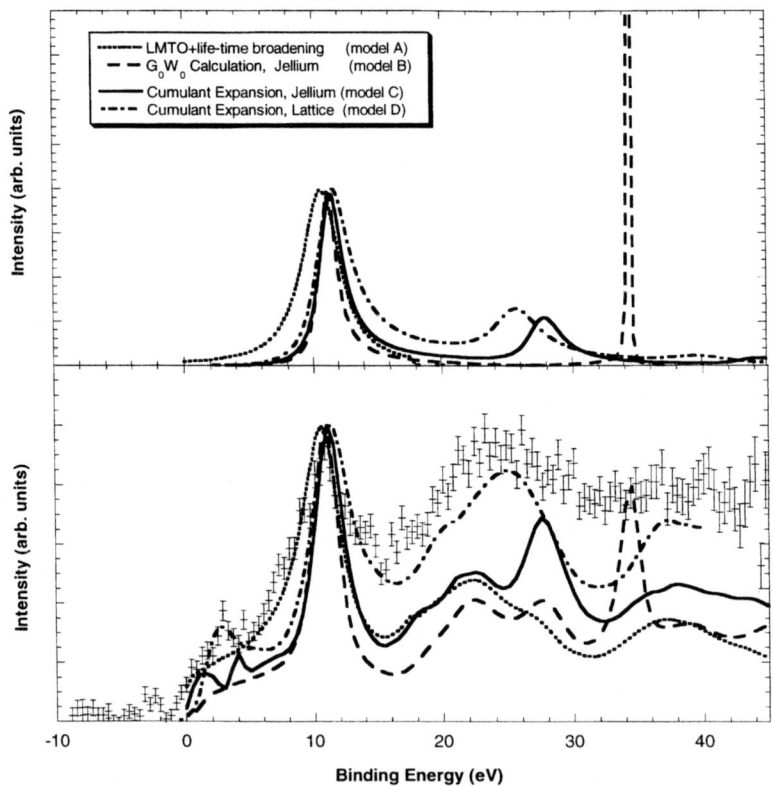

**FIGURE 13.** The valence spectral function of aluminum at zero momentum. In the top panel the different theoretical spectral functions as indicated. The lower panel shows the theories corrected for multiple-scattering effects compared with the EMS measurements. All have been normalised to the same quasiparticle peak height at $\epsilon = 11\text{eV}$. From ref [39], with permission from IOP Publishing Limited.

the lattice included (model D) gives the best description of the data. This is for both the lifetime broadening, ($\sim$ 3.5eV experimentally and $\sim$ 3eV for model D), and for the position and relative magnitude of the intrinsic satellite. The comparison at other momenta is shown in figure 14.

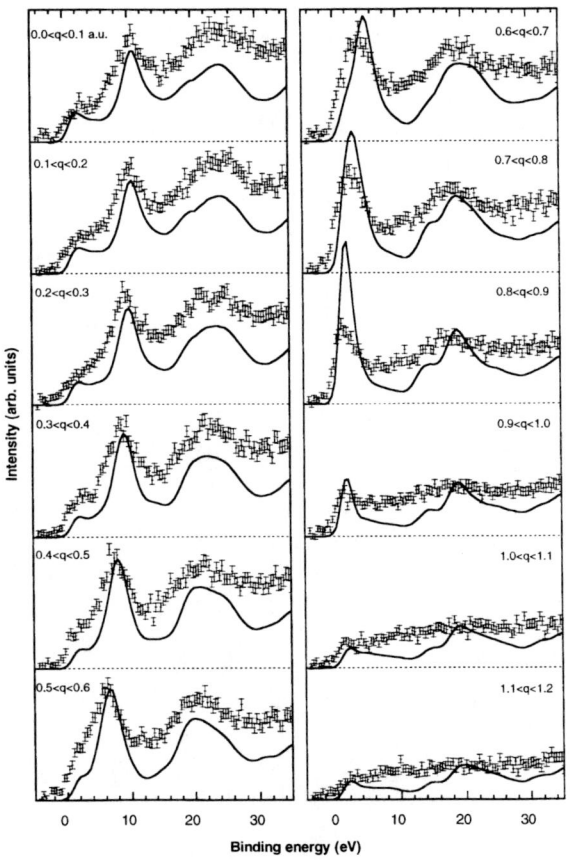

**FIGURE 14.** A comparison of the EMS measurements for the indicated momentum intervals with the spectral function based on the cumulant expansion including the crystal lattice (model D) (solid curves), including multiple-scattering corrections.

The data show that both electron-electron correlation and the lattice have to be considered to get a good description of the EMS measurement, although the band dispersion is actually well described by a free-electron parabola.

EMS seems to be an excellent probe for testing calculations of spectral functions, especially if multiple scattering effects can be reduced. This can be achieved simply by increasing the energies of the incoming and outgoing electrons, which leads to an increase in the mean free path of the electrons. A new high-energy spectrometer employing noncoplanar symmetric kinematics, with $E_o = 50\text{keV}$, $E_f = E_s = 25\text{keV}$, and $\theta_f = \theta_s$ variable around 44°, has been constructed at the Australian National University. With a 10nm target of amorphous carbon it has count rates in the range 1–10Hz with an energy resolution at present of around 1.5eV (which is being improved) and a coincidence-to-background count rate of around 1000:1. Preliminary data from this high-energy high-resolution spectrometer show that multiple-scattering effects are much reduced.

## VI  CONCLUDING REMARKS

Clearly EMS provides a powerful tool for investigating the electronic structure of atoms, molecules, and solids. It provides a direct measure of the spectral-momentum densities of the occupied states (bands) in matter. It is a sensitive probe of correlation effects in both initial and final states in atoms and molecules. Similarly in solids it provides a direct measure of the full spectral-momentum density function, including the band dispersion, intrinsic satellite structure and quasi-particle lifetime. Thus it provides a stringent test-bed for any many-body model of the electronic structure of the atom, molecule or solid.

Experimentally the developments are to go to higher energy with improved energy resolution using two-dimensional (energy and momentum) detection. Techniques have been developed for studying excited and oriented targets. Similarly polarised incident electrons have great potential in probing relativistic effects in atoms and molecules and magnetic effects in solids.

I am grateful for the many students and colleagues who have made much of the work reported here possible. In particular I must mention the major contributions of Ian McCarthy and Anatoly Kheifets on the theory aspects and Maarten Vos for carrying forward the experimental program.

## REFERENCES

1. Weigold E., and McCarthy I.E., *Electron Momentum Spectroscopy*, Kluwer Academic/Plenum Press, New York (1999).
2. McCarthy I.E., and Weigold E., *Rep. Prog. Phys.* **54**, 789 (1991).
3. Davidson E.R., *Can. J. Phys.* **74**, 757 (1996).
4. Duffy P., *Can. J. Phys.* **74**, 763 (1996).

5. Dey S., McCarthy I.E., Teubner P.J.O., and Weigold E., *Phys. Rev. Lett.* **34**, 782 (1975).
6. Lohmann B. and Weigold E., *Phys. Lett. A* **86**, 139 (1981).
7. Kheifets A.S., and Amusia M.Ya., *Phys. Rev. A* **46**, 1261 (1992).
8. McCarthy I.E., Pascual R., Storer P., and Weigold E., *Phys. Rev. A* **40**, 301 (1989).
9. Brunger M.J., McCarthy I.E., and Weigold E., *Phys. Rev. A* **59**, 1245 (1999).
10. Svensson S., Helenelund, K., and Gelius U., *Phys. Rev. Lett.* **58**, 1624 (1987).
11. Mitroy J., Amos K., and Morrison I., *J. Phys. B* **17**, 1659 (1984).
12. Hibbert A., and Hansen J.E., *J. Phys. B* **20**, L245 (1987).
13. Amusia M.Ya., and Kheifets A.S., *Aust. J. Phys.* **44**, 293 (1990).
14. Amusia M.Ya., and Kheifets A.S., *J. Phys. B* **18**, L679 (1985).
15. von Niessen, W. *Private communication.* (1989).
16. Cook J.P.D., McCarthy I.E., Stelbovics A.I., and Weigold E., *J. Phys. B* **17**, 2339 (1984).
17. Cook J.P.D., Mitroy J., and Weigold, E., *Phys. Rev. Lett.* **52**, 1116 (1984).
18. Dorn A., Elliot A., Lower J., Weigold, E., Berakdar J., Engelns, H. and Klar, H., *Phys. Rev. Lett.* **80**, 257 (1998).
19. Weigold E., Zhao K., and von Niessen W., *J. Chem. Phys.* **94**, 3468 (1991).
20. Bawagan A.O., Brion C.E., Davidson E.R. and Feller D., *Chem. Phys.* **113**, 19 (1987).
21. Cambi R., Ciullo G., Sgamellotti A., Brion C.E., Cook J.P.D., McCarthy I.E., and Weigold E., *Chem Phys.* **91**, 373 (1984).
22. Brion C.E., Hood S.T., Suzuki I.H., Weigold E., and Williams G.R.J., *J. Electron. Spectros.* **21**, 71 (1980).
23. Braidwood S.W., Brunger M.J., Konovalov D.A., and Weigold E., *J. Phys. B* **26**, 1655 (1993).
24. Snyder L.C., and Basch H., *Molecular Wave Functions and Properties*, Wiley, New York (1972).
25. Davidson E.R., Feller D., Boyle C.M., Adamowicz L., Clark S.A.C., and Brion C.E., *Chem. Phys.* **147**, 45 (1990).
26. Adcock W., Brunger M.J., Michalewicz M.T. and Winkler D.A., *Aust. J. Phys.* **51**, 707 (1998).
27. Adcock A., Brunger M.J., McCarthy I.E., Michalewicz M.T., von Niessen W., Wang F., Weigold E., and Winkler D.A., *J. Am. Chem. Soc.* (to be published).
28. Nicholson R.J.F., McCarthy I.E., and Brunger M.J., *Aust. J. Phys.* **51**, 691 (1998).
29. Lundqvist B.I., *Phys. Kondens. Mat.* **7**, 117 (1968).
30. Camilloni R., Giardini-Guidoni A., Tiribelli R., and Stefani G., *Phys. Rev. Lett.* **29**, 618 (1972).
31. Storer P.J., Caprari R.S., Clark S.A.C., Vos M., and Weigold E., *Rev. Sci. Instr.* **65**, 2214 (1994).
32. Fang Z., Guo X., Utteridge S., Canney S.A., McCarthy I.E., Vos M., and Weigold E., *Rev. Sci. Instr.* **68**, 4396 (1997).
33. Canney S., Brunger M.J., McCarthy I.E., Storer P.J., Utteridge S., Vos M., and Weigold E., *J. Electron. Spectrosc.* **83**, 65 (1997).
34. Vos M., Caprari R.S., Storer P.J., McCarthy I.E., and Weigold E., *Can. J. Phys.*

**74**, 829 (1996).
35. Vos M., Canney S.A., McCarthy I.E., Utteridge S., Michalewicz M.T., and Weigold E., *Phys. Rev. B* **56**, 1309 (1997).
36. Vos M., Fang Z., Canney S.A., Kheifets A.S., McCarthy I.E., and Weigold E., *Phys. Rev. B* **56**, 963 (1997).
37. Heske C., Treusch R., Himpsel F.J., Kakar S., Terminello L.J., Weyer H.J. and Shirley E.L., *Phys. Rev. B* **59**, 4680 (1999).
38. Vos M., and Bottema M., *Phys. Rev. B* **54**, 5964 (1996).
39. Vos M., Kheifets A.S., Weigold E., Canney S.A., Holm B., Aryasetiawan F. and Karlsson K., *J. Phys. Condens Matter* **11**, 3545 (1999).

# Photo- and charged-particle-ionization of He and $D_2$ studied with COLTRIMS

M.A. Abdallah [1], Matthias Achler[2], H.Braeuning [2], Angela Braeuning-Deminian[2], C.L.Cocke[1], Achim Czasch[2], R.Doerner [2], A. Landers[1,5], V.Mergel [2], T.Osipov[1], M.Prior [3], H.Schmidt-Boecking[2], M. Singh[1], T.Weber[2], W.Wolff [4], and H.E.Wolf [4]

[1]J.R.Macdonald Laboratory, Physics Department, Kansas State University, Manhattan, KS 66506;
[2]Institut fuer Kernphysik, Univ. Frankfurt, August-Euler-Str.6,D-60486 Frankfurt, Germany ;
[3]Lawrence Berkeley National Laboratory,Berkeley, CA 94720; [4]Instituto de Fisica, Universidade Federal do Rio de Janeiro Caixa Postal 68.528, 21945-970, Rio de Janeiro, Brazil;[5]Physics Dept., Western Michigan University, Kalamazoo, MI 49008

**Abstract.** The COLTRIMS (COLd Target Recoil Ion Momentum Spectroscopy) approach to final-state momentum imaging is now being widely used in at least a dozen accelerator and synchrotron-radiation laboratories in the world and its use is growing rapidly. The technique combines fast imaging detectors with a supersonically cooled gas target to allow the charged particles from a collision , including both recoil ions and electrons, to be collected with extremely high efficiency and with fully measured vector momenta. It allows the investigation of correlations between ejected momentum fragments and in some cases the identification of collective modes of disintegration. When molecular targets are used, it allows the *a posteriori* determination of the alignment of the molecule at the time of the collision. We will discuss the use of this approach to study the single and double ionization of He and $D_2$ by the impact of photons and of charged particles over a wide range of velocities.

## INTRODUCTION

The acronym COLTRIMS (COLd Target Recoil Momentum Spectroscopy) [1-3] , is sufficiently well established within the atomic collisions community that it has begun to be abused. Central to this experimental approach is the use of imaging detectors to register , in event-mode, the vector momenta of all charged ejecta from an ionizing collision except for that of the fast projectile (in the case of heavy particle projectiles). Unless neutral particles emerge from the reaction, conservation of energy and momentum allow the full kinematic final state of the system to be determined in this way. Implied by the acronym is also the use of a cooled gas target which allows the momentum of the recoil ion(s) to be measured with meaningful precision. However, the term is now occasionally applied to cases in which the full kinematics can be determined without knowledge of the recoil momentum (photo double ionization , for example, where two electron momenta are measured ) or for which the center-of-mass motion of the heavy fragments is not as important as the relative internal motion of these fragments (molecular fragmentation, for example). In this paper we will discuss single ionization of He and $D_2$ by charged particles and, briefly, photo-double ionization of $D_2$ by photons.

The ejection of slow electrons from neutral targets by fast charged particles is the central process whereby ions deposit energy in passing through matter. At least four decades of study of this problem have been driven by the need for understanding of this process for applications in the interaction of charged particles with biological materials, ion implantation, astrophysics, etc. and by the intellectual challenge of describing the dynamics of this electronic transition. What can possibly be new in such a time-honored field? One answer, that presented in this article is, *Momentum Imaging*. This approach adds at least two new ingredients to the mix: (1) The major part of the charged particle ionization cross section, typically more than 90 %, produces electrons with energies of less than 20 eV. Dispersive electron spectroscopy (DES), which excels for large electron energies and has far better resolution than imaging does for high energy, has always had severe experimental problems in the soft electron region. Only the bravest authors have published spectra measured this way below 10 eV. (2) Coupled with recoil momentum spectroscopy, momentum imaging allows the event-by-event determination of the momentum transfer to the electron (**k**), to the recoil (**q**) and to the projectile (**p**). For low beam velocity (v), this gives us the ability to select an impact parameter and collision plane; for high velocity, we can determine the momentum transfer **p** which appears in all perturbation theory treatments.

## EXPERIMENT

The basic COLTRIMS approach, which originated at the University of Frankfurt, has been discussed in many review articles [1-3]. For the reader's convenience, we show in fig.1, a schematic of the basic apparatus used in the present experiments. The target is a supersonic gas jet of He with an internal temperature below 1K in most experiments, flowing in the y-direction. The beam, travelling in the z-direction, intersects this beam and the electrons and recoiling ion fragments are collected by a static electric field of a few tens of volts/cm directed along the x-direction. More details are provided in refs. [1-6].

## RESULTS

### Low energy ionization of He and $D_2$

The ionization of neutral gaseous targets by charged particles whose velocity is much smaller than that of the target electrons has been a thorny problem for decades. For low v, where capture dominates, excitation is known to proceed via molecular

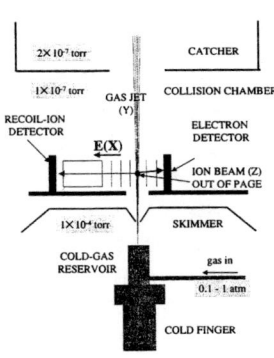

**FIGURE 1.** Schematic of apparatus.

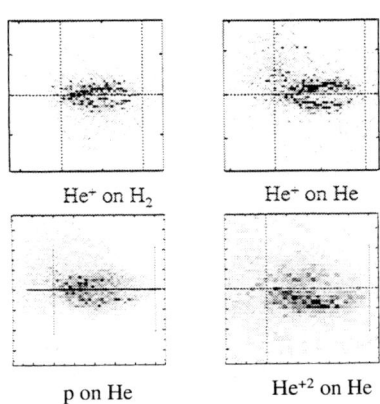

**FIGURE 2.** Electron momentum spectra for v=0.64 a.u. projectiles on He and $H_2$. The spectra are "topviews" (see text). The vertical lines indicate the positions in mometum space of the target and projectile (z direction), while the horizontal line indicates the zero for the y direction.

promotion mechanisms, as has been documented by much experimental work from nearly three decades ago [7-9]. As v increases, but remains short of matching velocity, the ionization cross section increases rapidly and the question arises of how the molecular excitation picture can be extended to describe the ejection of electrons into the continuum. COLTRIMS is perfectly suited to the study of this problem because it has a very high collection efficiency, low energy electrons can be studied with ease right down to zero energy, and it is possible to determine a collision plane and impact parameter for each event. This last property results from the fact that the transverse momentum transfer to the electron is very small, typically less than a few tenths of an atomic unit, while a major transverse momentum transfer of often tens of atomic units occurs between recoil and projectile. If the classical potential can be modeled, the impact parameter b can be calculated from **q** (or **p**).

In fig. 2 we show examples of electron momentum spectra taken for several projectiles singly ionizing He and $H_2$. These spectra are projections into the Y-Z plane of the momentum space distribution of electrons produced when recoil ions are selected to force the recoil to lie in the Y-Z plane. This view ("top view") represents a view of the electron distribution seen looking down on the collision plane from above. The corresponding "side view" images are dull: they show that the electrons are nearly confined to the collision plane, having typical transverse momenta with respect to this plane below 0.2 a.u.[6]. The longitudinal momentum transfer to the recoil has been used to either select target ionization of without excitation or to ensure that excitation contributions are small, while selection of the recoil charge state ensures that single target ionization has occurred (no transfer ionization, for example). Several features persist over all collision systems: first, the longitudinal electron momentum far exceeds the transverse momentum on the average, with an average value near the product of the electron mass and half of the projectile velocity. This behavior might be claimed as evidence for "saddle point" behavior were it not for the fact that the motion of the center does not move correctly with increasing projectile charge state [10]. Second, there is a two-fingered structure to all the spectra, with a quasi-nodal line along the Z axis, the ultimate internuclear axis of the departing heavy fragments. Slices across the spectra of fig. 2 show this feature more clearly [6]. The spectra resemble to some extent figures of π-orbitals drawn in configuration space in chemistry textbooks. We note that very similar spectra were first reported by Doerner et al. [11] and have very recently been observed in the transfer ionization channel for $He^{+2}$ on He by Schmidt et al. [12], suggesting that the ejection of one electron into the continuum follows the same basic mechanism regardless of the fate of the second one.

There are no theoretical calculations of these distributions, partially because even the basic ionization mechanisms are poorly known and partially because the calculation of continuum soft electron distributions are notoriously difficult. However, a mechanism suggested by Macek and Ovchinnikov [13] is illustrated in fig. 3, which shows the energy level scheme for $H_2^+$. The production of continuum distributions with π-symmetry is natural if the enabling process is the rotational coupling of 2pπ and 2pσ orbitals at small internuclear distances, followed by a saddle-like promotion

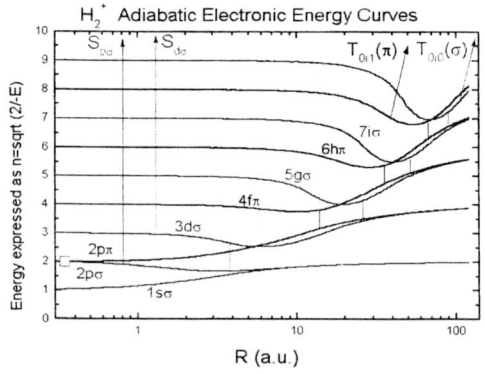

**FIGURE 3.** A subset of adiabatic electronic energy curves for the $H_2^+$ system.

($T_{01}$-promotion [13,14] in hidden variable language) of this orbital into the continuum. The slight asymmetries of the distribution with respect to the Z axis are attributed to the presence of an interfering sigma amplitude. Indeed, Doerner et al. [14] found that this asymmetry oscillates with v, and a quantitative analysis of the oscillation frequency was given by Macek and Ovchinikov [13] who concluded that the observed frequency was consistent with this interpretation.

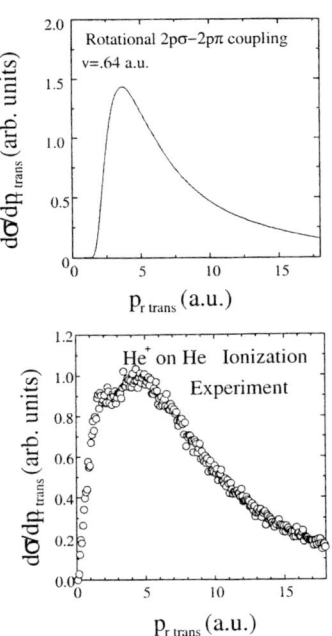

**FIGURE 4.** Comparison of the dependence on transverse momentum transfer expected for rotational $2p\pi$-$2p\sigma$ coupling with the experimental data for continuum electron production. The transverse momentum transfer for the model was calculated from the impact parameter using a classical scattering in the potential of two frozen $He^+$ cores.

If the rotational coupling is the dominant enabler, it should dominate the impact-parameter dependence of the process. Fig. 4 shows a comparison between the experimental transverse momentum dependence of continuum electron production from He by $He^+$ and the expected dependence calculated from the formulation by Taulbjerg et al.[15], using parameters for the $He_2^+$ system from Barat et al.[16]. A frozen charge $He^+$ on $He^+$ potential was used to convert the impact parameter dependence in the model into transverse recoil momentum. The agreement is remarkably good.

In spite of the apparent elegance of this qualitative interpretation, recent theoretical work by Sidky and Lin [17] presents a rather different explanation for the observed behavior. They have calculated the electron momentum spectra for slow p on H using a two center discretization of the continuum approach, solving the Schroedinger equation exactly within a discretized basis for the radial momentum functions. In their description, no mention of rotational coupling is made, although it is certainly included. The two-fingered structure is reproduced, but comes from target and projectile centered outgoing waves rather than from the symmetry properties of a promoted molecular orbital. Further work in progress by Sidky,

Illescas and Lin [18], using a CTMC approach, suggests that the two-fingered structure is really due to the exclusion of electrons from the internuclear axis (they would be captured by either target or projectile if they were there) and that saddle-point electrons promoted slowly on the outgoing trajectory are responsible for only a small fraction of the ionization cross section. That is, saddle point promotion is not a major ionizing mechanism in such collisions.

Further evidence that the whole story is not in is provided by the observation, in fig. 2, that a molecular hydrogen target provides spectrum very similar to that for He. This system does not have a united atom degeneracy of $2p\pi$ and $2p\sigma$ orbitals which exists for $He_2^+$, and one might expect that quite a different mechanism would prevail; nevertheless, the spectra are rather similar for the He and $H_2$ targets

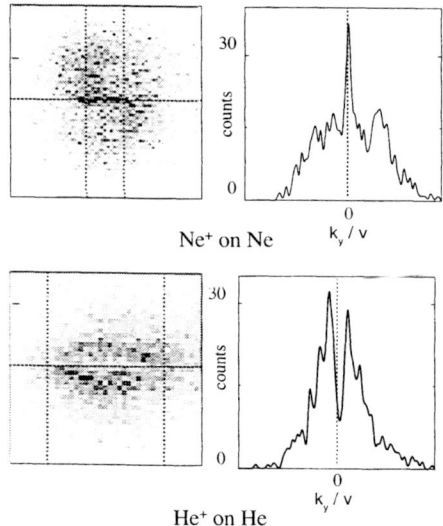

**FIGURE 5**. Top view electron momentum spectra for $Ne^+$ on Ne at v=0.45 a.u. and $He^+$ on He at v=0.9 a.u. The right hand figures are projections of a slice cut vertically across the center of each distribution, showing an anti-nodal peak for the Ne case and a nodal dip for the He case.

From fig. 2 one might wonder whether ALL collision systems give the two-fingered structure. Fig. 5 shows that this is not the case. For example, for $Ne^+$ on Ne, the orbital which would be expected to be promoted is the $3d\sigma$ one [19], and one might expect to see sigma character of the continuum electrons. The data show that the nodal line in this case is replaced by a anti-nodal peak along the Z axis, presumably a signature of the promotion of a sigma orbital, although the details are far from clear. Similar structure is also seen for $He^+$ on Ne and its sister $Ne^+$ on He. The similarity of the two last systems shows strongly that the properties of the promoted continuum are determined by those of the molecular system, independent of entrance channel. We summarize this section with the observation that COLTRIMS allows the observation of patterns in the continuum electron momentum distributions which we associate with the symmetries of promoted MO's, but that the problem of explaining these distributions theoretically is far from solved.

## High Velocity Ionization

For projectile velocities large compared to target electron velocities, the ionization becomes more impulsive in nature and approaches photoionization as v increases to the speed of light. It is in this velocity range where the most singles electron spectra have been measured with the DES method, and several recent review articles have summarized this venerable topic [20-22]. Ullrich, Moshammer and collaborators have recently emphasized that a very fast very highly charged ion applies a field pulse to the target which transfers considerable energy but little momentum to the target system, as is the case with photoionization [23-25]. Since COLTRIMS data are presented in a very different format (typically images in momentum space) from that of the DES data ($d\sigma/d\Omega dE$ at selected laboratory angles), it is not always obvious whether the newer format presents new information or old information in a new format. The major real advantages of the imaging approach is that it goes right down to zero electron energy and that it allows experimental control over the projectile momentum transfer. It also presents the results as one image instead of a series of separate graphs. In the high velocity region, the momentum transfers to projectile, recoil ion and electron are of roughly the same size and it is not possible to translate a particular projectile momentum transfer into a corresponding impact parameter even in principle.

Here we report recent results for single ionization of He over a very wide range of Z/v, the parameter which should measure the extent to which a perturbative treatment of the collision is valid. The experimental arrangement is that described above except that the Tandem-LINAC at KSU was used to provide a pulsed beam with a time resolution below 1 ns and the gas jet was only moderately supersonic, with a longitudinal momentum spread along the jet of approximately 0.5 a.u.

We begin our discussion with the statement that very little is learned from the COLTRIMS longitudinal momentum spectra which cannot be deduced from DES singles measurements of the electron spectra, except that the COLTRIMS extends the measurement itself to lower laboratory electron energies. The additional knowledge of the recoil momentum transfer for each collision event, measured directly with COLTRIMS, can also be deduced event-by-event from the experimental measurement of the electron energy and emission angle alone [26]. The momentum transfer to the projectile is given by $p_z = -(B+k^2/2)/v$, where B is the binding energy of the target and k is the electron momentum in the laboratory system (atomic units are used). The value of $q_z$ follows from conservation of longitudinal momentum: $q_z = -p_z - k_z$ and thus the z momentum state is completely determined by measurement of vector **k**. It is well established that, for Z/v less than unity, a first order Born treatment of single ionization (B1), with the final electron described as a Coulomb wave emerging from the target, gives a good description of the singles electron distributions for point projectiles. As Z/v is raised, the projectile charge plays an increasingly important role in the final state [20-22], sending the electrons more forward (and the recoil more backward) than B1 calculations predict. Replacing B1 by a CDW or CDW-EIS treatment cures this problem, and excellent agreement is usually found between experiment and theory even for very large values of Z/v. This is seen, for example, in the singles electron data of refs. [29] and [30] and in the COLTRIMS data of [23] with Z/v ranging up to 2.2. A sample comparison of theory and experiment for $F^{9+}$ on He from our data is given in fig. 6, and compared to a similar comparison in DES format from Tribedi et al.[29].

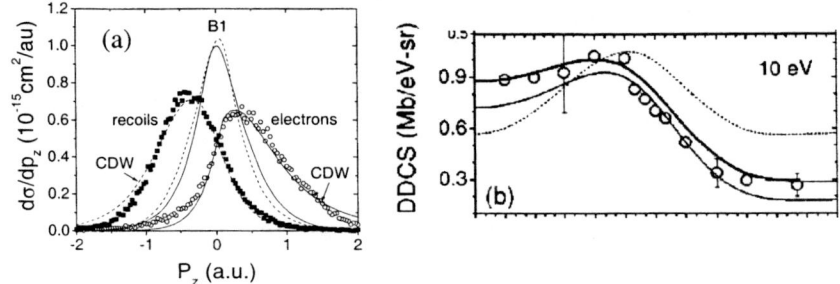

**FIGURE 6**. (a) Longitudinal distribution of electron and recoil momenta from single ionization of He by 1 MeV/u bare F from COLTRIMS measurements. The dashed lines and solid lines are theoretical B1 and CDW [30] results for recoils and electrons, respectively. (b) DES format electron angular distributions for bare C at 2.5 MeV/u on He. The dashed line and the heavy solid lines are B1 and CDW theoretical results [from Tribedi et al, ref. 29].

The real advantage of the COLTRIMS approach is that the projectile momentum transfer **p** is under experimental control. This means that, in principle, the Bethe-Born limit can be forced experimentally by selecting small **p** rather than having to assume that small **p** dominates. That is, the separation between photoionizationlike, or "three body ionization" [32], and binary-encounter, or "two body ionization", can in principle be made on this basis. Fig. 7 shows transverse momentum spectra for 5 Mev p on He. Fig. 7a shows electron momenta **k** plotted in a coordinate system in which **p** (the projectile momentum transfer) is defined as the y axis; in fig. 7b, **q** (the recoil) is the y axis. In fig. 7c, **p** is plotted in the **q** frame. It is seen from the strong correlation between **k** and **q**, and the much weaker one between **k** and **p**, that photoionizationlike events dominate over binary-encounter-like events, and that nuclear scattering between recoil and projectile is quite weak. Because the experimental resolution on **q** was much worse in the y (jet) direction than in the x( electric field or time) direction, due to the finite temperature of the supersonic jet, we can improve the correlation

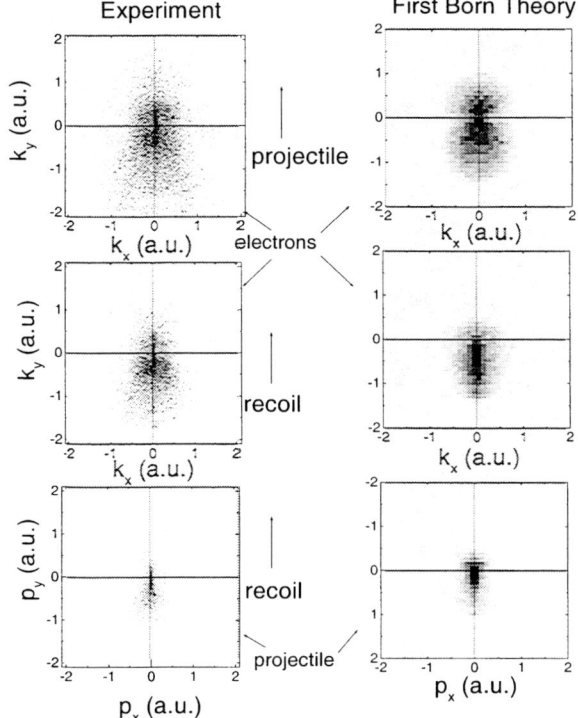

**FIGURE 7.** Left hand column: the top spectrum shows the transverse electron momenta from 5 MeV p on He , with the projectile momentum vector defined as the positive vertical axis. The middle spectrum is the same, but the recoil vector is defined as the positive vertical axis. The lowest figure is the transverse projectile momentum vector, with the recoil momentum vector defined as the positive vertical axis. Right hand column: First Born calculations for the same three spectra.

comparison by limiting ourselves to plots of $k_x$, $p_x$ and $q_x$ versus each other. This is done in fig. 8, where again the dominance of photoionizationlike events is clear.

These data can be compared with a B1 calculation , done in an imaging format by exploiting the closed form expression for the B1 amplitude for hydrogenic targets given in the literature in ref. [27] . Imaging theoretical spectra were produced by integrating this expression over the five-dimensional final momentum space and filling the desired image spectrum proportional to the cross section element for each element of this space. The results are shown in fig. 7 . The agreement between data and experiment is good. The B1 does not allow any momentum exchange between recoil and projectile due to the internuclear force, since this term does not contribute to ionization in first order. Transverse momentum exchange mediated through interactions with the electron is still possible. One can estimate the size of the error

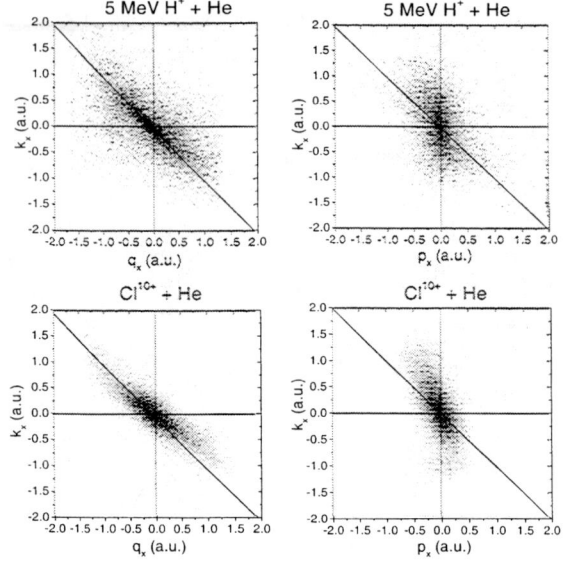

**FIGURE 8.** Plots of $k_x$ (electron momentum) versus $q_x$ (recoil momentum) and $p_x$ (projectile momentum transfer) for two collision systems.

due to the omission of the internuclear term using p(transverse, nuclear) ~ $Z_1 Z_2/bv$, which, for b=0.5 a.u., v=10, $Z_1$=1 and $Z_2$=1, yields only 0.2 a.u., comparable to the experimental resolution. Thus agreement of the data with B1 is therefore not surprising for this case.

The surprise comes when Z/v is raised, as is shown in fig. 8, going all the way to 2.3. Now the nuclear term might be expected to wash out the clear photoionizationlike behavior. It clearly does not: the correlation of $k_x$ with $q_x$ seen in fig. 8 for Cl on He is just as strong as for p on He. In fig. 9 a comparison of the x-momentum data for three systems with the B1 calculation is shown. Good agreement is found for both small and large Z/v. As was observed by Moshammer et al.[23-25], even a very highly charged projectile with a large Z/v still acts very much like a photon in ionizing the target. The Bethe-Born limit still dominates, even far from the perturbative region. Why? The answer probably lies at least partially in the fact that, by selecting single ionization, one selects non-perturbative collisions (large b), ones in which the He$^+$ recoil is not ionized further. Moshammer et al. [23] have suggested that the fast-ion-induced ionization "photographs" the initial state wave function. In fact, the PWBA transition amplitude is just proportional to the momentum space initial state wave function, evaluated at q, although the B1 amplitude is somewhat more complicated. When two or more electrons are removed in the same event, any

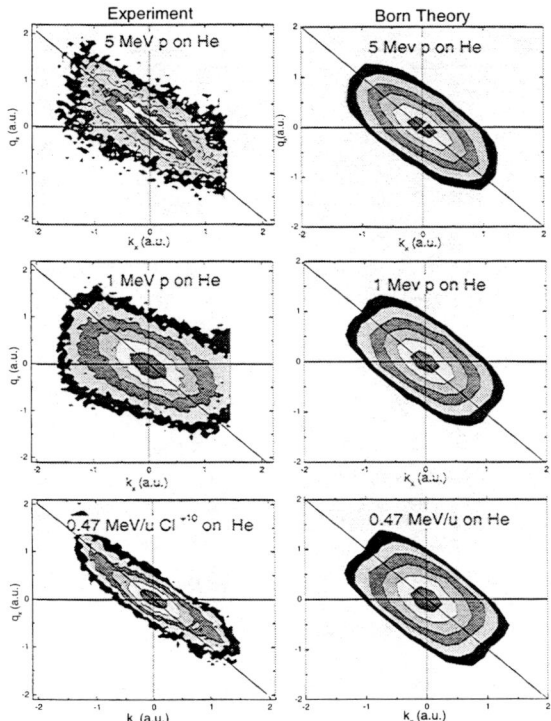

**FIGURE 9.** Comparison of $k_x$ versus $q_x$ with Born approximation results for different collision systems.

correlation of their relative motion is preserved in this amplitude, and for large Z/v such a process is very probable. Unfortunately, one probably loses somewhat the cleanliness of the interpretation for many electron transitions due to the influence of the internuclear term in highly non-perturbative collisions.

## Double Ionization of $D_2$

The double photoionization of a two-electron target takes place through the interaction between the two electrons. This interaction, referred to generally as correlation since it causes the electronic wave function to depart from a product-state form, can be in the initial state, final state or can occur during the ejection of the two electrons. The double ionization of the same target by a charged particle can occur through the same photoionizationlike process in perturbative collisions, and in this case should exhibit the same characteristics as photoionization. However, a second process can occur in the case of charged particle ionization which is absent in photoionization, namely ionization via the independent interactions of each electron with the projectile Coulomb field, a process roughly equivalent to the absorption of two photons in a single encounter of target with projectile.

For photo double ionization of an aligned molecule, the outgoing photoelectrons can experience anisotropic emission with respect to two different vectors, the polarization axis and the internuclear axis. Emission along the polarization axis is preferred in single ionization due to the dipole operator, and to a reduced extent this preference survives in double ionization. Anisotropic emission with respect to the molecular axis can be caused by interference between outgoing waves on the two centers and/or by scattering of the outgoing electrons in the anisotropic molecular potential. The first experimental data on differential photo double ionization of $H_2$ or $D_2$, without selection of the internuclear axis, appeared only recently [33,34]. We have performed measurements of double ionization of aligned $D_2$ by photons [35]. The alignment of the molecule is accomplished *a posteriori* by detecting the fragments from the dissociation in coincidence with the emitted electrons. The emission of the electrons is much faster than the dissociation so that the direction in which the fragments are emitted is a measure of the alignment of the molecule at time of ionization. A standard COLTRIMS geometry and jet is used with a high extraction field and close ion detector in order to capture the energetic fragments which emerge from the dissociation of the $D_2$. Fig. 10 shows the angular distribution of electrons from photo double ionization of He and $D_2$ taken at a photon energy of 58.8 eV, 7 eV above the Franck Condon threshold for double ionization.

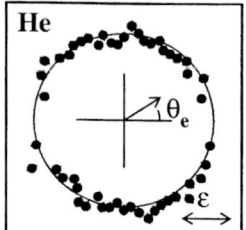 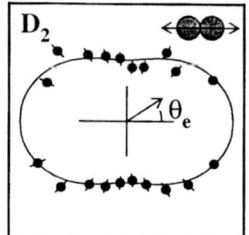 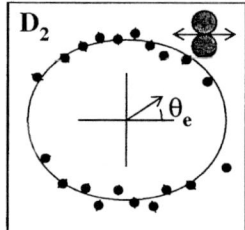

**FIGURE 10.** Polar plots of electron momentum from double ionization of He and $D_2$ by 58.8 eV photons. The polarization vector is horizontal. In the middle and right hand panels the molecular axis is parallel to and perpendicular to the polarization vector, respectively. From Doerner et al. ref. 35.

In all cases the polarization vector $\varepsilon$ is parallel to the x axis. The left hand panel is a control, for He, and shows that the dipole preference of emission along $\varepsilon$ is almost removed for photo double ionization. The middle panels show distributions for two different alignments of the $D_2$ molecule. It is seen that a weak but clear preference for emission along $\varepsilon$ exists, and furthermore that this preference is enhanced if the molecule is also aligned along this axis. The molecule seems to channel the outgoing electrons along the internuclear axis. Recent photo double ionization work on $D_2$ without molecular axis selection has shown that $D_2$ is very similar to He in many respects [33,34] . The first theoretical analyses of the $D_2$ case have just appeared [36], but no explanation of the behavior observed in fig. 10 has appeared.

Does a similar propensity occur for charged-particle impact? Fig. 11 shows a polar plot of soft electrons emitted from $D_2$ when it is bombarded by a fast (2.5 MeV/u) bare carbon beam. In this case both double ionization and ionization-excitation to dissociating states were included in the data . However, since both double ionization and ionization-excitation are two-electron processes, the physics is expected to be similar. The figure shows polar plots for various selections of the size of the radial electron momentum. The results are completely different from those for double photoionization: a strong preference for emission along the momentum transfer vector **p** (analagous to $\varepsilon$ ; **p** is mainly transverse for charged particle impact) is seen, especially for large k. A very small enhancement of this effect is perhaps seen when the internuclear axis lies transverse, but this effect is much weaker than for photoionization.

Why? The answer probably lies in the fact that double ionization in the charged particle case is almost certainly due to two independent ionization (or excitation) interactions ("photon-exchanges") with the projectile, and thus the two electron process being observed in this case is really a one electron ionization accompanied by an incidental second exciting/ionizing interaction with the other electron. Thus there is little reason to expect strong similarity between photo and charged particle

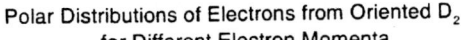
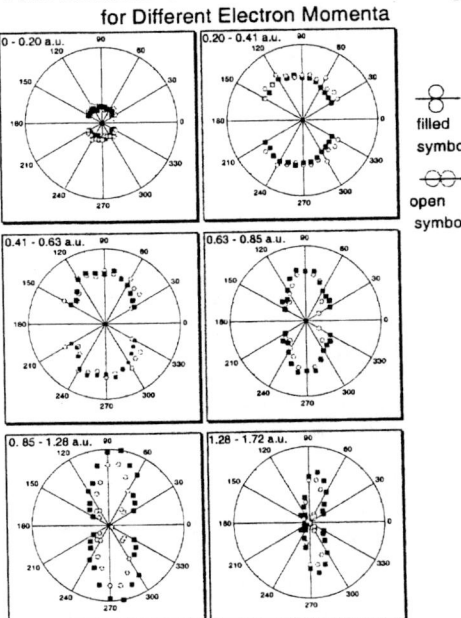

FIGURE 11. Polar distributions of electrons emitted from $D_2$ molecules bombarded by bare 2.5 MeV/u carbon ions. The open symbols are for $D_2$ molecules aligned along the beam axis, while the filled symbols have the $D_2$ molecule aligned perpendicular to the beam..

double ionization . However, this similarity might be expected to return if the charged particle projectile were sufficiently perturbative in nature (small Z/v).

## CONCLUSIONS AND SUMMARY

Momentum imaging is well suited for the study of the low energy products from atomic collisions, allow full correlations among the momenta of the products to be determined with high efficiency, and lead to a very compelling imaging display of the dominant features of physical processes. We have applied this to continuum electron production from light targets for charged-particle and photon projectiles. For the former case, one sees the progression of the soft electron patterns from promoted molecular orbitals at low velocity to the Bethe-Born photoionization-like patterns at high velocity. One surprising result is that the character of the transverse momentum transfers for small Z/v, which is well described by a first order Born approximation, changes little even for Z/v much above unity. In double ionization of $D_2$ by both photons, a propensity for electron emission along the internuclear axis is observed. This propensity is , at best, much weaker in the case of charged particle double ionization/excitation-ionization. We expect that the double ionization mechanisms are quite different for the two cases.

## ACKNOWLEDGMENTS

This work was supported by the Division of Chemical Sciences, Office of Basic Energy Sciences, Office of Science, U.S.Department of Energy. The photon work was performed at the Advanced Light Source at the Lawrence Berkeley Laboratory.

## REFERENCES

1. J.Ullrich,R.Doerner,V.Mergel,O.Jagutzki,L.Spielberger and H.Schmidt-Boecking,Comments At.Mol.Phys. 30,285 (1994).
2. J.Ullrich,R.Moshammer,R.Doerner,O.Jagutzki,V.Mergel,H.Schmidt-Boecking and L.Spielberger, J.Phys.B 30, 2917 (1997).
3. R.Doerner et al., Physics Reports (to be published, 1999).
4. M.A.Abdallah, C.L.Cocke,W.Wolff,H.Wolf,S.D.Kravis,M.Stockli and E.Kamber, Phys.Rev.Lett. 81, 3627 (1998).
5. M.A.Abdallah,A.Landers,M.Singh,W.Wolff,H.E.Wolf,E.Y.Kamber,M.Stockli and C.L.Cocke, Nucl.Inst.Meth. B 154 , 73 (1999).
6. M.A.Abdallah, W.Wolff,H.E.Wolf,C.L.Cocke, and M.Stockli, Phys.Rev. A 58, R3379 (1998).
7. J.C.Bernot,D.Dhuicq,J.P.Gauyacq,J.Pommier,V.Sidis,M.Barat and E.Pollack, Phys.Rev.A 11,1245(1975).
8. J.C.Bernot,D.Dhuicq,J.P.Gauyacq,J.Pommier,V.Sidis,M.Barat and E.Polllack, Phys.Rev.A 11,1933(1975).
9. M.Barat,D.Dhuicq,R.Francois,R.McCarroll,R.D.Piacentini and A.Salin, J.Phys. B 5,1343 (1972).
10. S.D.Kravis,M.A.Abdallah,C.L.Cocke,C.D.Lin,M.Stockli,B.Walch, D.Wang,R.E.Olson, V.D.Rodriguez,W.Wu,M.Pieksma and N.Watanabe,Phys.Rev.A54, 1394 (1996).
11. R.Doerner,H.Khemliche,M.H.Prior,C.O.Cocke,J.A.Gary,R.E.Olson,V.Mergel,J.Ullrich and H.Schmidt-Boecking, Phys.Rev.Lett.77,4520 (1996).

12. L.Schmidt, et al., Abstracts of Contributed papers, 21 Int. Conf. On Photonic, Electronic and Atomic Collisions, p. 482 (unpublished; Sendai, 1999).
13. J.H.Macek and S.Yu.Ovchinnikov, Phys.Rev.Lett. 80, 2298 (1998).
14. M.Pieksma,S.Y.Ovchinnikov,J.van Eck, W.B.Westerveld and A.Niehaus,Phys.Rev.Lett. 73, 46 (1994).
15. K.Taulbjerg,J.S.Briggs and J.Vaaben,J.Phys.B 9 , 1357 (1976).
16. M.Barat,D.Dhuicq,R.Francois,C.Lesech and R.McCarroll, J.Phys.B 6, 1206 (1973).
17. E.Y.Sidky and C.D.Lin, J.Phys.B31, 2949 (1998); E.Y.Sidky and C.D.Lin, Phys.Rev.A 60, 377 (1999).
18. E.Y.Sidky, C.Illescas and C.D.Lin, private communication (1999).
19. J.Eichler, U.Wille, B.Fastrup and K.Taulbjerg, Phys.Rev. A 14, 707 (1976).
20. Rudd, M.E., Y.-K.Kim, D.H.Madison and T.J.Gay, Rev.Mod.Phys. 64, 441 (1992).
21. M.W.Lucas,D.H.Jakubassa-Amundsen,M.Kuzel and K.O.Groeneveld, Int.J.Mod.Phys.A 12, 305 (1997).
22. N.Stolterfoht, R.D.DuBois and R.D.Rivarola, *Electron Emission in Heavy Ion-Atom Collisions*, Springer Series on Atoms +Plasmas #20 (Springer,Berlin Heidelberg, 1997).
23. R.Moshammer, J.Ullrich, H.Kollmus,W.Schmitt, M.Unverzagt,H.Schmidt-Boecking,C.J.Wood, R.E.Olson, Phys.Rev.A 56, (1997).
24. R.Moshammer et al., Phys.Rev.Lett. 79, 3621 (1997).
25. W.Schmitt et al., Phys.Rev.Lett. 81, 4337 (1998).
26. L.C.Tribedi, P.Richard, Y.Wang,C.D.Lin and R.E.Olson, Phys.Rev.Lett. 77, 3767 (1996).
27. D.S.F.Crothers and J.F.McCann, J.Phys.B 16, 3229 (1983).
28. P.D.Fainstein and R.D.Rivarola, J.Phys.B 20, 1285 (1987).
29. L.C.Tribedi et al., Phys.Rev. A 58, 3619 (1998).
30. N.H.Stolterfoht,H.Platten,G.Schiwietz,D.Schneider,L.Gulyas,P.D.Fainstein and A.Salin, Phys.Rev.A52, 3796 (195).
31. F.O'Rourke, private communication (1998).
32. N.Stolterfoht,J.-Y.Chesnel,M.Grether,B.Skogvall,F.Fremont,D.Lecler,D.Hennecart,X.Husson, J.P.Grandin,B.Sulik,L.Gulyas and J.A.Tanis,Phys.Rev.Lett. 80, 4649 (1998).
33. T.J.Reddish and J.M.Feagin, J.Phys.B 32, 2473 (1999).
34. N.Scherer, H.Loerch and V.Schmidt, J.Phys.B 31, L817 (1998).
35. R.Doerner et al., Phys.Rev.Lett. 81, 5776 (1998).
36. M.Walter and J.Briggs, J.Phys.B 32, 2487 (1999).

# Three-Electron Photo-Processes in Lithium

## Yoshiro Azuma

*Photon Factory, Institute of Materials Structure Science, KEK*
*1-1, Oho, Tsukuba, Ibaraki, Japan 305-0801*

Atomic lithium presents an ideal case for the investigations of electron correlation involving three electrons, since it possesses three and only three electrons. An overview as well as new results from a series of collaborative experiments utilizing the photoion time-of-flight methods with synchrotron radiation at the Photon Factory are presented. These include the photoexcitation resonances as well as photoionization into the continuum, with the direct participation of all three electrons.

## Introduction

Lithium has three and only three electrons. This leads to the possibility of a large variety of exotic correlational processes that find no analogue in the well studied helium atom. Nevertheless, lithium is the simplest neutral atom with the possibility of such processes. Some of the most interesting examples of these processes with photoexcitation include triple-photoexcitation (hollow lithium) together with its various decay mechanisms including one-step double-autoionization, two-step sequential autoionization, and two-down one-out processes. Furthermore, direct triple photoionization, as well as various types of satellite processes accompanying single and double photoionization are possible. It should also be noted that lithium is the simplest open-shell atom, and the simplest atom that exhibits intershell electron correlation, with one electron in the outershell. This outer shell electron can be excited easily be laser light, further increasing the variety of photo-processes that can be studied, including aligned or polarized targets. Photoabsorption processes with the direct participation of all three elctrons of the lithium atom came to be known during the last five years or so through the discovery of the triply photoexcited "hollow lithium" resonances. More recently, direct triple photoionization of lithium was observed for the first time by measuring $Li^{+++}$ photoion production.

# Total Photoion Yield Spectrum of Hollow Lithium Resonances

The first measurement of the $2s^2 2p$ "A" resonance was made by the Dublin group, utilizing the dual laser plasma technique (1). Subsequently, experiments with synchrotron radiation were done at the Photon Factory (2), as well as HASYLAB (3) and Super-ACO (4). The experiment at the Photon Factory was done at beamline BL3B by detecting the total photoion yield as the photon energy was scanned. Numerous new resonances were found above the lowest $2s^2 2p$ "A" (Fig. 1). However, no Rydberg series structure could be observed with this data and the apparent irregularity of the spectrum called for calculations in order to make any interpretation possible. The first calculation was performed by Koike utilizing the MCDF code combined iteratively with the CI method (2). This calculation, despite being somewhat less accurate compared to newer large scale calculations, provided important insight into the process through the analysis of orbital configurations involved.

1) The configurations were very severely mixed in an irregular manner. This revealed the dominance of electron correlation and the serious inadequacy of describing hollow lithium resonances with Hartree-Fock orbitals, let alone any assignment or labeling by them. Exceptions were found only in a few resonances including the lowest $2s^2 2p$ "A" for which a single configuration description could be a reasonable approximation.

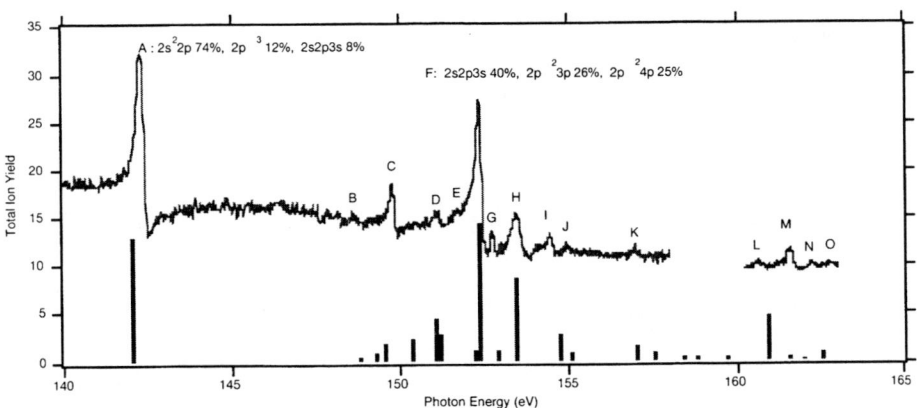

**FIGURE 1.** The first photoion spectrum of hollow lithium photoexcitation resonances (2). The bars indicate Koike's calculations by the MCDF-CI method.

2) The spectrum showed an irregular pattern with considerably more oscillator strength distributed to the resonances clustered around the "F" at ~150 eV than the lowest $2s^2 2p$ "A". This was explained as due to the severe shrinking of the n=2 orbitals caused by the sudden disappearance of core screening by the K electrons upon the photoexcitation of hollow lithium. This results in the greater overlap of n=3 orbitals with the initial ground state and thus leads to the redistribution of socillator strength to the resonance offset from the first by one unit of principle quantum number. An anology to this type of shift of oscillator strength by as much as one unit of principal quantum number due to the change in core shielding, can be found in the theoretical work on enhancement of photoelectron satellite lines in valence excited lithium targets (5). An important aspect of such photoexcitations involving a very big change in core shielding is that no common basis set can describe both the ground state and excited states adequately, and hollow lithium presents a very dramatic case. It should be noted that the framework of the total atomic structure changes completely before and after excitation, unlike photoexcitations in heavier atoms or solids where the rest of the framework may remain relatively unchanged.

Some photoion time of flight measurements were also made and revealed two types of decay processes (2). For resonances below the doubly excited singly charged $Li^+$ threshold the decay proceeded through one step double autoionization processes. For higher resonances, two step sequential autoionization processes via the doubly excited state of singly charged $Li^+$ became dominant. This was demonstrated in the rise of the peak-heights in the $Li^{++}$ and $Li^+$ channels for higher resonances by a factor of ~40. Actually this was analogous to the previous example of such increase of resonance strengths in the doubly charged ion channel, which was found in the case of barium by Hansen *et al.* (6).

## Partial Charge Yield Spectrum of Hollow Lithium Resonances

Since 1996, most of our experiments on hollow lithium were performed at the BL16B undulator beamline at the Photon Factory and Wulilleumier and cowokers pursued photoelectron measurements at the 9.0.1. undulator beamline of ALS (7,-,10). Measrurements at BL16B were done with the partial charge yield method, by gating the $Li^+$ peak and $Li^{++}$ peak in the ion time-of-flight spectrum taken at each step of the photon energy scan, so that the spectra in both final charge channels could be obtained at the same time. The much improved resolution and flux from insertion-device beamlines were instrumental to new findings including the (3,3,3) resonance with completley empty K and L shells (10,11) as well as the Rydberg series (9).

Questions were raised whether Rydberg series are formed in the hollow lithium photoexcitation spectrum (2,9). As confirmed on the high resolution photoion spectra (Fig. 2), the 2s2p $^3$P ns series and 2p$^2$ $^1$S np series form relatively unperturbed Rydberg series as stated by Diehl et al. (9). The quantum defects agree well with theory, and are relatively constant with increasing N. However, R-matrix calculations (12) indicate that, 2s$^2$ $^1$S np $^2$P$^o$ is strongly perturbed at low n(<7) as are 2p$^2$ $^3$Pnp, $^1$D np,nf, which are not truly "Rydberg" at all except at very high n (>>10). This is in contrast with the argument of Diehl et al. (9) for the existence of eight Rydberg series, and points instead to a chaotic pattern of interfering series of resonances converging on different thresholds.

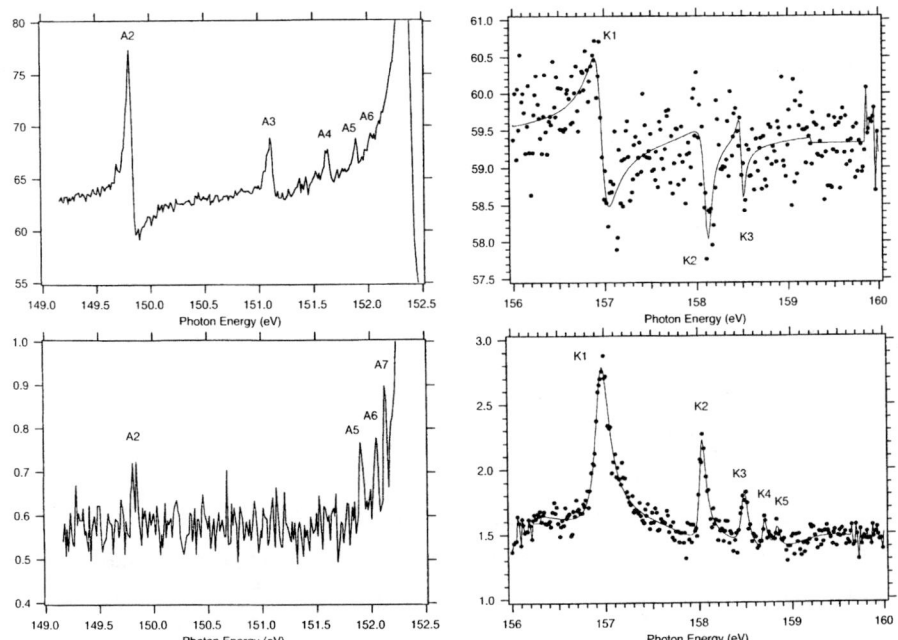

**FIGURE 2.** Partial charge yield spectra showing Rydberg series 2s2p $^3$P ns (left panels), and 2p$^2$ $^1$S np (right panels). The upper panels are Li$^+$ yields and the lower panels are the Li$^{++}$ yields. Note the sudden increase in the strength of the resonances in the Li$^{++}$ channel once the photon energy crosses the lowest doubly excited singly charged Li$^+$ threshold at 151.7 eV, as shown in the lower left panel.

Following Koike's first MCDF-CI calculation that provided significant physical insights (2) and the MCHF calculation that showed good agreement in energy positions if not in intensities (3), considerable developments in theoretical activities have taken place. A series of R-matrix calculations were performed by VoKy Lan and coworkers (7,-,10). Their reproduction of the spectral features has been impressive, although it is very hard (if not impossible in principle) to obtain any further physical interpretation from R-matrix calculations. More refined R-matrix calculations were published subsequently by Berrington and Nakazaki (18), VoKy Lan and coworkers (13,14) as well as Zhou et al.. (15). Chung has been publishing a series of very accurate calculations by the saddle-point complex rotation method (16,17) and Lin and co-workers employ the hyperspherical coordinates towards the goal of establishing new quantum numbers suitable for the description of this system (18,19).

## One-Step Double Autoionization of the $2s^22p$ "A" Resonance

Carefull partial charge yield measurements on the $2s^22p$ "A" resonance revealed asymmetry in the doubly charged $Li^{++}$ channel peak, which was previously reported to be symmetric (2). Hollow lithium can decay into a doubly charged $Li^{++}$ either by direct double autoionization or by two step sequential autoionization. However, for resonances below the first threshold for doubly excited singly charged lithium ion ($2s^2$) at 151.7 eV, one-step double-autoionization would be the only possibility. Therefore, the above asymmetry should be due solely to the interaction of the decay with the double-photoionization continuum. One-step double autoionization is of strong interest because they are processes that most directly measure the electron correlation involving all three electrons in the excited state. These cases present a theoretical challenge since the two ejected electrons will have a continuous distribution of energy sharing and calculations have to be made for all possible kinetic energies. Further experimental studies on angular correlation between the two ejected electrons should also be of interest. For higher photon energies above the doubly excited singly charged $Li^+$ thresholds, two-step sequential-autoionization becomes possible. This is usually the much stronger channel when it is available. The energy sharing will be limited to discrete values, and interaction with the continuum takes place only at those values.

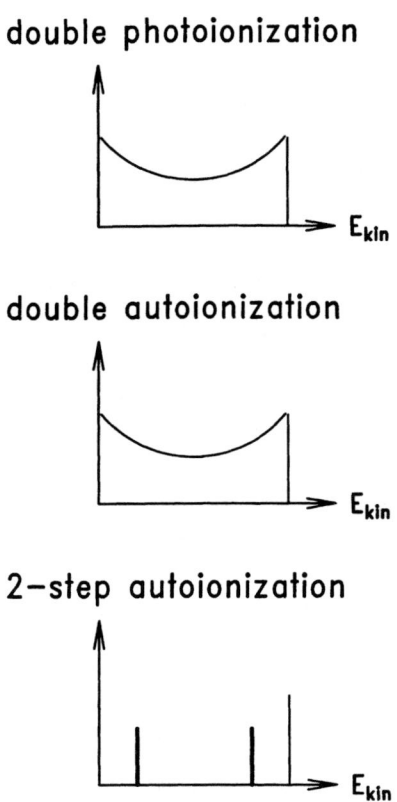

**FIGURE 3.** Energy sharing of the electrons for decay of excited states, interfering with the double photoionization continuum.

We have reported the resonance parameters and branching ratio recently (20). Theoretical calculation based on an R-matrix formulation using the "QB method" provides excellent agreement with experiment in the $Li^+$ channel and fairly good agreement regarding the q value in the $Li^{++}$ channel. However, agreement in the size of the resonance in $Li^{++}$ as well as the non-resonant double-to-single ionization ratio was less satisfactory. Considerable progress has been made on the analysis of two-step sequential processes by Wuilleumier and co-workers utilizing electron spectroscopy (21). Their studies also allowed the detailed mapping of doubly excited $Li^+$ ion states (22).

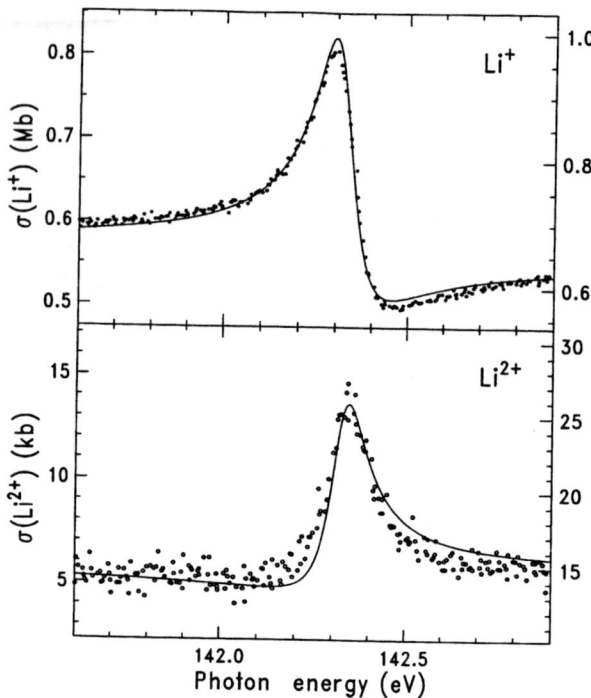

**FIGURE 4.** Partial charge yield spectra of the $2s^22p$ "A" resonance. The solid lines indicate R-matrix calculations.

## Partial Charge Yield Spectrum Over Broad Photon Energies

In parallel to the work on the high resolution spectrum of hollow lithium which focused on a rather narrow photon enrgy range, we also measured the double and single photoion yield from the double-ionization threshold (81 eV) up to 424 eV albeit with low resolution, in order to obtain a broader scope. Absolute cross-sections were derived by normalizing the data with a previous measurement by Mehlman et al at 103.3 eV (24). Figure 5 shows the total photoion yield (sum of $Li^+$ and $Li^{++}$ yields) in good agreement with the calculation by Reilman and Manson (25).

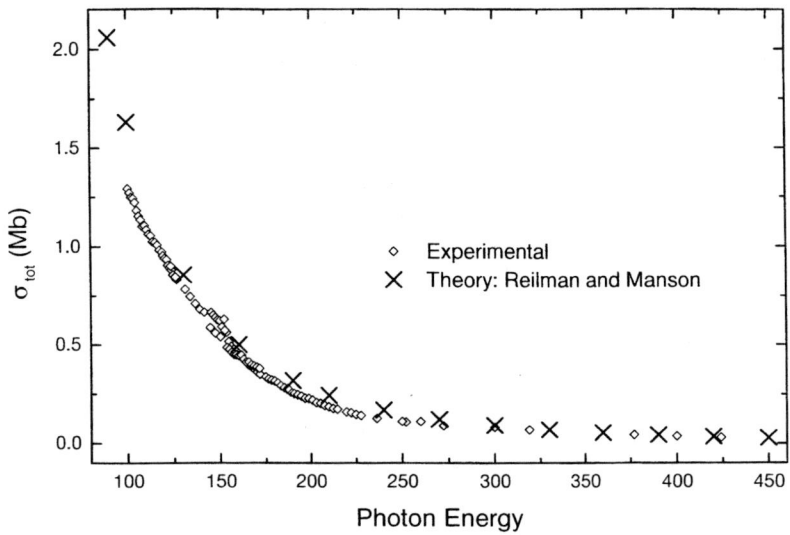

**FIGURE 5.** Total Li photoion yield and calculation by Reilman and Manson (25).


Figure 6 shows the ratio of double-to-single photoionization yields $\sigma^{++}/\sigma^+$ plotted against photon energy. Since both ion yields were taken simultaneously, changes in target density and photon flux cancel out. The ratio rises from zero at threshold to 1% above 100 eV and remains almost constant up to about 152 eV. This region corresponds to KL double photoionization. At higher photon energies, satellite processes due to photoionization followed by autoionization of doubly excited $Li^+$ become possible and cause the significant increase of doubly-charged ions formation. The region between 151.7 eV and 170 eV is strongly affected by two step sequential autoionization from the hollow lithium resonances and the double-to-single photoionization ratio fluctuates due to these resonances although the data was taken with low energy resolution. The second double-photoionization threshold (double K-shell photoionization) at 173 eV is not very obvious in the ratios which reach the maximum of 4% around 250 eV.

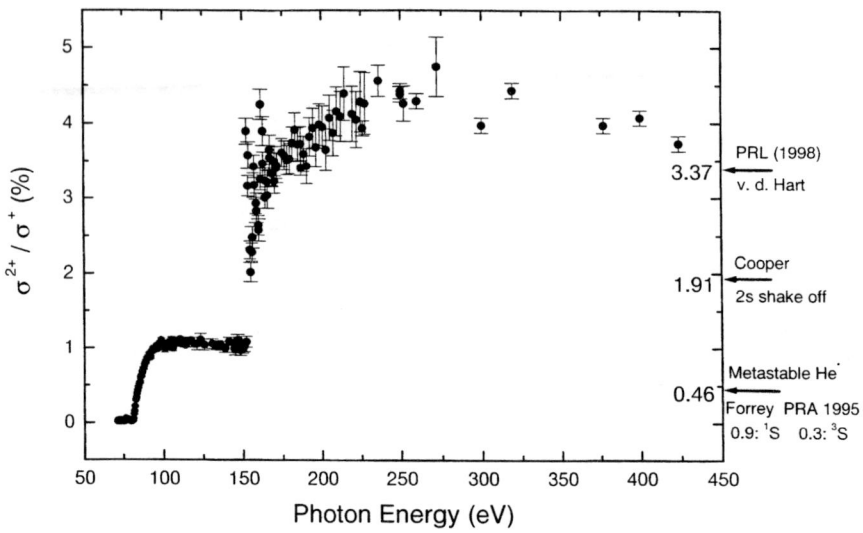

**FIGURE 6.** The $\sigma^{++}/\sigma^+$ ratio of lithium. Arrows to the right indicate asymptotic values with a new B-spline calculation by v.d. Hart et al (31), 2s shake-off calculation (23), and metastable helium (26).

An estimate of the $\sigma^{++}/\sigma^+$ ratio based on the photoionization of a 1s electron followed by 2s shakeoff with the other 1s electron remaining frozen, yielded the value of 1.91%, considerably larger than the 1.0% observed for the plateau like region near 150eV. Clearly, factors unaccounted for in a simple shake-off approximation are important. Since the other 1s electron of lithium does not directly participate for photon energies below 151.7 eV and probably just serves to screen the nuclear charge for the other two electrons the system should resemble metastable He (1s2s $^1S$, $^3S$). In a previous study by Forrey et al. (26), the asymptotic limits for $\sigma^{++}/\sigma^+$ of meta-stable He were calculated for the 1s2s $^1S$ and $^3S$ states using the asymptotic formulation of Dalgarno and Stewart with highly correlated Frankowski-Pekeris-type wave functions for the initial state. From their calculation, the asymptotic ratios are 0.9033% for He (1s2s $^1S$) and 0.3118% for He(1s2s $^1S$) were obtained. Since for lithium, the two ionized electrons can couple into either the $^1S$ or the $^3S$ states, comparsion should be made with metastable He data for 1/4He(1s2s$^1S$) + 3/4He(1s2s$^3S$) which yields 0.46%. This is considerably lower than the current measurement of 1.0% for lithium in the plateau region close to 150 eV. Most likely the asymptote is not reached there, but comparison in the higher photon energy region will be very much complicated by various processes other than simple KL double photoionization. In a recent paper by van der Hart et al. (27), the $\sigma^{++}/\sigma^+$ ratio of metastable helium were calculated using the R-matrix approach, up to 80 eV above the double photoionization threshold. An estimate made by van der Hart using 1/Z perturbation theory suggests that, to adequately compare these two cases, the energy axis of the He data needs to be scaled by a factor of 1.19 which is the ratio of the ionization potentials of the 2s electron of Li ($1s^22s$) and He(1s2s). Comparison is made with the calculated He data for 1/4He($1s2s^1S$) + 3/4He($1s2s^3S$), with the energy axis of He scaled by 1.19 and the agreement is very good as shown (Fig. 7). Since there is no experimental data available yet on double photoionization of metastable helium, perhaps our lithium data provides the best test of theory at present. The lithium $\sigma^{++}/\sigma^+$ data can also be compared to results from electron impact ionization of Li$^+$ ions, following the spirit of Samson's semi-empirical interpretation of $\sigma^{++}/\sigma^+$ in helium (29). It was argued that the He $\sigma^{++}/\sigma^+$ ratio should scale with the electron impact ionization cross-section of He$^+$ ion, based on the intuitive picture that the photon acts on one electron which on its way out, kicks out the other electron by "electron impact ionization". The agreement in the He case was reasonable up to everal hundred eV and failed for higher energies. In case of lithium, the agreement is fairly good within the photon energy region of our data, except that the structure from 100 eV to 150 eV is

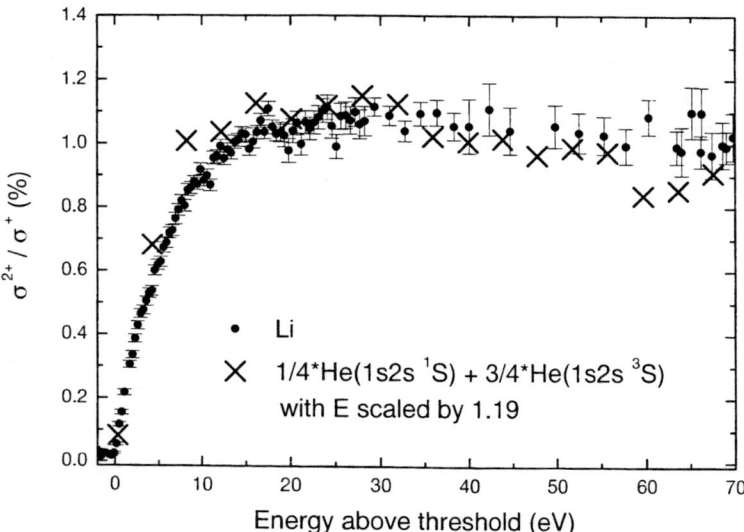

**FIGURE 7.** The KL double photoionization region compared to theory on metastable helium (27).

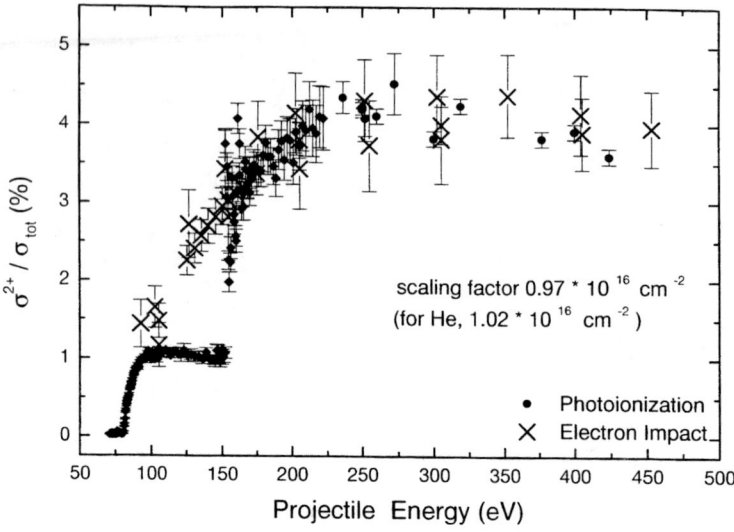

**FIGURE 8.** Comparison with scaled electron impact ionization data for Li$^+$, the "Samson Model".

smeared out in the electron impact ionization data. The value of the scaling factors are very close, $0.97 \times 10^{16}$ cm$^{-2}$ for Li, versus $1.02 \times 10^{16}$ cm$^{-2}$ for helium. We find these agreements rather surprising since the situation in lithium is quite different with the electron impact ionization analogy requiring the 2s electron to play the role of the projectile electron, whereas the 1s electron actually carries most of the photoionization cross-section.

## Triple Photoionization of Lithium

The most drastic of all three-electron photo-processes is perhaps the triple photoionization of lithium, in which all three electrons are ejected into the continuum in one step, leaving only the nucleus behind (bare lithium). The photoproduction of bare lithium ions was observed for the first time in the photoion time-of-flight (ToF) spectrum and the triple to single photoionization ratio from threshold to 424 eV (). Lithium is particularly well suited for the study of triple photoionization processes since Auger decay processes (or

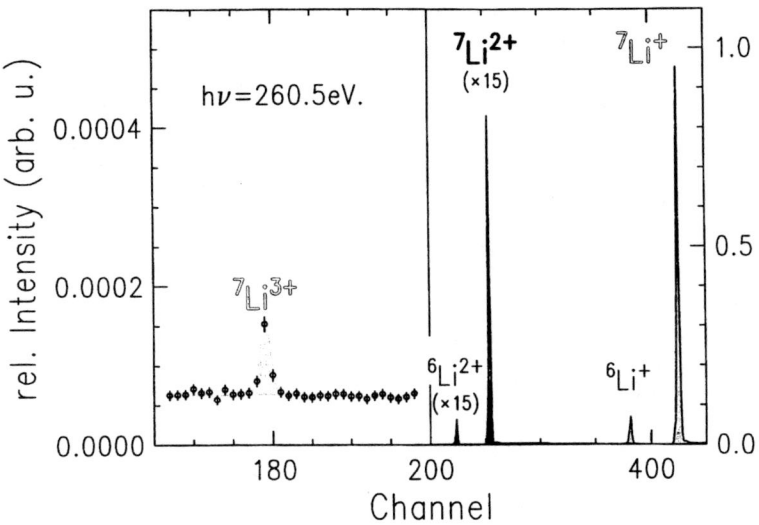

**FIGURE 9.** Photoion time-of-flight spectrum of lithium at hv = 260.5 eV, showing the formation of Li$^{+++}$.

autoionization), which leaves behind at least one bound electron cannot contribute to the triple-ionization cross section, and there is no added contribution from the relaxation or rearrangement of the rest of the atom that are important in heavier elements. However, a consequence of the fact that we have a pure, unambiguous direct triple-photoionization process in Li, is having to deal with a particularly cross section. It could be observed only after we moved the experiment to a powerful undulator beamline (BL16B), after earlier efforts without results at a bending magnet beamline (BL3B). Several ion time-of-flight spectra (Fig. 9) were taken over a range of photon energies. No theoretical calculation was available at the time of our first measurement. Our first estimate of the ratio for the high energy limit by a straight forward shake-off calculation yielded the value of $\sigma^{+++}/\sigma^{+} = 0.0015\%$, approximately a factor of 4 lower than the experimental value around 400 eV. Subsequently, a theoretical calculation was made using B-spline basis sets (31), which resulted in the high-energy limit of $\sigma^{+++}/\sigma^{+} = 0.00056\%$. Most recently, improved shake-off calculations (32) yielded asymptotic values of $(\sigma^{+++}+\sigma^{++})/\sigma^{+} = 1.81\%$ and $\sigma^{+++}/\sigma^{++} = 0.0465\%$ which should result in about $\sigma^{+++}/\sigma^{+} = 0.00084\%$. These predicted values for the high-energy limit are very much lower than the experimental data, possibly indicating that the photon energies for the current data points are way too low for meaningful comparison. Further measurements at higher photon energies are called for. So far there has been no calculation for photon energy dependences. The experimental result on the photon energy dependence of $\sigma^{+++}$ was found to be similar to the that of helium (33) with the energy axis multiplied by the factor $Z^2_{\text{eff}}(\text{He})/Z^2_{\text{eff}}(\text{Li}) = 4/9$ as suggested by Amusia et al. (34) and shifted according to the energy difference of the thresholds. Also, comparison was good with the theoretical curve for double photoionization of $\text{Li}^+$ ions (35,36). These suggest a model based on double photoionization of the K-shell followed by the shake-off of the loosely coupled 2s

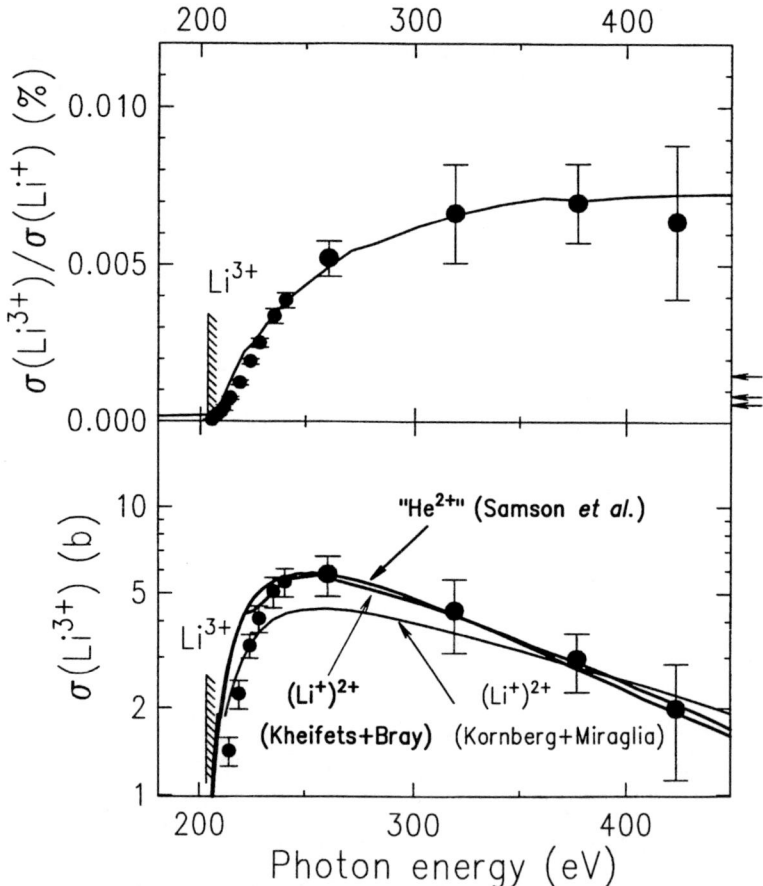

**FIGURE 10.** Triple photoionization of lithium. Ratio compared to single photoionization (upper panel), and derived absolute cross-section (lower panel). Comparisons are made with experimental results on helium double photoionization, as well as theory on double photoionization of lithium (see main text).

electron. Currently, analysis of some new data close to the triple-photoionization threshold region is in progress. This is very interesting since the triple photoionization threshold law at energies somewhat higher is still under much debate. Triple photoionization threshold studies were previously performed by Samson and Angel (37) for Ne and atomic oxygen. Their result indicates that there are two threshold laws, i.e. two different Wannier exponents depending on the energy region. Data on lithium promises to provide a cleaner test than other targets regarding the threshold laws and their range of validity. In fact, the threshold and above-threshold behaviour of lithium with an asymmetric electron configuration of two 1s electrons and a loosely coupled 2s electron could be different from that of Ne where the ejection of three equally coupled 2p electrons takes place. Future electron angular distribution, correlation measurements may shed light on these processes.

## ACKNOWLEDGEMENTS

Research on three-electron photo-processes in lithium at the Photon Factory has been pursued in collaboration with M.-T. Huang, R. Wehlitz, F. Koike, I.A. Sellin, T. Nagata, J. Cooper, T. Pattard, J. Burgdörfer, S. Hasegawa, B.D. DePaola, J.C. Levin, K. Berrington, S. Nakazaki, L. Pibida, E. Shigemasa, A. Yagishita, G. Kutluk, and M. Koide. Support has been provided by the Japanese Ministry of Education, Culture and Science (Monbusho), the Matsuo Foundation, the U.S. National Science Foundation through the Japan - U.S. Cooperative Science Program and the NSF Division of Physics.

## REFERENCES

1. L.M. Kiernan, E.T. Kennedy, J.-P. Mosnier, J.-T. Costello, and B.F. Sonntag, Phys. Rev. Lett. **72**, 2359 (1994).

2. Y. Azuma, S. Hasegawa, F. Koike, G. Kutluk, T. Nagata, A. Yagishita, and I.A. Sellin Phys. Rev. Lett. **74** 3768, (1995).

3. L.M. Kiernan, M.K. Lee, B.F. Sonntag, P. Sladeczek, P. Zimmerman, E.T. Kennedy, J.-P. Mosnier, and J.T. Costello, J. Phys. B **28** L161 (1995)

4. L. Journel, D. Cubaynes, J.-M. Bizau, S. Al Moussalami, B. Rouvellou, F.J. Wuilleumier, L. VoKy, P. Faucher, and A. Hibbert, Phys. Rev. Lett. **76**, 30 (1996)

5. Z. Felfi, and S.T. Manson, Phys. Rev. Lett. **68**, 1687 (1992).

6. J.E. Hansen, J. Phys. B **8**, L403 (1975).

7. S. Diehl, D. Cubaynes, J.-M. Bizau, L. Journel, B. Rouvellou, S. Al Moussalami, F.J. Wuilleumier, E.T. Kennedy, N. Berrah, C. Blancard, T.J. Morgan, J. Bozek, A.S. Schlachter, L. VoKy, P. Faucher, and A. Hibbert, Phys. Rev. Lett. **76**, 3915 (1996)

8. D. Cubaynes, S. Diehl, L. Journel, B. Rouvellou, J.-M.. Bizau, S. Al Moussalami, F.J. Wuilleumier, N. Berrah, L. VoKy, P. Faucher, A. Hibbert, C. Blancard, E.T. Kennedy, T.J. Morgan, J. Bozek, and A.S. Schlachter, Phys. Rev. Lett. **77**, 2194 (1996).

9. S. Diehl, D. Cubaynes, F.J. Wuilleumier, J.-M Bizau, L. Journel, E.T. Kennedy, C. Blancard, L. Voky, P. Faucher, A. Hibbert, N. Berrah, T.J. Morgan, J. Bozek, and A.S. Schlachter,
Phys. Rev. Lett. **79**, 1241 (1997)

10. S. Diehl, D. Cubaynes, K.T. Chung, F. J. Wuilleumier, E.T. Kennedy, J.-M. Bizau, L. Journel, C. Blancard, L. VoKy, P. Faucher, A. Hibbert, N. Berrah, T.J. Morgan, J. Bozek, and A.S. Schlachter, Phys. Rev. A, R1071 (1997)

11. Y. Azuma, F. Koike, J.W. Cooper, T. Nagata, G. Kutluk, E. Shigemasa, R.Wehlitz and I.A. Sellin Phys. Rev. Lett. **79** (13) 2419, (1997)

12. K. Berrington and S. Nakazaki, J. Phys. B **31**, 313 (1998).

13. L. VoKy, P. Faucher, A. Hibbert, J.-M. Li, Y.-Z. Qu, J. Yan, J.C. Chang, and F. Bely-Dubau, Phys. Rev. A **57**, 1045 (1998).

14. L. VoKy, P. Faucher, H.L. Zhou, A. Hibbert, Y.-Z. Qu, J.M. Li and
    F. Bely-Dubau, Phys. Rev. A **58**, 3688 (1998).

15. H.L. Zhou, S.T. Manson, L. VoKy, P. Faucher, F. Bely-Dubau, A. Hibbert, S. Diehl, D. Cubaynes, J.-M. Bizau, L. Journel, and F.J. Wuilleumier, Phys. Rev. A **59**, 462 (1999).

16. K.T. Chung and B.-C. Gou Phys. Rev. A **52** 3669 (1995).

17. K.T. Chung and B.-C. Gou Phys. Rev. A **53** 2189 (1996).

18. Xiazhou Yang, C.G. Bao, and C.D. Lin, Phys. Rev. Lett. **76** 3096 (1996).

19. Xiazhou Yang, C.G. Bao, and C.D. Lin, Phys. Rev. A **53** 3934 (1996).

20. R. Wehlitz, M.-T. Huang, K.A. Berrington, S. Nakazaki, and Y. Azuma
Phys. Rev. A **60** R17 (1999)

21. S. Diehl, C. Cubaynes, E.T. Kennedy, F.J. Wuilleumier, J.-M. Bizau, L. Journel, L. VoKy, P. Faucher, A. Hibbert, C. Blancard, N. Berrah, T.J. Morgan, J. Bozek, and A.S. Schlachter, J. Phys. B **30**, L595 (1997).

22. S. Diehl, C. Cubaynes, J.-M. Bizau, F.J. Wuilleumier, E.T. Kennedy, J.-P. Mosnier, and T.J. Morgan, J. Phys. B **32**, 4193 (1999).

23. M.-T. Huang, R. Wehlitz, Y. Azuma, L. Pibida, I.A. Sellin, J.W. Cooper, M. Koide, H. Ishijima, and T. Nagata, Phys. Rev. A **59** 3397 (1999).

24. G. Mehlman, J.W. Cooper, and E.B. Salomon, Phys. Rev. A **25**, 2113 (1982).

25. R.S. Reilman and S.T. Manson, Astrophys. J., Suppl. **40**, 815 (1979)

26. R.C. Forrey, H.R. Sadeghpour, J.D. Baker, J.D. Morgan III, and A. Dalgarno, Phys. Rev. A **51**, 2112 (1995) and references therein.

27. H.W. van der Hart, K.W. Meyer, and C.H. Greene, Phys. Rev. A **57**, 3641 (1998)

28. A. Müller, G. Hofmann, B. Weissbecker, M. Stenke, K. Tinschert, M. Wagner, and E. Salzborn, Phys. Rev. Lett. **63**, 758 (1989).

29. J.A.R. Samson, Phys. Rev. Lett. **65** 2861 (1990)

30. R. Wehlitz, M.-T. Huang, B.D. DePaola, J.C. Levin, I.A. Sellin, T. Nagata, J.W. Cooper, and Y. Azuma, Phys. Rev. Lett. **81**, 1813 (1998).

31. Hugo W. van der Hart and Chris H. Greene, Phys. Rev. Lett. **81**, 4333 (1998).

32. John.W. Cooper, Phys. Rev. A **59**, 4825 (1999).

33. J.A.R. Samson, W.C. Stolte, Z.-X. He, J.N. Cutler, Y. Lu, and R.J. Bartlett, Phys. Rev. A, **57**, 1906 (1998)

34. M.Ya Amusia, E.G. Drukarev, V.G. Gorschkov, and M.P. Kazachkov, J. Phys. B **8**, 1248 (1975).

35. M.A. Kornberg, and J.E. Miraglia, Phys. Rev. A **49**, 5120 (1994)

36. A. S. Kheifets and I. Bray, Phys. Rev. A **58**, 4501 (1998)

37. J.A.R. Samson and G.C. Angel, Phys. Rev. Lett. **61**, 1584 (1988).

# Unique Features of Photoelectron/Auger-Electron Coincidence Experiments

Volker Schmidt

*Fakultät für Physik Universität Freiburg*
*Hermann-Herder-Str. 3 D-79104 Freiburg, Germany*

**Abstract.** A comprehensive review of the multiply-differential cross section of sequential photo double-ionization is presented. The analytic form of this cross section allows controled studies to explore particular aspects of the three-body Coulomb continuum. Some of these are presented and illustrated by available experiments.

## INTRODUCTION

The study of multiply differential cross sections of photo double-ionization (PDI) in which two electrons are emitted after absorption of a single photon is a challenge to explore the three-body Coulomb continuum. Within the last few years important progress has been made to describe, interpret and predict PDI phenomena in helium (see review [1]). One reason for the fundamental role of helium is that only *direct* PDI is possible i.e. both electrons are emitted simultaneously. In other elements competing channels like *sequential* PDI can also contribute. There, the process goes through a well-defined intermediate resonance state with a characteristic lifetime [2,3]. Often this is described in terms of inner-shell photoionization with subsequent Auger decay. Although it is clear that in principle both processes cannot be separated, in favourable cases, and these are considered here, the direct channel is small enough to be neglected against the sequential process of interest.

The three-body Coulomb continuum without spin is characterized by the correlated motion of the escaping electrons. The interesting questions are how energy and angular momentum are shared and how do the electrons emerge from the doubly-charged ion. With known polarization of the incident light, this information is given by the differential cross section $d\sigma(\vec{k}_a, \vec{k}_b)$ in which the momenta $\vec{k}_a$, $\vec{k}_b$ of the ejected electrons are specified. Because of energy conservation, $E_a + E_b = \hbar\omega - E_I^{++}$ where $\hbar\omega$ is the photon energy and $E_I^{++}$ the double-ionization energy, the cross section is five-fold differential. However, in the present case the discussion is restricted to electron emission in a plane perpendicular to the photon beam direction and one has a triply differential cross section (TDCS). The experimental TDCS study then requires the coincident detection of both electrons with specified directions (angles $\theta_a, \theta_b$) and energies ($E_a, E_b$), taking into account energy conservation.

In the following we will concentrate on the TDCS of sequential PDI. In the first section parametrized forms of $d\sigma(\vec{k}_a, \vec{k}_b)$ will be presented for the simplest cases of direct and sequential PDI in helium and beryllium. From the given expressions remarkable analogies and differences can be elucidated. The main aspect is that sequential PDI leads to a fully analytical representation of the matrix element which is not possible for direct PDI. This superior property comes from the sequential process itself for which a special treatment of the three-body Coulomb continuum, known as *post − collision interaction* (PCI), can be applied (see review [4]). The analytic form for $d\sigma(\vec{k}_a, \vec{k}_b)$ allows unique studies of sequential PDI because one has control over the ingredients of the differential cross section i.e. one can make special selections and explore their impact on the observed features. For a convenient description of the resulting energy- and angle-dependences a new visualization of $d\sigma(\vec{k}_a, \vec{k}_b)$ in terms of two particular factors will be introduced. A demonstration of particular features of sequential PDI is then given in the second section. Two cases refer to large and small relative angles $\theta_{ab}$ of electron emission. For back-to-back emission and $E_a \approx E_b$ effects of electron exhange [5] dominate and PCI is small, but for emission at small relative angles and for $E_a \leq E_b$ effects of PCI are extremely important. Within the eikonal approximation [6] PCI leads, through energy sharing between the electrons, to modifications in the energy distribution which, however, preserve the emission probability. In a next example it will be demonstrated that beyond the eikonal approximation at very small relative angles the emission probability is reduced (see also [7]). Finally, the phenomenon of *circular dichroism* in the angular distribution (CDAD) of the emitted electrons will be discussed. For all cases available experimental data will complement the presentation.

# I  BASIC PROPERTIES

For a detailed discussion of PDI one needs a transparent representation of the differential cross section in terms of angular factors of the photon-atom interaction and internal factors of the process. In addition one needs a convenient way to present the energy- and angle-dependences. Examples for these points will be given for linearly polarized incident light, cross section $d\sigma_{lin}$, and observation of the emitted electrons in the plane perpendicular the the photon beam. The data are represented as TDCS *patterns* in which the angle $\theta_a$ of emission of one electron $e_a$ is kept fixed and the coincident intensity, i.e. the magnitude of the differential cross section, is plotted versus the angle $\theta_b$ of the second electron. Because of energy conservation it is sufficient to specify one energy only ($E_a$). The most interesting TDCS patterns are the two *complementary* patterns obtained for unequal energies with $E_a \ll E_b$ or $E_a \gg E_b$, and the *equal – energy* pattern with $E_a = E_b$.

## (a) Direct PDI in Helium

Within the dipole approximation $d\sigma(\vec{k}_a, \vec{k}_b)$ for PDI in helium is given by [1,8]

$$d\sigma_{lin} = |b_1 \cos\theta_a + b_2 \cos\theta_b|^2 \qquad (1)$$

or alternatively in the symmetrized form

$$d\sigma_{lin} = |a_g (\cos\theta_a + \cos\theta_b) + a_u (\cos\theta_a - \cos\theta_b)|^2 \qquad (2)$$

where the $a_i$ follow from the $b_i$ and obey the symmetry relations $a_g(E_a \leftrightarrow E_b) = +a_g$ and $a_u(E_a \leftrightarrow E_b) = -a_u$. The two $\cos\theta_i$ terms in eq.(1) are due to the photon interaction with either one of the two helium 1s-electrons, and the $b_i$ coefficients contain the dynamics of the process. In the high-energy limit it is reasonable to identify the fast electron with the "photo"-electron and the slow electron with a "shake-off"-electron and the complementary patterns carry this information (in some way even down to quite low energies, details are in [1]). In eq.(2) complementary patterns are simply related to each other by $a_u \leftrightarrow -a_u$. For equal energies one has $b_1 = b_2$ or $a_u = 0$. This means $d\sigma$ then factorizes into an *angular factor* from photoabsorption and a *correlation factor*, $|a_g|^2$, which reflects the Coulomb repulsion between the electrons (and contains the overall strength of the process). These properties will be also important for the sequential process.

## (b) Sequential PDI in Beryllium

Starting from the correct formulation of sequential PDI [2,3,9], taking into account the indistinguishability of the electrons [5] and incorporating effects of PCI [10] one gets for sequential PDI in beryllium i.e. 1s photoionization with subsequent $K - L_1 L_1$ Auger decay

$$d\sigma_{lin} = \left| \frac{R(a,b)}{E_P^0 - E_a + i\Gamma/2} \cos\theta_a + \frac{R(b,a)}{E_a - E_A^0 + i\Gamma/2} \cos\theta_b \right|^2 . \qquad (3)$$

The presence of the intermediate resonance-state manifests itself in the energy denominators which contain the level width $\Gamma$ and the nominal Auger energy $E_A^0$ (note $E_P^0 - E_a = E_b - E_A^0$). The factors $R(a,b)$, $R(b,a)$ describe PCI if they differ from unity. In the eikonal approximation they are given by

$$R(a,b)|_{eikonal} = \frac{e^{-\pi\xi_a/2} \, \Gamma(1+i\xi_a)}{(E_b - E_A^0 + i\Gamma/2)^{i\xi_a}} \qquad (4)$$

where $\xi$ is the characteristic PCI parameter, $\xi_{a/b} = 1/v_{ab} - 1/v_{a/b}$, with the electron velocities $v_a$ or $v_b$ and the relative velocity $v_{ab}$, and $\Gamma(1+i\xi)$ is the complex gamma function. Eq.(3) is valid for all electron energies, including $E_a \approx E_b$ where it is not possible to distinguish between the photoelectron and the Auger electron. Still, its structure reflects the sequential nature of the process: in the first amplitude electron $e_a$ takes the role of the photoelectron and electron $e_b$ that of the Auger electron, whereas these roles are interchanged in the second amplitude because of electron exchange.

The analogy to eq.(1) is obvious! As in helium, the two $\cos\theta_i$ terms are due to the photon interaction with either one of the two 1s-electrons. (Since the emitted Auger electron has s-symmetry no additional angle-dependences appear in the present case of beryllium.) But, as stated already above, the $b_i$ coefficients are known analytically for the sequential process. This property allows a systematic selection of particular features to be studied where each of these reflects a separate ingredient of the general PDI process. In this context another interesting feature shall be noted [11]. In principle both amplitudes in eq.(3) contribute to $d\sigma$. However, tuning the photon energy, or equivalently the electron energy $E_a$, one can explore rather distinct situations. For $E_a = E_b$ the matrix element in eq.(3) is governed by the full coherence of both contributions and remarkable destructive/constructive interference effects can be expected (see below). In contrast, for $E_a$ very different from $E_b$ only one $b_i$ term dominates (cross terms vanish) and the process can be described incoherently ($2-step\ model$, see the complementary patterns below). Within these limits any ratio between the $b_i$ coefficients is possible and will affect the observables. As a consequence of these many possibilities the appropriate parameter selection to get the best appearance of certain aspects in the PDI process is no simple task.

Even though eq.(3) is given for beryllium only and different aspects will be worked out for this case, the general structure is valid also for sequential processes in other elements. Modifications must be applied if the photon-atom interaction and the Auger decay is with electrons of higher orbital angular momentum, and possibly if the intermediate state has other symmetries i.e. other angular momenta $LSJ$ and/or parity $\Pi$ [12].

### (c) New Representation of Sequential PDI in Beryllium

Concentrating on TDCS patterns of beryllium the complementary patterns will be considered first for electron $e_a$ at fixed direction and $E_a$ very different from $E_b$. For reasons of simplicity, PCI effects shall be neglected, i.e. R(a,b) = R(b,a) = 1. From eq.(3) one verifies (i) that for $E_a \approx E_P^0$ the first amplitude dominates and leads to $d\sigma \propto \cos^2(\theta_a = \text{const}) = const$, and (ii) that for $E_a \approx E_A^0$ it is the second amplitude which dominates and gives $d\sigma \propto \cos^2(\theta_b = \text{varied})$. Both TDCS patterns are shown in Figure 1. Typical for these patterns is that their energy

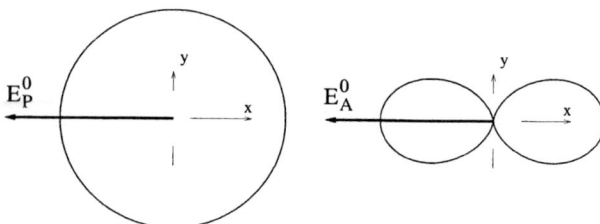

**FIGURE 1.** Complementary TDCS patterns in beryllium for $E_A^0 \ll E_P^0$. Left: $E_a = E_P^0$, right: $E_a = E_A^0$. The xyz-frame is defined by z = photon beam direction and x = electric field vector of linearly polarized light.

parameter is $E_a = E_P^0$ and $E_a = E_A^0$ i.e. the actual difference of $E_a$ around these nominal values is omitted. Justification for this comes from the fact that these dependences are simply described by Lorentzian functions and these do not affect the *shape* of the TDCS patterns. However, turning now to cases with $E_a \approx E_b$ such representations fail because the energy denominators in eq.(3) get equal weighting resulting in dramatic interference effects. In other words, the sequential TDCS patterns for $E_a \approx E_b$ will depend not only on the *nominal* values $E_P^0$, $E_A^0$ but also on the *actual* values $E_a$, $E_b$. This makes their representation more complicated than in the case of direct PDI where the energy dependence is smooth.

For a convenient representation of the energy- and angle-dependence in sequential PDI we will introduce the following factorization of the differential cross section

$$d\sigma(\theta_a, \theta_b, E_a) = P(\theta_a, \theta_b) \, [d\sigma/dE_a]_{coi} \, . \tag{5}$$

$P(\theta_a, \theta_b)$ is the *energy-integrated* angular correlation of both emitted electrons and $[d\sigma/dE_a]_{coi}$ the *coincident* energy distribution normalized to unit area. Both quantities follow from $d\sigma(\vec{k}_a, \vec{k}_b)$ calculated (or measured) for all sets of $\vec{k}_a$, $\vec{k}_b$. However, one has easier access to $P(\theta_a, \theta_b)$ provided there is only a redistribution of energies at given angles but no reduction of emission probability (see below). In such a case, the energy-integrated angular correlation is simply the sum of both complementary patterns, in the present case

$$P(\theta_a, \theta_b) = \cos^2 \theta_a + \cos^2 \theta_b \, . \tag{6}$$

This function is shown for $\theta_a = 180^0$ in Figure 2 together with the corresponding coincident energy distributions at four particular angles $\theta_b$. These distributions are in the present case isolated Lorentzian functions at $E_A^0$ and $E_P^0$, because for $E_A^0 \ll E_P^0$ interference and PCI effects are negligible. In the general case, however, these coincident energy distributions can be rather structured and reflect the interference and PCI effects. These general distributions, together with the energy-integrated angular distribution provide the full energy- and angle-dependence of the TDCS. In the present case, the two complementary patterns of Figure 1 are restored. To give

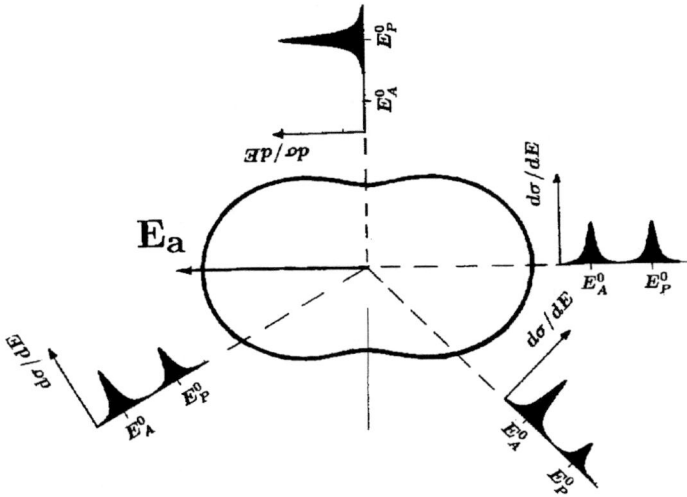

**FIGURE 2.** Energy-integrated angular correlation of beryllium, $P(\theta_a = 180^0, \theta_b) = 1 + \cos^2\theta_b$, together with some coincident energy distributions abbreviated here by $d\sigma/dE$ and calculated for $E_A^0 \ll E_P^0$.

one example, at $\theta_b = 210^0$ one has $P(\theta_a = 180^0, \theta_b = 120^0) = 1.75$ and obtains, with $[d\sigma/dE_a]_{coi}(E_a = E_A^0) = 0.57$ and $[d\sigma/dE_a]_{coi}(E_a = E_P^0) = 0.43$, the corresponding values $d\sigma = 1.0$ and $d\sigma = 0.75$ which agree with the respective complementary patterns at $\theta_b = 210^0$.

## II  DEMONSTRATION OF PARTICULAR FEATURES OF SEQUENTIAL PDI

Out of the many possibilities to probe particular facettes of sequential PDI some examples will be discussed. Instead of looking at the shapes of TDCS patterns we will concentrate on the coincident energy distributions at given angle-settings. These distributions can be explored in two ways, as a function of photon energy for preselected $E_a$, $E_b$, and for fixed photon energy (fixed $E_P^0$) as a function of $E_a$. The examples are arranged to illustrate interference effects at large relative angles, PCI effects at small relative angles, including results beyond the eikonal approximation, and circular dichroism in the angular distribution which can occur for circularly polarized incident light. In all these cases guidance is given by the simple beryllium example. (Some of these topics have been presented at a previous workshop [13].)

## (a) Interference Effects in Back-to-Back Electron Emission

For back-to-back electron emission, PCI effects can be neglected in first approximation, $R(a,b) = R(b,a) = 1$. Hence one can study interference effects between the direct and exhange amplitudes. Keeping $E_b = E_A^0$ while tuning $E_a = E_P^0$, one searches for changes in the Auger-line intensity when the coincident photoline is moved towards and across the Auger-line. Without interference one would naively expect that on-resonance ($E_a = E_P^0 = E_A^0 = E_b$) the signal increases by a factor of two, because there the electron spectrometers can detect either the photoelectron-Auger-electron-pair or the Auger-electron-photoelectron-pair (dashed curve in Figure 3). However, from the correct eq.(3) which allows for interference, one gets

$$d\sigma(\theta_{ab} = \pi, E_a = E_A^0)|_{E_P^0 \text{ tuned}} \propto \left| \frac{\cos\theta_a}{E_P^0 - E_A^0 + i\Gamma/2} + \frac{\cos(\pi - \theta_a)}{i\Gamma/2} \right|^2$$

$$\propto \cos^2\theta_a \, d\sigma_0 \, \frac{\Delta^2}{\Delta^2 + \Gamma^2/4} \quad (7)$$

where $\Delta = E_P^0 - E_A^0$ is the *coarse*-detuning parameter and $d\sigma_0$ the off-resonance signal. The result of eq.(7) is plotted as solid curve in Figure 3 and clearly demonstrates destructive interference effects: towards the resonance the coincident signal becomes reduced to a window function with a fwhm of $\Gamma$ and a zero on-resonance. In actual experiments the unavoidable finite instrumental resolutions will smear

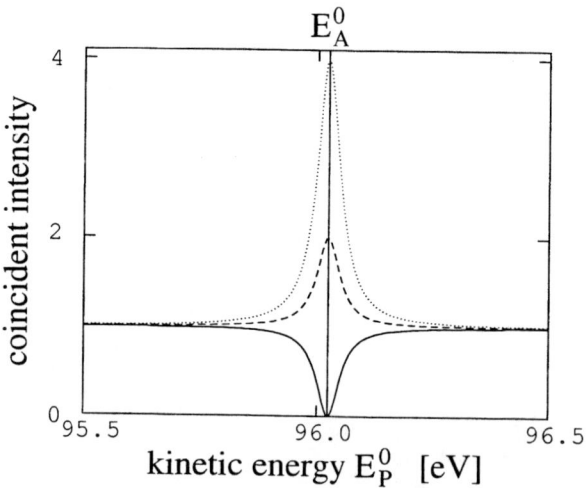

**FIGURE 3.** Coincident intensity for back-to-back electron emission in beryllium as a function of $E_a$ with $E_a = E_P^0$ and $E_b = E_A^0$. Solid: destructive interference, dashed: no interference, dotted: constructive interference. The position $E_A^0 = 96.03$ eV is indicated.

out the destructive interference effects. In spite of this difficulty, clear experimental verification is found [14-16].

The analysis shows that the destructive interference is connected to the $^2S^e$ symmetry of the intermediate resonance state considered so far and can be constructive in other cases [12]. In eq.(7) this constructive interference would require replacing the plus-sign between both amplitudes by a minus-sign. A convincing example for such a constructive interference where the photon-dependent intensity increases on-resonance to $4*d\sigma_0$ has been found for 2s photoionization in magnesium with subsequent $L_1-L_{2,3}M_1$ Coster-Kronig decay [17].

Concentrating now on the resonant coincident energy distribution ($E_P^0 = E_A^0$) one gets, for back-to-back electron emission, and neglecting again PCI effects,

$$d\sigma(\theta_{ab}=\pi, E_a)|_{E_P^0=E_A^0} \propto \left| \frac{\cos\theta_a}{E_b - E_A^0 + i\Gamma/2} + \frac{\cos(\pi - \theta_a)}{E_P^0 - E_a + i\Gamma/2} \right|^2$$

$$\propto \cos^2\theta_a \frac{4\delta^2}{(\delta^2 + \Gamma^2/4)^2} \quad (8)$$

with the *fine*-detuning parameter $\delta = E_A^0 - E_a$. The resulting dependence is shown in Figure 4 as solid curve (for comparison the incoherent distribution is plotted as dashed curve). According to the previous discussion $[d\sigma/dE_a]_{coi}$ is zero for $\delta = 0$, but with increasing $\pm\delta$ the distribution rises to a maximum for $\delta = \pm\Gamma/2$ and decreases afterwards. Again, finite instrumental energy resolutions will lead to a smearing out of this energy distribution. Here also, qualitative experimental confirmation has been found [16].

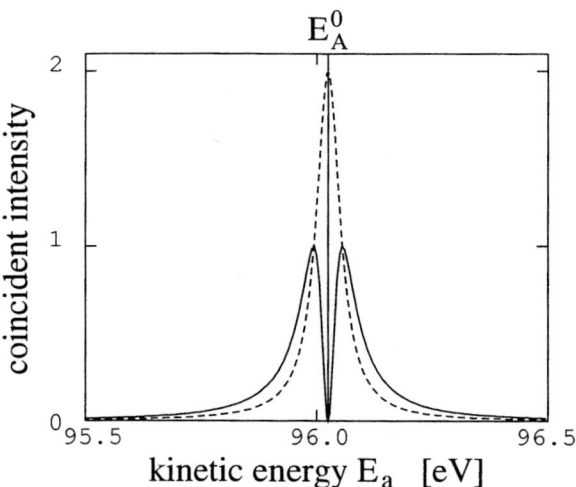

**FIGURE 4.** Coincident on-resonance energy distribution for back-to-back electron emission in beryllium. Solid: destructive interference, dashed: no interference.

## (b) PCI Effects at Small Relative Angles

For nearly parallel electron emission, PCI effects become very important [10]. They dominate the observed phenomena for all cases $E_P^0 < E_A^0$ and $E_P^0 \approx E_A^0$ where, in the latter, PCI leads also to an increase of the energy range where the interference between the direct and the exchange amplitude plays a role. As a first example the photon energy, and with it $E_P^0$, shall be tuned again, keeping the pass energies of the electron spectrometers at $E_a = E_b = E_{av} = (E_P^0 + E_A^0)/2$. From eq.(3) one gets

$$d\sigma(\theta_{ab}\, small, E_a = E_b = E_{av})|_{E_P^0\ tuned}$$
$$\propto \exp(-\pi\xi)\,|\Gamma(1+i\xi)|^2 \left|\frac{1}{(E_P^0 - E_A^0)/2 + i\Gamma/2)^{i\xi+1}}\right|^2 (\cos\theta_a + \cos\theta_b)^2. \quad (9)$$

The reason for this relative simple expression is the equivalence R(a,b) = R(b,a) which comes from equal values of $\xi_a$ and $\xi_b$, called $\xi$, and equal energy denominators, dependent on the detuning parameter $\Delta = E_P^0 - E_A^0$. Interestingly the common factors $\exp(-\pi\xi)$ and $|\Gamma(1+i\xi)|^2 = \pi\xi/\sinh(\pi\xi)$ are those which arise from the normalization factor of the electron-electron interaction in the three-Coulomb (3C) wavefunction [18] (frequently this Coulomb normalization factor is called a Gamov or Sommerfeld factor). Comparison of eq.(9) with eq.(2) for the equal-energy sharing case ($a_u = 0$) demonstrates that such a study probes the correlation factor $|a_g|^2$. The calculation gives

$$|a_g|^2 = \frac{\pi\xi}{\sinh\pi\xi}\,\exp(-2\xi\,\arctan(\frac{\Delta}{\Gamma}))\,\frac{4}{\Delta^2+\Gamma^2}\,. \quad (10)$$

As can be expected the correlation factor of the sequential process does also depend on the resonance parameters i.e. on the detuning $\Delta = E_P^0 - E_A^0$ and on $\Gamma$. In fact there is a competition between the Sommerfeld factor, lowering the intensity, and the resonance-dependent PCI factors, raising the intensity [10]. In summary, such an experiment probes the $|a_g|^2$ correlation factor of a sequential process. Its photon-energy dependence is shown by the solid curve in Figure 5 (for comparison, the dashed curve is without PCI effects). First experimental verification of this behaviour and confirmation of the underlying theoretical PCI model is presented in [11] for $4d_{5/2}$ photoionization in xenon with subsequent $N_5-O_{2,3}O_{2,3}$ $^1S_0$ Auger decay.

Another example is the shape of the coincident energy distribution at small relative angles, with fixed $E_P^0 < E_A^0$. Here in principle $\xi_a$ differs from $\xi_b$, but for a demonstration of the main effect it is still possible to keep the approximation $\xi \approx \xi_a \approx \xi_b$ where $\xi$ is calculated with the nominal values $E_P^0$ and $E_A^0$ (for the exact treatment see [10]). For fixed angles, $\xi$ and the Sommerfeld factors are independent of $E_a$, and the latter are the same for both amplitudes. Therefore the whole $E_a$-dependence comes from the two remaining energy denominators and it is possible to restrict the discussion to these i.e. to

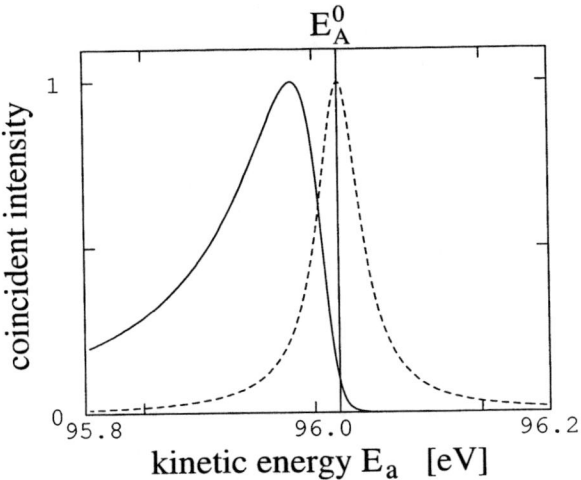

**FIGURE 5.** Coincident intensity for electron emission in beryllium at small relative angle, $\theta_{ab}$ = 10°, as a function of photon energy with $E_a = E_b = (E_P^0 + E_A^0)/2$. Solid: with PCI and interference, dashed: without PCI.

$$d\sigma(\theta_{ab}\text{small}, E_a)|_{E_P^0 < E_A^0} \propto \left| \frac{\cos\theta_a}{(E_P^0 - E_a + i\Gamma/2)^{i\xi+1}} + \frac{\cos\theta_b}{(E_a - E_A^0 + i\Gamma/2)^{i\xi+1}} \right|^2. \quad (11)$$

From the equation it can be seen that such an experiment probes interference effects with PCI modified amplitudes. These interferences will strongly depend on the phases $\varphi_{a/b}$ given by

$$\varphi_a = -\arctan\left(\frac{E_P^0 - E_a}{\Gamma/2}\right) + \pi/2, \quad \varphi_b = -\arctan\left(\frac{E_a - E_A^0}{\Gamma/2}\right) + \pi/2 \quad (12)$$

as can be seen from the cross-term of $d\sigma(\theta_{ab}\text{small}, E_a)|_{E_P^0 < E_A^0}$ which is given by

$$2\cos\theta_a \cos\theta_b \frac{\exp(\xi\varphi_a + \xi\varphi_b)}{\sqrt{(E_P^0 - E_a)^2 + \Gamma^2/4}\sqrt{(E_a - E_A^0)^2 + \Gamma^2/4}} \cos\Phi$$

$$\text{with} \quad \Phi = \varphi_b - \varphi_a + \xi \ln\sqrt{\frac{(E_a - E_A^0)^2 + \Gamma^2/4}{(E_P^0 - E_a)^2 + \Gamma^2/4}}. \quad (13)$$

The coincident energy distribution calculated in this way for the detuning $\Delta$ = −0.5eV is shown in Figure 6. One can note three peaks of equal importance, the outer ones are close to the nominal energies $E_P^0$, $E_A^0$, and the middle peak at $(E_P^0 + E_A^0)/2$. The intensity profile clearly demonstrates that any identification of structure in $[d\sigma/dE_a]_{coi}$ to a "photo"-electron or an "Auger"-electron, as done above for reasons of simplicity, becomes, strictly speaking, meaningless. Due to

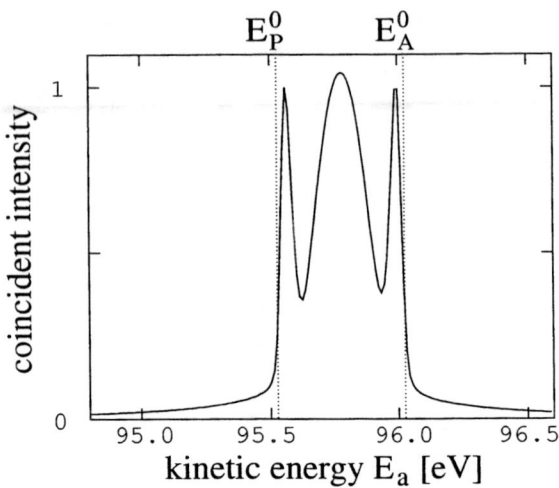

**FIGURE 6.** Coincident energy distribution for electron emission in beryllium at small relative angle, $\theta_{ab} = 10^0$, and for $E_P^0 < E_A^0$ (calculated from eq.(11), for exact results see [9]). The positions $E_P^0 = 95.53$ eV and $E_A^0 = 96.03$ eV are indicated.

experimental difficulties in performing coincidence experiments with two electrons of nearly the same direction and practically the same energy, and having good energy resolution, this experiment has not yet been performed.

### (c) PCI Beyond the Eikonal Approximation

As the relative angle $\theta_{ab}$ between the emitted electrons gets smaller, approaching zero, the eikonal approximation for describing PCI fails i.e. $R(a,b)$ and $R(b,a)$ of eq.(4) must be replaced by more complicated, but still analytical functions [19] which are given for PDI in [10]. Going beyond the eikonal approximation there are mainly two consequences. First, the on-resonance minimum of $[d\sigma/dE_a]_{coi}$ becomes zero because the Sommerfeld factor wins the competition against the resonance-dependent PCI factors. Second, the energy-integrated angular correlation $P(\theta_a, \theta_b)$ itself drops to zero. The latter aspect is shown in Figure 7 for the detuning parameter $\Delta = -0.5$ eV. This phenomenon is striking, because on energy grounds, the large detuning would allow identification of the two electrons as either the photoelectron or the Auger electron with the consequence that the photoelectron *must* have been emitted before the Auger electron could follow.

This 2-step description, however, would lead immediately to a conflict in explaining the reduced emission probability. The correct explanation must be that these two electrons, even though they are distinguishable in their energies, are still subject to their mutual Coulomb repulsion. The Sommerfeld factor reduces the emission probability if the electrons are ejected into the same direction and with nearly the same energies. First experimental verification of this effect and confirmation of the underlying theoretical PCI model is presented for $4d_{5/2}$ photoionization in xenon with subsequent $N_5 - O_{2,3}O_{2,3}$ $^1S_0$ Auger decay in [20]. (This should be compared to the counterpart of intensity enhancement in the forward direction of angle-dependent Auger transitions in ion-atom collisions [21–23].)

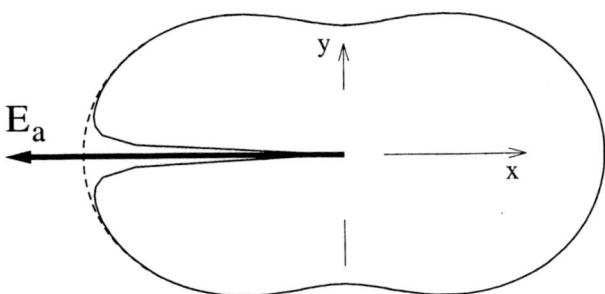

**FIGURE 7.** Energy-integrated angular correlation of beryllium for PCI described within (dashed; compare Figure (2)) and beyond (solid) the eikonal approximation, calculated for $E_P^0 = 95.53$ eV and $E_A^0 = 96.03$ eV.

### (d) Circular Dichroism in the Angular Distribution

Circular dichroism describes a dependence of the differential PDI cross section on the left or right circular polarization of the incident light, $d\sigma(\vec{k}_a, \vec{k}_b)_\ell \neq d\sigma(\vec{k}_a, \vec{k}_b)_r$, and a quantitative measure is provided by $CDAD = d\sigma_\ell - d\sigma_r$. On very general grounds, it can be shown that CDAD depends on a dynamical parameter $\Delta_{circ}$ and the triple product $\hat{k}_{photon} \cdot (\hat{k}_a \times \hat{k}_b)$. The latter describes the geometrical dependences and is maximum in the plane perpendicular to the photon beam direction $\hat{k}_{photon}$ where it depends on $\sin(\theta_b - \theta_a)$. Of particular interest is the dynamical parameter $\Delta_{circ}$. In direct PDI it yields information on the sine of the phase difference between the $a_g$, $a_u$ amplitudes of eq.(2) complementing the cosine information obtained from complementary TDCS patterns of linearly polarized light.

For sequential processes, compact expressions for $\Delta_{circ}$ have been derived with conditions for a non-vanishing signal: within the 2-step formulation of the process one must have $J_{intermediate\ state} > 1/2$ [24], but within the one-step formulation this condition does not apply [25]. It is interesting to describe $d\sigma(\vec{k}_a, \vec{k}_b)_{circ}$ in the same way as given above for linear polarization, because then the resulting CDAD

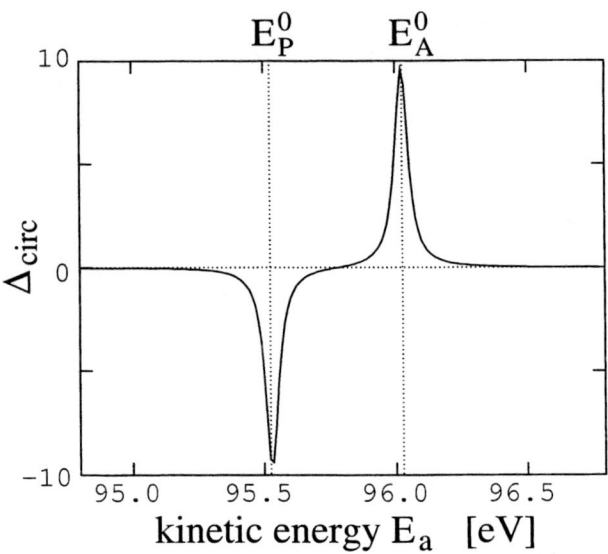

**FIGURE 8.** Energy-dependence of the CDAD parameter $\Delta_{circ}$ of beryllium, calculated for $E_P^0 < E_A^0$ and neglected PCI. The positions $E_P^0 = 95.53$ eV and $E_A^0 = 96.03$ eV are indicated.

with intermediate J = 1/2 can be illustrated clearly. In the perpendicular plane geometry one gets, neglecting for reasons of simplicity PCI modifications by setting R(a,b) = R(b,a) = 1,

$$d\sigma_{r/\ell} \propto \left| \frac{e^{\pm i\theta_a}}{E_P^0 - E_a + i\Gamma/2} + \frac{e^{\pm i\theta_b}}{E_a - E_A^0 + i\Gamma/2} \right|^2 \qquad (14)$$

where ± refer to right/left circular polarization. From this relation it can be seen that CDAD vanishes for $E_P^0$ very different from $E_A^0$ because there only one energy denominator dominates and no dependence on the light polarization remains. Also, CDAD vanishes for $E_a = E_b$ since then the energy denominators are equal (note $E_P^0 - E_a = E_b - E_A^0 \to E_a - E_A^0$) and the remaining $|e^{\pm i\theta_a} + e^{\pm i\theta_b}|^2$ gives $2(1+\cos(\theta_b - \theta_a))$ which also does not depend on the light polarization. In contrast, for $E_P^0 \approx E_A^0$ one gets

$$\Delta_{circ} = 2 \frac{1}{\sqrt{(E_P^0 - E_a)^2 + \Gamma^2/4}} \frac{1}{\sqrt{(E_a - E_A^0)^2 + \Gamma^2/4}} \sin(\varphi_a - \varphi_b) \qquad (15)$$

with $\varphi_{a/b}$ defined in eq.(12). This signal is plotted for $E_P^0 < E_A^0$ as a solid curve in Figure 8. One derives now a detailed understanding of the basic CDAD features. A non-vanishing signal depends on interfering amplitudes. In the present example these are the direct and exchange amplitudes, eq.(14). Eq.(15) then tells the details. The largest effect is for $E_a = E_P^0$ or $E_a = E_A^0$ where one of the two energy denominators is maximum. Note that this goes together with the other energy denominator which can be, depending on the detuning parameter $\Delta$, rather small. Therefore a visible CDAD signal can occur only in energy regions where *both* Lorentzian distributions overlap. (This is in contrast to CDAD produced by other mechanisms, for example by two interfering photoionization channels with $\epsilon(\ell-1)$ and $\epsilon(\ell+1)$.) Further, $\Delta_{circ}$ vanishes for $E_a = E_b = (E_P^0 + E_A^0)/2$ where $\varphi_a = \varphi_b$, as it must be. Finally, $\Delta_{circ}$ has a different sign for $E_a$ being lower or higher than this equal-energy value. This sign-change comes from the phases $\varphi_a$, $\varphi_b$ accumulated at $E_a = E_P^0$ and $E_a = E_A^0$: because of the negative detuning parameter one has at the first position $\varphi_a - \varphi_b = -\pi/2$ and at the second position $\varphi_a - \varphi_b = +\pi/2$. This implies $\sin(\varphi_a - \varphi_b) = -1$ and $\sin(\varphi_a - \varphi_b) = +1$ i.e. the $\Delta_{circ}$ signal at $E_a \approx E_P^0$ is negative and the $\Delta_{circ}$ signal at $E_a \approx E_A^0$ is positive. So far no experimental study of these particular phenomena has been made (for CDAD in direct PDI see [26,27]; for CDAD in a 2-step process see [28]).

## III  CONCLUSION

The fact that the implementation of the three-body Coulomb continuum in a sequential PDI process by the PCI model leads to a differential cross section which can be fully expressed by analytical functions gives the unique possibility to select and explore the impact of certain ingredients on the observed features. Guided by the simplest system, beryllium, which shows sequential PDI, some of these features have been represented and complemented by available experimental data. Obviously the discussion cannot be complete. Some aspects have not yet been worked out, others are even not yet thought of. Of particular interest might be the analysis of the sequential $|a_g|^2$ correlation factor with respect to its angle-dependence contained in the $\xi$-parameter. This might be accompanied by a search for particular features which arise for $E_a \neq E_b$ due to the non-vanishing $a_u$ amplitude and its interference with $a_g$. Another aspect might be modifications of the CDAD parameter $\Delta_{circ}$ when PCI is taken into account as well as in cases for which competing CDAD exists due to the presence of interfering photoionization channels.

These studies are stimulated by the analysis of the known theoretical expressions, but they must be complemented and tested by experiments. Such particular demands as the access to small (and large) relative angles, the necessary small or large energy acceptance and the adapted instrumental resolution, make these investigations more difficult than common electron-electron coincidence measurements. However with the experimental studies done so far, a demonstration has been made that this field can be explored for all desirable conditions.

The discussion has concentrated primarily on sequential PDI with an initial comparison with the direct PDI. Apart from these extreme cases, many processes lie between these limits and a unified treatment of both is needed. For a first example of this case, see [29] and the Poster B26 given at this conference.

**Acknowledgments.** I thank all members of my research group and S.A. Sheinerman and W. Mehlhorn for their contributions and stimulating discussions on the subject. The work is within the SFB 276, TP B5, and financial support by the Deutsche Forschungsgemeinschaft is gratefully acknowledged.

# REFERENCES

1. Briggs, J.S., and Schmidt, V., *J. Phys. B* submitted (1999).
2. Åberg, T., *Phys. Scripta* **21**, 495 (1980).
3. Åberg, T., and Howat, G., Theory of the Auger effect *Corpuscles and Radiation in Matter, Encyclopedia of Physics* vol. 31, ed. W. Mehlhorn (Berlin: Springer), pp. 469-619 (1982).
4. Kuchiev, M. Yu., and Sheinerman, S. A., *Sov. Phys.-Usp.*, 569 (1989).
5. Végh, L., and Macek, J. H., *Phys. Rev. A* **50**, 4031 (1994).
6. Kuchiev, M. Yu., and Sheinerman, S. A., *Sov. Phys. JETP* **63**, 986 (1986).
7. Kuchiev, M. Yu., and Sheinerman, S. A., *J. Phys. B* **27**, 2943 (1994).
8. Huetz, A., Selles, P., Waymel, D., and Mazeau, J., *J. Phys. B* **24**, 1917 (1991).
9. Tulkki, J., Armen, G. B., Åberg, T., Crasemann, B., and Chen, M. H., *Z. Phys. D* **8**, 241 (1997).
10. Sheinerman, S. A., and Schmidt, V., *J. Phys. B* **30**, 1677 (1997).
11. Scherer, N., Lörch, H., Kerkau, T, and Schmidt, V. *Phys. Rev. Lett.* **82**, 4615 (1999).
12. Selles, P., Mazeau, J., Lablanquie, P., Malegat, L. and Huetz, A. *J. Phys. B* **31**, L353 (1998).
13. Schmidt, V., *Atomic and Molecular Physics at High Brilliance Synchrotron Radiation Facilities*, Workshop in Harima Science Garden City, Hyogo, Japan (1998).
14. Schaphorst, S. J., Jean, A., Schwarzkopf, O., Lablanquie, P., Andric, L., Huetz, A., Mazeau, J., and Schmidt, V., *J. Phys. B* **29**, 1901 (1996).
15. Schwarzkopf, O., and Schmidt, V. *J. Phys. B* **29**, 3023 (1996).
16. Viefhaus, J., Snell, G., Hentges, R., Wiedenhöft, M., Heiser, F., Gešner, O., and Becker, U. *Phys. Rev. Lett.* **80**, 1618 (1998).
17. Lörch, H., et al., to be published (1999).
18. Brauner, M., Briggs, J. S., Klar, H., Broad, J. T., Rösel, T., Jung, K., and Ehrhardt, H. *J. Phys. B* **24**, 657 (1991).
19. Kuchiev, M. Yu., and Sheinerman, S. A., *J. Phys. B* **21**, 2027 (1988).
20. Scherer N., et al,. to be published (1999).
21. Swenson, J. K., Havener, C.C., Stolterfoht, N., Sommer, K., and Meyer, F. W., *Phys. Rev. Lett.* **63**, 35 (1989).
22. Barrachina, R. O., and Macek, J. H., *J. Phys. B* **22**, 2151 (1989).
23. Cordrey, I. L., and Macek, J. H., *Phys. Rev. A* **48**, 1264 (1993).
24. Kabachnik, N. M., and Schmidt, V., *J. Phys. B* **28**, 233 (1995).
25. Åberg, T., and Heinäsmäki, S., *Appl. Physics A* **65**, 131 (1997).
26. Viefhaus, J., Avaldi, L., Snell, G., Wiedenhöft, M., Hentges, R., Rüdel, A., Schäfers, F., Menke, D., Heinzmann, U., Engelns, A., Berakdar, J., Klar,H., and Becker, U., *Phys. Rev. Lett.* **77**, 3975 (1996).

27. Mergel, V., Achler, M., Dörner, R., Khayyat, Kh., Kambara, T., Awaya, Y., Zoran, V., Nyström, B., Spielberger, L., McGuire, J. H., Feagin, J., Berakdar, J., Azuma, Y., and Schmidt-Böcking. H., *Phys. Rev. Lett.* **80**, 5301 (1998).
28. Soejima, K., Shimbo, M., Danjo, A., Okuno, K., Shigemasa, E., and Yagishita, A., *J. Phys. B* **29**, L367 (1996).
29. Sheinerman, S. A., and Schmidt, V., submitted to *J. Phys. B* (1999) and contributed paper at this conference.

# Angular Distributions and Correlations in Auger Cascades of Atomic Argon Following 2p → 4s Excitation

K. Ueda*, Y. Shimizu*, H. Chiba*, Y. Sato*, M. Kitajima[†],
H. Tanaka[†], S.-M. Huttula[‡], H. Aksela[‡], I.P. Sazhina[§],
and N.M. Kabachnik*,[§],[¶]

*RISM, Tohoku University, Sendai 980-8577, Japan
[†]Department of Physics, Sophia University, Tokyo 102-8554, Japan
[‡]Department of Physical Science, University of Oulu, Oulu, Finland
[§]Institute of Nuclear Physics, Moscow State University, Moscow 119899, Russia
[¶]Fakultät für Physik, Universität Bielefeld, 33615 Bielefeld, Germany

**Abstract.** We have measured the angular distributions of the resonant Auger and the second-step Auger electrons ejected in the processes $Ar^*(2p^5 3s^2 3p^6 (^2P_{j_0})4s, J_0 = 1) \to Ar^{*+}(2p^6 3s 3p^5 (^1P_1) 4s\ ^2P_{1/2,3/2}) + e_1^-$ and $Ar^{*+}(2p^6 3s 3p^5 (^1P_1) 4s\ ^2P_{1/2,3/2}) \to Ar^{++}(2p^6 3s^2 3p^4\ ^3P_{0,1,2}) + e_2^-$, respectively, and the angular correlations between these two electrons. Based on these results, we discuss a key role of the lifetime interference effect in the angular distribution of the second-step Auger electrons and the extraction of the Auger amplitudes from the experimental data. The experimental results are compared with the multi-configuration Dirac-Fock calculations.

## INTRODUCTION

Tunability of the monochromatized synchrotron radiation allows one to excite an electron of a specific innershell to a specific unoccupied orbital. The innershell-excited atomic state thus created decays *via* the resonant Auger transition. A particularly interesting case is when the ion formed *via* the resonant Auger decay can decay further by emitting the second-step Auger electron. In this case, one can gain some dynamical information of the decay process of the core-excited state, as well as spectroscopic information of the states involved, from the angular distribution measurements for these resonant Auger and second-step Auger electrons and the angular correlation measurement between these two electrons.

As a specific example of such cascade processes, we consider here the following cascade in Ar photoexcited to the $2p_{j_0}^{-1} 4s$ state by linearly polarized light:

$$Ar(^1S_0) + h\nu \to Ar^*(2p^5 3s^2 3p^6 (^2P_{j_0})4s, J_0 = 1) \quad (1)$$

$$\to \text{Ar}^{*+}(2p^63s3p^5(^1P_1)4s\ ^2P_{J_1}) + e^-(l_1j_1) \qquad (2)$$
$$\to \text{Ar}^{++}(2p^63s^23p^4\ ^3P_{J_2}) + e^-(l_2j_2) \qquad (3)$$

where $j_0 = 1/2, 3/2$, $J_1 = 1/2, 3/2$, and $J_2 = 0, 1, 2$. The corresponding Auger lines for the transitions (2) and (3) were observed in electron-electron coincidence experiment [1]. The configuration of the intermediate state, $3s^{-1}3p^{-1}4s$, in (2) should be considered only as a label; this state is a mixture of this configuration with $3p^{-3}3d4s$ [1].

In the present paper, we discuss the following subjects using this cascade process: (1) the lifetime interference effect in the angular distribution of the second-step Auger electrons [2], (2) the extraction of the Auger amplitudes from the experimental data (*complete experiment*) [3], and (3) the comparison of the experimental anisotropy of Auger transitions with the results of multi-configuration Dirac-Fock (MCDF) calculations. The details of the experiment were described in our previous papers [2,3].

## LIFETIME INTERFERENCE EFFECT

Lifetime interference due to a coherent superposition of the Auger-initial states is expected to affect the angular distribution of the Auger electrons [4]. This effect is in general believed to diminish the anisotropy of the Auger emission due to cancellation of the various contributions from different substates. In this section, we show that the strong enhancement of the anisotropy in the Auger emission may also occur due to the lifetime interference.

We focus on the second-step Auger decay given by Eq. (3). The key point here is that the energy separation between the two spin-orbit states $2p^63s3p^5(^1P_1)4s\ ^2P_{1/2,3/2}$ is much smaller than the natural widths of these states and thus these two states are populated coherently *via* the resonant Auger decay (2). As a result the lifetime interference effect is expected in the second-step Auger emission (3).

To demonstrate the role of the lifetime interference, we have calculated the matrix elements describing Auger processes (2) and (3) with the simple models. We adopted the $LSJ$-coupling approximations to describe the final states of both Auger processes (2) and (3) and assumed that the spin-orbit interaction in the continuum can be neglected. Then only one partial wave ($p$-wave) is possible for the Auger emission (3). Furthermore, we assumed that, from two possible partial waves $s$ and $d$ of the resonant Auger emission, the $s$-wave strongly dominates. (We will discuss the $d$-wave contribution in the following section.)

Using the models described above and summing over all possible fine-structure states $J_2$ of the final multiplet $^3P$ we obtain the following expression for the angular distribution asymmetry parameter $\beta^{(2)}$ of the second-step Auger emission:

$$\beta^{(2)} = T_{int} - 1/18 \text{ for } 2p_{3/2} \to 4s\ ,\ \beta^{(2)} = T_{int} - 4/9 \text{ for } 2p_{1/2} \to 4s\ , \qquad (4)$$

where the term due to interference is $T_{int} = -4/9$ for $2p_{3/2} \to 4s$ and $T_{int} = 4/9$ for $2p_{1/2} \to 4s$. Without this interference term, the value of the asymmetry parameter is $\beta^{(2)} = -0.065$ for $2p_{3/2} \to 4s$ and $\beta^{(2)} = -0.444$ for $2p_{1/2} \to 4s$. Due to the interference term, however, the anisotropy is strongly enhanced, resulting in $\beta^{(2)} = -0.5$, for $2p_{3/2} \to 4s$, while it is strongly suppressed, resulting in $\beta^{(2)} = 0.0$, for $2p_{1/2} \to 4s$. Our experimental values, $\beta^{(2)} = -0.44 \pm 0.05$ for $2p_{3/2} \to 4s$ and $\beta^{(2)} = -.01 \pm 0.05$ for $2p_{1/2} \to 4s$, agree better with our theoretical predictions taking into account the lifetime interference than those neglecting the interference, though the agreement is only qualitative. The difference of the simple estimate from the experimental value is mostly due to the neglect of the $d$-wave contribution in the resonant Auger decay (2), as we discuss in the following section.

## DETERMINATION OF THE AUGER AMPLITUDES

The experiment for determining all of the dynamical parameters characterizing the atomic photoionization process, often called a *complete* experiment, has been well-established. A similar goal of complete characterization may be pursued for the Auger decay process. In this section we report one of the first attempts of the *complete* experiment for the Auger decay: experimental determination of the ratio of the amplitudes and their phase difference with the use of angular distribution and correlation measurements in Auger cascades [3].

We focus on the Auger emissions (2) and (3), following the excitation (1) with $j_0 = 3/2$ and extract the Auger amplitudes for the resonant Auger emission (2). In order to describe the angular distributions of the Auger emissions (2) and (3) and the angular correlation between them, we use almost the same models as those described in the previous section. Namely, we use the $LSJ$-coupling approximations to describe the final states of the Auger decay processes (2) and (3) and neglect the spin-orbit interaction in the continuum. We assume a complete interference between the intermediate two spin-orbit states $2p^6 3s 3p^5 (^1P_1) 4s\ ^2P_{1/2,3/2}$. The only difference from the models in the previous section is that we include here the contribution from both $s$ and $d$ waves in the resonant Auger emission (2).

Using these models we obtain the expression for the angular distribution parameters $\beta^{(1)}$ and $\beta^{(2)}$ for the resonant Auger emission (2) and the second-step Auger emission (3), respectively, in terms of the two amplitudes $M_s, M_d$ and their phase difference $\Delta$ which describe the resonant Auger decay (2):

$$\beta^{(1)} = \frac{M_d^2 - 2\sqrt{2} M_s M_d \cos\Delta}{2(M_s^2 + M_d^2)} \ , \ \beta^{(2)} = -\frac{M_s^2 + 0.1\, M_d^2}{2(M_s^2 + M_d^2)} \ . \quad (5)$$

In our experiment, the angular correlation between these two electrons was measured in the plane perpendicular to the photon beam, in such a way that the angular distribution of the resonant Auger emission (2) was recorded in coincidence with the second-step Auger emission (3) detected at the right angle (270°) to the photon polarization axis [3]. Then the angular correlation function can be expressed as:

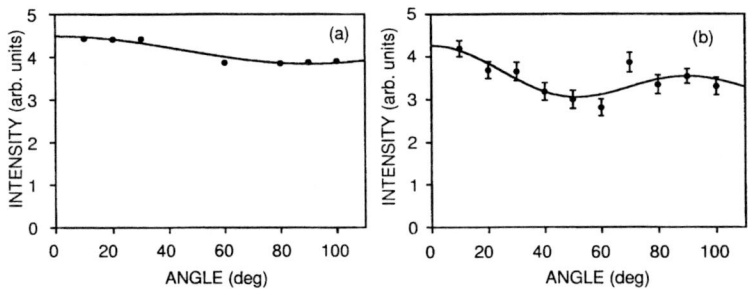

**FIGURE 1.** Angular distributions for the resonant Auger electrons ejected in the resonant decay of the Ar $2p \to 4s$ excitation; (a) without detecting the second-step Auger electrons and (b) with detecting in coincidence the second-step Auger electrons in the direction at $\theta = 270°$. The solid lines in (a) and (b) correspond to the result of the fit.

$$I(\theta) = A_0 + A_2 \cos 2\theta + A_4 \cos 4\theta ,  \quad (6)$$

where the coefficients $A_i$ depends on the amplitudes as

$$A_2/A_0 = \frac{48M_d^2 - 96\sqrt{2}M_s M_d \cos \Delta}{80M_s^2 - 16\sqrt{2}M_s M_d \cos \Delta + 61M_d^2} , \quad (7)$$

$$A_4/A_0 = \frac{27M_d^2}{80M_s^2 - 16\sqrt{2}M_s M_d \cos \Delta + 61M_d^2} . \quad (8)$$

Figure 1 (a) shows the angular distribution of the resonant Auger emission (2), whereas Fig. 1 (b) shows the angular distribution of the resonant Auger emission (2) recorded in coincidence with the second-step Auger emission (3). It is clear that the anisotropy of the resonant Auger emission is enhanced by detecting it in coincidence with the second-step Auger emission. Using Eqs. (5)–(8) and all the experimental data $\beta^{(1)}$, $\beta^{(2)}$, $A_2/A_0$, and $A_4/A_0$, we have obtained the ratio of the amplitudes and the phase difference: the resulting values are $M_d/M_s = 0.52 \pm 0.15$ and $\cos \Delta = 0.01 \pm 0.03$.

The experiment described above is the first complete experiment for the resonant Auger decay using the angular distribution and correlation measurements in the Auger cascade. The *completeness* however relies on our models: the non-relativistic $LSJ$-coupling approximations. In the following section, we lift these limitations and compare the experimental results with more general multi-configuration Dirac-Fock (MCDF) calculations.

## COMPARISON WITH MCDF CALCULATIONS

We have calculated the angular anisotropy parameters for the first and the second Auger transitions within the MCDF approach. Both initial state configuration

**TABLE 1.** Experimental and theoretical asymmetry parameters $\beta^{(1)}$ and $\beta^{(2)}$ for the resonant and second-step Auger transitions, respectively. A simple model was described in the second section, while MCDF is described in the fourth section.

|  | $2p_{3/2} \to 4s$ | | | $2p_{1/2} \to 4s$ | | |
| --- | --- | --- | --- | --- | --- | --- |
|  | Experiment | Simple model | MCDF | Experiment | Simple model | MCDF |
| $\beta^{(1)}$ | $0.09 \pm 0.05$ | 0.0 | $-0.061$ | $-0.03 \pm 0.09$ | 0.0 | 0.003 |
| $\beta^{(2)}$ | $-0.44 \pm 0.05$ | $-0.5$ | $-0.413$ | $-0.01 \pm 0.05$ | 0.0 | $-0.011$ |

interaction and final ionic state configuration interaction (FISCI) were taken into account. The bound orbitals optimized for the final states were used for both transitions. For the first-step Auger decay the initial resonant state has almost pure $2p^{-1}4s$ configuration with small admixture of the $2p^{-1}3d$ configuration. Strong FISCI between the $3s3p^54s$ and $3p^33d4s$ configurations results in considerable redistribution of the intensity of Auger transitions. However, the anisotropy parameter for the first Auger decay is not very sensitive to FISCI. The calculations agree reasonably well with the measurements. We note however that the calculated $d$-wave contribution is much lower than obtained in the previsous section. This discrepany may cast some doubts on the validity of the non-relativistic $LSJ$-coupling approximation adopted in the previous section, though we cannot draw the decisive conclusion yet. For the second-step Auger the lifetime interference effect in the intermediate state was taken into account using the formalism from [4]. The resulting $\beta^{(2)}$ parameter agrees well with the experiment (see Table 1).

## ACKNOWLEDGMENTS

This experiment was carried out with the approval of the Photon Factory Advisory Committee (Proposal Nos. 98G231). The authors are grateful to T. Hayaishi and the staff of the Photon Factory for their help in the course of the experiments, to R. Wehlitz and U. Becker for their contribution in the first stage of this work [2], and to J. Hansen, B. Sonntag, U. Becker, J. Viefhaus, and P. Lablanquie for discussions.

## REFERENCES

1. E. von Raven, M. Meyer, M. Pahler, and B. Sonntag, J. Electr. Spectrosc. Relat. Phenom. **52**, 677 (1990); E. von Raven *Thesis at Hamburg University* (1992).
2. K. Ueda, Y. Shimizu, N. M. Kabachnik, I. P. Sazhina, R. Wehlitz, U. Becker, M. Kitajima, and H. Tanaka, J. Phys. B **32**, L291 (1999).
3. K. Ueda, Y. Shimizu, H. Chiba, Y. Sato, M. Kitajima, H. Tanaka, and N. M. Kabachnik, to be published in Phys. Rev. Lett.
4. N. M. Kabachnik, J. Tulkki, H. Aksela, and S. Ricz, Phys. Rev. A **49**, 4653 (1994); and references cited therein.

# K-shell ionization and double-ionization of Au atoms with 1.33 MeV photons

A. Belkacem[1], D. Dauvergne[2], B. Feinberg[1], D. Ionescu[1], J. Maddi[1], and A. H. Sorensen[3].

[1] *Lawrence Berkeley National Laboratory, Berkeley, CA 94611*
[2] *Institut de Physique Nucleaire, Universite Claude Bernard, 69100 Villeurbanne Cedex*
[3] *Institut of Physics and Astornomy, Aarhus University, DK-8000 Aarhus C, Denmark*

**Abstract.** At relativistic energies, the cross section for the atomic photoelectric effect drops off as does the cross section for liberating any bound electron through Compton scattering. However, when the photon energy exceeds twice the rest mass of the electron, ionization may proceed via electron-positron pair creation. We used 1.33 MeV photons impinging on Au thin foils to study double K-shell ionization and vacuum-assisted photoionization. The preliminary results yield a ratio of vacuum-assisted photoionization and pair creation of $2 \times 10^{-3}$, a value that is substantially higher than the ratio of photo double ionization to single photoionization that is found to be $0.5-1 \times 10^{-4}$. Because of the difficulties and large error bars associated with the small cross sections additional measurements are needed to minimize systematic errors.

Inner-shell photoionization of an atom or an ion is one of the most basic processes in atomic collisions. Ionization may proceed through the photoelectric effect or through Compton scattering (1-2). At MeV photon energies and beyond, the cross sections associated with both processes decrease almost linearly with increasing photon energy (3-5). When the photon energy exceeds twice the rest mass of the electron, the negative-energy continuum will play an additional important role; photoionization of, say, the K-shell can now proceed through a new channel in which the excess of energy is taken by one of the negative-energy continuum electrons (6). The final result is the creation of the K-vacancy along with an electron-positron pair.

In the relativistic regime, an atom can be described as a many-body system containing Z electrons occupying discrete bound states, Z being the atomic charge number and, in the Dirac picture of the hole theory, an infinite number of electrons occupying the negative-energy continuum (7). The removal of the inner-shell electron through this process (called here "vacuum-assisted photoionization") will result in the creation of two vacancies, one in the inner-shell and the other in the negative-energy sea. This means that, from a theoretical point-of-view, vacuum-assisted photoionization can also be viewed as a double-ionization with a single photon. However, compared to the well-known photo double ionization (8-10), the situation is made somewhat more complicated by the presence of an extra lepton, the positron. The positron reflects the creation of a vacancy in the negative-energy continuum, and this vacancy can interact in the post-collision with the inner-shell electron.

Two different mechanisms contribute to vacuum-assisted photoionization. In the first, the photon converts into an electron-positron pair in the field of the nucleus and, subsequently, the bound electron (of the inner-shell) is ionized through an electron-electron or positron-electron encounter. In the second mechanism, the electron-positron pair is produced in the field of the inner-shell electron, that, in the process, takes enough recoil to be freed from the atom or the ion. If $E_B$ denotes the binding energy of the inner-shell electrons, vacuum-assisted photoionization occurs with a threshold of $\omega_{thr} = 2mc^2 + E_B$ for the first mechanism and $\omega_{thr} = 4mc^2 + E_B$ for the second mechanism, m being the rest mass of the electron.

Figure 1 shows the cross sections for creation of a K-vacancy for Pb. The details of the calculation can be found in ref. (6).

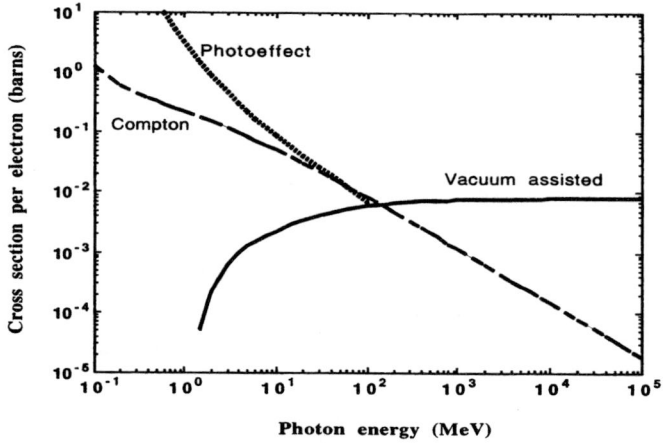

Figure 1

**Figure 1**: Cross section for creation of a K-shell vacancy for Pb (Z=82) for different photoionization processes as a function of the photon energy. The dashed-dotted line corresponds to the Compton effect and the dotted line to the photoelectric. Vacuum-assisted photoionization cross section is shown in a solid line.

The cross section for the creation of a K-vacancy by photons with energies below few MeV is dominated by the photoelectric effect contribution. However, with increasing photon energy, the photoelectric effect cross section decreases as a high negative power of the photon energy, and then, above about 10 MeV, as the inverse of the photon energy. In contrast, the contribution from vacuum-assisted photoionization increases with increasing photon energy, starting from a threshold of approximately 1

MeV. At high energies the photoionization cross section saturates at 7.5 mb per electron, a value that is due entirely to the contribution of vacuum-assisted photoionization.

We set up a first experiment to study this process at an intense Cobalt Source at the Lawrence Berkeley National Laboratory (LBNL). The 1500 Curies $^{60}$Co source emits two gammas at 1.17 MeV and 1.33 MeV, respectively. These energies are below the threshold for pair creation in the field of the inner-shell electron. Thus, the study reported in this paper focuses on the first mechanism, namely, pair creation in the field of the nucleus, with subsequent inner-shell ionization through electron-electron and electron-positron interaction. The photon energy of 1.33 MeV is very close to threshold, and its corresponding cross section is two orders of magnitude smaller than the cross section at saturation. Most of the contribution to photoionization comes from the photoelectric effect and Compton scattering, making the experiment difficult. The measurement of vacuum-assisted photoionization would have been somewhat easier at higher energies (above 100 MeV), but there are very few high energy photon facilities where such experiments can be undertaken.

A beam of $2 \times 10^8$ photons/second is produced through a 6-mm diameter collimator in a 25-cm thick lead shielding enclosing the source. We estimate the number of photons in the beam by measuring the rate of single K-shell vacancies created in a 3 mg/cm$^2$ Au target and comparing to tabulated values for the cross section.

A signature of single K-shell ionization is given by the detection of a $K_\alpha$ or $K_\beta$ photon emitted when a K-vacancy is filled. A signature of vacuum-assisted photoionization is given by the simultaneous detection of a K-vacancy and of a positron emitted. The positron is detected through its annihilation in a thin, low-Z foil set immediately downstream from the Au target. Two (7.5-cm diameter and 7.5-cm thick) NaI crystal detectors are set on opposite sides of the target to detect the 511-keV photons emitted back-to-back, characteristic of positron annihilation. The experiment is complicated by two backgrounds: First, a two-step process where the pair creation occurs on one atom and the K-shell vacancy on a different atom. The other false event can be simulated by two separate photons striking the target within the time window of the coincidence, one creating an electron positron pair and the other creating a K-vacancy. We set the coincidence window to 50 ns to minimize the latter. One of the difficulties of the experiment is to balance the need of a thick target for statistics and a thin target to minimize the two-step background. Unfortunately, because of the low (5%) geometrical detection efficiency of the germanium detectors, such an ideal target thickness is hard to achieve. To make our way around this problem, we perform a target-thickness-dependent measurement of vacuum assisted photoionization. A two-step process should vary as the square of the target thickness instead of linearly. Thus, a ratio of vacuum-assisted photoionization to pair creation should become independent of the target thickness for very thin targets.

Figure 2 shows the ratio of "vacuum-assisted photoionization" to pair production. A fit to the data yields a value of $2 \times 10^{-3}$, a value that is a factor of two to five larger than the theoretical value given in ref. (6). This preliminary number constitutes the first attempt to observe this new ionization process and measure its cross section relative to the pair creation cross section, which is well known at these energies. Because of the inherent difficulties associated with the small cross sections additional measurements are needed to minimize systematic errors. A series of follow up experiments to refine the measurement and make a more meaningfull comparison to theory are ongoing at the Lawrence Berkeley National Laboratory and also in planning at the European Synchrotron Radiation Facility (ESRF).

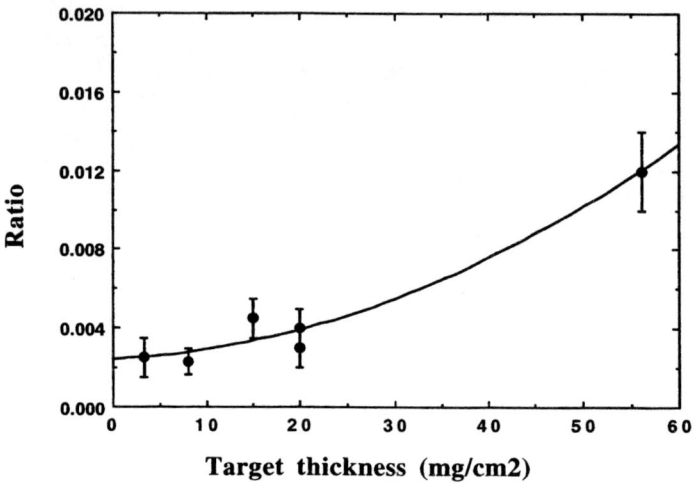

Figure 2

**Figure 2**: Ratio of the coincidence of "vacuum-assisted photoionization" cross section to the pair creation cross section as a function of the Au target thickness. The solid curve represents a least-square fit to the data used to extrapolate to vanishing thickness.

We used the same experimental set up to measure the ratio of double K-vacancy to single K-vacancy creation in Au. Because of the low count rate, we used a similar target dependent measurement of the ratio of double ionization to single ionization. Using an extrapolation to very thin targets, we obtain a value of $0.5\text{-}1 \times 10^{-4}$ for the ratio. This low value is at the limit of sensitivity of our current set up, and should be considered as an upper limit. This sensitivity could be improved by an order of magnitude by requiring a better timing between the germanium detector signals. The theoretical relativistic limit given in ref. (9) and ref. (10) differ by a factor of approximately two and are $0.54 \times 10^{-4}$ and $0.94 \times 10^{-4}$, respectively.

## ACKNOWLEDGEMENTS

This work is supported by the Director, Office of Energy Research, Office of Basic Energy Sciences, Chemical Sciences Division, of the U.S. Department of Energy (DOE) under Contract No. DE-AC-03-76SF00098. We acknowledge the support of this work by the France-Berkeley Fund. One of us (DD) thanks NATO for grant No 23C96FR of support of his sabbatical at Berkeley.

## REFERENCES

1. Lagarde P., Wuilleumier F. J., and Briand J. P., Special issue of J. Phys. (Paris) Colloq. 48, C9-48 (1987)
2. Drake G., *Atomic, Molecular and Optical Handbook*, AIP, Woodsbury, NY; see section written by Bernd Crasemann, 701 (1996).
3. Heitler W., *The Quantum Theory of Radiation*, Dover, New York, (1984).
4. Pratt R. H., Akiva Ron, and Tseng H. K., Rev. Mod. Phys. **45**, 273 (1973).
5. Hubbell J. H., Gimm H. A., and Overbo I., J. Phys. Ref. Data **9**, 1023 (1980).
6. Ionescu D. C., Sorensen A. H., and Belkacem A., Phys. Rev. A 59, 3527 (1999).
7. Greiner W., Muller B., and Rafelski J., Quantum Electrodynamics of Strong Fields, Springer-Verlag, Berlin, 1985.
8. McGuire J. H. et al., J. Phys. B **28** 913 (1995).
9. Mikhailov A. I., and Mihkailov I. A., JETP **87**, 833 (1998).
10. Drukarev E. G., and Karpeshin F. F., J. Phys. B **9**, 399 (1976).

# III. MOLECULES

# Dynamics of competing ultra-fast fragmentation and resonant Auger processes

O. Björneholm

*Department of Physics, Uppsala University, Box 530, S-751 21 Uppsala, Sweden*

When a molecule is core-excited to a dissociative state, dissociation may occur on the time scale of the core hole decay, i.e. a few femtoseconds. This is discussed in terms of the nuclear dynamics of the dissociation, and the electronic dynamics of the core hole decay. Examples demonstrating different aspects of these phenomena are presented. Semi-classical methods, which provide quantitative information about the dissociation dynamics, are outlined. More advanced theories, including time-dependent quantum mechanics and interference, as well as possible experimental developments, are briefly discussed.

## INTRODUCTION

When a molecule is excited to an anti-bonding core excited state, two different but inter-connected processes start, as schematically shown in Fig. 1. The nuclei will start to move apart, aiming to dissociation of the molecule into two neutral fragments, one of them core-excited. In parallel, the core hole will in a few femtoseconds (fs) be filled by an outer electron, often in an Auger-type decay. The timescale of the two processes is sometimes approximately equal, as first observed for HBr (1), and the rapid nature of the process has lead to it often being called ultra-fast dissociation. The same phenomenon has later been observed for many other hydrogen-containing molecules (2-12), and also other ones such as $O_2$ (13, 14). In the case of the hydrogen halides, the rapid nature of the dissociation may be qualitatively understood within the Z+1 approximation: the core excited halogen atom resembles the following rare gas atom, and the core excited molecule is thus created in a highly repulsive state. Similar, but less transparent, arguments may be used to obtain a simple understanding of the root of ultra-fast dissociation. The resulting Auger spectra reflects the interplay between the two competing processes, and may be used to study the dynamics of this ultra-fast dissociation on the low femtosecond timescale.

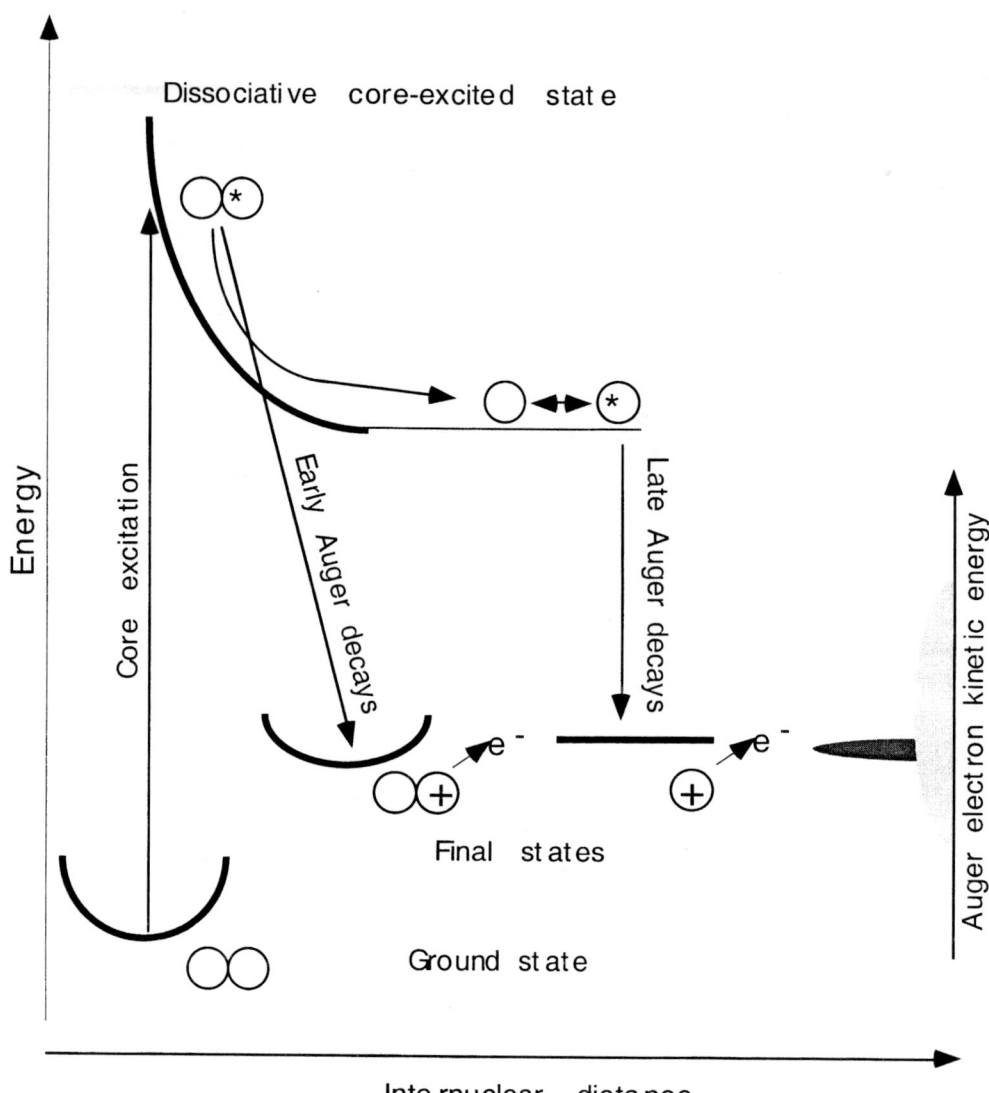

**FIGURE 1.** The two competing processes, Auger decay and dissociation. Early Auger decays are of "molecular" character, late Auger decays are of "fragment" character. * denotes a core excitation.

As indicated in Fig. 1, the "molecular" contribution due to the differently sloping potential energy curves forms mainly a broad background to the Auger spectra, whereas the "fragment" contribution is characterized by sharp structures. As a consequence of energy conservation, the two contributions exhibit different behaviours upon photon energy variations: the "molecular" contribution is characterized by constant binding energy, whereas the "fragment" contribution is characterized by constant kinetic energy. In the "molecular" case, the only outgoing particle is the Auger electron, which consequently will vary in kinetic energy with the exciting photon energy, resulting in a constant binding energy of the corresponding spectral features. In the "fragment" case the Auger decay takes place in the asymptotically flat region of the potential energy curve, resulting in constant kinetic energy of the Auger electron. Variations of the exciting photon energy will instead be reflected by the kinetic energy of the usually undetected neutral fragment. Connected to this are the different spectral widths: the width of "fragment" peaks in Auger decay is mainly given by the core hole life time width, and will thus be largely unaffected by variations in excitation bandwidth, in marked contrast to "molecular" features (7).

Ultra-fast core-excitation-induced dissociation processes produce not only Auger electrons but also ions. These will also reflect the processes, but in a different way, as the timescales of the two types of measurement are quite different. While Auger spectroscopy reflects the development in the system within the first few femtoseconds after excitation, ions may be produced both then but also much later in the often microsecond-long ion detection process. Ion yield spectroscopy may nevertheless provide information about ultra-fast dissociation processes (8, 9, 10).

In this paper, some examples illustrating different aspects of the nuclear and electronic dynamics of core-excitation-induced dissociation will be discussed. It is often useful to consider the nuclear dynamics of the dissociation process and the electronic dynamics of the core-decay processes separately (15). As dissociation is a dynamic process, it is tempting to use time-dependent descriptions. Conceptually simple semi-classical "core-hole clock" models (6, 11, 16), based on the life time of the core hole as an internal femtosecond stopwatch, will be outlined as mean of improving the understanding of these combined processes. Such models have of course limitations, and more advanced quantum models containing time-dependent elements will be discussed (17, 18, 19), as well as possibilities of obtaining more direct temporal information using pump-probe techniques.

## EXPERIMENTAL

The resonant Auger experiments discussed in this paper have been performed at beamlines 51 and I411 of MAX-LAB, the National Swedish Laboratory for synchrotron radiation (20, 21, 22). These consist of an undulator and a modified Zeiss SX-700 monochromator. The endstation was equipped with a Scienta SES-200 electron energy analyzer, which is possible to rotate so that measurements can be performed both parallel and perpendicular to the polarization direction of the light.

Ion yield spectroscopy has been performed with the PEPICO (PhotoElectron-PhotoIon Coincidence) technique. In this the charge-to mass ratio of the ions is determined using the time-of-flight technique, with a photo or Auger electron as start signal and the ion as stop signal (23). The measurements of this type discussed in this paper have been performed at the Brazilian national laboratory for synchrotron radiation, Laboratório Nacional de Luz Síncrotron (LNLS), in Campinas, Brazil (24). The time-of-flight PEPICO spectrometer was designed and built at the University of Brasilia, Brazil (25).

## DISSOCIATION VS CORE HOLE DECAY

Simple, semi-classical models can be used to derive some quantitative information about the dynamics of ultra-fast dissociation. A characteristic timescale for the ultra-fast dissociation may be obtained from resonant Auger measurements, as has been demonstrated using $H_2S$ (6).

The Auger spectra resulting from resonant excitation of a molecule such as $H_2S$ can be described as consisting of two components: one due to the Auger decay of neutral S*H fragments (* denotes a core excitation), and another one due to the decay taking place in H-S*H molecules in an intermediate, i.e. partially dissociated, geometry. The S*H contribution consists of sharp, well-defined transitions, indicating well-defined initial and final states. The H-S*H contribution, on the other hand, is observed as broad and structureless features. This can be seen as the result of Auger decays occurring during the dissociation process, thus consisting of a superposition of different H-S*H geometries, ranging from the ground state molecular geometry all the way to configurations where the outgoing H atom has only the weakest influence on the S*H fragment. This is consistent with an interplay between the two different processes, Auger decay and dissociation, occurring on the same time scale.

It is possible to apply the so called "core hole clock" method (16) to the dissociation of $H_2S$. The lifetime of the core excited state is used as an internal femtosecond stopwatch to monitor processes occurring on the same time scale as the Auger decay, such as electron transfer or nuclear motion. The Auger decay is well described with an exponential function, with the core hole life time $\tau$ as time constant. On the other hand, the H-S*H separation process follows a different time development. In the here discussed $H_2S$ case, it was possible to spectroscopically distinguish between

Auger decay processes taking place in the *dissociating* H-S*H species and the *dissociated* S*H species, but not between different H-S*H geometries. It is assumed that the actual electronic separation between H and S*H takes place in a very short time compared to the time scales of the dissociation and the core hole lifetime. As a first approximation, it is thus possible to use a model in which the molecule is described as H-S*H from the core excitation event at t=0 up to t=$t_D$, the dissociation time, and after $t_D$ as S*H. Letting $n_{"molecular"}$ denote the fraction of "molecular" Auger decays, the dissociation time $t_D$ may then be obtained as (6):

$$t_D = -\tau \ln(1 - n_{"molecular"}) \tag{1}$$

By fitting of the experimental spectra, the "molecular" decay fraction $n_{"molecular"}$ is determined to 0.43±0.1. An estimate for the core hole lifetime can be obtained from the width of high-resolution S 2p photoelectron or x-ray absorption spectra (26, 27). This yields a value for $\tau$ of 9.4 fs. Inserting the values of $n_{"molecular"}$ and $\tau$ into equation (1), a dissociation time of 5.3±1.5 fs for the S $2p_{3/2}^{-1}6a_1$ core excited state is obtained.

It is thus possible to obtain relatively detailed quantitative information about the dissociation process by using simple semi-classical models in the analysis of Auger spectra.

## ISOTOPIC SUBSTITUTION

The core-excitation-induced relaxation doesn't only produce Auger electrons, but also ions. These may have different charges, and both be of the same composition as the parent molecule or be fragments thereof. These will also reflect the ultra-fast dissociation process, but as the time scale of ion detection is much longer than the time scale probed by the Auger decay (a few fs), this is more indirectly manifested. One possibility for quantitative studies is offered by isotopic substitution, which allows a change of the nuclear dynamics, while the electronic dynamics is unaffected. This has been done in a PEPICO study of water (11).

From PEPICO spectra, it is immediately clear that a substantial number of molecules dissociate as a result of the core excitation and decay processes, which are illustrated schematically in Fig. 2. Starting from the top, the molecule is first core excited by absorption of a photon. The core-excited state will undergo Auger decay, the dynamics of which is well described by an exponential function. Some "early" decays will thus occur very quickly after core excitation into $H_2O^*$, producing mainly $H_2O^+$ in an essentially molecular geometry. But some decays, the "late" ones, occur after a

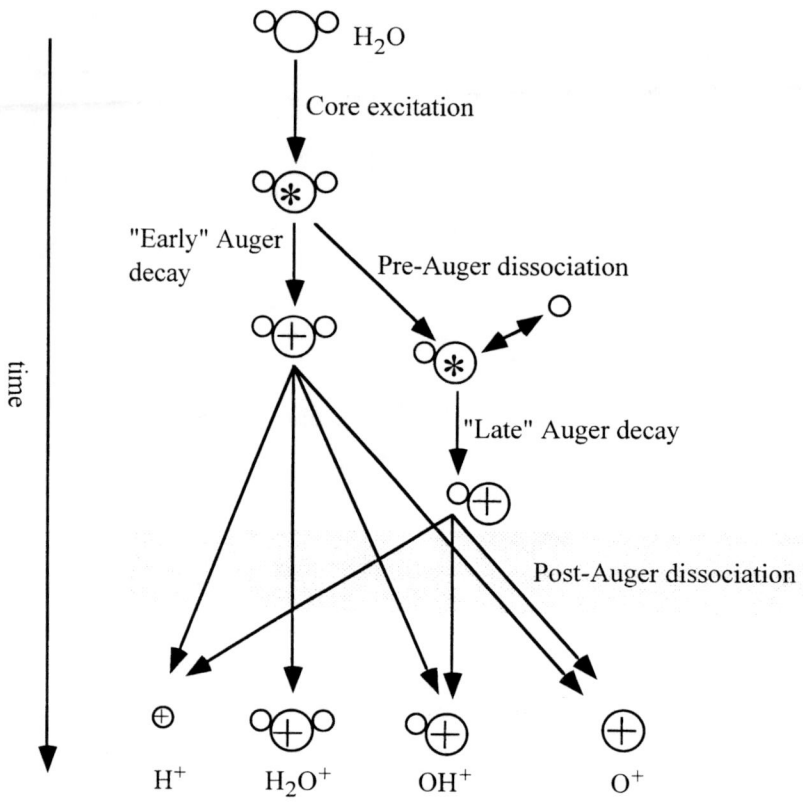

**FIGURE 2.** Pre-Auger and post-Auger dissociation of core excited water

much longer time, and there is a finite probability of dissociation of $H_2O^*$ into $HO^*$ + H before the Auger decay, an ultra-fast process which we denote pre-Auger dissociation. The neutral core excited $HO^*$ will subsequently undergo "late" Auger decay, producing mainly singly charged $HO^+$. In both "early" and "late" Auger decays there is also some production of doubly charged ions through shake-off processes accompanying the Auger decay. Finally, the $H_2O^+$ and $HO^+$ may dissociate into $HO^+$ and $O^+$, a process which we denote post-Auger dissociation.

PEPICO spectra for $H_2O$ and $D_2O$ reveal that the relative intensity of the unfragmented ion, $H_2O^+/O^+$ and $D_2O^+/O^+$ respectively, is stronger for $D_2O$ than for $H_2O$; in fact the same is true for the $H_2O^+/HO^+$ and $D_2O^+/DO^+$ ratios. The electronic structure and dynamics, such as energies of states, Auger decay rates and inter-atomic forces responsible for bonding or dissociation, are practically identical for $H_2O$ and $D_2O$. The nuclear dynamics, however, are not unaffected by the substitution of H with D, and all movements such as rotations, vibrations and dissociation, are changed. In this case we will restrict ourselves to dissociation. The repulsive force between the atoms is the same in the $H_2O$ and $D_2O$ cases, but the masses involved are larger for $D_2O$ than for $H_2O$, making dissociation of $D_2O$ slower than for $H_2O$. Let us consider the relative intensity of the unfragmented ions, $H_2O^+$ and $D_2O^+$. As seen in Fig. 2 these may only be produced via one single pathway. To increase the relative amount of the unfragmented ion, one or more of the bifurcations leading to dissociation must get a decreased branching ratio. As $H_2O$ and $D_2O$ are electronically identical, but nuclear motion is slower for $D_2O$, all branching ratios would be identical if the processes were not time limited. After the Auger decay the ions have practically unlimited time to dissociate, and there should thus not be any appreciable differences between $H_2O$ and $D_2O$ in the post-Auger dissociation pattern. However, the situation is different for the branching ratio of the first bifurcation, describing the possible ultra-fast pre-Auger dissociation in the core excited state. In this case the process is time limited by the finite lifetime of the O1s core hole, $\approx 3.3$ fs. As the core hole lifetime is the same for $H_2O^*$ and $D_2O^*$, fewer $D_2O^*$ than $H_2O^*$ will undergo pre-Auger dissociation. The observed difference in fragmentation patterns between $H_2O^+/HO^+$ and $D_2O^+/DO^+$ has thus been interpreted as evidence for the dissociation reaction $H_2O^*/D_2O^* \rightarrow HO^*/DO^* + H/D$ occurring on a time scale comparable to the core-hole lifetime of 3.3 femtoseconds. Similar conclusions are valid for fragmentation patterns between $H_2O^+/O^+$ and $D_2O^+/O^+$.

These arguments have moreover been quantified, leading to a simple model based on the lifetime of the core hole as an internal femtosecond stopwatch (the "core-hole clock") (11, 16). This yields an approximate quantitative value of 4.0 fs for the characteristic dissociation time of the core-excited water molecule. The model also yields quantitative information about the complete fragmentation pattern of water after core excitation. The results show that the outlined method to compare the fragmentation dynamics for "normal" and deuterated molecular species in combination with the short lifetime of core holes may be used for quantitative studies of dissociation dynamics on the low femtosecond time scale. Corresponding effects connected to isotopic substitution have also been observed by Auger spectroscopy (5).

## DOPPLER SPLITTING AND FRAGMENT VELOCITIES

A characteristic time scale is a first measure of a dynamic process. There are however other aspects of the ultra-fast dissociation process which may be interesting, such as what velocity the fragments acquire. While such data are routinely obtained for dissociation processes from time-of-flight ion techniques, such as PEPICO, these data do not selectively reflect the dynamics of the ultra-fast dissociation process due to the long time scale of ion detection and secondary processes. To study the fragment velocities on the low femtosecond time scale, Auger spectroscopy offers a possibility, as has been demonstrated for oxygen molecules (28).

O1s core electrons in oxygen molecules were selectively excited to the unoccupied $\sigma^*$ orbital, also refered to as the shape resonance. The $1s \to \sigma^*$ excitation probability depends on the orientation of the molecules as $\cos^2\theta$, where $\theta$ is the angle between the polarization vector and the molecular axis. Out of the randomly oriented gas phase molecules, the excitation thus selects a quasi-aligned subset with the molecular axes preferentially parallel to the polarization vector.

The core excited $1s^{-1}\sigma^*$ state has a lifetime of 3 fs, after which it typically will undergo Auger decay. The quasi-alignment will be preserved during the Auger decay, as rotational motion occurs on an approximately one hundred times longer time scale. The core excited $1s^{-1}\sigma^*$ state is also strongly repulsive, partially leading to dissociation on the time scale of the Auger decay (13, 14). The resulting resonant Auger spectrum will contain contributions from decay both in the molecule and the dissociated atoms.

The experimental Auger spectrum is shown in the upper part of fig. 3. The spectrum has been estimated to contain 10% contribution from "atomic" decays (13, 14), which together with the core hole lifetime of 3 fs and Eq. 1 may be used to estimate a characteristic dissociation time of 7 fs. An expanded view of one of the atomic Auger transitions, $1s^{-1}2p^5$ (3P) $\to 2s^{-2}p^4$ ($^2$D), is shown in the lower part of the figure. The spectrum has been measured with the analyzer at 0° and 90° relative to the polarization direction. In the 90° case, the velocities of the atomic dissociation fragments are mainly directed perpendicular to the measurement direction. The two dissociation directions are thus equivalent, and only one Auger peak is observed. In the 0° case, however, the velocities of the atomic dissociation fragments are mainly directed parallel to the measurement direction. The two dissociation directions, towards or away from the electron energy analyzer, are thus inequivalent, and the Auger peak is split into two components. This splitting is caused by a Doppler shift of the Auger electron due to the directed velocity acquired by the atomic fragment in the dissociation process.

The Doppler splitting, $\Delta E_{Doppler}$, depends on the velocity V of the outgoing fragment relative the molecular centre of mass (or alternatively the kinetic energy release $E_{KER}$ of the ultra-fast dissociation process), the kinetic energy $E_{Auger}$ of the Auger electron, and the masses of the parent molecule $M_{mol}$, of the core-excited fragment $M_{fragm}$ and of the electron, m. This may to a good approximation be treated classically, and as

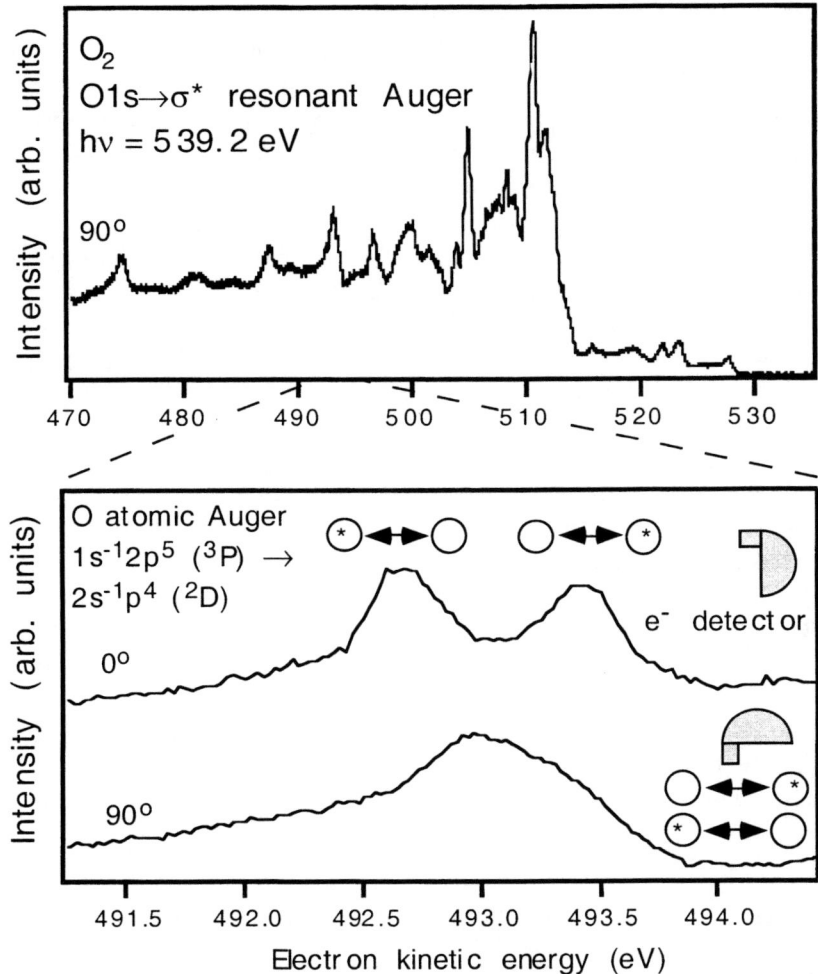

**FIGURE 3.** The resonant Auger spectrum resulting from excitation of $O_2$ to the $1s^{-1} \sigma^*$ state.

the ions are much heavier than the electrons, and $O_2$ is a homo-nuclear diatomic molecule this reduces to:

$$\Delta E_{Doppler} = 2 \, V \, ( \, 2 \, E_{Auger} \, m \, )^{1/2} = 2( \, 2 \, E_{Auger} \, E_{KER} \, m \, / \, M_{fragm} \, )^{1/2} \quad (2)$$

From the observed splitting, the velocity V may be determined to be 14 km/s, or 0.14 Å/fs. The distance between the two oxygen atoms thus increases with twice this value, 0.28 Å/fs. This may be compared to the equilibrium O-O distance of 1.21 Å and the characteristic dissociation time of 7 fs.

Apart from yielding such detailed quantitative information about the dissociation dynamics on the very low femtosecond time scale, the observed Doppler effect has consequences for core hole localization / de-localization. In the discussion so far, the core holes have been assumed to be localized to a certain atomic site. This is a good model for hetero-nuclear molecules, but has recently been challenged for molecules containing two or more identical atoms, which have been shown to exhibit signs of de-localized behaviour (29-33). One such example is the presently discussed $1s^{-1}\sigma^*$ state of oxygen molecules, for which x-ray emission measurements indicate that the core-excited intermediate state cannot be regarded as O-O*, but may have to be treated as a coherent superposition (O-O* + O*-O) (29). These and the Auger Doppler results indicate that the issue of core hole localization / de-localization may involve aspects of the measurement procedure.

One aspect of this could be that when the spectrometer is positioned parallel to the molecular axes, the two atomic fragments are non-identical with respect to the measurement process, forcing the wave function to collapse into either O-O* or O*-O. In the case of identical fragments, however, i.e. with the spectrometer positioned perpendicular to the molecular axes, the core-excited intermediate state then should be treated as a coherent superposition (O-O* + O*-O) (29). This has been predicted to give rise to interference effects, which would show up as peaks or valleys in the spectral profile (17, 34, 35). The width of these features may be narrower than both the monochromator resolution and the core hole lifetime width. No such features have been observed, but this may be due to the initially random arrangement of the molecules. Future experiments on spatially aligned molecules may unravel such effects.

## ELECTRONIC DYNAMICS AND DURATION TIME

In the discussion so far, the electronic dynamics, i.e. the core excitation and decay process, has been considered to be unchanged by the experimental conditions. Recent studies of core-hole decay, both using Auger and x-ray emission spectroscopy, have shown a strong dependence of the spectra on the exact primary excitation energy. This has been explained in a time-dependent model using an effective duration time of the combined core excitation-decay process, which has a maximum on resonance and decreases as the excitation energy is detuned from this energy (17). This may be exemplified by a study of the ultra-fast dissociation of HCl (9). In this study, Auger spectra of HCl after excitation of a Cl 2p electron to the dissociative σ* orbital were recorded. The energy width of the exciting radiation was substantially smaller than the width of the Cl 2p→σ* absorption resonance, which allowed selective excitation on top of the resonance as well as at various detuned off-resonance energies. The resulting Auger spectra were observed to be markedly altered

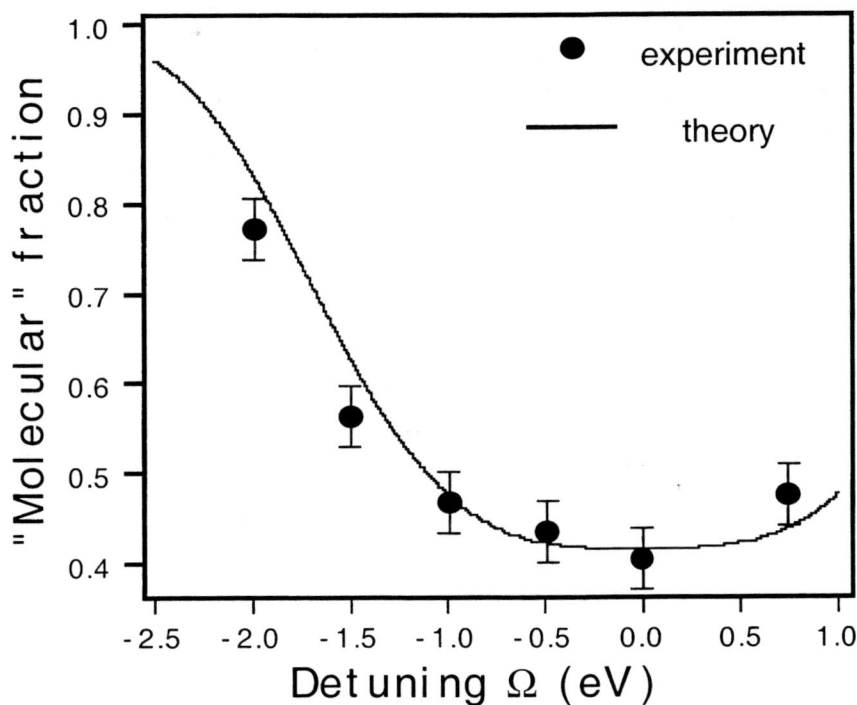

**FIGURE 4.** The "molecular" fraction to the Auger spectum of HCl after excitation to the $2p^{-1}\sigma^*$ state as function of detuning.

by this photon energy detuning: the broad features resulting from decay in the dissociating H-Cl* species were strongly enhanced relative to the sharp spectral lines from decay in the dissociated Cl* species as the exciting photon energy was detuned from resonance. This can be seen in fig. 4, where the molecular fraction obtained by fitting of the experimental spectra is shown as function of detuning.

These results have been interpreted in a time-dependent model, using a framework of a detuning-dependent effective duration time of the process as well as the involved potential energy curves (9). The characteristic duration time of the core-excitation-decay process, $\tau_c$, may be made shorter than the core hole life time width, $\Gamma$, by detuning of the excitation energy, $h\nu$, away from the resonance maximum $E_r$ (the detuning $\Omega = h\nu - E_r$) (17):

$$\tau_c = (\Omega^2 + \Gamma^2)^{-1/2} \qquad (3)$$

The detuning-dependent duration time may be seen as the the result of non-classical interference effect (17). Detuning gives the molecule less time to dissociate before the

Auger decay, thus increasing the intensity of the broad "molecular" features relative to the sharp "atomic" features. This has been quantified, and in Fig. 4 results of simulations based on the duration time concept are compared to the experimentally obtained molecular fraction.

Later studies of the same system have suggested a less symmetric dependence of the molecular fraction upon detuning (36). This may be partially connected to the different starting points on the potential energy surface connected to different excitation energies. Relative to on-resonance excitation, negative detuning goes to less steep sections of the potential energy curve, and leads to lower initial acceleration and consequently to slower dissociation. This would tend to increase the "molecular" fraction in the Auger spectra. Positive detuning would have the opposite effect, i.e. higher relative fragment acceleration, which would tend to make the dissociation faster, and the "molecular" fraction smaller. For negative detuning the effect of detuning-dependent starting points on the potential energy surface would thus add to the duration-time-related increase of the "molecular" fraction, whereas the two effects would counteract each other for positive detuning. Due to the complex nature of the HCl resonant Auger spectrum, caused by two partially overlapping spin-orbit split intermediate states ($2p_{3/2}^{-1}\sigma^*$ and $2p_{1/2}^{-1}\sigma^*$), the different effects are difficult to disentangle experimentally. In this respect, HF constitutes a more favourable system with only one intermedate state ($1s^{-1}\sigma^*$), and future experiments on this or some other more suitable system may provide more insight into the duration time concept.

## BEYOND SEMI-CLASSICAL PICTURES

Even though the detuning-dependent duration time in the example above was used in a semi-classical model, it should in itself be seen as the result of a non-classical interference effect. This points towards shortcomings of semi-classical models to describe these inherently quantum mechanical processes. The separation of the spectral features into "atomic" or "fragment" and "molecular" contributions is intuitively appealing but at the same time somewhat simplistic. The dissociation process is continuous, and even though very early Auger decays take place in a "molecular" environment and very late ones occur in an "atomic-like" or "fragment-like" environment, the range in between is more difficult to characterize. Theoretically, the core hole excitation, Auger decay and dissociation processes should not be seen as separate processes, but rather be treated in a one step picture. Theoretical treatments along these lines have lead to predictions of not yet observed interference effects (17, 18, 19, 34, 35). Apart from the already mentioned interference effects connected to the Doppler effect (17, 34, 35), simulations for HCl suggest for instance the existence of "spectral holes" caused by destructive interference between the "atomic" and "molecular" contributions to the Auger spectra when the photon energy is tuned in such a way as to make these contributions overlap (17, 19). In the case of HF, interference effects have been discussed in another picture (18). For positive detuning, i.e. photon energies slightly above the resonance maximum, the intermediate state wavepacket may either move outwards or inwards on the repulsive potential. In the latter case, the wavepacket is reflected by the potential, and starts moving outwards. These two possible pathways for the developing wavepacket could cause interference, which would be manifested as strong modulations of the intensity of the "molecular" spectral contribution (18). Future experiments may confirm these theoretical predictions, but also possibly reveal new effects.

## FUTURE EXPERIMENTAL DEVELOPMENTS

The "core hole clock" method discussed in this paper provides quantitative temporal information on the very low femtosecond timescale, and has even been used to study attosecond processes (37). By using short-lived core levels instead of long-lived ones, it would be possible to study processes on the time scale of a few hundred attoseconds. The information is however obtained in a rather indirect way, as the exponential nature of Auger decay constitutes a type of time integration of the dynamic process. The Auger spectrum thus reflects the complete development of the system from core excitation to infinite time, but the contributions from different times are exponentially weighed, with early decays giving the largest contributions. It would naturally be very interesting to obtain more direct temporal information. One promising technique for this is laser-based pump-probe spectroscopy, which is routinely used for time-resolved studies of the valence electronic structure in the 100 fs time range. Presently the pulse duration, and thus the ultimate time resolution, is approaching that of long-lived core holes, i.e. a few femtoseconds. At the same time the available photon energies are increasing. Techniques such as coherent harmonic generation now produce coherent radiation up to several hundred eV (38). A parallel development is taking place with free electron lasers (39). It may thus in the not too distant future be possible to have a photon source combining high photon energies, allowing excitation and/or probing of atomically localized core levels, and a time resolution in the order of the core hole life time. This would allow completely new studies with both high spatial and temporal control of ultra-fast dissociation and other types of dynamics in quantum systems.

## CONCLUSIONS

When a molecule is core excited to a repulsive state, dissociation of the molecule and Auger decay of the core excited state may proceed on a similar time scale. This leads to different contributions to the resulting Auger spectra, which contain both "atomic" or "fragment" and "molecular" contributions. On this level, semi-classical models are well suited to discuss the dynamics of the dissociation process. More advanced theoretical predictions point to quantum interference effects, which require a unified, one-step description of the core hole excitation, Auger decay and dissociation processes. Experimental investigations of such interference effects, e.g. the duration time concept, have been carried out, but new high performance beamlines dedicated to gas phase studies at third generation synchrotron radiation facilities are now opening completely new possibilities for studying fundamental quantum phenomena of these dynamic processes.

## ACKNOWLEDGEMENTS

The author gratefully acknowledges the contributions of A. Ausmees, F. Burmeister, M. Bässler, R. Feifel, R. F. Fink, F. Gel'mukhanov, I. Hjelte, A. B. Machado, R. T. Marinho, C. Miron, A. Naves de Brito, M. N. Piancastelli, L. Rosenquist, P. Salek, S. Sundin, S. Svensson, S. L. Sörensen, H. Wang, K. Wiesner and H. Ågren. This work has been supported by the Swedish Foundation for International Cooperation in Research and Higher education (STINT), the Swedish Natural Research Council (NFR) and CNPq (Brazil).

## REFERENCES

1. Morin, P., and Nenner I., *Phys. Rev. Lett.* **56**, 1913 (1986).
2. Aksela, H., Aksela, S., Ala-Korpela, M., Sairanen, O.-P., Hotokka, M., Bancroft, G. M., Tan, K. H., and Tulkki, J., *Phys. Rev.* **A41**, 6000 (1990).
3. Naves de Brito, A., and Ågren, H., *Phys. Rev.* **A45**, 7953 (1992).
4. Aksela, S., Aksela, H., Naves de Brito, A., Bancroft, G. M., and Tan, K. H., *Phys. Rev.* **A45**, 7948 (1992).
5. Kukk, E., Aksela, H., Sairanen, O.-P., Aksela, S., Kivimäki, A., Nõmmiste, E., Ausmees, A., Kikas, A., Osborne, S. J., and Svensson, S., *J. Chem. Phys.* **104**, 4475 (1996).
6. Naves de Brito, A., Naves de Brito, A., Björneholm, O., Neto, J., Machado, A., Svensson, S., Osborne, S. J., Ausmees, A., Sæthre, L., Aksela, H., Sairanen, O.-P., Kivimäki, A., Nõmmiste, E., and Aksela, S., *J. Mol. Str. (Theochem)* **394**, 135 (1997).
7. Kukk, E., Aksela, H., Aksela, S., Gel'mukhanov, F., Ågren, H., and Svensson, S., *Phys. Rev. Lett.* **76**, 3100 (1996).
8. Hansen, D. L., Arrasate, M. E., Cotter, J., Fisher, G. R., Leung, K. T., Levin, J. C., Martin, R., Neill, P., Perera, R. C. C., Sellin, I. A., Simon, M., Uehara, Y., Vanderford, B., Whitfield, S. B., and Lindle, D. W., *Phys. Rev.* **A57** 2608 (1998).
9. Björneholm, O., Sundin, S., Svensson, S., Marinho, R. R. T., Naves de Brito, A., Gel'mukhanov, F., and Ågren, H., *Phys. Rev. Lett.* **79**, 3150 (1997).
10. Piancastelli, M. N., Hempelmann, A., Heiser, F., Gessner, G. Rüdel, A., and Becker, U., *Phys. Rev.* **A59**, 300 (1999).
11. Naves de Brito, A., Feifel, R., Mocellin, A., Machado, A. B., Sundin, S., Hjelte, I., Sorensen, S. L. and Björneholm, O., *Chem. Phys. Lett.*, in press
12. Liu, Z. F., Bancroft, G. M., Tan, K. H., and Schlachter, M., *Phys. Rev.* **A48**, R4019 (1993).

13. Schaphorst, S. J., Caldwell, C. D., Krause, M. O., and Jimenez-Mier, J., *Chem. Phys. Lett.* **213**, 315 (1993).
14. Caldwell, C. D., Schaphorst, S. J., Krause, M. O., and Jimenez-Mier, J., *J. Electron. Spectrosc. Relat. Phenom.* **67**, 243 (1994).
15. Recent resonant Auger results for the B state in $N_2$ (Piancastelli, M. N., et al., submitted to *Phys. Rev. Lett.*) show that this may not always be a good approximation. It is however here assumed that the B state in $N_2$ is an intersting but rather special case.
16. Björneholm, O., Nilsson, A., Sandell, A., Hernnäs, B., and Mårtensson, N., *Phys. Rev. Lett.* **68**, 1892 (1992).
17. Gel'mukhanov, F., and Ågren, H., *Physics Reports* **312**, 87 (1999), and references therein
18. Pahl, E.,. Cederbaum, L. S., and Meyer, H.-D., *Phys. Rev. Lett.* **80**, 1865 (1998).
19. Goertel, Z., Teshima, R., and Menzel, D., *Phys. Rev A*, accepted for publication
20. Aksela, S., Kivimäki, A., Naves de Brito, A., Sairanen, O.-P., and Svensson, S., *Rev. Sci. Instrum.* **65**, 831 (1994).
21. Osborne, S. J., Ausmees, A., Forsell, J. O., Wannberg, B., Bray, G., Dantas, L. B., Svensson, S., Naves de Brito, A., Kivimäki, A., and Aksela, S., *Synchr. Rad. News* 7(1), 25 (1994).
22. Bässler, M., Forsell, J-.O., Björneholm, O., Feifel, R., Jurvansuu, M., Aksela, S., Sundin, S., Sorensen, S., Nyholm, R., Ausmees, A., and Svensson, S., *Journal of Electron Spectroscopy and Rel. Phen.*, **101-103**, 953-957 (1999)
23. See for instance Simon, M., LeBrun, T., Morin, P., Lavolleé, M. andMaréchal, J. L., *Nucl. Instr. & Meth.* **B62**, 167 (1991), and references therein.
24. Lira, A. C., Rodrigues, A. R. D., Rosa, A., Gonçalves da Silva, C.E.T., Pardine, C., Scorzato, C., Wisnivesky, D., Rafael, F., Franco, G. S., Tosin, G., Liu Lin, Jahnel, L., Ferreira, M. J., Tavares, P. F., Farias, R. H. A., and Neuenschwander, R. T., Proceedings of the EPAC, Stockholm (1998), and references therein.
25. Naves de Brito, A., et. al. to be published.
26. Svensson, S., Ausmees, A., Osborne, S. J., Bray, G., Gel'mukhanov, F., Ågren, H., Naves de Brito, A., Sairanen, O.-P., Kivimäki, A., Nõmmiste, E., Aksela, H., and Aksela, S., *Phys. Rev. Lett.* **72**, 3021 (1994).
27. The lifetime of the core excited $2p_{3/2}^{-1} 6a_1$ state need not be exactly identical to that of the $2p_{3/2}^{-1}$ state. Apart from possibly causing a somewhat different lifetime, the influence of the $6a_1$ electron on the decay would manifest itself in a resonant enhancement of the spectral features corresponding to single-hole valence final states, produced by participator-type decays. The absence of any significant such resonant enhancements strongly indicates that the influence of the $6a_1$ electron on the decay is very small, and consequently that the lifetimes of the $2p_{3/2}^{-1}$ and $2p_{3/2}^{-1} 6a_1$ states are practically identical at this level of accuracy.
28. Björneholm, O., Bässler, M., Ausmees, A., Hjelte, I., Feifel, R., Wang, H., Miron, C., Piancastelli, M. N., Svensson, S., Sorensen, S. L., Gel'mukhanov, F., and Ågren, H., submitted for publication in *Phys. Rev. Lett.*
29. Glans, P., Gunnelin, K., Skytt, P., Guo, J.-H., Wassdahl, N., Nordgren, J., Ågren, H., Gel'mukhanov, F., Warwick, T., and E. Rotenberg, *Phys. Rev. Lett.* **76**, 2448 (1996).
30. Glans, P., Skytt, P., Gunnelin, K., Guo, J.-H., and Nordgren, J., *J. Electron Spectrosc. and Relat. Phenom.* **82**, 193 (1996).
31. Kempgens, B., Köppel, H., Kivimäki, A., Neeb. M., Cederbaum, L. S., and Bradshaw, A. M., *Phys. Rev. Lett.* **79**, 3617 (1997).
32. Thomas, T. D., Sæthre, L. J., Sorensen, S. L., and Svensson, S., *J. Chem. Phys.* **109**, 1041 (1998).
33. Thomas, T. D., Berrah, N., Bozek, J., Carroll, T. X., Hahne, J., Karlsen, T., Kukk, E., and Sæthre, L. J., *Phys. Rev. Lett.* **82**, 1120 (1999).
34. Gel'mukhanov, F., Ågren, H., and Salek, P., *Phys. Rev.* **A57**, 2511 (1998).
35. Salek, P., Gel'mukhanov, F., and Ågren, H., *Phys. Rev.* **A59**, 1147 (1999).
36. Kukk, E., Will, A., Berrah, N., Langer, B., Bozek, J. D., Nayadin, O., Alsherhi, M., Farhat, A., and Cubaynes, D., *Phys. Rev.* **A57**, R1485 (1998).
37. Keller, C., Stichler, M., Comelli, G., Esch, F., Lizzit, S., Wurth, W., and Menzel, D., *Phys. Rev. Lett.* **80**, 1774 (1998).

38. See for instance Rundquist, A., Durfee, C. G. III., Zenghu-Chang, Herne, C., Backus, S., Murnane, M. M., Kapteyn, H. C., *Science* **280**, 1412 (1998), Lappas, D. G., and L'Huillier, A., *Phys. Rev.* **A58**, 4140 (1998), Gaarde, M. B. Antoine, P., Persson, A., Carre, B., L`Huillier, and A., Wahlström, C., G., *J. Phys.***B29**,.L163-8 (1996), and references therein.

39. See for instance Brinkmann, R., Materlik, G., Rossbach, J., Schneider, J. R., and Wiik, B.-H., *Nuclear Instruments and Methods in Physics Research*, A**393**, 86 (1997), and Tatchyn, R., et al., *Nuclear Instruments and Methods in Physics Research*, A**375**, 274 (1996), and references therin.

# Dynamical Effects and Selective Fragmentation after Inner Shell Excitation

Marc Simon*, Catalin Miron*[§], Renaud Guillemin*, Karine Le Guen*, Denis Ceolin*, Eiji Shigemasa*[£], Nicolas Leclercq* and Paul Morin*

*LURE, Bât. 209d, Université, BP34, 91898 Orsay CEDE and CEA/ DRECAM/ SPAM, France
[§] Department of Physics, Uppsala University, BOX 530, S-751 21 Uppsala, Sweden.
[£] Institute for Molecular Science, Myodaiji, Okazaki 444-8585, Japan.

**Abstract.** We have studied the fragmentation of core excited molecules probed by the Auger electron-ion coincidence technique. Our results illustrate site selective fragmentation by hexamethyldisiloxane ionized at the Si 2p and the C 1s thresholds. By combining high resolution electron spectroscopy with resonant Auger electron-ion coincidence, we reveal the role played by nuclear motion in the fragmentation of core excited $CO_2$ and $CF_4$ molecules.

## INTRODUCTION

The development of third generation synchrotron radiation facilities and well optimised electron analyzers has opened new opportunities for high resolution electron spectroscopy. With a photon bandwidth and an electron resolution narrower than the natural width of the core excited state, it is now possible to work in the sub-lifetime regime for resonant Auger spectroscopy (1). In this context, it has been observed that nuclear motion can play a significant role within the lifetime of the core excited states in some diatomic molecules (2-4). As an extreme case, faster fragmentation as compared to the resonant Auger decay has been observed with HBr (5), HCl (6-7), and $H_2S$ (8) and has been theoretically predicted for HF (9). Such phenomena have been found after deep inner shell excitation (1s) of the HCl molecule (10), where a lifetime as short as 1 fs has to be accounted for. Most of these studies have focused on diatomic systems. However, despite their increased complexity, the case of polyatomic systems, to which the present paper is addressed, represents a more general situation in this context, as shown on a previous study on $BF_3$ (11).

Mass spectrometry on polyatomic systems without electron energy selection has revealed that very fast nuclear motion is responsible for dramatic changes in the electron-ion-ion coincidence pattern of $N_2O$ (12) and $BF_3$ (13). These effects have been confirmed and quantitatively determined on $SO_2$ (14), with the help of an ion position sensitive detector, allowing vectorial correlations of the momenta.

Site specific excitation of polyatomic molecules can induce specific fragmentation channels (15-16). In most cases, complex polyatomic molecules are known to undergo fast internal energy conversion, leading to the loss of the memory of the initial excitation site (17). Further insight into the site-selective photochemistry has been obtained by the use of an energy analysed Auger-electron-ion experiment. The main advantage of this technique is that the detection of only one electron simultaneously determines the hole which was initially created and the binding energy of the doubly charged ion. By selecting the kinetic energy of Auger electrons in coincidence with fragment ions, strong site-selective fragmentation for the lowest states of the $ClCH_2Br^{2+}$ ion has been observed following selective Br 3d or Cl 2p core photoionization (18). At higher energy the selectivity decreases and is completely lost for the highest states.

## EXPERIMENTAL SET UP

Our energy selected-electron-ion coincidence results shown below have been obtained at LURE on the bending magnet beamline SA22 of the SuperACO storage ring with a resolving power of 2500 during the measurements. We have designed and constructed a high luminosity electron analyzer shown in figure 1, the full details of which have been published elsewhere (19). The electrons produced in the source volume are retarded and focused by a conical lens onto a circular entrance slit. The electrons are then dispersed by double toroidal electrodes and detected by a position sensitive detector, allowing us to simultaneously determine their energy and ejection angle.

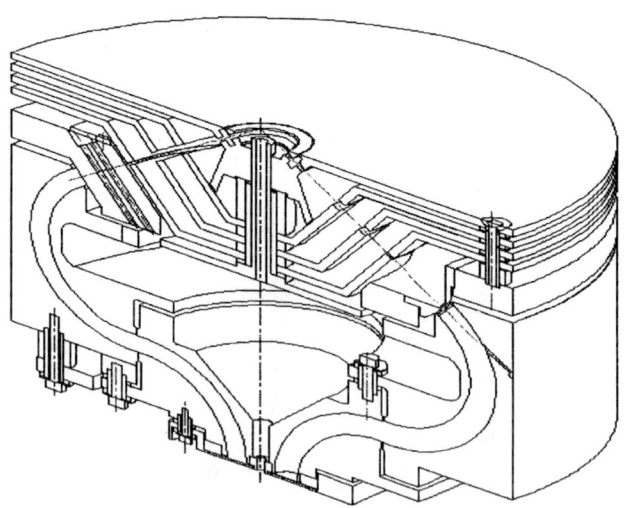

**FIGURE 1**. Sketch of our Double Toroidal Electron Analyser.

The performance of this double toroidal analyzer can be summarized as:

. large acceptance angle (5% of $4\pi$)

. parallel detection over 15% of the pass energy

. resolving power equal to 1% of the pass energy.

A short (12 cm) ion time of flight mass spectrometer faces the electron analyzer. The ionisation region is kept field free during the electron energy analysis to preserve high resolution performance. The electron signal triggers a high pulsed electrostatic field (1 kV/cm, rise time 50 ns) which extracts the ions toward the ion time of flight detector.

## SITE SELECTIVE FRAGMENTATION OF HEXAMETHYLDISILOXANE (HMDSO)

Investigations on site selective fragmentation have been reported for a large variety of molecules. The main advantage of the Auger electron-ion coincidences technique is that the detection of only one electron simultaneously determines the binding energy of the doubly charged ion and the hole which was initially created.

We have studied the hexamethyldisiloxane molecule $(CH_3)_6Si_2O$, comparing the fragmentation induced by Si 2p ionization and C1s ionization. A detailed analysis will be given in a forthcoming publication, and the main results are only briefly summarized here. The photon flux was reduced by closing the beamline slits to keep a low coincidence rate (25 Co/s), thus avoiding fortuitous coincidences. We used an analyzer pass energy of 80 eV, which corresponds to a resolution of about 1 eV. The binding energies of the core holes were experimentally determined and found to be 106 eV for the Si 2p threshold (the two spin orbit components were not resolved) and 290.5 eV for C 1s.

In Figure 2, we compare the mass spectra recorded in coincidence with electrons of 73 and 251 eV kinetic energy. The upper spectrum corresponds to the ions formed after Auger relaxation of the Si 2p hole, whereas the bottom corresponds to the C 1s hole relaxation. In both cases, we choose an electron energy corresponding to the lowest binding energies of the doubly charged ions. Note, in figure 2, that the Auger relaxation process produces very different doubly charged states, with a minimum of 33 eV of binding energy for the Si 2p hole, whereas the value is 39.5 eV for C 1s. Figure 2 clearly shows how different localization of the two valence holes created via the Auger decay can lead to different fragmentations. Due to electric noise during the rise time of the pulsed field, we apply a logical veto on the ion signal discriminator which forbids any detection of the $H^+$ ion.

Only the doubly charged fragment corresponding to the loss of two neutral methyl groups is produced via the Si 2p ionization threshold. Whereas, at the C 1s threshold, the internal energy of the doubly charged ion is much higher. This excess of energy is probably redistributed among the different degrees of freedom, leading to many different fragmentation channels, as expected from the statistical dissociation model.

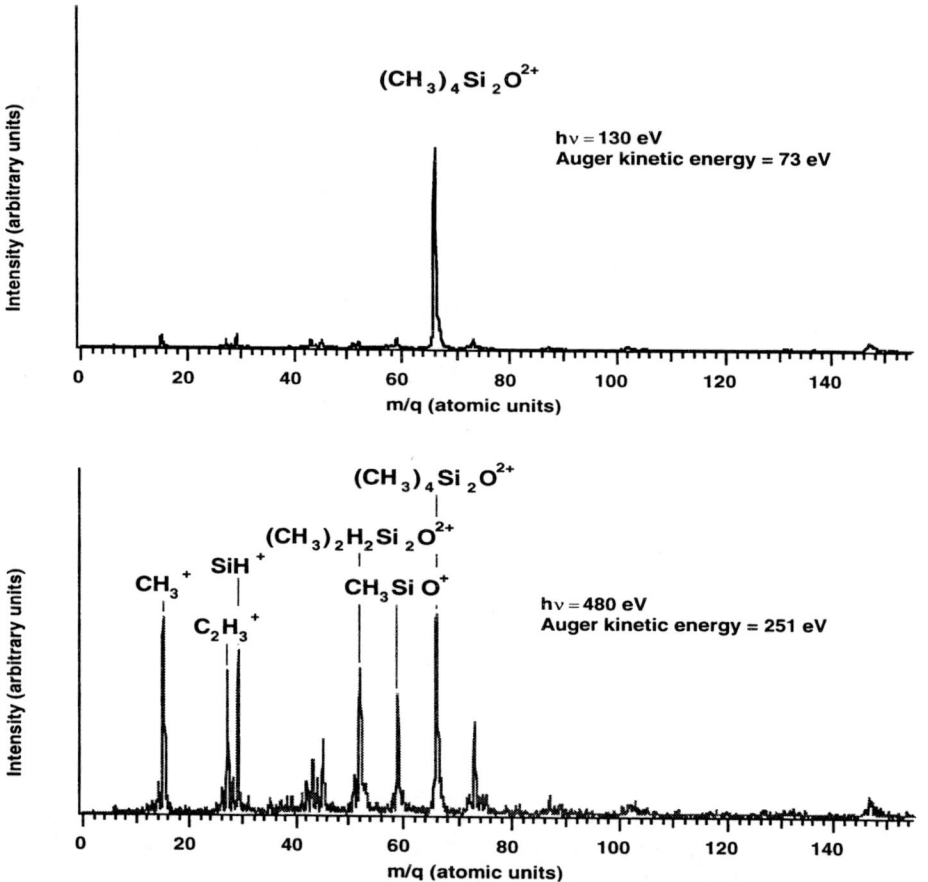

**FIGURE 2.** Auger-ion coincidence measurements obtained with the Fastest Available Auger electron after Si 2p ionisation (top) or C 1s ionisation (bottom).

The spectra emphasize a very pronounced site-selective photochemistry: the site-selective formation of a stable, doubly charged fragment. This kind of site-selective photochemistry, apart from its fundamental interest, should be linked to more applied fields like photodesorption, photoreactivity, or the stability of biological structures. There is no doubt that this effect could also have industrial applications.

With polyatomic molecules, it is often difficult to go beyond qualitative determination of the role played by the internal energy of the doubly charged ion, because of the high density of states, the high number of degrees of freedom, the IVR processes, and the lack of available theoretical calculations. The situation is completely different for resonant excitation below threshold with smaller molecules, for which many studies have been made experimentally and theoretically on the singly charged ion, allowing dynamical effects to be pointed out.

## DYNAMICAL EFFECTS : $CO_2$ AT THE CARBON K EDGE

The $CO_2$ molecule is an interesting case for the purpose of these studies. Indeed, starting with the initial state described in linear geometry, resonant core shell excitation can induce significant geometrical changes, which can be observed through electron and ion spectroscopy.

Here we focus on core C 1s excitation, as observed through a resonant Auger electron-ion coincidence experiment. Dissociative ionization has been studied previously in great detail, both experimentally (20-21) and theoretically (22-23), in the valence ionization regime. The results will only be summarized here and will be presented more extensively in a separate publication (24). We will only discuss here the relaxation of the linear $CO_2$ molecule after promotion of the C 1s electron to the first unoccupied molecular orbital which has $\pi^*$ character.

### Absorption spectrum

The core-equivalent model applied to the $CO_2$ molecule suggests that the molecular geometry of the core excited molecule is the same as the $NO_2$ molecule, i.e., bent. As a consequence of this bending, previous PEPIPICO experiments (25) on this molecule revealed that the central carbon atom is emitted with significant kinetic energy.

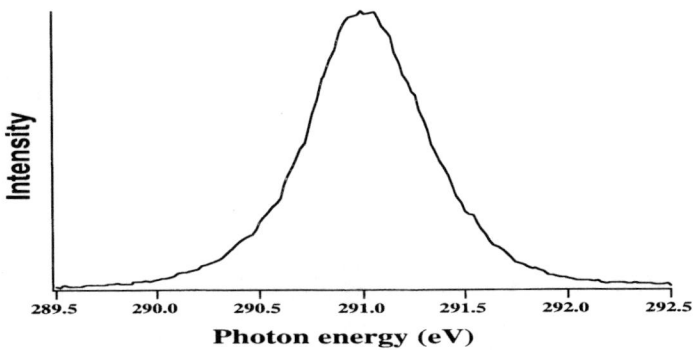

**FIGURE 3.** Total ion yield spectrum for the $CO_2$ molecule obtained with a 75 meV photon bandwidth.

The corresponding absorption (ion yield) spectrum is shown in figure 3. The molecular geometry change gives rise to a broad structure (0.69 eV FWHM). Symmetry-resolved-absorption spectra obtained on this molecule (26) have been rationalized using the Renner-Teller effect. It has been concluded that, due to the vibrationnal overlap, the left part of the resonance corresponds to a preferential excitation of bending modes and the right part to stretching excitation.

## Deexcitation Spectra

Resonant electron spectra have been recorded along this resonance at the ALS by Edwin Kukk *et al.* and will be the subject of a separate publication. They show that:

1) the A state of $CO_2^+$ is the dominant decay channel of the resonance, and

2) its vibrational progression is changes dramatically along the resonance profile.

Interestingly, they show that bending modes are efficiently excited at low energy.

## Mass Spectra

Figure 4 shows mass spectra recorded at three photon energies (Left, Top and Right) in coincidence with electrons of the same binding energy, 19.1 eV, in each case. This corresponds to the high vibrational modes of the A and B states.

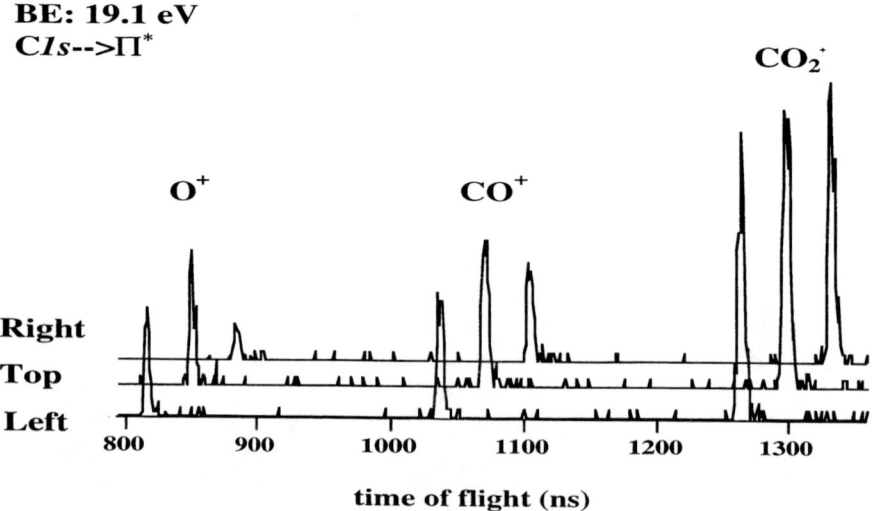

**FIGURE 4.** Coincidence mass spectra recorded with 19.1 eV electron binding energy along the $\pi^*$ resonance (Left, Top and Right as in Fig. 3)

The measurements have been performed with an energy resolution of 1.2 eV covering the A, B and C electronic states. We subtracted the C-state contribution, which is known to dissociate into $O^+ + CO$ (20). The B-state contribution is negligible. The most striking observation is that, despite the same internal energy selection, the $O^+$ ion production decreases for the Right excitation. For the A state, the only energetically accessible potential energy surface to produce $O + CO^+$ is via a spin-orbit coupling with the dissociative $^4\Pi_u$ state. Polak et al (23) have calculated a constant coupling independent of the bending angle. The explanation of the $O^+ + CO$ channel makes use of a conical intersection, which is open via the lowering of symmetry as a result of the bending (22-23).

This is a nice example of dissociation being indirectly mediated through bending motion. We show, with this example, that fragmentation of an ion depends not only on its internal energy, but also on the way it has been produced.

## DYNAMICAL EFFECTS : CF$_4$ AT THE CARBON K EDGE

As a third example, we investigate the CF$_4$ molecule at the carbon K edge. In this case, coupling between the electronic and nuclear motions is expected to be strong in the C $1s^{-1}\sigma^*$ core-excited state, as suggested by the broad resonance (see Fig. 5). On the other hand, dissociation after valence shell ionization has been well established (27). We only briefly review here the main results, which have been published in more detail (28).

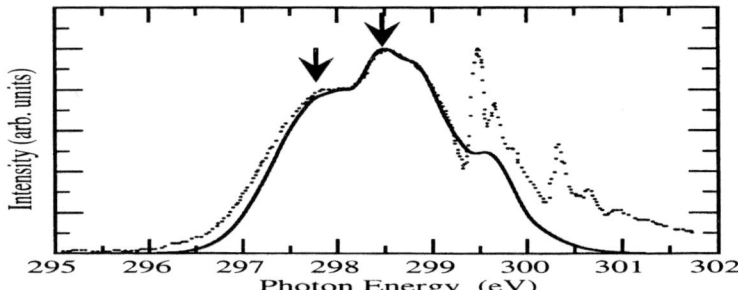

**FIGURE 5.** Total ion yield of CF$_4$ spectrum (dotted line) and theoretical spectrum calculated by Tanaka (see below).

The electronic configuration of the CF$_4$ molecule in its ground state is :

$(core)1a_1^2 1t_2^6 2a_1^2 2t_2^6 1e^4 3t_2^6 1t_1^6 ; 4t_2^0 3a_1^0$.

We will focus our interest on the $2t_2$ and $3t_2$ molecular orbitals. The $2t_2$ has a C-F bonding character, and the $3t_2$ is mostly a F 2p-like weakly bonding orbital. The electron emission spectra shown in Fig. 6 have been recorded at the C 1s → σ* resonance at the excitation energies indicated by arrows in Fig. 5.

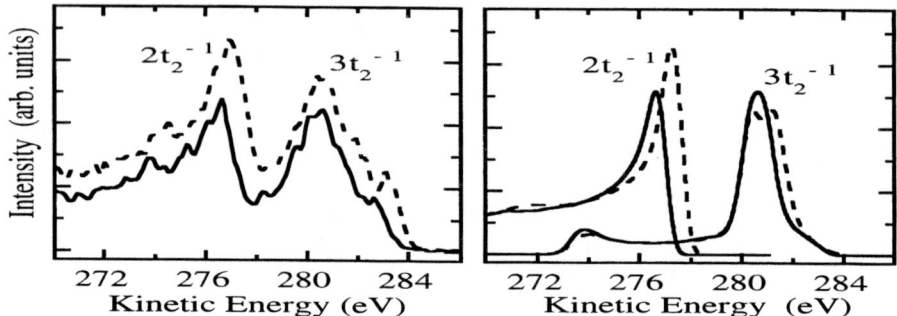

**FIGURE 6.** Experimental (left) and theoretical (right) resonant Auger spectra obtained at 297.7 eV (dashed line) and 298.5 eV (solid line).

We used the double toroidal electron analyzer at the SA22 bending magnet beamline. Figure 6 shows that the $2t_2$ line has a constant binding energy (dispersive) and a tail towards lower kinetic energy, whereas the $3t_2$ line has a constant kinetic energy (non-dispersive). This confirms the previous results obtained by Ueda *et al.* (29) and Neeb *et al.* (30). The calculations of Tanaka (28) reproduce these observations (both the dispersion and the tail).

The resonant Auger electron-ion coincidence technique has been used to investigate these states, in order to check an eventual role played by the core excitation, and the results are displayed in Fig. 7.

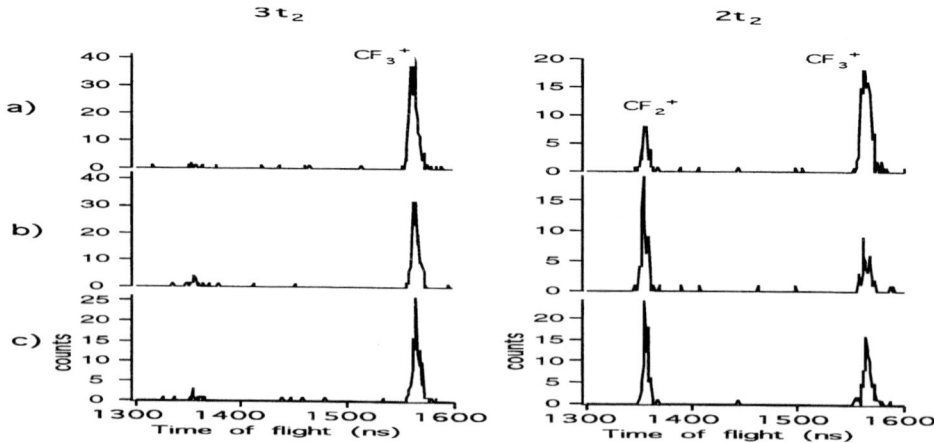

**FIGURE 7.** Ion time of flight mass spectra recorded in coincidence with $3t_2$ electrons (left) and $2t_2$ electrons (right) at three photon energies (a) 295 eV (below resonance), (b) 297.7 eV , and (c) 298.5 eV.

It is clear that the $3t_2$ channel leads to dissociation into $CF_3^+ + F$, independent of the photon energy, whereas the $2t_2$ channel dissociates into two fragmentation channels $CF_2^+ + F_2$ and $CF_3^+ + F$. These fragmentation channels have been previously identified by Creasey et al. (27). They concluded that two mechanisms are in competition: a radiative decay on the ns time scale leading to the X or A state of the ion, which dissociates into $CF_3^+ + F$, and a predissociation over a potential barrier, dissociating into $CF_2^+ + F_2$. Our results show that the $CF_2^+/CF_3^+$ branching ratio increases by a factor of 6 on the low energy part of the resonance, and we call this effect *core-excitation-induced-dissociation*.

A calculation has been made by Tanaka (28) with adjustable parameters to reproduce the absorption spectrum (see Fig. 5) and the resonant Auger decay (see Fig. 6). Adiabatic potential energy surfaces of the core excited state and of the final states have been computed and are reported in Fig. 8. The vibrational modes are classified as a symmetric stretching mode $Qa_1$ and an anti-symmetric $Qt_2$. The latter mode is interesting, because it leads to an anti-symmetric C-F molecular deformation. The linear coupling of this mode has been taken into account for the two core excited states C $1s^{-1}4t_2$ and C $1s^{-1}3a_1$. The Jahn-Teller effect has also been taken into account for the final states. The adiabatic potentials are plotted versus the Q[111] coordinate, for which a negative value denotes a particular elongated C-F chemical bond.

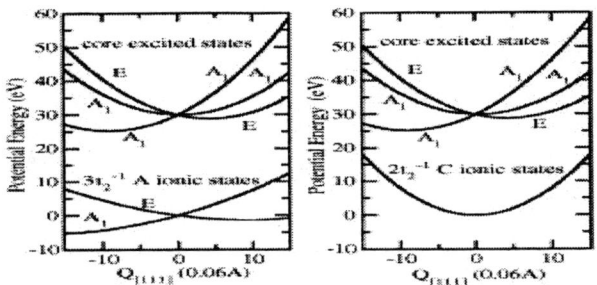

**FIGURE 8.** Adiabatic potential energy surfaces of the core excited state and the $3t_2^{-1}$ and $2t_2^{-1}$ final states plotted along the $Q_{[111]}$ coordinate (see text).

The shape of the potential energy surface shows that, in the core excited state, the molecule starts to move along the potential toward the $A_1$ curve with an elongating C-F bond. The $3t_2^{-1}$ potential curve shows the ion is dissociating with an elongated C-F bond, leading to $CF_3^+ + F$, as already experimentally observed. The core excited state and the $3t_2^{-1}$ final state have almost parallel curves, which explains the non dispersive character of the photoelectron line. In contrast, the $2t_2^{-1}$ final state is stable along the $Q_{[111]}$ coordinate. At time t = 0 of the excitation, the molecule starts at $Q_{[111]} = 0$ and starts to move along the core excited state potential until the Auger decay. This explains why the kinetic energy of the electron decreases with time. A C-F bond is elongated for a long time, which could explain how it assists the molecule in passing through the potential barrier to predissociate into $CF_2^+ + F$.

## CONCLUSION

The Auger-ion coincidence technique has been shown to be a unique tool for probing selective fragmentation and nuclear dynamics. Indeed, in this kind of study, it is essential to know the internal energy stored in the ion before it undergoes dissociation.

The HMDSO molecule provides a spectacular demonstration of this technique. Its fragmentation totally depends on the final state.

Our results on $CO_2$ and $CF_4$ concerning nuclear dynamics show how crucial is the role of wave packet evolution within the resonant state lifetime. This indicates that such kinds of measurements should be of great help in the future to understand very fast processes, like relaxation of adsorbed molecules on surfaces, for instance.

## ACKNOWLEDGMENT

We warmly thank K. Ueda, S. Tanaka, N. Berrah, J. Bozek and E. Kukk for their collaborative work.

## REFERENCES

1. Kivimäki, A., Naves de Brito, A., Aksela, S., Aksela, H., Sairanen, O. P., Ausmees, A., Osborne, S. J., Dantas, L. B., Svensson, Phys. Rev. Letters **71**, 4307-4310 (1993)

2. Neeb, M., Rubensson, J.E., Biermann, M., Eberhardt, W., J. Electr. Spectrosc. Relat. Phenom. **67**, 261 (1994)

3. Gel'mukhanov, F. and Ågren, H., Phys. Rev. A **54**, 379 (1996)

4. Sundin, S., Gel'mukhanov, F., Ågren, H., Osborne, S., Kikas, A., Bjorneholm, O., Ausmees, A. and Svensson, S., Phys. Rev. Letters **79**, 1451-1453 (1997)

5. Morin, P. and Nenner, I., Phys. Rev. Letters **56**, 1913-1916 (1986)

6. Kukk, E., Aksela, H., Aksela, S., Gel'mukhanov, F., Ågren, H. and Svensson, S. Phys. Rev. Letters **76** 3100-3103 (1996)

7. Björneholm, O., Sundin, S., Svensson, S., Marinho, R. R. T., Naves de Brito, A., Gel'mukhanov, F. and Ågren, H., Phys. Rev. Letters **79**, 3150-3153 (1997)

8. Naves de Brito, A., Björneholm, O., Naves de Brito, Al., Neto, J., Machado, A., Svensson, S., Osborne, S., Ausmees, A., Saethre, L., Aksela, H., Sairanen, O.-P., Kivimäki, A., Nommiste, E. and Aksela, S., Theo. Chem. Phys., **394** 135 (1997)

9. Pahl, E., Cederbaum, L.S., Meyer, H.-D. and Tarantelli, F., Phys. Rev. Letters **80** 1865-1868 (1998)

10  Hansen, D. ,Arrasate, M., Cotter, J., Fisher, G., Leung, K., Levin, J., Martin, R., Neil, P., Perera, R., Sellin, I., Simon, M., Uehara, Y., Vanderford, B., Whitfield, S. and Lindle, D., Phys. Rev. A **57** 2608-2611 (1998)

11  Simon, M., Miron, C., Leclercq, N., Morin, P., Ueda, K., Sato, Y., Tanaka, S. and Kayanuma, Y., Phys. Rev. Letters **79** 3857-3860 (1997)

12  Lebrun, T., Lavollée, M., Simon, M. and Morin, P., J. Chem. Phys. **98** 2534 (1993)

13  Simon, M., Morin, P., Lablanquie, P., Lavollée, M., Ueda, K. and Kosugi, N., Chemical Physics Letters **238** 42 (1995)

14  Lavollée, M. and Brems, V., J. Chem. Phys. **110** (1999)

15  Schmelz, H., Reynaud, C., Simon, M. and Nenner, I., J. Chem. Phys. **101** 3742 (1994)

16  Nenner, I., Reynaud, C., Schmelz, H., Ferrand-Tanaka, L., Simon, M. and Morin, P., Zeitschrift für Physik D **195** 43-63 (1996)

17  Simon, M., Lavollée, M., Morin, P. and Nenner, I., J. Phys. Chem. **99** 1733 (1995)

18  Miron, C., Simon, M., Leclercq, N., Hansen, D. and Morin, P., Phys. Rev. Letters **81** 4104-4107 (1998)

19  Miron, C., Simon, M., Leclercq, N. and Morin, P., Rev. Sci. Instrum. **68** 3728-3737 (1997)

20  Richard-Viard, M., Dutuit, O., Amarkhodja, A. and Guyon P.M. in *Photophysics and Photochemistry above 6 eV*, p153, Lahmani, F. editor, Elsesevier, Amsterdam (1985)

21  Locht, R. and Davister, M., Int. J. Mass Spec. **144** 105 (1995)

22  Praet, M. T, Lorquet, J. C. and Raseev, G., J. Chem. Phys. **77** 4611 (1982)

23  Polak, R., Hochlaf, M., Levinas, M., Chambaud, G. and Rosmus, P., Spectrochemica Acta Part A (1998)

24  Morin, P., Simon, M., Miron, C., Leclercq, N., Kukk, E., Bozek, J. D. and Berrah, N., submitted for publication

25  Morin, P., Lavollée, M. Meyer, M. and Simon, M., ICPEAC conf. Proc. AIP 295, p139, Andersen et al eds, NY 1993

26  Adachi, J., Kosugi, N., Shigemasa, E. and Yagishita, A., J. Chem. Phys. **102** 7369 (1995)

27  Creasey, J. C., Jones, H. M., Smith, D. M., Tuckett, R. P., Hatherly P. A. and Codling, K., Chem. Phys. **174** 441 (1993)

28  Ueda, K., Simon, M., Miron, C., Leclercq, N., Guillemin, R., Morin, P. and Tanaka, S., accepted at Phys. Rev. Letters (99)

29  Ueda, K., Shimizu, Y., Chiba, H., Okunishi, M., Ohmori, K., Sato, Y., Shigemasa, E. and Yagishita, A., J. El. Spectrosc. Relat. Phenom. **79** 441 (1996)

30  Neeb, M., Kivimäki, A., Kempgens B., Köppe H. M. and Bradshaw, A. M., J. Phys. B **30** 93 (1997)

# High-Resolution Molecular Inner-Shell Electron Spectroscopies

J.D. Bozek*, N. Berrah[♀], E. Kukk[♀], T.D. Thomas[♋], T.X. Carroll[δ], L.J. Saethre[⌶], J.A. Sheehy[⚘] and P.W. Langhoff[♒]

* Lawrence Berkeley National Laboratory, 1 Cyclotron Rd., Berkeley, CA 94720
[♀] Physics Department, Western Michigan University, Kalamazoo, MI 49008
[♋] Department of Chemistry, Oregon State University, Corvallis, OR 97331
[δ] Keuka College, Keuka Park, NY 14478
[⌶] Department of Chemistry, University of Bergen, N-5007 Bergen, Norway
[⚘] Air Force Research Laboratory, AFRL/PRS, Edwards AFB, CA 93524
[♒] Department of Chemistry, Indiana University, Bloomington, IN 47405

**Abstract.** High-resolution inner-shell photoelectron and resonant Auger electron spectroscopies are examined as probes of the local chemical environment of specific atoms within a molecule. The C 1s spectra of $CH_4$ are reported at an experimental resolution better than the natural line width. The spectra were analyzed to extract the basic physical information contained in the line width, vibrational spacings, and vibrational intensities. Spectra of $C_2H_2$ were measured at a range of photon energies above the C 1s ionization threshold. The spectra were measured at the highest possible resolution to obtain intensity ratios for the symmetry split C 1s photoelectron lines. A definitive assignment of a shape resonance in the $k\sigma_u$ photoionization channel was obtained from these results. Photoelectron spectra of propyne, $HC\equiv CCH_3$, were measured at high resolution and a definitive assignment of the three peaks was obtained from spectra of the model compounds ethane and ethyne ($CH_3CH_3$ and $HC\equiv CH$), and theoretical calculations of the vibrational structure. Angle-resolved molecular-field split S 2p photoelectron spectra of COS are reported and the methods used to extract body-frame information from these spectra described. Resonant Auger electron spectra of CO measured at the three vibrational levels of the C $1s^{-1}2\pi^*$ inner-shell excited state were obtained at about half the intrinsic line width of the inner-shell hole state. The spectra are shown to be a sensitive probe of the geometry of the intermediate excited state and allow access to portions of the final state (one valence hole) potential energy surfaces not open to Franck-Condon transitions from the ground-state neutral molecule.

## INTRODUCTION

Electron spectroscopy of inner-shell electrons in atoms and molecules is experiencing a renaissance in the last decade due to the availability of photon sources and electron energy analyzers with resolutions significantly better than the natural line width of most inner-shell hole states. Progress in high-resolution electron spectroscopy has become more rapid with the recent commercial availability of high energy resolution electron energy analyzers such as Scienta's SES-200 and high-flux, high-resolution photon sources on undulators at electron storage rings. The more recent development of third generation light sources (such as the ALS, MAX-II, ELLETRA) and their accompanying beamlines with very-high resolution (SGM's and PGM's) have now made it possible to routinely measure photoelectron spectra of inner-shell levels with a resolution better than the natural line width.

Photoelectron spectra measured at these high resolutions reveal features that are not apparent in low-resolution spectra measured using line sources. Effects due to vibrational excitation of the final state, symmetry splitting of chemically equivalent atoms, small energy shifts in chemically inequivalent atoms, anisotropic molecular fields, and changes in the molecular geometry in multi-step processes all become apparent in photoelectron spectra measured at high resolution. It has also become possible to study the intrinsic shape of the inner-shell photoelectron line with confidence when the experimental resolution is smaller than the line width. The line shape contains line width ($\Gamma$) contributions from the lifetime ($\tau$) of the core-hole state according to:

$$\Gamma = \frac{\hbar}{\tau}. \tag{1}$$

The line shape of inner-shell photoelectron lines can also be modified by post-collision interaction, PCI, when the kinetic energy of the photoelectron is less than the energy of one or more of the Auger lines. PCI results in an asymmetric shape for the inner-shell photoelectron line with a tail to the low kinetic energy side. The effects of PCI are greatest at threshold and typically decrease with increasing kinetic energy of the photoelectron. The effects of PCI will not be studied explicitly here, but rather modeled in the fits to the experimental data using the functions of van der Straten *et al.* (1).

Several illustrative examples have been chosen from our recent results to highlight the phenomena described above. Vibrationally resolved C 1s photoelectron spectra of methane, $CH_4$, are presented to clearly show the effects of vibrational structure on the experimental photoelectron spectra (2). Methane also serves as a benchmark for theoretical calculations because of its relative simplicity. Analysis of the vibrational structure provides a stringent test for the theoretical methods used to predict the molecular structure of the ionic state and hence the photoelectron spectra. Owing to the large vibrational spacing, the natural line shape including the inherent line width and PCI effects are clearly evident in the C *1s* spectrum of methane, providing a demanding test for the line shape models.

High-resolution C *1s* photoelectron spectra of ethyne, HC≡CH, commonly known as acetylene, have been shown to exhibit a 105 meV splitting between the $1\sigma_u^{-1}$ and $1\sigma_g^{-1}$ final states (3). This splitting, which can only be resolved in photoelectron spectra measured at high resolution, makes it possible to investigate the question of the existence and symmetry of a shape resonance in the C *1s* continuum at ≈310 eV.

The use of shape resonances as bond length "rulers" within a group of related molecules was proposed and raised a heated debate about the assignment of such features (4,5). Photoionization cross section results for ethyne are relevant to this debate.

Propyne, HC≡CCH$_3$, is representative of a large class of molecules (hydrocarbons), which have not been successfully analyzed by the analytical technique of x-ray photoelectron spectroscopy (XPS) up until now. XPS is a powerful analytical tool owing to the relationship between the chemical shift (the change in core ionization energy between an atom in one environment and the same atom in a different environment) of an inner-shell photoelectron line with the electron density at the atom being studied. In silicon oxides, for example, the binding energy of a Si *2p* photoelectron line is related to the oxidation state of the atom (the binding energy increases with increased oxidation). It is also sensitive to the number and electronegativity of the ligands bound to the atom. In the extremely important group of molecules composed of only carbon and hydrogen however, the shift in the C *1s* binding energy as a function of its chemical environment is small compared to the experimental resolution available until recently (6). The high resolution available at the ALS using the Scienta SES-200 electron spectrometer makes it possible to uniquely identify the peaks in the C *1s* photoelectron spectrum due to each of the chemically different carbon atoms.

An anisotropic molecular field can also induce splitting in inner-shell photoelectron spectra. This splitting, known as molecular-field, ligand-field, or crystal field splitting occurs only for degenerate inner-shell levels such as $p_{3/2}$, *d* and *f* levels. Molecular-field effects are well-understood and theoretical calculations available (7,8). Combined with measurements of angular distributions, the molecular-field split core levels can provide body frame information in the absence of explicit sample alignment. We have measured angle resolved S *2p* photoelectron spectra of COS at 25 meV experimental resolution, sufficient to completely resolve the spin-orbit splitting, vibrational splitting and molecular-field splitting.

Resonant Auger spectroscopy provides a means to examine inner-shell spectra with sub-natural line width resolution. A bound core-hole excited state is photoexcited and the subsequent electron emission spectrum recorded when performing this technique. In the one-electron picture, two distinct processes can be identified in the resonant Auger spectrum; participator decay, where the photoexcited electron takes part in the Auger decay; and spectator decay, where the excited electron does not take part in the de-excitation. The final states for participator decays, which determine the energies of the peaks in the electron spectra, are the single valence hole states accessed in direct valence photoionization. Spectator Auger decays result in two-hole one- (excited) electron final states, which are present in the valence photoelectron spectra as shake-up peaks.

Carbon monoxide, with its simple molecular and electronic structure, has been studied extensively in the past by resonant Auger spectroscopy (9) and the resulting spectra are well understood. As a benchmark system, however, it is useful to revisit the resonant Auger spectra with the highest available experimental resolution. We report here on the electron emission spectra of CO following photoexcitation of a C $1s$ electron to the $2\pi^*$ first unoccupied molecular orbital. The C $1s \rightarrow 2\pi^*$ excitation in the photoabsorption spectrum exhibits a vibrational series with three components (10). Vibrationally resolved resonant Auger electron spectra measured at the peaks of each of the three vibrational levels of the C $1s^{-1}2\pi^*$ state are reported.

## EXPERIMENTAL

All of the spectra reported here were measured using the High-Resolution Atomic and Molecular Electron Spectrometer (HIRAMES) end station located on beamline 10.0.1 at the Advanced Light Source (ALS) of Lawrence Berkeley National Laboratory (11). The HiRAMES end station consists of a Scienta SES-200 electron energy analyzer mounted on a rotatable vacuum chamber. The analyzer can be oriented at any angle between 0° and 90° relative to the polarization vector of the 100% linearly polarized undulator radiation orthogonal to the beam propagation direction. A gas cell with openings for the photon beam and ejected electrons was used to differentially pump the sample. The gas cell incorporates field compensation electrodes (12) to achieve the highest resolution. Spectra of the Xe $5p$ photolines have been measured with total line widths of approximately 5 meV using this spectrometer and beamline combination.

Beamline 10.0.1 at the ALS is optimized to deliver high photon energy resolution over the photon energy range from 20 – 340 eV. A 4.5m long 10cm period undulator in the low emittance ALS electron storage ring provides a bright source of radiation in the desired range. A vertical focusing mirror demagnifies the photon beam onto the entrance slit of the monochromator. The spherical grating monochromator contains three gratings to cover the entire photon energy range of the beamline. Only the 2100 line/mm high-energy grating was used for the results reported here. Refocusing optics provide a beamsize of approximately 0.5mm in the interaction region of the HiRAMES spectrometer.

A variety of experimental resolutions, with contributions from the photon bandpass and electron energy analyzer resolution, were used for these experiments. The optimal settings for a given experiment were determined from the compromise between photon flux and resolution and the luminosity of the electron spectrometer at different combinations of pass energy and slit settings. Typically photon energy resolutions of 30-40 meV were used at photon energy of 330 eV. Contributions from the electron spectrometer at the settings employed were estimated from measurements of the Xe $5s$ line at several photon energies with very high photon energy resolution and found to be approximately 30 meV. The total instrumental resolution, obtained by combining the two contributions, was in the range of 40-50 meV.

The transmission of the electron energy analyzer, which varies as a function of kinetic energy, was determined using the method of Jauhiainen et al. (13). Xenon NOO Auger and $4d$ photoelectron spectra were measured as a function of photon energy and ratios of intensities used to determine the transmission. The transmission was found to very nearly follow the expected inverse of the kinetic energy and spectral intensities were corrected appropriately.

## RESULTS AND DISCUSSION

### C *1s* Photoelectron Spectra of Methane

The C *1s* photoelectron spectrum of methane was the first to be measured with a resolution sufficient to resolve the vibrational structure (14). Using monochromatized Al $K\alpha$ radiation, Asplund and coworkers were able to report the methane C *1s* photoelectron spectrum with 0.22 eV resolution (15) in 1985. Using monochromatized undulator radiation, several groups have recently reported $CH_4$ C *1s* photoelectron spectra with a resolution comparable to the natural line width of about 95 meV (16,17). We report here on measurements made with a resolution about half the natural line width (2) and therefore obtain a clearer view of the vibrational structure and intrinsic line shape of the C *1s* photoelectron line.

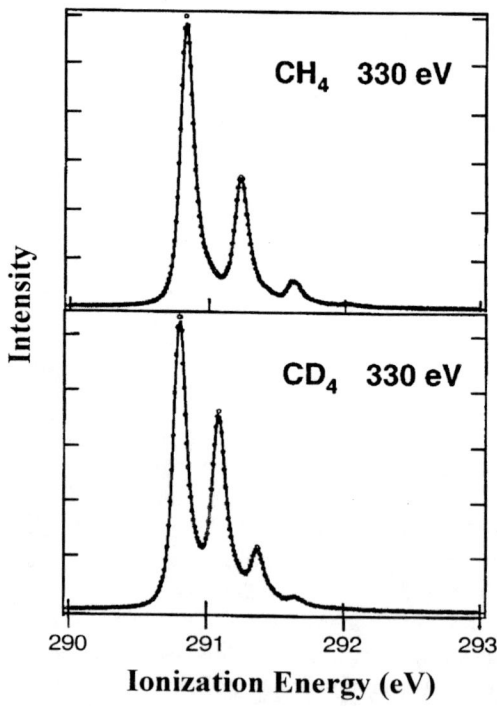

**FIGURE 1.** Carbon *1s* photoelectron spectra of methane and deuteromethane measured at 330 eV. The solid lines show results of least squares fit to the data using the model described in the text. Figure adapted from Ref. (2).

Spectra of the C $1s$ lines of $CH_4$ and $CD_4$ were measured at several photon energies and representative spectra are shown in Fig. 1. The experimental data points are fit with the PCI line shape function of van der Straten et al. (1) convoluted with the experimental resolution function and a Gaussian function to account for the Doppler broadening. The fits included parameters for the position and height of the v = 0 peak, the width of the Lorentzian line used to describe the lifetime limited line width, the relative positions and intensities of the higher vibrational peaks, and a constant background. The fit, represented by the solid line through the data points, is seen to be in very good agreement with the experimental data.

The vibrational progression, which results from a change in the C–H bond length upon removal of the C $1s$ electron can be analyzed to provide information about the difference in the bond length between the ionized and ground-state molecule. Using the peak positions determined in the fitting process, the fundamental (difference in energy between the energies of the v=1 and 0 states) stretching energies were found to be 397.1±0.9 and 284.2±0.7 meV for $CH_4$ and $CD_4$ respectively at 330 eV. To evaluate the possibility of anharmonic vibrational motion, the peak positions were fit to a quadratic function of v+½, where v is the quantum number of the symmetric stretching mode. The results indicate no conclusive sign of anharmonicity (2).

Franck-Condon factors, describing the overlap between the wavefunction for the ground vibrational state of the ground electronic state with the wavefunctions for the various vibrational levels of the core hole ionic state, have been extracted from the intensities of the various vibrational levels. The results are shown in Table 1 for both $CH_4$ and $CD_4$ at 330 eV. The average vibrational energy (which is the product of the spacing from the v=0 level and the relative intensity) for both $CH_4$ and $CD_4$ at 330 eV is 156 meV. The intensities of the different vibrational levels can be modeled using a harmonic oscillator model. The only variable parameter in the model is the change in the C–H bond length upon ionization, and using the ratio of the v=0 to 1 intensities in the spectrum to determine the best fit, absolute values for the bond length changes of 4.94±0.01 and 4.96±0.01 pm result for $CH_4$ and $CD_4$. Anharmonicity can also be included in this calculation and it results in a better fit to the intensities of the higher vibrational levels and a slightly different bond length change. Interested readers are directed to a more detailed examination of the methane spectra (2).

The Lorentzian widths extracted from the fitting process contain fundamental

**TABLE 1.** Intensities of the vibrational structure in core ionized methane. Normalized to 1 for the v=0 transition. Uncertainties are shown in parenthesis.

| v | $CH_4$ | $CD_4$ |
|---|--------|--------|
| 0 | 1 | 1 |
| 1 | 0.424(2) | 0.618(3) |
| 2 | 0.0748(9) | 0.161(1) |
| 3 | 0.0063(4) | 0.0238(7) |
| 4 |  | 0.0013(4) |

information about the lifetime of the core-hole state. At 330 eV the Lorentzian widths for $CH_4$ and $CD_4$ are 95.4 and 95.0 meV respectively. The widths change with different photon energies getting larger as the photon energy is decreased. This might be a result of changes in lifetime with photon energy but is more likely related to a failure of the PCI function to adequately describe the shape of the experimentally observed photoelectron lines close to threshold. Uncertainty in the function used to describe the effects of PCI on the line shape adds a degree of uncertainty to the line width. At higher energies, however, the spectra are well fit by the function and the lifetimes derived from these spectra are expected to be within 2 meV of the true value. The lifetime width for methane is therefore estimated to be between 93 and 95 meV. A more detailed evaluation of the failure of the PCI line shape function can be found in Ref. (2).

## Symmetry Resolved Carbon 1s Cross-Sections for Ethyne

The C 1s photoelectron spectrum of ethyne, HC≡CH, was recently reported to exhibit a 105 meV splitting between the $1\sigma_u^{-1}$ and $1\sigma_g^{-1}$ states (3). The splitting arises

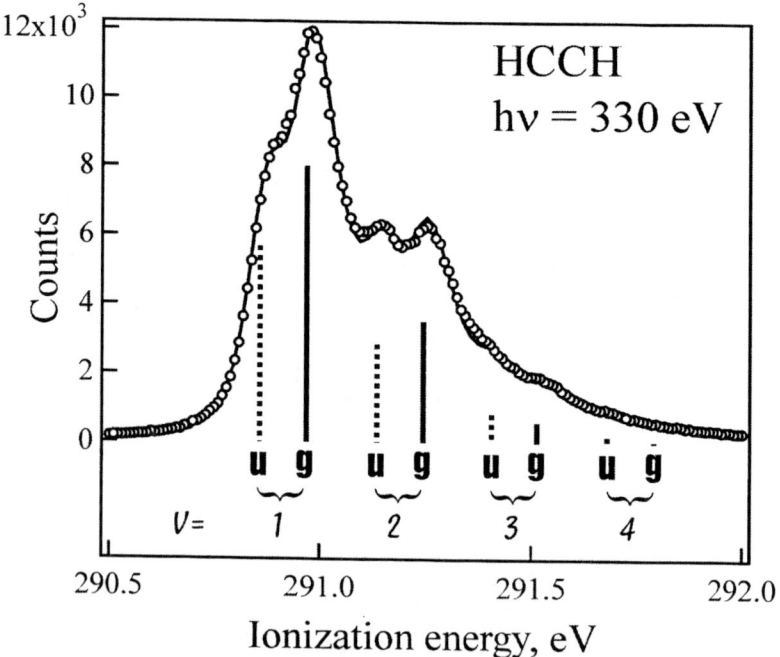

**FIGURE 2.** Carbon 1s photoelectron spectrum of HCCH measured at a photon energy of 330 eV. The points are experimental data and the solid line a fit assuming that only C–C stretching contributes to the vibrational structure. Adapted from Ref. (17).

from the overlap of the two C 1s wave functions owing to the short carbon-carbon triple bond and resulting in molecular orbitals that are symmetric and antisymmetric linear combinations of the atomic wavefunctions. This splitting, which is similar to the intrinsic width of the C 1s line allows the two final states to be resolved in the photoelectron spectrum.

A C 1s spectrum of HCCH measured at an angle of 54.7° and a photon energy of 330 eV is shown in Fig. 2. The spectrum contains a great deal of structure due to the symmetry splitting as well as vibrational structure assigned to C–C stretching (18). The splitting of the C 1s line into $1\sigma_u^{-1}$ and $1\sigma_g^{-1}$ peaks is identified by the letters u and g on the figure, and the four vibrational levels are also labeled. It is readily apparent that at 330 eV the $1\sigma_u$ cross section is less than that of the $1\sigma_g$ level. The spectra were fit using a 32 meV Gaussian monochromator function, a 105 meV Lorentzian for the lifetime width for both the $1\sigma_u^{-1}$ and $1\sigma_g^{-1}$ lines and the PCI line shape function used above (1). Only the carbon-carbon stretching mode (and not the

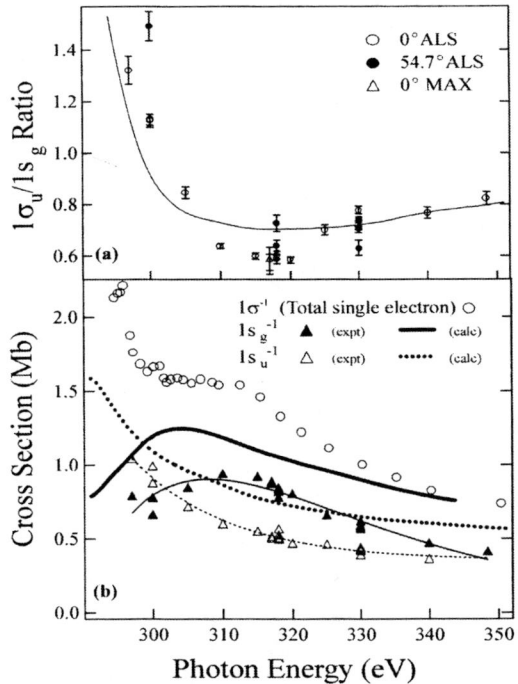

**FIGURE 3.** (a) The ratio of the cross sections for ionization of the $1\sigma_u$ and $1\sigma_g$ electrons of ethyne as a function of photon. The solid line shows a theoretical calculation from results given in Ref. (19). (b) Cross sections for symmetry resolved C 1s photoionization. Figure from Ref. (18).

C–H stretching mode or any bending modes) is included in the vibrational manifold. The u-g splitting and vibrational frequency from the fit are in agreement with previous results (3). The intensities of the v=2 and 3 levels will probably be in error due to the omission of the C-H stretching, but since only the intensities of the u and g levels of the v=0 band are of interest, they are not significant.

The ratio of intensities for the $1\sigma_u^{-1}$ and $1\sigma_g^{-1}$ bands of the first vibrational level (v=0) is the quantity of interest and it is 0.71 for the spectrum in Fig. 2. Intensity ratios were similarly extracted from a series of spectra measured at photon energies over 297 – 348 eV and both 54.7° and 0° for HCCH and the deuterated version of the molecule DCCD. No evidence was found of an angular dependence for the $1\sigma_u^{-1}/1\sigma_g^{-1}$ intensity ratio above photon energies of 300 eV (18) and therefore, no significant error is expected from mixing results for the two different angles.

The photon energy dependence of the $1\sigma_u^{-1}/1\sigma_g^{-1}$ intensity ratio is shown in Fig. 3a. Data from previous work (17) at 317 eV are also included. The solid line results from theoretical calculations of the symmetry resolved cross sections (19). The ratio exhibits a shallow minimum at 315-320 eV, which corresponds to a region of enhanced C $1s$ single hole cross section as determined by Kempgens et al. (3). The single hole cross sections for the symmetry resolved $1\sigma_u^{-1}$ and $1\sigma_g^{-1}$ lines have been determined by taking the product of the intensity ratio and the total single hole cross section and are illustrated in Fig. 3b with the thin solid and dashed curves drawn through the data for clarity. The thick lines show ab initio calculated values for the partial photoionization cross sections (19). The cross section for photoionization of the $1\sigma_u$ orbital shows a monotonic decrease while that for the $1\sigma_g$ orbital has a maximum at around 310 eV. The position of the peak is consistent with previous assignment of a shape resonance at 310 eV (4) but we are now able to experimentally assign the symmetry of the resonance to the $k\sigma_u$ photoionization channel.

## Carbon $1s$ Photoelectron Spectrum of Propyne

High experimental resolution can also be used to elucidate chemical information about individual atoms in a molecule. Chemical shifts are related to the electron density at the atom being probed, providing a means of studying the chemical reactivity of a molecule at the site of a specific atom. We have chosen to apply our high-resolution capabilities to the C $1s$ spectrum of propyne, a prototypical aliphatic alkyne molecule, to gain insight into the electron distribution within the molecule. The three different carbon atoms in propyne are chemically distinct, with the CH unit at the basic end of the molecule and the $CH_3$ group at the acidic end. Low-resolution (1.1 eV) spectra were previously available but they exhibited only a broad asymmetric peak (6). Definitive assignment of the unresolved spectrum was not possible.

The C $1s$ photoelectron spectrum of propyne measured with approximately 45 meV experimental resolution at 330 eV photon energy is shown in Fig. 4. The spectrum consists of three major peaks of equal intensity with distinctive vibrational

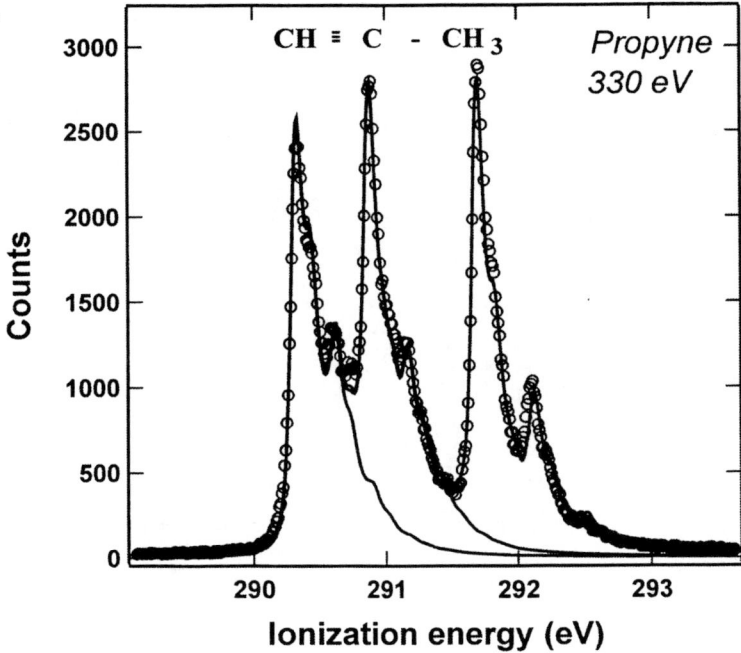

**FIGURE 4.** Carbon $1s$ photoelectron spectrum of propyne. Circles indicate the experimental data and the solid lines show the results of theoretical calculations of the vibrational structure.

structure. In order to assist in the assignment of the propyne spectrum, C $1s$ photoelectron spectra of ethyne (HCCH) and ethane (H$_3$CCH$_3$) were measured under approximately the same conditions, albeit with slightly degraded resolution. Each spectrum was measured simultaneously with CF$_4$ to provide an internal calibration of the relative binding energies of the different compounds. The spectra of the three related compounds are shown in Fig. 5.

The assignment of the methyl peak in propyne is greatly simplified by the comparison of the vibrational structure of the three peaks with the line shape for ethane. The vibrational structure in ethane is due primarily to excitation of the C-H stretching vibration, resulting in a strong peak 400 meV above the main line. The shoulder on the main peak is a result of HCH bending (17). The vibrational structure of the ethane peak is remarkably similar to that of the rightmost peak in the propyne spectrum in Fig 4. The assignment, made on the basis of this resemblance, is in agreement with theoretical calculations.

The other two peaks are obviously related to the acetylenic group, both by the similarity in their shapes to the ethyne peak and by process of elimination. Other methods need to be used to assign each peak to an individual carbon atom in the molecule. We choose to use the comparison of the line shapes with the theoretical

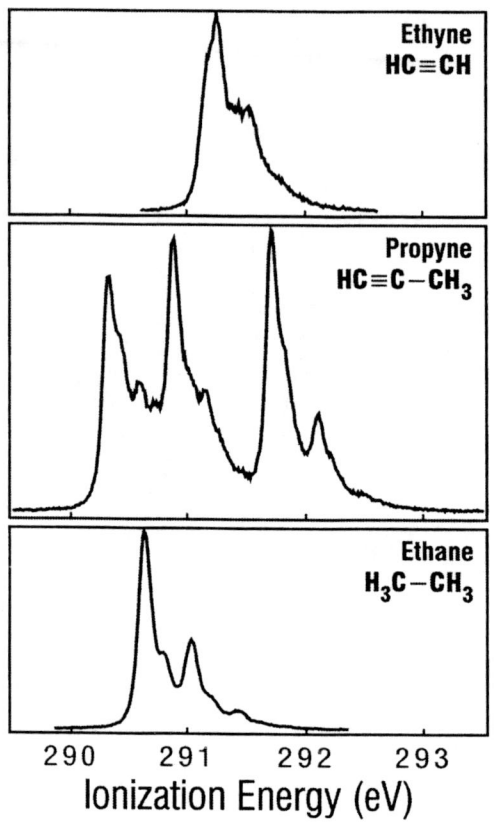

**FIGURE 5.** Carbon *1s* photoelectron spectra of ethyne, propyne and ethane measured at 330 eV.

vibrational structure obtained from *ab initio* calculations of the vibrational energies and molecular geometry (which determines the Franck-Condon factors) of the core hole state as described elsewhere (17). The details of the calculations and approximations will be published elsewhere, but the results are shown by the solid line in Fig. 4. The line is a result of the *ab initio* calculation, and is not "fit" to the data in any way other than by the position and intensity of the v=0 peak. The resulting assignment, with the leftmost peak assigned to the terminal CH carbon and the middle peak to the central carbon is shown on Fig. 4. Reversing the assignment results in a much poorer fit to the experimental data.

The ionization energies are somewhat surprising when compared with the model compounds. The ionization energy of the methyl group is 1.08 eV higher in propyne than it is in ethane and that for the CH group is 0.87 eV lower than in ethyne.

Considering the relationship between the chemical shift and electron density, this means that the acetylinic group withdraws electron density from the methyl group. Its electronegativity can be compared with those of the halogen atoms by comparing the ionization energy of the methyl carbon in propyne with the ionization energies of halogen substituted methane molecules, $CH_3X$. Based on this comparison, the electronegativity of the acetylenic group, HC≡C, falls between that of bromine and iodine.

## Angle-Resolved Molecular Field Split S *2p* Photoelectron Spectra of COS

The anisotropic field present in molecules with non-cubic symmetry can result in additional fine structure in the inner-shell electron spectra of degenerate levels. The S 2p levels of carbonyl sulfide, COS, which are shown in Fig. 6 at three different angles, exhibit this phenomenon with additional structure on the S $2p_{3/2}$ line that is not seen on the S $2p_{1/2}$ line. Molecular-field splitting has previously been observed for *d* and *p* levels of many different molecules and is generally well understood (7,8,20).

Angular distributions of photoelectrons probe the potential field experienced

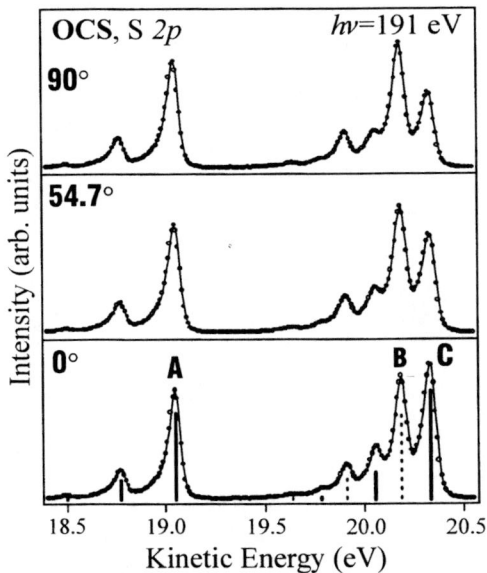

**FIGURE 6.** Sulfur 2p photoelectron spectra of the COS molecule measured angles of 90°, 54.7°, and 0° relative to the polarization vector of the light at 191 eV. The solid lines indicate the results of non-linear least squares fits to the data as discussed in the text (23).

by electron as it departs the ionized atom. For an ensemble of randomly oriented molecules, the angular distributions of the electrons can be represented by an anisotropy parameter, $\beta(h\nu)$, in the dipole approximation. Determination of anisotropy parameters for molecular field split-levels can provide a means to access body frame anisotropies and transition matrix elements that can otherwise only be determined from oriented molecules. In order to measure electron anisotropies from oriented molecules, however, low rate coincidence experiments must generally be performed (21). Anisotropy parameters from fully spin orbit and molecular field-split photoelectron spectra can provide an alternative to these coincidence techniques.

Sulfur 2p spectra of the COS molecule have been measured at photon energies between 180 and 300 eV and at three angles, 0°, 54.7°, and 90°, relative to the polarization direction of the 100% linearly polarized light. Examples obtained at 191 eV are shown in Fig. 6. The spectra were fit using a Gaussian width of 35 meV to represent the instrumental width, the PCI line shape (1), and a Lorentzian line with a variable width (65±5 meV) to represent the intrinsic lifetime limited line width. The positions and intensities of the peaks were included as parameters in the fit.

The three peaks at lower kinetic energy result from a vibrational progression of the S $2p_{1/2}$ level. The structures at higher kinetic energy correspond to overlapping vibrational manifolds for the two molecular-field split S $2p_{3/2}$ levels. The molecular-field splitting is determined to be 145±1 meV from the fit and is in good agreement with previous theoretical values (7). The areas of the three v=0 peaks, labeled A, B, and C on the figure, are approximately equal in the spectra measured at the magic angle for all photon energies. The spectra in Fig. 6 show a significant change in the relative intensities of the three peaks at the three different angles, however, indicating different angular distributions for these three lines. Experimental anisotropy parameters ($\beta(h\nu)$) were determined from the angle-resolved photoelectron spectra using a method described by Kivimaki et al. (22).

In order to understand the origins of the spectral features in Fig. 6, wavefunctions can be constructed from linear combinations of the $2p\sigma^{-1}$ and $2p\pi^{-1}$ body-frame molecular hole states as described in greater detail elsewhere (23). Expressions related to experimentally accessible anisotropy parameters can be derived for the linear combinations of the body-frame anisotropy parameters:

$$\beta_{B+C}(h\nu) \rightarrow \frac{1 - 2\rho(h\nu)\cos\delta(h\nu)}{1 + \frac{1}{2}\rho^2(h\nu)} \tag{2}$$

$$\beta_C(h\nu) - \beta_B(h\nu) \rightarrow \frac{2\rho(h\nu)\{\cos\Delta_B(h\nu) - \cos\Delta_C(h\nu)\}}{1 + \frac{1}{2}\rho^2(h\nu)}$$
$$\approx 2\rho(h\nu)\sin\delta(h\nu)(\Delta_{BC}), \tag{3}$$

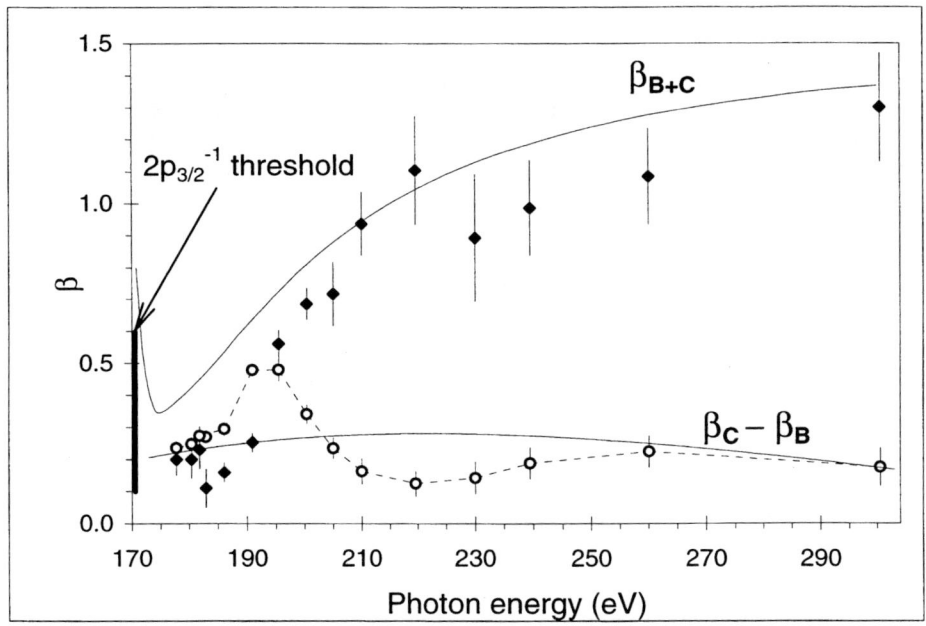

**FIGURE 7.** Experimental anisotropy parameters obtained from the PES measurements. Solid lines show the values obtained from the theory discussed in the text (23).

from the wavefunctions using appropriate continuum states of the dipole allowed transitions to construct final-state wavefunctions (see Ref. (23) for the details). Experimental $\beta_{B+C}(h\nu)$ and $[\beta_C(h\nu)-\beta_B(h\nu)]$ values are shown in Fig. 7 and compared with values calculated using the above equations. The average anisotropy values $\beta_{B+C}(h\nu)$ in Fig. 7 exhibit a photon energy dependence characteristic of atomic $2p$ shells (23) as predicted by Eq. 2. Body-frame anisotropy parameters and transition moment-phase shift products can be extracted from this data as described in Ref. (23). Body-frame information can thus be extracted from high-resolution non-coincident molecular field split photoelectron spectra.

## Resonant Auger Spectra of CO

Auger spectroscopy is another powerful technique useful for investigating inner-shell processes in molecules. Resonance Auger spectroscopy further offers the opportunity to utilize the high-resolution capabilities of third generation synchrotron light sources and modern electron spectrometers. Line widths are limited by the experimental resolution rather than the lifetime of the core hole state in resonant Auger spectroscopy assuming that the final state is stable.

We have measured resonant Auger spectra of CO upon excitation of the three vibrational levels of the C $1s^{-1}2\pi^*$ bound state resonance at around 287 eV using an experimental resolution of 45 meV, about half the intrinsic width of the C $1s$ photoelectron line (24). The resonant Auger spectra of CO have been studied extensively in the past (9) and the general structure is understood. The spectra are worth revisiting, however, to highlight the advantages of high spectral resolution with high flux such as is available in our experimental setup. Experimental electron energy spectra of CO measured at photon energies corresponding to the maxima of the C $1s \rightarrow 2\pi^*$ (v=0,1,2) absorption features are shown in Fig. 8 along with a normal photoelectron spectrum measured with a photon energy below the absorption resonance (non-resonant photoionization). The spectra were obtained at 54.7° relative to the polarization of the linearly polarized light. We have also obtained angle-resolved spectra and extracted anisotropy parameters for each of the peaks in the electron energy spectra at each of the excitation energies (24). Those results will not be discussed here, but interested readers are directed to the reference indicated. Compared with previously published spectra, the present results have significantly greater experimental resolution and improved signal. The spectra reported here were measured with a total resolution of 80 meV except for the nonresonant spectrum that

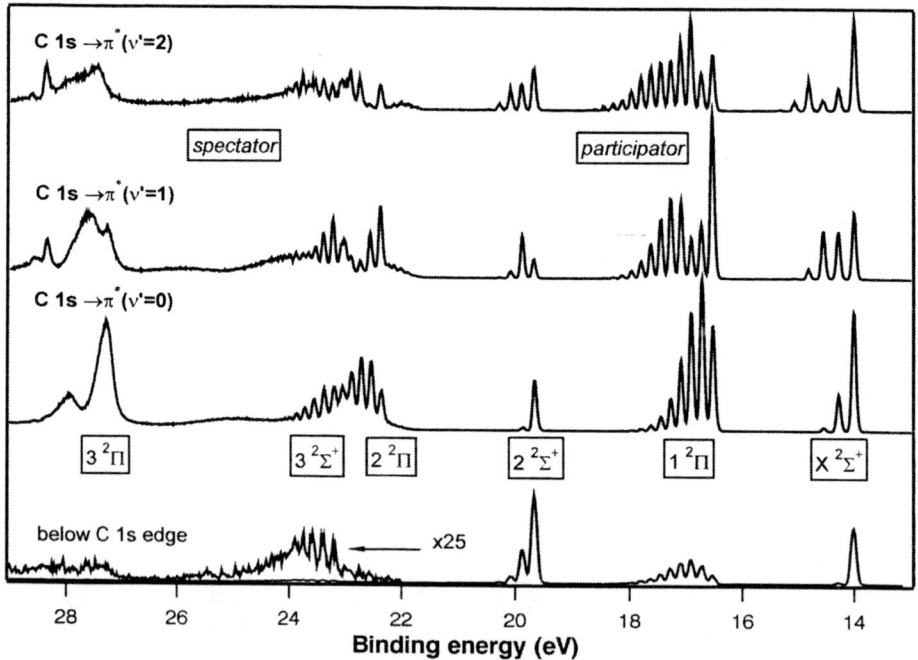

**FIGURE 8.** Experimental Auger electron spectra of the decay of the C $1s^{-1}2\pi^*$, v=0,1,2 states taken at 54.7°. A nonresonant spectrum measured at 280 eV is shown in the bottom panel for comparison.

was obtained with 140 meV resolution to compensate for the much lower intensity of the signal.

Spectra have been obtained at 52 meV resolution for the strong v=0 level of the excited state, and while they are slightly better resolved, the spectrum is not significantly different. The experimental line width of the strong $1^2\Sigma^+$ lines for these higher resolution spectra is 75 meV, suggesting the presence of an additional contribution of 40 meV to the width, which could arise from the excitation of rotational levels in the final state of the ion.

Three participator Auger bands are seen with full vibrational resolution of the final single valence hole state in the binding energy range from 14 to 20 eV. The vibrational manifold of these final states changes dramatically with the vibrational level of the initial excited state. All three of the final states exhibit a considerably more extensive vibrational structure than is found for the direct valence photoionization. The complex structure in the spectrum between 21 and 29 eV arises from spectator Auger decay of the excited state. The high resolution of these spectra combined with *ab initio* theoretical calculations have allowed us to unambiguously assign the experimental spectrum to overlapping $2^2\Pi$ and $3^2\Sigma^+$ states in the 22-25 eV binding energy range (24).

## SUMMARY

High resolution inner-shell photoelectron and resonant Auger electron spectra have been presented for a number of simple molecules in the gas phase to highlight the wealth of chemical information that can be obtained from such spectra. Experimental resolution is crucial to resolve the many overlapping peaks in the lifetime broadened inner-shell photoelectron spectra. Such high resolution depends on the development of high flux/high resolution beamlines on undulator sources at third generation light sources and the availability of modern high-resolution electron energy analyzers.

Carbon *1s* photoelectron spectra of the benchmark carbon compound methane were investigated to probe the understanding of vibration excitation accompanying inner shell ionization. The well-resolved vibrational lines also offered the opportunity to probe the line broadening mechanisms of core hole lifetimes and postcollision interaction. Applying high resolution to the C *1s* spectra of ethyne allowed symmetry resolved cross sections to be determined resulting in a definitive assignment of a shape resonance in the $k\sigma_u$ photoionization channel. Atom specific chemical information was obtained from the high-resolution photoelectron spectrum of propyne where vibration band profiles were used to assign the chemical shifts of the individual carbon atoms. Angle resolved photoelectron measurements of the S *2p* lines of COS were shown to provide access to molecular frame anisotropy parameters and transition moment phase shift products in the absence of explicit alignment of the target molecule, therefore foregoing low rate coincidence experiments. Finally, vibrationally resolved resonant Auger spectra of CO measured at the C $1s^{-1}2\pi^*$ excitation were reported and shown to provide a sensitive probe of the dynamics of the excitation/deexcitation processes.

## ACKNOWLEDGEMENTS

This work was supported by the Divisions of Chemical Sciences and Material Sciences of the Office of Energy Research of the U.S. Department of Energy. T.D.T. and T.X.C. acknowledge support by the National Science Foundation, and L.J.S. thanks the Research Council of Norway (NFR) for support. We are also indebted to our many colleagues for their support and suggestions.

## REFERENCES

1. P. van der Straten, R. Morgenstern, and A. Niehaus, Z. Phys. D **8**, 35 (1988).
2. T.X. Carroll et al., Phys. Rev. A **59**, 3386 (1999).
3. B. Kempgens et al., Phys. Rev. Lett. **79**, 35 (1997); J. Chem. Phys. *107*, 4219 (1997).
4. A.P Hitchcock et al., J. Chem. Phys. **80**, 3927 (1984); J. Stöhr, F. Sette, and A.L. Johnson, Phys. Rev. Lett. **53**, 1684 (1984); F. Sette, J. Stöhr, and A.P. Hitchcock, J. Chem. Phys. **81**, 4906 (1984).
5. M.N. Piancastelli, D.W. Lindle, T.A. Ferrett, and D.A. Shirley, J. Chem. Phys. **86**, 2765 (1987); A.P. Hitchcock and J. Stöhr, J. Chem. Phys. **87**, 3253 (1987); M.N. Piancastelli, D.W. Lindle, T.A. Ferrett, and D.A. Shirley, J. Chem. Phys. **87**, 3255 (1987).
6. R.G. Cavell, J. Electron Spectrosc. Relat. Phenom. **6**, 281 (1975).
7. K.J. Borve, Chem. Phys. Lett. **262**, 801 (1996).
8. K. Ellingsen, T. Saue, H. Aksela, and O. Gropen, Phys. Rev. A **55**, 2743 (1997).
9. M.N. Piancastelli, et al., J. Phys. B **30**, 5677 (1997) and references therein.
10. M. Domke et al., Chem. Phys. Lett. 173, 122 (1990).
11. N. Berrah et al., J. Electron Spectrosc. Relat. Phenom. **101-103**, 1 (1999).
12. P. Baltzer, B. Wannberg, M. Carlsson Gothe, Rev. Sci. Intrum. **62**, 643 (1991); P. Baltzer, L. Karlsson, M. Lundqvist, and B. Wannberg, Rev. Sci. Instrum. **64**, 2179 (1993).
13. J. Jauhiainen et al., J. Electron Spectrosc. Relat. Phenom. **69**, 181 (1994).
14. U. Gelius, J. Electron Spectrosc. Relat. Phenom. **5**, 985 (1974).
15. L. Asplund et al., J. Phys. B **18**, 1569 (1985).
16. H.M. Koppe et al., J. Chin. Chem. Soc. (Taipei) **42**, 255 (1995); H.M. Koppe et al., Phys. Rev. A **53**, 4120 (1996); S.J. Osborne et al., J. Chem. Phys. **106**, 1661 (1997); L.J. Saethre et al., Phys. Rev. A **55**, 2748 (1997).
17. T.D. Thomas et al., J. Chem. Phys. **109**, 1041 (1998).
18. T.D. Thomas et al., Phys. Rev. Lett. **82**, 1120 (1999).
19. R.E. Farren, Ph.D. thesis, Indiana University, Bloomington, 1989.
20. J.N. Cutler, G.M. Bancroft, and K.H. Tan, J. Phys. B **24**, 4897 (1991); M.R.F. Siggel et al., J. Chem. Phys **105**, 9035 (1996), and references therein.
21. E. Shigemasa, J. Adachi, M. Oura, and A. Yagishita, Phys. Rev. Lett. **74**, 359 (1995); N. Watanabe et al., Phys. Rev. Lett. **78**, 4910 (1997); F. Heiser et al., Phys. Rev. Lett. **79**, 2435 (1997).
22. A. Kivimaki et al., Phys. Rev. A **57**, 2724 (1998).
23. E. Kukk, J.D. Bozek, N. Berrah, J.A. Sheehy, and P.W. Langhoff, J. Phys. B, *submitted*.
24. E. Kukk, J.D. Bozek, W.-T. Cheng, R.F. Fink, A.A. Wills, and N. Berrah, J. Chem. Phys., *in press*.

# Photoelectron Emission from Oriented Molecules

Uwe Becker

*Fritz-Haber-Institut der Max-Planck-Gesellschaft, Faradayweg 4-6, D-14195 Berlin, Germany*

**Abstract.** The development of a new kind of coincidence spectroscopy – angle resolved photoelectron-photoion coincidence spectroscopy (ARPEPICO) – made inner-shell photoelectron spectroscopy on oriented molecules in the gas phase possible. For the first time, molecule-frame photoelectron angular distributions can be investigated; dipole matrix elements and relative phases can be derived. Furthermore, the study of electron-ion momentum vector correlations opens a new field in molecular photoemission, intramolecular scattering and, interference phenomena.

## INTRODUCTION

Photoelectron spectroscopy probes the initial and final states of the photoionization process, i. e. the ground state and continuum state charge distributions, in the simplest approximation their energy-dependent mutual overlap. The particular charge distribution of molecular orbitals gives rise to rather different photoelectron angular distributions due to the loss of spherical symmetry compared to atoms. This richer structure, however, is not revealed in the molecular photoelectron angular distribution as long as the target molecules are randomly oriented; their angular distributions can be completely described by the same formula including the same angular distribution asymmetry parameter β as used for atoms because the unobserved molecular axis gives rise to the spherical symmetry of the problem as given in atoms [1]. Therefore, any molecule specific information beyond the pure comparison with ab-initio β-calculations relies on orientation or alignment of the molecules, which are ionized by photon impact. In many cases, molecules are spatially oriented when adsorbed on a surface; this fact has been exploited in photoelectron and Auger spectroscopy of adsorbates to obtain additional information from the orientation of the molecular axis with respect to the propagation direction or electric vector of the incoming light [2]. But in the gas phase there is no such natural orientation. Brute force methods such as applying electric fields would require in most cases prohibitively high field strengths for angle-resolved photoelectron spectroscopy. Other methods such as the use of molecular beam techniques still suffer from the low degree of orientation they yield. Because of these difficulties only very few and specific experiments succeeded in angle resolved photoelectron measurements on oriented molecules [3]. Considering the alternative of preparing an aligned sample of molecules, e. g. by excitation with polarized radiation, rotational resolution has to be achieved in order to analyze properly the angular distribution of the photolines. First experiments of this kind were performed at energies accessible for high-resolution laser spectroscopy [4]. Because of the limited resolution of the light sources available for higher photon energies, this approach is not yet feasible for inner-shell ionization. But another method, which rather probes than causes the orientation of a molecule, has been developed. It requires probing the spatial orientation of a molecule in a sample of randomly oriented molecules at the moment of photoionization. Such a probe includes all ionization events resulting in ionic fragmentation of the molecule, and hence their angle resolved detection in coincidence with the associated photoelectron, which is emitted in a certain solid angle. We may call this method angle resolved photoelectron-photoion coincidence spectroscopy (ARPEPICO) [5]. Since it correlates the momentum vector of the photoion with the corresponding vector of the photoelectron it is sometimes also called electron–ion recoil vector correlation method. There is a condition for its successful application to photoelectron studies on oriented molecules, namely the so-called axial recoil approximation, which means that the rotation of the molecule has to be much slower than its fragmentation velocity. This approximation is well fulfilled in inner-shell ionization processes, which result mostly in energetic fragmentation of the molecular ions [5-7]. It works also well for many valence ionization processes, although clear deviations have been observed in near-threshold photoionization [8]. For several years now, this method has been successfully applied to study molecule-frame photoelectron angular distributions for a variety of molecules.

## CORE LEVEL PHOTOELECTRON EMISSION

Core electrons are highly localized; this makes site specific excitation and ionization in hetero-nuclear molecules possible. Thus, core level photoelectron emission from oriented molecules is inherently affected by the topology of the molecule. In a sense, extended absorption fine structure (EXAFS) oscillations have even been observed for randomly oriented molecules. The trapping of photoelectrons in the centrifugal barrier they experience after scattering from the neighboring atoms causes pronounced shape resonances in the 1s photoionization cross section of small molecules. One may consider these resonances as the low frequency part of an EXAFS oscillation. Shape resonances have attracted considerable interest over the years [9]. Already twenty years ago, theoretical calculations based on a multiple scattering model of molecular photoelectron emission predicted that $f$-wave enhancement would predominantly contribute to the $\sigma^*$-shape resonance in molecules like $N_2$ and CO, for example [10]. This prediction, however, could be experimentally proved only five years ago when the first inner-shell gas phase photoionization experiment on oriented molecules was carried out by Shigemasa et al. [11]. They used an ARPEPICO setup similar to that used by Golovin [5]. They derived from their measurement for the first time molecule-frame photoelectron angular distributions, which clearly exhibited the $f$-wave character of the σ-type photoelectrons ejected on top of the shape resonance – a beautiful proof of the theoretical predictions. It also opened the door to a completely new class of photoionization experiments on molecules; i. e. partial wave resolved photoelectron spectroscopy. Such spectroscopy is the basis of complete photoionization experiments on molecules, similar to that on atoms.

The coincident photoelectron emission pattern of homo-nuclear diatomic molecules is always symmetric because the core hole sides are indistinguishable. The molecule-frame angular distributions of hetero-nuclear molecules, on the other hand, should show pronounced forward-backward asymmetries with respect to their two distinguishable atomic sides. A showcase for such behavior is CO since many calculational efforts have been spent on this molecule. Crucial preconditions of this experiment are the mass selectivity as well as the angular resolution within the fragment ion detection. Time-of-flight mass spectrometry combined with position sensitive detection is well suited for both tasks. Figure 1 shows a scheme of such an experimental setup.

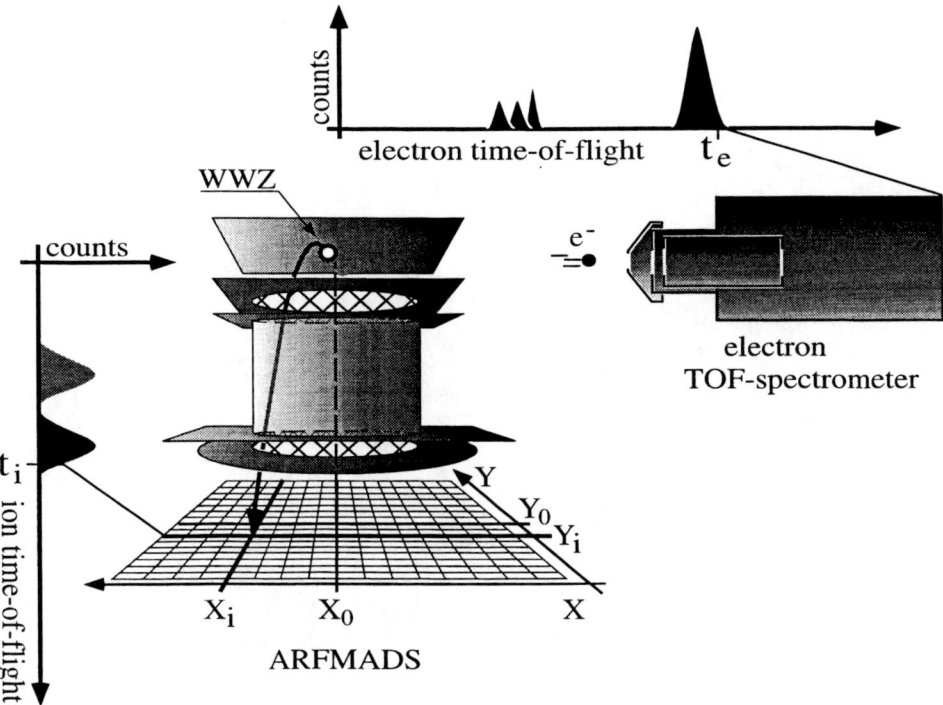

**FIGURE 1** Scheme of the experimental setup for angle resolved photoelectron – photoion coincidence spectroscopy (ARPEPICO). The ion time-of-flight spectrometer is equipped with a position sensitive anode. Each ion detector event delivers three data: the ion time of flight $t_i$, the hit coordinate $X_i$ and the hit coordinate $Y_i$. The ion is measured in coincidence with an electron characterized by its time of flight and the detection angle. The simultaneously accumulated data represent molecular axis distributions for a fixed electron emission direction.

Because of the great interest CO has attracted over the years, it was the first molecule on which such an experiment was carried out [12]. The data from ARPEPICO experiments give basically two-dimensional spectra of photoions and photoelectrons. Both are angle-resolved, but while all emission directions of the ionic fragments are detected simultaneously by the position-sensitive anode, electron emission is measured in this experiment in one direction only. In principle, the number of electron detectors positioned at different angles can be increased until they cover nearly the entire solid angle, but in the first experiment only one rotatable electron detector was used. In order to measure complete molecule-frame angular distributions one has to make many measurements under different angles at the same photon energy. Figure 2 shows a two-dimensional ARPEPICO spectrum of the C(1s) photoionization of CO at a photon energy of 320 eV.

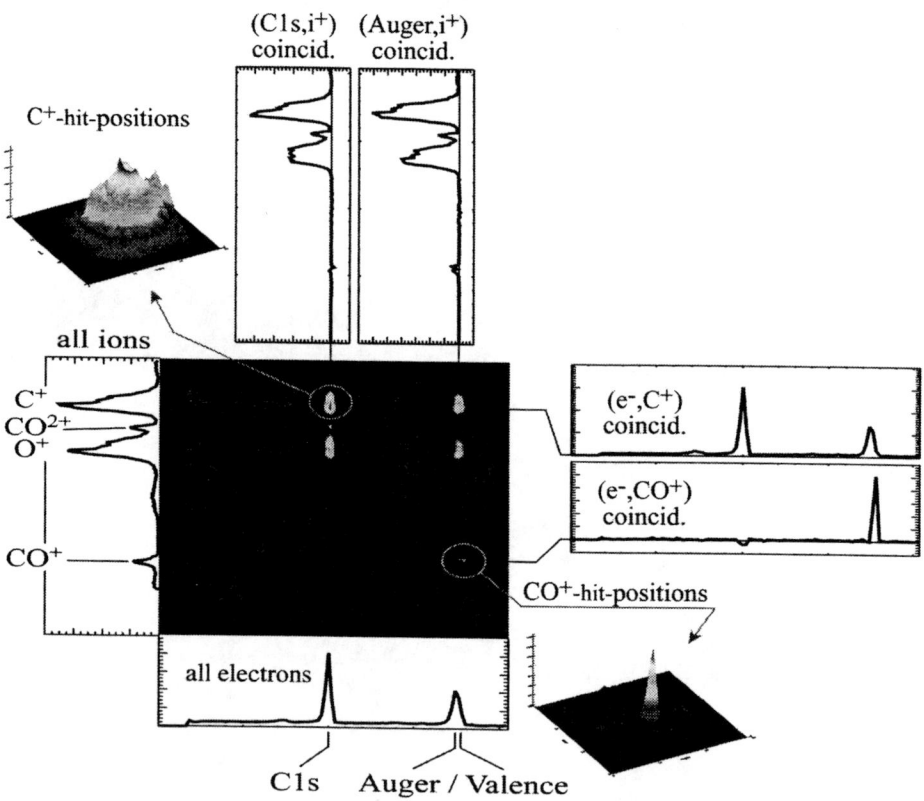

**FIGURE 2** Two-dimensional ion – electron coincidence spectrum for the C(1s) photoionization of CO. The intensities along the X-axis sum up to the electron spectra, those along the Y-axis to the fragment and molecular ion spectra. Selected coincidence rates are plotted above and to the right of the two-dimensional spectra. Intensity distributions on the position-sensitive anode are shown for two cases: [$C^+$, C(1s)] and [$CO^+$, valence].

A closer inspection of this figure shows that even the angle integrated coincident ion intensities exhibit a large forward–backward asymmetry of the electron emission into the direction of the different fragment ions. It illustrates that the intramolecular scattering of the photoelectron on its way out is a significant effect. In order to obtain the complete coincident photoelectron angular pattern of parallel and perpendicular transitions with respect to the electric vector one has to analyze spectra which are taken under a variety of electron ejection angles.

Figure 3 shows the corresponding angular patterns taken at a photon energy of $h\nu = 320$ eV in two polar diagrams. The parallel transition exhibits the most distinct forward-backward asymmetry. The intensity of the photoelectron emitted into the direction of the $C^+$ ions is three times higher than that into the direction of $O^+$. This is a clear indication for strong intramolecular scattering and interference. There is also some forward-backward asymmetry in the perpendicular transition but it is much less pronounced. Fitting an expansion of Legendre polynomials truncated at a certain value of the angular momentum $l$ of the partial waves results in principle in a complete set of matrix elements and phases.

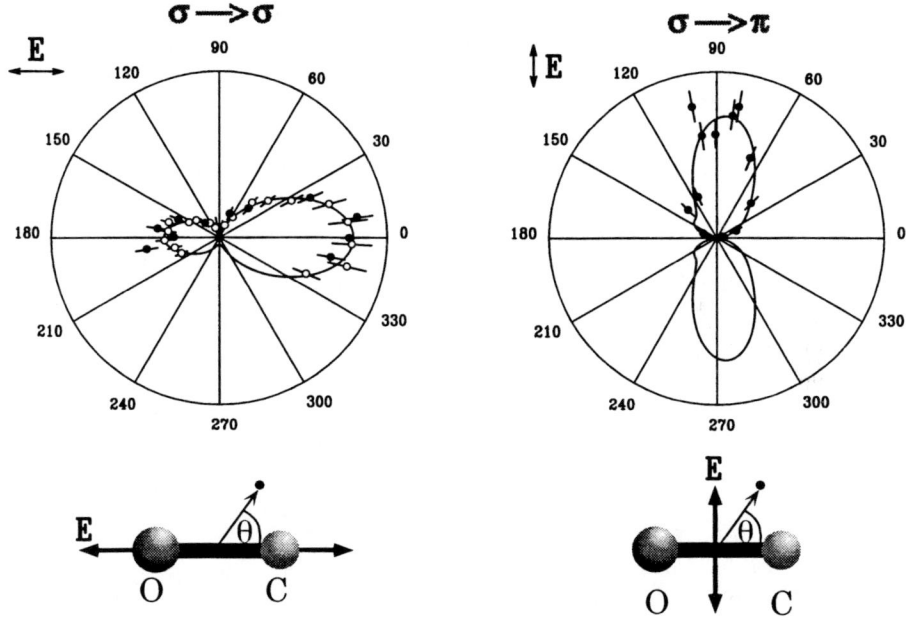

**FIGURE 3** Coincident photoelectron angular distribution for fragment ion ejection along and perpendicular to the electric vector. These two coincident angular patterns represent molecule-frame photoelectron angular distributions for the $\sigma \to \sigma$ and $\sigma \to \pi$ transitions of the C(1s) photoelectron ejection in CO. The closed symbols are from [12], the open ones from [13]. The solid curve illustrates a fit to the joint data set.

In atomic photoionization such a procedure yields unambiguous results, at least under the assumption of LS-coupled continuum waves [14]. This does not apply to molecular photoionization because the number of partial waves necessary for a consistent result is not well defined, not to say arbitrary. Hence, the experimental data have to be of very good quality to prove if expansions of different sizes converge to the same main components of the partial wave expansion. Nevertheless, the first corresponding experiments were performed and analyzed, and the results were compared with ab-initio calculations [15]. A still open question is the size of the phase shift between the $\sigma$ and $\pi$ components of the continuum wave. The phase shift is assumed to be zero outside of resonances, since there are no particular forces that could cause it. But in the vicinity of a $\sigma$-shape resonance the analysis of transitions other than purely parallel or perpendicular ones could be a sensitive probe. The data obtained from the position sensitive ion detector provide the information needed to explore the sensitivity of the various coincident angular patterns with respect to the dipole matrix elements and to examine their various relative phase shifts including the phase shift between continua of different molecular symmetries.

In addition to the dynamical aspects of molecular photoionization it is also possible to study structural aspects, in the simplest case the bond lengths of a diatomic molecule, via photoelectron scattering in the free molecules. This method known as photoelectron diffraction is frequently used to study the geometrical properties of surfaces and adsorbates see e. g. [16]. Here again as mentioned in the beginning, a fixed orientation and periodic structure of the substrate support the effectiveness of the method, but for randomly oriented targets the effect is washed out. Therefore, one has to use spatially oriented molecules as a target when looking for pronounced photoelectron diffraction effects. They have an optimum geometry for a diffraction experiment when all three quantization axes – electric vector, molecular axis and electron detection direction – are parallel. In this case, the largest amplitude should be found in the differential cross section modulation [17]. Due to the small bond lengths of small molecules the expected cross section modulations extend over a kinetic energy range of several hundreds of electron volts.

The use of a beamline behind a bending magnet is still a good choice for such an experiment although the overall photon flux is lower than that of insertion devices. Hence, the first experiment of this kind was performed at beamline 9.3.2 of the Advanced Light Source (ALS) in Berkeley. Figure 4 shows the first preliminary results [18].

There is indeed a large amplitude of the oscillation, even though the error bars are large due to the low coincident count rate. Compared to gas phase EXAFS a much greater sensitivity to relative changes in the differential cross sections can be expected. For a first fit of the data we used a program that takes diffraction and multiple scattering of the photoelectron on its way out into account. The reasonable result is represented by the solid curve in Figure 4. The derived bond lengths are in good agreement with the literature values for CO. This is a convincing demonstration of photoelectron scattering effects in free molecules. In future, the method may be applied in topological studies of excited molecules.

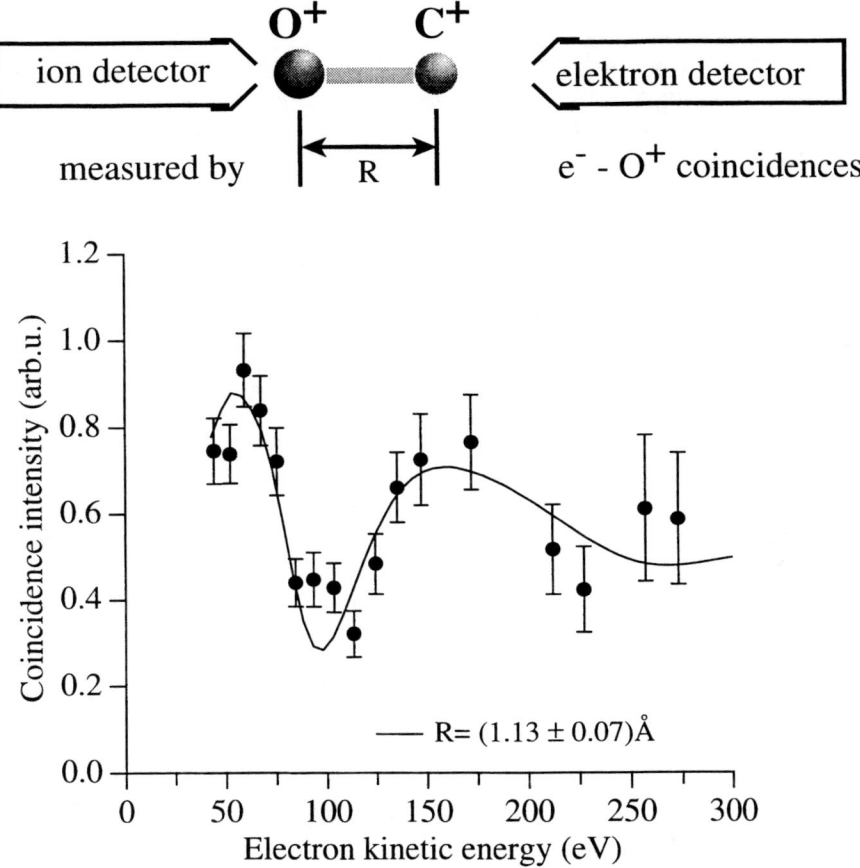

**FIGURE 4** Coincident photoelectron intensities for ejection angles parallel to the molecular axis and electric vector. The coincident intensities are normalized with respect to the non-coincident intensities, and plotted versus electron kinetic energy. The solid curve represents a fit of the data using a program that takes the diffraction of the photoelectron by the neighboring atom on its way out into account.

## VALENCE PHOTOELECTRON EMISSION

The scattering process of localized core electrons on the various atomic sides of its molecular surrounding is well understood, above all because of its numerous surface and solid state applications. In contrast, the photoemission of outer valence electrons is less well understood since these electrons are delocalized and form molecular orbitals in free molecules or a quasi-free electron gas in metals. Their photoemission properties rather reflect their momentum distribution in the solid than topological features. Valence photoemission from molecules should reflect the orbital structure of the valence electrons rather than its topology. Until very recently, intramolecular scattering effects have not been observed in valence photoemission.

In order to perform ARPEPICO experiments with valence shell electrons, one has to look for repulsive final states, which cause ionic fragmentation, because otherwise there are no fragment ions to be detected in coincidence with electrons. This precondition reduces the number of potential candidates for such studies. Significant examples of ionic fragmentation are Rydberg excited states in $O_2$. They can dissociate into ionic fragments by autoionization [19]. The low energy of the emitted autoionizing electrons suggests using a threshold electron detector instead of a regular electron time-of-flight detector, which has a typically small acceptance angle of less than 1 percent of the total solid angle. The advantage of the threshold electron detector is a $2\pi$ detection efficiency, its disadvantage a practically not existing angular resolution. It just distinguishes the electrons moving towards the detector from those travelling away from it. Because of the high detection efficiency in both the ion and electron detection channels this property is sufficient to reveal possible scattering effects shown by forward-backward asymmetries.

Figure 5 shows a scheme of an experimental arrangement (left side) together with two schematic spectra (right side). The electrons are moving upwards and are detected with a MCP, while the ions are moving downwards and hit the position sensitive detector. On the right side there is a low kinetic energy electron spectrum shown (top), together with the spectrum of the fragment ions (bottom). The electron peak with no particular designation has an energy of 0.16 eV ('zero-volt electrons') and corresponds to the autoionizing line in question. It is clearly split into a „forward" ($v^-$) and „backward" ($v^+$) peak with respect to the electron detector position, because the electrons travelling towards the electron detector will arrive earlier at the MCP than those travelling in the reverse direction. With this setup it is possible to measure all ionic fragments in coincidence with either forward or backward electrons.

Figure 6 visualizes that the process under investigation evolves over three stages. First (fig. 6a), a Rydberg electron is coherently excited on both atomic sites of the oxygen molecule. Then (fig. 6b), the excited molecule starts to dissociate with a fifty-to-fifty probability of having the Rydberg electron coherently on either side. Now, the autoionization destroys the coherence, thus localizing the emission side by production of an observable ion. On its way out the autoionizing electron may then be scattered by the other still neutral oxygen atom. When the dissociation is complete (fig. 6c) both, the fragment ion and the autoionizing electron travel toward their detectors. The measured coincident ion angular distributions are shown at the right end of figure 6c) [20]. These ion angular distributions for fixed electron emission direction are in case of an autoionizing electron equivalent to coincident electron angular distributions for fixed fragment ion directions.

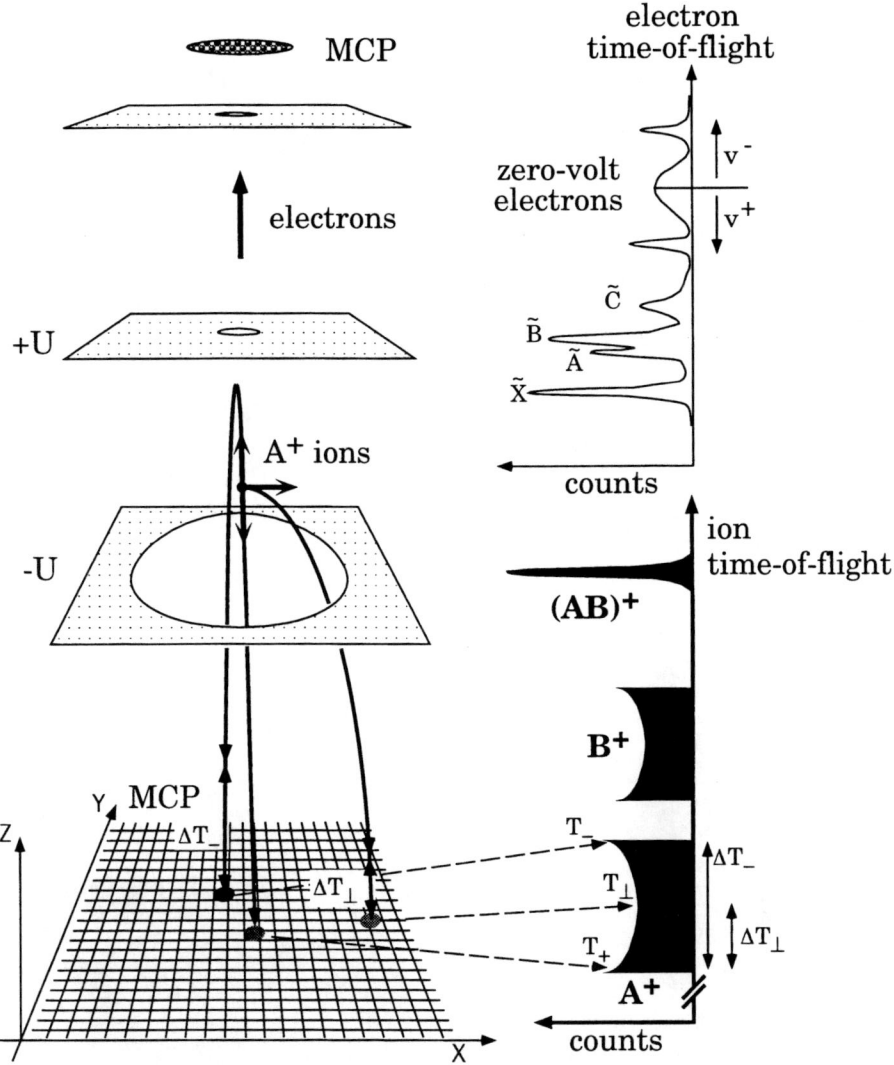

**FIGURE 5** Scheme of the experimental setup used for ARPEPICO measurements with low energy electrons. On the right side, there is a scheme of an electron time-of-flight and an ion time-of-flight spectrum, respectively. Note the particular line shapes of the fragment ions, which reveal information on the angular distribution already in the time domain.

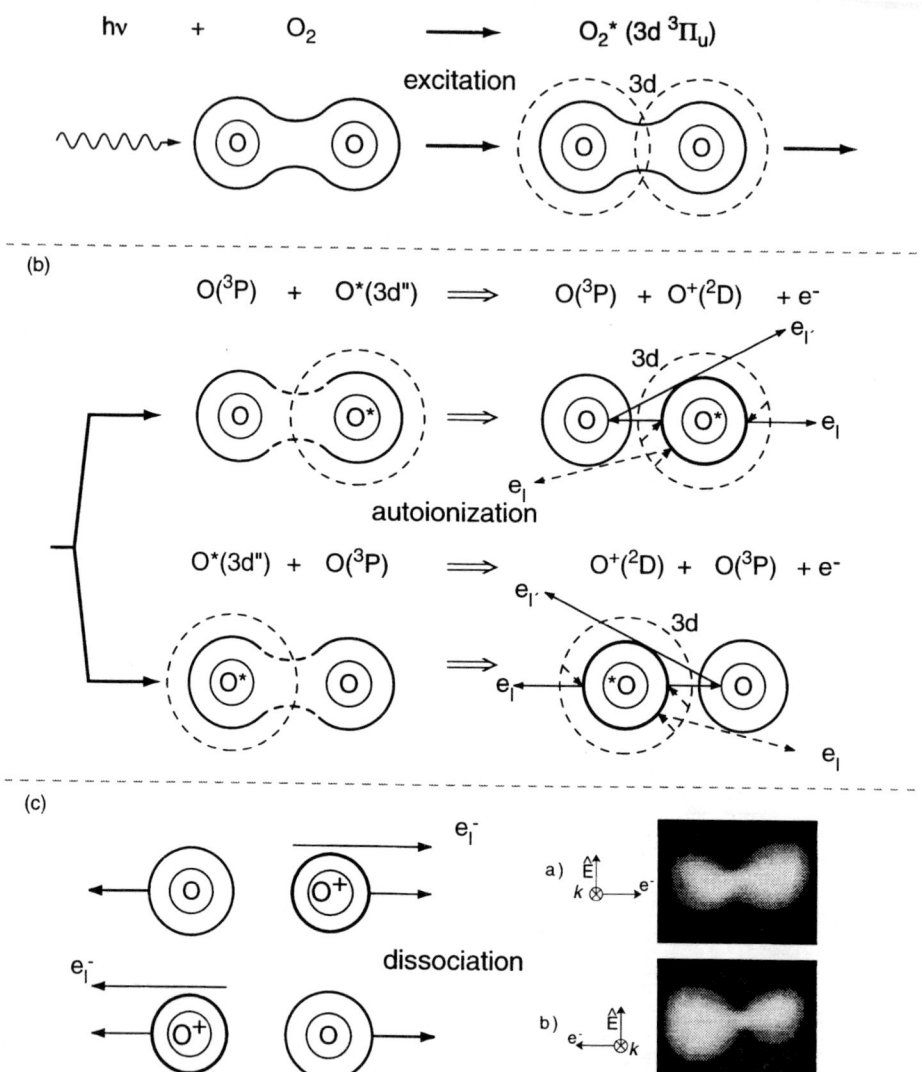

**FIGURE 6** a) Visualization of the coherent excitation of a 3d Rydberg electron at both atomic sides of $O_2$. b) Onset of the dissociation of the two atoms; both are equally predetermined to carry the Rydberg electron. During fragmentation, autoionization occurs on one side of the departing atoms. Backscattering from the other atoms causes more electrons to be ejected along with the opposite ionic fragment, if intramolecular scattering still plays a role. c) Schematic time evolution of the electron emission and fragmentation along with experimentally proved intensity distributions. The forward-backward asymmetries are clearly exhibited in the ionic fragment angular distribution for a fixed half-space electron emission direction.

They clearly reveal forward-backward asymmetries, which become reversed when electrons, travelling in the other direction, are selected from the threshold electron spectrum. For the first time, there is a clear indication of intramolecular scattering of electrons ejected from outer shells. Similar experiments were performed with NO using photoelectrons instead of autoionizing electrons. They, too, show asymmetries [21]. More interestingly however, when using circularly polarized light during the photoionization process, evidence for circular dichroism in the coincident angular pattern was seen for the first time very recently at the synchrotron radiation laboratory BESSY in Berlin [22].

These studies open a wide field for complete experiments in molecular photoionization, on the one hand, and for probing the coherence and localization properties of Rydberg excited molecules, on the other hand.

## SUMMARY

The recent development of new coincidence methods in molecular photoionization opened the door to angle resolved photoelectron spectroscopy on oriented molecules in the VUV and soft x-ray regimes. The angular patterns thus obtained provide information on the dynamics of molecular photoionization, particularly on dipole matrix elements and their relative phase shifts. Additionally, topological parameters e.g. bond lengths, as well as information on intramolecular scattering and interference phenomena may be obtained.

## ACKNOWLEDGEMENT

The author is indebted to the Deutsche Forschungsgemeinschaft for financial support. He likes to thank A. Hempelmann for critical reading of the manuscript.

## REFERENCES

[1]   Yang, C. N., Phys. Rev., **74**, 764 (1948).
[2]   Umbach, E. and Hussain, Z., Phys. Rev. Lett., **52**, 457 (1984).
[3]   Kaesdorf, S., Schönhense, G., and Heinzman, U., Phys. Rev. Lett., **54**, 885 (1985).
[4]   Reid, K. L., Leahy, D. J., and Zare, R. N., Phys. Rev. Lett., **68**, 3527 (1992).
[5]   Golovin, A. V., Opt. Spectrosc. (USSR), **71**, 537 (1991).
[6]   Saito, N. and Suzuki, I. H., Phys. Rev. Lett., **61**, 2740 (1988).
[7]   Yagishita, A., Maezawa, H., Ukai, M., and Shigemasa, E., Phys. Rev. Lett., **62**, 36 (1989).
[8]   Downie, P. and Powis, I., Phys. Rev. Lett., **82**, 2864 (1999).
[9]   Dehmer, J. L. and Dill, D., Phys. Rev. Lett., **35**, 213 (1975).
[10]  Dehmer, J. L. and Dill, D., J. Chem. Phys., **65**, 5327 (1976).
[11]  Shigemasa, E., Adachi, J., Oura, M., and Yagishita, A., Phys. Rev. Lett., **74**, 359 (1995).
[12]  Heiser, F., Geßner, O., Viefhaus, J., Wieliczek, K., Hentges, R., and Becker, U., Phys. Rev. Lett., **79**, 2435 (1997).

[13]  Shigemasa, E., Adachi, J., Soejima, K., Watanabe, N., Yagishita, A., and Cherepkov, N. A., Phys. Rev. Lett., **80**, 1622 (1998).
[14]  Becker, U., J. Electr. Spectrosc. Relat. Phenom., **96**, 105 (1998).
[15]  Motoki, S., Adachi, J., Hikosaka, Y., Sano, M., Shigemasa, E., Soejima, K., Yagishita, A., Ito, K., Raseev, G., and Cherepkov, N. A., *XXI-ICPEAC, Abstracts of Contributed Papers,* Sendai, Japan, 1999, p. 102.
[16]  Schaff, O. and Bradshaw, A. M., Phys. Bl., **52**, 997 (1996).
[17]  Gessner, O., Heiser, F., Cherepkov, N. A., Zimmermann, B., and Becker, U., J. Electr. Spectrosc. Relat. Phenom., **101-103**, 113 (1999).
[18]  Gessner, O., Heiser, F., Moler, E. J., Hussain, Z., Shirley, D. A., and Becker, U., *12th International Conference on Vacuum and Ultraviolet Radiation Physics, Program and Abstracts,* LBNL, San Francisco, 1998, p. We020.
[19]  Guyon, P., Golovin, A., Quayle, C., Vervloet, M., and Richard-Viard, M., Phys. Rev. Lett., **76**, 600 (1996).
[20]  Golovin, A. V., Heiser, F., Quayle, C. J. K., Morin, P., Simon, M., Geßner, O., Guyon, P. M., and Becker, U., Phys. Rev. Lett., **79**, 4554 (1997).
[21]  Eland, J. H. D. and Duerr, E. J., Chem. Phys., **229**, 1 (1998).
[22]  Gessner, O., Hempelmann, A., Guyon, P. M., and Becker, U., to be published.

# Oscillating Partial Cross Sections in $C_{60}$: Evidence for a Beating Frequency

A. Rüdel, R. Hentges, and U. Becker

*Fritz-Haber-Institut der Max-Planck-Gesellschaft, Faradayweg 4-6, D-14195 Berlin, Germany*

**Abstract.** The partial photoionization cross sections of $C_{60}$ exhibit strong modulations of the HOMO and HOMO-1 photoelectron line intensities over a wide range of photon energies. These modulations were interpreted as quantum oscillations resulting from the size of the short range potential in which the valence electrons are bound. More recent measurements revealed an additional modulation superimposed on the main oscillations. This low frequency modulation can be understood as a beating frequency reflecting the finite width of the spherical-shell like potential. First evidence for such a behavior is reported and discussed.

## INTRODUCTION

Strong oscillations of the photoelectron line intensities of the highest occupied molecular orbital (HOMO) and the next lower molecular orbital HOMO-1 were first observed for solid $C_{60}$. Large variations in the density of states of empty odd and even final states were suggested by P. Benning et al. as a possible explanation [1]. However, after the observation of such oscillations in the partial photoionization cross section of free $C_{60}$ molecules, an interpretation based on the specific geometry of the fullerenes, i.e., the nearly spherical cage structure of the $C_{60}$ molecule, seemed to be more appropriate. So far, very few experiments have been performed on this subject and the available data sets cover only a few selected energies (T. Liebsch et al. [2], D. Lichtenberger et al. [3], M. Biermann et al. [4]). Only recently a systematic study of the cross section oscillations in free $C_{60}$ and $C_{70}$ within the energy range from $h\nu = 19$ eV to $h\nu = 125$ eV has been completed by T. Liebsch et al. [5] and A. Rüdel et al. [6,7].

## EXPERIMENT

The experiments were performed at the synchrotron radiation facility HASYLAB at DESY in Hamburg under single bunch conditions using the electron time-of-flight method. $C_{60}$ molecules were evaporated by a resistively heated oven running at about 600°C and ionized by monochromatic synchrotron radiation from the undulator beamline BW3 equipped with a plane grating monochromator (SX 700). The time-of-flight photoelectron spectra were recorded at the magic angle, 54.7° with respect to the polarization vector of the synchrotron light. In order to calibrate the photon energy of the monochromator, we recorded Ne $2s$, $2p$ photoelectron spectra in the valence ionization region and total-yield electron spectra of CO and $N_2$ in the core ionization region. Further details about the experimental setup are given in U. Becker et al. [8].

## RESULTS AND DISCUSSION

The branching ratios of the HOMO and HOMO-1 line intensities are compared with the result of a semiempirical calculation based on the following simplified model by Y. B. Xu et al. [9]: The observed intensity oscillations can be visualized by the formation of "standing spherical waves" of the final state electron inside the hollow fullerene molecule depending on the wave number and thus on the final state energy. One can imagine the skeleton of a fullerene molecule as a three-dimensional quantum sphere formed by the initial wave functions of delocalized valence electrons distributed at this shell with its relatively sharp boundaries. The "standing waves" are the eigenstates of the three-dimensional box potential giving rise to "quantum

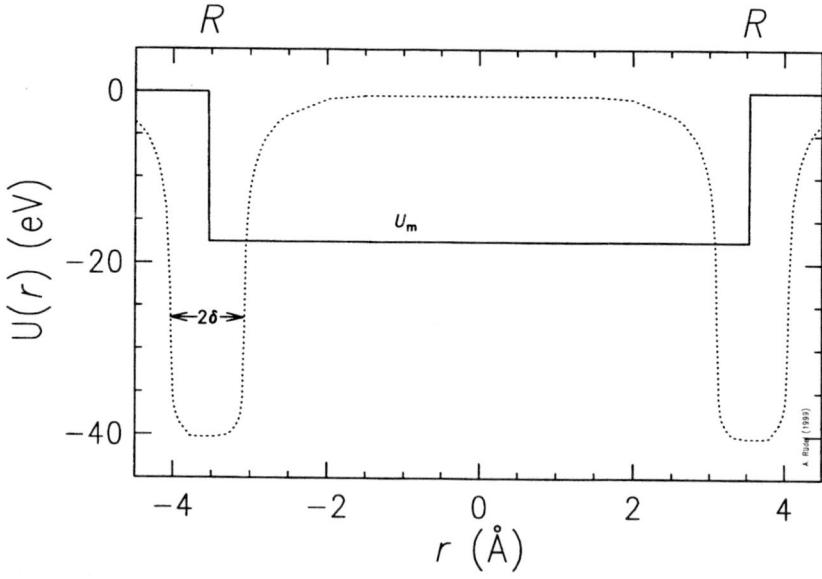

**FIGURE 1.** Simplified potential energy of the initial state electron as a function of the radial coordinate. The solid line represents a simple box potential with a radius of $R = 0.354$ nm and a potential depth of $U_m = 17.5$ eV. The dotted line shows a more realistic jellium potential of M. Puska and N. Nieminen [10].

oscillations" which are equivalent to intensity variations observed in photoelectron diffraction. Figure 1 shows a two-dimensional representation of such a box potential with a radius of $R = 0.354$ nm and a potential depth of $U_m = 17.5$ eV. This potential is equivalent to a jellium potential of a hollow sphere with a thickness of $2\delta = 0.015$ nm and a depth of approximately $U_m = 41$ eV (as given, for example, by M. Puska and N. Nieminen [10]) because in both potentials the electron has the same kinetic energy before leaving the molecule. The semiempirical curves for the case of $C_{60}$ derived from this model agree very well with experimental data as shown in figure 2. Similar results have been obtained for the case of $C_{70}$.

Further measurements on both gas phase and solid state samples by S. Hasegawa et al. [11] corroborated the strong oscillations in the two outermost valence orbitals of $C_{60}$. In fact these oscillations are not restricted to these two orbitals only, rather they occur in all other photolines as well. However, in groups of many unresolved lines the corresponding oscillations from lines with alternating odd and even angular momenta tend to cancel each other's modulations. Therefore our extensive study concentrated on the behavior of the HOMO and HOMO-1 lines and in particular their branching ratio, which is more sensitive for the observed oscillations because no photon flux calibration is required for the branching ratio measurements.

In the discussion of the interpretation and origin of the observed oscillations a new aspect was brought in by O. Frank and J.-M. Rost [12]. They argued that

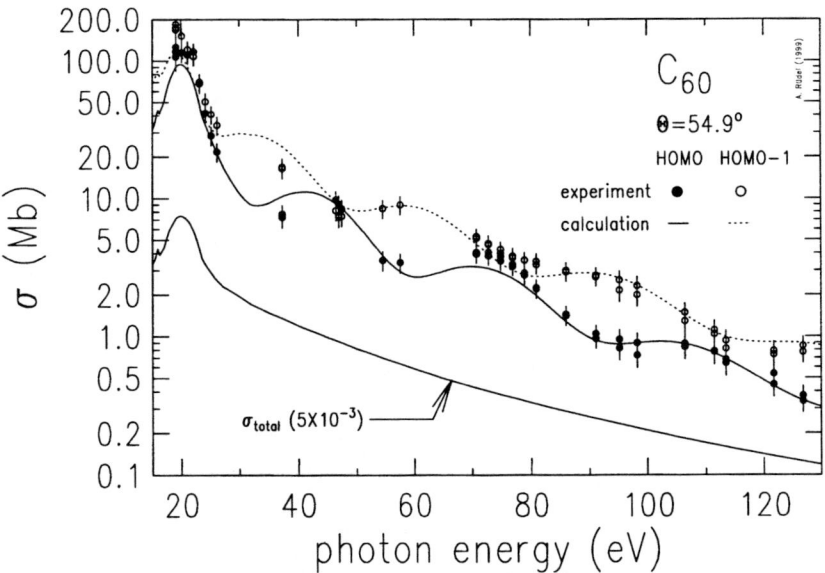

**FIGURE 2.** Partial cross sections of the highest occupied molecular orbitals HOMO and HOMO-1 of $C_{60}$ with a total photoionization cross section curve normalized by the sum rule of Thomas-Reiche-Kuhn.

the model proposed by Y. B. Xu et al. [9] may have to be extended by the effect of the finite width of the spherical shell potential on the cross section oscillations. Frank and Rost substantiate their proposal by an explicit calculation using the potential of M. Puska and N. Nieminen [10]. The calculation shows a pronounced beating frequency superimposed on the basic modulation. The effect of this beating frequency on the partial cross section behavior was very prominent because the amplitudes of the oscillations were taken to be constant. In reality the oscillations are on a nonoscillating background and are damped towards higher energies. Figure 3 shows the branching ratio from a variety of measurements, including solid state results reproduced as a shaded area at lower energies and as symbols at higher energies. The tendency of damping in the amplitudes of the oscillations is clearly exhibited by the experimental data. However, the scarce data points at higher

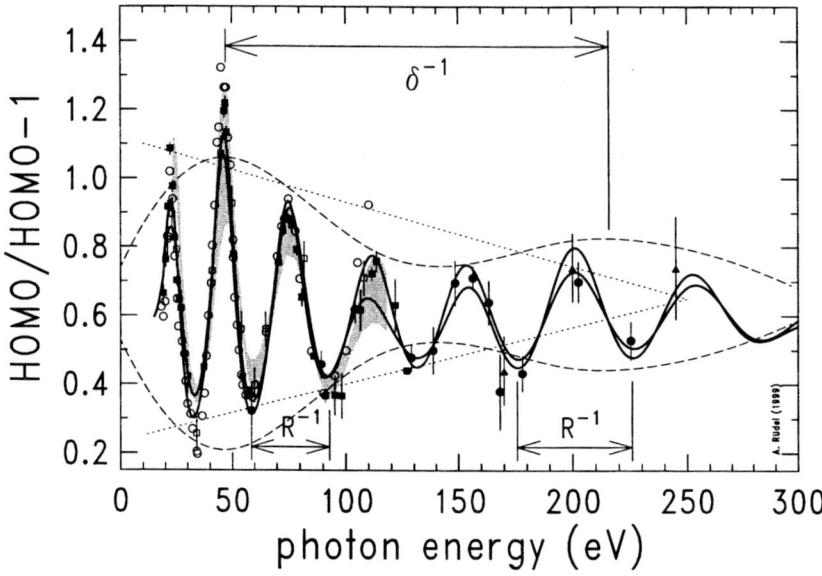

**FIGURE 3.** The branching ratio between the two partial cross sections exhibits more pronounced oscillations due to the alternating modulation in the different partial cross sections. The experimental data are from T. Liebsch et al. [2] (open rectangles), A. Rüdel et al. [6,7] (filled rectangles, filled circles), D. Lichtenberger et al. [3] (filled triangles), M. Biermann et al. [4] (filled triangles) and S. Hasegawa et al. [11] (open circles). The two solid lines represents two fits with different parameter settings through the data using an extension of model 'B' of Y. B. Xu et al. [9]. The shaded area indicates a variety of solid state measurements, particularly the data of P. Benning et al. [1]. The dotted lines represent a linear damping of the oscillations whereas the dashed curves show the behavior of a possible beating frequency. While the basic oscillation is related tp the radius $R$ of the spherical shell, the superimposed low frequency modulation depends on the finite width $2\delta$ of the potential.

energies did not allow a conclusive decision concerning the existence of a beating frequency in the partial cross section behavior.

For this purpose we have performed extensive partial cross section and branching ratio measurements during two beamtimes at the undulator beamline BW3 at HASYLAB (A. Rüdel et al. [13]). The results confirm the existence of such a frequency. However, their effect is much reduced compared to the theoretical prediction due to the damping of the main oscillation and the existence of an incoherent contribution which is not affected by the beating frequency. The preliminary results of these measurements (not shown in the figure) are consistent with the dashed curve plotted together with the former data.

## ACKNOWLEDGEMENT

This work was supported by the BMBF and the DPG. The authors would like to thank Thomas Möller for his excellent support during the beamtimes.

## REFERENCES

1. Benning P., Poirier D., Troullier N., Martins J., Weaver J., Haufler R., Chibante L., and Smalley R., *Phys. Rev. B* **44**, 1962 (1991).
2. Liebsch T., Plotzke O., Heiser F., Hergenhahn U., Hemmers O., Wehlitz R., Viefhaus J., Langer B., Whitfield S. B., and Becker U., *Phys. Rev. A* **52**, 457 (1995).
3. Lichtenberger D., Jatcko M., Nebesny K., Ray C., Huffman D., and Lamb L., in *Cluster and Cluster-Assembled Materials*, Vol. 206, edited by Averback R., Nelson D., and Bernholc J. (Mat. Res. Soc., Pittsburgh, 1991), p. 673.
4. Biermann M., Neeb M., Johnen F., and Krummacher S., in *Proceedings of ECS Symposium on "Fullerenes: Chemistry, Physics and New Directions VI"*, Vol. 94-24, edited by Kadish K. M. and Ruoff R. S. (The Electrochemical Society, Inc., San Francisco, 1994), p. 952.
5. Liebsch T., Hentges R., Rüdel A., Viefhaus J., Becker U., and Schlögl R., *Chem. Phys. Lett.* **279**, 197 (1997).
6. Rüdel A., Hempelmann A., Hergenhahn U., Prümper G., Viefhaus J., Liebsch T., and Becker U., HASYLAB at DESY annual report p. 231 (1996).
7. Rüdel A., Hentges R., Hergenhahn U., Kempgens B., and Becker U., HASYLAB at DESY annual report p. 197 (1997).
8. Becker U., Szostak D., Kerkhoff H., Kupsch M., Langer B., Wehlitz R., Yagishita A., and Hayashi T., *Phys. Rev. A* **39**, 3902 (1989).
9. Xu Y. B., Tan M. Q., and Becker U., *Phys. Rev. Lett.* **76**, 3538 (1996).
10. Puska M. and Nieminen N., *Phys. Rev. A* **47**, 1181 (1993).
11. Hasegawa S., Miyamae T., Yakushi K., Inokuchi H., Seki K., and Ueno N., *Phys. Rev. B* **58**, 4927 (1998).
12. Frank O. and Rost J.-M., *Chem. Phys. Lett.* **271**, 367 (1997).
13. Rüdel A., Hentges R., and Becker U., HASYLAB at DESY annual report p. 48 (1998).

# Beyond the Dipole Approximation: Angular-Distribution Effects in the 1s Photoemission from Small Molecules

O. A. Hemmers, H. Wang, D. W. Lindle, P. Focke,[a] I. A. Sellin,[a] J. D. Mills,[b] J. A. Sheehy,[b] and P. W. Langhoff[c]

*Department of Chemistry, University of Nevada, Las Vegas, Nevada, 89154-4003, USA*
[a]*Department of Physics, University of Tennessee, Knoxville, TN 37996*
[b]*Air Force Research Laboratory, AFRL/PRS, Edwards, AFB, CA 93524-7680*
[c]*Department of Chemistry, Indiana University, Bloomington, IN 47405*

**Abstract.** Over the past two decades, the dipole approximation has facilitated a basic understanding of the photoionization process in atoms and molecules. Recent experiments on the 1s inner shells of small molecules at relatively low photon energies ($\leq 1000$ eV) show strong nondipole effects. They are significant and measurable at energies close to threshold, in conflict with a common assumption that the dipole approximation is valid for photon energies below 1 keV.

## INTRODUCTION

The electric-dipole (**E1**) approximation [1], applied to photoionization, leads to the well-known expression for the differential cross section [2],

$$\frac{d\sigma}{d\Omega} = \frac{\sigma}{4\pi}\left[1 + \frac{\beta}{2}\left(3\cos^2\theta - 1\right)\right] \tag{1}$$

which describes the angular distribution of photoelectrons from a randomly oriented sample created by 100% linearly polarized light. Here, $\sigma$ is the partial photoionization cross section, and $\theta$ is the angle between the vector of the outgoing electron and the vector of linear polarization. The parameter $\beta$ completely describes the angular distribution of photoelectrons, within the dipole approximation. In this approximation, all higher-order interactions, such as electric-quadrupole (**E2**) and magnetic-dipole (**M1**), are neglected. This assumption is justified by the argument that the strengths of the E2 and M1 interactions relative to electric-dipole effects are approximately equal to the ratio of the photoelectron's velocity to the speed of light [3], a ratio which is small except at very high energies.

Over the past two decades, the dipole approximation has facilitated a basic understanding of the photoionization process in atoms and molecules [2], as well as the application of photoelectron spectroscopy to a wide variety of condensed-phase systems. The first hint of deviations from the dipole approximation was provided by Krause [4] in measurements using unpolarized x-rays [5]. A small deviation from the expected dipolar angular distribution at photon energies between 1 and 2 keV was observed and attributed to the influence of **E2** and **M1** interactions. These lowest-order, non-electric-dipole corrections to the dipole approximation lead to so-called *nondipole* effects in the angular distributions of photoelectrons, described by [6]

$$\frac{d\sigma}{d\Omega} = \frac{\sigma}{4\pi}\left[1 + \frac{\beta}{2}(3\cos^2\theta - 1) + (\delta + \gamma\cos^2\theta)\sin\theta\cos\phi\right] \quad (2)$$

for 100% linearly polarized light. The nondipole angular-distribution parameters $\gamma$ and $\delta$ are attributable to interference terms between electric-dipole and electric-quadrupole interactions. Figure 1 describes the geometry and the angles $\theta$ and $\phi$.

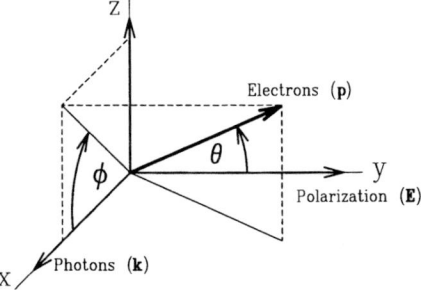

**Figure 1.** Geometry applicable to photoelectron angular-distribution measurements using polarized light. $\theta$ is the polar angle between the photon polarization vector $\varepsilon$ and the momentum vector **p** of the photoelectron. $\phi$ is the azimuthal angle defined by the photon propagation vector **k** and the projection of **p** into the x-z plane.

More-recent measurements [7,8], focussing on noble-gas core levels (Ar $K$ and Kr $L$) and photon energies above 2 keV, have begun to investigate nondipole effects in photoelectron angular distributions in more detail. In contrast, the present experiment concentrates on the $N_2$ $N1s$ and CO $C1s$ inner shells at relatively low photon energies (300 to 700 eV). Nondipole effects are observed to be large and highly energy dependent in this region, especially close to core-level thresholds, in conflict with a common assumption in applications of photoelectron spectroscopy; namely, that the dipole approximation is valid for photon energies below 1 keV. The potential significance of these findings is nicely illustrated by comparison of the present results for the $N_2$ and the CO $\gamma_{1s}$ parameters with theories for atomic nitrogen and atomic carbon [9], where the influence of nondipole effects are expected to be much smaller.

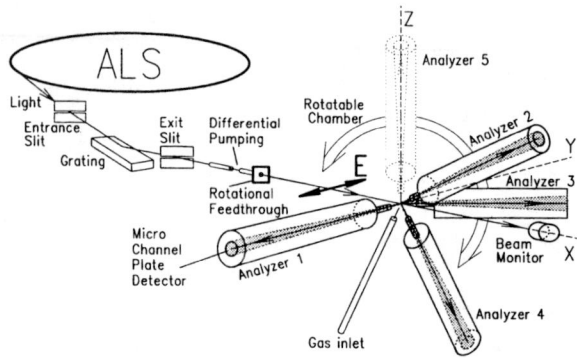

**Figure 2.** Experimental schematic of the electron time-of-flight system. Light from the ALS storage ring passes through beamline optics into a differential-pumping section. The chamber and analyzers can rotate around the photon beam for more accurate electron angular-distribution measurements.

## EXPERIMENT

The experiments were performed on undulator beamline 8.0, [10], which covers the 100-1500 eV photon-energy range. The monochromator entrance slit was set to 70 μm and the exit slit to 100 μm yielding very high flux, because high photon resolution was not needed. During the measurements the ALS operated at 1.9 GeV in two-bunch mode with a photon pulse every 328 ns. Four time-of-flight (TOF) electron analyzers, equipped with microchannel plates for electron detection, collect spectra simultaneously at different angles. The total electron flight paths are 437.5 mm, and the analyzers have a full cone acceptance angle of 5.4°.

The interaction region is formed by an effusive gas jet intersecting the photon beam which has a diameter of less than 1 mm. Energy resolution of the TOF analyzers with a focus size of 1 mm is 1% of the electron kinetic energy. Each spectrum was collected for about 600 s. The gas samples were obtained either commercially (CO) or directly from ambient air ($N_2$). A mixture of the sample with xenon was used sometimes because Xe has an abundance of Auger lines below 100 eV kinetic energy, which provide excellent internal calibration for each spectrum.

## RESULTS

Figure 3 shows two superimposed spectra, both taken at the magic angle θ=54.7°, but at different φ angles. The spectra were measured close to the C1s threshold (296 eV) and are scaled to the area of the Xe NOO Auger lines and the obvious intensity differences between the CO C1s peaks in the two spectra are due entirely to nondipole

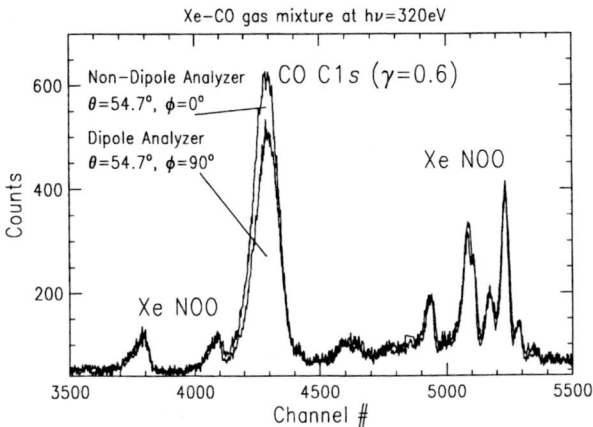

**Figure 3.** Photoelectron spectra of a CO-Xe mixture measured at a photon energy of 320 eV. One spectrum was taken with the dipole magic-angle analyzer and the other spectrum with the nondipole analyzer. The spectra are normalized to the Xe NOO Auger lines. The intensity differences in the CO C1s lines between the two analyzers is due entirely to nondipole effects.

effects because both spectra are at the magic angle where the β parameter has no influence.

For the dipole magic-angle analyzer the differential cross section in Eq. (2) reduces to the partial cross section; **E2** and **M1** effects vanish in the $\phi=90°$ plane even if relativistic effects are included [11]. For the nondipole analyzer,

$$\frac{d\sigma}{d\Omega} = \frac{\sigma}{4\pi}\left[1+\sqrt{\frac{2}{27}}(3\delta+\gamma)\right] = \frac{\sigma}{4\pi}\left[1+\sqrt{\frac{2}{27}}\zeta\right] \tag{3}$$

which simplifies further for $s$ subshells [6,12] in the non-relativistic approach where $\delta$ vanishes. We are using $\zeta=3\delta+\gamma$ for measurements that don't resolve the $\delta$ and $\gamma$ parameters of the angular distributions. In the case of molecular effects it is not clear if the $\delta$ parameter for $s$-shells vanishes near threshold.

With our experimental geometry, it is possible to measure the $\zeta$ parameter for $s$ subshells directly, if the degree of linear polarization is known, by using the two magic angle analyzers. The data points for CO and $N_2$ in Figures 4 and 5 show strong nondipole contributions with maxima of $\zeta=1.2$.

The difference between CO and $N_2$ lies in the position of the maxima. For $N_2$ the maximum is about 60 eV above the $N_2$ 1s ionization threshold much higher than the maximum of the dipole shape resonance at about 420 eV. The maximum of the $\zeta$ parameter for the CO C1s is close to the maximum of the dipole shape resonance at 305 eV.

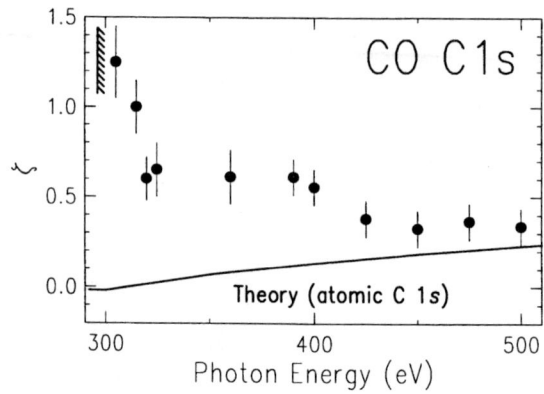

**Figure 4.** Electron angular anisotropy parameter $\zeta$ for the CO C1s photoline from threshold to hv= 500 eV. The theoretical curve for atomic carbon is from Lajohn and Pratt [9].

A qualitative explanation for the behavior of $\zeta$ can be obtained from the following model. Just as molecular $\beta$ values can change rapidly with photon energy for large differences in polarization components for ionization along and perpendicular to a molecular axis (due to a resonance, for example), so also $\zeta$ values can behave similarly but with greater sensitivity to the difference in polarization components because of the higher power of the transition moment coordinate involved. Thus, the observed molecular $\zeta$ effects may be universal.

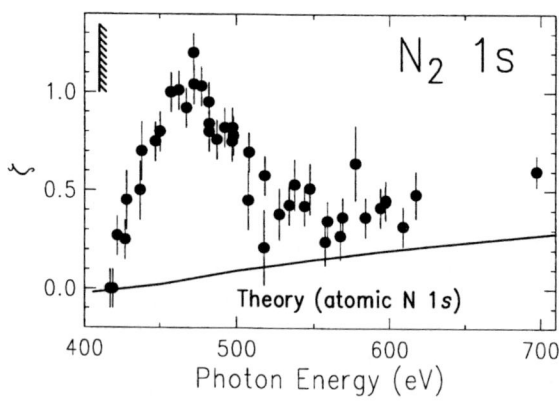

**Figure 5.** Electron angular anisotropy parameter $\zeta$ for the $N_2$ 1s photoline from threshold to hv= 700 eV. The theoretical curve for atomic nitrogen is from Lajohn and Pratt [9].

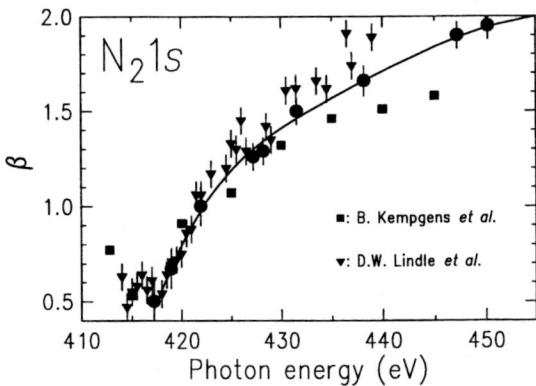

**Figure 6.** Electron angular anisotropy parameter β for the $N_2$ 1s photoline from threshold to hv= 450 eV. Measurements by Kempgens [14] and Lindle [15] did not take non-dipolar effects into account and deviate from our measurements.

The variations of β (and ζ) with photon energy in atoms are due to the interference of different partial waves ($p \rightarrow s$ and $d$, for example) whereas in molecules this can be due to the interference of the polarization components (1s → $p\sigma$ and $p\pi$ in $N_2$, for example).

One consequence of these strong molecular nondipole effects near threshold is the possibility of influences on previous measurements of β-parameters, as demonstrated in Fig. 6. If the β-parameter is not measured in the plane perpendicular to the direction of the light (and linear polarized light) the intensities used to determine β are influenced by the forward/backward intensities of ζ. These intensities reduce or increase the β values as shown in Fig. 6 for Kempgens and Lindle. Larger values of ζ lead to larger deviations in β.

The present results illustrate that any photoemission experiment, whether on gases, solids, or surfaces, can be influenced by nondipole effects at relatively low photon energies, pointing to a general need for caution in interpreting angle-resolved photoemission data.

## ACKNOWLEDGMENTS

The authors thank the staff of the ALS and the IBM, LBNL, LLNL, the University of Tennessee, and Tulane University collaboration for their support. This research is funded by the NSF (PHY-9303915), the DOE Nevada EPSCoR. The ALS is supported by the U.S. DOE through the Materials Science Division, Office of Basic Energy Sciences, Office of Energy Research at the Lawrence Berkeley National Laboratory under contract No. DE-AC03-76SF00098.

# REFERENCES

1. Bethe, H. A. and Salpeter, E. E., *Quantum Mechanics of One- and Two-Electron Atoms*, Berlin: Springer-Verlag, 1957.
2. Manson, S. T., and Dill, D., *Electron Spectroscopy: Theory, Techniques, and Applications*, New York: Academic, 1978, Vol. 2, edited by Brundle, C. R. and Baker, A. D. (Academic, New York, 1978).
3. Cooper, J. and Zare, R. N., *J. Chem. Phys.* **48**, 942 (1968).
4. Krause, M. O., *Phys. Rev.* **177**, 151 (1969).
5. For unpolarized incident light, $\beta/2$ is replaced by $\beta/4$ in Eq. (1), and $\theta$ is measured between the propagation vectors of the photon and the photoelectron. Otherwise, the essential physics is the same.
6. Cooper, J. W., *Phys. Rev. A* **42**, 6942 (1990); **45**, 3362 (1992); **47**, 1841 (1993).
7. Krässig, B., Jung, M., Gemmell, D. S., Kanter, E. P., LeBrun, T., Southworth, S. H., and Young, L., *Phys. Rev. Lett.* **75**, 4736 (1995).
8. Jung, M., Krässig, B., Gemmell, D. S., Kanter, E. P., LeBrun, T., Southworth, S. H., and Young, L., *Phys. Rev. A* **54**, 2127 (1996).
9. Lajohn, L. and Pratt, R. H., (private communication).
10. Perera, R. C. C., *Nucl. Instrum. Methods* **A319**, 277 (1992).
11. Scofield, J. H., *Phys. Rev. A* **40**, 3054 (1989); *Phys. Scripta* **41**, 59 (1990).
12. Amusia, M. Ya., Arifov, P. U., Baltenkov, A. S., Grinberg, A. A., and Shapiro, S. G., *Phys. Lett.* **47A**, 66 (1974); Amusia, M. Ya., Baltenkov, A. S., Grinberg, A. A., and Shapiro, S. G., *Sov. Phys.-JETP* **41**, 14 (1975); Amusia, M. Ya. and Cherepkov, N. A., *Case Studies in Atomic Physics*, Amsterdam: North-Holland, 1975, Vol. 5.
13. Hemmers, O., Whitfield, S. B., Glans, P., Wang, H., Lindle, D. W., Wehlitz, R., and Sellin, I. A., *Rev. Sci. Instrum.* **69**, 3809 (1998).
14. Kempgens, B., Kivimäki, A., Neeb, M., Köppe, H.M., Bradshaw, A.M., and Feldhaus, J., *J. Phys. B* 29 (1996).
15. Lindle, D. W., Truesdale, C.M., Kobrin, P.H., Ferrett, T.A., Heimann, P.A., Becker, U., Kerkhoff, H.G., and Shirley, D.A., *J. Chem. Phys.* 81 (1984).

# IV. SOLIDS AND SURFACES

# Atomic Effects seen in Solid Phases

B. Sonntag

*II. Institut für Experimentalphysik der Universität Hamburg, Luruper Chaussee 149,
22761 Hamburg, Germany*

**Abstract.** The importance of atomic effects in solids is demonstrated by the multiple inner-shell excitations, the threshold dynamics of Auger lines, the multiplet and lifetime effects in inner-shell absorption and photoelectron spectra and the dichroism in inner-shell photoelectron spectra. The comparison of the solid state spectra with the corresponding spectra of the free atoms is emphasized.

## INTRODUCTION

Atoms and ions are the basic building blocks of solids. Their properties determine the geometric and electronic structure of solids. An a priori calculation of the mechanical, thermal, electrical, magnetical and optical properties of solids solely based on the properties of the atomic/ ionic constituents still poses a formidable task. Solids are many-body systems and the interaction between and within the atoms/ ions can only be described by models based on various approximations ranging from single particle models to self-consistent many-body approaches. Experimental data are crucial for the development and testing of these models. Photons are a powerful means for probing solid state properties. In this article we will concentrate on photoabsorption, photonemission and electronemission processes in which transitions from or between inner shells of atoms/ions are involved. These core spectra, due to the characteristic atomic/ ionic absorption edges, photon and electron emission lines are element specific probes. The often considerable changes of the core spectra, brought about by the interaction with the other atoms/ ions in the solid, contain detailed information on the electronic and geometric solid state structure. In order to extract the information from the spectra one has to cope with both, the local intraatomic/ ionic – and the non-local, interatomic/ ionic interactions. Especially in transition metal – and rare earths compounds the balance between intraatomic/ ionic and interatomic/ ionic effects is very delicate. Giant resonances, multiplet splitting and term dependent lifetime broadening for example are clear atomic/ ionic signatures. Besides relying on models, like the modified local density models, the Anderson impurity model or the ligand field multiplet model there is a very attractive experimental method to assess the relative importance of intraatomic/ ionic and interatomic/ ionic interactions. This approach, based on the comparison of the spectra of free atoms/ ions with the corresponding spectra of atoms/ ions bound in solids or on surfaces will be the focus of the following paragraphs.

## ABSORPTION

Except for the energy region close to inner shell thresholds the gross features of solid state absorption spectra are well described by atomic models taking many electron correlation and relativistic effects into account. The spectra display the giant resonances, the Cooper minima and the smooth decrease of the photoionization cross section towards higher photon energies encountered in the corresponding spectra of the free atoms (see e. g. 1 – 4 and references therein). Since the core states are confined within a very narrow region around the nucleus the atomic character of these states is preserved. This also holds for the innermost part of the continuum state wave function mainly accessed by transitions from core levels. Solid state effects give rise to the x-ray absorption fine structure (XAFS) superimposed on the atomic cross section (5, 6).

## Multielectron excitations in inner-shell photoabsorption spectra

The simultaneous excitations of a deep inner-shell electron and an outer shell electron, e. g. Ar 1s3p; Kr 1s4p, 1s3d; Xe 2p4d; 2p5p in free atoms have been studied in great detail experimentally and theoretically (7 – 9). For solids it is often difficult to separate the multielectron photoexcitations (MPE) from the x-ray absorption fine structure (XAFS). Recently Gomilsek et al. (10) succeeded to clearly identify the 1s3p MPE in the x-ray absorption spectra of 4p elements by combining the information from different compounds of the same element and by studying the systematic changes of the MPE along a series of neighbouring
elements. The Ge 1s3p MPE spectra of various Ge compounds displayed in Figure 1 have been obtained by subtracting the XAFS signal and the smooth atomic background from the experimental spectra.

The MPE spectra given in Figure 2 for the series of elements from Ga to Kr have been determined in the same way. The good agreement of the MPE spectra of various Ge compounds (Figure 1) and the systematic change of the MPE spectra along the series of elements Ga-Kr corroborate the interpretation. The spectra can be well described by theoretical atomic cross sections.

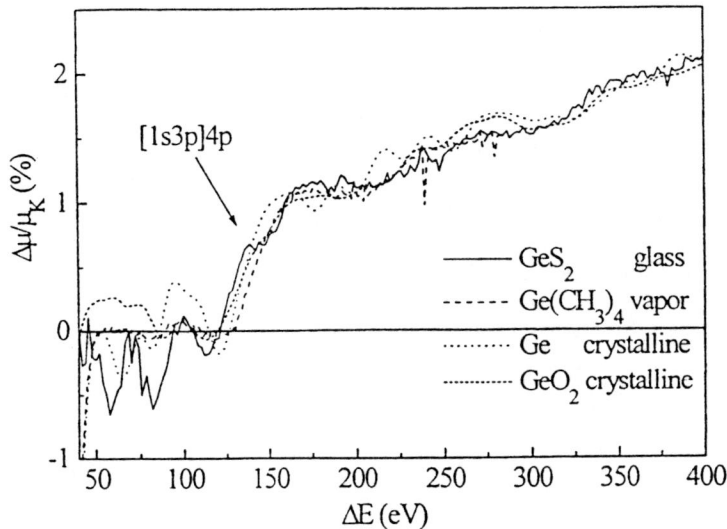

**FIGURE 1.** The 1s3p MPE contribution to the Ge photoabsorption of different Ge samples. The origin of the energy scale is set at the 1s threshold (from Ref. 10).

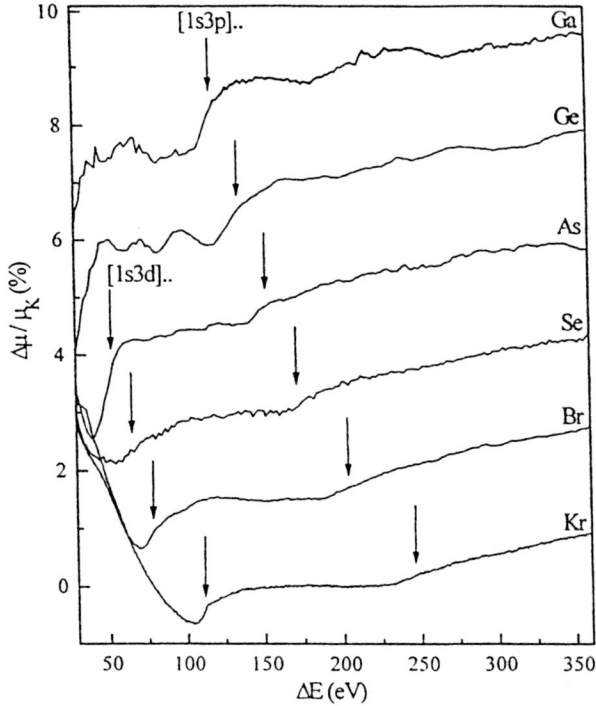

**FIGURE 2.** The contribution of the 1s3p MPE to the photoabsorption of the elements from Ga to Kr. The HF energy estimates of the lowermost resonance are indicated by arrows. For the elements from As to Kr the 1s3d MPE contribution is also resolved. A relative energy scale as in Figure 1 is used (from Ref. 10).

# Double K-vacancy production by x-ray photoionization

Multiple photoexcitations are caused by electron-electron interaction and therefore the observation of these processes serves as an extremely sensitive probe of electron correlation. The double photoionization of atomic He has been the focus of a great number of experimental and theoretical studies (11 – 14 and references therein). Recently Kanter et al. (15) succeeded to observe the double K-vacancy production in x-ray photoionization of molybdenum. Irradiating a thin Mo film by powerful 50 keV undulator radiation they were able to detect the $K^{-2} \rightarrow K^{-1} L^{-1}$ x-ray emission hypersatellites (16). The K-fluorescence emitted normal to the x-ray beam was registered by two Si (Li) detectors facing each other. The K-fluorescence spectra in each detector for the K-x-rays detected in coincidence with K-x-rays in the other detector are shown in Figure 3. For the "delayed" accidental coincidences the spectra display the normal Mo $K_{\alpha,\beta}$ diagram lines at 17,5 and 19,6 keV. The prompt coincidence spectra show a clear excess on the high-energy sides of each peak corresponding to the hypersatellite emission. The hypersatellites, in agreement with theoretical prediction (17) are shifted up in energy by 465 and 594 eV respectively.

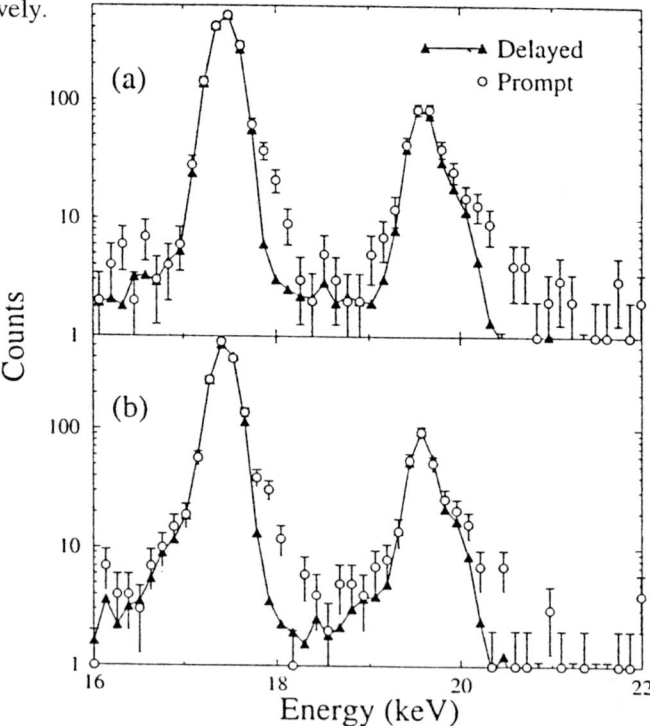

**FIGURE 3.** Energy spectra in each detector for coincidences with $K_{\alpha,\beta}$ x-rays in the other detector Spectrum for detector "1" is in (a) while that for "2" is in (b). Incident photon energy was 50 KeV. Data points (open circles) are for prompt coincidences while the filled triangles connected with the solid line correspond to delayed coincidences (from Ref. 15).

Based on the approximation that the fluorescence yield for the Mo K hypersatellites is the same as for the K-diagram lines (18) Kanter et al. determined for the ratio R of double/single ionization a value $R = 3{,}4\ (6) \cdot 10^{-4}$. Since the exciting photon energy of 50 keV is well below the asymptotic regime it is not surprising that this ratio is larger than the asymptotic ratio estimated to be an order of magnitude smaller.

## THRESHOLD DYNAMICS OF AUGER LINES

Auger decays only involving inner shells preserve their atomic character in solids. In addition to the atom to solid energy shifts solid state effects influence the satellite emission and the threshold dynamics of the Auger lines (19).

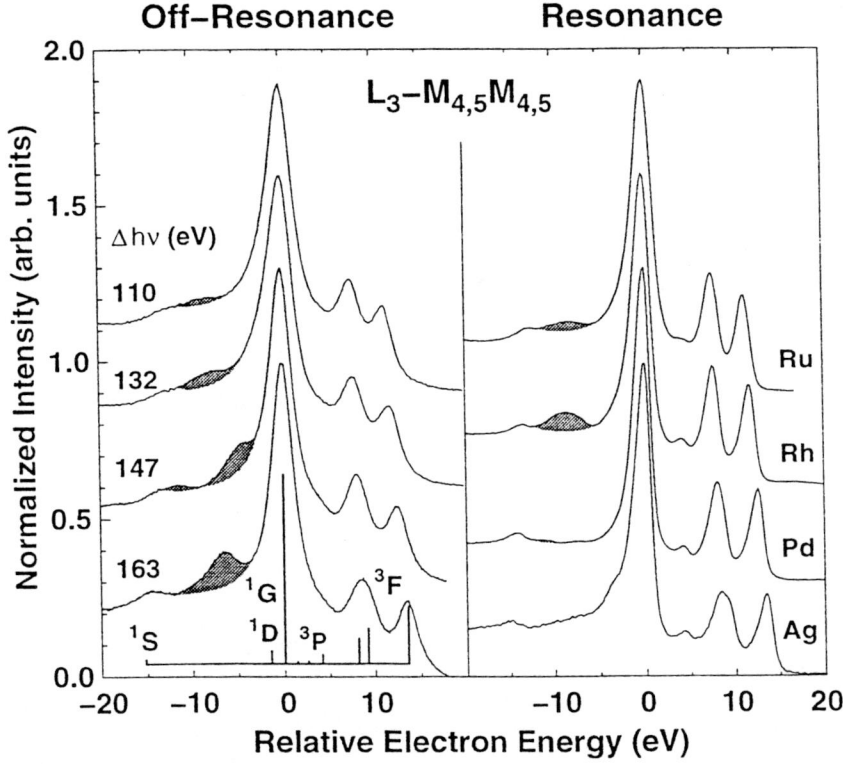

**FIGURE 4.** $L_3$-$M_{4,5}M_{4,5}$ spectra for the studied 4d metals excited far above (left) and close to (right) the $L_3$-threshold. The figure shows raw data, a constant background has been subtracted. The electron energies are aligned relative to the prominent $^1G$ line, peak maxima are normalized to unity. The $^1G$ kinetic energy ranges from 2250 eV (Ru) to 2577 eV (Ag). Note the significant line narrowing for resonant excitation and the markedly different behavior of the satellite structures below the main line. A calculation for atomic Ag is shown for comparison (from Ref. 20).

Combining the intense x-ray radiation of a multiple wiggler monochromatized by a Si (111) double crystal monochromator with a high-energy photoelectron spectrometer Drube et al. (20) succeeded to study the resonant $L_3$-$M_{4,5}$ $M_{4,5}$ Auger emission of the 4d metals Ru, Rh, Pd and Ag. The x-ray photon energy (spectral bandpass 0,5 eV) was tuned through the $L_3$ thresholds. Since the resolution of the electron energy analyzer was set to 0,2 eV the total instrumental width was smaller than the lifetime (~ 2 eV) of the $L_3$ hole state. This is essential for the observation of the gradual transition from non-radiative resonant Raman-like scattering to characteristic Auger electron emission (19, 21, 22). Figure 4 shows the $L_3$-$M_{4,5}$ $M_{4,5}$ Auger spectra of Ru, Rh, Pd and Ag metals taken at a photon energy above the $L_3$ edge (off-resonance) and (on-resonance).

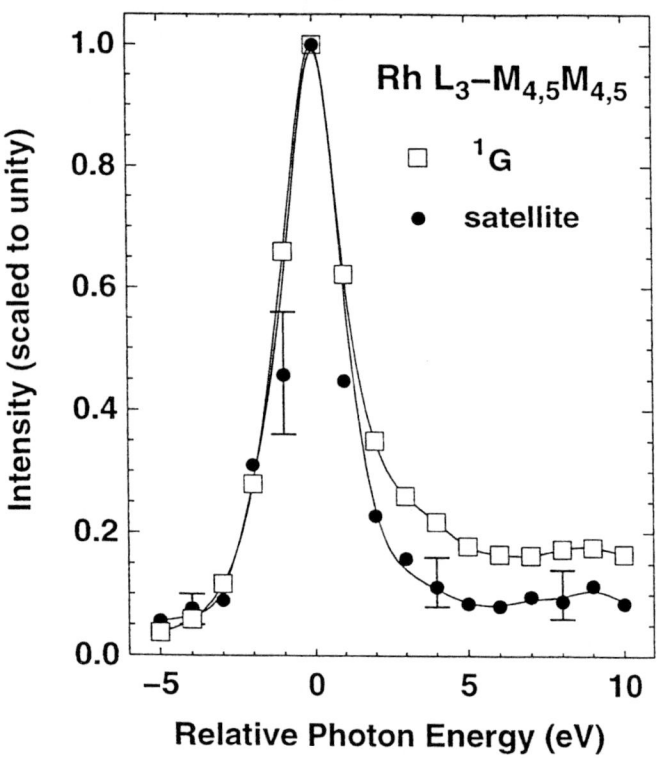

**FIGURE 5.** Threshold resonance behavior of the Rh - 8.6 eV satellite intensity (circles) compared to $^1G$ line (squares). The photon energy is measured relative to the peak of the $L_3$ absorption (from Ref. 20).

Based on calculations for atomic Ag the main $L_3$-$M_{4,5}$ $M_{4,5}$ Auger lines have been assigned (19, 20) to different terms of the final state $3d^8$ multiplet. The calculated Ag spectrum is included in Figure 4. The sub-lifetime narrowing of the Auger lines for the spectra taken on resonance proves the one-step nature of the process. Except for the resonant Auger spectra of Pd and Ag all spectra show prominent satellite lines (shaded peaks in Figure 4). Tuning the photon energy through the threshold region the intensity of the Rh satellite line closely replicates the resonance of the main $^1$G Auger line. Both resonance curves are given in Figure 5. The Ru satellite displays the same behavior. Consequently the satellites, like the main Augerlines, are resonantly enhanced by excitations of a 2p electron into localized unoccupied 4d states. Drube et al. (20) associated these Ru and Rh satellites with additional final state multiplet splitting involving the open 3d and 4d shells. The prominent satellites in the spectra of Pd and Ag behave quite differently. Their intensity is almost zero at resonance and increases when tuning the photon energy above resonance. Drube et al. (20) invoked shake-up and shake-off processes for the assignment of these satellites.

# WHITE LINES AT THRESHOLD

## Rare earths

The absorption spectra of atomic and solid rare earths display prominent white lines at the $L_2$ and $L_3$ thresholds which are attributed to the $2p_{1/2,3/2} \to 5d$ transitions (23). The $L_3$ absorption spectra of atomic and metallic Ce, Sm, Gd and Er are presented in Figure 6. In the solids interatomic interactions cause a considerable broadening of the almost Lorentzian atomic lines. This reflects the hybridization of the 5d states in the solid. An interesting feature is the shift of the white line towards higher energies by about 8 eV for Sm and Er which undergo a change of the $4f^n$ configuration on solidification (Sm $4f^6 \to 4f^5$; Er $4f^{12} \to 4f^{11}$). This shift provides a sensitive probe for valence changes. Transitions of a 2p electron into contracted 5d final state orbitals have been invoked for the explanation of the circular dichroism, especially the branching ratio at the rare earths $L_{2,3}$ edges. The 5d – 4f interactions are responsible for the radial contraction of the final state 5d orbital (24).

In contrast to the 5d orbitals the 4f orbitals of the rare earths are fully localized within the atomic core (25). Therefore the strong white lines at the 3d absorption thresholds can be well described by atomic $3d^{10} 4f^n \to 3d^9 4f^{n+1}$ transitions (26 and references therein). The complex structure of the absorption lines is due to the $3d^9 4f^{n+1}$ multiplet. Atomic calculations in intermediate coupling based on the usual reduction of the Slater F and G integrals to ~ 80 % give good agreement with the experimental solid state spectra.

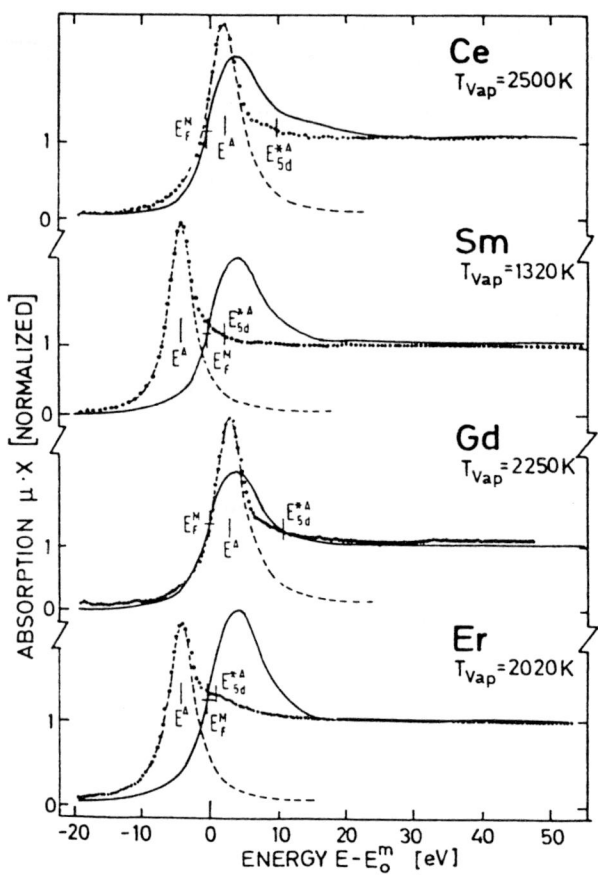

**FIGURE 6.** $L_3$ absorption spectra of atomic (dotted lines) and metallic (solid lines) Ce, Sm, Gd Er. A background, determined by extrapolating the absorption spectrum below the 2p threshold towards higher energies by a straight line, has been subtracted. The positions of the Fermi levels $E_F^M$, the atomic 5d excitation energies $E^A$, and the ionization energies $E_{5d}^{*A}$ are indicated. The dashed line gives a Lorentzian absorption profile. The edge heights are normalized. The inflection point $E_o^m$ of the absorption edge of the solid-state spectrum is chosen as the origin of the energy scale (from 23).

## 3d transition metals

Due to the partly local, partly itinerant character of the 3d electrons the situation is much more complicated for the 3d metals and 3d metal compounds (25). The 2p absorption spectra of the solid 3d-transition-elements and their compounds have been studied by electron energy loss spectroscopy (27 – 30 and references therein) and by x-ray absorption spectroscopy (31 – 35). The spectra of the metals cannot be described by the density of unoccupied d states. The shape of the dominant lines at the $L_{2,3}$ thresholds markedly deviates from that expected from the density of states. Furthermore there is a strong deviation from the statistical ratio of the $L_2$ to $L_3$ near-edge structure intensities. Only if the interaction of the 3d electrons with the 2p-hole is taken into account together with solid state band-structure effects reasonable agreement between the calculated and the experimental spectra can be achieved. It is interesting to note that only the monopole $F^0$ (2p-3d) Coulomb interaction is strongly screened in the solid whereas the higher multipole (F and G) interactions are only reduced by 20 – 30 % from the free-atom values (29, 30). The branching ratio of the lines in the absorption spectra could be well explained by an atomic approach, including crystal or ligand field effects (36). Only for Cr and Mn the 2p-absorption spectra for the free atoms have been determined (37, 38).

In figure 7 the $L_{2,3}$ absorption spectrum of manganese vapor after background subtraction is shown. The spectrum is dominated by two prominent groups of resonances corresponding to $2p_{1/2,3/2}$ excitations, which are separated by the spin-orbit interaction of the 2p level. Besides this spin-orbit interaction the direct and exchange Coulomb interaction of the $2p^5$ core with the $3d^6$ electrons and between the $3d^6$ electrons determine the atomic $2p^5\ 3s^2\ 3p^6\ 3d^6\ 4s^2$ multiplet.

The experimental absorption spectrum is compared to the results of a Hartree-Fock (HF) calculation for the excitations from the Mn$1s^2\ 2s^2\ 2p^6\ 3s^2\ 3p^6\ 3d^5\ 4s^2$ $^6S_{5/2}$ ground-state. The calculation was performed as a single configuration calculation in intermediate coupling.

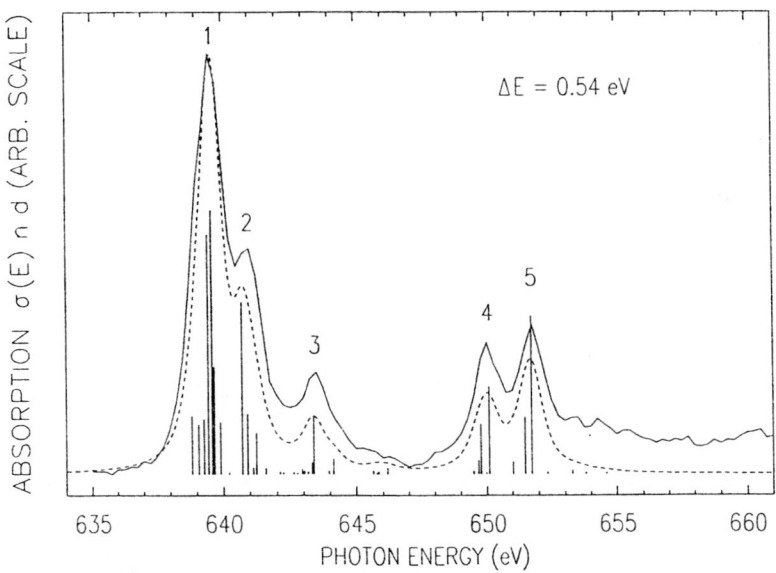

**Figure 7.** Comparison of the experimental and calculated 2p-spectra of atomic Mn. The dashed spectrum results from a convolution of the calculated lines with a Lorentzian and a Gaussian in order to take lifetime and instrumental broadening into account (from 38).

In the upper part of figure 8 the experimental atomic Mn $L_{2,3}$ absorption spectrum is presented together with the electron energy loss spectra of manganese impurities in Ag and Cu and of solid Mn (38, 28). There is a shift towards higher energies and a broadening that increase from the spectrum of the atom to that of the metal but the gross features of the spectra are similar corroborating the importance of intraatomic interactions. The bottom of figure 8 shows the Mn $L_{2,3}$ absorption spectra for 0,9 and 10 monolayers of Mn deposited on an Ag (001) (39). Very similar spectra have been reported for 15 monolayers of Mn on Cu (001), for the Mn on Cu (001) c (2x2) surface alloy (40) for 0,75 and 0,2 monolayers of Mn on Cu (110) (41). The sub-monolayer spectra are very close to the spectrum of the free atom while the spectra for higher coverages are similar to the spectrum of the bulk metal.

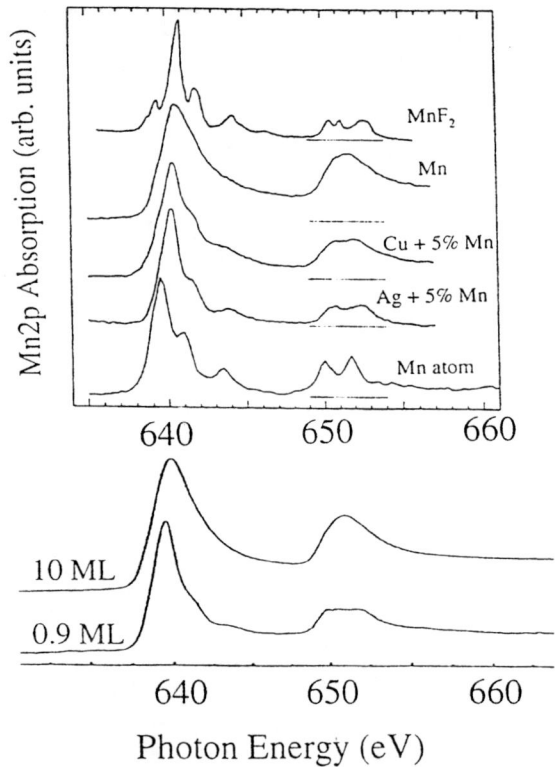

**Figure 8.** Upper part: experimental $L_{2,3}$ absorption spectra of atomic Mn, Mn impurities in Ag and Cu, Mn-metal and Mn $F_2$ (from 38).
Lower Part: Mn $L_{2,3}$ absorption spectra for 0,9 and 10 monolayers of Mn on Ag (001) (from 39).

The rich structure in the Mn $L_{2,3}$ spectrum of Mn $F_2$ is well reproduced by an approach based on ionic excitations modified by the cubic crystal field (34).

# DICHROISM

## X-ray absorption magnetic circular dichroism

The x-ray absorption and dichroism of transition metals and their compounds has recently been reviewed by de Groot (41). In this review emphasis was given to the 2p core spectra. J. Stöhr focussed in his review (42) on the x-ray magnetic circular dichroism (X-MCD) spectroscopy of transition metal thin films. The field of x-ray magnetic circular dichroism within the last decades has attracted much interest because it promises element specific information on the spin and orbital moments. Various sum rules have been derived with the help of which the values of the moments can be extracted from the spectra (43 – 47 and references therein). Experimentally the difference in absorption of a magnetized sample between left and right circularly polarized x-rays is determined. The x-ray absorption and magnetic-circular-dichroism spectra for 0,25 monolayers of Mn on a Fe/ Cu (110) substrate reported by Dürr et al. (48) is presented in figure 9.

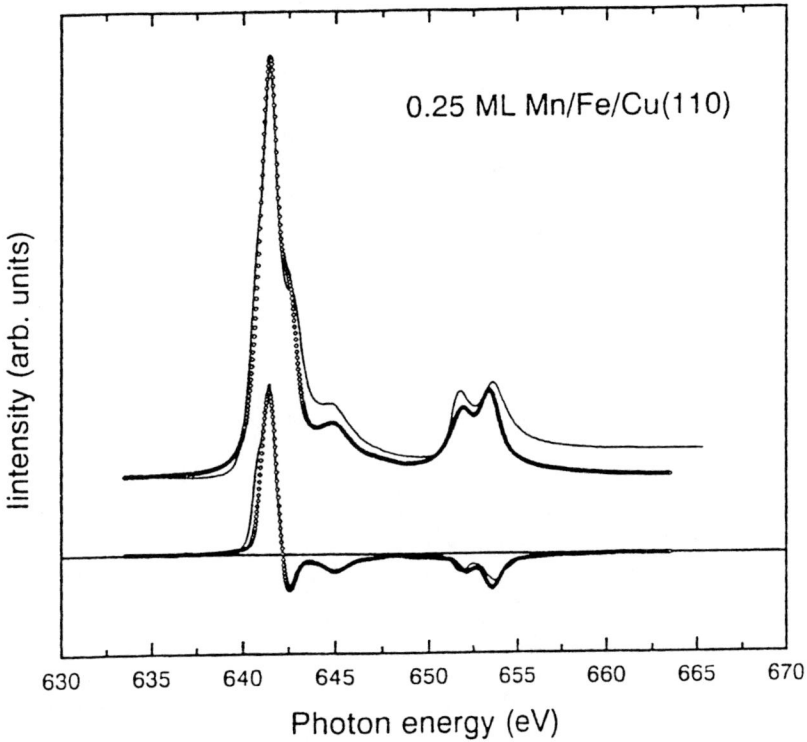

**FIGURE 9.** X-ray-absorption (top) and magnetic-circular-dichroism (bottom) spectra for 0,25 ML Mn/ Fe/ Cu (110). Experimental spectra (solid lines) are compared to an Anderson impurity model calculation (symbols) (from 48).

The absorption spectrum closely resembles the 2p-absorption spectrum of free Mn atoms given in figures 7 and 8.

Dürr et al. were able to reproduce all the multiplet structures in the absorption and the x-MCD spectrum by an Andersen impurity model calculation for a ground state with 95% $3d^5$ and 5% $3d^6$ configurations. The spin magnetic moment of 4,5 ± 0,2 $\mu_B$ is very close to the maximum moment possible for the pure $3d^5$ high-spin ground state.

## Dichroism in the inner-shell photoelectron spectra

The role of atomlike excitations in the photoelectron spectra of transition metals and their compounds has been addressed by L. C. Davis in his excellent review (49). The pioneering experiments on the circular magnetic dichroism in the 2p-photoelectron spectra of Fe-metal (50) were quickly followed by a series of energy, angle and spin resolved 2p photoelectron studies using linearly or circularly polarized synchrotron radiation (51 – 54 and references therein). Various models have been used to describe the dichroism detected in the photoelectron spectra. Single particle band-structure or atomic approaches contrast many-body descriptions including initial state configuration mixing and final state screening and multiplet splitting caused by the interaction of the core hole with the valence electrons. Since parameters enter into almost all models the agreement between the experimental and theoretical spectra is not a guarantee for the correct description of the emission of an electron from a core level of an atom/ ion embedded in a solid or bound to a surface. Especially for the latter case the comparison with the corresponding spectra of free atoms can help.

Free Cr-atoms in an atomic beam can be prepared in an oriented ground state by optical pumping with circularly polarized CW laser radiation tuned to the resonance transition

$$\text{Cr } 3d^5 \, 4s \, ^7S_3 \rightarrow 3d^5 \, 4p \, ^7P_2$$

at $\lambda = 429,09$ nm. The experimental arrangement used for the core photoelectron spectroscopy of laser oriented atoms (55 - 57) is presented in figure 10.

The oriented Cr atoms were ionized

$$\text{Cr } 2p^6 \, 3s^2 \, 3p^6 \, 3d^5 \, 4s \, ^7S_3 \rightarrow \text{Cr}^+ \, 2p^5 \, 3s^2 \, 3p^6 \, 3d^5 \, 4s \, \varepsilon s, \varepsilon d$$

by monochromatized linearly polarized undulator radiation. The counter-propagating laser and undulator beams interact with the atoms in the source volume of a Scienta SES-200 electron energy analyzer.

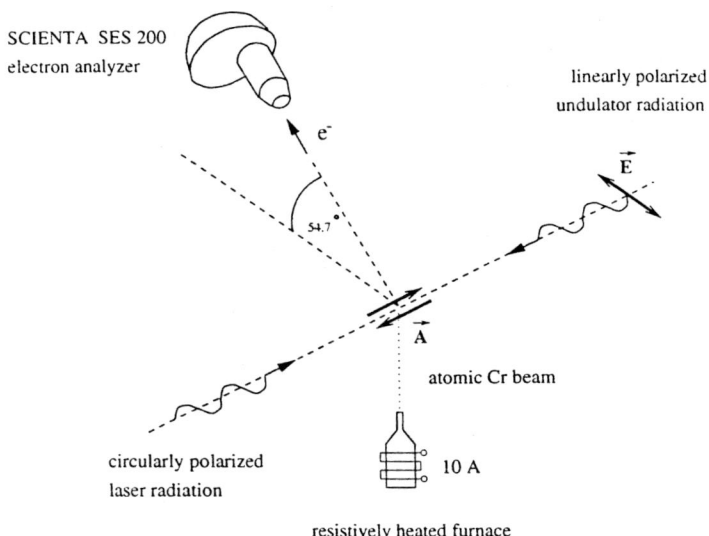

**FIGURE 10.** Experimental arrangement for inner-shell photoelectron spectroscopy on laser oriented free atoms

The linear magnetic dichroism in the angular distribution (LMDAD) is defined as the difference between two spectra recorded for two opposite orientations $\pm A_{10}$ produced alternately by right and left circularly polarized laser radiation. The orientation $A_{10}$ describes the non-statistical population of the magnetic sublevels of the Cr ground state prepared by the laser. Normalizing the LMDAD to the cross section leads to the relation

$$\beta_{LMDAD} \cdot A_{10} = \frac{I(\uparrow\downarrow) - I(\uparrow\uparrow)}{I(\uparrow\downarrow) + I(\uparrow\uparrow)}$$

for the asymmetry parameter $\beta_{LMDAD}$.

Assuming LS-coupling for the initial state and jK-coupling for the final state and neglecting configuration interaction and spin-orbit coupling in the continuum the asymmetry parameter is given by

$$\beta_{LMDAD} = C_{J,j} \sin(\delta_s - \delta_d) \frac{D_s D_d}{D_s^2 + D_d^2}$$

$D_s$ and $D_d$ are the reduced dipole amplitudes for the $2p \to \varepsilon s, \varepsilon d$ transitions and $\delta_s$ and $\delta_d$ are the relative phases of the outgoing $\varepsilon s, \varepsilon d$ electron waves. For a detailed description of the sum rules and the spectral patterns of the dichroism in inner-shell photoelectron spectra the reader is referred to a recent article by Verweyen et al. (58). The Cr 2p spin-orbit interaction ($\zeta$ (2p) = 5,7 eV) mainly determines the separation of the two peaks observed in the photoelectron spectra. Since the (2p 3d) and (3d 3d) Slater parameters are of similar magnitude jK-coupling is a reasonable approximation. Inserting HF-values for the parameters in the equation for ß$_{LMDAD}$ results in a satisfying agreement between experiment and theory. In figure 11 the experimental LMDAD spectra of free Cr atoms (57) and a magnetized Cr surface layer on a Fe crystal (59) are presented. The close similarity of both spectra corroborates the importance of atomic effects in the 2p ionization of the Cr-surface layer.

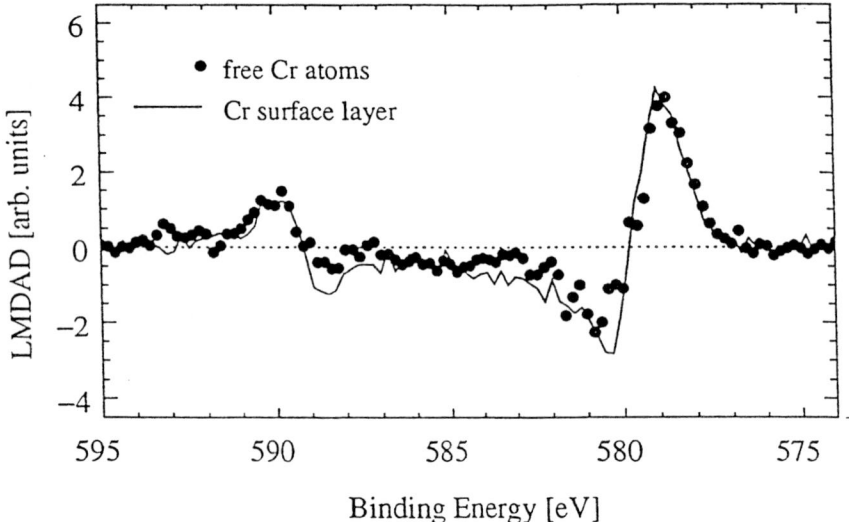

**Figure 11.** LMDAD of free oriental Cr atoms and of a magnetized Cr surface layer (from 57, 59).

A similar agreement has been found for the Cr-3p spectra (55). For the Cr 3p ionization the (3p 3d) Slater parameters are much larger than the 3p spin-orbit interaction. The states are well represented in LS-coupling. A description in terms of spin-orbit split $3p_{1/2,3/2}$ single particle states is inadequate.

## TERM DEPENDENT LIFETIME BROADENING

Term dependent decay rates can strongly influence the inner-shell photoelectron spectra (60 – 63 and references therein). As an example the Mn 3p photoelectron spectra of atomic Mn (64) and Mn $F_2$ (63) are discussed. The experimental 3p photoelectron spectrum of atomic Mn is given in the upper part of figure 12. The sharp and prominent $3p^5 3d^5 (^6S) 4s^2\ ^7P$ lines at low binding energies (A) lie 18 eV below the $3p^5 3d^5 (^6S) 4s^2\ ^5P$ lines with the highest binding energy (F). The 3p – 3d exchange interaction causes this separation of these $^7P$ and $^5P$ states where the total 3p spin is parallel and anti-parallel to the total 3d spin, respectively. Recoupling of the 3d electrons give rise to the $^5P$ lines B and C. The Z + 1 model (see upper part of figure 3) predicts shake up lines in the binding energy range of lines D and E. The calculated $3p^5 3d^5 (^6S) 4s^2\ ^7P, ^5P$ photoelectron lines are given by the bar diagram in the lower part of figure 12. The obvious discrepancy between the experiment and theory rests with the relative amplitude of the lines. These amplitudes are strongly influenced by the term dependent line width.

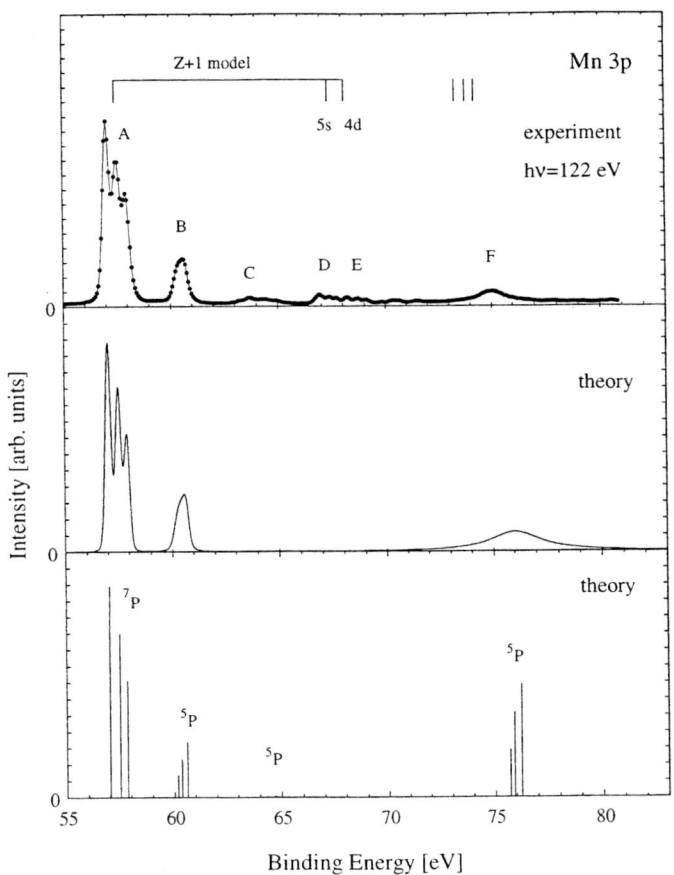

**FIGURE 12.** Upper part: Experimental 3p photoelectron spectrum of Mn atoms taken at a photon energy of 122 eV. The energy positions of the Mn II $3p^5 3d^5$ 4s nl shake-up states predicted by the Z + 1 model are indicated.
Center and lower part: Calculated Mn II $3p^5 3d^5 4s^2\ ^7P, ^5P$ photoelectron lines. The strength of the lines is given by the length of the bars in the lower part. Convoluting this bar spectrum with Lorentzian and Gaussian profiles the spectrum in the center part was obtained (from 64).

The Super-Coster-Kronig decay, forbidden for the low energy $^7$P lines, gives rise to a dramatic broadening of the high energy $^5$P lines. In order to facilitate the comparison with the experimental spectrum we convoluted the bar spectrum (lower part of figure 12) with a Lorentzian and Gaussian profile. Choosing the width of the Lorentzians equal to the calculated line width and the width of the Gaussian equal to the instrumental resolution (0,3 eV) the spectrum of the center part of figure 12 was generated. This spectrum describes the main features of the experimental spectrum reasonably well, especially the dramatic reduction of the amplitude of the high energy $^5$P lines. The characteristic structures of the atomic Mn 3p spectrum are clearly to be recognized in the photoelectron spectrum of MnF$_2$ displayed in figure 13 (63). The Mn F$_2$ consists of two main peaks with energy separation of about 17 eV. The peaks with the low and high binding energies correspond to $^7$P and $^5$P final states, where the total 3p spin is parallel and anti-parallel to the total 3d spin, respectively. The line spectrum, representing the various 3p excitations, has been calculated by Taguchi et al. (63) in a Mn F$_6$ cluster model taking into account the intraatomic multiplet coupling and the interatomic hybridization. The solid curve was obtained by convoluting the lines with a Gaussian, representing the instrumental broadening, and Lorentzians with term-dependent width. There is a good agreement between experiment and theory.

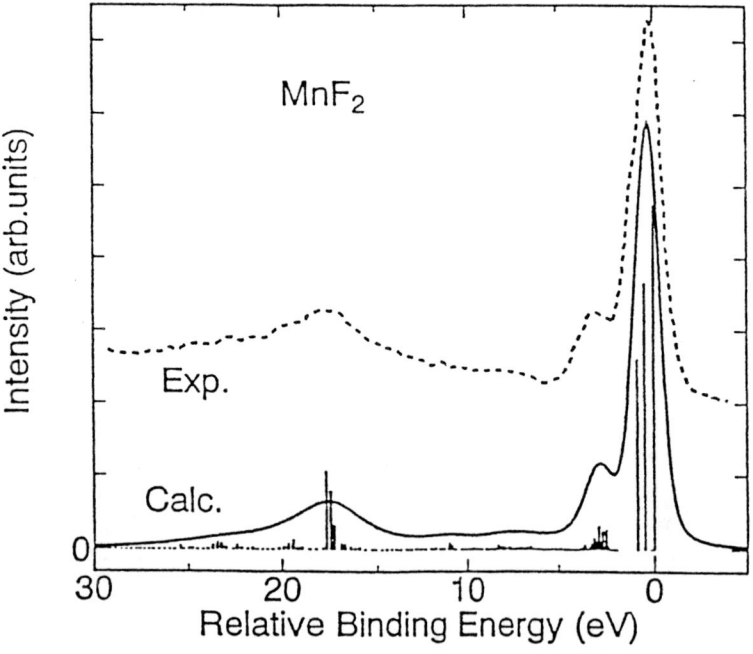

**FIGURE 13.** Theoretical and experimental Mn 3p-XPS for Mn F$_2$. The theoretical results are obtained using term-dependent $\Gamma_M$ (from 63).

In the light of these findings it is no surprise that the 3p-photoelectron spectra of Cr and Mn films only display one asymmetric line which corresponds to the dominant $3p^5 3d^5\ ^7P$ in the spectra of the free atoms (55, 64 – 66).

## Acknowledgements

Discussion and support by many students and colleagues is gratefully acknowledged.

## REFERENCES

1. Sonntag, B. *Journal de Physique*, Colloque C4, supplément au n° 7, 39, 4 (1978)

2. Jitschin, W. *Progress in Atomic Spectroscopy*, Part D, edited by H. J. Beyer and H. Kleinpoppen, Plenum Publishing Corporation, 1987, pp. 295

3. Starace, A. F. *Handbuch der Physik*, Vol. 31, ed. W. Mehlhorn, Springer Verlag, 1982, pp. 1

4. Sonntag, B., and Zimmermann, P., *Rep. Prog. Phys.* 55, 911 (1992)

5. Yacoby, Y., and Stern, A. E. "Achievements and Prospects in X-ray Absorption Spectroscopy, in Proceedings of the 17$^{th}$ International Conference on X-ray and Inner-shell Processes", AIP Conference Proceedings 389, 1997, pp. 535

6. Proceedings of the 10$^{th}$ International Conference on X-ray Absorption Fine Structure, *Journal of Synchrotron Radiation* 6, 121-807 (1999)

7. Deslattes, R. D., La Villa, R. E., Cowan, P. L., and Henins, A., *Phys. Rev.* A 27, 923 (1983)

8. Schaphorst, S. J., Kodre, A. F., Ruscheinski, J., Crasemann, B., Aberg, T., Tulkki, J.,Chen, M. H., Azuma, Y., and Brown, G. S., *Phys. Rev.* A 47, 1953 (1993)

9. Arčon, I., Kodre, A., Štuhec, M., Glavič-Cindro, D., and Drube, W., *Phys. Rev.* A 51, 147 (1995)

10. Gomlišek, J. P., Kodre, A., Arcon, I., Loireau-Lozach, A. M., and Bénazeth, S., *Phys. Rev.* A 59, 3078 (1999)

11. Qiu, Y., and Burgdörfer, J., *Phys. Rev.* A 59, 2738 (1999)

12. Mergel, V., Achler, M., Dörner, R., Khayyat, Kh., Kambara, T., Awaya, Y., Zoran, V., Nyström, B., Spielberger, L., McGuire, J. H., Feagin, J., Berakdar, J., Azuma, Y., and Schmidt-Böcking, H., *Phys. Rev. Lett.* 80, 5301 (1998)

13. Kheifets, A. S., and Bray, J., *J. Phys.* B31, L447 (1998)

14. Samson, J. A. R., Stolte, W. C., He, Z. X., Cutler, J. N., Lu, Y., and Bartlett, R. J., *Phys. Rev.* A 57, 1906 (1998)

15. Kanter, E. P., Dunford, R. W., Krässig, B., and Southworth, S. H., *Phys. Rev. Lett.* 83, 508 (1999)

16. Briand, J. P., Chevallier, P., Tavernier, M., and Rozet, J. P., *Phys. Rev. Lett.* 27, 777 (1971)

17. Chen, M. H., Chasemann, B., and Mark, H., *Phys. Rev.* A 25, 391 (1988)

18. Chen, M. H., *Phys. Rev.* A 44, 239 (1991)

19. Drube, W. Treusch, R., and Materlik, G., *Phys. Rev. Lett* 74, 42 (1995)

20. Drube, W., Grehk, T. M., Treusch, R., Materlik, G., Hansen, J. E., and Aberg, T., *Phys. Rev. B* in print

21. Aberg, T., *Phys. Scr.* T 41, 71 (1992)

22. Aberg, T., and Crasemann, B., "Resonant Anomalous X-ray scattering ", eds. Materlik, G., Sparks, C. G., and Fischer, K., Amsterdam, North-Holland, 1994, pp. 430

23. Materlik, G., Sonntag, B., and Tausch, M., *Phys. Rev. Lett.* 51, 1300 (1983)

24. van Veenendaal, M., Goedkoop, J. B., and Thole, B. T., *Phys. Rev. Lett.* 78, 1162 (1997)

25. van der Marel, D. and Sawatzky, G. A., *Phys. Rev.* B 37, 10674 (1988)

26. Thole, B. T., van der Laan, G., Fuggle, J. C., Sawatzky, G. A., Karnatak, R. C., and Esteva, J. M., *Phys. Rev.* B 32, 5107 (1985)

27. Leapman, R. D., Grunes L. A., and Fejes, P. L., *Phys. Rev.* B 26, 614 (1982)

28. Thole, B. T., Cowan, R. D., Sawatzky, G. A., Fink, J., and Fuggle, J. C., *Phys. Rev.* B 31, 6856 (1985)

29. Fink, J., Müller-Heinzerling, Th., Scheerer, B., Speier, W., Hillebrecht, U., Fuggle, J. C., Zaanen, J., and Sawatzky, G. A., *Phys. Rev.* B 32, 4899 (1985)

30. Zaanen, J., Sawatzky, G. A., Fink, J., Speier, W., and Fuggle, J. C., *Phys. Rev.* B 32, 4905 (1985)

31. van der Laan, G., Zaanen, J., Sawatzky, G. A., Karnatak, R., and Esteva, J. M., *Phys. Rev.* B 33, 4253 (1986)

32. Grioni, M., Goedkoop, J. B., Schoorl, R., de Groot, F. M. F., Fuggle, J. C., Schäfers, F., Koch, E. E., Rossi, G., Esteva, J. M., and Karnatak, R. C., *Phys. Rev.* B 39, 1541 (1989)

33. del Grande, N. K., *Physica Sc.* 41, 110 (1990)

34. de Groot, F. M. F., Fuggle, J. C., Thole, B. T., and Sawatzky, G. A., *Phys. Rev.* B 42, 5459 (1990)

35. Abbate, M., de Groot, F. M. F., Fuggle, J. C., Ma, Y. J., Chen, C. T., Sette, F., Fujimori, A., Ueda, Y., and Kosuge, K., *Phys. Rev.* B 43, 7263 (1991)

36. Thole, B. T., and van der Laan, G., *Phys. Rev.* B 38, 3158 (1988)

37. Arp, U., Iemura, K., Kutlak, G., Nayata, T., Yagi, S., and Yagishita, A., *J. Phys.* B 28, 225 (1995)

38. Arp, U., Federmann, F., Källne, E., Sonntag, B., and Sorensen, S. L., *J. Phys.* B 25, 3747 (1992)

39. Schieffer, P., Tuilier, M. H., Krembel, C., Hanf, M. C., Gewinner, G., Chandesris, D., Magnan, H., and Hricorini, K., *J. Synchrotron Rad.* 6, 784 (1999)

40. O'Brien, W. L., and Tonner, B. P., *Phys. Rev.* B 51, 617 (1995)

41. de Groot, F. M. F., *J. Electron Spectroscopy Relat. Phenom.* 67, 529 (1994)

42. Stöhr, J., *J. Electron Spectrosc. Relat. Phenom.* 75, 253 (1995)

43. Schütz, G., Wagner, W., Wilhelm, W., Kienle, P., Zeller, R., Frahm, R., and Materlik, G., *Phys. Rev. Lett.* 58, 737 (1987)

44. van der Laan, G., and Thole, B. T., *Phys. Rev.* B 43, 13401 (1991)

45. Carra, P., Thole, B. T., Altarelli, M., and Wang, X., *Phys. Rev. Lett.* 70, 694 (1993)

46. van der Laan, G., *Phys. Rev.* B 55, 8086 (1997)

47. Schmitz, D., Charton, C., Scholl, A., Carbone, C., and Eberhardt, E., *Phys. Rev.* B 59, 4327 (1999)

48. Dürr, H. A., van der Laan, G., Spanke, D., Hillebrecht, F. U., and Brookes, N. B., *Phys. Rev.* B 56, 8156 (1997)

49. Davis, L. C., *J. Appl. Phys.* 59, R 25 (1986)

50. Baumgarten, L., Schneider, C. M., Petersen, H., Schäfers, F., and Kirschner, J., *Phys. Rev. Lett.* 65, 492 (1990)

51. Ebert, H., Baumgarten, L., Schneider, C. M., and Kirschner, J. *Phys. Rev.* B 44, 4406 (1991)

52. Menchero, J. G., *Phys. Rev. Lett.* 76, 3208 (1996)

53. Hillebrecht, F. U., Roth, Ch., Rose, H. B., Park, W. G., Kisker, E., and Cherepkov, N. A., *Phys. Rev.* B 53, 12182 (1996)

54. Knabben, D., Koop, Th., Dürr, H. A., Hillebrecht, F. U., and van der Laan, G., *J. Electron Spectrosc. Relat. Phenom.* 86, 201 (1997)

55. von dem Borne, A., Dohrmann, T., Verweyen, A., Sonntag, B., Godehusen, K., and Zimmermann, P., *Phys. Rev. Lett.* 78, 4019 (1997)

56. Verweyen, A., Wernet, Ph., Sonntag, B., Godehusen, K., and Zimmermann, P., *J. Electron Spectrosc. Relat. Phenom.* 101-103, 179 (1999)

57. Wernet, Ph., Schulz, J., Sonntag, B., Godehusen, K., and Zimmermann, P., to be published

58. Verweyen, A., Grum-Grzhimailo, A. N., and Kabachnik, N. M., *Phys. Rev.* A, in print

59. Hillebrecht, F. U. (private communication)

60. Okada, K., Kotani, A., Ogasawara, H., Seino, Y., and Thole, B. T., *Phys. Rev.* B 47, 6203 (1993)

61. Ogasawara, H., Kotani, A., and Thole, B. T., *Phys. Rev.* B 50, 12332 (1994)

62. van der Laan, G., Arenholz, E., Navas, E., Bauer, A., and Kaindl, G., *Phys. Rev.* B 53, R5998 (1996)

63. Taguchi, M., Uozumi, T., and Kotani, A., *J. Phys. Soc. Japan* 66, 247 (1997)

64. von dem Borne, A., Johnson, R. L., Sonntag, B., Talkenber, M., Verweyen, A., Wernet, Ph., Schulz, J., Gerth, Ch., Obst, B., Tiedke, K., Zimmermann, P., and Hansen, J. E., to be published

65. Hillebrecht, F. U., Roth, Ch., Jungblut, R., Kisker, E., and Bringer, A., *Europhysics Lett.* 19, 711 (1992)

66. Roth, Ch., Kleeman, Th., Hillebrecht, F. U., and Kisker, E., *Phys. Rev.* B 52, R 15691 (1995)

# Multi-Atom Resonant Photoemission

*†Charles S. Fadley, †Elke Arenholz, *†Alex W. Kay,
†#Javier Garcia de Abajo, *†Bongjin S. Mun, †See-Hun Yang,
‡Zahid Hussain, and †Michel Van Hove

*Department of Physics, University of California, Davis, Davis, CA 95616 USA
†Materials Sciences Division, Lawrence Berkeley National Laboratory,
Berkeley, CA 94720 USA
‡Advanced Light Source, Lawrence Berkeley National Laboratory, Berkeley, CA 94720 USA
#Permanent address: Departamento de CCIA, Centro Misto CSIC-UPV/EHU,
San Sebastian, Spain

**Abstract** We report here on the first measurements and theoretical considerations of an interatomic multi-atom resonant photoemission (MARPE) effect that can enhance photoelectron intensities by as much as 100% and appears to be generally observable in solid materials. MARPE occurs when the photon energy is tuned to a core-level absorption edge of an atom neighboring the atom from which the photoelectron is being emitted, with the emitting level having a lower binding energy than the resonant level. Large peak intensity enhancements of 30-100% and energy-integrated effects of 10-30% have been seen by our group in various metal oxides and in a metallic system, as well as by other groups now in metal halides and an adsorbate system. The effect has also been observed in solids via the secondary decay processes of Auger emission and fluorescent x-ray emission. Weaker effects also appear to be present in gas-phase electron emission experiments. The range of the effect is so far estimated from both experiment and theory to be about 2-3 nm, with further work needed on this aspect. MARPE should thus provide a new and broadly applicable spectroscopic probe of matter in which the atomic identities and other properties (e.g. magnetic order) of atoms neighboring a given atomic type should be directly derivable. Such interatomic resonance effects also may influence normal x-ray absorption experiments, and in some cases, they may require a consideration of the degree of x-ray beam coherence for their quantitative analysis.

## INTRODUCTION

In this paper, we discuss a fundamentally new type of resonant photoemission or photoexcitation process that has been termed multi-atom resonant photoemission or MARPE [1]. In this interatomic effect, a photoelectron is ejected from a given core level of a first atom "A" and the photon energy is tuned through an absorption threshold for a core level on a neighboring atom "B". Although at first sight such an interatomic resonant photoexcitation process might be thought to be rather weak, and to depend critically on the overlap of electronic orbitals on the two atoms involved, neither of these things is true: the photoemission from A can be enhanced by as much as 100% (i.e. by a factor of two) and a first theoretical examination of the effect shows that orbital overlap between A and B is not required for a B atom to contribute to excitation from A [1,2]. In fact, the effect extends well beyond nearest-neighbor contributions, having a currently estimated range of 2-3 nanometers [2,3].

MARPE is thus quite different from the well-known intraatomic single-atom resonant photoemission, which we will denote as SARPE for clarity, a phenomenon which has been studied and used for many years to enhance photoemission intensities by an order of magnitude or more, and to study interesting many-electron aspects of the photoexcitation process [4,5]. A classic set of SARPE results due to Krause and co-workers [5] is shown in Fig. 1. Here, as shown in the energy-level diagram of Fig. 1(a), the excitation of a photoelectron from Mn 3d in a gas-phase Mn atom is followed as the photon energy is increased and reaches ~50 eV, an energy that is just sufficient to excite a deeper-lying Mn 3p electron in the same atom up to the first dipole-allowed bound excited state, which will also be of Mn 3d character. When this occurs, the very strong resonant Mn 3p-to-Mn 3d excitation can be considered to decay immediately so as to produce a free electron at the same energy as the photoelectron directly excited from Mn 3d at the same photon energy, as shown by the dashed arrows in Fig. 1(a). This resonant process, which is formally related to Auger electron emission, but more properly termed autoionization [4], is simultaneous with and coherent with the usual direct excitation of the photoelectron, and it can lead to significant increases and decreases in intensity, depending on the relative phases of the direct and resonant channels. The variation of the intensity of the Mn 3d photoelectrons has been measured as photon energy passes over this particular resonance region and the resulting data [5a] are shown in Figs. 1(b)--a typical spectrum, and 1(c)--the energy dependence of the Mn 3d intensity. Note the change in sign of the resonance effect as photon energy varies due to a change in phase between the direct and resonant channels, with the resonance curve generally following a Fano profile [4]. Simple one-electron Hartree-Fock (HF) theory is unable to predict this phenomenon, whereas many-body perturbation theory (MBPT) is [5b]. The peak intensity is increased by a maximum of a factor of about 7 for this case. As another overall measure of the strength of the resonance, the positive and negative effects of the resonance (darker grey-shaded areas) can be integrated over the full energy range over which the effect causes significant differences in intensity from the simple non-resonant Hartree-Fock (HF) theory [5b] and compared to the estimated non-resonant intensity (lighter grey-shaded areas); this leads to an overall effect of about 63%.

The core-level interatomic effect to be discussed in this paper is closely related to, but qualitatively distinct from, an intermediate type of interatomic resonant photoemission in which excitation of a valence electronic level of a system that is primarily associated with one atom (e.g. a valence level on one side of an interface [6,7] or a lone-pair molecular orbital primarily localized on one atom [8]) is enhanced as the photon energy passes through a core absorption threshold of another atom in the system (an atom on the other side of an interface or another atom in the same molecule, respectively). In what follows, we will focus on measurements on solid samples in which two well-localized core levels on A and B are involved, although other measurements of the intermediate type are certainly of interest for the future. Beyond observations of such core-level MARPE by our group and its

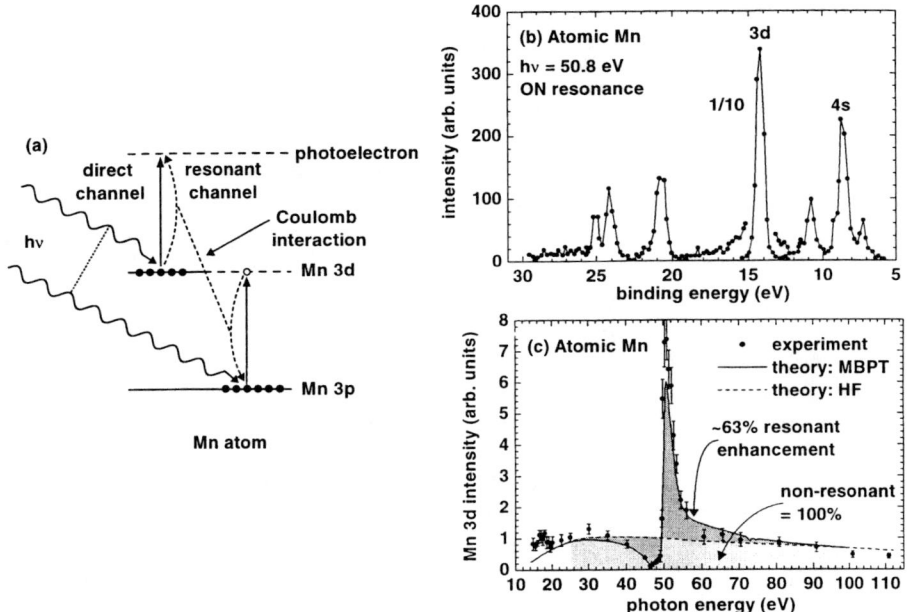

**FIGURE 1** Single-atom resonant photoemission (SARPE) for the case of Mn 3d emission from atomic Mn [5]: (a) The energy level diagram, with the resonance occurring via the Mn3p-to-Mn3d excitation. The direct and resonant excitations are indicated by the solid arrows, and the dashed arrows indicate the autoionization decay via the Coulomb interaction. (b) A photoelectron spectrum covering the Mn 3d region, with the photon energy set on resonance [5a]. (c) The measured variation of the Mn 3d intensity with photon energy, together with two types of theoretical calculation: Hartree-Fock = HF and many-body perturbation theory = MBPT [5b]. The region shaded dark grey represents the effect of the resonance; the overlapping region shaded light grey underneath the smooth HF curve represents the estimated non-resonant intensity, with relative areas as integrated over energy indicated in percent.

collaborators that will be discussed below [1,3,9], the effect has also by now been seen by others in other solid compounds [10], in an adsorbate sitting on a metallic substrate [11], and with reduced amplitude and more indirectly in photoemission from gas phase molecules [12].

Beyond representing an interesting new aspect of the x-ray absorption process, MARPE appears to constitute a new probe of matter in which emission/excitation of one atom A can be directly used to sense the presence of other atoms B near to it, at least on the nanometer scale that is commonly discussed in many current materials science developments. Adding magnetic sensitivity via magnetic circular or linear dichroism (as will be discussed below) should also permit selectively monitoring the magnetic order of those atoms B that are near to A. This projected capability can be compared to several current characterization techniques based on x-rays. Several methods presently permit determining the bulk atomic structures of solids, including x-ray diffraction and extended x-ray absorption fine structure (EXAFS) [13]. EXAFS is also element-specific via core-level electronic excitations, allowing the local structure around each atomic type to be determined, at least as to the radial positions of shells of neighboring atoms. If the atomic structure near solid surfaces is to be probed, one can add to this list low energy electron diffraction (LEED) [14], and photoelectron diffraction (PD) [15], with the latter also being element-specific via core excitation. However, as powerful and widely used as the foregoing methods are, none of them permits <u>directly</u> determining the type of atom that neighbors a given atom. That is, some of these techniques (e.g. EXAFS and PD) may be element-specific for the central atom in the structure, but there is no simple way to determine the near-neighbor atomic identities (atomic numbers) from them. In some cases, use can be made of the differences in electron-atom scattering strengths between different atoms in these last two methods, and then comparing experiment with model calculations, but this is only unambiguous when atomic numbers are relatively far apart, and even then, this procedure often provides only a semi-quantitative distinction of atomic identity. From this brief overview, it thus appears that MARPE should add a significant new aspect to the array of x-ray based methods for characterizing matter. More broadly, the fact that all atoms beyond He have a core level that can be used as one side of a MARPE experiment indicates that it could have broader applicability across the periodic table than two other powerful spectroscopic probes, Mössbauer spectroscopy and nuclear magnetic resonance, which require the use of certain nuclei or nuclear pairs.

We now turn to a more detailed consideration of the experimental aspects of measuring such multi-atom resonant photoemission effects, provide some examples of data obtained to date, and summarize the systems measured to date. An outline of the theory being developed to describe the effect [2] will also be given, and its relationship to other phenomena such as the Förster effect [16-19] and normal x-ray absorption spectroscopy will also be discussed. Finally, we discuss some implications, applications, and future perspectives.

## EXPERIMENTAL METHODOLOGY AND RESULTS

### Observation via Photoelectrons

The most direct method of detecting MARPE is simply by measuring the intensity of photoelectrons emitted in a given peak, with our first example being O 1s emission from MnO, as the photon energy is scanned through the Mn 2p absorption thresholds. The measurement thus requires scanning photon energy continuously over such thresholds, and a careful determination of the elastic or no-loss peak intensities above a suitably subtracted inelastic background. A variable-energy synchrotron radiation source is thus required. For the inelastic background, we have chosen to use the iteratively-derived Shirley background [20], but other forms could be used as well. All intensities were also normalized to allow for the time-dependent decay of the electron current in the Advanced Light Source storage ring at which all measurements were performed; the time between fills during these measurements was 4-5 hours, with current dropping to about 40-50% of its initial value just before a refill.

Our photoemission experiments (and Auger experiments to be discussed below) have been carried out at bend-magnet beamline 9.3.2 of the Advanced Light Source (ALS) in Berkeley [21], in particular using the Advanced Photoelectron Spectrometer/Diffractometer situated there [15,22]. This beamline permits scanning photon energy continuously over the energy range from 30 to 900 eV, thus spanning the resonant energies of interest here. The experimental geometry is shown in Fig. 2(a); the light was linearly polarized, with polarization vector $\hat{\varepsilon}$ lying in the plane of the figure. The higher brightness of the ALS (a third-generation synchrotron radiation source), coupled with the higher transmission and resolution of the rotateable Scienta ES200 electron spectrometer used to measure photoelectron spectra [15,22], permits determining photoelectron intensities rapidly enough to study any resonant effects observed as a function of the key experimental parameters of photon energy hv, photon incidence angle ($\theta_{hv}$), and photoelectron emission direction ($\theta,\phi$). The ability to rotate the electron spectrometer over approximately 60° in the plane of the storage ring, together with the polar ($\theta$) rotation of the sample goniometer, also permits varying $\theta$ and $\theta_{hv}$ independently, a feature that is important for distinguishing resonant effects from more simple effects due to the onset of significant x-ray reflection and refraction effects at the surface (as discussed further below).

## *Manganese oxide-MnO*

As a first example of experimental photoemission results, we show in Figs. 2(b)-(e) various aspects of the interatomic enhancement of O 1s emission from MnO, as photon energy crosses the Mn 2p thresholds. The O 1s level is a deep-lying core electron with approximately 530 eV binding energy, and the deeper-lying Mn $2p_{3/2}$ and $2p_{1/2}$ thresholds (i.e. peaks in the x-ray absorption curve) occur at approximately 640 and 652 eV, respectively [23,24]. The photoelectron kinetic energy is thus in the range of 110-120 eV in crossing these thresholds. The autoionization decay that would occur for the Mn $2p_{3/2}$ resonance is indicated by the dashed arrows in Fig. 2(b): it involves the simultaneous deexcitation of Mn 3d to Mn $2p_{3/2}$ and excitation of O 1s to a free-electron state at the photoelectron energy. Fig. 2(c) further illustrates that the peak intensity has to be measured carefully via inelastic background subtraction, since the background increases substantially as the photon energy goes above the 2p thresholds, primarily due to inelastic scattering of Auger electrons that are produced in the decay of the Mn $2p_{3/2}\rightarrow 3d$ excitations via secondary Auger processes that are different from the resonant autoionization channel. We have in fact used the photon-energy dependence of this inelastic background intensity to measure the x-ray absorption coefficient of MnO in our

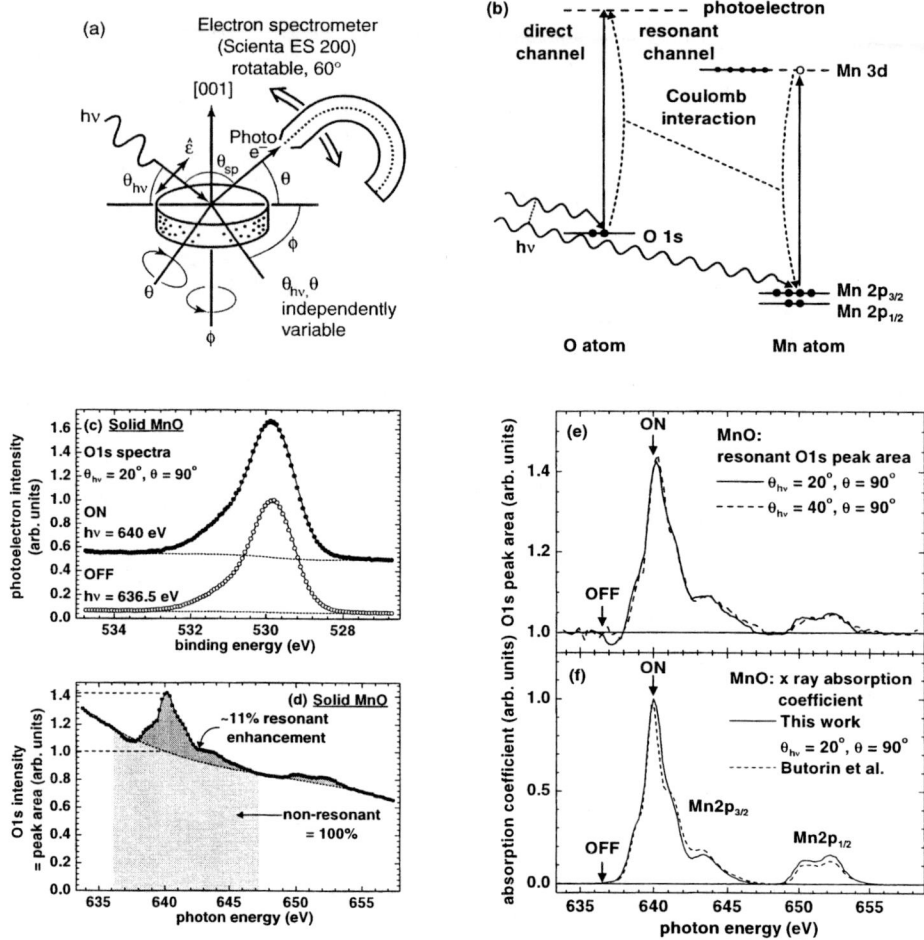

**FIGURE 2** Multi-atom resonant photoemission (MARPE) for the case of O 1s emission from MnO(001) [1]: (a) The experimental geometry, with various angles defined. (b) The energy level diagram, as in Fig. 1(a), but with the resonance occurring via the Mn2p-to-Mn3d excitation. (c) O 1s spectra on and off resonance, showing the energy-dependent inelastic background that can be used to measure the x-ray absorption coefficient. (d) MARPE enhancement of the O 1s intensity above background, with $\theta_{hv} = 20°$ and $\theta = 90°$ (normal emission), and the same shadings and percentage indications as in Fig. 1(c). (e) MARPE enhancement of the O 1s intensity with the non-resonant intensity subtracted and set equal to unity, for two different x-ray incidence angles of $\theta_{hv} = 20°$ and 40°. (f) The x-ray absorption coefficient of MnO over the Mn 2p region, as measured in this work (solid curve) and in a previous study [24].

experiment, and this is shown in Fig. 2(f) in direct comparison to an x-ray absorption spectrum measured in a prior study [24]; there is excellent agreement between the two curves. The final O 1s intensity data for a 20° x-ray incidence angle in Fig. 2(d) show a clear enhancement of the O 1s signal in crossing the Mn $2p_{3/2}$ and $2p_{1/2}$

absorption peaks: the peak intensity increases by about 42% at the $2p_{3/2}$ peak, and the effect as integrated relative to a smooth background over the $2p_{3/2}$ region is about 11%. Thus, although the overall enhancement in intensity is about an order of magnitude less than seen in the intraatomic case of Fig. 1, it is still pronounced and easily measurable. Also, the energy-integrated effect is only reduced by a factor of 1/5-1/6 compared to the more dramatic intraatomic case.

In Fig. 2(e), we have also overplotted background-normalized resonance effects for two different x-ray incidence angles of $\theta_{h\nu} = 20°$ (solid curve) and 40° (dashed curve), but with the electron exit fixed along the normal at $\theta = 90°$; this was achieved by rotating the sample and the electron spectrometer by the same amount with respect to the light beam between the two measurements. These two curves are essentially identical, and these results provide crucial evidence that these effects in MnO are in fact due to multi-atom resonant photoemission, rather than any kind of variation in the exciting near-surface x-ray flux. For example, an increase in reflectivity and reduction in x-ray penetration depth due to the strong Mn 2p absorption could yield higher electric field strengths near the surface [25], thus resulting in enhanced O 1s photoelectron intensities due to their short electron inelastic attenuation lengths that can be estimated to be in the 5-7 Å range [26]. But such phenomena can be ruled out in the present case, because the intensity enhancement does not change when $\theta_{h\nu}$ changes and because the exponential x-ray penetration depths are in any case still expected to be much larger than the electron escape depths, at approximately 32 Å and 61 Å for $\theta_{h\nu} = 20°$ and 40°, respectively. These latter numbers come from a detailed consideration of the MnO absorption coefficient, including Kramers-Kronig analysis of the experimental data in Fig. 2(f) to derive the real and imaginary parts of the dielectric constant ($\delta$ and $\beta$, respectively) and a subsequent x-ray optical calculation of the reflection coefficients and detailed profiles of the squared electric field below the surface, as shown in Fig. 3 [27,28]. The maximum reflectivities of the x-rays occur at the Mn $2p_{3/2}$ resonance and are calculated to be ~0.003 and ~0.00001 at $\theta_{h\nu} = 20°$ and 40°, respectively, thus indicating that the electric field strength near the surface cannot change significantly due to the onset of reflection. The detailed calculations shown in Fig. 3 confirm this, although we do note that a standing wave of about ±8% in strength is expected at the Mn $2p_{3/2}$ resonance and for $\theta_{h\nu} = 20°$, roughly what is expected from the square root of reflectivity, which would give ±5.4%. As a final comment, the electron inelastic attenuation length $\Lambda_e$ cannot change significantly from off-resonance to on-resonance, since the overall electronic structure and electronic states involved in inelastic scattering are not significantly perturbed by the x-rays used in even third-generation measurements [26b]. However, these combined experimental and theoretical results make it clear that the effects of enhanced reflectivity, refraction, and reduced x-ray penetration must be allowed for or ruled out by working at high enough incidence angles if MARPE effects distinct from them are to be clearly distinguished.

The forms of the MARPE enhancement is also found to be rather closely linked to the normal x-ray absorption coefficient. In Figs. 2(e), we have normalized the O 1s

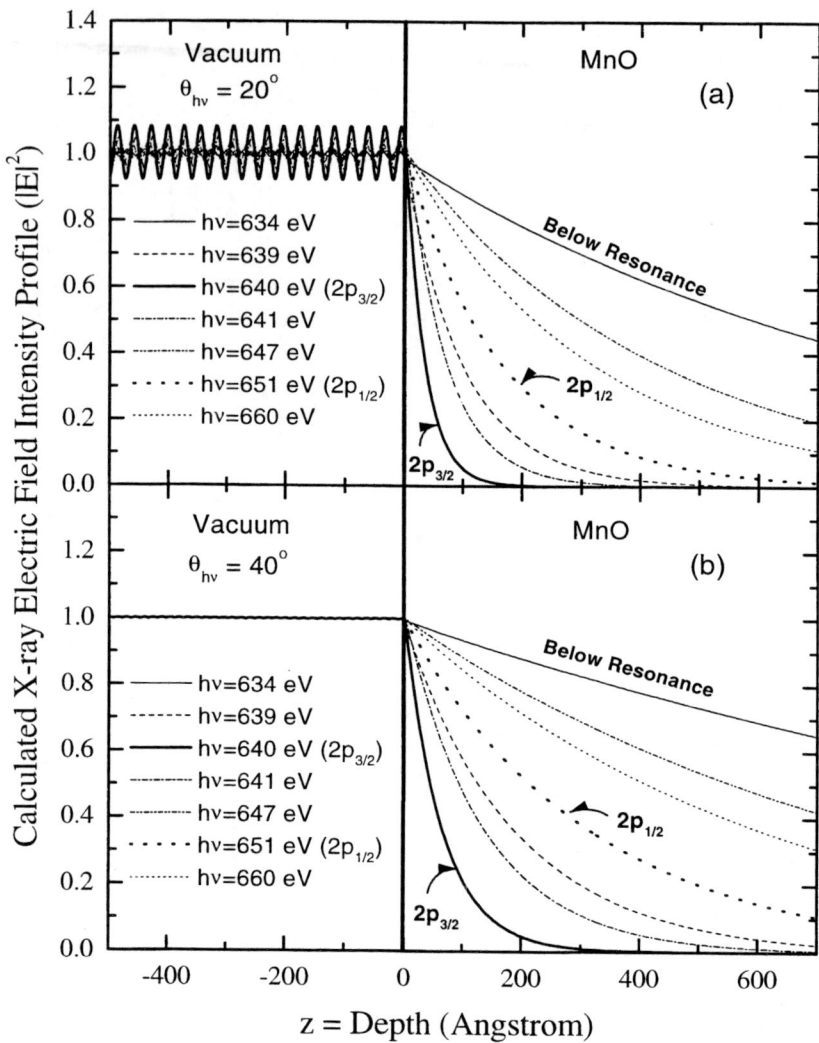

**FIGURE 3** Macroscopic x-ray optical calculation of the photon-energy-dependent field intensity $|E|^2$ as a function of depth for MnO [28]: (a) x-ray incidence angle of $\theta_{hv} = 20°$, and (b) x-ray incidence angle of $\theta_{hv} = 40°$.

data by subtracting off the smooth non-resonant background and setting its value equal to unity. This permits a more direct comparison with the x-ray absorption coefficient of MnO in Fig. 2(f), as measured previously [24] and repeated in our work, as described above. There is a strong degree of similarity between the normalized MARPE enhancement and the x-ray absorption coefficient that has by now been observed by our group in several metal oxides [1-3,9], metallic alloys [3], and metallic bilayers [3].

As a final comment on our study of MnO, we note that the resonant process can also be reversed in direction, in particular by choosing to emit photoelectrons from the multiplet-split Mn 3s levels (~85 eV binding energy), and looking for a resonance with the O 1s levels (with absorption peaks over ~530-542 eV). Strong MARPE effects are also seen here in both members of the spin-split Mn 3s doublet, with peak intensity enhancements of about 40% and energy-integrated effects of about 26% [3]. The MARPE curve here is similar to the O 1s absorption curve, but not nearly as close to it as for the case of O 1s-Mn 2p case shown in Figs. 2(e) and 2(f).

## *Other Transition-Metal Oxides*

We have also observed MARPE effects in several other inorganic transition-metal compounds, with all being studied as single crystals or epitaxial films with certain low-index surfaces exposed in ultrahigh vacuum: $Fe_2O_3(001)$, $NiO(001)$, and $La_{0.7}Sr_{0.3}MnO_3(001)$, with the last being a collosal magnetoresistive oxide containing mixed $Mn^{3+}$ and $Mn^{4+}$ oxidation states. Oxygen is bound directly to Fe or Ni or Mn in all three of these materials. A high degree of surface crystalline order in all four samples was verified by doing photoelectron diffraction measurements based on O 1s intensities, although this kind of order is not essential for observing the effect. In each case, we have studied the resonance of O 1s with the $2p_{3/2}$ and $2p_{1/2}$ levels of the transition-metal atom that is known to be bound directly to the oxygen, as shown previously for the first case of MnO in Fig. 2(b). In the perovskite-derived lattice of $La_{0.7}Sr_{0.3}MnO_3$, we have also studied resonances between O 1s and La $3d_{5/2,3/2}$, and between Mn 2p and La $3d_{5/2,3/2}$. For reference, the binding energies of the levels involved (relative to the Fermi energy) are: O 1s--532 eV, Mn $2p_{3/2,1/2}$--644 eV, 656 eV, Ni $2p_{3/2,1/2}$--869 eV, 882 eV, and La $3d_{5/2,3/2}$--837 eV, 854 eV. Any level can in principle resonate with another level at higher binding energy, so some of the pairs possible here are: O 1s with Mn 2p (resonant photoelectron kinetic energy = k.e. ≈ 110 eV), O 1s with Ni 2p (k.e. ≈ 344 eV), O 1s with La 3d (k.e. ≈ 302 eV), and Mn 2p with La 3d (k.e. ≈ 184 eV).

Figs. 4 and 5 summarize data from two of these oxides, $Fe_2O_3$ and $La_{0.7}Sr_{0.3}MnO_3$, respectively. Several resonances are observed, with the peak enhancements and energy-integrated effects being indicated directly on each plot. Several resonant enhancements are shown here: O 1s with Fe 2p in $Fe_2O_3$, and O 1s with Mn 2p, O 1s with La 3d, and Mn 2p with La 3d in $La_{0.7}Sr_{0.3}MnO_3$. Our measured x-ray absorption coefficient for $Fe_2O_3$ is shown in Fig. 4(d), and it is in excellent agreement with a prior measurement at higher resolution [29]. In Fig. 4(b), the resonant O 1s peak area as measured over the Fe $2p_{3/2,1/2}$ region and for

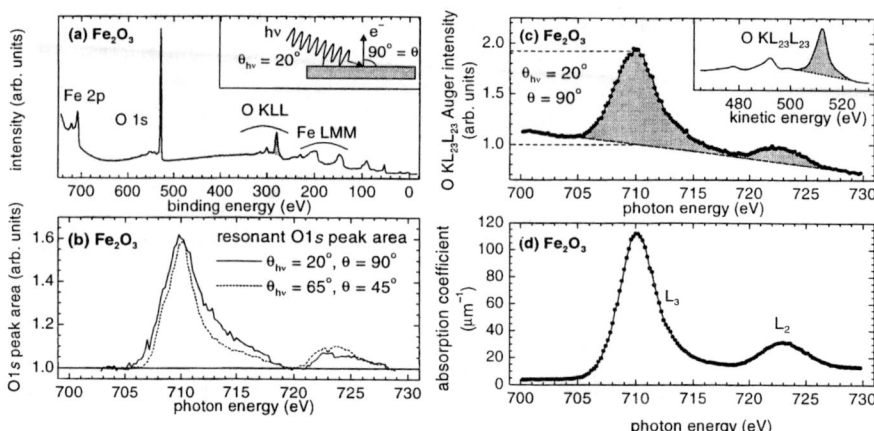

**FIGURE 4** MARPE effects on O 1s photoemission and O $KL_{23}L_{23}$ Auger emission from $Fe_2O_3(001)$ [9]. The resonance here is O 1s with Fe 2p: (a) An overall electron spectrum, with the photon energy well above the Fe 2p edges at hv = 800 eV. The inset shows the experimental geometry. (b) Resonant enhancement for the case of O1s photoemission, and with two different experimental geometries. (c) Resonant enhancement for the case of O $KL_{23}L_{23}$ emission in the geometry of (a), with the Auger peak used to measure intensity shown in the inset. (d) The x-ray absorption coefficient of $Fe_2O_3$ over the same energy region as in (c), measured by partial electron yield.

photoelectron emission normal to the $Fe_2O_3(001)$ surface is shown as the solid curve, with the smoothly varying non-resonant intensity underlying it again determined and used for normalization. Again, we see a significant resonance effect that for this case exhibits a maximum 62% peak intensity enhancement over the smoothly varying intensity without the resonance. The resonant intensity as energy-integrated over the $2p_{3/2}$ region of 705 to 720 eV for this case yields a 24% effect. The resonance enhancement also again follows very closely the x-ray absorption curve in Fig. 4(d), but is not identical to it. Also in Fig. 4(b) we show the normalized resonant intensity for an electron takeoff angle of 45° as the dashed curve. This curve is again similar to the x-ray absorption coefficient, and perhaps closer to it than for normal emission; the energy-integrated effect is here 17%. There is also a significant difference between the curves for the two emission directions, indicating some angular dependence--albeit small--of MARPE effects. For $La_{0.7}Sr_{0.3}MnO_3(001)$, we show an overall high-energy electron spectrum in Fig. 5(a), with the various resonances observed indicated by dashed arrows. In Fig. 5(b) is our measurement of the x-ray absorption coefficient over the Mn $2p_{3/2,1/2}$ region, a curve which agrees very well with prior data [30a], and in Fig. 5(c) our measurement of the x-ray absorption coefficient over the La $4d_{5/2,3/2}$ region, which agrees well with prior x-ray absorption data for La metal [30b]. Figs. 5(d)-5(f) show the MARPE enhancements observed. In 5(d), the normalized O 1s intensity in resonance with

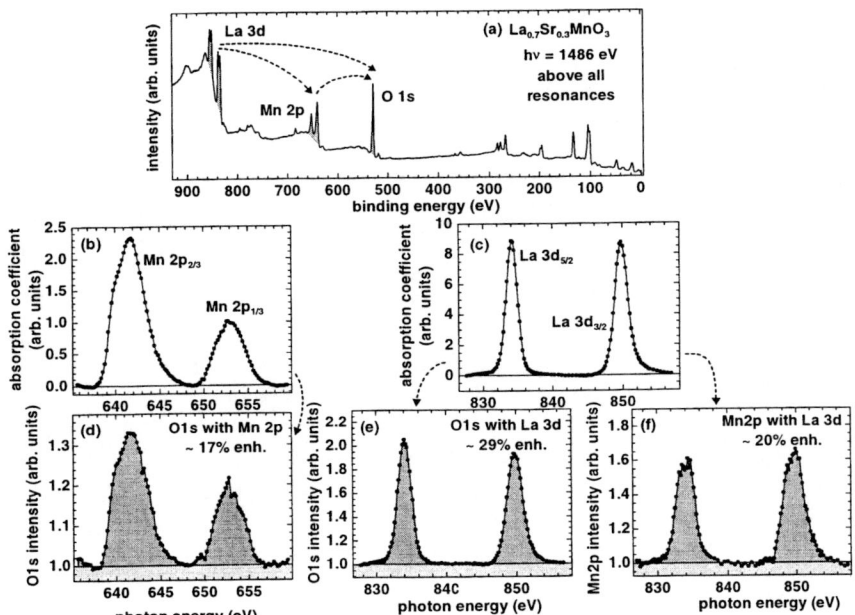

**FIGURE 5** MARPE effects on O 1s and Mn 2p emission from $La_{0.7}Sr_{0.3}MnO_3(001)$ [expanded from ref. 1]. (a) An overall electron spectrum for a photon energy well above all resonances studied at hν = 1487 eV. (b),(c) X-ray absorption coefficients over the Mn 2p and La 3d regions, respectively, as measured via the inelastic background in photoelectron spectra. (d) Enhancement of O 1s in resonance with Mn 2p. (e) Enhancement of O 1s in resonance with La 3d. (f) Enhancement of Mn 2p in resonance with La 3d. Percentage resonance effects as energy-integrated over the absorption region are again shown.

Mn 2p is shown. There is a resonant peak increase at the $2p_{3/2}$ position of about 33%, and an energy-integrated effect over $2p_{3/2}$ of 17%. The resonance is again seen to follow very closely the Mn 2p absorption curve in Fig. 5(b) for this material [30a]. The O 1s intensity is also found to resonate with the La $3d_{5/2,3/2}$ levels, as shown in Fig. 5(e), with very large peak enhancements of up to 92-105%, an energy-integrated effect over La $3d_{5/2}$ of 29%, and a form that follows very closely the x-ray absorption coefficient in Fig. 5(c). Finally, Fig. 5(f) shows an analogous normalized resonant enhancement of the Mn 2p intensity with the La $3d_{5/2,3/2}$ levels; with a peak enhancement of 60-70%, a 20% energy-integrated effect, and a form that again follows closely what we measure for the La 3d xay absorption coefficient in Fig. 5(c). Although the surface of this last sample was in no way treated after insertion into ultrahigh vacuum, we nonetheless are able to detect three strong distinct resonance effects among its constituents: O with Mn and La, and Mn with La.

## Other solids and systems

We here briefly note that similar MARPE effects have by now been observed by our group for Cr 2p emission in resonance with Fe 2p from Cr/Fe alloys [3] and

Ce/Fe bilayers [3]; by Kikas et al. for Cl 2p emission in resonance with transition-metal 2p in $MnCl_2$, $VCl_2$, $CrCl_2$ [10]; by Garnier et al. for N 1s emission in resonance with Ni 2p from $N_2$ adsorbed on Ni(001) [11]; and by Hemmers et al. more subtly through non-dipole effects on Auger emission in CO [12]. In all solid compounds and the adsorbate system, the effects are of similar magnitude to those reported here (~10-40% in peak intensity), and the MARPE enhancements are also found to rather closely follow the x-ray absorption profile of the resonating level. There is thus little doubt that this is a generally observable phenomenon. Our Cr/Fe bilayer work [3] has also permitted directly estimating the phenomenological falloff with distance from the emitter in the MARPE intensity enhancement, with an exponential decay constant of about 15-25 Å. This Cr/Fe work has also permitted using MARPE effects to study compositional clustering and phase separation in epitaxial Cr/Fe alloy films [3], and demonstrated that magnetic circular dichroism can be seen in the effect [3], thus providing a new probe of the local magnetic order (e.g. of Fe atoms that are known to be within a few nm of Cr).

## Observation via Secondary Decay Processes

### *Auger electron emission*

It is also of interest to ask whether the enhanced probability of creating a given core hole as photon energy is scanned across a MARPE region can be detected from the secondary decay processes involved, specifically, Auger electron emission and fluorescent x-ray emission. Observation of the effect via these secondary decay channels not only further verifies its origin, but also provides other experimental methods for detecting it [9].

As one example of detection via Auger processes, we show in Fig. 4(c) the O $KL_{23}L_{23}$ Auger intensity from $Fe_2O_3$, measured using the same sort of inelastic background subtraction as for photoelectron intensities. There is a clear enhancement of this intensity on passing through the Fe 2p absorption curve, and it again follows rather closely the absorption coefficient of the material, as shown in Fig. 4(d) [9].

Although this is the only case studied to date, it seems clear that MARPE also influences Auger decay as a secondary process to the initial photoexcitation.

### *Flourescent x-ray emission*

The influence of multi-atom resonance on x-ray fluorescence intensities has been studied for the case of MnO [9]. These measurements were performed on undulator beamline 8.0.1 of the Advanced Light Source, using a special high-resolution grating monochromator situated there [31]. The O $K_\alpha$ emission is resolvable from any other peak, as shown in the spectrum of Fig. 6(a). The area of this peak, with a

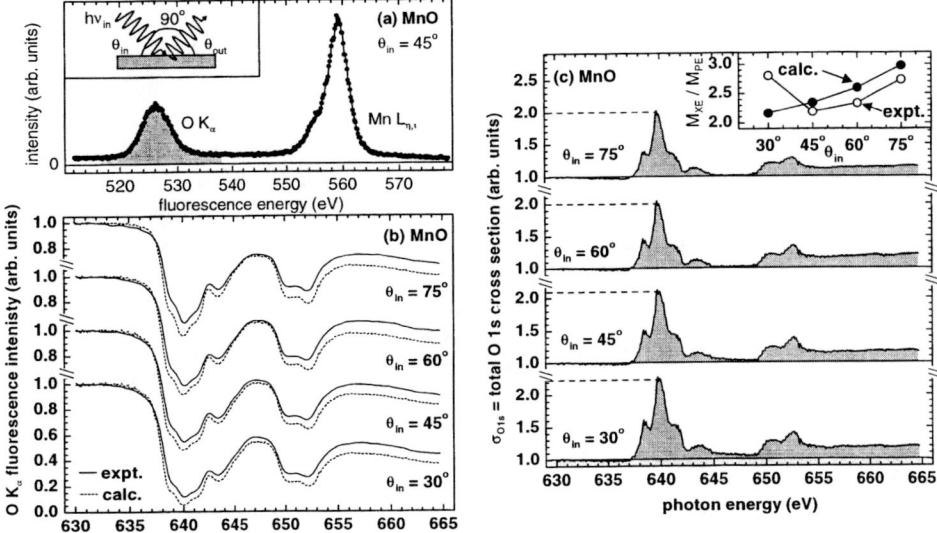

**FIGURE 6** MARPE effects on O $K_\alpha$ x-ray fluorescence from MnO [9]: (a) Fluorescence spectrum of MnO recorded at hv = 640 eV. The inset shows the experimental geometry, with a fixed angle of 90° between x ray incidence and exit. (b) Energy dependence of the O $K_\alpha$ fluorescence intensity over the Mn $2p_{3/2,1/2}$ edges for different angles of x-ray incidence $\theta_{in}$ and x-ray exit $\theta_{out} = 90° - \theta_{in}$. Curves are shown for both the experimental data (solid curves) and for calculations based on Eq. 1 (dashed curves). (c) The normalized total O 1s cross section in MnO, including interatomic resonance effects, as derived from dividing the two sets of curves shown in (b). The inset shows the experimental (open symbols) and calculated (filled symbols) ratio between the resonant enhancement at the Mn $2p_{3/2}$ edge as detected in the O $K_\alpha$ fluorescence intensity ($M_{XE}$), and in the O 1s photoelectron intensity ($M_{PE}$) for the different experimental geometries.

small background due to scattered light subtracted, has been measured as a function of photon energy over the Mn 2p region and for different experimental geometries, as shown in the inset in Fig. 6(a). Both the exciting x-ray incidence angle, $\theta_{in}$ and the outgoing $K_\alpha$ x-ray exit angle were varied over four values, with a fixed angle of 90° between them. The experimental O $K_\alpha$ intensity is shown in Fig. 6(b), and it exhibits strong and well-known self-absorption effects which actually <u>decrease</u> the emission significantly as one passes through the Mn 2p absorption spectrum. However, the energy-dependent x-ray absorption coefficient in MnO can be accurately determined from the data in Fig. 2(f) and a careful matching of it to tabulated off-resonance values at lower and higher energies [24]. With $\mu_{MnO}(hv)$ thus determined, we can allow for self-absorption via the standard expression [32]:

$$I_{OK_\alpha}(hv) \propto I_0(hv_{in})\sigma^0_{O1s}(hv_{in})\varepsilon_{O1s}(hv_{in})\left[\frac{\mu_{MnO}(hv_{in})}{\sin\theta_{in}} + \frac{\mu_{MnO}(OK_\alpha)}{\sin\theta_{out}}\right]^{-1} \quad (1)$$

where $I_0(h\nu_{in})$ is the incident x-ray flux at energy $h\nu_{in}$, $\sigma^0_{O1s}$ is the cross section for producing an O 1s hole at this energy (which we take to exclude MARPE effects and thus be described by a smooth curve which varies little over the Mn 2p region), and $\varepsilon_{O1s}(h\nu_{in})$ is the fluorescence yield at this energy (which we assume to be constant over the small Mn 2p energy range of our measurements). Using this equation now yields the four dashed curves shown in Fig. 6(b), which are consistently below the experimental data when all curves are normalized to unity well below the Mn 2p region. We can now estimate the MARPE effects on this data simple by taking the ratio of each pair of curves, and this give the results shown in Fig. 6(c) for the four experimental geometries studied. That is, this ratio we expect to be a measure of $\sigma_{O1s}/\sigma^0_{O1s}$, where $\sigma_{O1s}$ is the true cross section for O 1s hole production, including any MARPE effects. The four curves shown are quite self consistent in all showing a peak enhancement of 100-140%, and very similar shapes over the entire energy range. These data also add another new element to the phenomenon by suggesting that there is enhancement well above the Mn $2p_{1/2}$ threshold, although this needs further experimental and theoretical confirmation. Finally, the fact that the effects here are over twice as large as the approximately 40% seen in the direct photoemission process from MnO (cf. Fig. 2) can be understood by noting that a typical x-ray fluorescence emitter is well below the surface, and so can experience resonant effects from atoms in a spherical cluster around itself, whereas in the case of surface-sensitive photoemission, this cluster is essentially a hemisphere extending inward from the surface. Quantitative calculations of the ratio of MARPE peak enhancement on and off the $2p_{3/2}$ resonance in x-ray emission ($M_{XE}$) and in photoemission ($M_{PE}$) confirm this qualitative explanation, as shown by the comparison for experiment and theory in the inset of Fig. 6(c) [9].

Thus, although allowance for self-absorption effects will be critical in measuring MARPE via x-ray fluorescence, the procedure outlined here appears to provide a reliable method for doing this. Fluorescence detection also has the very desirable feature of making the measurement more bulk sensitive, and thus applicable to a broader range of systems.

## THEORETICAL MODELING OF PHOTOEMISSION

### Basic Theory

In order to further confirm our assertion of the interatomic nature of these resonant enhancements, as well as to understand their origins and approximate spatial extents, let us discuss some of the ingredients needed in the theoretical analysis for the O 1s-Mn $2p_{3/2,1/2}$ absorption region in MnO(001) [2]. The relevant interactions and quantum-mechanical matrix elements involved are indicated in Fig. 7. Specializing to the MARPE case, and writing all expressions specifically for O 1s-Mn $2p_{3/2}$ for clarity, we have:

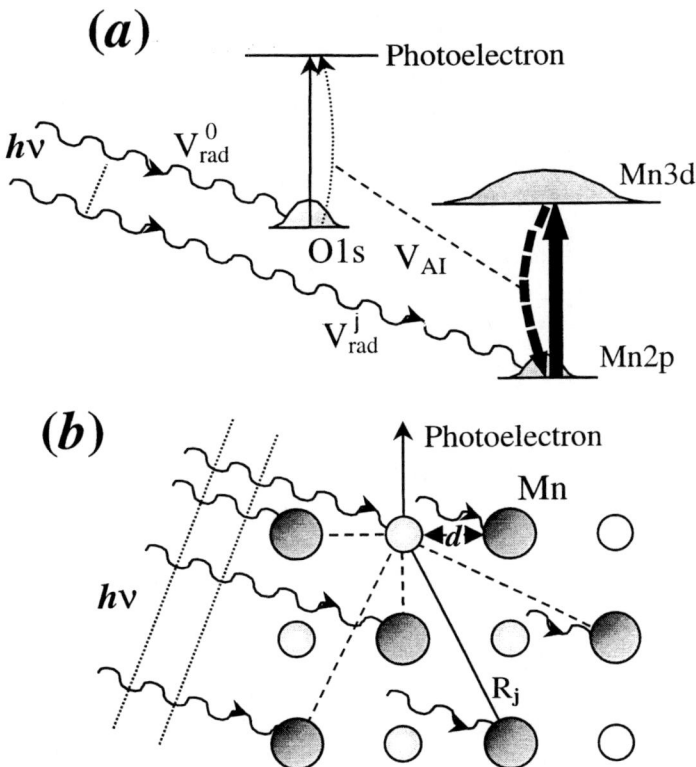

**FIGURE 7** Schematic representation of the energy levels involved in O1s MARPE in MnO [ref. 2]. (a) O1s photoemission takes place via two different interfering channels: direct emission (thin solid arrow) and emission assisted by the excitation of Mn2p to Mn3d (thick solid arrow), with subsequent interatomic autoionization (broken arrows) produced by electron-electron interaction (dashed line). Symbols for the various interactions are also indicated here. (b) The resonance involves coherent addition of effects over several Mn sites in the solid situated at $R_j$ with respect to the emitter O atom.

- The dipole excitation by light with polarization vector $\hat{\varepsilon}$ of an O 1s electron from a given oxygen atom to a photoelectron with kinetic energy E, angular momentum $\ell = 1 = p$ as dipole-allowed, and magnetic quantum number $m_f = 0, \pm 1$:

$$V_{rad}^0 = <Ep_{m_f} | \vec{\varepsilon} \cdot \vec{r} | O1s >; \qquad (2)$$

- The dipole excitation of a Mn $2p_{3/2}$ electron with magnetic quantum number m on the jth Mn atom located at position $\vec{R}_j$ in a cluster surrounding the emitter O atom to the first unoccupied Mn 3d level with magnetic quantum number m' on that atom:

$$V_{rad}^j = <Mn_j 3d_{m'}| e^{i\vec{k}_{h\nu} \cdot \vec{R}_j} \hat{\varepsilon} \cdot \vec{r} | Mn_j 2p_{3/2,m}>, \qquad (3)$$

where we have included the effects of retardation via the exponential pre-factor in the photon wave vector $\vec{k}_{h\nu}$ and $\vec{R}_j$, since the typical photon wavelength at the Mn$2p_{3/2}$ threshold of 640 eV energy of about 19 Å, coupled with the minimum interatomic O-Mn distance of d = 2.23 Å leads to a phase of 42° or 0.73 radians that can quickly accumulate to a non-negligible value in a sum over a cluster of several times this interatomic distance in radius; and

- The autoionization of the Mn 3d state via a Coulomb interaction coupling the Mn and O electrons:

$$V_{AI}^j = < Ep_{m_f}, Mn2p_{3/2,m} | \hat{V}_{Coul} | O1s, Mn3d_{m'} >, \qquad (4)$$

where the Coulomb interaction written here as $\hat{V}_{Coul}$ is an abbreviation for the retarded form including both charge-charge and current-current interactions and often referred to as the Møller expression [33,34]:

$$\hat{V}_{Coul} = \frac{e^2}{|\vec{R}_j + \vec{r}_1 - \vec{r}_2|} e^{ik_{h\nu}|\vec{R}_j + \vec{r}_2 - \vec{r}_1|} (1 - k_{h\nu}\vec{r}_1 \cdot \vec{r}_2), \qquad (5)$$

with this form having been used previously in high-energy Auger theory. Eq. (5) includes both dipolar charge-charge interactions (the "1" in parentheses) and dipolar current-current interactions (the "$-k_{h\nu}\vec{r}_1 \cdot \vec{r}_2$" in parentheses), as well as appropriate retardation.

The enhancement observed in MARPE originates in the excitation produced by the external radiation in every Mn atom j via the interaction described by Eq. (3) and its subsequent de-excitation via Eq. (5). The effective spatial range of this interaction is then determined by the sum over j of the phase factors appearing in Eqs. (3) and (5), together with the inherent distance dependence of the O-Mn interactions and possible dielectric screening effects. But considering only the sum over phase factors, far-distance contributions to this sum suffer phase cancellations, so that they play a minor role. Actually, a good approximation is already obtained when only Mn atoms up to 10 times the nearest-neighbor distance away from the O emitter are considered. Therefore, the range of MARPE in MnO can be estimated to be approximately 22 Å.

As a final comment on the theoretical picture of MARPE, we should point out its relationship to the inverse form of x-ray fluorescence holography [32,35], also referred to as multi-energy x-ray holography (MEXH). In this off-resonance measurement usually done at much higher energies of 6-15 keV, an incoming x-ray beam scatters from various neighbors around a given fluorescent emitter, with the relative phases of the different scattered-wave components thus encoding the local atomic geometry. The intensity variation as a function of x-ray incidence angle is thus a hologram, which can be inverted mathematically so as so directly produce an image of the local atomic structure [32,35]. MEXH involves the scattering of virtual photons from the various atoms surrounding the fluorescent emitter so as to add an induced component to the local electric field that produces fluorescence. However, when this picture is incorporated at the resonant edges in the 0.5-1.0 keV energy regime utilized in MARPE, particularly for the case of MnO, the resulting induced electric field produces a <u>decrease</u> rather than an increase in the magnitude of the total electric field, largely due to the inelastic part of the x-ray scattering factor [36]. Therefore, the resonant nature of MARPE involves other processes not simply accounted for by the simple scattering via virtual photons. This aspect of MARPE is currently under investigation.

## Relationship to the Förster Effect

As a final comment on the theoretical interpretation of MARPE, we contrast it with the well-known type of intermolecular or interatomic energy transfer involved in the Förster effect [16,18,19], which is also directly related to what has been described as sensitized luminescence [17]. In this effect, a low-energy excitation of the order of 1 eV that is very long-lived with respect to the core decay involved in MARPE is produced on one atom/molecule. This excitation then propagates from the "donor" atom or molecule to an "acceptor" atom or molecule with a matching energy-level system, with the propagation time being much shorter than the lifetime of the initial excited state. The relationship of these times dictates that coherent quantum interference effects can be neglected, and the process is well described by a transfer rate equation in which energy diffuses from one site to another via real photons in multiple-step cascades. The interaction is again fundamentally Coulombic, and it reduces due to the long wavelength of the excitation to a non-retarded $1/r^3$ for typical donor/acceptor distances of the order of 10 nm. The final transition rates scale as the square of the potential or $1/r^6$.

MARPE is thus fundamentally different in several respects. It involves much higher excitation energies of 100-1000 eV, much shorter relaxation times leading to coherent interference effects, and much shorter wavelengths that can require considering retardation effects at the high end of the energy scale given above. A rate equation like that in the Förster effect is thus not appropriate, and quantum interference with the direct process is then critical. MARPE thus involves virtual photons and a single-step process, compared to real photons and a multiple-step process in the Förster effect. Both processes can however be considered to involve multiple scattering of photons, whether real or virtual, and inclusion of this in MARPE theory is also currently underway.

# IMPLICATIONS, POTENTIAL APPLICATIONS, AND FUTURE DIRECTIONS

## Implications for X-ray Absorption Spectroscopy

We conclude by first pointing out some potential implications of the multi-atom resonant photoemission effect on other measurements. It seems likely that MARPE will have an effect on simple x-ray absorption spectroscopy, even though the effect will be masked in that type of measurement by simple contributing to an already-large increase in absorption in crossing a given core threshold. For example, in MnO we estimate from our experimental data that the net change in the strength of the O 1s excitation process due to Mn $2p_{3/2}$ resonance is about 1.5% of the strength of the primary Mn $2p_{3/2}$ absorption peak. But even at this kind of level, one can imagine a process involving the type of inter-atomic interaction that makes MARPE possible, but occurring between the <u>same</u> type of atoms in a sample and that this interaction could affect the main absorption resonance on something like that magnitude scale. In this case, one might refer to multi-atom resonant <u>photoexcitation</u> instead of photoemission. Thus, in view of the strength with which such resonant effects are found in photoemission from compounds, it seems quite possible that they would represent a measurable correction to an absorption coefficient. One might also expect such effects to be important in resonant x-ray scattering experiments, whether elastic or inelastic.

## Requirements on x-ray beam coherence

A further consequence of our discussion of MARPE (cf. Fig. 7(b)) and this line of reasoning is that the incoming x-ray beam must be coherent over the full volume necessary to excite all atoms j that can contribute significantly to the multi-atom resonant excitation. Thus, the coherence volume of the x-ray beam becomes important in order to "saturate" these effects. With an x-ray wavelength of $\lambda_x$, a wavelength resolution of $\Delta\lambda_x$, and a beam divergence of $\theta_{div}$, we can for the example of beamline 9.3.2 at the ALS estimate the transverse coherence length from the formula $d_{trans} = \dfrac{\lambda_x}{2\pi\theta_{div}}$ to be about 1200 Å and the longitudinal coherence length from the formula $d_{long} = \dfrac{\lambda_x^2}{2(\Delta\lambda_x)}$ to be about 9700 Å. These values are sufficiently large with respect to our present estimated range for the MARPE interactions of a few nanometers or 20-30 Å, that we do not believe our results have been influenced by beam coherence effects. Similar statements are probably true for most third-generation spectroscopy beamlines with state-of-the-art optics. However, differences in x-ray absorption strengths for different beamlines and experimental conditions have recently observed, and these may be linked to such beam coherence effects [37].

## Possible applications and future directions

Beyond their possible influence on x-ray absorption, or perhaps also on resonant x-ray scattering phenomena, multi-atom resonant photoemission or photoexcitation (MARPE) effects as we have discussed them here via photoelectron, Auger electron, or x-ray fluorescence detection appear to have broad potential applications in physics and chemistry, as well as in the materials and environmental sciences, and perhaps also the life sciences. From data obtained by our group and now also others, these effects are expected to be generally observable in condensed-phase materials, provided that suitable core level spacings can be found in the two atoms to be studied. The best estimates of the range of the effect to date are as noted a few nanometers, with nearest-neighbor atoms that are in more intimate bonding interaction with the emitter expected to exhibit especially strong contributions [3]. However, it is also important to note that there may be several hundred resonating atoms within a few nanometers of a given emitter in a typical solid. For many types of systems exhibiting special bonding sites and/or heterogeneity on a nanometer scale, MARPE measurements should permit determining the local composition, bonding, and magnetic environment (e.g. via magnetic circular dichroism), as well as the existence or non-existence of compositional clustering or phase separation [3], in a non-destructive and unique way.

Further studies exploring the precise sensitivity of MARPE to different interatomic distances and/or different pairs of core levels with varying energy spacings are certainly called for. Further theoretical analysis to explain the observed magnitudes of the effect and the low degree of angular sensitivity seen in experiment and to more quantitatively determine nearest-neighbor effects and the effective range of the effect are also needed. But even without more theoretical development, calibrating such interatomic resonant enhancements against standard compounds with the same or varying types of chemical bonding and/or heterogeneity may permit estimating the number of near neighbors of a given type for unknown cases. The well-known chemical shifts in photoelectron kinetic energies [15,22] should also permit measuring resonances separately around different chemically-distinct species, something which has already been done for $N_2$/Ni(001) [11]. Measuring photoelectron diffraction patterns [15,22] as differences on and off resonance may in addition permit focussing on scattering processes associated with a given type of near-neighbor atom, thereby leading to a more element-specific structural probe.

Although we have used single crystals here to better define these first measurements of the effects, there is no general requirement of a single-crystal specimen; however, with truly random atomic positions in an amorphous or glassy material, the range of the effect--and thus its magnitude--might be expected to be reduced to something like the wavelength of the radiation due to phase cancellation effects for distances larger than this. Further studies of such interatomic effects in free molecules are also called for, with first observations showing effects in the 1% range or smaller [12]. With soft-x-ray detection, as illustrated in Fig. 6, the experiment becomes more bulk sensitive, considerably widening its applicability to include perhaps the environment around active sites in systems of biological interest, although a careful allowance for the enhanced absorption of the *exciting* flux at resonance would be needed to unambiguously detect the resonant enhancement. With photoelectron, x-ray, or Auger detection, the identities of neighbors to atoms adsorbed at surfaces, around impurity atoms in solids, or around atoms at buried interfaces should be detectable. Finally, exciting with circularly-polarized radiation on magnetic samples should lead to resonant photoelectron spin polarization and/or magnetic dichroism effects (e.g. by subtracting intensities for right- and left- handed excitation, as already observed in Cr/Fe alloys [3]) that should be particularly sensitive to the short-range magnetic order of near neighbors. Future studies exploring and exploiting these various possibilities, as well as more accurate theoretical modeling of all of these effects, are thus clearly of interest.

## ACKNOWLEDGEMENTS

We are indebted to M.P. Klein and J.B. Kortright for helpful discussions, to E.J. Moler for assistance with beamline tuning and maintenance, to R. Denecke for help with initial experiments, and to S.A. Chambers, K. Krishnan, and J. Stohr for providing samples. The expert assistance of M.W. West and M.J. Press with various experimental aspects is also much appreciated. We also gratefully acknowledge a collaboration with M.M. Grush, T.A. Callcott, D.L. Ederer, and C. Heske on x-ray fluorescence experiments. This work was supported by the U.S. Department of Energy, Office of Science, Office of Basic Energy Sciences, Materials Sciences Division, under Contract No. DE-AC03-76SF00098. Additional support was provided by the Miller Institute (E. A.), as well as the Basque Government and the Fulbright Foundation (J. G. de A.).

## REFERENCES

1. A. Kay, E. Arenholz, S. Mun, F.J. Garcia de Abajo, C.S. Fadley, R. Denecke, Z. Hussain, and M.A. Van Hove, *Science* **281**, 679(1998).
2. F.J. Garcia de Abajo, C.S. Fadley, and M.A. Van Hove, *Phys. Rev. Lett.* **82**, 4126 (1999).
3. E. Arenholz, A. Kay, C.S. Fadley, to be published.
4. U. Fano, Phys. Rev. **124**, 1866 (1961); U. Fano and J. W. Cooper, *Rev. Mod. Phys.* **40**, 493 (1968).
5. (a) M.O. Krause, T.A. Carlson, and A. Fahlman, *Phys. Rev.* **A30**, 1316 (1984) and refs. therein; (b) L.J. Garvin, E.R. Brown, S.L. Carter, and H.P. Kelly, *J. Phys.* **B16**, L269 (1983).
6. K.L.I. Kobayashi, N. Watanabe, H. Nakashima, M. Kubota, H. Daimon, and Y. Murata, *Phys. Rev. Lett.* **52**, 160 (1984).

7. P. Pervan, M. Milun, and D.P. Woodruff, *Phys. Rev. Lett.* **81**, 4995 (1999).
8. Y.F. Hu, G.M. Bancroft, and K.H. Tan, private communication.
9. E. Arenholz, A.W. Kay, C.S. Fadley, M.M. Grush, T.A. Callcott, D.L. Ederer, C. Heske, and Z. Hussain, to be published.
10. A. Kikas, A. Maiste, E. Nommiste, R. Ruus, and A. Saar, "Multi-Atom Resonant Photoemission in Transition Metal Chlorides", presented at the X99 Conference, Chicago, IL, August, 1999.
11. M.G. Garnier, N. Witkowski, R. Denecke, D. Nordlund, A. Nilsson, M. Nagasono, and N. Mårtensson, and A. Föhlisch, private communication.
12. O. Hemmers, H.H. Wang, D. Lindle, and co-workers, private communication.
13. E.A. Stern, in *X-Ray Absorption: Principles, Applications, and Techniques of EXAFS, SEXAFS, and XANES*, D.C. Konigsberger and R. Prins, Eds., New York: Wiley, 1988, pp. 3-51; E.A. Stern, M. Newville, B. Ravel, Y. Yacoby, and D. Haskel, Physica B, **208-209**, 117 (1995); M. Newville et al., Physica B, **208-209**, 154 (1995); and refs. therein.
14. M.A. Van Hove, W.H. Weinberg, and C. M. Chan, *Low Energy Electron Diffraction*, Heidelberg: Springer, 1986.
15. C.S. Fadley, Y. Chen, R.E. Couch, H. Daimon, R. Denecke, J.D. Denlinger, H. Galloway, Z. Hussain, A.P. Kaduwela, Y.J, Kim, P.M. Len, J. Liesegang, J. Menchero, J. Morais, J. Palomares, S.D. Ruebush, E. Rotenberg, M. B. Salmeron, R. Scalettar, W. Schattke, R. Singh, S. Thevuthasan, E.D. Tober, M.A. Van Hove, Z. Wang, and R.X. Ynzunza, *Prog. in Surf. Sci.* **54**, 341 (1997) and refs. therein.
16. T. Förster, *Ann. Phys. (Leipzig)* **2**, 55 (1948).
17. D.L. Dexter, *J. Chem. Phys.* **21**, 836 (1953).
18. L. Van Hove, *Physica* **21**, 517 (1955).
19. P.T. Rieger, S. P. Palese, and R. J. Dwayne Miller, *Chem. Phys.* **221**, 85 (1997) and refs. therein.
20. D.A. Shirley, *Phys. Rev. B* **55**, 4709 (1972).
21. Z. Hussain, W.R.A. Huff, S.A. Kellar, E.J. Moler, P.A. Heimann, W. McKinney, H.A. Padmore, C.S. Fadley, and D.A. Shirley, *J. Electron Spectrosc. Relat. Phenom.* **80**, 401 (1996).
22. C.S. Fadley M.A. Van Hove, Z. Hussain, and A.P. Kaduwela, *J. Electron Spectrosc. Relat. Phenom.*, **75**, 273 (1995).
23. A. Tanaka and T. Jo, *J. Phys. Soc. Jpn.* **63**, 2788 (1994) and refs. therein.
24. S.M. Butorin, J.-H. Guo, M. Magnuson, P. Kuiper, and J. Nordgren, *Phys. Rev. B* **54**, 4405 (1996).
25. B.L. Henke, *Phys. Rev. A* **6**, 94 (1972); M. Mehta and C. S. Fadley, *Phys. Letters* **55A**, 59 (1975).
26. Electron escape depths estimated from (a) C.J. Powell, A. Jabloski, S. Tanuma, and D.R. Penn, *J. Electron Spectrosc. Relat. Phenom.* **68**, 605 (1994), and (b) C.J. Powell, private comm.
27. X-ray optical constants, penetration depths and reflectivities off resonance have been estimated from published sources: B.L Henke, E.M. Gullikson, and J.C. Davis, *Atomic Data and Nuc. Data Tables* **54**, 181 (1993) and E.M. Gullikson, Center for X-ray Optics, Lawrence Berkeley National Laboratory, public access program available at http://www.cxro-lbl.gov.
28. S.-H. Yang, B.S. Mun, S.K. Kim, J.B. Kortright, J.H. Underwood, Z. Hussain, and C.S. Fadley, to be published.
29. P. Kuiper, B.G. Searle, P. Rudolf, L.H. Tjeng, and C.T. Chen, *Phys. Rev. Lett.* **70**, 1549 (1993).
30. (a) M.A. Brewer et al., *Advanced Light Source Compendium of User Abstracts and Technical Reports*, Report LBNL-39981, UC-411 (Lawrence Berkeley National Laboratory, Berkeley, CA, 1997) p. 411; and K. Krishnan, private communication; (b) B.T. Thole et al., *Phys. Rev. B* **32**, 5107(1985).
31. J.J. Jia, T.A. Callcott, J. Yurkas, A.W. Ellis, F.J. Himpsel, M.G. Samant, J. Stohr, D.L. Ederer, J.A. Carlisle, E.A. Hudson, L.J. Terminello, D.K. Shuh, and R.C. Perera, *Rev. Sci. Instrum.* **66**, 1394 (1995).
32. G. Faigel and M. Tegze, Repts. Prog. in Phys. **62**, 355-393 (1999).
33. N.F. Mott and I.N. Sneddon, *Wave Mechanics and its Applications*, London, Oxford University Press, 1948), pp. 338-339.
34. J.P. Desclaux, in *Relativistic Effects in Atoms, Molecules, and Solids*, edited by G.L. Malli, New York, Plenum Press, 1981, pp. 115-143.

35. T. Gog, P.M. Len, D. Bahr, C.S. Fadley, G. Materlik, and C. Sanchez-Hanke, *Phys. Rev. Lett.* **76**, 3132 (1996); P.M. Len, C.S. Fadley, and G. Materlik, in *X-ray and Inner-Shell Processes: 17th International Conference*, R.L. Johnson, H. Schmidt-Boecking, and B.F. Sonntag, Eds., New York, AIP, 1997, American Institute of Physics Conference Proceedings, No. 389 pp. 295-319.
36. M. Labute, S.-H. Yang, E. Arenholz, P.M. Len, J.F. Garcia de Abajo, C.S. Fadley, and M.A. Van Hove, to be published.
37. J. Hunter Dunn, D. Arvanitis, R. Carr, and N. Mårtensson, to appear in Phys. Rev. Letters.

# X-ray Resonant Raman Scattering

Paolo Carra

*European Synchrotron Radiation Facility,*
*B.P. 220, F-38043 Grenoble Cédex, France.*

**Abstract.** A theoretical analysis of x-ray resonant Raman scattering in solids is outlined, with emphasis on symmetry properties of the scattering amplitude. Experimental configurations are also discussed to identify the electronic states probed by different energy scans.

## INTRODUCTION

For energies $\hbar\omega < 100$ keV, x rays interact weakly with bulk electrons, and their scattering may be accurately treated in the lowest Born approximation. The double differential cross section is then obtained by considering the usual semi-classical coupling

$$H = \frac{1}{2m}\sum_j \left(\mathbf{p}_j + \frac{e}{c}\mathbf{A}(\mathbf{r}_j)\right),$$

where $\mathbf{A}$ denotes the vector potential of the electromagnetic field; $m$ and $\mathbf{p}$ identify electron mass and momentum, respectively. (Relativistic corrections, which are of order $\hbar\omega/mc^2$, will not be considered in the current work.)

Away from resonance, the scattering is controlled by the $\mathbf{A}^2$ term (treated to first order), giving rise to a variety of phenomena as reviewed, for example, by Shülke [1].

Near an absorption edge, the $\mathbf{p}\cdot\mathbf{A}$ term (treated to second order) can result in significant contributions to the scattering amplitude. These effects have been thoroughly investigated in the elastic and inelastic limits, probing magnetic and anisotropic (low point-group symmetry) systems [2–4]. Particular attention has been given to resonant inelastic events characterised by the presence of a core hole in the final state: *the x-ray resonant Raman scattering* (XRRS) [5,6].

Theoretically, the process has been studied by applying spherical-tensor methods [7,8], previously developed for x-ray dichroism [9,10] and resonant diffraction [11]. Under a certain approximation, namely for fast collisions, simple expressions for the resonant amplitude are obtained: effective charge and magnetic (spin and orbital) operators are found, which describe the scattering in the allowed polarisation channels.

These developments provide an opportunity to outline here a comprehensive analysis of XRRS, with emphasis on symmetry properties and experimental configurations.

# THEORETICAL FRAMEWORK

This section discusses general features for x-ray resonant inelastic scattering in solids. The formalism we introduce will be specialised to XRRS in the remaining part of the paper.

## Response Function

Consider an x-ray resonant inelastic scattering event, where an energy $\hbar\omega = \hbar(\omega_\mathbf{k} - \omega_{\mathbf{k}'})$, and a momentum $\hbar\mathbf{q} = \hbar(\mathbf{k} - \mathbf{k}')$ are transferred to the target. The scattering rate is provided by Fermi's golden rule:

$$P(i \to f) = \frac{2\pi}{\hbar} \mid U_{i \to f} \mid^2 \delta(E_f + \hbar\omega_{\mathbf{k}'} - E_i - \hbar\omega_\mathbf{k}),$$

with the resonant transition amplitude $U_{i \to f}$ given by

$$U_{i \to f} = \sum_I N_\mathbf{k} N_{\mathbf{k}'} \sqrt{n_{\epsilon\mathbf{k}}} \frac{\langle \psi_f \mid \hat{H}^\dagger_{\epsilon'\mathbf{k}'} \mid \psi_n \rangle \langle \psi_n \mid \hat{H}_{\epsilon\mathbf{k}} \mid \psi_i \rangle}{E_i + \hbar\omega_\mathbf{k} - E_n + i\Gamma_n/2}, \quad (1)$$

with $N_\mathbf{k} = \sqrt{2\pi\hbar c^2/V\omega_\mathbf{k}}$. The operators $\hat{H}_{\epsilon\mathbf{k}}$ can be written as:

$$\hat{H}_{\epsilon\mathbf{k}} = \sum_{\boldsymbol{\kappa}\mathbf{G}} \sum_{l_z\sigma m} M_{l_z\sigma;jm}(\boldsymbol{\epsilon}; \mathbf{k}, \boldsymbol{\kappa}) l^\dagger_{l_z\sigma,\boldsymbol{\kappa}+\mathbf{k}+\mathbf{G}} c_{jm,\boldsymbol{\kappa}}, \quad (2)$$

with $\mathbf{G}$ a reciprocal lattice vector, and

$$M_{l_z\sigma;jm}(\boldsymbol{\epsilon}, \mathbf{k}, \boldsymbol{\kappa}) = \frac{e}{mc} \sum_\mathbf{R} e^{-i(\boldsymbol{\kappa}+\mathbf{k})\mathbf{R}} \int d\mathbf{x} \varphi^*_{l_z\sigma}(\mathbf{x}, \mathbf{R}) e^{i\mathbf{k}\cdot\mathbf{x}} \boldsymbol{\epsilon} \cdot \mathbf{p} \varphi_{cjm}(\mathbf{x}, 0). \quad (3)$$

Expressions (1)-(3) are derived by inserting the plane-wave expansion for the vector potential,

$$\mathbf{A}(\mathbf{r}) = \sum_{\mathbf{k}\epsilon} \sqrt{\frac{2\pi\hbar c^2}{V\omega_\mathbf{k}}} \left(\boldsymbol{\epsilon} a_\mathbf{k} e^{i\mathbf{k}\cdot\mathbf{r}} + \text{h.c.}\right),$$

into the interaction hamiltonian:

$$\hat{H}_{int} = \frac{e}{mc} \int d\mathbf{r} \mathbf{A}(\mathbf{r}) \cdot \Psi^\dagger(\mathbf{r}) \mathbf{p} \Psi(\mathbf{r}).$$

The electron field $\Psi(\mathbf{r})$ is taken to be of the form

$$\Psi(\mathbf{x}) = \sum_{\boldsymbol{\kappa}}^{BZ} \left[ \sum_{l_z\sigma} l_{l_z\sigma,\boldsymbol{\kappa}} \psi_{l_z\sigma,\boldsymbol{\kappa}}(\mathbf{x}) + \sum_m c_{jm,\boldsymbol{\kappa}} \psi_{cjm,\boldsymbol{\kappa}}(\mathbf{x}) \right], \qquad (4)$$

with the Bloch waves expanded using Wannier functions

$$\psi_{l_z\sigma,\boldsymbol{\kappa}}(\mathbf{x}) = \frac{1}{N^{\frac{1}{2}}} \sum_{\mathbf{R}} e^{i\boldsymbol{\kappa}\cdot\mathbf{R}} \varphi_{l_z\sigma}(\mathbf{x}-\mathbf{R}).$$

Expression (4) amounts to assuming that, for $\hbar\omega_\mathbf{k}$ near an absorption threshold, the transition amplitude $U_{i\to f}$ is dominated by the processes in which the ingoing photon, of momentum $\mathbf{k}$ and polarisation $\boldsymbol{\epsilon}$, is first absorbed by promotion of a core electron into the conduction band; the core-hole is then annihilated by creation of a hole in the conduction band and emission of a photon of momentum $\mathbf{k}'$ and polarization $\boldsymbol{\epsilon}'$.

The notation is as follows: $\mathbf{R}$ identifies a lattice site, and $N$ the number of sites; $l_{l_z\sigma,\boldsymbol{\kappa}}$ and $c_{jm,\boldsymbol{\kappa}}$ denote annihilation operators for conduction and core electrons, respectively. The core electron Wannier functions are labelled by atomic quantum numbers $c$ (orbital angular momentum), $j = c \pm \frac{1}{2}$ (total angular momentum), and $m$ (its projection on the quantisation axis); the spin-orbit coupling of the conduction band is neglected, thus justifying the labelling by a set of orbital quantum numbers $l_z$ and spin $\sigma$.

In expression (1), the intermediate states $|\psi_n\rangle$ are eigenstates of the system Hamiltonian with energies $E_n$; they are assumed to be of the form: $|\psi_n\rangle = l^\dagger c |\psi_i\rangle$. This amounts to neglecting many core-hole decay processes, e.g. non-radiative (Auger) decays, which might occur before the core-hole is filled by a conduction electron. In order not to completely disregard these events, a finite core-hole lifetime, $\tau_n = \hbar/\Gamma_n$, has been introduced.

The total scattering rate is obtained by summing $P_{i\to f}$ over all final states allowed by energy conservation; it is proportional to the response function (dynamic structure factor)

$$S(\mathbf{q},\omega) = \int_{-\infty}^{\infty} dt\, e^{i\omega t} \langle \psi_i | \hat{O}^\dagger(\mathbf{q},t) \hat{O}(\mathbf{q}) | \psi_i \rangle, \qquad (5)$$

with $\hat{O}(\mathbf{q},t) = e^{-i\hat{H}/\hbar} \hat{O}(\mathbf{q}) e^{i\hat{H}/\hbar}$, and

$$\hat{O}(\mathbf{q}) = \sum_n \frac{\hat{H}_{\text{int}} | \psi_n \rangle \langle \psi_n | \hat{H}_{\text{int}}}{E_i + \hbar\omega_\mathbf{k} - E_n + i\Gamma_n/2}.$$

The double differential cross-section is obtained by multiplying the total scattering rate by the density of states of the outgoing photon, $V\omega_{\mathbf{k}'}^2/\hbar c^3 (2\pi)^3$, and dividing by the incident flux, $cn_{\boldsymbol{\epsilon}\mathbf{k}}/V$.

## Multipolar Expansion

X-ray resonant scattering proceeds from the excitation of electric dipole and quadrupole transitions; an appropriate formulation is obtained by performing a multipolar expansion of the electron-photon interaction [12].

The $\mathbf{p} \cdot \mathbf{A}$ coupling between x rays and matter expands into spherical Bessel functions $g_l(k_i r)$ and spherical harmonics of $\mathbf{k}_i$ and $\mathbf{r}$:

$$\mathbf{p} \cdot \mathbf{A} \to \mathbf{p} \cdot \mathbf{e}_i \, g_l(k_i r) \sum_m Y_m^{l*}(\hat{\mathbf{k}}_i) Y_m^l(\hat{\mathbf{r}}).$$

Recoupling $\mathbf{p}$ and $Y^l(\hat{\mathbf{r}})$ to a total L yields the term

$$\sum_L \left[ [\mathbf{e}_i Y^l(\hat{\mathbf{k}}_i)]^L [\mathbf{p} Y^l(\hat{\mathbf{r}})]^L \right]^0 g_l(k_i r),$$

with the couplings defined by

$$[T^{L'} F^{L''}]_l^L = \sum_{l'l''} T_{l'}^{L'} F_{l''}^{L''} \langle L'L''l'l'' \mid Ll \rangle.$$

The product $g_l(k_i r)[\mathbf{p}, Y^l(\hat{\mathbf{r}})]_M^L$ determines the structure of the paramagnetic coupling that promotes an inner-shell electron to an empty valence orbital. The factor $[\mathbf{e}_i, Y^l(\hat{\mathbf{k}}_i)]_M^L$ describes the geometry of x-ray scattering. For $L = 0$, we have: $[\mathbf{e}_i, Y^1(\hat{\mathbf{k}}_i)]^0 = -(1/\sqrt{3})\mathbf{e}_i \cdot \hat{\mathbf{k}}_i = 0$. For $L > 0$, three values of $l$ contribute. In the long wavelength limit $k_i r \ll 1$, $g_l(k_i r) \sim (k_i r)^l$, and the lowest value $l = L - 1$ dominates. We obtain

$$[\mathbf{p}, Y^{L-1}(\hat{r})]_M^L r^{L-1} = \frac{1}{\sqrt{L(2L+1)}} \mathbf{p} \cdot \nabla(r^L Y_M^L),$$

yielding

$$\langle \psi_2 \mid \mathbf{p} \cdot \mathbf{A} + \mathbf{A} \cdot \mathbf{p} \mid \psi_1 \rangle \sim \frac{2m(E_2 - E_1)}{i\hbar \sqrt{L(2L+1)}} \langle \psi_2 \mid r^L Y_M^L(\hat{\mathbf{r}}) \mid \psi_1 \rangle,$$

i.e., an electric multipole matrix element. Here, for simplicity, we have dealt only with the electric part of the photon wave-function expansion in angular momentum and parity eigenstates. Full details, including magnetic multipoles, can be found in the book by Akhiezer and Berestetsky [13].

## Localised Model

By leaving a localised hole, x-ray resonant scattering selects a specific site in the solid. A local process is thus expected to control the excitation, to leading order. Additional contributions should emerge when the remaining sites in the lattice are taken into account, that is, when electron delocalisation is included. In the following, only the local contribution to the scattering will be retained, i.e., we will assume that the multipolar matrix element

$$\sum_{\mathbf{R}} e^{-i(\boldsymbol{\kappa}+\mathbf{k})\mathbf{R}} \int d\mathbf{x} \varphi^*_{\lambda\sigma}(\mathbf{x},\mathbf{R}) r^L Y^L_q(\mathbf{x}) \varphi_{cjm}(\mathbf{x},0)$$

is non-negligible only for $\mathbf{R}=0$. (The validity of this approximation has been discussed by Benoist and co-workers [14], using a minimal set of orthonormal linear muffin-tin orbitals.) Furthermore, Wannier functions with an atomic-like symmetry will be considered.

## Fast Collision Approximation

In the following section we will consider the case where the intermediate state lifetime, determined by the smaller of $\tau_n$ and $\hbar/\mid \hbar\omega_\mathbf{k} - E_n + E_i \mid$, is so small that photon absorption and emission are practically simultaneous (no core-hole propagation). Experimentally, this amounts to having

$$\text{Max}(\mid \hbar\omega_\mathbf{k} - E_n + E_i \mid, \Gamma_n) \gg D,$$

with $D = \langle (E_n - \langle E_n \rangle)^2 \rangle^{\frac{1}{2}}$ of the order of the conduction bandwidth. In a metal, the propagation of the core hole would give rise to the well-known infrared singularities [15]; such effects can be disregarded, if the above condition is fulfilled. The above assumption is equivalent to neglecting the dispersion of the intermediate states; $E_n$ and $\Gamma_n$ can be taken as constants, and the expansion for the resonant denominator

$$(E_n - E_i - \hbar\omega_\mathbf{k} - i\frac{\Gamma_n}{2})^{-1} = (\langle E_n \rangle - E_i - \hbar\omega_\mathbf{k} - i\frac{\langle \Gamma_n \rangle}{2})^{-1}$$
$$\sum_{n=0}^{\infty} \left( \frac{\langle E_n - i\frac{\Gamma_n}{2}\rangle - E_n + i\frac{\Gamma_n}{2}}{\langle E_n \rangle - E_i - \hbar\omega_\mathbf{k} - i\frac{\langle \Gamma_n \rangle}{2}} \right)^n, \quad (6)$$

truncated at $n=0$, thus reducing the resonant denominator to

$$\langle G(\omega_\mathbf{k}) \rangle = (E_g + \hbar\omega_\mathbf{k} - \langle E_n \rangle + i\Gamma_n/2)^{-1}.$$

It is known as the fast collision approximation [11,16]. When collisions are fast, the sum over the intermediate states in expression (1) can be removed by completeness.

# RESONANT RAMAN SCATTERING

## Symmetry analysis

XRRS amounts to having a final state $\mid \psi_f \rangle$ with an extra electron in the conduction (valence) band and a hole in a core level. (In a rare-earth system, this would correspond, for example, to a $2p \to 4f$ excitation followed by a $3d \to 2p$ decay.) In the fast collision approximation, the double-differential cross section can be given the form

$$\frac{d^2\sigma}{d\Omega d\hbar\omega_{\mathbf{k}'}} = \frac{\lambda}{\lambda'} | \langle G(\omega_{\mathbf{k}}) \rangle |^2 \sum_f \left| \sum_{z,\zeta} T_\zeta^{(z)*}(\boldsymbol{\epsilon}'^*, \mathbf{k}'; \boldsymbol{\epsilon}, \mathbf{k}) \langle \psi_f | F_\zeta^{(z)}(k, k') | \psi_i \rangle \right|^2$$
$$\delta(E_f + \hbar\omega_{\mathbf{k}'} - E_i - \hbar\omega_{\mathbf{k}}), \tag{7}$$

with

$$T_\zeta^{(z)*}(\boldsymbol{\epsilon}'^*, \mathbf{k}'; \boldsymbol{\epsilon}, \mathbf{k}) = \frac{[z]^{\frac{1}{2}}}{[L]^{\frac{1}{2}}} \sum_{M,M'} C^{LM}_{L'M';z\zeta} \left[ \boldsymbol{\epsilon} \cdot \mathbf{Y}^*_{LM}(\hat{\mathbf{k}}) \right] \left[ \boldsymbol{\epsilon}'^* \cdot \mathbf{Y}_{L'M'}(\hat{\mathbf{k}}') \right]. \tag{8}$$

The transition operator, $F^{(z)}$, is defined by

$$F_\zeta^{(z)}(\lambda, \lambda') = \sum_{l_z, \sigma, \text{all } m} S_\zeta^{(z)}(L\lambda, L'\lambda') c_{j_2 m_2} l^\dagger_{l_z \sigma}, \tag{9}$$

showing that, for fast collisions, the scattering is described by standard particle-hole creation operators. The angular part, $S_\zeta^{(z)}(L\lambda, L'\lambda')$, takes the form

$$S_\zeta^{(z)}(L\lambda, L'\lambda') = R^{L'\lambda'}_{L\lambda}(c_1, l; c_2, c_1) \frac{[z]^{\frac{1}{2}}}{[L]^{\frac{1}{2}}} \sum_{\substack{M,M' \\ \gamma_1 \gamma_2 \gamma'_1 \sigma'}} (-)^{-M'} C^{LM}_{L'M';z\zeta} \tag{10}$$

$$C^{j_1 m_1}_{c_1 \gamma'_1; \frac{1}{2}\sigma'} C^{j_2 m_2}_{c_2 \gamma_2; \frac{1}{2}\sigma'} C^{j_1 m_1}_{c_1 \gamma_1; \frac{1}{2}\sigma} C^{c_1 \gamma'_1}_{c_2 \gamma_2; L'-M'} C^{l l_z}_{c_1 \gamma_1; LM}.$$

Recoupling with use of standard theorems of angular momentum theory [17], we find

$$S_\zeta^{(z)}(L\lambda, L'\lambda') = R^{L'\lambda'}_{L\lambda}(c_1, l; c_2, c_1)(-1)^{j_1 + j_2 + 1}[j_1][j_2 c_1 z l]^{\frac{1}{2}}$$
$$\begin{Bmatrix} j_1 & j_2 & L' \\ c_2 & c_1 & \frac{1}{2} \end{Bmatrix} \sum_{jm} [j] \begin{Bmatrix} j_1 & j & L \\ l & c_1 & \frac{1}{2} \end{Bmatrix} \begin{Bmatrix} L & L' & z \\ j_2 & j & j_1 \end{Bmatrix} \tag{11}$$
$$\begin{pmatrix} \frac{1}{2} & l & j \\ \sigma & l_z & -m \end{pmatrix} \begin{pmatrix} j & z & j_2 \\ m & -\zeta & -m_2 \end{pmatrix},$$

with $[a \cdots b] = (2a+1) \cdots (2b+1)$. This result is particularly instructive. The last $3j$ symbol in expression (11) implies that, for fast collisions, XRRS can be viewed as a direct $2^z$-pole transition between the core level $j_2 m_2$ and the valence empty state $(l\frac{1}{2})jm$; the value of $z$ results from the coupling between ingoing and outgoing photons. In contrast to real absorption, however, the effective $2^z$-pole transition operator is not purely orbital; it also displays a spin dependence, as the spin-orbit coupling in the intermediate state (the $j_1$ level) allows for spin transitions, even in the absence of spin-orbit interaction in the ground and final states.

The foregoing results lead to the following form for the double-differential scattering cross-section

$$\frac{d^2\sigma}{d\Omega_{\mathbf{k}'}d\hbar\omega_{\mathbf{k}'}} = 8\pi\lambda^2 \int_{-\infty}^{\infty} dt\, e^{i(\omega_{\mathbf{k}}-\omega_{\mathbf{k}'})t} \sum_r \sum_{zz'} \mathcal{T}_0^{(zz')r} \langle g | \mathcal{O}_0^{(zz')r}(t) | g \rangle, \quad (12)$$

with the scattering geometry, $\mathcal{T}_0^{(zz')r}$, given by:

$$\mathcal{T}_0^{(zz')r} = \sum_{\zeta\zeta'} C^{r0}_{z\zeta;z'\zeta'} T^z_\zeta T^{z'}_{\zeta'}. \quad (13)$$

The scattering operator is given by

$$\mathcal{O}_0^{(zz')r}(t) = |\langle G(\omega_{\mathbf{k}}) \rangle|^2 \sum_{\zeta\zeta'} C^{r0}_{z\zeta;z'\zeta'} \sum_{m_2 m'_2 l_z l'_z \sigma\sigma'} S^{(z)\dagger}_\zeta S^{(z')}_{\zeta'} c^\dagger_{j_2 m'_2}(t) l_{l'_z \sigma'}(t) l^\dagger_{l_z \sigma} c_{j_2 m_2}. \quad (14)$$

As observed, Eq. (11) describes an absorption process, induced by an effective photon of energy $\hbar(\omega_{\mathbf{k}} - \omega_{\mathbf{k}'})$. Integrating over $\hbar\omega_{\mathbf{k}'}$, i.e summing over the final states, leads to an expansion for the integrated cross-section in terms of simple spin and orbital operators. In expression (13), ingoing and outgoing photons are coupled together. In some circumstances, namely when dealing with ingoing and/or outgoing isotropic photons, it is advantageous to work with each photon coupled to itself. This amounts to a re-definition of the geometric factors

$$\mathcal{T}_0^{(zz')r} \to \tilde{\mathcal{T}}_0^{(zz')r} = \sum_{\zeta\zeta'} C^{r0}_{z-\zeta;z'-\zeta'} \tilde{T}^z_\zeta(L) \tilde{T}^{z'}_{\zeta'}(L'),$$

with

$$\tilde{T}^z_\zeta(L) = [z]^{\frac{1}{2}}[L]^{-\frac{1}{2}} \sum_{M,M'} C^{LM}_{LM';z\zeta} \left[\boldsymbol{\epsilon} \cdot \mathbf{Y}^*_{LM}(\hat{\mathbf{k}})\right]\left[\boldsymbol{\epsilon}^* \cdot \mathbf{Y}_{LM'}(\hat{\mathbf{k}})\right]. \quad (15)$$

The scattering operator needs to be transformed accordingly: $\mathcal{O}_0^{(zz')r} \to \tilde{\mathcal{O}}_0^{(zz')r}$. Then, applying standard diagrammatic methods [17], one has

$$\frac{d\sigma}{d\Omega_{\mathbf{k}'}} \cong 8\pi\lambda^2 \sum_{zz'r} \tilde{\mathcal{T}}_0^{(zz')r}(L,L') \langle g | \tilde{\mathcal{O}}_0^{(zz')r}(0) | g \rangle \quad (16)$$

with [7]

$$\tilde{\mathcal{O}}_0^{(zz')r}(0) = |\langle G(\omega_{\mathbf{k}}) \rangle R^{L'\lambda'}_{L\lambda}(c_1, l; c_2, c_1)|^2 (-1)^{j_1+j_2+c_1+c_2}$$

$$[j_1]^2[j_2][zz']^{\frac{1}{2}} \begin{Bmatrix} j_1 & j_2 & L' \\ c_2 & c_1 & \frac{1}{2} \end{Bmatrix}^2 \begin{Bmatrix} L' & L' & z' \\ j_1 & j_1 & j_2 \end{Bmatrix} \sum_{ab}[ab]^{\frac{1}{2}} \quad (17)$$

$$\left[\sum_x [x] \begin{Bmatrix} a & b & r \\ z & z' & x \end{Bmatrix} \begin{Bmatrix} c_1 & \frac{1}{2} & j_1 \\ c_1 & \frac{1}{2} & j_1 \\ x & a & z' \end{Bmatrix} \begin{Bmatrix} L & l & c_1 \\ L & l & c_1 \\ z & b & x \end{Bmatrix}\right] W_0^{(ab)r},$$

that is, the cross-section expressed as a linear combination of *hole* double-tensor operators [11,18]

$$W_0^{(ab)r} = -\frac{[ab]^{\frac{1}{2}}}{2[l]^{\frac{1}{2}}} \sum_{\alpha\beta} C_{a-\alpha;b-\beta}^{r0} \sum_{l_z l'_z,\sigma\sigma'} C_{\frac{1}{2}\sigma';a\alpha}^{\frac{1}{2}\sigma} C_{ll'_z;b\beta}^{ll_z} l^\dagger_{l'_z\sigma'} l_{l_z\sigma},$$

describing the multipole moments of the charge and magnetic distributions of the valence $l$-electrons. (The complete derivation of these results is given in Ref. [8]). One has : $W^{(00)} \sim n_h$ (number of holes), $W^{(11)0} \sim \sum_i \mathbf{s}_i \cdot \mathbf{l}_i$ (spin-orbit), $W^{(01)} \sim \mathbf{L}$ (orbital angular momentum), $W^{(10)} \sim \mathbf{S}$ (spin); higher-order tensors are discussed in Ref. [19].

Different values of the tensor rank $r$ are selected by the order of the transitions $(L, L')$, and by photon polarisations, as determined by $\tilde{T}_0^{(zz')r}(L, L')$.

## Experimental configurations

XRRS experiments can be performed in different configurations (energy scans); different electronic states are probed accordingly, as described below.

Consider the Kramers-Heisenberg formula

$$\frac{d^2\sigma}{d\Omega_{\mathbf{k}'}d\hbar\omega_{\mathbf{k}'}} = \frac{\omega_{\mathbf{k}'}}{\omega_{\mathbf{k}}} \sum_{qq'} \sum_f \left| \sum_n \frac{\langle f|D_{q'}^{(1)}|n\rangle\langle n|D_q^{(2)}|i\rangle}{E_i + \hbar\omega_{\mathbf{k}} - E_n + i\Gamma_n/2} \right|^2$$
$$\delta(E_f + \hbar\omega_{\mathbf{k}'} - E_i - \hbar\omega_{\mathbf{k}}), \qquad (18)$$

where $D_q^{(L)}$, with $q = -L, ..., L$, denotes the components of the electric $2^L$-pole operator. A two-dimensional analysis of the spectrum, associated with expression (18), exhibits a variety of effects. A scan parallel to the $\hbar\omega_{\mathbf{k}}$-axis amounts to moving through the intermediate-state structure. The spectrum has resolution $\Gamma_n$, and is obtained by retaining only one final state (selected by the given energy transfer, up to $\gamma_f$) in Eq. (18). (To account for the finite lifetime of the final states, we broaden the energy-conservation $\delta$ function in the Kramers-Heisenberg formula with a Lorentian of width $\gamma_f$.)

Parallel to the $\hbar\omega = \hbar(\omega_{\mathbf{k}} - \omega_{\mathbf{k}'})$-axis, the final-state structure is scanned, with resolution $\gamma_f$. In this case, only a $\Gamma_n$-wide set of intermediate states (selected by the fixed ingoing energy) contribute in Eq. (18). For systems with incomplete $4f$ and $5d$ shells, two groups of final states are observed: (i) the $3d^94f^{n+1}$ multiplet reached from the intermediate states $2p^54f^{n+1}$, as probed by Krisch and co-workers [6]; for a suitable choice of photon polarisation, the spectrum is equivalent to a $3d \to 4f$ dipolar absorption, as discussed below. (ii) the $3d^94f^n5d^{m+1}$ final states, reached from the $2p^54f^n5d^{m+1}$ intermediate states, formally equivalent to a $3d \to 5d$ monopole transition, not observable in absorption.

Keeping $\hbar\omega_{\mathbf{k}'}$ fixed, while varying $\hbar\omega_{\mathbf{k}}$, as performed by Hämäläinen *et al.* [5], amounts to moving along the 45° line in the $\hbar\omega_{\mathbf{k}}, \hbar\omega$ plane. In this case, a superposition of the full set of intermediate and final states is recorded. Considerably different lifetimes, $\Gamma_n$ and $\gamma_f$, determine the particular structure. When one resolution is rather broad, it is still possible to observe narrow structures; these, however, are entirely due to the other, better resolved set of states. (When $\Gamma_n \gg \gamma_f$, the spectral resolution is controlled by $\gamma_f$.)

In the case of a $2p$-electron excitation into a broad $5d$ band, the interaction with the core-hole is believed to be small, and the one-electron picture to provide a good description of the electron density of states; general features of the spectra can then be derived by simple arguments. Consider the resonance: $5d^0 \to 2p^5 5d^1 \to 3d^9 5d^1$, as described by Eq. (18), with matrix element: $\langle f|D^{(1)}|n\rangle\langle n|D^{(1)}|g\rangle$; intermediate and final state energies can be defined as: $E_n = \epsilon_{dn} + \epsilon_{2p}$ and $E_f = \epsilon_{df} + \epsilon_{3d}$, with $\epsilon_{dn}$ and $\epsilon_{df}$ the one-electron energies of the $5d$ continuum (it could also denote the energy of a state of the $d^{m+1}$ configuration, in a multiplet picture). As $\langle f|D^{(1)}|n\rangle \sim \langle 2p|D^{(1)}|3d\rangle\langle \varepsilon_{dn}|\varepsilon_{df}\rangle$, the band structure $|\langle n|D^{(1)}|g\rangle|^2$ will appear only along the diagonal in the $\hbar\omega_\mathbf{k}$, $\hbar\omega$ plane. ($\varepsilon_{dn}$ changes along $\hbar\omega_\mathbf{k}$, whereas $\varepsilon_{df}$ changes along $\hbar\omega$, with $\langle \varepsilon_{dn}|\varepsilon_{df}\rangle = \delta_{dn,df}$ yielding intensity only on the diagonal.)

We note, in passing, that detection of off-diagonal spectral weight yields information on core-hole interactions in transitions to broad bands, therefore providing a stringent test of one-electron (band structure) calculations of x-ray spectra.

## CONCLUDING REMARKS

The previous sections have provided a summary of results for XRRS. It has been shown that, when the intermediate-state core hole is extremely short-lived, the scattering amplitude can be expressed in terms of elementary charge, spin and orbital effective operators, leading to a standard two-particle form for the dynamic structure factor. Experimental configurations have also been discussed to identify the electronic states probed by different energy scans.

## REFERENCES

1. W. Schülke, in *Handbook on Synchrotron Radiation*, G. Brown and D. E. Moncton Eds., Elsevier Science Publishers, Amsterdam (1991).
2. D. Gibbs, D. R. Harshman, E. D. Isaacs, D. B. McWhan, D. Mills, and C. Vettier, Phys. Rev. Lett. **61**, 1241 (1988).
3. J.-E. Rubensson, D. Mueller, R. Shuker, D. L. Ederer, C. H. Zhang, J. Jia, and T. A. Callcott, Phys. Rev. Lett. **64**, 1047 (1990).
4. K. D. Finkelstein, Q. Shen, and S. Shastri, Phys. Rev. Lett. **69**, 1612 (1992).
5. K. Hämäläinen, D. P. Siddons, J. B. Hastings, and L. E. Berman, Phys. Rev. Lett. **67**, 2850 (1991).
6. M. H. Krisch, C. C. Kao, F. Sette, W. A. Caliebe, K. Hämäläinen, and J. B. Hastings, Phys. Rev. Lett. **74**, 4931 (1995).
7. P. Carra, M. Fabrizio, and B. T. Thole, Phys. Rev. Lett. **74**, 3700 (1995).
8. M. van Veenendaal, P. Carra, and B. T. Thole, Phys. Rev. B **54**, 16010 (1996).

9. B. T. Thole, P. Carra, F. Sette, and G. van der Laan, Phys. Rev. Lett. **68**, 1943 (1992).
10. P. Carra, B. T. Thole, M. Altarelli, and X. Wang, Phys. Rev. Lett. **70**, 694 (1993).
11. J. Luo, G. T. Trammell and J. P. Hannon, Phys. Rev. Lett. **71**, 287 (1993).
12. P. Carra and B. T. Thole, Rev. Mod. Phys. **66**, 1509 (1994).
13. A. I. Akhiezer and V. B. Berestetsky, *Quantum Electrodynamics*, Consultants Bureau, New York, 1957.
14. R. Benoist, P. Carra, and O. K. Andersen, unpublished.
15. G. D. Mahan, *Many-Particle Physics*, Plenum Press, New York, 1991.
16. P. Carra and M. Fabrizio, in *Core-Level Spectroscopies for Magnetic Phenomena: Theory and Experiment*, P. S. Bagus et al. Eds., Plenum Press, New York, 1995.
17. D. A. Varshalovich, A. N. Moskalev, and V. K. Khersonskii, *Quantum Theory of Angular Momemtum* (World Scientific Publishing, Singapore, 1988).
18. B. R. Judd, *Second Quantisation in Atomic Spectroscopy* (Johns Hopkins University Press, Baltimore, 1967).
19. P. Carra, H. König, B. T. Thole, and M. Altarelli, Physica B **192**, 182 (1993).

# A band-structure-based approach to modeling x-ray absorption, fluorescence, and resonant inelastic scattering

Eric L. Shirley,[1] J. A. Carlisle,[2] S. R. Blankenship,[1] R. N. Smith,[2] L. J. Terminello,[3] J. J. Jia,[4] T. A. Callcott,[4] and D. L. Ederer[5]

[1]*National Institute of Standards and Technology, Gaithersburg, MD 20899*
[2]*Virginia Commonwealth University, Richmond, VA 23284*
[3]*Lawrence Livermore National Laboratory, Livermore, CA 94551*
[4]*University of Tennessee, Knoxville, TN 37996*
[5]*Tulane University, New Orleans, LA 70118*

X-ray optical processes in solids—absorption, fluorescence and resonant scattering—are modeled within a band-structure-based approach to describe electron states. The theory goes beyond a simple one-electron treatment by considering self-energy corrections for electron states, the perturbation acting on electron states because of a core hole, and, to a lesser degree, lifetime effects for high-energy electron states in solids. Aspects of the work that suggest extensions to be made in the future are discussed. Theoretical results for all three types of x-ray spectra (absorption, fluorescence and resonant scattering) are presented and compared to experiment. Results are presented for diamond, graphite, NaF and Al.

## INTRODUCTION

The availability of high-brightness, tunable x-ray light sources, sophisticated x-ray instrumentation, and current electronic structure calculation techniques suggests the potential for understanding x-ray optical processes in solids—absorption, fluorescence and resonant scattering—at an unprecedented level of quantitative detail. The past several years have witnessed several breakthroughs in the collective understanding of x-ray physics, many of which are documented in this Proceedings Volume. Because of the breadth of the field, this article touches on only a relatively narrow subset thereof, namely the x-ray optical processes of some of the simplest covalent, ionic and metallic solids. Only "s-p bonded" (that is, not transition-metal or rare-earth) systems are considered here, and x-ray fluorescence, absorption and resonant scattering are all considered briefly.

In a sense, x-ray optical processes are well understood on the level of a "renormalized" but essentially one-electron picture. X-ray absorption, fluorescence and resonant scattering in simple s-p solids probe electron states in valence and conduction bands, and these three spectroscopies can be used to analyze wave functions and energies of such states. A current theoretical description of the electronic structure of solids can easily account for band energies and wave functions of electron states, including corrections because of dynamical electron self-energy effects. In addition, perturbations acting on electron band states because of a core hole created by x-ray absorption are treatable, but it is hard to predict the screening effects that weaken such perturbations with sufficient accuracy in the solid.

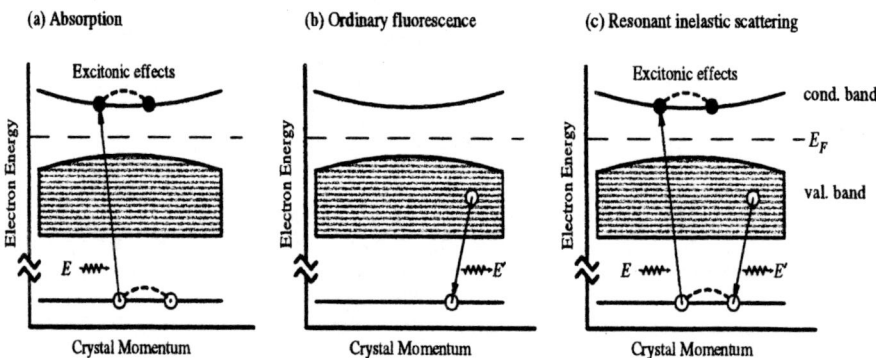

**FIGURE 1.** Three x-ray optical processes are schematically illustrated: absorption (a), ordinary fluorescence (b), and resonant inelastic scattering (c). Interaction (excitonic) effects that influence absorption and resonant scattering are indicated.

The physical processes of x-ray absorption, fluorescence and resonant inelastic scattering (specifically, electronic resonant Raman scattering) are indicated in Fig. 1. (Non-resonant scattering and elastic scattering of x-rays are not addressed here.) In the simplest interpretation, x-ray absorption permits one to probe unoccupied electron band states, perturbed by a core-hole potential. X-ray fluorescence permits one to probe occupied electron valence-band states, essentially without a core-hole perturbation. In the ordinary x-ray fluorescence processes considered here, a core hole decays by radiative recombination of the hole with a valence-band electron. Because the final state of the solid involves only a valence hole, the emitted x-ray photon energy is the difference between the valence hole's band energy and the core level energy.

In resonant inelastic x-ray scattering, conservation of total (electron plus hole) crystal momentum must be considered, because resonant scattering is a coherent absorption/emission event. The resonant scattering process leaves the system in a final state, in which an electron (promoted from a core level during absorption) and valence hole (created during subsequent radiative decay of the core hole) have crystal momenta related by the difference between the momenta of the absorbed and emitted photons. In the case of soft x-rays, e.g. near the carbon $K$ edge, photon momenta can be negligible. In the case of hard x-rays, inelastic scattering may be probed in a way that varies substantially the total momentum of the final-state electron-valence-hole pairs that are studied. In either case resonant inelastic x-ray scattering can be used to study both occupied and unoccupied electron states. Stokes shifts of scattered x-rays correspond roughly to energies of inter-band transitions achieved in the final state, and the incident photon energy can be tuned to favorably enhance transitions involving different unoccupied electron states.

On a deeper level, effects that are not considered within a "renormalized, essentially one-electron" picture might require consideration. Of the above three spectroscopies, fluorescence is probably treated the most completely within such a picture. It is straightforward to describe the final state with a valence hole using a quasiparticle model based on the electron Dyson equation. However, complications arising from the perturbation of all occupied electron states prior to photon emission could still be considered. That is, one could consider complications because of the x-ray edge singularity (1).

X-ray absorption spectroscopy (XAS) has long been addressed theoretically within a variety of contexts. The acronyms EXAFS (extended x-ray absorption fine structure), NEXAFS (near-edge x-ray absorption fine structure), and XANES (x-ray absorption near-edge structure) all indicate important aspects of spectra that are studied by XAS. With sufficiently quantitative treatments of unoccupied electron states, it seems clear that the fine structure of spectral features near an absorption edge (up to ≈20 eV above the edge) can be largely accounted for. Furthermore, much of the structure up to ≈100 eV can also be accounted for, except that it is not obvious how electron lifetime effects can be predicted with high reliability. In the present work, a conventional treatment (2) of lifetime effects accounts nicely for broadening of spectral features in metallic Al, which is a nearly-free-electron metal, but grossly overestimates such effects in NaF and diamond.

Regarding resonant inelastic x-ray scattering, the above-mentioned difficulties that can arise when modeling x-ray absorption also hamper the modeling of resonant inelastic scattering. Additional complications arise as well, because such scattering is a coherent process in which several coherent intermediate channels can lead to the same final state. Effects that introduce "dephasing" or "incoherence" (phonons and multi-electron excitations in cases of metals) remain largely untreated. Currently, interactions between the core hole and the promoted electron are considered, but calculations that include both these and final-state valence-hole-electron interactions remain to be performed. Because effects of the latter interactions have been included for problems such as ordinary optical absorption (3), it is reasonable to expect that the same effects will be included in the modeling of resonant inelastic x-ray scattering in the near future.

In what follows, a band-structure-based approach to modeling the above three x-ray spectroscopies is presented. Many details of its implementation have already been provided in other references, as is indicated, and some newer innovations of potential practical significance are also discussed here. Thereafter, x-ray absorption results for several solids are presented, with emphasis laid on the recently established capacity for treating absorption up to $\approx 100$ eV above an edge within a band-structure-based approach. Resonant inelastic x-ray scattering in graphite, for incident photon energies being tuned through the $\pi^*$ and $\sigma^*$ resonances, is discussed next, followed by x-ray fluorescence results for graphite and diamond. It is clear that several many-body effects that arise in absorption, fluorescence and resonant scattering are included in the theory. However, outstanding issues such as some details of the screening of core holes, electron lifetimes, and incorporation of "dephasing" or "incoherence" involving phonons or multi-electron excitations may not be amenable to treatments within the present "renormalized, essentially one-electron picture."

# THEORETICAL MODEL

Presentation of the approach is organized as follows. First, the treatments of one-electron states, their lifetimes, and photoabsorption matrix elements are discussed. This information alone is adequate for describing "ordinary" x-ray fluorescence, which involves radiative decay of an "ordinary" core hole, and is not the fluorescence associated with a resonant scattering process, resulting from decay of a coherent superposition of selected core-excited states. The perturbing potential that electrons experience because of a core hole is discussed next, completing the description of what is needed to treat x-ray absorption. Afterwards, resonant inelastic x-ray scattering is modeled by considering possible radiative decay channels for the core-excited states achieved during an absorption event.

## Electron states, lifetimes, photoabsorption matrix elements

### Electron-state wave functions and energies

Core states were evaluated within a Hartree-Fock atomic program, while electron band states were computed within a pseudopotential plane-wave approach (4). Using a plane-wave approach led to a relatively large number of degrees of freedom in electron wave functions, but the approach could be used to describe electron states that were occupied as well as states up to over 100 eV above the Fermi level or band gap. The local-density approximation (LDA, Ref. 5) was used when evaluating wave functions for the band states, but LDA band energies were corrected according to the known results of self-energy calculations done within the *GW* approximation (6) using the Hybertsen-Louie method (7).

In the LDA, an electron wave function $\psi_{nk}$ in band $n$ at crystal momentum **k** obeys the Kohn-Sham equation (5),

$$[\ \mathbf{p}^2/(2m) + V(\mathbf{r}) + V_{xc}(\mathbf{r})\ ]\ \psi_{nk}(\mathbf{r}) = \varepsilon_{nk}^{LDA}\ \psi_{nk}(\mathbf{r}), \tag{1}$$

where $\varepsilon_{nk}^{LDA}$ is an LDA band energy. Obviously, this equation strongly resembles the one-electron Schrödinger equation. The potential term $V$ accounts for electron-ion interactions and the average electrostatic potential because of the electron charge density. The potential term $V_{xc}$, which depends locally on the average electron density, accounts for exchange and correlation (self-energy) effects in a manner akin to Slater's local treatment of exchange (8).

In reality, electron band states are the solution of the electron Dyson equation, and the solution of that equation for each state can be written in simplified form as

$$[\ \mathbf{p}^2/(2m)+V(\mathbf{r})]\ \psi_{nk}(\mathbf{r}) + \int d^3\mathbf{r'}\ \Sigma(\mathbf{r}, \mathbf{r'}; \varepsilon_{nk})\ \psi_{nk}(\mathbf{r'}) = \varepsilon_{nk}\ \psi_{nk}(\mathbf{r}), \tag{2}$$

which is similar to the Kohn-Sham equation, except that the non-local, non-Hermitian, energy-dependent self-energy operator $\Sigma$ replaces $V_{xc}$. The self-energy may be expanded using diagrammatic perturbation theory in an efficient way in terms of the electron Green's Function $G$, which is the electron propagator and in this work is approximated by

$$G(\mathbf{r}, \mathbf{r'}; E\ ) = \Sigma_{nk}\ \psi_{nk}(\mathbf{r})\ \psi_{nk}^*(\mathbf{r'})/[E - (\varepsilon_{nk} + i\Gamma_{nk})], \tag{3}$$

and the dynamically screened Coulomb interaction, $W = \varepsilon^{-1}v$, where $\varepsilon$ is the dielectric function and $v$ is the bare Coulomb interaction. The first term in a diagrammatic expansion for the self-energy is

$$\Sigma(\mathbf{r}, \mathbf{r'};\ E\ ) = +i\ (2\pi)^{-1} \int d\omega \exp(i\eta\omega)\ G(\mathbf{r}, \mathbf{r'}; E+\omega)\ W(\mathbf{r}, \mathbf{r'}; \omega), \tag{4}$$

The $\eta$ parameter is a positive infinitesimal. Stopping at this term is known for reasons of its form as the *GW* approximation.

The electron lifetime parameter $\Gamma_{nk}$ is positive for occupied electron states and negative for unoccupied states. Having the same effect as an imaginary potential, $i\Gamma_{nk}$ approximately accounts for the imaginary part of $\Sigma$. In this work, relevant effects of replacing LDA band energies with *GW* energies are widening of the graphite (9) and diamond (10) valence bands and enhancement of the widths of the lowest unoccupied bands in graphite.

Given such a description of electron states, band states of greatest importance here were those lying within occupied or partially occupied bands and as many unoccupied bands as were of interest. Each band was sampled by considering states located on a uniform k-point mesh that filled the Brillouin zone. Despite their initial evaluation within reciprocal space, these states could be transformed readily into real space by Fourier techniques. Linear combinations of Bloch states would transform to wave packets moving within a large super-cell having Born-von Karman boundary conditions, and the k-point mesh density (or super-cell size) was increased to control k-point-sampling or super-cell-periodicity effects. Rapid evaluation of Bloch states at up to 4096 k-point and/or for up to 80 bands was done using basis sets optimized for calculations requiring such sampling (11). The basis functions themselves were constructed using a plane wave representation, so electron wave functions could also be expressed in the plane wave representation by carrying out appropriate linear transformations. Core states considered in this work were readily modeled as localized states, because site-to-site hopping of core holes was a negligible effect. Such localization was a useful conceptual tool, but during a coherent scattering event one cannot neglect possible interference between translationally equivalent scattering channels involving core excitations in different unit cells. This fact remains true, even if the intermediate-state channels are not coupled to each other.

## *Electron-state lifetime effects*

Lifetime effects are substantial for electron states far from the band gap or Fermi level, and in this work states that are the most affected are high-energy (free-electron-like) states in unoccupied bands. Lifetime effects influence x-ray absorption spectra by broadening (and flattening) spectral features resulting from variations in the core-hole-perturbed density of states probed by photoabsorption. For a high-energy electron state, an important decay process is that of the electron scattering into a lower-energy state, while energy and momentum are conserved by the creation of an electron-hole pair, or, more likely, a plasmon. It appears difficult at present to predict lifetime effects in every case with satisfactory accuracy, although lifetime effects do become easy to describe at very high energies. Therefore, there is a range of energies, roughly between the Fermi level plus the plasmon energy and the Fermi energy plus ≈75 eV, over which the precise onset of lifetime-broadening effects is both significant and difficult to predict. (Lifetime effects may be inferred from the broadening effects that are apparent in measured absorption spectra.)

Lifetime effects can be computed directly from the imaginary part of the self-energy, as discussed, for instance, by Tanuma, Penn, and Powell (2). In this work, a single-pole model for the dielectric function $\varepsilon(q,\omega)$, with

$$\mathrm{Im}\, \varepsilon(q,\omega) = A_q [\, \delta(\omega - \omega_q) + \delta(\omega + \omega_q)\,] , \qquad (5)$$

and a homogeneous-electron-gas model for electron states were considered for computing Im $\Sigma$ in the *GW* approximation. For each value of $q$, the pole position $\omega_q$ and pole strength $A_q$ were chosen to satisfy the *f*-sum rule for the $\omega^1$ frequency moment of Im $\varepsilon(q,\omega)$ and to force $\varepsilon(q,\omega=0)$ to equal the Levine-Louie dielectric function (12). The average valence electron density is required as input for such a single-pole model. In metallic Al, graphite and diamond, it is clear how to distinguish core and valence electrons. In an insulator such as NaF, such a distinction is less clear, because of the semi-core Na 2s and Na 2p states that are deeply bound but participate substantially in solid-state screening effects, especially at high frequencies such as the bulk plasma frequency. This difficulty remains to be addressed, probably by considering screening models that explicitly incorporate the screening effects of valence and shallow core electrons.

To apply the electron-gas model *GW* results, lifetime effects on an electron in the solid were equated to analogous effects on a fictitious electron in the gas. The fictitious electron was assumed to be in a state with an energy equal to the electron-gas Fermi energy plus the difference between the real electron's assumed energy and the energy of the lowest unoccupied band states. (Free-electron dispersion was assumed in the gas.) This procedure implicitly assumes that an electron excited by x-ray absorption virtually occupies only states with energies close or equal to its true energy. To the degree that this assumption is not valid, one should actually reconsider evaluation of $\Sigma$ "off-shell," that is, for combinations of energy and momentum for which no quasiparticle states actually exist (13). In the future, it is also clearly desirable to study lifetime effects without neglecting band-structure effects on electron states, because such refinements might substantially improve the agreement between theoretical and observed lifetime-broadening effects.

## *Photoabsorption matrix elements*

It should first be noted that photon-emission matrix elements are related to photoabsorption matrix elements by interchanging electron initial and final states. Photoabsorption matrix elements of interest were those of the type,

$$M_{C\tau nk} = <\psi_{nk}(\mathbf{r})|\mathbf{p}\cdot\mathbf{A}|\psi_C(\mathbf{r}-\mathbf{\tau})>. \tag{6}$$

Here, $\psi_C(\mathbf{r}-\mathbf{\tau})$ is the wave function for a core level $C$ on site $\tau$. Our method to evaluate such a matrix element has evolved somewhat in the past, but the current, most versatile approach will be discussed first. Evaluation of such a matrix element is complicated by the fact that $\psi_{nk}(\mathbf{r})$ is directly evaluated only for the pseudopotential version of a band-state wave function, $\psi^{ps}_{nk}(\mathbf{r})$, whereas matrix elements are desired for true wave functions. To overcome this difficulty, $\psi^{ps}_{nk}(\mathbf{r})$ was expressed near $\tau$ as a linear combination of partial waves:

$$\psi^{ps}{}_{nk}(\tau+s)=\Sigma_{vlm}\, C_{vlm}(C\tau nk)Y_{lm}(s/|s|)\, F^{ps}{}_{vl}(s). \tag{7}$$

Here, $l$ and $m$ are standard angular-momentum quantum numbers, and the index $v$ specifies partial waves with different scattering energies for a given $l,m$ channel. The $C$'s are weight parameters, $F^{ps}{}_{vl}(s)$ gives the radial dependence of a partial wave in the pseudopotential case, and $s$ should be taken as small. Once the $C$'s are known, the true band-state wave function can be expressed as a linear combination of the true counterparts of pseudopotential partial waves:

$$\psi_{nk}(\tau+s)=\Sigma_{vlm}\, C_{vlm}(C\tau nk)Y_{lm}(s/|s|)\, F_{vl}(s). \tag{8}$$

Corresponding pseudopotential and true versions of partial waves were all derived within the same atomic program. Once a true wave function was expressed as a linear combination of partial waves, photoabsorption matrix elements were evaluated using standard atomic calculation techniques.

In past work involving only x-ray $K$ edges, it was reasonable to assume that a photoabsorption matrix element was proportional to the gradient of $\psi^{ps}{}_{nk}(\mathbf{r})$ at $\mathbf{r}=\tau$. This assumption has meanwhile become untenable, because of the large energy ranges for unoccupied states considered when modeling x-ray absorption. When one considers electron states that span a large energy range, the radial de Broglie wavelengths of partial waves can vary greatly, and the above expression involving multiple partial waves per $l,m$ channel becomes necessary. In addition, the authors are presently also studying $L_2$ and $L_3$ edges, and a proportionality relationship is not sufficient when partial waves for more than one value of $l$ can contribute to a matrix element.

## "Ordinary" x-ray fluorescence processes

If a core level $C$ on site $\tau$ is vacant, an x-ray fluorescence spectrum is given approximately by

$$S_{em}(E') \sim \Sigma_{nk} \Sigma_{e'} |<\psi_C(\mathbf{r}-\tau)|\, \mathbf{p}\cdot\mathbf{e'}^* \,|\, \psi_{nk}(\mathbf{r})>|^2\, \delta\,(\varepsilon_{nk} - E_C - E'). \tag{9}$$

Here, $E'$ and $E_C$ denote photon and core-level energies, respectively, $\mathbf{e'}$ indicates the emitted photon polarizations that are considered, and $S_{em}(E')$ is the fluorescence spectrum.

## Core-hole effects on electrons

During the existence of a core hole at some lattice site, band electrons feel the perturbation of the core hole's potential that results from the vacancy of a core state ordinarily filled by an electron with charge $-e$. This "core-hole" perturbation is screened by the dynamical dielectric response of the solid, but core-hole lifetimes and the over-all strength of the perturbation are small enough to make a static-screening assumption plausible. The strength of the interaction is relevant here, because it governs the possible energy difference between unperturbed electron band states that become significantly admixed because of the core-hole perturbation. If states that are sufficiently far apart in energy are strongly coupled, plasmons that help mediate the electron-core-hole interaction might exist in more than a virtual sense, which undermines a static-screening picture. An additional breakdown of a simple static-screening picture might arise in the case of photoabsorption near threshold. There, interference between the couplings of a core hole and the ejected electron to the dielectric response of the solid may require consideration, in regards to both screening and lifetime effects (14).

Screening effects arise within the atomic core containing the hole, because of core electrons, but screening also occurs because of valence electrons, which have a band character (versus atomic character) in the solid state. While it is straightforward to deduce that the core-hole's potential is screened, quantifying the degree of screening is another matter. Yet because of the utility of experimental techniques such as EXAFS studies, improving predictive models for the screening would be a worthwhile endeavor. Features very close to or even at x-ray edges appear to be remarkably sensitive to the details of the screening, and the ultimate realizable quality of theoretical absorption spectra based on non-empirical screening models is not known. In earlier work (15), the Levine-Louie-Hybertsen (12,16) approach was used to treat the screening, but theoretical x-ray absorption spectra can be substantially improved at the cost of using one empirical parameter per spectrum.

To treat screening in this work, the following empirical approach has been developed in our latest modeling, whereas some results presented here rely on earlier, nearly equivalent, empirical screening models. The current approach is described as follows. A central, local potential is used for the electron-core-hole interaction. Non-spherical screening effects should be of secondary importance, and they are neglected here. Also, a local potential should account only for the "direct" part of the electron-core-hole interaction, which is attractive, whereas a repulsive "exchange" part should also be considered (15). In some cases, this term was explicitly treated (15,17-19), but for simplicity the local potential is now adjusted to account for the entire electron-core-hole interaction, including direct and exchange parts.

Regarding the local potential, screening effects should force the potential to have the long-range behavior of $-e/(\varepsilon_\infty r)$, where $r$ denotes distance from the core-hole's site, and $\varepsilon_\infty$ is the static value of the dielectric constant, neglecting phonon contributions. At short range, the potential should involve effects of atomic and solid-state screening, so that screening by core electrons would occur much like in the isolated-atom/ion limit, whereas screening by valence electrons may be stronger or weaker in the solid than in the isolated-atom/ion limit.

Four potentials and a "screening function" were considered when constructing the local potential, and the combination of these five objects was done in a way that permitted one empirical parameter to adjust the strength of screening effects. One potential was the spherical part of the Hartree potential, in the isolated-atom/ion limit, of the core electrons in the case of no valence electrons being present. The second potential was the spherical part of the analogous Hartree potential, but with an electron removed from the appropriate core level. The third and fourth potentials were analogously derived, except that the valence electrons were present according to an element's neutral, ground-state configuration in the isolated-atom limit. These Hartree potentials were all obtained using self-consistent Hartree-Fock calculations. The difference between the first two potentials, henceforth called $V_c$, is the core-hole's potential including screening only because of core electrons. The difference between the last two potentials, henceforth called $V_v$, is the core-hole's potential including screening because of core electrons and valence electrons in the isolated-atom/ion limit. The screening function, $f(r)$, was essentially the ratio, of the spherically averaged, Levine-Louie-Hybertsen (16) screened potential for a test charge, located on the core site, to the bare Coulomb potential for the same test charge. The local potential, $W(r)$, was then given by

$$W(r) = f(r) \{ V_c(r) + x [ V_v(r) - V_c(r) ] \}, \tag{10}$$

and $x$ was adjusted empirically to maximize agreement between theoretical and experimental x-ray absorption spectra. Effects of adjusting $x$ were mainly changes in the relative strengths of the very lowest feature or features in spectra, whereas the relative strengths of higher-energy features were almost unaffected. Binding energies of core-hole excitons were also weakly influenced.

## X-ray absorption

To obtain x-ray absorption spectra, it was necessary to consider solution of the Hamiltonian appropriate for an electron-core-hole pair,

$$H_{\text{pair}} = H_e - E_C + V_{\text{int}}. \tag{11}$$

The three terms indicated account for the electron's band energy, the core-level energy, and the interaction between the electron and core hole. For states with the core hole in state $C$ on site $\tau$, a complete set of electron-core-hole pair-states could be enumerated by considering all values of the electron band index $n$ and crystal momentum $\mathbf{k}$. If we then specify such a *basis* state by $|C\, \tau n\mathbf{k}\rangle$ and a stationary solution of $H_{pair}$ and its energy by $|P\rangle$ and $E_P$, we may write

$$S_{ab}(E) \sim \Sigma_P |\langle P| \mathbf{p}\cdot\mathbf{e} |0\rangle|^2 \, D(E-E_P, \Gamma(E)) \tag{12}$$

or

$$S_{ab}(E) \sim -\pi^{-1} \operatorname{Im} \langle 0| \mathbf{p}\cdot\mathbf{e}^* \, (E-H_{pair} + i\,\Gamma(E))^{-1} \mathbf{p}\cdot\mathbf{e} |0\rangle. \tag{13}$$

The function $D$ is a Lorentzian with a half-width at half-maximum equal to the energy-dependent lifetime parameter, $\Gamma(E)$, and $|0\rangle$ denotes the ground state (which has no electron-core-hole pair). The first expression for an absorption spectrum uses Fermi's golden rule, and this expression requires solution of $H_{pair}$ for all states $\{|P\rangle\}$. In contrast, the second expression requires only evaluating the indicated expectation value (for the indicated state, $\mathbf{p}\cdot\mathbf{e}\,|0\rangle$) at each energy $E$. This can be done efficiently using the Haydock recursion method (20), which requires only evaluating the action of $H_{pair}$ on a certain sequence of states, beginning with $\mathbf{p}\cdot\mathbf{e}\,|0\rangle$, but does not require complete solution or inversion of $H_{pair}$. For this reason, the Haydock recursion method was used in this work, mainly because of the large secular order of $H_{pair}$ (up to $\approx 10^5$). Further details of calculations are provided elsewhere (15,17-19).

## Resonant inelastic x-ray scattering

Resonant inelastic x-ray scattering is described using the resonant term in the Kramers-Heisenberg formula for the scattering of light, so that the scattering rate for incident and emitted photons with energies $E$ and $E'$, momenta $\mathbf{q}$ and $\mathbf{q}'$, and polarizations $\mathbf{e}$ and $\mathbf{e}'$ is given by

$$S_{scat}(E,E') \sim \Sigma_{f,\mathbf{e}'} |\Sigma_m \langle f| \mathbf{p}\cdot\mathbf{e}'^*|m\rangle\langle m|\mathbf{p}\cdot\mathbf{e}|0\rangle/(E-E_m + i\,\Gamma)|^2 \delta(E-E'+E_0-E_f), \tag{14}$$

where states $\{|f\rangle\}$ are final states with energies $\{E_f\}$, states $\{|m\rangle\}$ are intermediate states with energies $\{E_m\}$ and a typical lifetime parameter $\Gamma$, and $E_0$ is the ground-state energy. The delta-function insures that energy lost by the photon upon scattering is the energy gained by the solid. Completeness of states $\{|m\rangle\}$ allows one to write

$$S_{scat}(E,E') \sim \Sigma_{f,\mathbf{e}'} |\langle f| \mathbf{p}\cdot\mathbf{e}'^*|x(E)\rangle|^2 \, \delta(E-E'+E_0-E_f), \tag{15}$$

where we have

$$|x(E)\rangle = (E - H_{pair} + i\Gamma)^{-1} \mathbf{p} \cdot \mathbf{e} |0\rangle, \qquad (16)$$

and $|x(E)\rangle$ is a combination of states $\{|C\,\tau n\mathbf{k}\rangle\}$. Unlike in the case of absorption, it is now important to note that $|x(E)\rangle$ is built up from combinations of states of the Bloch-sum form,

$$\Sigma_{\mathbf{R}} \exp[i(\mathbf{q}-\mathbf{k}) \cdot \mathbf{R}]|C(\mathbf{R}+\tau)n\mathbf{k}\rangle. \qquad (17)$$

Although states for different values of **R** are not coupled significantly, the coherent excitation of a solid into such translationally equivalent states forces a valence electron that radiatively recombines with the core hole during emission to have crystal momentum $\mathbf{k}+\mathbf{q}'-\mathbf{q}+\mathbf{G}$, where **G** is a reciprocal-lattice vector. If final-state interactions between the electron that was promoted and the valence hole resulting from recombination are neglected, one may deduce that there is one final state for each possible electron-valence-hole pair having total crystal momentum $\mathbf{q}-\mathbf{q}'+\mathbf{G}$. Such a state would have a relative energy of $\varepsilon_{n\mathbf{k}} - \varepsilon_{n'\mathbf{k}+\mathbf{q}'-\mathbf{q}+\mathbf{G}}$ compared to the ground state, where $n'$ is the hole's band index. Such "non-interacting" final states are assumed in this work, i.e., final-state, electron-valence-hole interactions are neglected. Because such interactions have been included in problems such as ordinary optical absorption (3), it would be worthwhile in future work to incorporate these final-state interactions into models for resonant inelastic x-ray scattering.

If $V_{int}$ could be neglected, $|x(E)\rangle$ would weigh unoccupied band states with $\varepsilon_{n\mathbf{k}}$ closest to $E_C+E$ most heavily. Tuning $E$ would vary the unoccupied band states most frequently involved in scattering, and emission spectra would peak at values of $E'$ for which $E-E'$ corresponds to inter-band transitions into the portions of the Brillouin zone that contain the frequently accessed unoccupied states. Ma and co-workers (21) first realized that this effect offers the potential for using resonant inelastic x-ray scattering as a probe of band structure, and results for resonant scattering involving excitation near the graphite $\pi^*$ and $\sigma^*$ resonances are presented in this work.

Matters are somewhat more complicated because of $V_{int}$, and the authors have found the following guidelines helpful for anticipating the effects of $V_{int}$ on scattering spectra. First, while total scattering yields depend on the magnitude of $|x(E)\rangle$, the relative strengths of emission features are affected only by the shape of $|x(E)\rangle$. Next, one may write $|x(E)\rangle$ as

$$|x(E)\rangle = (E - H_e + E_C + i\Gamma)^{-1} [1 + V_{int}(E - H_{pair} + i\Gamma)^{-1}] \mathbf{p} \cdot \mathbf{e} |0\rangle. \qquad (18)$$

In this way, effects of $V_{int}$ arise precisely from the bracketed quantity differing from the identity operator. Note that $\mathbf{p} \cdot \mathbf{e} |0\rangle$ results from photoejection of an electron from an atomic core, and that within $\mathbf{p} \cdot \mathbf{e} |0\rangle$ the electron should exist within a fairly localized wave packet. Likewise, the action of $V_{int}$ is strongest at short range, so that

$(E - H_e + E_C + i\,\Gamma)|x(E)\rangle$ is generally highly localized. (In the immediate context, "localization" denotes the relative localization of the electron and core hole.) The operator, $(E - H_e + E_C + i\,\Gamma)^{-1}$, weighs the constituents of such a localized state according to electron band energies only, and not in any way that depends on $V_{int}$. Obviously, the resulting $|x(E)\rangle$ can be affected strongly by $V_{int}$ in regards to its magnitude, but the shape of $|x(E)\rangle$ should be strongly affected only if there is more than one way to construct a localized wave packet for the electron using band states that lie within a given energy range. In particular, only one partial-wave channel (belonging to C 2p orbitals pointing normal to the carbon sheets) is relevant when probing states at the bottom of the graphite $\pi^*$ band, and $V_{int}$ should not change much the shape of $|x(E)\rangle$. To illustrate this within a real-space multiple-scattering picture, note that the core-hole potential essentially adds phase shifts to the ejected electron's outgoing and scattered waves. However, such phase shifts can only renormalize the photoabsorption matrix element when only one partial-wave channel is relevant, so that the *magnitude* of $|x(E)\rangle$ is changed, though not its *shape*. When probing at the bottom of the $\sigma^*$ band, however, $\pi^*$ and $\sigma^*$ band states exist in the same energy range, so $V_{int}$ could modify relative weights of $\pi^*$ and $\sigma^*$ portions of $|x(E)\rangle$. As shown later, features in theoretical scattering spectra are not affected strongly by $V_{int}$ for $E$ near the $\pi^*$ resonance, but they are affected strongly for $E$ near the $\sigma^*$ resonance. Further details of calculations are provided elsewhere (15,18,19).

## X-RAY ABSORPTION SPECTROSCOPY RESULTS

X-ray absorption spectra are presented for graphite (Fig. 2, experiment taken from Ref. 19), diamond (Fig. 3, experiment taken from Ref. 22), aluminum (Fig. 4, experiment taken from Ref. 23), and the Na and F edges in NaF (Fig. 5, experiment taken from Ref. 24). There is satisfactory agreement between theory and experiment regarding many spectral features, with results being the best in NaF and the poorest in covalently bonded systems. Electron lifetime effects based on model *GW* calculations were included for Al, but the model *GW* results appeared to overestimate lifetime effects in other systems and were ignored. However, the lifetime parameter was always set to at least a non-zero value (between 0.25 eV and 0.5 eV) to simulate core-hole-lifetime, phonon-broadening, and experimental resolution effects, as well as continuous Brillouin-zone sampling.

The present results may be compared to current EXAFS results obtained using multiple-scattering theory (25). Clearly, difficulties remain in describing absorption, regarding the strength of $V_{int}$ and lifetime effects for high-energy electron states. However, the present capacity to model observed spectral features with such detail suggests that one-electron aspects of x-ray absorption are well understood, and many-body effects should now be addressed. It may also be good to use band-structure-based and multiple-scattering treatments as mutual quality checks.

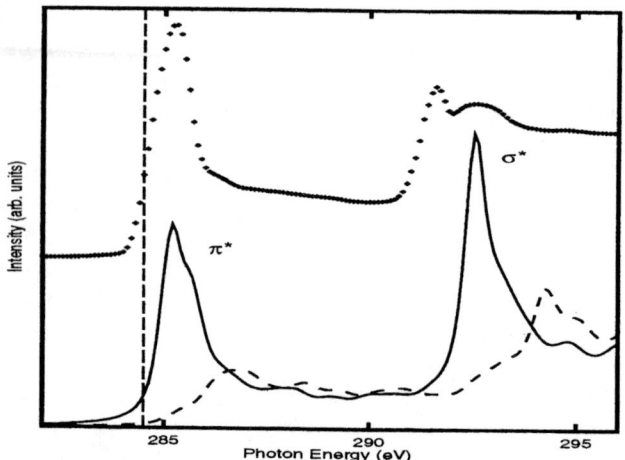

**FIGURE 2.** X-ray absorption spectra at the carbon $K$ edge in graphite near the $\pi^*$ and $\sigma^*$ resonances. Points are experiment, the solid curve is theory that includes the core-hole perturbation on electron states, and the dashed curve is theory omitting the perturbation.

## RESONANT INELASTIC X-RAY SCATTERING RESULTS

In Figs. 6, 7 and 8, a transposed band structure of graphite and resonant x-ray scattering spectra for $E$ tuned through the $\pi^*$ and $\sigma^*$ resonances are presented. Experimental results in Figs. 7 and 8 were taken from Ref. 26. In Figs. 7b and 8b, theoretical spectra either include (solid) or omit (dashed) the effects of $V_{int}$, except for the fact that the spectra were subsequently normalized in a similar fashion. This normalization removes visible effects of $V_{int}$ for $E$ near the $\pi^*$ resonance but not the $\sigma^*$ resonance, for reasons discussed earlier involving the number of relevant partial-wave channels at the core-hole site. As discussed elsewhere (15,18,19,26,27), the evolution of emission features in Fig. 7 results from core-electron excitation near the $K$ point at lower photon energies evolving into excitation near the $M$ point at higher photon energies because of the dispersion of the $\pi^*$ band. Emission because of $\pi$ electrons recombining with the core hole moves downwards from the x-ray edge with increasing $E$. Meanwhile, emission near $E' = 276$ eV increases rapidly in strength with increasing $E$ because of strong emission from $\sigma$ electrons recombining with the core hole near the $M$ point.

In Fig. 8, there is enhancement and subsequent lessening of emission near $E' = 281$ eV, 276 eV and 269 eV as $E$ is tuned through the $\sigma^*$ resonance [291.5 eV as measured, 292.5 eV in the calculations (28)]. This emission is emphasized in the theory only when effects of $V_{int}$ are considered. The emission features correspond to energies of occupied bands in the vicinity of the $M$ point, which is also the location of the $\sigma^*$ band minimum and hence near the electron states most important for the

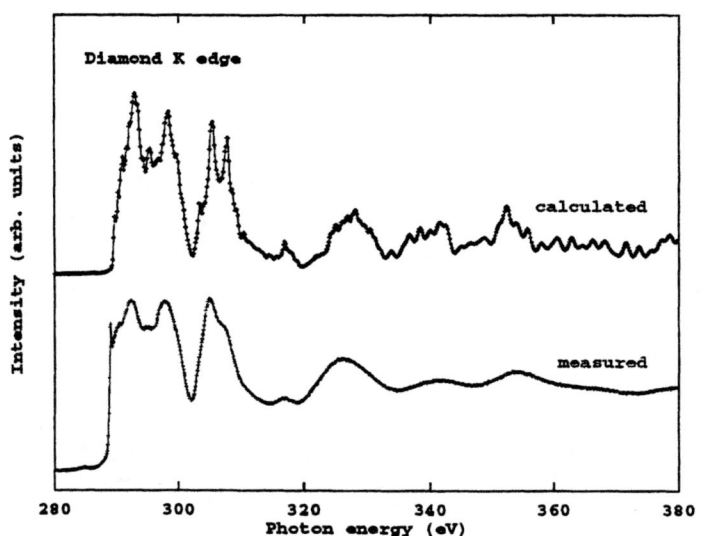

**FIGURE 3.** X-ray absorption spectra at the carbon $K$ edge in diamond, as calculated and measured (data provided by M. Jaouen).

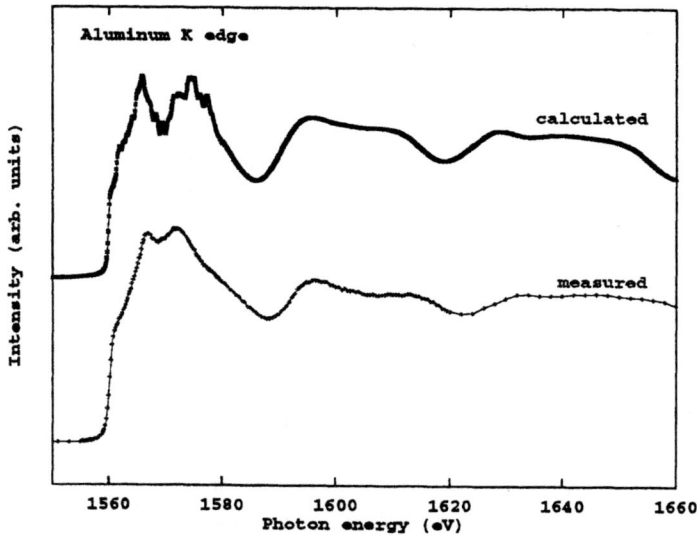

**FIGURE 4.** X-ray absorption spectra at the aluminum $K$ edge in Al, as calculated and measured (data provided by G. N. George).

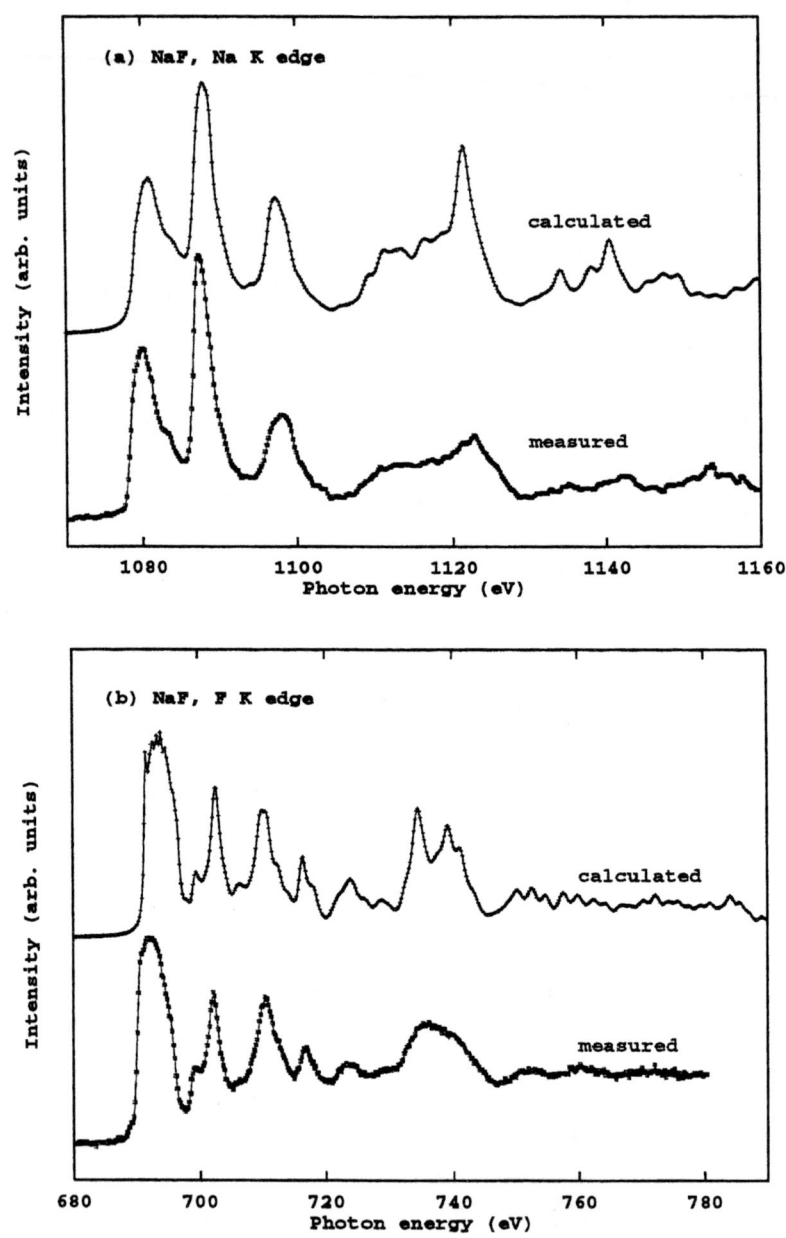

**FIGURE 5.** X-ray absorption spectra at the sodium (a) and fluorine (b) *K* edges in NaF, as calculated and measured (data for fluorine provided by E. A. Hudson, data for sodium taken from Ref. 24).

$\sigma^*$ core-hole exciton. Our interpretation of the behavior of the above emission features is that they result from radiative decay of the core hole in the $\sigma^*$ core-hole exciton, whose relative oscillator strength in absorption is reasonably computed only if $V_{int}$ is considered. As a cautionary note, others (29) have found stronger effects because of $V_{int}$ even for $E$ near the $\pi^*$ resonance, though the present authors have not obtained the same results (15,18,19). Also, several outstanding difficulties regarding the present theoretical treatment [e.g., the complete neglect of phonons, treated in one work by others on cBN (30), signatures of dephasing in the scattering, and certain selection-rule effects] are discussed elsewhere (15,18,19).

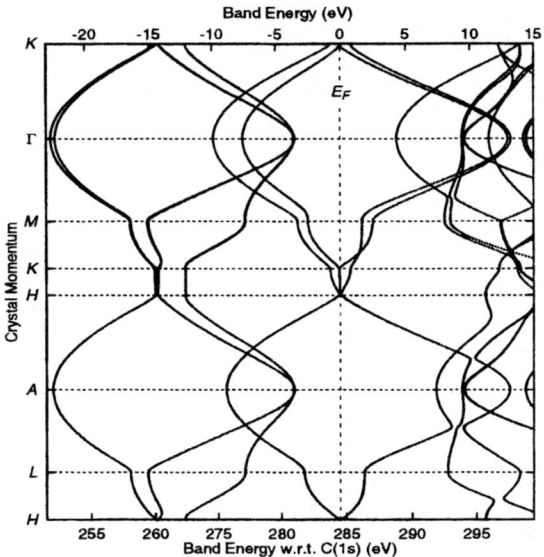

**FIGURE 6.** Transposed band structure of graphite. Electron state energies are indicated referenced to the Fermi level $E_F$ (numbers on top axis) and core level (numbers on bottom axis).

## "Ordinary" x-ray fluorescence

As a by-product of resonant x-ray scattering experiments, one may obtain essentially ordinary x-ray fluorescence spectra when $E$ is tuned far above an x-ray edge. The microscopic understanding for how such seemingly ordinary fluorescence results from "off-resonant" x-ray excitation is far from complete, but it is our intent here to point out the practical utility of ordinary fluorescence as a probe of partial densities of occupied electron states. One advantage of ordinary x-ray fluorescence over photoemission for analyzing densities of states is the relative simplicity of the transition matrix elements involved. Disadvantages also exist, such as poorer typical energy resolutions. In Fig. 9, ordinary x-ray fluorescence spectra are presented for graphite and diamond, both as measured and computed.

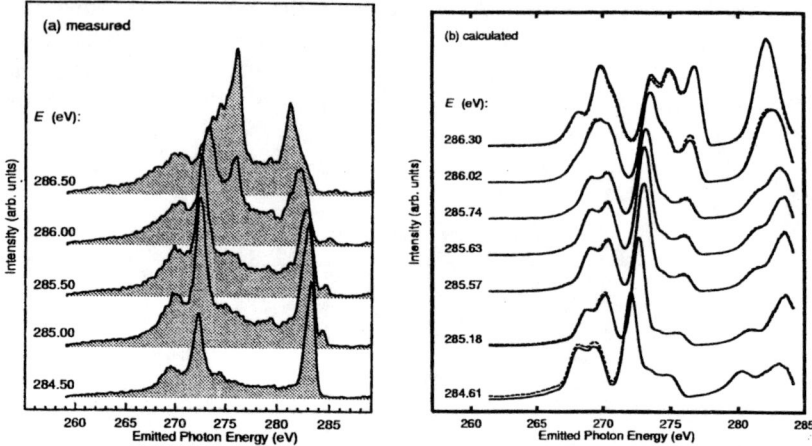

**FIGURE 7.** Measured (left) and theoretical (right) resonant inelastic scattering spectra for graphite for $E$ near the $\pi^*$ resonance. Dashed and solid lines respectively indicate theoretical spectra omitting and including core hole perturbations. Important emission features occur around 272 eV, 276 eV and between 281 eV and 284 eV. Note different emission-energy scales. Measured and simulated $E$'s were chosen slightly differently to illustrate analogous steps in emission spectrum evolution.

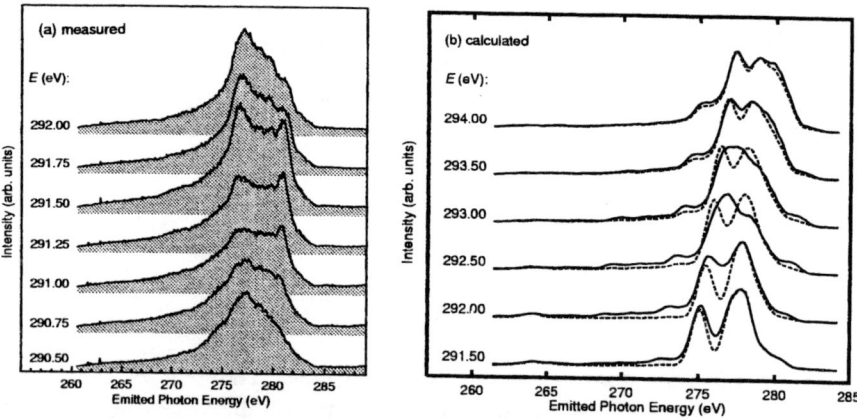

**FIGURE 8.** Measured (left) and theoretical (right) resonant inelastic scattering spectra for graphite for $E$ near the $\sigma^*$ resonance. Dashed and solid lines respectively indicate theoretical spectra omitting and including core hole perturbations. Important emission features occur around 270 eV, 276 eV and 281 eV when $E$ is tuned near the resonance (291.5 eV as measured, but 292.5 eV as simulated). Note different emission-energy scales. Measured and simulated $E$'s were chosen slightly differently to illustrate analogous steps in emission spectrum evolution.

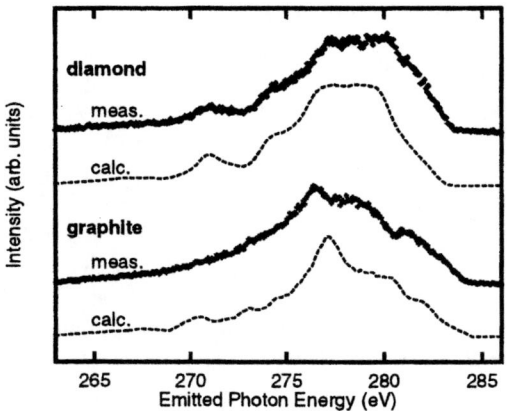

**FIGURE 9.** Measured off-resonance fluorescence spectra (meas.) for C 1s decay in diamond and graphite and calculated ordinary fluorescence spectra (calc.).

## CONCLUSIONS

Current experimental and theoretical capabilities offer potential for understanding the x-ray optical processes in solids—absorption, fluorescence, and resonant scattering—at an unprecedented level of quantitative detail. This work presents a band-structure-based approach to modeling these processes. The approach may, in a limited sense, be called a "renormalized, essentially one-electron approach," because it requires one to compute the dynamics of one or at most two particles at a time. Many-body effects that are included in this work include electron self-energy contributions to band energies, core-hole perturbations acting on electron states, and, to a limited degree, broadening effects because of core-hole decay and the decay of high-energy electron states. Results presented here have dealt with x-ray absorption in graphite, diamond, Al and NaF, x-ray fluorescence in graphite and diamond, and resonant inelastic x-ray scattering spectra in graphite at the $\pi^*$ and $\sigma^*$ resonances. In the future, it would be useful to eliminate empirical aspects of the treatment of core-hole perturbations and refine the treatment of electron lifetime effects. Meanwhile, the present accounting for many observed spectral features is encouraging.

## ACKNOWLEDGMENTS

We have benefited greatly from discussions with J. J. Rehr, J. A. Soininen, C. J. Powell, D. R. Penn, and D. A. Shirley. We also gratefully acknowledge permission from G. N. George, M. Jaouen and E. A. Hudson to present their x-ray absorption data. J. A. Carlisle acknowledges a Research Corporation Cottrell College Science Award No. CC4526, a Jeffress Trust Memorial Foundation Grant No. J-424, and a Faculty Grant-in-Aid at his institution. The work presented was also supported by the Division of Materials Science, Office of Basic Energy Sciences, and performed under the auspices of the U.S. Department of Energy (DOE) by the Lawrence Livermore National Laboratory under Contract No. W-7405-ENG-48, by National Science Foundation Grants, Nos. DMR-9420425 and DMR-9801084, and by the Louisiana Education Quality Support Fund and DOE-EPSCOR Grant No. LEQSF (93-95)-03 at Tulane University. Experimental data presented in Figs. 2, 7, 8 and 9 were obtained at the Advanced Light Source, which is also supported by the Office of Basic Energy Sciences, U. S. Department of Energy, under Contract No. DE-AC03-76SF00098.

## REFERENCES

1. Mahan, G. D., Phys. Rev. **163**, 612 (1967); Nozières, P. and De Dominicis, C. T., Phys. Rev. **178**, 1097 (1969).
2. Tanuma, S., Penn, D. R., and Powell, C. J., Surface and Interface Analysis **11**, 577 (1988); *ibid.* **17**, 911 (1991).
3. Benedict, L. X., Shirley, E. L., and Bohn, R. B., Phys. Rev. Lett. **80**, 4514 (1998), Phys. Rev. B **57**, R9385 (1998); Benedict, L. X. and Shirley, E. L., Phys. Rev. B **59**, 5441 (1999); Albrecht, S., Reining, L. Del Sole, R., and Onida, G., Phys. Rev. Lett. **80**, 4510 (1998); Rohlfing, M. and Louie, S. G., Phys. Rev. Lett. **81**, 2312 (1998).
4. For a review, see Pickett, W. E., Comp. Phys. Rep. **9**, 115 (1989).
5. Hohenberg, P. and Kohn, W., Phys. Rev. **136**, 864 (1964); Kohn, W. and Sham, L. J., Phys. Rev. **140**, 1133 (1965).
6. Hedin, L., Phys. Rev. **139**, 796 (1965); Hedin, L. and Lunqvist, S., in *Solid State Physics*, Vol. 23, edited by Ehrenreich, H., Seitz, F., and Turnbull, D., New York: Academic, New York, 1969, p. 1.
7. Hybertsen, M. S. and Louie, S. G., Phys. Rev. Lett. **55**, 1418 (1985), Phys. Rev. B **34**, 5390 (1986).
8. Slater, J. C., Phys. Rev. **81**, 385 (1951).
9. Zhu, X. and Louie, S. G., as cited in Louie, S. G., in *Topics in Computational Materials Science*, edited by Fong, C. Y., Singapore: World Scientific, 1997, p. 96; Heske, C., Treusch, R., Himpsel, F. J., Kakar, S., Terminello, L. J., Weyer, H. J., and Shirley, E. L., Phys. Rev. B **59**, 4680 (1999).
10. See Ref. 7, as well as Rohlfing, M., Kruger, P., and Pollman, J., Phys. Rev. B **48**, 17791 (1993) and Jiménez, I., Terminello, L. J., Sutherland, D. G. J., Carlisle, J. A., Shirley, E. L., and Himpsel, F. J., Phys. Rev. B **56**, 7215 (1997).
11. Shirley, E. L., Phys. Rev. B **54**, 16464 (1996).
12. Levine, Z. H. and Louie, S. G., Phys. Rev. B **25**, 6310 (1982).
13. Lundqvist, B. I., Phys. kondens. Materie **6**, 193 (1967), *ibid.*, **6**, 206 (1967), *ibid.*, **7**, 117 (1968), Phys. Stat. Sol. **32**, 273 (1969).
14. Rehr, J. J., Bardyszewski, W., and Hedin, L., J. Phys. IV (France) **7**, 97 (1997).
15. Shirley, E. L., Phys. Rev. Lett. **80**, 794 (1998).
16. Hybertsen, M. S. and Louie, S. G., Phys. Rev. B **37**, 2733 (1988).

17. Shirley, E. L., in *Proceedings of the Seventh International Symposium on Physics and Chemistry of Luminescent Materials*, edited by Struck, C. W., Mishra, K. C., and Di Bartolo, B., Pennington, NJ: The Electrochemical Society, Inc., 1999, p. 36.
18. Shirley, E. L., J. El. Spect. and Rel. Phenom. (in press).
19. Carlisle, J. A., Shirley, E. L., Terminello, L. J., Jia, J. J., Callcott, T. A., Ederer, D. L., Perera, R. C. C., and Himpsel, F. J., Phys. Rev. B **59**, 7433 (1999).
20. Haydock, R., Comput. Phys. Commun. **20**, 11 (1980).
21. Ma, Y., Phys. Rev. B **49**, 5799 (1994); Johnson, P. D. and Ma, Y., Phys. Rev. B **49**, 5024 (1994).
22. Jaouen, M., private communication.
23. George, G. N., private communication.
24. Hudson, E. A., private communication; Hudson, E. A., Moler, E., Zheng, Y., Kellar, S., Heimann, P., Hussain, Z., and Shirley, D. A., Phys. Rev. B **49**, 3701 (1994).
25. Ankudinov, A. L., Ravel, B., Rehr, J. J., and Conradson, S. D., Phys. Rev. B **58**, 7565 (1998); Rehr, J. J. and Albers, R. C., Phys. Rev. B **41**, 8139 (1990).
26. Carlisle, J. A., Shirley, E. L., Blankenship, S. R., Smith, R. N., Terminello, L. J., Jia, J. J., Callcott, T. A., and Ederer, D. L., J. El. Spect. and Rel. Phenom., **101-103**, 839 (1999); reprinted from this reference in a slightly changed form with permission from Elsevier Science.
27. Carlisle, J. A., Shirley, E. L., Hudson, E. A., Terminello, L. J., Callcott, T. A., Jia, J. J., Ederer, D. L., Perera, R. C. C., and Himpsel, F. J., Phys. Rev. Lett. **74**, 1234 (1995).
28. For a discussion about difficulties with the theoretical energy, see Refs. 9b, 18 and 19.
29. van Veenendaal, M. and Carra P., Phys. Rev. Lett. **78**, 2839 (1997).
30. Minami, T. and Nasu, K., Phys. Rev. B **57**, 12084 (1998).

# Resonant inelastic scattering at the 3d and 4d resonances of LaAlO$_3$

### A. Moewes

*Center for Advanced Microstructures and Devices, CAMD at Louisiana State University,*
*6980 Jefferson Hwy, Baton Rouge, LA 70806*

**Abstract.** Soft x-ray emission through the 4d-4f and 3d-4f resonance of LaAlO$_3$ is studied with monochromatic photon excitation. At the 4d-4f resonance strong mixing of intermediate states over an extended energy range of up to 21 eV is observed. The resonant emission is fundamentally different for selective excitation energies. Inelastic scattering due to charge-transfer excitations from the valence band to 4f and 5p to 4f is observed. At excitation energies above the $M_V$ edge we observe Coster-Kronig enhanced fluorescence that refills the 3d hole via 5p and 4f electrons. The experimental data are in good agreement with our calculations.

## INTRODUCTION

The electronic excitation of a solid with monochromatic synchrotron radiation is described as a first order optical process and the excitation can occur either to a bound state – as studied in x-ray absorption spectroscopy (XAS) or to the continuum level as described in photoemission specroscopy (PES). Soft x-ray emission (XES) is the radiative decay of an excited state and it is described as a second order optical process. The second order optical process is weaker and it can provide more detailed information than the first order process. Depending on whether the electronic excitation occurs to a bound state or the continuum level, it is common to distinguish the resonant (RXES) from the non-resonant x-ray emission (NXES), the latter one often being referred to as fluorescence. The intermediate states of RXES and NXES are identical to the final states of the XAS and PES processes, respectively. XES inherently suffers from low intensities because it competes with non-radiative decay mechanisms such as Autoionization, Auger and Coster-Kronig processes. This is even more of a problem in below threshold excitation where the resonant inelastic scattering (RIXS) takes place. Experiments have improved drastically with the advent of the more brilliant third generation of synchrotron sources.

The electronic properties of the d and f electrons, as in the transition metals and the Lanthanides, give rise to such phenomena as magnetism, Kondo resonances and mixed valency. In order to understand these properties, it is important to study the electronic states of the d and f electrons. X-ray emission spectra (XES) of Lanthanides excited near the 3d-4f and 4d-4f thresholds show a variety of resonance phenomena revealing specific information about the electronic structure. Resonant inelastic soft x-ray scattering (RIXS) [1] has been used to elucidate charge-transfer and inner-shell excitations in correlated systems [2-8]. *Only* in the presence of mixed valency such as in CeO$_2$ [2] and PrO$_2$ [3], the one-electron approach fails to interpret the emission spectra. In this case the *f*-states may then be treated alternatively within the framework of the Anderson impurity model [9] as degenerated impurity levels hybridized with the valence band [10]. This approach, however, does not explain all observed spectral features and the fitting procedure usually includes variation of several free parameters, which does not allow for estimation of the contribution of atomic and band-like effects in the formation of XES.

Lanthanum does not have any 4f electrons in the ground state and in the elements following Lanthanum in the periodic table the 4f shell is being filled. As a result the decay process in La can directly provide information about the character of the 4f (and 5p) wave functions. La is also an ideal system to study the charge transfer to empty 4f states. In the present study we address the decay mechanisms of the 3d and 4d core holes of La.

## EXPERIMENT

The experiments were performed at Beamline 8.0 of the Advanced Light Source at Lawrence Berkeley National Laboratory. Spherical gratings monochromatize the undulator radiation. The resolving power of the monochromator is set to about $E/\Delta E = 300$ (at 97 and 850 eV) for our measurements. The fluorescence end station [11] is equipped with a Rowland circle grating spectrometer that provides a resolving power of about 700 (at 97 eV) and 440 (for 850 eV). In order to lower the amount of elastically scattered radiation (for excitation at the 4d-4f resonance) the incident angle of monochromatic beam is about 20° to the sample normal. The plane of incidence is the plane of polarization. The sample is a single crystal of $LaAlO_3$ (a cubic perovskite with tilted $O_6$ octahedra, see Ref. 12).

## MEASUREMENTS AND DISCUSSION

Fig. 1 shows three experimental soft x-ray emission spectra (XES) of $LaAlO_3$. The excitation energy is tuned to the three absorption features (97.4, 101.7 and 119 eV) that correspond to excitations of 4d electrons to the 4f-shell ($4d^{10}4f^0 \rightarrow 4d^9 4f^1$). Arrows in the partial fluorescence spectrum (shown in the insert) indicate the excitation energies selected for the XES spectra. The emission features show dramatic differences in energy and intensity for selected excitation energies.

Calculated spectra are shown below the experimental data. The intensity I of inelastic and elastic scattering as well as of resonant fluorescence is described by the Kramers-Heisenberg formula [13]:

$$I(h\nu_{in}, h\nu_{out}) \propto \sum_f \left| \sum_m \frac{\langle f | p \cdot A | m \rangle \langle m | p \cdot A | i \rangle}{E_m - E_i - h\nu_{in} - i\Gamma/2} \right|^2 \cdot \delta(E_f + h\nu_{out} - E_i - h\nu_{in}) \quad (1)$$

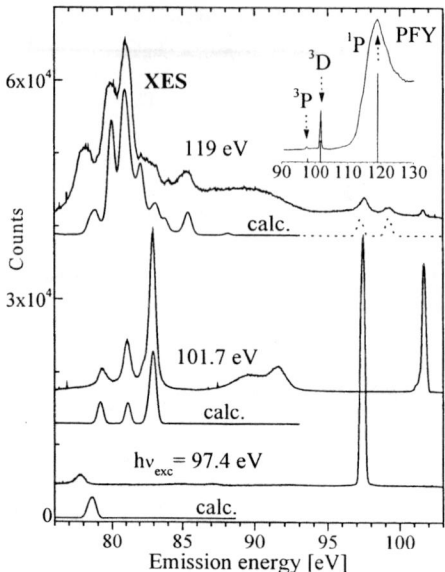

**FIGURE 1.** Soft x-ray emission spectra of LaAlO$_3$. The excitation energy is given above each experimental spectrum. It is tuned to the three features of the partial fluorescence spectrum, PFY, (shown in top right insert) corresponding to weakly dipole allowed $^3P_1$ (at 97.4 eV) and $^3D_1$ (at 101.7 eV) and dipole allowed $^1P_1$ (at 119 eV) terms of the intermediate configuration 4d$^9$4f$^1$. Calculations are shown below the experimental data.

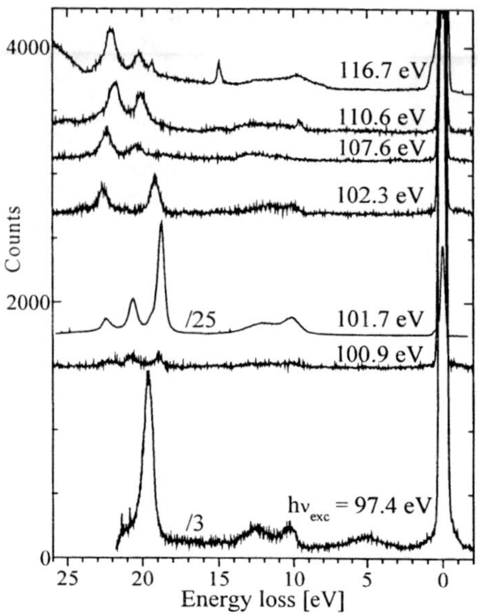

**FIGURE 2.** Energy loss spectra of LaAlO$_3$ for excitation energies near the 4d-4f resonance. Two groups of loss features at 10 to 13 eV and 19 to 23 eV are prominent. For clarity the spectra in both Figures have been displaced along the y axis. In order to display all spectra on the same scale, the spectrum for 101.7 eV has been divided by a factor of 25 and the spectrum at 97.4 eV by a factor of 3.

In this equation, p·A is the dipole operator, $|i\rangle$ is the initial state of the system with energy $E_i$, $|m\rangle$ and $E_m$ describe the intermediate state, $|f\rangle$ and $E_f$ stand for the final state, and $\Gamma$ is the life-time broadening in the intermediate state. $h\nu_{in}$ and $h\nu_{out}$ are the energies of the incoming and emitted photons respectively. In order to calculate the emission intensity, the transition rates were computed relativistically for initial to intermediate and intermediate to final state using Cowan's code [14]. The PFY spectrum was used to determine the set of parameters used for the calculations. The scaling factor for the Slater integrals was set to 80% [15] and the life-time broadening $\Gamma$ to 0.5 eV [16]. The transition rates then are summed incoherently in Eq. 1 and broadened with a Gaussian profile of 0.4 eV, which is approximeately the resolution of the spectrometer.

When exciting at 97.4 eV, the elastic peak dominates the spectrum. The 5p-4d decay of the $^3P_1$ term of the 4d$^9$5p$^6$4f$^1$ configuration results in the single peak at 78 eV. At an excitation energy of 101.7 eV (intermediate state $^3D_1$) the emission is strongly enhanced and shows three features that are due to 5p-4d transitions to different final states ($^3F_2$, $^1D_2$ and $^3D_{1,2}$). For this excitation energy it is necessary only to include the $^3D$ states with J=1 in order to obtain good agreement between calculations and measurements, because only the terms with J=1 are populated in the absorption process. When tuning the excitation energy to 119 eV two groups of emission features in the range at 96 to 102 eV and much stronger ones at 77 to 85 eV are observed. The 5p-4d emission through only the $^1P_1$ intermediate state is shown in Fig. 1 as the dotted part of the calculation below the measured spectrum for 119 eV. The much stronger additional features dominate the emission in the range from 77 to 85 eV, which is about 18 eV below the features resulting from the $^1P_1$ intermediate states. In order to reproduce this part of the measured spectrum by the calculations, the matrix elements of *all* intermediate terms ($4d^9 5p^6 4f^1$ configuration) to all final state terms ($4d^{10} 5p^5 4f^1$) have been taken into account. The fact that all terms of the intermediate state rather than only the $^1P_1$ terms have to be taken into account shows the strong mixing of terms that are separated by about 17 eV (more than 15% of the 4d ionization energy). The strong hybridization of the intermediate terms is also emphasized by the fact that (weak) 4d-4f emission from the $^3P_1$ and $^3D_1$ intermediate states is observed (97.4 and 101.7 eV respectively) when exciting at the $^1P_1$ term (119 eV). The lower one of these two features overlaps in energy with the 5p-4f emission through the $^1P_1$ intermediate state though. Non-resonant 4d-4f emission excited by photons is observed only in La. Usually the emission of the Lanthanides near the giant resonance is dominated by inelastic scattering due to 5p-4f net transitions [4] and 4f inner-shell excitations [5,7] as well as elastic scattering. The peak at 85.4 eV is due to 5p-4d emission from the four times ionized atom $La^{4+}$. This peak appears at an excitation of 119 eV only because the ionization thresholds for 4d electrons lie in the range of 110 to 117 eV (for La atoms) [17].

Generally speaking, XES spectra can exhibit normal fluorescence emission, in which case the energetic position is fixed as a function of excitation energy and inelastic and elastic features, usually observed for excitation energies near absorption thresholds. In the latter case the energy positions of the spectral features follow changes in the excitation energy and have a fixed energy difference relative to the excitation energy. In order to demonstrate the resonant character of the emission, soft x-ray emission spectra for various excitation energies are displayed in Fig. 2. The energy loss is obtained by subtracting the emission energy for each spectrum from the excitation energy. The emission intensity is highly resonant when tuning the excitation energy to the sharp absorption feature that corresponds to the $^3D_1$ term of the $4d^9\ 4f^1$ configuration (at 101.7 eV). The intensity of this emission spectrum is divided by a factor of 25 in order to display all of the spectra on the same scale. Detuning of the excitation energy by only 0.6 eV above or 0.8 eV below the $^3D$ resonance (101.7 eV) leads to a drop in emission intensity by two orders of magnitude.

Two groups of loss features are found: A double-peak energy loss feature is found in the range of 18 to 23 eV and another weaker double-peak loss structure appears around 10 and 12.5 eV below the excitation energy. The energy losses in the range 18 to 23 eV correspond to net transitions in which a 5p electron is promoted to the 4f level. These are usually the strongest energy loss mechanisms at the 3d and 4d thresholds and they have been observed for La (at 3d threshold) [6], Nd (at 3d and 4d) [7] and Gd (4d) [4]. The weaker band-like loss features around 10 and 12.5 eV is assigned to charge transfer transitions in which electrons from the O 2p valence bands are promoted to empty 4f states through the intermediate state configuration $4d^9 5p^6 4f^1 VB$. This is described in detail elsewhere [18].

We now focus on the decay of the 3d core hole. Fig. 3 displays the x-ray emission spectrum (XES) of LaAlO$_3$ obtained at an excitation energy of 10 eV above the 3d-4f threshold (858.9 eV). Four peaks are resolved and labeled A through D.

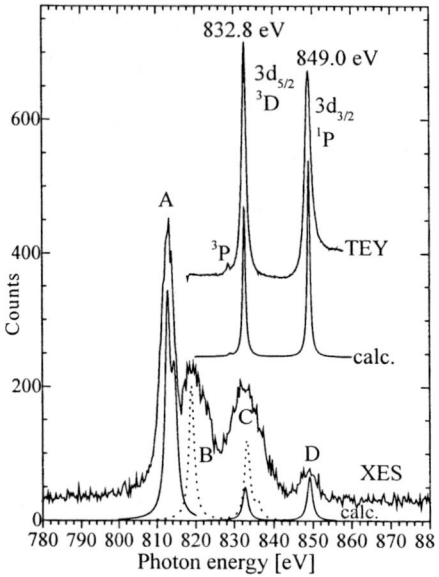

**FIGURE 3.** Total electron yield spectrum (TEY) and soft x-ray emission spectrum (XES) excited above M$_{3,4}$ threshold of LaAlO$_3$ and calculated transitions for La from the refill of the 3d hole. For details regarding the calculations see text. The XES spectrum is taken at an excitation energy of 858.9 eV and the same one as displayed in Fig. 2.

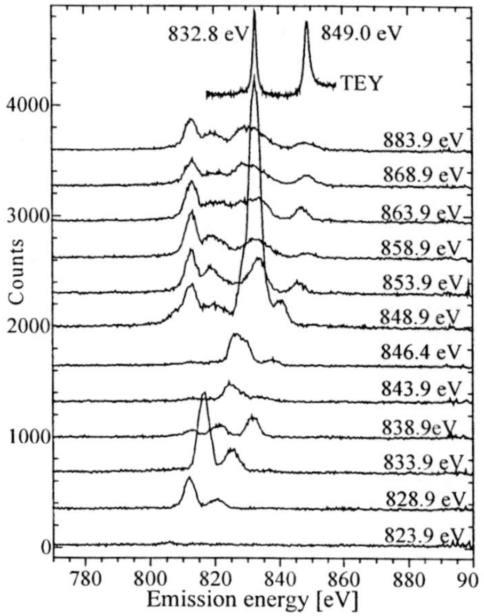

**FIGURE 4.** Soft x-ray emission spectra and absorption spectrum (TEY) of LaAlO$_3$ in the region through the 3d-4f threshold. The excitation energy for the spectra is given above each spectrum. For clarity, the spectra have been displaced along the y axis.

We have calculated the atomic transition rates to fill the hole in the M$_{IV}$ shell for La by using Cowan's code [14]. The scaling factor for the Slater integrals was 80% and the calculated spectra have been broadened by 2 eV (full width-half maximum), which is about the resolution of the spectrometer in this energy range. The results are shown below the XES spectrum. Both **L•S** and j-j notation will be used for the terms of the configurations in this paper. Peaks A and B are due to emission from transitions of 5p electrons to the 3d$_{5/2}$ shell with (A: 3d$_{5/2}$ 4f$^1$ $\underline{L}$ → 5p$_{3/2}$ 4f$^1$ $\underline{L}$) and without (B: 3d$_{5/2}$ → 5p$_{3/2}$) the 4f spectator electron. The charge transfer from O 2p valence band to the 4f leaves a hole in the ligands which is denoted as $\underline{L}$. Peak C shares its intensity with 3d$_{3/2}$ → 5p$_{3/2,1/2}$ and 3d$_{5/2}$ 4f$^1$ $\underline{L}$ → $\underline{L}$. Peak D is the fluorescence from the transition: 3d$_{3/2}$ 4f$^1$ $\underline{L}$ → $\underline{L}$. In Fig. 3 calculated fluorescence involving charge-transfer are presented as solid lines while the non charge-transfer transitions (3d$_{5/2}$ → 5p$_{1/2,3/2}$) are given as the dotted curve. The peak assignment in this emission spectrum is consistent with the interpretation of Jouda et al. [19]. When exciting above the 3d threshold, charge-transfer processes provide occupancy of the 4f level: electrons from the oxygen valence band can shake down to the 4f levels and then refill the 3d hole. According to Jouda et al. [19] and Okusawa et al. [20], who measured the La emission spectrum with electron beam excitation, peaks C and D were suppressed by self-absorption.

The absorption of $LaAlO_3$ in the region of the 3d threshold has been measured by detecting the total electron yield (TEY). The data are displayed above the XES spectrum. The features arise from transitions between the $3d^{10}4f^0$ ($^1S_0$) ground state to the $^3P$, $^3D$ and $^1P$ terms of the $3d^94f^1$ configuration. Spin-orbit splitting as well as the energy positions at 832.8 eV for the transitions $3d^{10}4f^0$ ($^1S_0$) → $3d^94f^1$ ($^3D$ or $3d_{5/2}$) and at 849 eV and for $3d^{10}4f^0$ ($^1S_0$) → $3d^94f^1$ ($^1P$ or $3d_{3/2}$,) are almost the same as those reported for La metal (834.8 eV and 850.7 eV, see Ref. 20).

In Fig. 4 the measured emission spectra for $LaAlO_3$ are shown with the TEY spectrum (top) for various photon excitation energies. Above the $M_{IV}$ threshold the emission spectra consists of four fluorescence peaks. The fluorescence evolves into inelastic scattering at excitation energies below the $M_{IV,V}$ absorption thresholds. As soon as the excitation energy drops below the $M_{IV}$ absorption threshold at 849 eV, the emission from lanthanum is completely dominated by inelastic scattering.

When exciting at the $M_{IV}$ excitation threshold, the emission corresponding to peak C shows strong resonant behavior. Emission, corresponding to peaks A and B, suddenly appears from the refill of the $3d_{5/2}$ hole. This is due to Coster-Kronig (C-K) enhanced fluorescence from the decay of the $3d_{5/2}$ hole (peaks A, B, and C) that commences at the $3d_{3/2}$ threshold. The C-K transition $^1P$ ($3d_{3/2}$) → $^3D$ ($3d_{5/2}$) promotes a charge-transfer 4f L creating the $3d_{5/2}$ and $3d_{5/2}4f$ L intermediate states that result in fluorescence to final states of the type $5p^5$ 4f L (peak A), $5p^5$ (peak B and C) and L (peak C). There is only Raman scattering that involves 4f L $3d_{3/2}$ intermediate states and L final states (peak D), because the resonant energy (849.0 eV) is below the threshold to produce the 4f L charge-transfer and excite a $3d_{3/2}$ subshell electron. Furthermore, the rates for $M_{IV}M_VN_{VI,VII}$ C-K transitions are about two orders of magnitude greater than the rates for $M_{IV}M_VO_{II,III}$ C-K transitions [16]. This is why no final states of the type $5p^44f$ are observed. Auger processes dominate photon absorption in the spectral range between the $M_V$ and the $M_{IV}$ absorption thresholds. Very weak fluorescence from the decay of the $3d_{5/2}$ hole is observed even when exciting at the $3d^{10}4f^0$ $^1S$ → $3d_{5/2}4f^1$ $^3D$ absorption resonance.

In Fig. 5 two emission spectra (XES) from Fig. 4 excited at resonances associated with the transition of 3d electrons to $3d_{5/2}$ 4f and $3d_{3/2}$ 4f excited states are displayed as energy loss spectra. The calculated energy loss from Eq. (1) with a life-time broadening of 1 eV is shown below the measurements. In our calculation the initial state is $3d^{10}$ $5p^6$ $4f^0$ ($^1S$), intermediate states are $3d^9$ $5p^6$ $4f^1$, ($^3D$, $^1P$) and the final states are $3d^{10}$ $5p^5$ $4f^1$ ($^3F$, $^1D$, $^3D$). The 16.3 eV energy loss corresponds to a net transition of a 5p electron to a 4f orbit. The good agreement of model and experiment indicates that the intermediate states are not strongly coupled in the presence of a 3d core hole. The two peaks at higher loss energies in the spectrum excited at threshold (848.9 eV) are fluorescence peaks A and B as calculated in Fig. 3. The energy loss at about 9 eV is due to a charge transfer of an O 2p electron to the 4f level as seen in the 4d-4f spectra as well. It is discussed in more detail in Ref. 6.

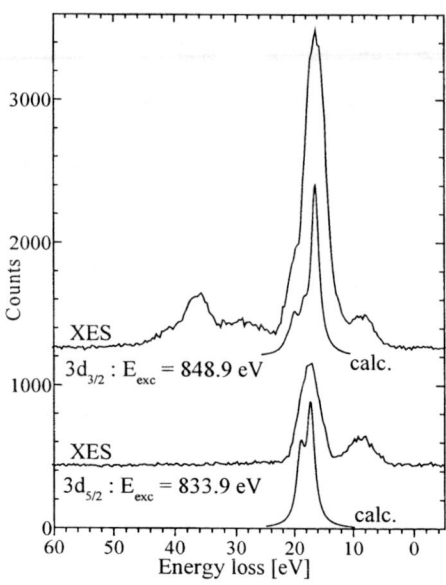

**Figure 5.** Energy loss spectra of LaAlO$_3$ for excitation at $^3$D and $^1$P threshold and atomic calculations of the 16.3 eV loss feature. The energy loss is due to the net transition 5p-4f.

## SUMMARY

We have selectively excited LaAlO$_3$ through the 4d-4f and 3d-4f resonances with monochromatic synchrotron radiation. At the giant resonance we observe strong mixing of intermediate states over an unusual extended energy range of 15 to 20 eV along with strong dispersion and strong resonant inelastic scattering. Charge-transfer transitions from the O 2p valence band to the 4f as well as energy losses due to excitations within the La atom (5p→4f net transitions) are observed. While both the 5p-4f energy loss and the 4f $\underline{L}$ charge transfer are present at the 3d-4f resonance as well, strong Coster-Kronig transitions (between M$_{IV}$ and M$_V$ levels) are apparent at the 3d-4f resonance only.

## ACKNOWLEDGMENTS

This work was supported by National Science Foundation grant DMR-9801804 and the Science Alliance Center for Excellence Grant from the University of

Tennessee. The Advanced Light Source is supported by the office of Basic Energy Sciences, U.S. Department of Energy, under contract no. DE-AC03-76SF00098. CAMD is supported by the state of Louisiana.

## REFERENCES

[1] W. Eberhardt (ed.), *Resonant Inelastic Soft X-ray Scattering (RIXS)*, Special issue of Appl. Phys. A **65** (1997).

[2] S.M. Butorin *et al.*, Phys. Rev. Lett. **77**, 574 (1996).

[3] S.M. Butorin *et al.*, J. Phys.: Condens. Matter **9**, 8155 (1997).

[4] J.-J. Gallet *et al.*, Phys. Rev. B **54**, 14238 (1996).

[5] A. Moewes *et al.*, Phys. Rev B **57**, R8059 (1998).

[6] A. Moewes *et al.*, Phys. Rev B **58**, R15951 (1998).

[7] A. Moewes *et al.*, Phys. Rev B **59**, 5452 (1999).

[8] J.-E. Rubensson *et al.*, Appl. Phys. A **65**, 91 (1997).

[9] O. Gunnarson and K. Schönhammer, Phys. Rev. B **28**, 4315 (1983).

[10] S. Tanaka *et al.*, J. Phys. Soc. Jpn. **61**, 4212 (1992).

[11] J.J. Jia *et al.*, Rev. Sci. Instrum. **66** (2), 1394 (1995).

[12] C. deRango *et al.*, Comp. Rend. Acad. Sc. **263** ser. C, 64 (1966).

[13] T. Aberg, Phys. Scr. **21**, 495 (1980).

[14] R.D. Cowan, *The theory of atomic structure and spectra*, Berkeley: University of California press, 1981.

[15] H. Ogasawara *et al.*, Sol. State Comm. **81**, 645 (1992).

[16] E.J. McGuire, Phys. Rev. A **5**, 1052 (1972).

[17] M. Richter *et al.*, Phys. Rev. A **11**, 5666 (1989).

[18] A. Moewes *et al.*, submitted.

[19] K. Jouda *et al.*, J. Phys. Soc. Jpn **64**, 192 (1995).

[20] M. Okusawa *et al.*, Phys. Rev. B **35**, 478 (1987).

# Resonant Inelastic X-ray Scattering from Transition Metal Oxides

## J.P. Hill

*Department of Physics, Brookhaven National Laboratory, Upton, NY 11973*[1]

**Abstract.** Recent developments in hard x-ray resonant inelastic x-ray scattering as a probe of strongly correlated systems are reviewed. Particular attention is paid to studies of $Nd_2CuO_4$. A charge transfer excitation is observed when the incident photon energy is tuned in the vicinity of the copper K-edge. It is shown that the presence of resonant enhancements is controlled by the polarization dependence of the excitation process and by the overlap between a given intermediate state and the particular excitation being studied. This latter observation has shed light on the non-local effects present in certain intermediate states.

## INTRODUCTION

Strongly correlated electron systems, and in particular the transition metal oxides, are amongst the most actively investigated systems in condensed matter physics today. This interest is driven both by the wide variety of phenomena displayed by these materials and by the fundamental issues raised in trying to understand them. Notable examples in this class include, the high-$T_C$ superconductors and the colossal magnetoresistance manganates.

Strongly correlated systems are characterized by electronic behavior intermediate between the highly localized behavior of ionic insulators and the completely delocalized behavior of simple metals and as a result a complete theoretical description remains problematic. The strong correlations preclude the possibility of successful band structure calculations, and a variety of numerical approaches have therefore been applied, utilizing small clusters of ions for which on-site interactions can be treated explicitly. These models are typically local, that is the translational symmetry of the lattice is neglected. However, as is increasingly recognized, solid state (non-local) effects can be important and recently calculations have been performed for a number of clusters for which some degree of translational symmetry has been restored.

On the experimental front, of crucial importance are measurements of the electronic structure and dynamics, to provide insights into the behavior and to test the

---

[1] Work supported by the US D.O.E. under contract no. DE-AC02-98CH10886

various theoretical approaches. In this endeavor, inelastic x-ray scattering in the hard x-ray regime offers a number of potential advantages over other techniques. In particular, it probes bulk-like properties and is a "photon-in, photon-out" process. This ensures that the experiments do not suffer from so-called final state effects, in which the interpretation of the observed spectra is complicated by the presence of a core-hole (as is the case in, for example, a photoemission process). In addition, the technique can be applied to insulators, without problems due to electrostatic charging, and in principle experiments can also be performed in applied magnetic and electric fields.

In the non-resonant regime, far from any absorption edges, the cross-section is well understood and is given by the dynamic structure factor, $S(\mathbf{q}, \omega)$. This quantity is proportional to the Fourier transform of the charge density-density correlation function and may be directly related to important quantities, such as the complex dielectric function [1].

Unfortunately, to date this technique has been limited largely to studies of low-Z materials and is thus not applicable to the majority of interesting transition metal oxides. This is because the inelastic cross-section is extremely small and only by studying samples for which the x-ray absorption length is large, resulting in relatively large sample volumes, can this limitation be overcome and reasonable signals be obtained.

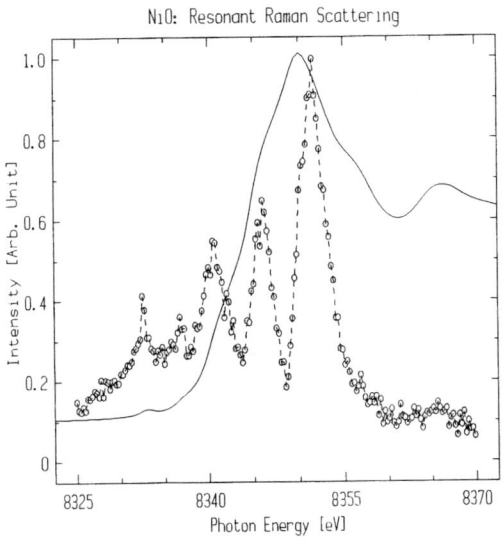

**FIGURE 1.** Incident photon energy dependence of a 5 eV charge transfer excitations, as observed in NiO, with the incident energy being tuned through the Ni K-edge (open circles). The solid line shows the Ni K-edge absorption spectrum for comparison. Figure taken from ref. [2].

Recently, however, Kao and co-workers discovered resonant enhancements in the inelastic scattering cross-section which offer the potential for overcoming this limitation [2], and it is these enhancements which are the subject of this review. Note this discovery of enhancements in the inelastic scattering of valence band excitations is in a certain sense an extension of the more familiar resonant Raman scattering from emission lines (see e.g. ref. [3]) to very small energy transfers.

Kao et al. observed the large increase in the inelastic scattering in NiO when the incident x-ray energy was tuned through the Ni K-edge. In particular, they found excitations at 5 eV and at 8 eV that were only observable when the incident photon energy was tuned to the vicinity of the Ni K-edge. The intensity of each was found to be strongly dependent on the incident energy (while the position of the excitations moved only slightly with incident energy). Figure 1 shows this incident energy dependence for the 5 eV excitation. A large number of sharp resonances are observed in the scattered intensity, each of which appears to correspond to a feature in the absorption spectrum (solid line), albeit better resolved in the inelastic scattering.

On the basis of a configuration-interaction cluster model calculation, Kao et al. identified these features as charge transfer-type excitations in which the system is excited from a $3d^8$ ground state to a $3d^9\underline{L}$ excited state, where the $\underline{L}$ represents a hole on the ligand site. Of perhaps greatest significance was the size of the resonant enhancement. By comparing non-resonant x-ray data to electron scattering data taken at the same momentum transfer, Kao et al. estimated that the resonant inelastic cross-section to be about 100 times larger than the non-resonant cross-section.

While a number of questions concerning the detailed understanding of the resonant cross-section arose as a result of the NiO work (and still remain to some extent), the potential for expanding the applicability of inelastic x-ray scattering to materials containing heavy elements, and specifically the high-$T_C$ cuprates, was immediately apparent. Furthermore, the technique offers the added advantage of element specificity allowing, for example, only those excitations associated with the copper orbitals to be probed by tuning to the Cu K-edge.

Note that this new work in the hard x-ray regime extends and complements a larger body of work utilizing resonant inelastic scattering in the soft x-ray regime. This latter effort includes the early work of Ma and others (see eg. [4,5]) in studies of diamond, graphite and other systems and extends to the recent work on strongly correlated systems (e.g. [6–9] and references therein). Theoretical efforts have accompanied these advances in soft x-ray techniques and include [10–12] and references therein.

In this paper, we review a series of experiments carried out to elucidate the details of the resonant inelastic cross-section. These experiments, primarily carried out on $Nd_2CuO_4$ observed a similar charge transfer process to that seen in NiO, and demonstrate the feasibility of using these techniques to study the high-$T_C$ cuprates. Measurements of the incident energy and polarization dependence of the cross-section illustrate the importance of the intermediate (highly excited) states in determining the resonant lineshape, and in particular the sensitivity of that lineshape to non-local effects in the intermediate state. In addition, the momentum dependence of the cross-section will be briefly discussed. Portions of this work have appeared previously [13–16].

# I RESONANT INELASTIC SCATTERING IN $ND_2CUO_4$

$Nd_2CuO_4$ is the parent compound of the electron doped high-$T_C$ family, $Nd_{2-x}Ce_xCuO_4$. Its crystal structure is body centered tetragonal of the $T'$ type and consists of $CuO_4$ plaquettes connected in a two dimensional corner-sharing network. Unlike the $La_2CuO_4$ series, there are no apical oxygens. The crystal used in this work was of high crystallographic quality (mosaic 0.014° FWHM), which minimizes the elastic scattering background, and was in the form of a platelet, 20mm x 10mm x 0.1mm with a c-axis surface normal.

## A Experimental Details

The $Nd_2CuO_4$ experiments were performed at two wiggler sources at the National Synchrotron Light Source, beamlines X21 and X25. The initial work [13] was performed at X21 and utilized a four bounce monochromator comprised of two channel cut Si(220) crystals. The incident resolution was 0.2 eV and the flux delivered on the sample was $5 \times 10^{10}$ photons s$^{-1}$. The horizontally scattered radiation was collected by a spherically bent Si(553) analyzer and the overall energy resolution was a Gaussian of 1.9 eV full-width-at-half-maximum (FWHM), as measured by the quasi-elastic scattering from the sample. The sample was oriented such that the momentum transfer was along the c-axis and the incident polarization was therefore largely perpendicular to the copper oxide planes [13]. A second set of measurements were performed at X25, for which a vertical scattering geometry was utilized [15]. In this case, with the momentum transfer again along the $\hat{c}$-axis, the incident polarization was entirely within the plane of the CuO sheets. and the $1s \rightarrow 4p_\sigma$ transition was utilized. For these latter measurements, a double crystal Si(111) monochromator was used which provided an incident energy resolution of 2.2 eV (FWHM) and delivered a flux of $5 \times 10^{11}$ photons s$^{-1}$. The same analyzer set-up as in the X21 experiments was used, and the overall energy resolution was 2.3 eV (FWHM). All results discussed here were taken at room temperature at a momentum transfer of $q = 4.6 Å^{-1}$. The same sample was used in all measurements.

The experiments on $CuGeO_3$ [14,17] were performed on X21 with an identical set-up to that outlined above.

# B  Charge Transfer Excitation

The scattered intensity observed in $Nd_2CuO_4$ is plotted as a function of energy loss $(E_i - E_f)$ for a series of incident photon energies around the Cu K-edge in figure 2. Each data set took approximately 24 hours to collect. The excitation at 6 eV is observed only for incident energies around 8990 eV, and remains at approximately fixed energy loss for all incident energies.

The amplitude of the 6eV excitation was extracted by fitting the energy loss scans to a Gaussian peak on a sloping background (to account for the tails of the nearby $K\beta_5$ emission line) and the results are plotted in figure 3 as a function of incident photon energy. Resonant behavior was observed, with a peak at 8990 eV. Even at resonance, the count rates remain small, due to the large absorption of the Nd, as well as that of the copper. Count rates were on order of a few counts per minute.

Also shown in figure 3 is the absorption of $Nd_2CuO_4$ powder, in which features were observed at 8983, 8990, 8995, and 9002 eV (see also refs. [19,20]). These have been interpreted as arising from two sets of dipole transitions [19–21]. The first two (lower energy pair) are the $1s \rightarrow 4p_\pi$ transition (4p orbitals perpendicular to the CuO planes) and the second two, the $1s \rightarrow 4p_\sigma$ transition (in-plane 4p orbitals). Figure 3 shows that the resonant enhancement is associated with the $1s \leftrightarrow 4p_\pi$

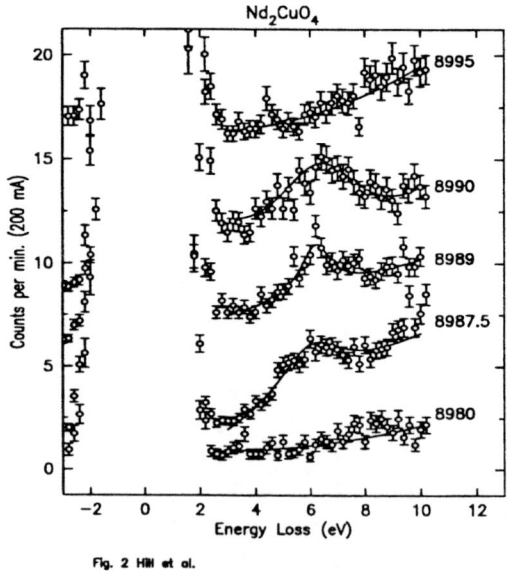

Fig. 2 Hill et al.

**FIGURE 2.** Scattered intensity as a function of energy loss ($E_i$ - $E_f$) for a number of incident energies in the vicinity of the Cu K-edge. Data are offset vertically for clarity and solid lines are guides to the eye. Figure taken from ref. [13].

transition. There was no apparent enhancement at the $1s \leftrightarrow 4p_\sigma$ transition. The absence of this latter transition is a result of the polarization dependence of the excitation process, as discussed below.

In order to identify the character of the excitation, numerical calculations of the electronic structure of $Nd_2CuO_4$ were performed within the Anderson impurity model [22,23,13]. The intermediate states of the scattering process were then calculated within the same model including on-site interactions between both the $4p$ and the core hole ($U_{pc}$=5 eV) and the $4p$ and $3d$ states($U_{dp}$= 3 eV) and a finite $4p_\pi$ bandwidth of 2 eV. The scattering was then treated as a coherent second order (dipole) process between these states [22]:

$$F(\omega_f, \omega_i) = \sum_f \left| \sum_n \frac{<f|T|n><n|T|g>}{E_g + \omega_i - E_n - i\Gamma} \right|^2 \delta(E_g + E_i - E_f - \omega_f), \quad (1)$$

where $|g>$ is the ground state of the Anderson Hamiltonian with energy $E_g$, and $|n>$ and $|f>$ are the intermediate and final states with energies, $E_n$, and $E_f$, respectively. $T$ is the dipole transition operator and $\Gamma$ is the lifetime of the core hole in the intermediate state.

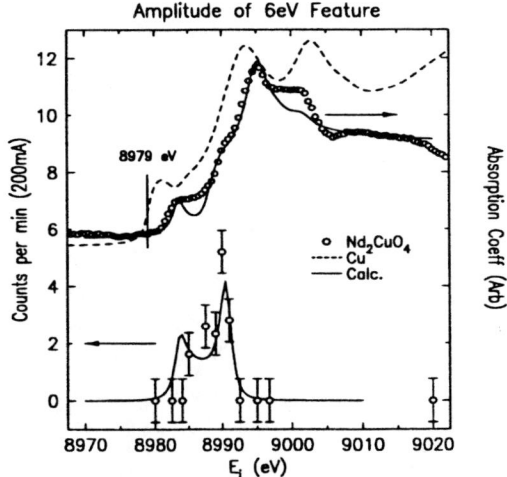

**FIGURE 3.** The amplitude of the 6eV excitation, as a function of incident energy, showing a resonance in the vicinity of 8990 eV (open circles). The solid line is the result of the Anderson impurity model calculation described in the text. The top half of the figure displays the absorption measured for powdered $Nd_2CuO_4$ (open circles) together with a calculation using the same parameters as for the scattered intensity. The absorption from a Cu foil (dashed line) is shown for reference. Figure taken from ref. [13].

In the CuO planes, the Cu $3d^9$ configuration hybridizes with $3d^{10}\underline{L}$, where $\underline{L}$ represents an O $2p$ ligand hole of finite bandwidth. Within the Anderson impurity model, this results in discrete bonding and anti-bonding states composed of a mixture of $3d^9$ and $3d^{10}\underline{L}$ configurations, with a continuous band between them (figure 4). The ground state is then the bonding state, with about 60% $3d^9$. The lowest edge of the continuous band (charge transfer gap) is about 2 eV above this and the discrete anti-bonding state is $\sim 6$ eV above the ground state.

In the intermediate state of the resonant scattering process, a Cu $1s$ electron is excited to the (for example) Cu $4p_\pi$ band, and the core hole potential reverses the balance between the $3d^9$ and $3d^{10}\underline{L}$ configurations. The lowest energy state is then predominately $\underline{1s}3d^{10}\underline{L}4p_\pi$ and is about 7 eV lower than the anti-bonding state, $\underline{1s}3d^9 4p_\pi$ (fig. 4). These states form the 8983 and 8990 eV features of the Cu K-edge XAS. (The $1s \rightarrow 4p_\sigma$ is similarly split.)

Energy loss spectra calculated within this model show a single excitation at 5.4 eV, when the incident photon energy is 8990 eV. (Note, higher quality data taken at X25 placed the experimental value for the excitation at 5.7 eV). The peak results from the decay of the intermediate state into the anti-bonding excited state. A single peak is observed because, for the $\underline{1s}3d^9 4p_\pi$ intermediate state (which is excited at 8990 eV), there is no significant overlap with the continuous band, which is predominately $3d^{10}\underline{L}$. Decays into this band are therefore suppressed, so that only

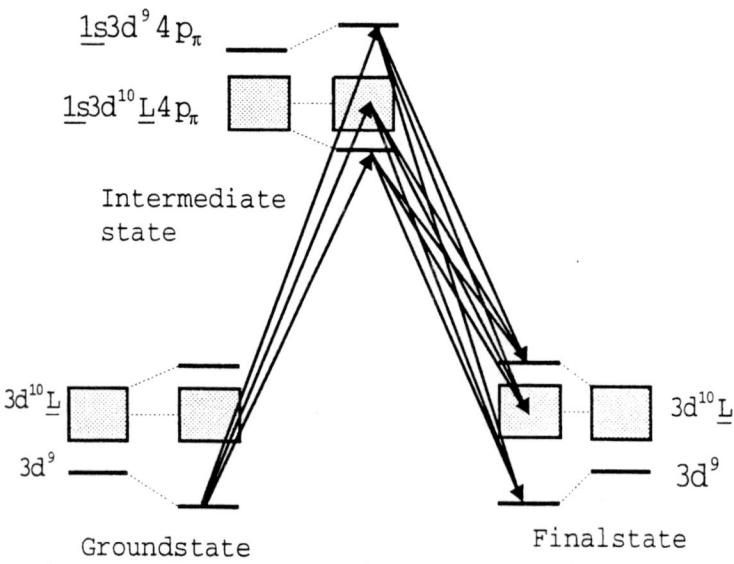

**FIGURE 4.** Schematic energy level diagram for inelastic scattering from a copper site. Arrows indicate processes summed over in the calculation of the scattered intensity. Figure taken from ref. [13].

the discrete anti-bonding state is observed. As the incident photon energy is moved away from 8990 eV, the intensity of the 6 eV feature decreases, correctly reproducing the resonant behavior of this excitation. Thus, the calculations showed good agreement with experimental observations in the vicinity of 8990 eV and therefore the excitation was identified as a charge transfer excitation into an antibonding state.

In addition to $Nd_2CuO_4$, two other cuprates have also been studied $CuGeO_3$ and $YBa_2Cu_3O_7$ [14,17], and in each case, similar excitations were observed, at 6.5 eV and $\approx$ 5 eV, respectively. The similarity in excitation energies suggests that the important structure in determining the electronic properties is the $CuO_4$ plaquette, which these three materials have in common. This is consistent with the charge transfer identification of the excitation.

## C  Polarization Dependence

As noted above, in the original work on $Nd_2CuO_4$, there were no resonances associated with the higher energy transitions associated with the $4p_\sigma$ band. In ref. [13], it was hypothesised that this was connected with polarization effects. In order

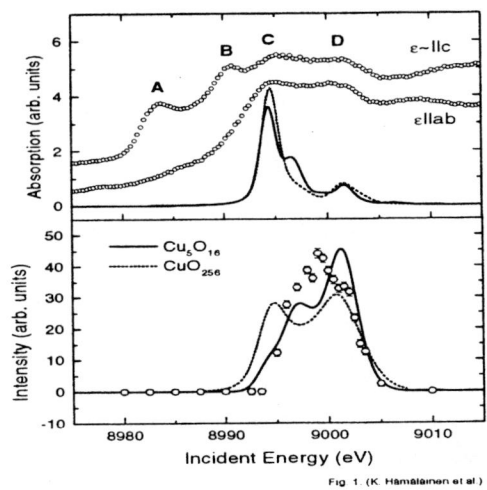

**FIGURE 5.** Top panel: Polarization dependent absorption measured via monitoring the fluorescence yield. Lower panel: The intensity of the 6 eV excitation extracted by fitting the individual energy loss scans as a function of incident energy (open symbols). The solid and dashed lines in both panels are the results of numerical calculations for the $1s \rightarrow 4p_\sigma$ transitions described in the text. Figure taken from ref. [15].

to test this, measurements were carried out with the incident polarization aligned in the CuO planes ($\epsilon||ab$ plane), that is about 90° to the earlier work ($\epsilon \approx \perp ab$ plane). The momentum transfer was perpendicular to the CuO sheets ($q||c$), as previously [15,16].

As in the previous experiment a 6 eV excitation is observed only at certain, resonant, incident energies. The amplitude of the 6 eV excitation as a function of incident energy was extracted by fitting a series of scans with a Gaussian peak on a linear background. The resulting amplitudes are shown in the bottom half of Fig. 5 (open circles). The peak positions and widths were almost constant showing only very slight variations (in fact, the fits were essentially unchanged if the peak position and width were held fixed). A single resonance was observed at 8999.5 eV, of width $\approx$ 7 eV FWHM. Thus, in this geometry, a resonant enhancement is observed at a $4p_\sigma$ transition and not a $4p_\pi$ excitation.

Polarized x-ray absorption spectra, shown in top half of Fig. 5, provide a natural explanation for these results. These data were obtained by monitoring the intensity of the Cu $K\alpha$ fluorescence from the sample as the incident energy was tuned through the $K$-edge and were not corrected for self absorption effects [24].

Features A and B (fig. 5) are associated with the transitions to the $4p_\pi$ orbitals and they are preferentially excited when the incident polarization is along $c$ direction (top panel, fig 5), and aligned with the excited state orbitals, maximizing the dipole matrix element for the excitation process. Conversely, C and D are associated with $1s \rightarrow 4p_\sigma$ transitions and, correspondingly, these were most prominent when the incident polarization is in the $ab$ plane and parallel with the $4p_\sigma$ orbitals.

The significance of these data is that the *final* state of the XAS process is the *intermediate* state in the resonant scattering process. For a significant resonant enhancement to occur, it is necessary that the intermediate state be strongly excited. The polarized XAS results show that the $4p_\sigma$ intermediate states are strongly excited (and the excitation transitions $1s \rightarrow 4p_\pi$ strongly suppressed) when the polarization is in the CuO planes. It was this polarization dependence of the excitation process which was responsible for the absence of resonant enhancements at the $4p_\sigma$ position in figure 3 and for the absence of a $4p_\pi$ resonance in figure 5.

A strong excitation to the intermediate state, however, is only a necessary condition for observing a resonant enhancement, it is not a sufficient one. This is demonstrated by the absence of resonances associated with the $\underline{1s}3d^{10}\underline{L}$ intermediate states (features A and C) in either geometry. An explanation for this absence is presented in the following section.

## D  Resonant Lineshape and Non-Local Effects

Calculations of electronic structure and the resonant inelastic x-ray scattering process, based on a single copper site, correctly predict the incident polarization dependence discussed above. However, they fail to account for the systematic absence of the $\underline{1s}3d^{10}\underline{L}$ resonances (features A and C); specifically in both cases (solid line, figure 3, dashed line, figure 5) a lower energy resonance is also predicted. This failure of the single site calculation led to speculation [13] that the suppression of the lower energy resonance was due to non-local effects arising from the fact that the real system does not consist of a single isolated $CuO_4$ cluster, but rather forms a continuous CuO network, and therefore more complicated, solid state, effects may be required to accurately describe the resonance process.

One explanation for the discrepancy came from non-local screening effects, which are active in the case of a $3d^{10}\underline{L}$ intermediate state, of the type first proposed by Veenendaal, Eskes and Sawatzky [25], In this scenario, it was proposed that in the $3d^{10}\underline{L}$ intermediate state, the $\underline{L}$ hole is repelled by the $\underline{1s}$ on the copper site and moves off elsewhere in the copper oxide sheet, to form a Zhang-Rice singlet bound state [27] with a $3d^9$ copper site. The binding energy of this singlet stabilizes such a process. This non-local intermediate state cannot then simply decay into the charge-transfer antibonding state (the "6 eV" feature), which is a local excitation; the transition probability between states with and without a Zhang-Rice singlet is small, and thus no resonance would be observed. The solid line of fig. 3 is the result of calculations performed for a single copper site for which there are no such non-local effects, since the $\underline{L}$ hole of necessity must remain in the immediate vicinity of the central core hole. This latter state has significant overlap with the antibonding state (and indeed also with the continuous band of states in fig 3) and therefore a resonance appears at 8983 eV in the calculations.

In order to explore explicitly such non-local screening effects, the earlier calculations were improved upon, and the resonant inelastic scattering was modeled with a $Cu_5O_{16}$ cluster, in which five $CuO_4$ plaquettes were arranged in the $ab$ plane with $D_{4h}$ symmetry (see e.g. [26]). A Cu $3d_{x^2-y^2}$ orbital was placed at each Cu site, and an O $2p_x$ or $2p_y$ orbital at each O site, hybridized with the neighboring Cu $3d_{x^2-y^2}$ orbital. Parameter values used in the cluster model were; the charge transfer energy $\Delta = 2.5$ eV, the electron hopping energy between neighboring Cu $3d$ and O $2p$ orbitals, $T_{pd} = 1.21$ eV, that between the neighboring O $2p$ orbitals $T_{pp} = 0.55$ eV, the on-site Coulomb interaction between Cu $3d$ electrons $U_{dd} = 8.8$ eV, and for the Cu $1s$ core hole potential acting on the Cu $3d$ hole $U_{dc} = 8.0$ eV. For the Cu $4p$ orbitals, nearest neighbor hopping was considered, as represented by the Slater-Koster parameters $(pp\sigma) = 0.24$ eV and $(pp\pi) = -0.8$ eV, the on-site Coulomb interaction between Cu $4p$ and $3d$ states $U_{pd} = 3.0$ eV, and the on-site Coulomb potential of the core hole acting on the $4p$ electron $-U_{pc} = -4.0$ eV. The resonant inelastic spectra were again calculated with the coherent second order optical formula following an exact diagonalization of electronic states of the cluster.

The calculated result for the 6 eV amplitude at the $1s \rightarrow 4p_\sigma$ transition is shown in the bottom panel of Fig. 5 (solid curve), after broadening by incident and final resolutions of 2.0 eV. A strong resonance is seen only at the higher energy XAS feature ($\underline{1s}3d^9$), in broad agreement with the experimental results. For comparison, the results of the Anderson impurity model calculation for a large cluster with a *single* Cu site and many O sites (256) are also shown (dashed curve). This exhibits resonances at both the $\underline{1s}3d^9$ and $\underline{1s}3d^{10}\underline{L}$ energies. Taken together, these calculations demonstrated that the suppression of the resonant enhancement associated with the $1s3d^{10}\underline{L}$ intermediate states results from the presence of multi-Cu sites and reflects non-local effects, as originally speculated.

In addition, the higher intensity obtained at the $4p_\sigma$ resonance (at least partially due to the higher flux at beamline X25) allowed a more accurate determination of the excitation energy. It was found to be $E_{CT} = 5.7$ eV. The single site calculation estimated $E_{CT} = 5.2$ eV and the multi-site calculation, $E_{CT} = 5.7$ eV. Thus it appears that non-local effects are also manifest in the position of the excitation as well as the incident energy dependence. Such quantitative comparisons highlight the potential for resonant inelastic x-ray scattering measurements to provide sensitive tests of electronic structure calculations for these materials.

The non-local hypothesis was also tested experimentally, by studying samples for which the crystal structure is such that the connectivity of the $CuO_4$ plaquettes is radically different. In this regard $CuGeO_3$ provided a good test case since the $CuO_4$ units are connected in a 1 dimensional "edge-sharing" chain in contrast to the two dimensional "corner-sharing" network of $Nd_2CuO_4$. If the non-local ideas are correct, one would expect significant differences in the resonant lineshapes for the two materials.

The incident energy dependence of the inelastic scattering in $CuGeO_3$ is shown in fig. 6, together with the absorption as measured in a powder sample [14,17]. In line with the expectations outlined above, the resonance lineshape was found to be

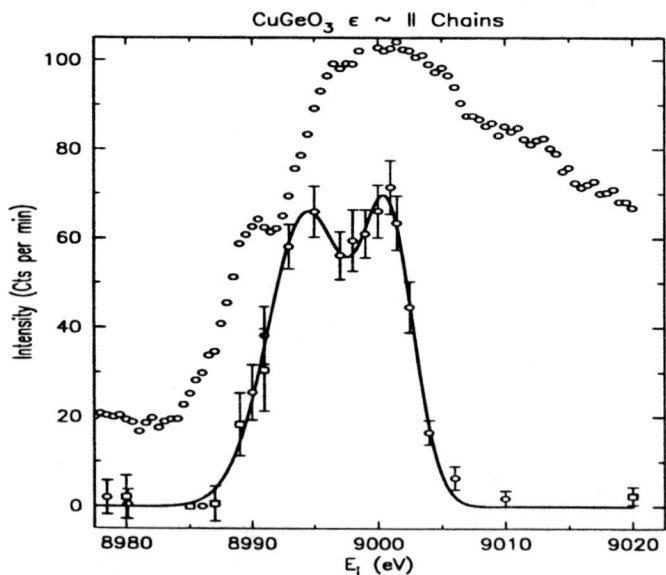

**FIGURE 6.** Incident energy dependence of the inelastic scattering in $CuGeO_3$. Solid line is a guide to the eye. The absorption of powdered $CuGeO_3$ is also shown. Figure taken from ref. [14].

quite different, with a width three times that observed in $Nd_2CuO_4$ and a strong suggestion of two resonances being present. While these results have yet to be analyzed in the context of detailed electronic structure calculations, they are certainly consistent with the idea that non-local effects are important in determining the resonance phenomena.

## E  Momentum Dependence

The $q$-dependence of the charge transfer scattering was also investigated. Data were taken in a vertical geometry, i.e. with the polarization in the $ab$ plane, and at the peak of the $4p_\sigma$ resonance ($E_1 = 8999.5$ eV). The excitation was measured for momentum transfers in the range 3.5 - 7.9 Å$^{-1}$. The observed intensities were normalized by the intensity of the $K\beta_1$ emission line, as measured with the incident photon energy set well above the edge, to account for any variations in the beam footprint at different scattering angles. No dispersion in the peak position, width, or intensity was observed.

As emphasized above, this charge transfer excitation is a localized excitation and thus would not be expected to exhibit much dispersion. It was therefore not possible to comment from these data on the speculations of ref.s [28,29] that q-dependent information can be obtained with the resonant technique. However, these data do rule out any strong angular intensity dependence in the resonant cross-section in this geometry. Future work will focus on the 2 eV charge transfer gap excitation for which predictions of dispersion have been made [18].

# II  OTHER WORKS

Finally, there has been recent work on other copper oxides, namely that of Abbamonte et al. [28] on $La_2CuO_4$ and $Sr_2CuO_2Cl_2$. These experiments were carried out at X21 and at 3-ID, an undulator beamline at the Advanced Photon Source. Broadly, the results obtained in these materials are consistent with those discussed above. In particular, a feature was observed around 6 eV which is attributed to charge transfer excitations between the copper and oxygen. In addition, the $Sr_2CuO_2Cl_2$ experiments were performed at 0.45 eV resolution and were able to resolve a feature at 2 eV energy loss, which was attributed to the charge transfer gap (that is to transitions to the bottom of the continuous band, figure 4).

However, in contrast to the approach outlined above, these authors chose to analyze their data following the ideas put forth by Platzman and Isaacs [29]. Specifically, they follow a perturbative approach in treating the core hole-valence electron interaction and argue that near a sharp dipole allowed transition, the dominant term occurs in third order and has the form,

$$M = \sum_{1s,4p} \frac{M_{em} M_{Coul} M_{abs}}{(\omega_f - E_{1s,4p} + i\gamma_K)(\omega_i - E_{1s,4p} + i\gamma_K)} \quad (2)$$

Where $M_{em}$ ($M_{abs}$) are the dipole emission (absorption) matrix elements, $M_{Coul}$ is the Coulomb interaction between the core and valence states, $E_{1s,4p}$ is the energy of the virtual exciton-like intermediate state and $\gamma_K$ is its lifetime. Physically, this expression represents a "shakeup" process, whereby a virtual $\underline{1s}4p$ exciton is created, scatters off the valence electrons, leaving them in an excited state, and decays. By assuming that all the intermediate states are degenerate, of energy $E_K$, then the energy denominators can be pulled out of the sum, leaving

$$\frac{d\sigma}{d\Omega d\omega} \propto \frac{S_K(\mathbf{q},\omega)}{[(\omega_f - E_K)^2 + \gamma_K^2][(\omega_i - E_K)^2 + \gamma_K^2)]} \quad (3)$$

with

$$S_K(\mathbf{q},\omega) = \frac{2\pi}{\hbar} \sum_f \left| \sum_{\underline{1s},4p} M_{em} M_{Coul} M_{abs} \right|^2 \delta(\omega - E_f + E_i) \quad (4)$$

The quantity, $S_K(\mathbf{q},\omega)$ depends only on the difference quantities $\mathbf{q} = \mathbf{k}_i - \mathbf{k}_f$ and $\omega = \omega_i - \omega_f$. Since it does not depend explicitly on the incident energy, it is a fundamental property of every spectrum. Abbamonte et al. show that their $La_2CuO_4$ data exhibit the correct scaling predicted by such arguments, for a reasonable choice of $E_K$ and $\gamma_K$. They further show that $S_K(\mathbf{q},\omega)$ is related to a particular weighted sum of $S(\mathbf{q} + \mathbf{G}, \omega)$ where $S(\mathbf{q},\omega)$ is the dynamic structure factor of the valence electrons.

Note one of the features of equation (3) is that it predicts that the position of the energy loss feature will depend on the incident excitation energy. Such dispersion was seen in the $La_2CuO_4$ data (and perhaps in the NiO work [2]), but not in the $Nd_2CuO_4$ and $CuGeO_3$ experiments. However, the size of the dispersion is a sensitive function of the parameters and small changes in, for example the excitation lifetime, can greatly reduce the any dispersion observed in the energy loss feature.

It remains to be seen under what conditions the approximations used to obtain equation (3) are valid. If it is generally applicable, the importance of the ability to relate measured quantities to the dynamic structure factor cannot be overstated.

## III  SUMMARY

Recent progress in the use of resonant inelastic x-ray scattering in the hard x-ray regime to study transition metal oxides was reviewed. Particular emphasis was placed on the work on $Nd_2CuO_4$ in which the details of the resonant process was explored. A charge transfer excitation to the antibonding excited state is seen at 5.7 eV when the incident photon energy is tuned to certain resonances in the vicinity of the copper K-edge. It was shown that the systematic absence of resonant enhancements associated with certain transitions is caused by the polarization dependence of the excitation process into the intermediate state. Specifically, the relative orientation of the incident polarization and the particular $4p$ intermediate state. In addition, in order that there be a resonant enhancement of the inelastic signal, there needs to be a significant overlap between the intermediate states and the particular excited state being studied. In the case of the antibonding excitation in $Nd_2CuO_4$, this was manifest in the absence of any resonant enhancement at transitions corresponding to well screened intermediate states. Experiments on other materials and calculations are consistent with explanations based on these intermediate states being extended, non-local objects with little overlap with the localized excitation being studied. These results suggest that such experiments may provide a sensitive measurement of the breadown of the localized picture of the electron structure, and of the resulting importance of non-local effects.

The important remaining question in the field concerns the interpretation of the inelastic spectra themselves, as measured on resonance, and in particular what, if any, connection can be made to the dynamic structure factor of the valence electrons. In this regard, q-dependent measurements to be carried out in the near future will be very important in answering such questions.

## IV ACKNOWLEDGEMENTS

I would particularly like to acknowledge the contributions of C.-C. Kao, with whom all of this work was performed. In addition it is a great pleasure to acknowledge my other collaborators in this work, including, L.E. Berman, W.A.L. Caliebe, R.L. Greene, K. Hämäläinen, K. Hirota, S. Huotari, T. Idé. A. Kotani, T. Masuda, M. Matsubara, J.L. Peng, I. Tsukada, K. Uchinokura, and M. v Zimmermann. This work was supported by the U.S. Department of Energy, Division of Materials Science under contract no. DE-AC02-98CH10886.

## REFERENCES

1. W. Schülke *Handbook on Synchrotron Radiation*, Ed.s D.E. Moncton, and G.S. Brown, North Holland, 1991, pp 565-639.
2. C.-C. Kao, W.A.L. Caliebe, J.B. Hastings and J.M. Gillet, Phys. Rev. B **54**, 16361 (1996).
3. P. Eisenberger, P.M. Platzman, and H. Winick, Phys. Rev. B **13** 2377, (1976).
4. Y. Ma et al., Phys. Rev. Lett., **69**, 2598 (1992).
5. J.J. Jia, T. A. Callcott, E. L. Shirley, J.A. Carlisle, L.J. Terminello, A. Asfaw, E.L. Ederer, F.J. Himpsel and R.C.C. Perera, Phys. Rev. Lett., **76**, 4054 (1996).
6. S.M. Butorin et al., Phys. Rev. Lett. **77**, 574 (1996).
7. S.M. Butorin et al., Phys. Rev. B **55**, 4242 (1997).
8. P. Kuiper, J.-H. Guo, C. Sathe, L.-C. Duda, J. Nordgren, J.J.M. Potuizen, F.M.F. de Groot, and G.A. Sawatzky, Phys. Rev. Lett. **80**, 5204 (1998).
9. A. Moewes, S. Stadler, R.P. Winarksi, D.L. Ederer, M.M. Grush and T.A. Callcott , Phys. Rev. B. **58**, R15951 (1998).

10. F. Gel'mukhanov and H. Agren, Phys. Rev. B **49**, 4378, (1994).
11. M. v Veenendaal and P. Carra, Phys. Rev. Lett., **78**, 2839 (1997).
12. E. Shirley, Phys. Rev. Lett., **80**, 794 (1998).
13. J.P. Hill, C.-C. Kao, W.A.L. Caliebe, M. Matsubara, A. Kotani, J.L. Peng and R.L. Greene, Phys. Rev. Lett. **80**, 4967 (1998).
14. J.P. Hill, C.-C. Kao, M. v Zimmermann, K. Hämäläinen, S. Huotari, L.E. Berman, W.A.L. Caliebe, K. Hirota, M. Matsubara, A. Kotani, T. Masuda, I. Tusukada and K. Uchinokura, J.L. Peng and R.L. Greene, to appear in Jap. J. Appl. Phys. (1999).
15. K. Hämäläinen, J.P. Hill, S. Huotari, C.-C. Kao, L.E. Berman, A. Kotani, J.L. Peng, and R.L. Greene, submitted to Phys. Rev. B.
16. J.P. Hill, C.-C. Kao, K. Hämäläinen, S. Huotari, L.E. Berman, W.A.L. Caliebe, M. Matsubara, A. Kotani, J.L. Peng and R.L. Greene, to appear in J. Phys. Chem. of Solids. (1999).
17. J.P. Hill, C.-C. Kao, M. v Zimmermann, K. Hirota, A. Kotani, T. Masuda, I. Tusukada and K. Uchinokura, in preparation
18. K. Tsutsui, T. Tohyama and S. Maekawa, LANL Cond. Mat/9905372
19. Z. Tan, J.I. Budnick, C.E. Bouldin, J.C. Woicik, S.W. Cheong, A.S. Cooper, G.P. Espinoza and Z. Fisk, Phys. Rev. B **42**, 1037 (1990).
20. J.M. Tranquada, S.M. Heald, W. Kunnmann, A.R. Moodenbaugh, S.L. Qiu, Y. Xu, and P.K. Davies, Phys. Rev. B **44**, 5176 (1991).
21. Z. Wu, M. Benfatto and C.R. Natoli, Phys. Rev. B **54**, 13409 (1996).
22. S. Tanaka, K. Okada and A. Kotani, J. Phys. Soc. Jpn. **60**, 3893 (1991).
23. S. Tanaka and A. Kotani, J. Phys. Soc. Jpn. **62**, 464 (1993).
24. Note that while the sample geometry made it possible to place the incident polarization solely in the $ab$ plane (vertical scattering geometry), in the horizontal scattering geometry the polarization was in fact 30° off the $c$ direction.
25. M.A. Veenendaal, H. Eskes and G.A. Sawatzky, Phys. Rev B. **47**, 11462 (1993).
26. K. Okada and A. Kotani, J. Electron Spectrosc. Relat. Phenom. **86**, 119 (1997).
27. F. Zhang and T. Rice, Phys. Rev. B **37**, 3759 (1988).
28. P. Abbamonte, C.A. Burns, E.D. Isaacs, P.M. Platzman, L.L. Miller, S.W. Cheong, M.V. Klein, Phys. Rev. Lett., **83**, 860 (1999).
29. P.M. Platzman and E.D. Isaacs, Phys. Rev. B. **57**, 11107 (1998).

# Electronic States of Metals and Alloys Investigated by High-Resolution Bloch-$\vec{k}$ selective X-ray Raman Scattering

A. Kaprolat[1], H. Enkisch[1], M. H. Krisch[2] and W. Schülke[1]

[1] *University of Dortmund, Inst. of Physics E1b, D 44221 Dortmund, Germany*
[2] *ESRF, Exp. Division, BP 220, F 38043 Grenoble, France*

**Abstract.** The shape of resonantly excited fluorescence spectra from valence bands of single crystal solids strongly depends on both the exact value of the excitation energy and the scattering angle, the latter determining the momentum transferred in resonant scattering experiments. This non-isotropic behaviour can be explained by considering the absorption of the incident photon and the re-emission of the fluorescence photon as one single resonant inelastic scattering process which underlies a law of Bloch-$\vec{k}$ momentum conservation thus revealing a strong influence of the properties of the electronic band structure on the spectral shape. By measuring resonantly excited valence fluorescence emission spectra of single crystal NiAl for incident energies around the Ni 1s binding energy (8,333 eV) and comparing the obtained spectra to calculations based on the Bloch-$\vec{k}$ selectivity and a LAPW electronic band structure, we find a good agreement.

## I INTRODUCTION

Inelastic X-ray scattering in its various forms is a well established technique to investigate properties of the electrons in solids [1]. For the nonresonant case, the experiment directly yields the *dynamical structure factor* $S(\vec{q},\omega)$, revealing the dynamics of the valence electrons. Choosing the incident energy to be close to an internal transition energy of the scattering sample leads to so-called resonant scattering processes involving lower lying core electrons. During these processes, the incoming photon excites a core electron into the continuum or an empty state close above the Fermi level (intermediate state). This core hole is then filled by an electron of a higher core level or a valence electron, emitting the outgoing photon. With incident energies around an absorption edge of the system under investigation, these processes, although of second order, will be dominant due to their resonant nature.

# II RESONANT INELASTIC SCATTERING FROM VALENCE BANDS

To account for resonant scattering processes in the calculation of the double differential cross section, the Hamiltonian describing the interaction of the photon field with the sample electrons

$$H = \sum_j \frac{1}{2m} \left( \vec{p}_j - \frac{e}{c} \vec{A}_j \right)^2 + V(\vec{r}_j) \tag{1}$$

($\vec{A}$ denoting the vector potential of the photon field, $V$ being the potential seen by the electrons in the unperturbed system and $\vec{p}_j$, $\vec{r}_j$ momentum and position of the electron $j$) has to be treated in $2^{nd}$ order perturbation theory.

Besides the term quadratic in the vector potential (so-called $\vec{A}^2$-term leading to Thomson scattering) arising from the $1^{st}$-order-treatment, in $2^{nd}$ order the $\vec{p}\cdot\vec{A}$-term contributes leading to resonant terms in the cross section described by the Kramers-Heisenberg-Formula [2]

$$\frac{d^2\sigma}{d\omega_2 d\Omega} \sim \sum_f \left| \sum_m \frac{\langle f|\vec{p}\cdot\vec{A}|m\rangle\langle m|\vec{p}\cdot\vec{A}|i\rangle}{E_m - E_i - \hbar\omega_1 - i\Gamma_m/2} \right|^2 \delta(E_f - E_i - \hbar\omega_1 + \hbar\omega_2) \tag{2}$$

The resonant scattering process can be understood as developing as follows (see Fig. 1)

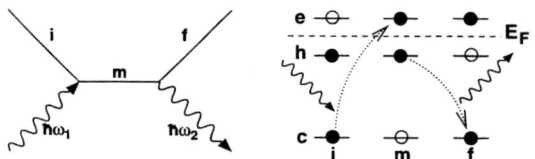

**FIGURE 1.** The resonant inelastic scattering process

The incoming photon ($\hbar\omega_1$) excites a core electron into an energy level above the Fermi energy, transferring the system from the ground state $|i\rangle$ into the intermediate state $|m\rangle$, the core state of which is then reoccupied by a valence electron emitting the outgoing fluorescence photon ($\hbar\omega_2$). $E_{i,m,f}$ are the respective energies of the electron states and $\Gamma_m$ denotes the energy width of the intermediate state due to its finite life time. Rewriting Eq.2 in these terms yields

$$\frac{d^2\sigma}{d\omega_2 d\Omega} \sim \sum_{e,h} \left| \sum_c \frac{\langle c|\vec{p}\cdot\vec{A}_2|h\rangle\langle e|\vec{p}\cdot\vec{A}_1|c\rangle}{E_e - E_c - \hbar\omega_1 - i\Gamma_m/2} \right|^2$$
$$\times \delta(E_e - E_h - \hbar\omega_1 + \hbar\omega_2) \tag{3}$$

Calculation of Eq. 3 for the case of valence emission was first carried out by Ma et al. [3] using Bloch states for the conduction and valence band electronic states (described by Bloch-$\vec{k}$ vectors $\vec{k}_1$ and $\vec{k}_2$ respectively), plane waves for incident and outgoing photon (wave-vectors $\vec{q}_1$ and $\vec{q}_2$ respectively) and a tight-binding approach to describe the core state involved. This single particle treatment results in

$$\frac{d^2\sigma}{d\omega_2 d\Omega} \sim \sum_{\vec{k}_1,\vec{k}_2} |M_{\vec{k}_1,c}|^2 \; \delta(E(\vec{k}_1)-E_c-\hbar\omega_1)$$
$$\times \; \delta_{\vec{G},(\vec{k}_1-\vec{k}_2+\vec{q}_1-\vec{q}_2)} \; |M_{\vec{k}_2,c}|^2 \; \delta(E(\vec{k}_2)-E_c-\hbar\omega_2). \tag{4}$$

$M_{\vec{k}_{(1,2)},c}$ denote the transition matrix element of the absorption and the emission process respectively. Writing $c$ instead of $\vec{k}_c$ for the core state stresses the fact that these are not functions of $\vec{k}$ as the core state is localized in direct space. If the energy transfer is small compared to the incident energy, the momentum transferred during the scattering process $\hbar\vec{q} = \hbar(\vec{q}_1-\vec{q}_2)$ is connected to the scattering angle $\vartheta$ via $|\vec{q}| \approx 2|\vec{q}_1|\sin\frac{\vartheta}{2}$. Thus, one can set $\vec{q}$ to any desired value by choosing the scattering angle and the orientation of the sample with respect to the directions of $\vec{k}_1$ and $\vec{k}_2$ appropriately.

It is important to note that this treatment of the resonant inelastic scattering process as *one coherent process* yields a law of Bloch-$\vec{k}$ momentum expressed in the Kronecker-$\delta$

$$\delta_{\vec{G},(\vec{k}_1-\vec{k}_2+\vec{q}_1-\vec{q}_2)} \tag{5}$$

(using the conventional two-step model of absorption followed by emission would have yielded Eq. 4 without the Kronecker-$\delta$). It states that the momentum transferred to the electron system of the sample by the scattered photon must equal the difference of the Bloch-$\vec{k}$ vectors $\vec{k}_1$ and $\vec{k}_2$ of the two electrons involved in the scattering process modulo any reciprocal lattice vector $\vec{G}$.

Eq. 4 describes the influence of the properties of the electronic band structure on the shape of the valence emission spectra as a function of incident energy and momentum transfer. This is illustrated in Fig. 2, showing the electronic band structure from valence electrons of Ni in NiAl. The radii of the circles are proportional to the Ni-p partial charge of the respective state.

The first delta function of Eq.4 describes energy conservation for the absorption process of the incoming photon, thus determining the allowed energy eigenvalues for the excited electron (horizontally shaded area). The conduction band structure determines, which Bloch-$\vec{k}$ vectors $\vec{k}_1$ can be occupied by the excited electron (namely those, for which energy bands $E(\vec{k})$ with considerable DOS cross the shaded area). The Kronecker-$\delta$ then determines the Bloch-$\vec{k}$ vectors $\vec{k}_2$ of the valence states that are allowed to reoccupy the core hole (vertically shaded area). Therefore, only a small part of the Brillouin zone contributes to the fluorescence process. The fluorescence spectrum hence changes shape, when $\hbar\omega_1$ or $\vec{q}$ is changed. In particular, if $\hbar\omega_1$ and thus $\vec{k}_1$ is kept fixed, one can scan the emission points $\vec{k}_2$ within the

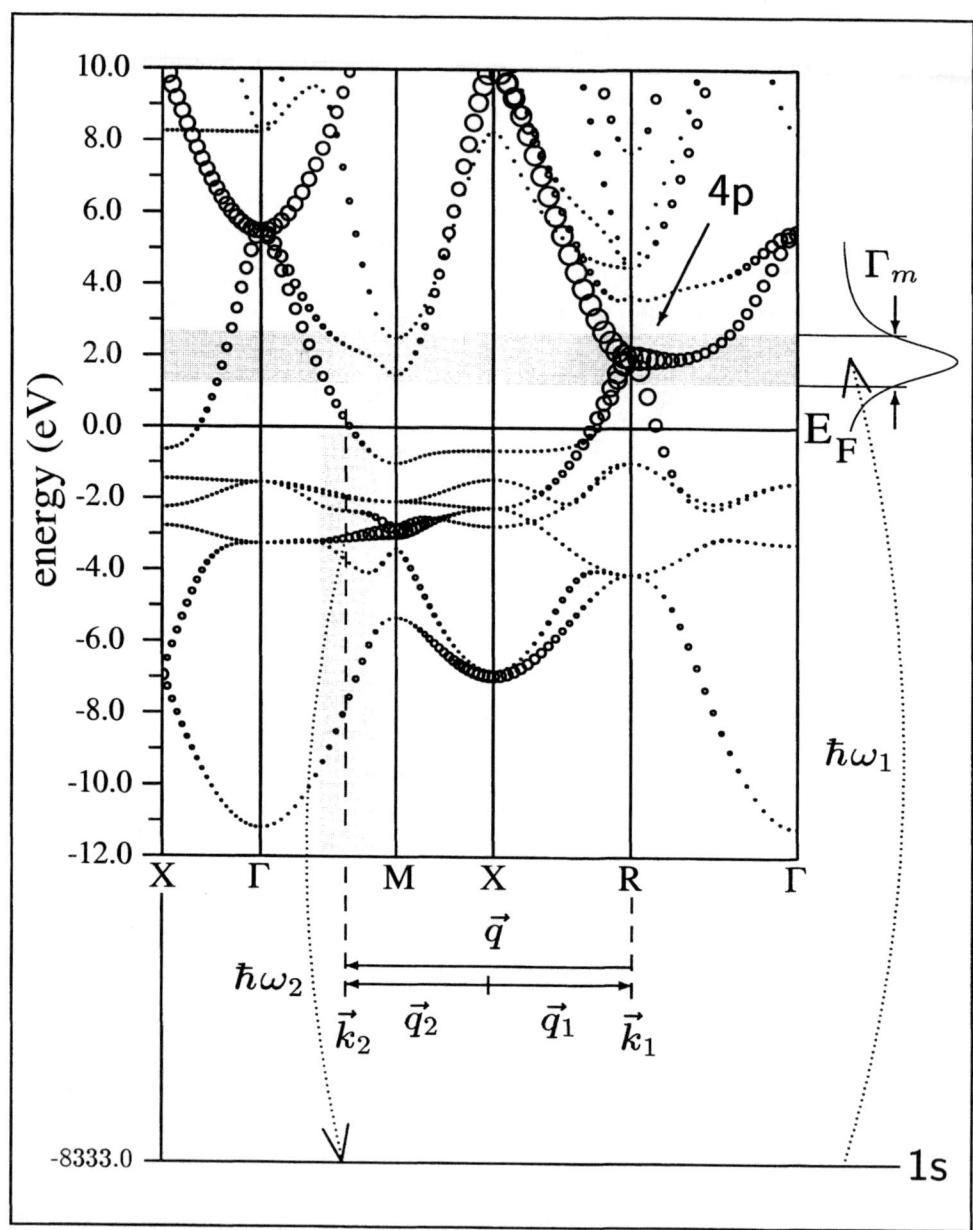

**FIGURE 2.** resonant valence emission from NiAl at the Ni K-edge

irreducible wedge of the first Brillouin zone by changing $\vec{q}$ and thus trace dispersive valence bands.

As the excitation process is of resonant nature due to the participation of a core state, this mechanism is element specific. In addition, the transition matrix elements $M_{\vec{k}_{(1,2)},c}$ introduce certain symmetry selection rules. Therefore one can in principle trace dispersive bands of certain symmetry originating from a certain element in a compound sample.

However, the strict connection of the Bloch-$\vec{k}$ momenta via Eq. 5 can be spoiled if there exist concurrent decay processes for the intermediate state, i.e. excitonic or phononic excitations [4], [5]. The probability for these processes is inversely proportional to the lifetime of the intermediate state. Unfortunately, a small lifetime of the intermediate state means a large energy width $\Gamma_m$, smearing out the well defined region of allowed $\vec{k}_1$ for absorption and $\vec{k}_2$ for emission respectively.

There exist a variety of studies using the described mechanism within the soft X-ray region [6-14]. It was our goal to extend this technique to the regime of hard X-rays (around 10 keV). The use of hard X-rays has the advantage that the length of the photon wave vector is comparable to or even larger than the first Brillouin zone. Therefore, the momentum transfer $\vec{q}$ is not restricted by experimental conditions but can be varied freely allowing one to scan the whole Brillouin zone without changing $\hbar\omega_1$. For soft X-rays, on the contrary, since $|\vec{q}| \ll$ any reciprocal lattice vector $|\vec{G}|$ and therefore $\vec{k}_1 \approx \vec{k}_2$, the variation of $\hbar\omega_1$ is the only way to scan the first Brillouin zone.

It must be emphasized that a direct reconstruction of the valence band structure out of resonant inelastic valence fluorescence spectra is not feasible, as the Bloch-$\vec{k}$ vectors $\vec{k}_2$ contributing to a fluorescence spectrum for certain values of $\vec{q}$ and $\hbar\omega_1$ are distributed over a rather large part of the Brillouin zone. A possible way to get direct band structure information that is contained redundantly in a set of fluorescence spectra would be the use of a selfconsistent iterative reconstruction procedure that refines a given initial band structure.

## III  EXPERIMENT

To prove the validity of the Bloch-$\vec{k}$ selectivity together with the element specificity of the resonant inelastic scattering process, we chose as sample a single crystal of the ordered stoichiometric alloy $Ni_1Al_1$ and excited the system using incident energies $\hbar\omega_1$ close to the Ni-K-edge.

The experiment has been carried out at beamline ID28 of the ESRF using a 1m spherical crystal Rowland spectrometer [16] with an energy resolution of about 0.5 eV. The diffraction plane of the analyzer was horizontal. To monochromatize the incident photons, the premonochromator of ID 28 (cryogenically cooled Si 111 double crystal) was used together with an additional Si 333 channel cut crystal. The horizontal slits were set in a way to achieve an overall energy resolution of about 1.0 eV FWHM, determined from the width of the quasi-elastic (Rayleigh)

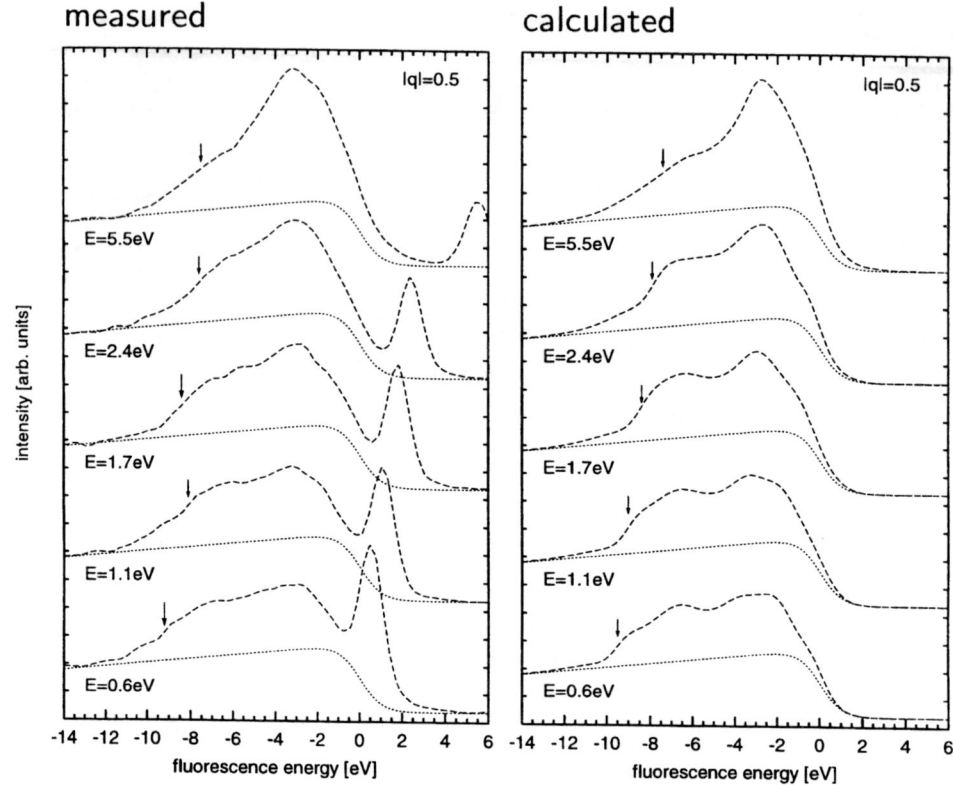

**FIGURE 3.** resonantly excited valence emission spectra of NiAl for constant momentum transfer $\vec{q} = 0.5 \cdot (1,1,0)\frac{2\pi}{a}$

line. The achieved $|\vec{q}|$-resolution roughly corresponds to 8 per cent of the length $\Gamma$ to $X$ in reciprocal space.

The experimentally chosen values of the momentum transfer $\vec{q}$ were of 0.1, 0.25, 0.4 and 0.5 in units of $|\langle 110 \rangle| \cdot \frac{2\pi}{a}$, $a$ being the lattice constant, whereas for each $\vec{q}$ a spectrum with excitation energy of 0.6, 1.1, 2.4 and 5.5 eV above the Ni 1s binding energy was measured.

A sample series of measured valence fluorescence spectra either for fixed momentum transfer $\vec{q}$ or fixed incident energy $\hbar\omega_1$ are shown in Fig. 3 (left) and Fig. 4 (left), respectively.

It can be seen from the figures, that the valence emission spectra lie on top of a certain background, the so-called radiative Auger background or shakeup satellite to the valence fluorescence line. This satellite originates from the excitation of another valence electron into the conduction band during the emission process, leading to an energy loss of the emitted photon. The energy transferred by this

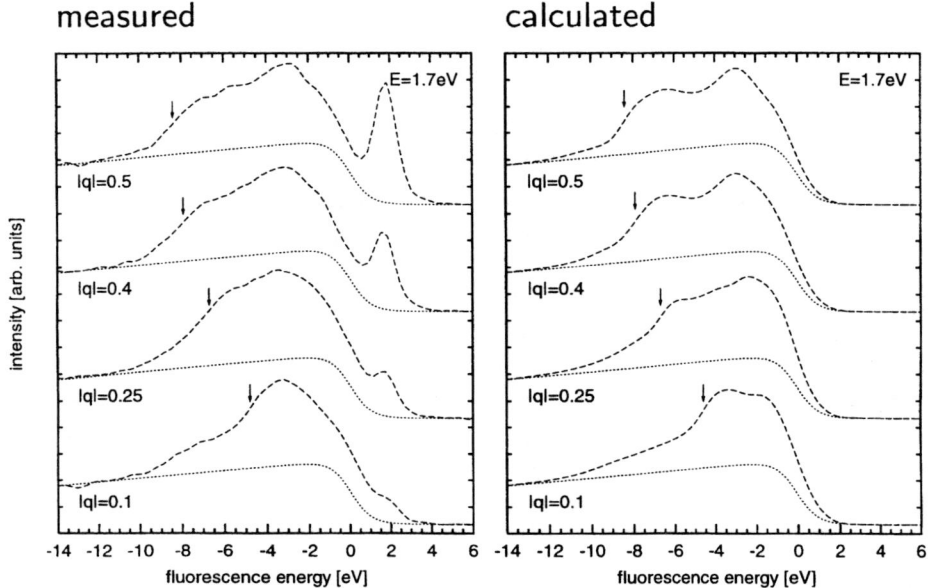

**FIGURE 4.** resonantly excited valence emission spectra of NiAl for constant excitation energy $\hbar\omega_1 = 1.7eV$ above Ni 1s binding energy

process can in principle take any value from zero up to the fluorescence energy, therefore the satellite will show a steep drop on its high energy side (close to the Fermi level) and a slowly decreasing tail on its low energy side. To our knowledge, no calculation of the shape of this satellite is available and a separate experimental determination is impossible, so we chose to model the satellite by a linear function multiplied by a Fermi function as indicated by dotted lines in Figs 3 and 4.

To verify the validity of the Bloch-$\vec{k}$ selectivity within the resonant inelastic scattering process, we calculated the spectra obtained experimentally following the arguments of the $\hbar\omega_1$- and $\vec{q}$-dependence of the fluorescence spectra given in Sec. II. The calculation is based on a LAPW band structure computed by the WIEN97 package [15]. Energy eigenvalues and band character at about 9000 points within the irreducible wedge of the first Brillouin zone were used. The calculation accounts for the energy resolution of both monochromator and analyzer, the lifetime broadening of the core state as well as the energy dependence of the dipole transition matrix element of the Ni 1s to Ni 4p transition and a selfabsorption correction. The strength of quadrupolar transitions was estimated from measurements of Ge valence fluorescence, where p and d emission lines are nicely separated, to be two orders of magnitude smaller. These calculated spectra are shown in Fig. 3 (right) and Fig. 4 (right). It has to be mentioned that electron correlations are not yet taken into account in these calculations.

It can be seen from theses figures that there is considerable agreement between measured and calculated spectra. In both series of experimental spectra, a low-energy shoulder is visible and follows the same dispersion as in the calculated spectra. For the series for constant $\vec{q}$, an increase in strength of a peak at -3 eV fluorescence energy is seen in the experimental as well as in the calculated spectra. These features find their explanation in the NiAl electronic band structure. For the spectrum with constant $\hbar\omega_1 = 1.7$eV above Fermi level, the possible Bloch-$\vec{k}$ vectors $\vec{k}_1$ are distributed around the R-point (see Fig. 2). With increasing $|\vec{q}|$ the allowed Bloch-$\vec{k}$ vectors $\vec{k}_2$ are shifted parallel to the 110 direction from R to X and therefore the lowest p-band around X causes this shoulder. Similarly, the growth of the peak at -3 eV (see Fig. 4) originates from p-states around the M-point. Calculated emission spectra from Al-3p orbitals show no agreement with the experiment. From this one can conclude that for resonant inelastic scattering from valence electrons the element selectivity holds.

## IV CONCLUSION

In conclusion, measurements of resonantly excited valence emission spectra from a stoichiometric NiAl single crystal sample prove the Bloch-$\vec{k}$ selectivity of the resonant inelastic scattering process and therefore the influence of the electronic band structure on the shape of the fluorescence spectra. The measurements show the symmetry and element selectivity as predicted. Features within the spectra can be assigned to distinct features of the electronic band structure. To reconstruct parts of the band structure, selfconsistent methods to extract the redundantly spread information from sets of fluorescence spectra for different $\hbar\omega_1$ and $\vec{q}$ have to be established.

## REFERENCES

1. W. Schülke, *Inelastic Scattering by Electronic Excitations* in G. Brown and D. Moncton (Ed.), Handbook of Synchrotron Radiation, Vol.3, p. 565, Elsevier, Amsterdam (1991).
2. see for example M. Blume, J. Appl. Phys. **57**, 3615 (1985).
3. Y. Ma, Phys. Rev. **B 49**, 5799 (1994).
4. C.-O. Almbladh, Phys. Rev. **B 10**, 4343 (1977).
5. M. van Veenendaal, P. Carra, Phys. Rev. **B 78**, 2839 (1997).
6. J. E. Rubensson, D. Mueller, R. Shuker, D. L. Ederer, C. H. Zhang, J. Jia, and T. A. Callcatt, Phys. Rev. Lett **64**, 1047 (1990).
7. Y. Ma, N. Wassdahl, P. Skytt, J. Guo, J. Nordgren, P. D. Johnson, J.-E. Rubensson, T. Boske, W. Eberhardt, and S. D. Kevan, Phys. Rev. Lett **69**, 2598 (1993).
8. K.E. Miyano, D. L. Ederer, T. A. Callcott, W. L. O'Brine, J. J. Jia, L. Zhou, Q.-Y. Dong, Y. Ma, J. C. Woicik, and D. R. Mueller, Phys. Rev. **B 48**, 1918 (1993).
9. P. D. Johnson, Y. Ma, Phys. Rev. **B 49**, 5024 (1994).

10. Y. Ma, K. E. Miyano, P. L. Cowan, Y. Aglitzkiy, and B. A. Karlin, Phys. Rev. Lett. **74**, 478 (1995).
11. A. Carlisle, E. L. Shirley, E. A. Hudson, L. J. Terminello, T. A. Callcott, J. J. Jia, D. L. Ederer, R. C. C. Perera, and F. J. Himpsel, Phys. Rev. Lett **74**, 1234 (1995).
12. J. J. Jia, T. A. Callcott, E. L. Shirley, A. Carlisle, L. J. Terminello, A. Asfaw, D. L. Ederer, F. J. Himpsel, and R. C. C. Perera, Phys. Rev. Lett **76**, 4054 (1996).
13. J. Lüning, J.-E. Rubensson, C. Ellmers, S. Eisebitt, and W. Eberhardt, Phys. Rev. **B 56**, 13147 (1997).
14. E. L. Shirley, Phys. Rev. Lett. **80**, 794 (1998).
15. P. Blaha, K. Schwarz and J. Luitz, **WIEN 97**, Vienna University of Technology, Vienna 1997, Updated version of P. Blaha, K. Schwarz, P. Sorantin, and S. B. Trickey, Comp. Phys. Commun. **59**, 399 (1990).
16. C. C. Kao, W. A. Caliebe, J. B. Hastings, K. Hämäläinen, and M. H. Krisch, Rev. Sci. Instrum. **67**, 1 (1996).

# X-Ray Magnetic Circular Dichroism Of Model Heisenberg Ferromagnets.

Andrei Rogalev and José Goulon

*European Synchrotron Radiation Facility (E.S.R.F.)*
*B.P.-220, F-38043 Grenoble Cedex, France*

**Abstract.** We review first the principles of the X-ray magnetic circular dichroism (XMCD) which will be used to investigate the magnetic properties of Europium Chalcogenides. These compounds are good models of magnetism in solids and were selected to illustrate the potentialities and limitations XMCD at high excitation energy. A thorough analysis of the $Eu$ $L$-edges XMCD spectra allowed us to identify several effects which, if they are not properly taken into account, may cause the failure of the magneto-optical sum rules at the $L$-edge of rare-earths. These effects include: (i) the exchange splitting of the $2p$ core states; (ii) the crystal field splitting of the $5d$ final states, and (iii) the strong $2p \to 4f$ quadrupolar contributions to the XMCD spectra. It is also shown that Magnetic EXAFS measurements at the $Eu$ $L$-edges combined with XMCD studies at the Chalcogen $K$-edges produce strong evidence that the highly localized $4f$ moments of $Eu$ are indirectly coupled *via* the magnetically polarized $p$-states of the anions. This is the first time that such magnetic interactions mediated by the anions are unambiguously detected by X-ray absorption spectroscopy.

## I   INTRODUCTION

In these recent years, the list of the applications of X-ray Absorption spectroscopy was continuously growing as illustrated by the discovery of a variety of new experimental techniques associated with the use of circularly polarized synchrotron radiation. The existence of X-ray magnetic circular dichroism (XMCD) was first anticipated by Erskine and Stern [1] and the reality of this effect was demonstrated many years later by Schütz and coworkers in the spectral ranges of both XANES [2] and EXAFS [3]. XMCD is now a well established technique with an extensive list of applications concerning technologically important materials. The quantity of interest in an XMCD experiment is the difference in the X-ray absorption absorption cross sections of spectra recorded with left and right circularly polarized photons while the sample magnetization is kept parallel or antiparallel to the wavevector of the incident X-ray beam. The interest in XMCD increased very rapidly when it was realized that, with the help of the magneto-optical sum rules [4,5], one could extract from these spectra *orbital-projected* magnetic moments both in magnitude and direction with the full benefit of the element selectivity that is inherent to X-ray absorption spectroscopy: this established XMCD as an attractive probe of the spin-dependent electronic structure of ferro(i)- or paramagnetic compounds. XMCD is now commonly used for element-specific magnetometry in heteromagnetic systems including magnetic multilayers with giant magnetoresistance. Magnetic circular dichroism can also be recorded in the spectral region of EXAFS region: this effect most often referred to as *Magnetic EXAFS* (MEXAFS), is sensitive to the magnetic moments carried by the atoms surrounding the absorbing center and could become a very valuable probe of the local magnetic environment of the absorbing atom.

Even though the first XMCD experiments were performed at rather high excitation energy, the applications of this technique were most spectacular in the soft X-ray spectral range [6]. The soft X-ray region is of considerable interest because the $3d$-states of the transition metals and $4f$-states of the rare-earths are directly accessible *via* strong dipolar transitions from $2p$ ($L_{II,III}$-edges) and $3d$ ($M_{IV,V}$-edges) core states, respectively. At higher energy, *i.e.* above 2 keV, XMCD remains a very valuable tool to detect *induced* magnetism in metallic multilayers and intermetallic alloys because "non-magnetic" atoms such as $Pd$ [7], $Pt$ [8] or even $S$ [9], can acquire a significant magnetic moment in a magnetic environment. $M_{IV,V}$-edges of acinides are also accessible and make it possible to investigate the manifold magnetic behavior of the $5f$-electrons [10,11]. Finally, XMCD spectra recorded at the $L_{II,III}$-edges of rare-earths could shed some light on the role played by the $5d$-electrons which are magnetically coupled to the strongly localized $4f$ electrons [12]. High energy X-ray photons are indeed most appropriate for MEXAFS experiments: due to the large spin-orbit splitting of the $2p$ core states, the overlap between $L_{II}$ and $L_{III}$ edge EXAFS oscillations is less dramatic than in the soft X-ray range and this makes the analysis of the MEXAFS signals a lot easier and more reliable.

In this paper, we will report on recent XMCD experiments performed at the ESRF on the beamline ID12A [13] which was operated so-far above 2 keV. In order to illustrate the potentiality of the method and the reliability of our instrumentation, we have selected the case of the Europium Chalcogenides ($EuS$ and $EuSe$), which belong to an important class of magnetic semiconductors with a highly symmetric $NaCl$ structure. These compounds are often given as good model compounds for the study of the Heisenberg exchange in solids [14]. Their magnetism originates from the spin-only $4f$ magnetic moments carried by the $Eu^{2+}$ cations which are in the $^8S_{7/2}$ spectroscopic ground state. According to the Heisenberg model, the magnetic properties of these compounds should result basically from the competition between two magnetic interactions: ($i$) a direct ferromagnetic coupling of the localized $Eu^{2+}$ spin with the twelve nearest $Eu^{2+}$ neighbors; ($ii$) an antiferromagnetic exchange with six $Eu^{2+}$ ions mediated by the anions. The first interaction is thought to dominate in $EuS$ in which a ferromagnetic order is observed below $T_C = 16.57K$ [15]. $EuSe$ has the unique feature that the two exchange interactions nearly cancel each other and several magnetic phases have been observed below the ordering temperature ($4.6K$) [16]. The situation becomes different in the presence of an external magnetic field: it was shown that at $T = 4.2K$, a

$1.5T$ magnetic field is enough to align all moments ferromagnetically with a net magnetization that is ca. 0.8 of saturation. The Heisenberg picture does not offer a detailed description of the underlying mechanisms of magnetic interactions. Moreover, it has been argued on the basis of Mößbauer and NMR-studies [17] that the range of the magnetic interactions may not be limited to the second nearest $Eu^{2+}$ neighbors but could extend well beyond. Still recently, neutron diffraction studies of $EuS$ under high pressure [18] suggested in a rather obscure way that the valence $p$ electrons of Sulphur should contribute to the ferromagnetic interaction in $EuS$. We will show below that XMCD and MEXAFS experiments confirm unambiguously this conclusion which is also supported by band structure calculations [19] : cation-anion $4f - np$ and $5d - np$ interactions may induce a polarization of the anion $p$ band and promote a ferromagnetic polarization exchange coupling between the $Eu^{2+}$ cations.

## II PHYSICAL CONTENT OF XMCD AND OF THE SUM RULES.

The physical origin of XMCD can be most easily explained with the so-called *two step model* first proposed by G. Schütz et. al. [2] and revisited later by J. Stöhr and R. Nakajima [20]. The first step describes the excitation by a circularly polarized X-ray photon of a core electron. One should simply remember that a circularly polarized photon carries an angular momentum ($+\hbar$ for a right-handed photon and $-\hbar$ for a left-handed photon), the corresponding *helicity* vector being parallel (right) or antiparallel (left) to the propagation direction. As a consequence of the conservation of angular momentum in the absorption process, the angular momentum carried by the X-ray photon will be entirely transferred to the excited photoelectron. Depending on the initial core state and on the nature of the interactions, both the orbital moment and the spin of the photoelectron are concerned with this momentum transfer. Let us first assume that the photoelectron has been excited from a spin-orbit-split core level (as this is the case for $L_{II,III}$ absorption edges): then part of the angular momentum carried by the photon will be converted into spin *via* spin-orbit coupling. This is basically the origin of the well-known Fano effect [21]. In the spin-orbit-split core level, there are always two electrons that have opposite spin-orbit coupling ($l + s$ and $l - s$) whilst the helicity of left and right circularly polarized photons is characterized by opposite signs: thus 2 types of photoelectrons will be excited with antiparallel spins and which are currently referred as "spin-up" or "spin-down" photoelectrons depending on the helicity of the incident X-ray photon which may be parallel (right) or antiparallel (left) to the wavevector. In the absence of any spin-orbit coupling (as in the case of the excitation of atomic $s$ core electrons), the angular momentum of the photon is entirely converted into $+\hbar$ ($-\hbar$) orbital moments and there is no spin polarization of the photoelectron.

The magnetic properties of the sample are driving the second step. In the XANES region, X-ray absorption spectra primarily reflect the density of empty states for an angular momentum $l$ given by the symmetry of the initial core state and dipole selection rules. XMCD spectra simply reflect the difference in the density of states with different spin or orbital moments. In the case of spin-orbit split edges, the excited photoelectron carries both a spin and an orbital momentum and any imbalance in either spin or orbital momentum in the final states will immediately cause a dichroic effect. However, if we *sum* the contributions of all electrons that can be excited from a split core level (taking properly into account their degeneracy), the result can only reflect a difference in the *orbital moments* of the final states because there is always a pair of photoelectron with opposite spin polarization created at spin-orbit split absorption edges. This is basically what the first sum rule is telling us. In contrast, the *difference* in the dichroism intensity measured at both edges will reflect a *spin* imbalance in the empty states because the orbital momentum transferred to the photoelectron has the same sign at the both edges. This is precisely the content of the second sum rule. If there is no orbital moment in the final states, then the ratio of the dichroic signal at the two spin-orbit split edges should be negative and inversely proportional to the ratio of the corresponding degeneracies. We would like to emphasize that the summation over two spin-orbit split edges is equivalent to what can be measured for a core level with no spin-orbit interaction. This implies that a dichroism effect at the $K$-edge can only be due to the orbital moments of $p$ electrons in the valence shell.

Eventhough the two-step model can help us to understand qualitatively the underlying physics, the discrimination between the two steps is purely artificial. The derivation of the magneto-optical sum rules by Thole and Carra [4,5] in the case of spin-orbit split edges does not require any artificial trick such as a two step model. The same conclusion also hold true for a recent discussion of the XMCD spectra at the $K$-edge given by Igarashi et al. [22]. For the relevant absorption edges, the sum rules relate the integrated absorption $\mu(E)$ and the integrated XMCD signal $\Delta\mu(E)$ to the *ground state* expectation values of the operators $\langle\sigma_z\rangle$ and $\langle l_z\rangle$ of the probed level of the absorbing atom. At the $L_{II,III}$-edges and for $d$-electrons, the sum rules can be expressed as

$$\langle l_z\rangle_d = -\frac{2n_{hole}}{N}\int[\Delta\mu_{L_{III}}(E)+\Delta\mu_{L_{II}}(E)]dE \qquad (1)$$

$$2\langle\sigma_z\rangle_d + 7\langle T_z\rangle_d = -\frac{3n_{hole}}{N}\int[\Delta\mu_{L_{III}}(E)-2\Delta\mu_{L_{II}}(E)]dE, \qquad (2)$$

where $\langle T_z\rangle$ is the expectation value of the magnetic dipole operator which measures the asphericity of the spin magnetization. At the $K$-edge and for $p$-electrons there is only one sum rule which is given by:

$$\langle l_z\rangle_p = \frac{n_{hole}}{3N}\int\Delta\mu_K(E)\,dE. \qquad (3)$$

Here $N$ denotes the integrated polarization averaged absorption spectrum corresponding to the transitions to particular final states, e.g. $5d$ or $3p$, and $n_{hole}$ is the number of holes in these final states. Thus, the magneto-optical sum rules provide us with a very valuable tool to measure separately the spin and orbital components of the total magnetic moment at a given absorbing center. It should be underlined, however, that their derivation relies on various approximations which, in practice, appear to be valid for the $L_{II,III}$-edges of $3d$ [23,24], $4d$ [7] and $5d$ [8] transition metals as well as for the $M_{IV,V}$-edges of actinides [10] provided that reasonable estimates can be obtained for $N$ and $n_{hole}$. Unfortunately, the case of the $L_{II,III}$-edges of the rare-earths is far more complicate and the sum rules can yield erroneous conclusions [25] if one is not paying enough attention to the following limitations:
- the sum rules are restricted to dipole allowed transitions
- they neglect the exchange splitting of the core levels
- they neglect the difference between $5d_{3/2}$- and $5d_{5/2}$-wavefunctions
- they neglect the energy dependence of the matrix elements.

## III EXPERIMENTAL.

Figure 1 reproduces schematically the experimental arrangement of the ESRF beamline ID12A which is essentially dedicated to polarisation dependent X-ray absorption spectroscopy [13]. In the experiments which we report below, circularly (elliptically) polarized X-rays were produced by an helical undulator (Helios-II) that has the capability to flip the helicity of the emitted photons [26]. The typical energy range covered by Helios-II is 2.05 - 18 $keV$ depending on which harmonic is used and on how closed are the half gaps defining the horizontal and vertical magnetic fields ($B_x$ and $B_z$). For our experiments the fixed-exit double crystal monochromator was equipped with a pair of $Si <111>$ crystals cooled down to $-140°C$. Due to the ultra-low emittance of the source, the energy resolution was always found very close to the theoretical limits: $0.35eV$ at the $S$ $K$-edge; $1.1eV$ at the $Eu$ $L_{II,III}$-edges and $2.0eV$ at the $Se$ $K$-edge. In all cases, the instrumental resolution was smaller than the core hole life-time broadening. Given the fact that the absorption edges of interest for the present study ($S$ $K$-edge; $Eu$ $L_{II,III}$-edges; $Se$ $K$-edge) are very distant in energy one from each other, three different experimental configurations were required. Experiments at the Sulfur K-edge ($2.47keV$) were performed using the fundamental harmonic of the undulator spectrum (FWHM $\approx 100eV$). The higher order harmonics were cut off very efficiently using two SiC mirrors operated at 8 $mrad$ angle of incidence. These mirrors located upstream with respect to the monochromator also concurred to minimize the heat load on the first crystal. We found more appropriate to use the second harmonic of the undulator to record our spectra at the $L_{II,III}$-edges of Europium (6.9 and 7.6$keV$). Even though the spectral bandwidth of the $2^{nd}$ harmonic is rather broad (FWHM $\approx 500eV$), it remains still too narrow to record a whole EXAFS spectrum covering both $L_{III}$ and $L_{II}$- edges. This is why MEXAFS spectra were recorded using the

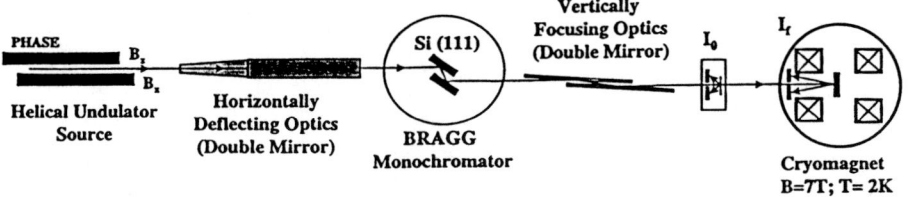

**FIGURE 1.** Layout of the ESRF beamline ID12A.

so-called "gap-scan technique" [27]. The key advantage of this method is in that, during an EXAFS scan, the undulator is permanently tuned to the maximum of its emission peak in order to maximize the photon statistics whereas the circular polarization rate can be kept almost constant over the whole energy range. In order to get rid of the higher order harmonics, the mirrors were set at 4 $mrad$ angle of incidence, whilst the intensity of $1^{st}$ harmonic of the undulator was significantly attenuated by inserting a $0.9\mu m$ thick aluminium foil before the monochromator: this concurred again to reduce the heat load on the first crystal and improved the long term stability of the spectrometer. Finally, we found most advantageous to use the $3^{rd}$ harmonic to record XMCD spectra at the Selenium $K$-edge ($12.6 keV$) even though its spectral bandwidth is quite narrow (FWHM $\approx 100 eV$). For the latter experiment, the emitted light was elliptically polarized since the horizontal and vertical magnetic fields of undulator could not be kept equal anymore. The lower order harmonics were again attenuated by inserting a $25\mu m$ thick aluminium foil before the monochromator. We decided not to use any mirror at such high energies because the very weak $6^{th}$ harmonic of the undulator has anyhow a very low transmission with a $Si < 111 >$ double crystal monochromator whereas the intensity of the $9^{th}$ harmonic of an helical undulator can be taken as negligible. Polarimetric studies were performed using a quarter wave plate [28] and allowed us to check that, under the experimental conditions described above, the circular polarization rate of the *monochromatic* beam was of the order of ca. 85% at the $Eu$ $L$-edges, ca. 78% at the $Se$ $K$-edge but only 16% at the $S$ $K$-edge. Given the very low polarization rate at the S K-edge, we thought initially that it would be a real "tour de force" to measure XMCD at the Sulfur $K$-edge but the experiment turned out to be nevertheless very successful. The sample was a pellet of polycrystalline powder (purity grade: 99.9%) mixed with boron nitride in order to ensure the temperature homogeneity inside the sample. The pellet was attached to the cold finger of a liquid helium cryostat inserted in a split superconducting coil. All measurements were performed at $T = 2K$ and with a magnetic field of $1.5T$ applied parallel or antiparallel to the direction of the X-ray beam propagation. It was carefully checked by simply monitoring the amplitude of the XMCD signal at the $Eu$ $L_{II}$-edge that the field was strong enough to reach a complete magnetic

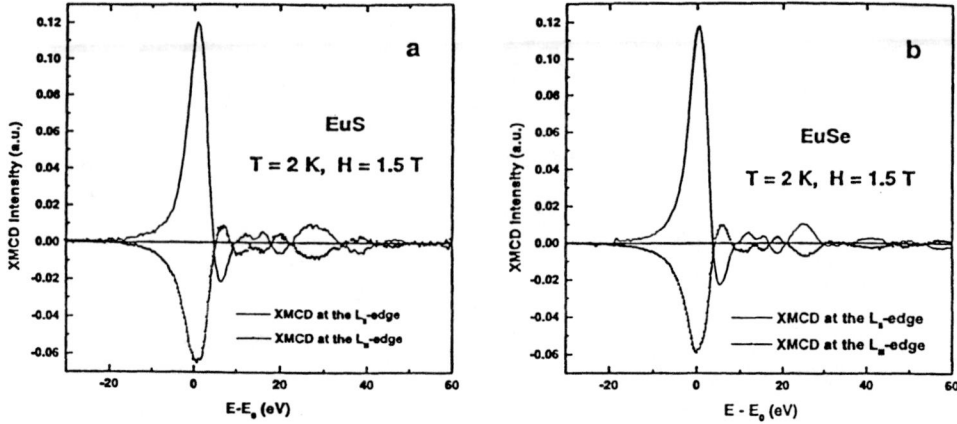

**FIGURE 2.** XMCD spectra at the $Eu$ $L_{III}$- and $L_{II}$-edges in $EuS$ (a) and in $EuSe$ (b).

saturation of the sample. All spectra were recorded in the total fluorescence yield detection mode. The rough spectra were carefully corrected for self-absorption using an homographic transform described elsewhere [29]. The dichroism spectra displayed below are spectral differences between consecutive scans recorded with opposite photon helicity taking advantage of optimized sequences minimizing the residual effects of slow angular drifts of the undulator beam. We checked that strictly the same dichroism spectra were obtained on flipping the direction of the magnetic field while keeping the photon helicity unchanged.

## IV RESULTS AND DISCUSSIONS

### A XMCD at the L-edges of Europium

We have reproduced in Figure 2 the XMCD spectra recorded at the $Eu$ $L_{II,III}$-edges for $EuS$ (Fig.2a) and $EuSe$ (Fig. 2b). The absorption spectra were first preset to zero before the edge and the atomic background was renormalized to 2 above the $L_{III}$-edge but only to unity above the $L_{II}$-edge. Such a renormalization procedure is already taking into account *a priori* the expected statistical correction for the branching ratio. For the sake of comparison, the origin of the energy scale was selected at the inflection point of the white lines. For both compounds, the $L_{III}$ ($L_{II}$) XMCD spectra consist of a strong negative (positive) peak with the ratio of their amplitudes $\approx -2$. This result was rather unexpected for the $Eu^{2+}$ ion if we assume a $5d^0$ ground state for which the magneto-optical sum rules would predict $\langle l_z \rangle = \langle \sigma_z \rangle = \langle T_z \rangle = 0$. Note that the shape and the amplitude of the XMCD spectra are fairly similar for both $EuS$ and $EuSe$. This observation implies that the nature of the anion has very little influence on the $Eu$ $L_{II,III}$ near edge XMCD spectra which seem to be mainly controlled by intraatomic exchange interactions. Among the various interatomic interactions expected in the rare-earth cations, the $4f-2p$ exchange is certainly quite insensitive to chemical bonding. One may expect this interaction to induce a sizeable splitting of the $2p$ core states resulting in a quite significant dichroism effect. In this case, the XMCD spectrum should exhibit the same characteristic lineshape as the first derivative of the absorption spectrum whereas its amplitude would depend both on the energy of the exchange splitting and on the core-hole lifetime. On the other hand, the exchange splitting should be different for $2p_{1/2}$ and $2p_{3/2}$ core levels with the consequence that the amplitude of the XMCD signal should also be different at the corresponding absorption edges. It is therefore our interpretation that the $4f-2p$ exchange interaction is at least one of the primary origins of the unexpected behaviour of the $Eu$ $L_{II,III}$-XMCD in europium chalcogenides.

Another mechanism which deserves attention is the spin-orbit interaction in the final $5d$-states of Europium. In europium chalcogenides the $5d$ states form a narrow conduction band which is split by the cubic crystal field into: (i) a low energy $t_{2g}$ subband; (ii) a high energy $e_g$ subband. The energy separation between these two subbands is $\approx 2.2 eV$ in $EuS$ and $\approx 1.9 eV$ in $EuSe$ [30], the width of each subband being of the order of $0.6 eV$. According to group theory, no spin-orbit interaction is expected in a $e_g$ subband, whereas the spin-orbit interaction in the $t_{2g}$ subband was estimated to be $\geq 0.3 eV$ and should cause an additional splitting. Since the spin-orbit coupling in the final $5d$ states is comparable with the subband width, it can modify significantly the branching ratio at the $L$-edges: a similar interpretation has already been formulated for transition metals [31]. Unfortunately, there is no hope to resolve the $t_{2g}-e_g$ splitting in XMCD or in XANES spectra in the presence of the large $2p$ core-hole life time broadening ($\approx 4 eV$).

A further complication is related to the significant contribution of quadrupolar transitions ($2p \to 4f$) that have been identified at the pre-edge region of the $L_{III}$ spectra [32–34]. The quadrupolar contributions are usually small in XANES spectra [35] but they are strongly enchanced in XMCD spectra because as a consequence of the large magnetic moment carried by the $4f$ electrons. However, there has been no clear evidence reported yet that quadrupolar transitions could contribute to the XMCD signal at the $L_{II}$-edge and this remains an open question. In the $L_{III}$-XMCD spectra of rare earths, the quadrupolar contributions are typically located $\leq 10 eV$ below the main dipolar peak and have the opposite sign with respect to the dipolar contribution [12]. Suprisingly, our experimental spectra (see fig.2) do not reveal any signature that could be assigned to quadrupolar transitions neither at the $L_{II}$ nor at the $L_{III}$ edges of $Eu^{2+}$ in both EuS and EuSe. What is even more puzzling is that the $Eu^{2+}$ ion has exactly the same spectroscopic ground state as $Gd^{3+}$, for which sizeable quadrupolar contributions were predicted at the $L_{III}$-edge [32]. Since we were interested in refining the energy of the quadrupolar features in our experimental spectra, we have performed atomic calculations for both $Eu^{2+}$ and $Gd^{3+}$ using the Cowan's atomic Hartree-Fock program with relativistic corrections

[36]. The difference between the average energies of the final state configurations $2p^54f^85d^0$ and $2p^54f^75d^1$ which is relevant for the present problem should take into account a strong attractive $2p$ core-hole potential. The calculated value for $Gd^{3+}$ ($\approx -8eV$) agrees well with the experiment [12] and with previous calculations [32], whereas for $Eu^{2+}$ the calculated difference is of the order of $+2eV$. It means that in the present example, the quadrupolar contributions at the L-edges of $Eu^{2+}$ are expected to overlap completely with strong dipolar transitions with the practical consequence that they cannot be resolved experimentally. This calculation also suggest that quadripolar transition could contribute to a modification of the XMCD branching ratio because the quadrupolar transitions should have different amplitudes at the $L_{III}$- and $L_{II}$-edges respectively.

## B  Magnetic EXAFS at the L-edges of Europium.

The similarity found in the near edge XMCD spectra of $EuS$ and $EuSe$ cannot hold true in the EXAFS regime : this is because EXAFS and MEXAFS oscillations depend on the interatomic distances and on the nature of the backscattering atoms which are obviously different in both compounds. Whereas the branching ratio was found to deviate appreciably from the statistical value in the near edge region, this is no longer the case in the MEXAFS regime. It can be checked from fig. 2 that the ratio of the amplitudes of the MEXAFS oscillations recorded at the $L_{III}$- and the $L_{II}$-edges is very close to $-1$. The striking similarity which appears between the $L_{II}$- and the $L_{III}$-MEXAFS spectra reproduced in Fig.2 clearly indicates that the effects responsible for the anomalies detected in the near edge XMCD spectra play only a minor role in the EXAFS region. This is certainly true for the quadrupolar contributions and for the crystal field effects. In contrast, the exchange splitting of the core states could generate additional problems in the EXAFS regime as well. However, this contribution should be less perceptible in MEXAFS spectra than in the near edge region because we expect a signal that should be proportional to the first derivative of the absorption spectrum which is weak beyond the edge. Thus, if we refer to the interpretation given in the previous section, we do not find any fundamental contradiction in the fact that the measured MEXAFS spectra satisfy the statistical branching ratio whereas this is not the case for the XMCD spectra recorded in the near edge region.

MEXAFS spectra are usually analysed *via* direct Fourier transformation and the FT spectra are simply compared to the FT spectra of the polarization averaged EXAFS oscillations. However, there is the prerequisite that both spectra should be properly normalized. In practice, the direct difference $(\mu^+(E) - \mu^-(E))$ is not normalized to the edge jump. We have found that a convenient way to circumvent this difficulty is to exploit the normalized quantity $\frac{\mu^+(E)}{mu^-(E)} - 1$. This approximation was checked to introduce an error of less than 5% in the magnitude of MEXAFS oscillations [37]. We have compared in figure 3 the FT spectra of the spin averaged and spin polarized EXAFS oscillations recorded at the Eu $L_{III}$-edge for $EuS$ (fig. 3a)

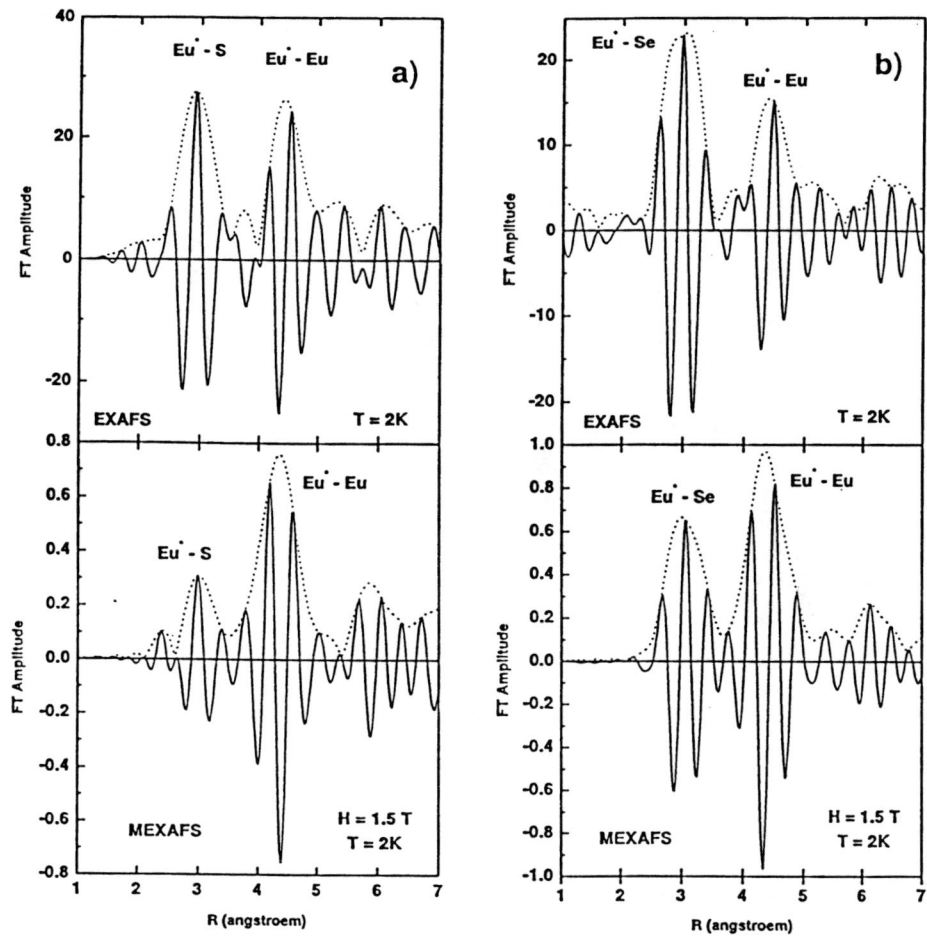

**FIGURE 3.** FT spectra of spin-averaged (upper plots) and magnetic (lower plots) EXAFS recorded at the $Eu\ L_{III}$-edge in $EuS$ (a) and in $EuSe$ (b).

and $EuSe$ (fig. 3b). For the sake of clarity, the FT spectra were systematically corrected for the phase-shifts and scattering amplitudes of the [Absorber...Scatterer] pair calculated with the FEFF code [38] for the first shell, i. e. the $Eu...S$ and $Eu...Se$ pairs, respectively. As expected, the most intense peaks in the spin polarized FT spectra are due to the $Eu...Eu$ pairs. What is most remarkable is the persistence of the signature assigned to the $Eu...S$ or $Eu...Se$ coordination shells eventhough their amplitudes are heavily reduced. A more quantitative analysis would yield an amplitude reduction of factor of 30 for the $Eu...Eu$ signal, while the amplitude of the $Eu...Anion$ signal is less than 1% of its value in the spin-averaged spectrum. If one would accept the assertion by Knulle et al. [39] that the amplitude of the MEXAFS oscillations is directly proportional to the spin moment carried by the backscattering atom, then one would end up with the completely unrealistic conclusion that the spin moment carried by the anions in europium chalcogenides should be ca. $1.2\mu_B$. Thus, there is a clear evidence from our spectra that the MEXAFS signals do not scale linearly with the magnetic moments carried by the neighbouring atoms. Note that it was already pointed out in [40] that such a crude assumption was totally inappropriate. We wish also to emphasize that our data reveal quite significant differences between the phase-shifts of the EXAFS and MEXAFS signals for the same scattering atoms. Such phase differences are more perceptible for the $Eu...Eu$ shells whereas the phase of the $Eu...Anion$ signals looks fairly similar in the EXAFS and MEXAFS spectra.

There is an additional point which deserves special attention. It has been established in the multiple scattering theory of XMCD [41] that, at the L-edges, MEXAFS oscillations are due not only to the magnetically polarized neighbours but also to the spin polarization of the absorbing atom. Following this model MEXAFS spectra can be observed even when the backscattering atoms are not magnetically polarized because one expects a difference in the central atom phase shifts for the "spin-up" and "spin-down" photoelectrons. However, the corresponding MEXAFS signal *should be in quadrature* with the spin averaged EXAFS oscillations. A direct inspection of the phase of the Imaginary parts of the FT spectra (Fig. 3) immediately shows that this contribution cannot be dominant in our experiment. It is therefore our conclusion that the observation of the signal assigned to the $Eu...Chalcogen$ pair in the MEXAFS spectra is a strong argument to support the view that the anions are carrying a magnetic moment in ferromagnetic europium chalcogenides.

## C  XMCD at the anion K-edges.

If our interpretation of the MEXAFS spectra is correct, then one should be able to detect also some XMCD signal at the $S$ K-edge in $EuS$ or at the $Se$ K-edge in $EuSe$. This stimulated us to try such challenging experiments. We have reproduced in Fig. 4 the measured XMCD spectra together with the corresponding XANES spectra. Since the XANES spectra were systematically normalized with

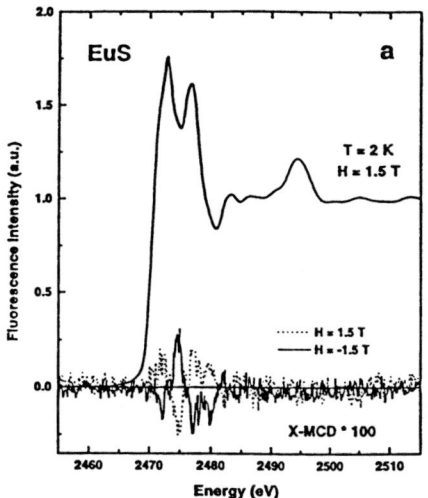 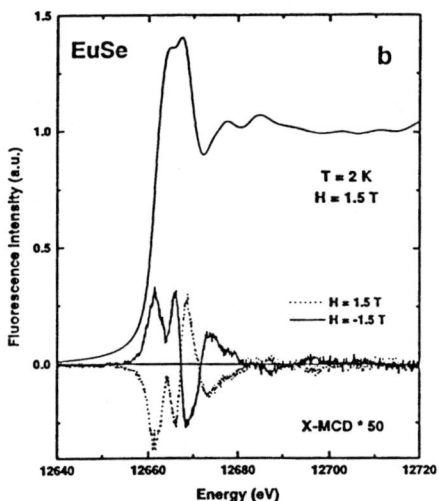

FIGURE 4. a). $S$ $K$-edge XANES and XMCD spectra of ferromagnetic $EuS$. b). $Se$ $K$-edge XANES and XMCD spectra of ferromagnetic $EuSe$. Raw XNCD spectra recorded with two directions of the magnetic field are shown.

respect to the edge jump, it appears that the intensity of the XMCD signals is unexpectedly large : ca. 1.2% at the $Se$ $K$-edge and nearly 2% at the $S$ $K$-edge once the XMCD amplitudes are corrected for the polarization transfer function of the monochromator. Note that the corresponding XMCD signals are even much larger than what is commonly measured at the $K$-edge of $3d$ transition metals. For the sake of comparison, let us recall that a XMCD signal of the order of $8 \cdot 10^{-3}$ has already been reported at the $Ni$ $K$-edge for a ferromagnetic $TbNi_5$ single crystal [42]. We feel nevertheless very confident that the XMCD signals which were measured at the $Se$ and $S$ $K$-edges were quite real and free of artefacts mainly because they could be nicely inverted on reversing the direction of the magnetic field. Moreover, for both $EuX$ compounds, the XMCD spectra exhibit a characteristic structure which seems to be well correlated with the splitting of the white line in the XANES spectra. The K-edge magneto-optical sum rules [22] tell us that the observed XMCD signals are to be ascribed to the orbital polarization of the $S$ $(3p)$ and $Se(4p)$ final states. Thus, our results produce strong experimental evidence that the anions $p$ band in the europium chalcogenides is involved in the magnetic interactions in these semiconductors. This is possible only if we admit that there is a strong hybridization either with the highly polarized $4f$ or $5d$ states of Europium as this was suggested by recent band structure calculations [9]. Moreover, this hybridization has been shown to be precisely at the origin of the splitting the white line and of the structure in the $K$-edge XMCD spectra. This hybridization could indeed explain the existence of small induced moments carried by the chalcogen anions and these small moments could promote some further polarization exchange interaction between the strongly localized $4f$ magnetic moments of the Europium atoms bridged by the chalcogen anions.

We wish to point out here that our results regarding the polarization of the anion $p$ band are complementary with respect to the results derived from previous NMR studies because a comparison of the hyperfine fields of the $S^{33}$ and $Se^{77}$ nuclei led to the conclusion that the hyperfine fields arose essentially from the spin polarization of the outer $s$ electrons [43]. In our opinion, the experimental observation of the $K$-edge XMCD spectra of the chalcogen anions yields a very strong and fairly direct argument supporting the suggestions drawn from a recent neutron diffraction study performed under very high pressure on $EuS$ crystals [18].

## V  CONCLUSIONS

In conclusion, the present study has illustrated the remarkable complementarity of the near-edge XMCD and Magnetic EXAFS experiments. It was indeed demonstrated that very valuable information could be extracted from X-ray absorption spectroscopy regarding the magnetic properties of ferromagnetic compounds. It was shown that the XMCD signal at the $L_{II,III}$-edges of rare-earths was controlled to some extent by the *intraatomic* $4f - 2p$ exchange interaction which should be nearly independent of the nature of the surrounding atoms. The contributions of quadrupolar transition to the XMCD signal and intricate effects the crystal field along with exchange splitting of the $2p$ core states were identified as possible explanations for the failure of the magneto-optical sum rules when applied to the $L$-edges spectra of rare-earths.

High quality MEXAFS spectra were also recorded at the $L$ edge of $Eu$ and a careful analysis of these spectra gave a strong indication that the chalcogen anions could carry a magnetic moment. This was independently confirmed by the detection of a fairly intense XMCD signal at the $K$-edge of Sulfur and Selenium. The magnetic polarization of the anion $p$ band can be explained its hybridization with the strongly polarized $4f$ and $5d$ states of $Eu$. This interpretation is also consistent with the observation nicely resolved structures in the XMCD spectra recorded at the $K$-edge of Sulfur and Selenium. This hybridization may explain as well the presence on the chalcogen atoms of induced moments which could promote the indirect exchange coupling between the $Eu$ moments. This is the first time that magnetic interactions mediated by chalcogen anions can be unambiguously established by X-ray absorption spectroscopy.

## VI  ACKNOWLEDGEMENTS

The authors are indebted to Ch. Brouder for enlightening discussions and to Y. Petroff for his constant support and permanent interest in this project.

# REFERENCES

1. Erskine J.L. and Stern E.A., *Phys. Rev.* **B12**, 5016 (1975).
2. Schütz G., Wagner W., Wilhelm W., Kienle P., Zeller R., Frahm R., Materlik G., *Phys. Rev. Lett.* **58**, 737 (1987).
3. Schütz G., Frahm R., Mautner P., Wienke R., Wagner W., Wilhelm W. and Kienle P., *Phys. Rev. Lett.* **62**, 2620 (1989).
4. Thole B.T., Carra P., Sette F. and van der Laan G., *Phys. Rev. Lett.* **68**, 1943 (1992).
5. Carra P., Tole B.T., Altarelli M., and Wang X., *Phys. Rev. Lett.* **70**, 694 (1993).
6. Chen C.T., Sette F., Ma Y., Modesti S., *Phys. Rev.* **B42**, 7262 (1990).
7. Vogel J., Fontaine A., Cros V., Petroff F., Kappler J.P., Krill G., Rogalev A. and Goulon J., *Phys. Rev.* **B55**, 3663 (1997).
8. Grange W., Maret M., Kappler J.P., Vogel J., Fontaine A., Petroff F., Krill G., Rogalev A., Goulon J., Finazzi M. and Brooks N.B., *Phys. Rev.* **B58**, 6298 (1998).
9. Rogalev A., Goulon J. and Brouder Ch., *J. Phys.: Cond. Matter.* **11**, 1115 (1999).
10. Finazzi M., Sainctavit Ph., Dias A.M., Kappler J.P., Krill G., Sanchez J.P., Dalmas de Reotier P., Yaouanc A., Rogalev A. and Goulon J., *Phys. Rev.* **B55**, 3010 (1997).
11. Grange W., Finazzi M., Kappler J.P., Delobbe A., Krill G., Sainctavit Ph., Sanchez J.P., Rogalev A. and Goulon J., *J. Alloys and Compounds* **275-277**, 583 (1998).
12. Neumann C., Hoogenboom B.W., Rogalev A. and Goedkoop J.B., *Solid State Comm.* **110**, 375 (1999).
13. Goulon J., Rogalev A., Gauthier Ch., Goulon-Ginet Ch., Paste S., Signorato R., Neumann C., Varga L., and Malgrange C., *J. Synch. Rad.*, **5**, 232 (1998).
14. Wachter P., in: *Handbook on the Physics and Chemistry of Rare Earth*, eds. by K.A. Gschneider, Jr. and L. Eyring, (North-Holland Publ. Co), **Ch. 19**, 507 (1979).
15. Passel L., Dietrich O.W. and Als-Nielsen J., *Phys. Rev.* **B14**, 4897 (1976).
16. Argyle B.E., Suits J.C. and Freiser M.J., *Phys. Rev. Lett.* **15**, 822 (1965).
17. Zinn W., *J. Magn. Magn. Mat.* **3**, 23 (1976).
18. Goncharenko I.N. and Mirabeau I., *Phys. Rev. Lett.* **80**, 1082 (1998).
19. Lee V.-C. and Liu L., *Solid State Comm.* **48**, 795 (1983).
20. Stöhr J. and Nakajima R., *IBM J. Res. Develop.* **42**, 73 (1998).
21. Fano U., *Phys. Rev.* **178**, 131 (1969).
22. Igarashi J. and Hirai K., *Phys. Rev.* **B50**, 17820 (1994).
23. Wu R., Wang D. and Freeman A.J., *Phys. Rev. Lett.* **71**, 3581 (1993).
24. Chen C.T., Idzerda Y.U., Lin H.-J., Smith N.V., Meigs G., Chaban E., Ho G.H., Pellegrin E. and Sette F., *Phys. Rev. Lett.* **75**, 152 (1995).
25. Dartyge E., Baudelet F., Giorgetti C. and Odin S., *J. Alloys and Compounds* **275-277**, 526 (1998).
26. Elleaume P., *J. Synch. Rad.* **1**, 19 (1994).
27. Rogalev A., Gotte V., Goulon J., Gauthier Ch., Chavanne J. and Elleaume P., *J. Synchrotron Rad.* **5**, 989 (1998).
28. Varga L., Giles C., Neumann C., Rogalev A., Malgrange C., Goulon J. and de Bergevin F., *J. Phys. IV France* **7**, C2-309 (1997).
29. Loos M., Ascone I., Goulon-Ginet C., Goulon J., Guillard C., Lacroix M., Breysse M., Faure D. and Descourieres T. *Physica B* **158**, 145 (1989).

30. Dimmock J.O., Hanus J. and Feinleib J., *J. Appl. Phys.* **41**, 1088 (1970).
31. de Groot F.M.F., Hu Z.H., Lopez M.F., Kaindl G., Guillot F. and Trone M., *J. Chem. Phys.* **101**, 6570 (1994).
32. Carra P. and Altarelli M., *Phys. Rev. Lett.* **64**, 1286 (1990).
33. Lang J.C., Srajer G., Detlefs C., Goldman A.I., König H., Wang X., Harmon B.N. and McCallum R.W., *Phys. Rev. Lett.* **74**, 4935 (1995).
34. Bartolomé F., Tonnerre J.M., Sève L., Raoux D., Lorenzo J.E., Chaboy J., García L.M., Bartolomé J., Krisch M., Rogalev A., Serimaa R., Kao C.C., Cibin G., and Marcelli A., *J. Appl. Phys.* **83**, 7091 (1998).
35. Brouder Ch., *J. Phys.: Condens. Matter* **2**, 701 (1990).
36. Cowan R.D., *The Theory of Atomic Structure and Spectra* (Univ. of California Press, Berkeley, 1981)
37. Srivastava P., Lemke L., Wende H., Chauvistré R., Haack N., Baberschke K., Hunter-Dunn J., Arvanitis D., Mårtensson N., Ankudinov A. and Rehr J.J., *J. Appl. Phys.* **83**, 7025 (1998).
38. Ankudinov A. and Rehr J.J., *Phys. Rev.* **B52**, 10214 (1996).
39. Knülle M., Ahlers D. and Schütz G., *Solid State Comm.* **94**, 267 (1995).
40. Chakarian V., Idzerda Y.U., Kemner K.M., Park J.H., Meigs G. and Chen C.T., *J. Appl. Phys.* **79**, 6493 (1996).
41. Brouder Ch. and Hikam M., *Phys. Rev.* **B43**, 3809 (1991).
42. Galera R.M. and Rogalev A., *J. Synch. Rad.* **6**, 691 (1999).
43. Suzuki H., Komaru T., Hihara T. and Koi Y., *J. Phys. Soc. Japan* **30**, 288 (1971).

# Diffracted Anomalous Fine Structure (DAFS) to Study Surfaces and Interfaces

M. Benfatto [a], R. Felici [b] and F. Comin [c]

[a] Istituto Nazionale Fisica Nucleare, Via E.Fermi 40, I-00044 Frascati, Italy

[b] Istituto Nazionale Fisica della Materia - Operative Group in Grenoble
c/o ESRF, BP 220, F-38043 Grenoble Cedex 9, France

[c] European Synchrotron Radiation Facility, BP 220, F-38043, Grenoble Cedex 9, France

**Abstract.** The structural determination of surfaces is crucial for the understanding of their properties. In this article we will outline a general treatment of the anomalous correction to the atomic scattering factor which can be used for structural determination in any system and in particular in the case of surfaces. Data collected on bulk Ge will be compared with theoretical results, while the data measured on reciprocal lattice points belonging to surface reconstruction rods will be interpreted using the described theory.

## INTRODUCTION

The structural characterization of reconstructed surfaces is of fundamental importance for the determination of their properties. Several techniques, which employ an x-ray beam, are routinely employed for this purpose as Grazing Incidence X-Ray Diffraction (GIXRD), Surface Extended X-ray Absorption Fine Structure (SEXAFS) and X-ray Standing Waves (XSW). A different approach, based on the measurements of the scattered intensity as a function of the energy near and above an absorption edge usually referred as Diffracted Absorption Fine Structure (DAFS), has been proposed and used to determine bulk structure (1). In principle this technique joins the advantages of the x-ray diffraction and EXAFS techniques and, for this reason, can become one of the most powerful structural techniques. However very few examples of such experiments appeared in the literature because of difficulties either in the experimental setup and in the data interpretation. The application of DAFS to the case of surfaces is of fundamental importance because, in the case of a clean reconstructed surface, it will allow to extract the EXAFS signal of the surface atoms only, separating it from the bulk ones, while, in the case of absorption of different atomic species, it should permit to separate the signal coming from the atoms which are in ordered positions with respect to the others.

These possibilities have focused the attention of several research groups in this field (2), but the data interpretation is still far from being fully understood. The data are usually analyzed by separating the form factor in a smooth part and an oscillating contribution, which is supposed to be equal to the EXAFS signal (3). This is an approximation which is, in general, not valid. In the next paragraph we will theoretically derive a more general expression for the anomalous corrections to the atomic scattering factors.

## THEORY

The intensity of x-rays diffracted by crystals depends on the structure factor F and then on the atomic scattering factors $f$ (4). In this paragraph we will deal with the corrections arising when the energy is close to an absorption edge. This dispersion correction, usually referred as anomalous correction, can be described by a complex quantity, $\Delta f = f' + if''$. Because of the causality theorem, the real and imaginary terms of the anomalous correction are related by a Kramers-Kronig transformation. In the forward scattering limit, $f''$ is proportional to the absorption cross-section. In the following we shall derive, in the dipole approximation, a more general expression for the anomalous correction to the atomic scattering factor which are valid for any condition of scattering.

In the dipole approximation the resonant correction $\Delta f$ to the Thompson scattering amplitude $f_0$ may be written as:

$$\Delta f = m\omega^2 \sum_n \frac{\langle \Psi_i | \hat{\varepsilon}_i \cdot \mathbf{r} | \Psi_n \rangle \langle \Psi_n | \hat{\varepsilon}_s \cdot \mathbf{r}' | \Psi_i \rangle}{E_i - E_n + \hbar\omega + i\Gamma/2} + C.C. \tag{1}$$

where $\hat{\varepsilon}_i$ ($\hat{\varepsilon}_s$) is the polarization vector of the incident (outgoing) photon of energy $E=\hbar\omega$, m is the electron mass and C.C. stands for complex conjugate. The sum is over all the intermediate states $\Psi_n$ with energy greater than the Fermi energy. For simplicity $\Gamma$ does not depend on n. We consider only elastic scattering and photon energies close to the absorption edge. In this case the complex conjugate is very small and can be neglected.

Using the Multiple Scattering (MS) approach, following Vedrinskii et al. (5), it is convenient to introduce the Green function $G(\mathbf{r},\mathbf{r}';E)$ of the system. In this way the anomalous term becomes:

$$\Delta f = \frac{(\hbar\omega)^2}{2} \langle 1s | \hat{\varepsilon}_i \cdot \mathbf{r} G(\mathbf{r},\mathbf{r}';E) \hat{\varepsilon}_s \cdot \mathbf{r}' | 1s \rangle \tag{2}$$

using atomic units and neglecting the smooth background contribution coming from the integration over the occupied states below the Fermi energy (4). For $\mathbf{r}$ and $\mathbf{r}'$ close to the origin, where the initial $1s$ state is localized, the Green function is given by:

$$G(\mathbf{r},\mathbf{r}';E) = k \sum_{L,L'} R_L^o(\mathbf{r}) \tau_{L,L'}^{oo} R_{L'}^o(\mathbf{r}') - k \sum_L R_L^o(\mathbf{r}_<) S_L^o(\mathbf{r}_>) \tag{3}$$

where $\tau_{L,L'}^{00}$ is the scattering path operator of the MS theory. The other symbols have been introduced elsewhere (6) and here have the same meaning.

The anomalous term, using atomic units, becomes:

$$\Delta f = \frac{(\hbar\omega)^2}{2} M_{01}^2(k) \sum_{\mu,\mu'} Y_{1\mu}(\hat{\varepsilon}_i) \tau_{\mu,\mu'}^{00} Y_{1\mu'}^*(\hat{\varepsilon}_s)$$

$$M_{01}(k) = \tfrac{2}{3}\sqrt{k\pi} \int \varphi_{1s}(r) R_1(k,r) r^3 dr \qquad (4)$$

$$k = \sqrt{\hbar\omega + E_{1s}}$$

where $Y_{l,m}$ are complex spherical harmonics. $\Delta f$ depends on both the polarization vectors $\hat{\varepsilon}_i$ and $\hat{\varepsilon}_s$ of the incident and outgoing photons, although, in the dipole approximation, a direct dependence on the corresponding photon wavevectors $\mathbf{k}_i$ and $\mathbf{k}_s$ is not present. For comparison the total absorption correction depends only on the polarization vector of the incident photon, $\mathbf{k}_i$. The anomalous scattering polarization dependence is generally different from the corresponding absorption polarization dependence. By expanding the scattering path operator in spherical tensor components (7), the anomalous term finally becomes:

$$\Delta f(E,\hat{\varepsilon}_i,\hat{\varepsilon}_s) = A(\omega)\left[\frac{\sqrt{3}}{4\pi}\tau(11;00)\hat{\varepsilon}_i\cdot\hat{\varepsilon}_s + \sum_{c\gamma\neq 0}\tau(11;c\gamma)B_{11}^{c\gamma}(\hat{\varepsilon}_i,\hat{\varepsilon}_s)\right] \qquad (5)$$

with:

$$A(\omega) = \tfrac{1}{2}(\hbar\omega)^2 M_{01}^2(k)$$

$$B_{11}^{c\gamma}(\hat{\varepsilon}_i,\hat{\varepsilon}_s) = \sum_{\mu,\mu'}(-1)^{1-\mu}C_{1-\mu 1\mu'}^{c\gamma} Y_{1\mu}(\hat{\varepsilon}_i) Y_{1\mu'}^*(\hat{\varepsilon}_s) \qquad (6)$$

where $C_{1-\mu 1\mu'}^{c\gamma}$ are the Clebsch-Gordon coefficients and $\tau(11;c\gamma)$ is a $(c,\gamma)$ spherical tensor component.

In this last expression we have separated the spherical part from the anisotropic contribution in analogy with the general structure of the polarization dependent absorption cross section (7). We note that the Thompson scattering amplitude $f_0$ has the same polarization vector dependence of the spherical term and, therefore, can be simply added to the $\tau(11;00)$ term to calculate the elastic scattering cross section. In the atomic limit only the $c=\gamma=0$ term survives and the usual expression for the anomalous term can be recovered.

# RESULTS

Measurements of diffracted intensities, I(E), as a function of incident photon energies have been performed at the European Synchrotron Radiation Source of Grenoble using the ID32, windowless, SEXAFS and XSW beamline. The beam is monochromatized by a double Si(111) monochromator. The first crystal is cooled to liquid nitrogen temperature to remove all the heat load while the second crystal can be bent to perform a horizontal focusing of the beam. The beamline is also equipped with a high quality diffractometer coupled with a UHV chamber where a clean surface can be prepared and GIXRD experiments can be carried out.

We measured I(E) in the so called "top mode". In this procedure the intensity of a reciprocal space point is measured as a function of the energy and varying all the diffractometer angles. At the end the sample is rotated, maintaining then the same scattering geometry, and the same scan in this new point of the reciprocal space is performed for background subtraction. We choose this procedure with respect to the more classical peak integration because it is less time consuming. The two procedures differ only if the width of the peaks changes with energy but in our cases we never observed this phenomenon. This method is not practical when the diffraction peaks are very narrow (less than 0.01°) because it is very difficult to align the sample with the necessary accuracy.

We are dealing with absolute intensities and then several corrections have to be considered: absorption, surface area, Lorentz and polarization corrections. The absorption is particularly important in the case of bulk measurements because introduces oscillations in the measured intensities. Moreover we also need to consider the different energy efficiencies of the ionization chamber used as a monitor and of the scintillation detector.

All the above corrections, but not the absorption one, have a smooth dependence on the energy and they can be easily simulated by a simple three parameters function which was determined by comparison with the theoretical atomic intensity. Because we do not consider "forbidden reflections", the intensity is dominated by the real part of the structure factor. In this case it is straightforward to extract the real part of the anomalous corrections being negligible the oscillating term of the imaginary part.

First we discuss on the measurements on the bulk which present two problems: the absorption correction and very narrow diffraction peaks. To overcome both of these problems, we performed our measurements using as sample a thin Ge film grown by MBE technique on a Si(001) substrate. The film thickness was 700 nm, well above the critical thickness, resulting in a completely relaxed structure, but much shorter than the absorption length which is of several microns. Because of the large mismatch existing between Ge and the Si substrate a large quantity of defects was present in the structure increasing the width of the diffraction peaks of about one order of magnitude compared to a pure Ge single crystal values. The peaks result large enough to perform "top-mode" energy scans without particular problems.

Because of the symmetry of the Ge structure the second term of eq. 5 gives an imaginary term equal to the $f''$ measured in an absorption experiment. $f'$ can then be calculated starting from absorption data and using the Kramers-Kronig transformation. Experimentally we determined $f'$ by measuring the intensity of the (111), (224) and (113) peak as function of energy. After a proper normalization to the atomic scattering form factor the three experimental $f'(h,k,l)$ were identical within the statistical noise and we averaged them to obtain the experimental $f'$. In fig. 1a we show the experimental $f'$ (points) and the Kramers-Kronig transformation of the $f''$ (continuos line) that has been measured on the same sample with the same experimental setup by simply rotating the sample of 2 degrees in the surface plane in a off-Bragg position. The agreement is excellent and experimentally confirms that in the case of germanium $f'$ is simply related to $f''$ measured in an absorption experiment. In fig. 1b we report a comparison between the experimental data and a theoretical curve calculated by first principles modifying the CONTINUUM program (8,9), already widely used for absorption analysis, and modified to calculate $\Delta f$ as defined in eq. 5. The agreement, considering that we calculated $f'$ from first principles and without including any thermal coefficient is more than satisfactory. The analysis of DAFS spectra can then be carried out using the same MS programs developed for EXAFS. In the case of absorption only the imaginary part is used, while in the diffraction case both the real and imaginary terms have to be considered.

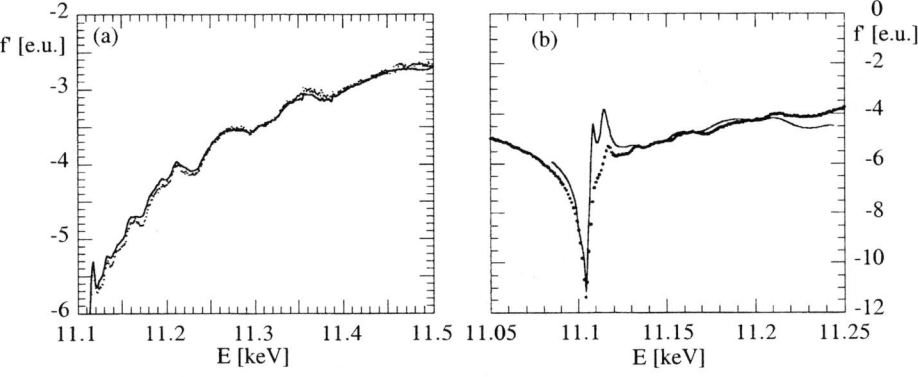

**FIGURE 1.** Experimentally determined real part of the anomalous correction, $f'$, for Ge (points) compared with Kramers-Kronig transformation of the absorption data, continuos line in (a), and with a theoretical calculation carried out using eq. 5, continuos line in (b).

In the case of the clean Ge(001) surface reconstructed 2x1 the system does not have anymore the symmetry properties of the bulk and we expect $f''(E,\hat{\varepsilon}_i,\hat{\varepsilon}_s)$ to be different from the absorption case. This system is then ideal to understand the modifications induced in the anomalous part of the scattering factor by the 2nd term of eq. 5. The structure of Ge(001)2x1 surface has been widely studied by x-ray diffraction (10,11). Buckled Ge dimers, 4% longer than the Ge bulk bonds, terminate the structure. We have measured the intensities of several points of surface reconstruction rods as a function of the incident photon energy. In fig. 2a we show the real part of the anomalous scattering factor measured at one of these points: the (0.5,0,0.03). In this case we used incidence and exit angles smaller than the critical value. Under these conditions the signal is dominated by the dimer atoms and we are mainly sensitive to their local structure. Two features are evident: an oscillatory term similar to the standard EXAFS and a rapidly oscillating function near the edge. A first shell analysis of the EXAFS-like signal can be carried out and it shows the presence of a frequency corresponding to near neighbor distances about $3 \pm 1\%$ longer than the bulk. This value is in excellent agreement with the local structure of dimer atoms determined by x-ray diffraction where each dimer atom has three different bonds whose average value is 2.53 Å, 3% longer than the bulk bond distance of 2.45 Å.

The near edge structure is instead due to the second term of eq. 5 where terms with a $(kr)^2$, or higher power, are present. This term is sensitive to both $\hat{\varepsilon}_i$, $\hat{\varepsilon}_s$ and to their orientation with respect to the lattice unit cell. For this reason it depends strongly on the geometrical scattering angles.

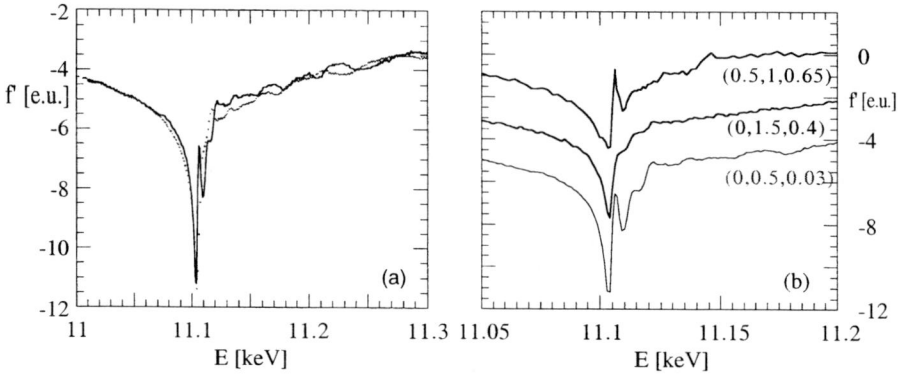

**FIGURE 2.** Real part of the anomalous corrections determined on points of surface reconstruction rods corresponding to the clean Ge(001) 2x1 reconstruction. In subset (a), we compare $f'$ measured at the reciprocal point (0,0.5,0.03) (continuos line) with the bulk experimental $f'$ (points). In subset (b), we show the experimental $f'$ for points of different surface reconstruction rods showing the strong dependence of the near edge structure on the scattering conditions. The two upper curves are shifted of +2 and +4 for clarity reasons.

However as soon as we brake the grazing incidence and grazing exit conditions the diffracted intensity presents contribution from deeper layer Ge atoms with a "bulk-like" local structure. The oscillating term is then a weighted product of the different terms corresponding to the different atoms of the surface unit cell with different frequency components. Their average results on the disappearance of the EXAFS-like oscillations for large energy transfers (E>50 eV). The near edge structure is instead always present, being due mainly to the two dimer atoms, and depends strongly on the geometrical scattering conditions as it appears in fig. 2b. This part still contains structural information and we will try to calculate it using the modified CONTINUUM program.

In conclusions we have outlined a general treatment for the corrections to the atomic scattering factor and showed how programs developed using the MS theory to analyze absorption data can be easily modified to interpret DAFS spectra. This is of particular importance in the case of surfaces where the corrections to the scattering factors cannot be generally derived by absorption measurements.

## ACKNOWLEDGMENTS

We would like to express our thanks to C.R. Natoli for the helpful discussions and to all the staff of the ID32 beamline for the technical support. We also thanks the ESRF for the travel support given to one of us (M.B.).

## REFERENCES

1. Stragier, H., Cross, J.O., and Rehr, J.J., *Phys. Rev. Lett.* **69**, 3064 (1992)
2. Materlik, G, Sparks, C.J., and Fisher, K., *Resonant Anomalous X-ray Scattering*, Amsterdam, North-Holland, 1994, pp. 365-388; Iwasawa, Y., *X-ray Absorption Fine Structure for catalysts and Surfaces*, Singapore, World Scientific, 1995, pp. 372-382
3. Pickering, I.J., Sansone, M., Marsch, J., and George, G.N., *J.Am.Chem.Soc.* **115**, 6302 (1993)
4. Warren, B.E., *X-ray Diffraction*, New York, Dover Scientific, 1969
5. Vedrinskii, R.V., Kraizman, V.L., Novakovich, A.A., and Machavariani, V.Sh., *J.Phys.Cond.Matt.* **4**, 6155 (1992)
6. Natoli, C.R., and Benfatto, M., *Jou. de Physique* **47**, C8-11 (1986)
7. Brouder, C., Ruiz Lopez, M.F., Pettifer, R.F., Benfatto, M., and Natoli, C.R., *Phys. Rev.* **B39**, 1488 (1989)
8. Tyson, T.A., Hodgson, K.O., Natoli, C.R., and Benfatto, M., *Phys. Rev.* **B46**, 5997 (1992)
9. Benfatto, M., Solera, J.A., Chaboy, J., Proietti, M.G., and Garcia, J., *Phys. Rev.* **B56**, 2447 (1997)
10. Rossmann, R., Meyerheim, H.L., Jahns, V., Wever, V.J., Moritz, W., Wolf, D., Dornish, D., and Shult H., *Surf. Sci.* **279**, 199 (1992)
11. Torrelles, X., van der Vegt, H.A., Etgens, V.H., Fajardo, P., Alvarez, J., and Ferrer, S., *Surf. Sci.* **364**, 242 (1996)

# Charge Stripes Formation by X-ray Illumination in High Tc Superconductors

Ginestra Bianconi,
*Department of Physics, Notre Dame University, 46556 Indiana*

Daniele Di Castro, Naurang L. Saini, Antonio Bianconi
*Unità INFM, Dipartimento di Fisica, Università di Roma La Sapienza,
P. Aldo Moro 2, 00185 Roma Italy*

Marcello Colapietro
*Dipartimento di Chimica, Università di Roma "La Sapienza", 00185 Roma, Italy*

Augusto Pifferi
*Istituto Strutturistica Chimica, Area della Ricerca, CNR, Monterotondo Stazione, Italy*

**Abstract.** Charge ordering of the doped holes in $La_2CuO_{4.1}$ with $T_c$=40 K has been investigated. The formation of a 1D anharmonic incommensurate charge density wave (ICDW), i.e., stripes, in the $CuO_2$ plane has been detected by x-ray diffraction using high photon flux at the third generation synchrotron radiation source, Elettra. We have identified the ICDW peaks that are found to increase by photodoping at low temperature, T<180K, where the lifetime of the photoexcited electron-hole pairs is very long and the mobility of interstitial oxygens is frozen. The 3D long range oxygen ordering is identified and a photo-assisted transition from an incommensurate ordered phase O1, to a second phase O2 is observed at T=300 K.

## INTRODUCTION

The coexistence of a one-dimensional (1D) incommensurate charge density wave (ICDW) and itinerant carriers in high Tc superconductors has been found by joint x-ray diffraction and x-ray absorption spectroscopy (1-7) in 1990-1992. This coexistence provides a key to understand the metallic phase showing high $T_c$ superconductivity (HTCS). In fact the ICDW, not driven by nesting vectors on the Fermi surface, is expected to be a characteristic feature of an electron gas close to a metal-insulator transition (MIT) driven by a long range Coulomb interaction Q (i.e., for a non-negligible ratio of Q on the Fermi energy), so called a generalized Wigner MIT (8-14).

Experimental detection of the ICDW driven by the long range Coulomb interaction has been difficult due to the fact that the signal is usually very weak, broad, with a short coherence length. Moreover, in different materials the charge fluctuations are found to have largely different time scales and evolve differently with doping. The coexistence of a 1D ICDW and the metallic superconducting phase, first observed in $Bi_2Sr_2CaCu_2O_{8+\delta}$ (Bi2212) (1-7, 15-17), has been confirmed by several experiments in $La_2CuO_{4+\delta}$ by nuclear magnetic resonance (NMR) (18), $YBa_2Cu_3O_{7-x}$ (YBCO) by scanning tunnelling microscopy (STM) (19), $La_{2-x}Sr_xCuO_4$ (LSCO) by high resolution extended x-ray absorption fine structure (EXAFS) (20), and recently in $La_{1.6-x}Nd_{0.4}Sr_xCuO_4$ (LNSCO) by neutron diffraction (21) and LSCO by NMR (22). The suppression of spectral weight at the spots of the Fermi surface connected by the second harmonic of the ICDW wavevector has been observed by angle scanning photoemission in Bi2212 (23,24). This result shows that nesting effects are not the origin of the ICDW.

It is known that a metallic phase close to the Wigner metal-to-insulator phase transition, called a polarized electron gas, exhibits an anomalous dielectric function $\varepsilon(\omega,q)$ that gives ICDW, and spin density waves (SDW) (25). In this regime it is possible to have superconducting pairing driven by electronic fluctuations (26). This particular superconducting phase is characterized by the shortest possible Pippard coherence length $\xi_0$ i.e, as short as the average distance between the charges (27).

The critical temperature $T_c$ for an electron gas with Fermi temperature $T_F$, Fermi wavevector $k_F = 2\pi/\lambda_F$, gap energy $\Delta_0$, is given by:

$$T_c = 0.36 \, T_F / (f \, k_F \, \xi_0)$$

where f is a measure of the deviation from the weak coupling limit $2\Delta_0/K_B T_c = 3.52f$. For a 2D electron gas $\rho = 1/\pi(r_s a_B)^2$, $k_F = \sqrt{2}/(r_s a_B)$ and $E_F(Ry) = 2/(m^* r_s^2)$. The coherence length $\xi_0$ of the superconducting phase of a polarized electron gas, where pairing is mediated by charge fluctuations, i.e., with $\varepsilon(0,q) \to 0$, is given by $\xi_0 = 2 r_s a_B$ (27). Therefore the critical temperature is given by:

$$T_c \, (K) = (35440.8/f) \, \rho/m^*$$

where $\rho$ is measured in $Å^{-2}$, and $m^*$ is the effective mass (8,12). Thus in this limit the critical temperature depends only on the ratio $\rho/m^*$.

The discovery of Uemura et al. (28) and Keller et al. (29) that *$T_c$ is proportional to $\rho/m^*$* and that of Bianconi et al. (1-6) that *1D ICDW coexists with superconductivity* have been interpreted as a direct experimental evidence that the 2D electron gas in the HTCS is close to an MIT driven by the long range Coulomb interaction at the second Phase Separation workshop at Cottbus in Sept. 1993 (8). It was proposed (9-14) that the pairing mechanism active in the high temperature superconductors was mediated by charge fluctuations in a polarized electron gas. The particular generalized Wigner MIT was described as driven by both electronic correlations (the enhanced electron effective mass resulting by doping an Hubbard insulator is estimated by the slave boson approximation) and polaronic electron lattice interactions (the enhanced electron effective mass is estimated in the polaronic weak coupling limit). The observed Wigner MIT was found to be not far from a critical point, since we observed the coexistence of a 1D-ICDW and superconductivity i.e., a so called striped phase. The universal curve of the critical temperature as a function of doping $T_c(\delta)$ calculated for the particular striped phase determined in the experiments considering a pairing mechanisms mediated by charge fluctuations was found to be in agreement with experimental data (12).

The coexistence of the 1D ICDW and the metallic superconducting phase was first seen in 1990 (1) and characterized in the following two years (2-4). In 1994 and 1995 Löw et al. (30) and C, Castellani et al. (31) included the long range Coulomb interaction Q as an intrinsic property of the metallic phase of cuprate superconductors, i.e., a 2D electron gas with a non negligible ratio $Q/E_F$. Both theoretical papers show that the effect of introducing Q in a correlated electron gas is to trigger 1D ICDW with finite wavevector determined by the intrinsic charge density fluctuations, in agreement with the experimental findings (1-6).

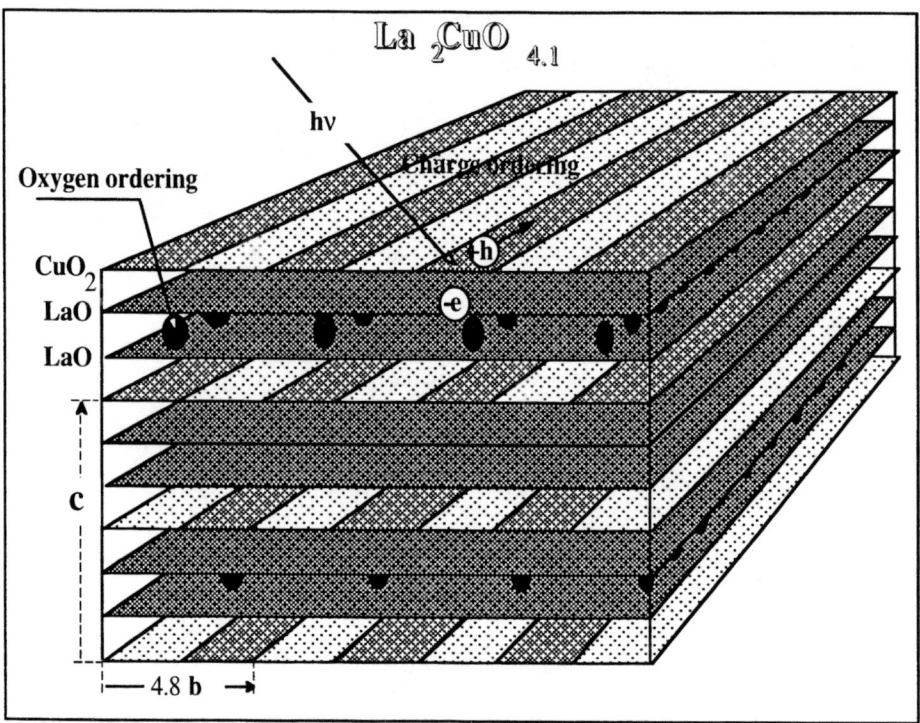

**FIGURE 1.** Pictorial view of the $La_2CuO_{4+\delta}$ crystal with interstitial oxygens (black circles) between the LaO layers and photoinduced electron (in LaO block layers) and hole (in the $CuO_2$ plane) pairs. The charge ordering in the $CuO_2$ plane, i.e., the 1D ICDW, is indicated by stripes of larger and lower charge density.

The coexistence of charge ordering, ICDW, and superconductivity in the $CuO_2$ planes could suppress the critical temperature. In fact the Fermi surface is broken by the ICDW, with the formation of missing points, due to the coupling of electrons with the wavevector of the ICDW. This effect generally induces a decrease of the density of states at $E_F$ and a suppression of the critical temperature. It was shown that for particular values of the chemical potential and the shape of the anharmonic 1D-ICDW as that found experimentally in cuprates at optimum doping (see Fig. 1,\ it is possible to amplify the critical temperature. The process to amplify the critical temperature has been disclosed in a patent, with priority date 7 Dec. 1993, (32) and following papers (33-37). The 1D ICDW modulation forms a superlattice of quantum stripes as shown in Fig. 1 and the chemical potential is tuned at a shape resonance of the superlattice. Under this resonance condition a large amplification of the critical temperature is obtained (33-37).

The stripes scenario described above "*charges that move freely mainly in one direction, like the water running in grooves of a corrugated iron foil*", as introduced for the first time in 1992 (3), has been followed by other models for the striped phases considering the stripes formed by doped charges in presence of antiferromagnetic correlations. Theoretical models and experiments on stripes have discussed at the first and second international conferences on "Stripes and High $T_C$ superconductivity" held in Rome in 1996 (38) and 1998 (39). Recently experimental research in HTCS has been focused to the detection of charge stripes by novel methods (15-24).

In this work we have explored a new experimental approach to identify the ICDW modulation associated with doped charges using x-ray diffraction that is a direct probe to the charge fluctuations associated with the 1D-ICDW. The idea is based on the use of the high intensity x-ray flux, available at the new third generation synchrotron radiation x-ray sources, to create a relevant number of photodoped holes at low temperature in a surface layer of thickness H, determined by the x-ray penetration depth, and to probe the charge ordering in the same slab by x-ray diffuse scattering using a fast 2D detection. Photodoping of cuprate superconductors using a high laser light photon flux has been shown to be an effective experimental way to promote persistent photoconductivity and the insulator to metal transition (40-42). In fact the photons, with a threshold energy of about 2 eV, create electron-hole pairs with a long life time at temperature lower than 200 K. The electrons are ejected from the $CuO_2$ plane, leaving an itinerant hole, to the block layers as shown in Fig. 1, and in $La_2CuO_4$ is

$$(La_2O_2)^{++}(CuO_2)^{--} + h\nu \rightarrow (La_2O_2)^{(2-y)} (CuO_2)^{-(2-y)}.$$

Photodoping by using x-rays has allowed us, for the first time, to add y itinerant holes in the $CuO_2$ plane and y electrons in the LaO planes, and to probe their ordering by x-ray diffraction.

We have focused our interest to the superconducting $La_2CuO_{4+\delta}$ system (43-52) where the metallic stripe phase coexists with superconductivity (19, 53-58). The interstitial oxygen ions enter between two LaO layers of $La_2CuO_{4+\delta}$ (49) shown in Fig. 1, and they have a high mobility for temperatures larger than about 200 K (59). The relation between the oxygen doping and the actual number of holes per Cu ions is object of discussion since the samples are not homogenous and a large part of the doped holes are localized (60-62). A phase separation below $T^*_{ps}=250$ K into an antiferromagnetic phase with oxygen content $\delta=0.01$ and a first superconducting phase with $\delta_1=0.055$ has been found. In the superconducting and metallic phase with $T_c \sim 31$ K a diffuse x-ray scattering peak associated with modulated charge and tilts of the $CuO_6$ octahedra has been detected with wavevector $q_{cdw}=0.09$ **a**\*+0.153 **c**\* typical of diagonal charge stripes in the $CuO_2$ plane and a stage 6 in the c axis direction (55). The onset temperature for the staging is 250 K, while the ordering temperature $T_{CO}$ for the 1D charge stripes is 180 K. This sample shows the characteristic four diffuse inelastic incommensurate magnetic scattering peaks of the vertical and horizontal spin density waves as in other cuprate superconductors with $q_{sdw}=\pm0.105\mathbf{a_t}$ and $q_{sdw}=\pm0.105\mathbf{b_t}$ (63) below 70 K. The spin ordering at the higher oxygen concentration, where stages 4, 3 and 2 appear, has been recently observed: the coexistence of inelastic diffuse incommensurate peaks below 70K and elastic sharp peaks below $T_c=40K$ has been detected (64).

In this work we show two coexisting modulations associated with charge ordering and oxygen ordering in a sample with $\delta=0.1$ ($T_c=40$ K). We observe first a 3D long range oxygen ordering as in $La_2NiO_{4+\delta}$ for $\delta>0.125$ (65), and a diffuse, short range charge ordering of diagonal stripes with a wavevector as in Bi2212 at optimum doping. We discuss the relation between the charge stripes observed in this work with a recent magnetic scattering experiment on a similar crystal. (64).

## EXPERIMENTAL

The single crystal of $La_2CuO_{4.1}$ was grown first as $La_2CuO_4$ by flux method and then doped by electrochemical oxidation (47,48). The measurements were made on a crystal of size $3\times2\times0.5$ mm$^3$ showing a sharp superconducting transition at $T_c=40$ K as measured by surface resistivity in the radiofrequency region. The temperature dependent diffraction data were collected on the crystallography beamline at the Elettra storage ring. The X-ray beam emitted by the wiggler source on the Elettra 2 GeV electron storage ring at Trieste, was monochromatized by a Si(111) double crystal monochromator, and focused on the sample. The temperature of the crystal was monitored with an accuracy of $\pm1K$).

We have collected the data in the K geometry, with a photon energy of 12.4 KeV, wavelength $\lambda=1$ Å, using an imaging plate as a 2D detector. The sample oscillation around the **b** axis was in a range $0<\phi<30^o$, where $\phi$ is the angle between the direction of the photon beam and the **a** axis. We have investigated a portion of the reciprocal space up to 0.6 Å$^{-1}$ momentum transfer i.e., recording the diffraction spots up to the maximum indexes 3, 3, 19 in the **a***, **b***, **c*** direction respectively. Using the high flux X-ray beam available at the third generation synchrotron radiation sources with a maximum flux of $1.6\ 10^{12}$ photons/(sec.mm$^2$), and photon energy of 12.4 KeV, we have illuminated the sample with a maximum power of $2\times10^{16}$ eV/(sec mm$^2$).

The surface density of Cu ions in a surface slab of thickness 1.5 μm, determined by the X-ray penetration depth, is $1.5\ 10^{16}$ Cu/mm$^2$. Considering that the energy required to create an electron-hole pair in the cuprates is 2 eV and taking into account an efficiency for quantum conversion of the order of $10^{-4}$, $10^{-5}$, we can create a detectable number of photo-doped holes as was obtained before by laser irradiation (41,42) and to observe ordering of the mobile photo-doped holes and photo-assisted oxygen ordering (41) driven by photo-induced oxygen diffusion (42).

## RESULTS AND DISCUSSION

The orthorhombic lattice parameters of our crystal are a=5.351 Å, b=5.418 Å, c=13.171 Å. Thanks to synchrotron radiation it has been possible to record a large number of weak superstructure spots around the main peaks of the average structure. The indexing of the superstructure has been conducted taking into account the twinning of the crystal and coexistence of 3 incommensurate modulations (Fig. 2). The first two superstructures are characterized by 3D ordering and narrow, resolution limited, diffraction peaks.

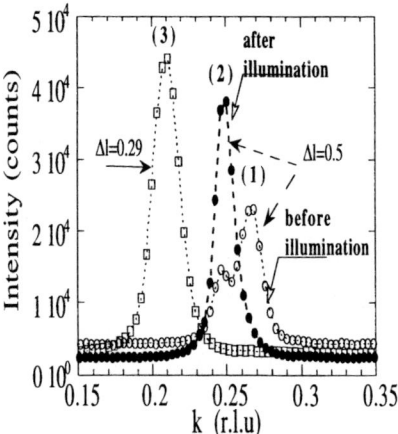

**FIGURE 2.** Scans along the Q= (0,k,6+0.29) due to the diffuse scattering peaks of stage 3.5 superstructure (3), squares, and along the Q= (0,k,6+0.5) squares due to the superstructure (2) and (1) before and after illumination at T=300K.

$q_1$= 0.049 (±0.0014) $a^*$, 0.268 (±0.0014) $b^*$, 0.490 (±0.0039) $c^*$

$q_2$=0.089 (±0.0031) $a^*$, 0.244 (±0.0024) $b^*$, 0.495 (±0.0046) $c^*$

that coexist with a pattern of diffuse spots due to a third superstructure, with a coherence length of about 350 Å.

$q_3$= 0.2080 (±0.0016) $b^*$, 0.290(±0.0055) $c^*$,

The modulations can be classified according to the different modulation along the **c** axis. The wavevectors $q_1$ and $q_2$ are associated with two phases, namely O1 and O2, and classified as stage 2 oxygen ordered phases.

The third wavevector $q_3$ is associated to in plane charge ordering with a period of 3.5 **c** in the axial direction. The in plane modulation along the **b**-axis of the superstructure $q_3$ is the same to the one found in Bi2212 and characterized by diagonal stripes with wavevector of $0.21b^*$.

We have investigated temperature dependence of the superstructures in the region around 100-300K and explored the effect of illumination dose. The charge ordering, or stripe formation, is expected in the range 100-180K. In fact in this temperature range the polaron formation has been observed in this system by EXAFS (56,57). On the contrary oxygen ordering, such as the stage formation, is expected to be in the range 300-200K where the mobility of doped oxygen ions is high.

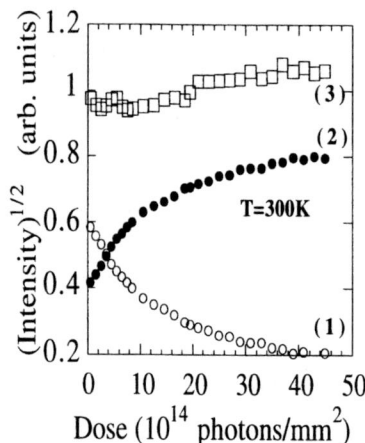

**FIGURE 3.** Effect of the photo illumination on the intensity of the peaks (1) (2) and (3) at T=300K. We have used a flux of 4 $10^{11}$ photons/sec. Evidence for photo-assisted phase transition from the coexisting 3D oxygen ordered phases O1 and O2 to a the phase O2.

At room temperature we observe two coexisting phases O2 and O1. The larger probability of the phase O1 (60%) indicates that the free energy of the phase O1 at room temperature is larger than that of the phase O2. The free energy for the long range oxygen ordering is determined by both elastic crystalline strains and charge interactions therefore it can change by decreasing the temperature. Under x-ray illumination we have observed a photon assisted phase transition from the phase O1 to the phase O2. We observe a disappearance of the phase O1 under illumination, while the superstructure (3) remains nearly constant (Fig. 3).

The photo-induced oxygen diffusion under the local electric field suppresses the O1 phase with an increase of the O2 phase, while their sum remains constant. Since the free energy of the phase O1 is lower we observe a slow decay of the phase O2 towards the phase O1 with a decay rate decreasing with the temperature, going from 300 K to 240 K, as expected for a temperature activated process.

The kinetics of the photo-assisted oxygen ordering in the range 180<T<300K under a photon flux $\phi$ from the ordered phase O1 to the phase O2 is given by

$$\frac{\partial O1}{\partial t} = -k_i O1 + k_d O2$$

where $k_i$ is the photo-assisted ordering rate under extended illumination, that is proportional to the photon flux and $k_d(T)$ is the rate of the decay of the ordered phase O2 toward the O1 phase.

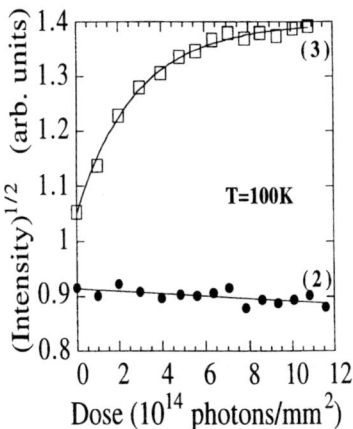

**FIGURE 4.** Effect of the x-ray illumination on the intensity of the charge ordering modulation (3) by photo doping at T=100K, while the integrated intensity of the peaks (3) increases the peaks of the 3D oxygen ordering superstructure (2) remains constant.

The sum O1(t)+O2(t) remains constant, the probability of the two phases are P2=O2/(O2+O1) and P1 = 1- P2. If we assume that $k_i$ and $k_d$ independent on the populations of the two phases O1(t) and O2(t), we can easily integrate the equation

$$\frac{\partial P_1}{\partial t} = -P_1(k_i + k_d) + k_d \; ; \; P_1(t) = \frac{k_d + k_i e^{-(k_i + k_d)t}}{k_i + k_d}$$

The probability at the saturation for $t \to \infty$, is a measure of the ratio of the two rates $k_d/k_i = \frac{P_1(\infty)}{1 - P_1(\infty)}$. The data in Fig. 3 show the photo-assisted phase transition at room temperature as a function of the x-ray dose, D= $\phi t$ where $\phi$ is the photon flux and t the time. The fit provides a measure of the dose D* needed for the transition D*=1.9 $10^{15}$ photons/mm², i.e., 0.12 photons per Cu ion, corresponding to an energy density E*=2.3 $10^{21}$ eV/cm².

From these data we have found a lifetime, $1/k_d$ = 1.1 $10^4$ sec for the photo-induced phase O2 during the illumination at 300K. Nevertheless, the assumption that $k_d$ is independent on the population O2(t) has to be considered only as a first approximation since further data show that $1/k_d$ increases with the population of this phase. In fact when the transition is completed the lifetime of the phase O2 becomes

2.5 $10^4$ sec. The lifetime of the phase O2 becomes much longer below 250 K, where it seems to be the stable phase. This result can shed light on many reports of an increase of the superconducting critical temperature by annealing around 180-200 K in oxygen doped $La_2CuO_4$.

The time scale of this metastable phase above 240 K is of the same order of magnitude as the persistent photoinduced conductivity observed in YBaCuO samples (41,42) under continuos laser illumination. Here we would like to remark that this phase transition induced by the polarized synchrotron radiation x-rays illumination is similar to the alignment of nematic liquid crystals obtained by polarized laser light (66). Therefore we describe this phenomenon as a photo-assisted oxygen ordering (41); in fact this process is possible only at high temperature where the oxygen ions are mobile (59).

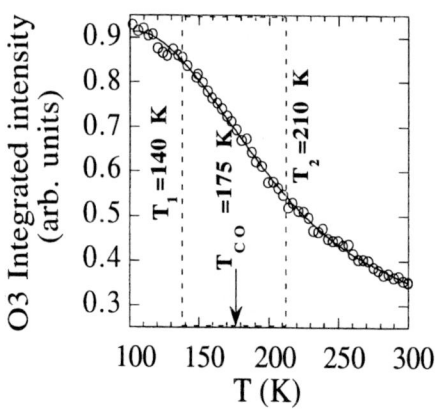

**FIGURE 5.** Effect of the temperature on the intensity of the ordered photo-doped charges with wavevector q3. The experiment shows the increase of the intensity of the diffraction peaks (3) by cooling the sample under a continuos photon flux. A crossover from 210 K to 140 K around the characteristic temperature of $T_{co}$ =180 K for the ordering of photodoped charges is found.

At low temperature, T=100K, we have studied the effect of charge ordering without oxygen motion. Indeed we have observed that below 170K, when the oxygen ions are not mobile the x-ray illumination has no effect on the 3D oxygen ordered phase O2 as shown in Fig. 4. We have observed an increase of the charge modulation (3), indicated by the increase of the intensity of diffraction satellites due to the short range, diffuse, superstructure $q_3$ that is shown in Fig. 4. The data have been collected with photon flux of 2.4 $10^{11}$ photons/sec and the characteristic dose for the increase of the $q_3$ modulation is D*=3.3 $10^{14}$ photons/mm$^2$.

The temperature dependence of this effect is shown in Fig. 5. We have plotted the variation of the diffraction intensity due to ordered photo-induced charge stripes formed under continuos illumination. This experiment provides direct evidence that, below 180K, the photo-induced electron hole pairs give an increase of the total diffraction intensity, i.e., the photo-induced holes get ordered, with the wavevector $q_3$. Therefore $q_3$ can be assigned to the charge ordering. It is relevant to mention that this wavevector has the same magnitude and direction as in Bi2212 at optimum doping. Thus the results show a similarity between the superlattice of stripes in Bi2212 and in oxygen doped $La_2CuO_4$. This assignment for the modulation $q_3$ is also supported by the temperature dependence of the intensity of the peaks as a function of temperature under a high x-ray flux shown in Fig. 5.

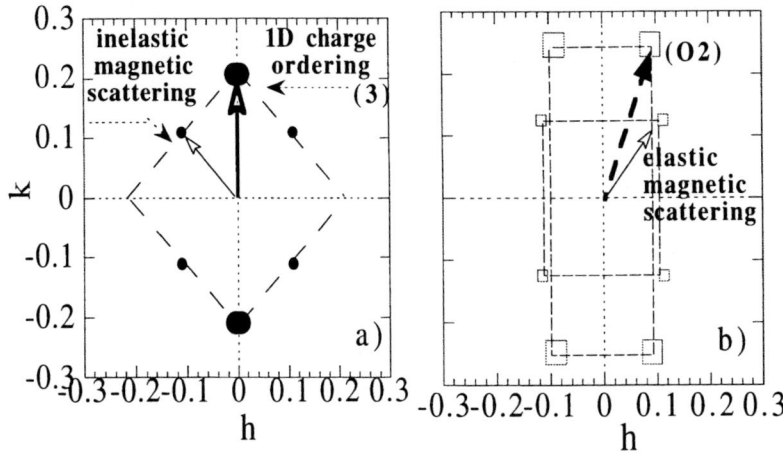

**FIGURE 6.** The projection on the ab plane of the wavevector of the short range charge ordering, superstructure (3), determined by x-ray diffraction, compared with the short range inelastic magnetic diffraction wavevector determined in ref. 64 below 70 K (panel a) and the ab projection of the wavevector of the 3D long range oxygen order modulation (2) compared with the elastic magnetic scattering wavevectors detected below 40 K (64).

In conclusion this work provides a direct measure of the stripe charge ordering in an optimally doped superconductor. We show that there is a universal charge ordering wavevector at the optimum doping in cuprate superconductors related with the characteristic density of doped holes at the optimum doping. Also, we shown that the charge stripes are diagonal at optimum doping. The dynamic and static magnetic scattering on an oxygen doped crystal, very similar to the sample studied here, with $T_C = 40$ K and with both stage 2 and stage 3 modulations, has been studied recently by neutron scattering by Lee et al. (64). We have compared in Fig. 6 the results of their work with the charge modulation vector determined in this work. The vertical and horizontal dynamic spin modulations indicated by the inelastic diffuse magnetic scattering peaks appearing below 70K are associated with the 1D charge ordering wavevector, i.e., the wavector $q_3$ of the 1D ICDW determined by the long range Coulomb interaction in the 2D metallic and superconducting phase, shown in panel (a} of Fig. 6. The magnetic and charge modulation wavevectors are not parallel, in fact the 4 magnetic peaks are at 45 degree from the 1D charge modulation $q_3$. The long range static incommensurate magnetic order observed below $T_c$ is associated with long range oxygen ordering with wavevector $q_2$ in panel (b) of Fig. 6.

## ACKNOWLEDGEMENTS

We would like to thank D. C. Johnston and P. C. Hammel for growing and characterizing the crystal, A. Lanzara, P. Radaelli, P. Bordet and A. Valletta for help in the early stage of the work at ESRF. This research has been supported by Istituto Nazionale di Fisica della Materia (INFM) in the frame of the progetto PAIS "Stripes", the Ministero dell'Università e della Ricerca Scientifica (MURST) Programmi di Ricerca Scientifica di Rilevante Interesse Nazionale coordinated by R. Ferro, e Progetto 5% Superconduttvità del Consiglio Nazionale delle Ricerche (CNR).

## REFERENCES

1. A. Bianconi, P. Castrucci, M. De Simone, A. Di Cicco, A. Fabrizi, C. Li, M. Pompa, D. Udron, A. M. Flank, P. Lagarde, and G. Calestani in Proc. of the 3rd SATT meeting, Genova Feb. 12-14, 1990 *"High Temperature Superconductivity"* edited by C. Ferdeghini and A. S. Siri, World Scientific Co, Singapore, pp. 144-148 (1990); A. Bianconi in Proc. of the *"International Conference on Superconductivity"*, held in Bangalore India Jan 10-14, 1990, edited by S. K. Joshi CNR Rao, & S. V. Subramanyam, World Scientific Singapore pag. 448 (1990).
2. A. Bianconi, S. Della Longa, M. Missori, I. Pettiti, and M. Pompa Proc. of the Int. Conf. held at Santa Fe Jan. 13-15, 1992, in *Lattice Effects in High-$T_c$ Superconductors,* edited by Y. Bar-Yam, World et al. Scientific Pub., Singapore, pp. 65-76 (1992).
3. A. Bianconi, in Proc. of the workshop, held at Erice May 6-12,1992, in *"Phase Separation in Cuprate Superconductors,* edited by K. A. Muller, and G. Benedek, World Scientific Pub., Singapore, pp. 125-138 and at pag. 352 (1993).
4. A. Bianconi, S. della Longa, M. Missori, C. Li, M. Pompa, A. Soldatov, S. Turtu, and S. Pagliuca Proc. of the int. Conf. BHTSC '92 held at Beijing in *High Temperature Superconductivity* edited by Z. Z. Gan, S. S. Xie and Z. X. Zhao World Scientific Singapore pp. 147-155 (1993).
5. A. Bianconi S. della Longa M. Missori, I. Pettiti, M. Pompa, and A. Soldatov *Jpn. J. Appl. Phys.* **32** Suppl. 32-2, 578-580 (1993).
6. A. Bianconi M. Pompa, S. Turtu, S. Pagliuca, P. Lagarde, A. M. Flank, and C. Li *Jpn. J. Appl. Phys.* **32** (Suppl. 32-2 pp. 581-583 (1993).
7. M. Missori, A. Bianconi, H. Oyanagi, and H. Yamaguchi *Physica C* **235-240,** 1245-1246 (1994).
8. A. Bianconi and M. Missori,in Proc. of the Workshop held at Cottbus, Germany Sept 4-10, 1993 *"Phase Separation in Cuprate Superconductors"* edited by E. Sigmund and K. A. Müller, Springer Verlag, Berlin-Heidelberg, pp. 272-289 (1994).
9. A. Bianconi, M. Missori, H. Oyanagi, H. Yamaguchi, D. H. Ha, Y. Nishihara (Proc. of *ETL Workshop on High Temperature Superconductors* (Tsukuba, Ibaraki, Japan Dec. 6-8 1993) *Bulletin of the Electrotechnical Laboratory* Vol. **8** n. 6 pp. 16, (1994)..
10 A. Bianconi, M. Missori, H. Oyanagi and H. Yamaguchi, Proc. of the SPIE Conference, Los Angeles, California, USA, January 24-28, 1994, in *Oxide Superconductors Physics and Nano-Engineering*, D. Pavuna Ed., SPIE, Washington, Vol. 2158 pp. 78-85, (1994).

11. A. Bianconi, M. Missori *J. Phys. I (France)* **4**, 361 (1994).
12. A. Bianconi *Sol. State Commun.* **91**, 1 (1994).
13. A. Bianconi, M. Missori *Sol. State Commun.* **91**, 287 (1994).
14. A. Bianconi. in Proc. of the Int. Conf. "Materials and Mechanisms of Superconductivity, High Temperature Superconductors" ($M^2$S-HTSC IV) Grenoble, France, July 5-9 (1994) *Physica C* **235-240**, 269-272 (1994).
15. A. Bianconi, M. Missori, H. Oyanagi, H. Yamaguchi, D. H. Ha, Y. Nishiara and S. Della Longa *Europhys. Lett.* **31**, 411-415 (1995).
16. A. Bianconi, N. L. Saini, T. Rossetti, A. Lanzara, A. Perali, M. Missori, H. Oyanagi, H. Yamaguchi, Y. Nishihara and D. H. Ha, *Phys. Rev. B* **54**, 12018 (1996).
17. A. Bianconi, M. Lusignoli, N. L. Saini, P. Bordet, Å. Kvick, P. G. Radaelli *Phys. Rev. B* **54** 4310 (1996).
18. B. W. Statt, P. C. Hammel, Z. Fisk, S. W. Cheong, F. C. Chou, D. C. Johnston and J. E. Schirber, *Phys. Rev.* B **52**, 15575 (1995).
19. H. L. Edwards, D. J. Derro, A. L. Barr, J. T. Market, and A. L. de Lozanne *Phys. Rev. Lett.* **75**, 1387 (1995).
20. A. Bianconi N. L Saini, A. Lanzara, M. Missori, T. Rossetti, H. Oyanagi, H. Yamaguchi, K. Oka, T. Ito *Phys. Rev. Lett* **76**, 3412-3415 (1996).
21. J. Tranquada J. D. Axe, N. Ichikawa, A. R. Moodenbaugh, Y. Nakamura, and S. Uchida *Phys. Rev. Lett.* **78**, 338 (1997); N. Ichikawa, S. Uchida, J. M. Tranquada, T. Niemöller, P. M. Gehring, S. H. Lee, J. R. Schneider Cond -mat/9910037 4 Oct 1999
22. A. W. Hunt, P. M. Singer, K. R. Thurber, and T. Imai *Phys. Rev. Lett.* **82**, 4300 (1999) and P. M. Singer, A. W. Hunt, and T. Imai *Phys. Rev. B* (in press).
23. N. L. Saini, J. Avila, M. C. Asensio, S. Tajima, G. D. Gu, N. Koshizuka, A. Valletta, A. Lanzara, A. Bianconi *Z. Phys.* B **104**, 703-706 (1997) ).
24. N. L. Saini, J. Avila, A. Bianconi, A. Lanzara, M. C. Asenzio, S. Tajima, G. D. Gu, N. Koshizuka *Phys. Rev. Lett.* **79**, 3467-3470 (1997); *Phys. Rev. Lett.* **82**, 2619 (1999).
25. B. Tanar, and D.M. Ceperly *Phys. Rev. B* **39**, 5005 (1989).
26. O. V. Dolgov, D. A. Kirzhnitz, E. G. Maksimov in *Superconductivity, Superdiamagnetism, Superfluidity* Edited by V. L. Ginzburg, Mir Publishers Moscow, pp. 18-68 (1987)..
27. Y. Takada *J. Phys. Soc. Japan* **45**, 786-794 (1978); *J. Phys. Soc. Japan* **61**, 3849 (1992).
28. Y. J. Uemura et al. *Phys. Rev. Letters* **62**, 2317 (1989); *Phys. Rev. B* **38**, 909 (1988); *Phys. Rev. Lett.* **66**, 2665 (1991);*Nature* **364**, 605 (1993).
29. H. Keller, et al. *Physica* (Amsterdam) **185-189C**, 1089 (1991); T. Schneider & H. Keller *Phys. Rev. Lett.* **69**, 3374 (1992); Ch. Niedermayer et al. *Phys. Rev. Lett.* **71**, 1764 (1993).
30. U. Löw, V. J. Emery, K. Fabricius, and S. A. Kivelson *Phys. Rev. Lett.* **72**, 1918 (1994).
31. C. Castellani, C. Di Castro and M. Grilli *Phys. Rev. Lett.* **75**, 4650 (1995).
32. A. Bianconi, European Patent N. 0733271 "High $T_c$ superconductors made by metal heterostuctures at the atomic limit" ; (priority date 7 Dec 1993), published in *European Patent Bulletin* **98/22**, (May 27 1998).
33. A. Bianconi *Sol. State Commun.* **89**, 933 (1994).
34. A. Bianconi, M. Missori, N. L. Saini, H. Oyanagi, H. Yamaguchi, D. H. Ha, and Y. Nishiara (Proc. of the workshop on high Tc superconductivity, Miami Jan 5-11 1995) *Journal of Superconductivity* **8** 545-548 (1995).
35. A. Perali, A. Bianconi, A. Lanzara, and N. L. Saini *Solid State Communications* **100**, 181-186 (1966).
36. A. Valletta, G. Bardelloni, M. Brunelli, A. Lanzara, A. Bianconi and N. L. Saini *J. Superconductivity* **10**, 383-387 (1997).
37. A. Bianconi, A. Valletta, A. Perali, and N. L. Saini *Physica C* **296**, 269-280 (1998).
38. Proc. of the First Inter. Conference on *Stripes, Lattice instabilities and high Tc Superconductivity* edited by A. Bianconi and N. L. Saini, special issue of *J. of Superconductivity* Vol. **10**, No. 4 August (1997).
39. Proc. of the Second Inter. Conference on Strepes and High Tc superconductivity published in *"Stripes and Related Phenomena"* edited by A. Bianconi and N. L. Saini, Plenum Press, New York (1999).

40. G. Yu, C.H. Lee, and A. J. Heeger in Proc. of the Erice workshop, May 6-12,1992, *"Phase Separation in Cuprate Superconductors,* edited by K. A. Muller, and G. Benedek, World Scientific Pub., Singapore pp. 17-35 (1993).
41. O. Osquiguil, M. Maenhoudt, B. Bruynseraede, D. Lederman, and I. K. Schuller *Phys. Rev. B* **49** 3675-3678 (1994).
42. V. I. Kudinov, I. L. Chaplygin, A. I. Kirilyuk, and N. M. Kreines, R. Laiho, E. Lähderanta, and C. Ayache *Phys. Rev.* **B 47**, 9017 –9028 (1993)
43. P. C. Hammel, et al. in Proc. of the Erice workshop, May 6-12,1992 *"Phase Separation in Cuprate Superconductors,* edited by K. A. Muller, and G. Benedek, World Scientific Pub., Singapore pp. 139-157 (1993).
44. D. C. Johnston et al. in Proc. of the Workshop held at Cottbus, Germany Sept 4-10, 1993 *"Phase Separation in Cuprate Superconductors"* edited by E. Sigmund and K. A. Müller, Springer Verlag, Berlin-Heidelberg, pp. 82-100 (1994).
45. J. D. Jorgensen, B. Dabrowski, S. Pei, D. G. Hinks, L. Soderholm, B. Morosin, E. L. Venturini and D. S. Ginley, *Phys. Rev. B* **38** 11337 (1988); P. G. Radaelli, J. D. Jorgensen, R. Kleb, B. A. Hunter, F. C. Chou and D. C. Johnston, *Phys. Rev.* **B 49**, 6239 (1994); B. Dabrowski, Z. Wang, K. Rogacki, J. D. Jorgensen, R. L. Hitterman, J. L. Wagner, B. A. Hunter, P. G. Radaelli and D. G. Hinks, *Phys. Rev. Lett.* **76**, 1348 (1996).
46. P. C. Hammel, A. P. Reyes, S. -W Cheong, Z. Fisk and J. E. Schriber *Phys. Rev. Letters* **71**, 440 (1993); P. C. Hammel, A. P. Reyes, Z. Fisk, M. Takigawa, J. D. Thompson, R. H. Heffner, S. Cheong, and J. E. Schirber *Phys. Rev.* **B 42**, 6781 (1990); E. T. Ahrens, A. P. Reyes, P. C. Hammel, J. D. Thompson, P. C. Canfield, Z. Fisk, J. E. Schirber, *Physica C* **212**, 317 (1993).
47. F. C. Chou, J. H. Cho, D. C. Johnston, *Physica* C **197**, 303 (1992).
48. F. C. Chou, D. C. Johnston *Phys. Rev. B* **54**, 572-583 (1996).
49. C. Chaillout, J. Chenavas, S. W. Cheong, Z. Fisk, M. Marezio, B. Morosin and E. Schirber, *Physica* C,**170**, 87 (1990).
50. J. C. Grenier, N. Lagueyte, A. Wattiaux, J. P. Doumerc, P. Dordor, J. Etourneau, M. Pouchard, J. B. Goodenough and J. S. Zhou, *Physica* C **202**, 209 (1992); A. Demourges, F. Weill, B. Darriet, A. Wattiaux, J. C. Grenier, P. Gravereau, and M. Pouchard, *J. Sol. State Chem.* **106**, 330 (1993).
51. M. Itoh, T. Huang, J. D. Yu, Y. Inaguna and T. Nakamura, *Phys. Rev. B* **51**, 1286 (1995); T. Kyomen, M. Oguni, M. Itoh and J. D. Yu, *Phys. Rev. B* **51**, 3181 (1995).
52. R. K. Kremer, E. Sigmund, V. Hizhnyakov, F. Hentsch, A. Simon, K. A. Müller and M. Mehring, *Z. Phys. B* 86, **319** (1992); R. K. Kremer, V. Hizhnyakov, E. Sigmund, A. Simon, and K. A. Müller, *Z. Phys. B* **91**, 169 (1993)
53. N. Lagueyte, F. Weill, A. Wattiaux and J. C. Grenier, *Eur. J. Solid State Inorg. Chem.* **30**, 859-869 (1993).
54. P. G. Radaelli, J. D. Jorgensen, A. J. Schultz, B. A. Hunter, J. L. Wagner, F. C. Chou and D. C. Johnston, *Phys. Rev. B* **48**, 499 (1993).
55. X. Xiong, P. Wochner, S. C. Moss, Y. Cao, K. Koga and N. Fujita, *Phys. Rev. Letters* **76**, 2997 (1996).
56. A. Lanzara, N. L. Saini, A. Bianconi, J. L. Haemann,, Y. Soldo, F. C. Chou, and DC. Johnston *Phys. Rev. B* **55**, 9120-9124 (1997)
57. N. L. Saini, A. Lanzara, A. Bianconi, F. C. Chou, and D. C. Johnston, *Journal of Physical Society of Japan* **67**, 16-19 (1998)
58. X. L. Dong, Z. F. Dong, B. R. Zhao, Z. X. Zhao, X. F. Duan, L. -M. Peng, W. W. Huang, B. Xu, Y. Z. Zhang, S. Q. Guo, L. H. Zhao and L. Li *Phys. Rev. Letters* **80**, 2701-2704 (1998).
59. F. Cordero and R. Cantelli *Physica C* **312**, 213-224 (1999).
60. P. Blakeslee, R. J. Birgenau, F. C. Chou, R. Christianson, M. A. Kastner, Y. S. Lee, and B. O. Wells *Phys. Rev. B* **57**, 13915-13921 (1998).
61. P. C. Hammel, B. W. Statt, R. L. Martin, F, C. Chou, D. C. Johnston, and S. -W. Cheong *Phys. Rev. B* ( 1998).
62. Z. G. Li, H. H. Feng, Z. Y. Yang, A. Hamed, S. T. Ting, and P. H. Hor *Phys. Rev. Letters* **77**, 5413-5416 (1996).
63. B. O. Wells, Y. S. Lee, M. A. Kastner, R. J. Christianson, R. J. Birgeneau, K. Yamada, Y. Endoh, and G. Shirane *Science* **277**, 1067-1071 (1997).

64. Y. S. Lee, R. J. Birgeneau, M. A. Kastner, Y. Endoh, S. Wakimoto, K. Yamada, R. W. Erwin, S. -H. Lee and G. Shirane Phys. Rev. B **60**, 3643-3654 (1999).
65. J. M. Tranquada, D. J. Buttrey, V. Sachan, and J. E. Lorenzo *Phys. Rev. Lett.* **73**, 1003-1006 (1994); J. M. Tranquada, P. Wochner, and D. J. Buttrey *Phys. Rev. Lett.* **79**, 2133-2136 (1997).
66. W. M. Gibbons, P. J. Shannon, S. –T. Sun, and B. Swetlin *Nature* **351**, 49-50 (1991).

# Charge Transfer at Surfaces on Femtosecond Timescales: New Information from Electron Spectroscopies

D. Menzel and W. Wurth

*Physik-Department E20, Technische Universität München, D-85748 Garching, Germany*

Charge transfer (CT) between an adsorbed atom or molecule and its substrate is of direct importance for the understanding of photochemical surface processes and more generally of the adsorbate-substrate coupling. A direct measurement of its timescales is difficult as it is extremely fast (from less than a fs to some or some tens of fs, as deduced from indirect evidence). New methods have become possible with the availability of synchrotron light at very high resolution and intensity, which utilize core hole excitations to determine timescales for fast electron transfer processes. These techniques are based on the use of the core hole lifetime as an internal time standard. This approach can be applied in the time domain as well as in the frequency domain.

This survey describes mainly the former approach which consists of the study of the decay spectra after resonant excitation of an adsorbate core hole, induced with excitation band widths below the core hole lifetime width (so called Auger resonant Raman conditions). Tuning through the resonance allows one to separate those parts in the decay spectra which correspond to decay before, vs. after, transfer of the locally excited electron from the adsorbate into the substrate. The ratio of intensities of these two types of spectra is connected to the ratio of the timescales of CT and of core hole decay via simple rate equations. Since the core hole lifetime is known, the CT time can be calculated from this spectral branching. For adsorbate resonances between the Fermi and the vacuum level of the substrate, values from below 1 fs (strong chemisorption), via the range of some fs to some 10 fs (physisorption with graded coupling), to above 50 fs (decoupled condensate) have been found; their dependence on the actual excitation energy can also be determined.

The CT times of electrons screening a core hole on a chemisorbate - a process which happens at the Fermi level - on the other hand are even shorter, namely fast even on the timescale of sudden photoionization. This is demonstrated by the possibility to measure vibrationally resolved XPS peaks even for chemisorbed species (indicating the dominance of a final state of core ionisation which is made neutral by CT). The observed total and partial suppression of the PCI (post collision interaction) effect for chemisorbates and physisorbates, respectively, is consistent with such timescales of screening.

## INTRODUCTION

The transfer of charge between an adsorbate and a substrate is at the heart of their interaction. In the ground state the hopping of electrons between them determines the strength of their interaction which expresses itself in the adsorbate bond. As an equilibrium property, it cannot be measured directly but only be assessed theoretically; it can be quantified as the hopping matrix element of the electron in the bond.

Charge transfer in nonequilibrium cases can be observed directly and, given suitable methods, their time scale can be measured. Since these processes are the excited state equivalent for electron hopping in the ground state, they also give a feeling for this basis of the surface chemical bond. They are of direct decisive importance for surface photochemistry, i.e. the modification of photochemical processes by the coupling of the excited species to a surface. Semiquantitative information on these time scales has been derived quite early from the processes usually referred to as DIET (Desorption Induced by Electronic Transitions), where strong modifications of electronically induced bond breaking is found by such surface coupling of the species concerned (1). On a metal surface, one usually observes strong selective quenching in particular of ionic fragmentation channels which is most distinct for single excitations (quenching more probable than survival by factors $10^3$ or more) and becomes weaker the more complex - and therefore more localized by correlations - the primary excitation is. In the frame of the generalized MGR (Menzel-Gomer Redhead) model (2) this is due to fast charge and energy exchange between adsorbate and substrate following the primary localized excitation. To explain the observed order of magnitude of these modifications, charge transfer (CT) times in the range of 1 fs or below must be assumed for chemisorbed species on metal surfaces. However, exact CT times cannot be derived in this way. More direct information is therefore very desirable.

Two principal possibilities exist: the measurements can be done either in the frequency or in the time domain. As to the first, lifetimes of holes created in occupied states of the combined system, or of excited electrons in the empty range, can be extracted from the respective widths of the corresponding spectral lines, provided inhomogeneous broadening and the contributions of coupling to other degrees of freedom (such as vibrations or electron-hole excitations) can be excluded or accurately corrected for. Since this is usually not possible reliably, the extracted widths are upper limits, or the derived times lower bounds. However, there are cases when one can draw semiquantitative information from spectral widths and shapes, and we will come back below to recent examples of this type.

Measurements in the time domain are superior in principle, and especially laser pump-probe techniques appear attractive. The problem is that for the interesting types of systems such as strongly or intermediately coupled chemisorbates or even weakly coupled physisorbates, the respective times are too short to be accessible to presently available technology (for an informative review of recent work, see (3)). While progress is being made towards shorter times (4), other methods with faster time response are very desirable. In the vein of this conference which concentrates on the applications of X-ray photons, we will survey in the following what knowledge about these time scales can be obtained by X-ray induced core electron spectroscopies. The particular usefulness of these types of spectroscopies is that they contain an internal time standard of suitable magnitude for comparison which is the core hole lifetime, so that we apply what has aptly been called the core hole clock (5). On the other hand, the existence of the core hole - which at first glance seems to change the situation drastically compared to the ground state - can be readily accounted for by application of the equivalent core approximation. It will be shown that one can use this clock in the frequency domain, as well as directly in the time domain, though for the latter in a somewhat different way from the usual pump-probe experiment.

# INFORMATION IN THE TIME DOMAIN

The principle of this measurement (6,7) is simple (Fig. 1): An electron is taken from a core level of an adsorbate atom to an empty bound state on this adsorbate by resonant absorption of a suitable photon. If the excited electron stays localized on the adsorbate for the core hole lifetime, then the final states of core hole decay are one-hole and two-hole/one-electron states, and the decay spectra consist of participant and spectator lines. If the excited electron has been transferred to the substrate before the core hole is filled again, two-hole decay spectra result which are equal to the normal Auger spectra.

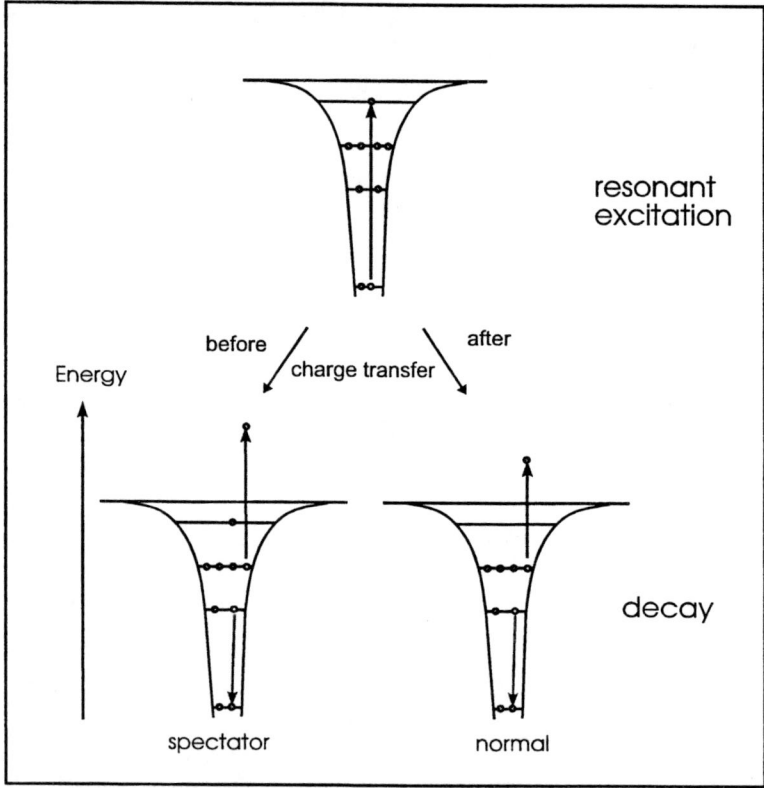

**FIGURE 1:** A schematic depiction of the resonant core hole excitation and decay processes for an adsorbate on a metal substrate. The kinetic energy of the outgoing Auger electron in a spectator decay process (before CT) is higher than in a normal Auger process (after CT).

The two types of spectra are shifted against each other because of the Coulomb interaction of the final state holes with the excited electron. (To be sure, in both cases the interaction with the substrate leads to screening, but there is still a clear difference between the results of the two decay modes. Also, the excited adsorbate state is not a purely atomic or molecular state but may contain admixture of substrate states; it is still a localized excited state which can decay into a delocalized substrate state by CT). In a simple rate approach ((5-7); see below) the ratio of the intensities of those two types of decay spectra gives the relation between the CT time and the core hole lifetime; as the latter is known, the former can be calculated.

A basically equivalent type of experiment has been done already some time ago by comparing the decay spectra resulting after resonant, and after continuum excitation, respectively (in spectroscopic terms: decay of NEXAFS and XPS final states). It has been found that for a chemisorbed species on a metal surface the two types of spectra are identical (8,9) which must mean that CT of the excited electron is much faster than core hole decay; a quantification of the ratio was not possible. For a physisorbed species, both types of spectra were found superimposed, meaning that CT and core hole decay have comparable time scales (10,11). However, the separation of the spectral contributions is not easy even here because of the many features in both types of decay spectra which often strongly overlap. In both cases of similar and extremely different time scales the application of Auger resonant Raman conditions (exciting photon band width narrower than the core hole lifetime width which is contained in the resonance width, see above) brings a definite improvement. Under these conditions the excitation takes longer than the decay (12), so that a coherent one-step picture has to be used for the total excitation/decay process, provided there are no coupled processes which destroy the coherence - like the CT process of interest here. By tuning the narrow excitation band through the resonance and observing the resulting decay spectra, the ratio of the integrated intensities of "Raman-type" and of "Auger-type" decays can be mapped out: for the first the electron kinetic energies increase in parallel with the increase of the photon energy (i.e. the binding energy stays constant), while for the second the kinetic energies are independent of photon energy. The separation of the two channels is made easy by this different reaction to changes of the photon energy. Here the Raman-type spectra indicate the decay channel for which the excited electron is still localized on the adsorbate at the instance of core hole decay; the overall process is then coherent as in the isolated molecule, and energy conservation demands the kinetic energy of the decay electrons to increase in the same way as the exciting photon energy. The Auger-type spectra, on the other hand, indicate decays for which the excited electron has been transferred into the substrate *before* core hole decay, so that energy is *not* conserved in the observed channel (the transferred electron takes with it any surplus energy provided by tuning), and the kinetic energy of decay electrons stays constant. Because transfer of the excited electron leads to loss of the primary energy information (the electron in the substrate cannot be observed), it breaks the coherence of the overall process. The two types of spectra can now be easily separated because of their different behavior with photon energy, constant difference of photon energy and electron kinetic energy - i.e. constant binding energy - for "core hole decay before CT" (the socalled Raman fraction), and constant kinetic energy for the Auger fraction (CT faster than core hole decay). A more detailed analysis (13 ) shows that the distinction is not just in the preservation of the energy information, but also that of the excitation phase, as generally in the features of the Auger resonant Raman effect (12); nevertheless the ratio of the two time constants is not influenced by coherent detuning effects. The important aspect in the present context is that this different behavior of the spectra upon tuning the narrow photon bandwidth through the core-to-bound resonance makes it easy to unequivocally separate the two contributions and determine their intensity ratio. Since the two decay modes correspond to two competing channels for the disappearance of the primary excited state, both of which have simple exponential decay characteristics, the ratio of the two decay rate constants, $k_i$, (or the inverse ratio of the decay time constants, $\tau_i$) is proportional to

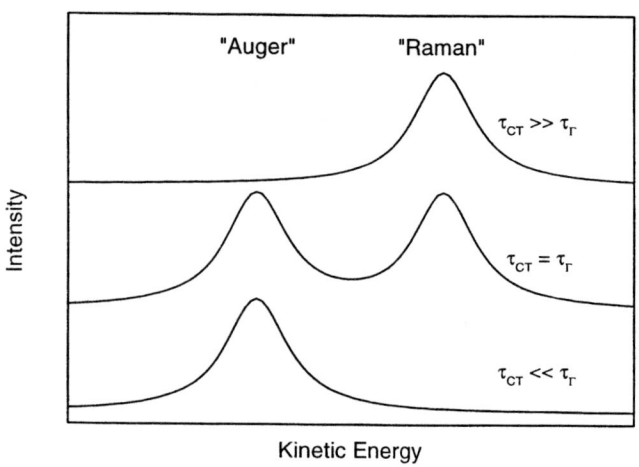

**FIGURE 2:** A schematic depiction of the intensity ratios between the two decay channels, the Raman and the Auger channel, for extreme ratios between the time constants governing the two underlying processes. For details, see the text.

the ratio of the total intensities, $I_i$, accumulated in each of the two channels after all excitations have decayed: $(I_{Raman}/I_{Auger}) = (k_{Raman}/k_{Auger}) = (\tau_{Auger}/\tau_{Raman})$; the somewhat different formula found in the cited work (5-7) is due to their use of branching ratios (e.g. the "Raman fraction" $f_R = I_{Raman}/(I_{Raman} + I_{Auger})$), instead of direct intensity ratios. $\tau_{Raman}$ is identical to the intrinsic core hole decay time, $\tau_\Gamma$, because if no other processes are possible, this is the limiting time constant. Since this time constant varies little for slightly different valence configurations of the same core hole, it can be assumed constant and taken from literature. On the other hand the fraction of Auger decays is connected to the transfer of the excited electron to the bulk, so that $\tau_{Auger}$ is equal to the CT time, $\tau_{CT}$. Thus the latter can be calculated from the measured intensity ratios between Raman- and Auger-type spectra, using either their integrals or (if their shapes are unchanged by tuning) the amplitude of a representative peak. In this way one can quantify the CT times, even for chemisorbates, as sketched above. One can

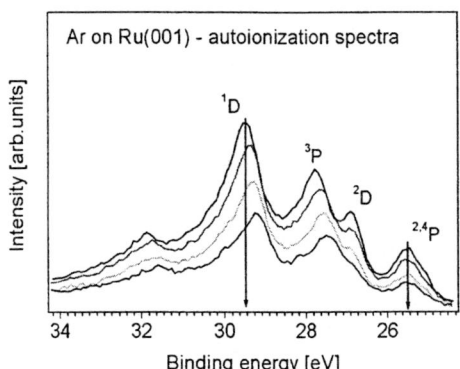

**FIGURE 3:** Depicted are Ar $L_3M_{2,3}M_{2,3}$ decay spectra obtained after resonant excitation at the Ar $2p_{3/2}^{-1}4s^{+1}$ resonance, directly on resonance and in steps of 0.1 eV to lower photon energies, respectively. The arrows emphasize the different reaction of the peaks to changes of the photon energy: the peaks $^1D$, $^3P$ belong to normal Auger channels (decay of ionic core excitations), while $^2D$, and $^{2,4}P$ are Raman channels (decay of neutral core excitations).

then make comparative measurements for adsorbates with graded interaction strength, and even study the dependence of the CT time on the exact energy of the excitation, i.e. on the energetic position of the excited electron with respect to the Fermi level.

For an example see Fig. 3 which shows a few of the decay spectra measured in this way tuning through the $2p_{3/2} \gg 4s$ resonance of Ar adsorbed on the Ru(001) surface. Qualitatively, the different behavior of the peaks with changing photon energy can be discerned. From an intensity analysis of the spectra we determine a Raman fraction (see above) of about 20 %. Comparison to the lifetime of the $2p_{3/2}$ hole of about 6 fs (assumed to be unchanged from the atomic value) leads to the CT time. Interestingly, the Raman to Auger fraction (and therefore the CT time) varies with photon energy, i.e. with the position within the resonance, between 1 and 2 fs (14). Furthermore, very different CT times are obtained for Ar with graded coupling to the Ru(001) surface; these can be produced by different spacer layers (chemisorbed O or CO, physisorbed Xe, or Xe+Ar) (7). For three different examples Fig. 4 shows that the Raman fraction can vary from quite small (strongest coupling, for Ar directly on the metal surface) to totally predominant, so that only a lower bound on the CT time can be given.

On the other extreme is the spectral ratio for excitation of a chemisorbed species, as already indicated. For CO on Ru(001) as example (15), only a small part of the Raman-type spectrum is observable because a large part is covered up by the overwhelming contribution of the Auger-type spectrum. Nevertheless it is possible to extract a CT

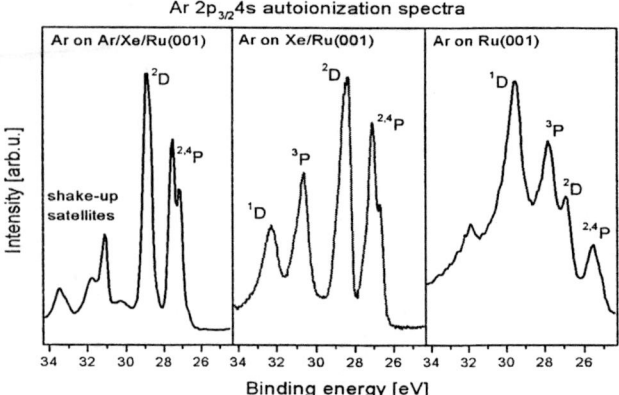

**FIGURE 4:** Autoionisation spectra after resonant Ar $2p_{3/2} \gg 4s$ excitation for three different adsorption situations of Ar. The right spectrum represents Ar adsorbed on the clean Ru(001) surface, the spectrum in the middle shows the result for Ar adsorbed on Xe/Ru(001), while the left spectrum represents the case of Ar on top of two monoatomic rare gas layers (Ar/Xe on Ru(001). The very different weighting of the two types of peaks is clearly seen (Peaks are labelled as in fig. 3. Compare, e.g. the $^1$D Auger peaks).

time of the order of 0.6 fs for CT of an electron excited from the O1s to the $2\pi$ level of the adsorbate. While the absolute magnitude of the CT time in this case is not very accurate due to the overlap of the spectra, its change with the excitation energy can be measured with better accuracy (15). The observed trends are qualitatively similar to the Ar case (14).

Obviously, the total range of the CT time which can be observed depends on the relative intensity of the smaller spectral contribution which can still be seen besides the larger one. While this depends to some extent on the shapes of the spectra, a conservative estimate (7) is based on a range of detectable intensity ratios - and thus of the ratios of time constants - from 10:1 to 1:10, i.e. a total range of 100. A compilation of the measured values is given in Fig. 5. To first order core-excited Ar can be approximated by a K atom in the location of the adsorbed Ar, and the CT time is then the time for tunneling of the K 4s electron into the metal (16,17). As discussed in ref. (7), the main contribution for the changes of this tunneling time by spacer layers is most likely the change of the tunneling barrier's width and height, although contributions from electronic states in the spacer layer cannot always be excluded (the resonance positions relative to the Fermi level are different for the various situations). Furthermore, the tunneling rate is also determined by the substrate states into which the excited electron has to tunnel, i.e. the empty surface band structure. This must also be the decisive factor in the mentioned changes of CT time when tuning through the resonances. Detailed theoretical analyses utilizing several approaches are under way.

## Timescales for charge transfer relaxation

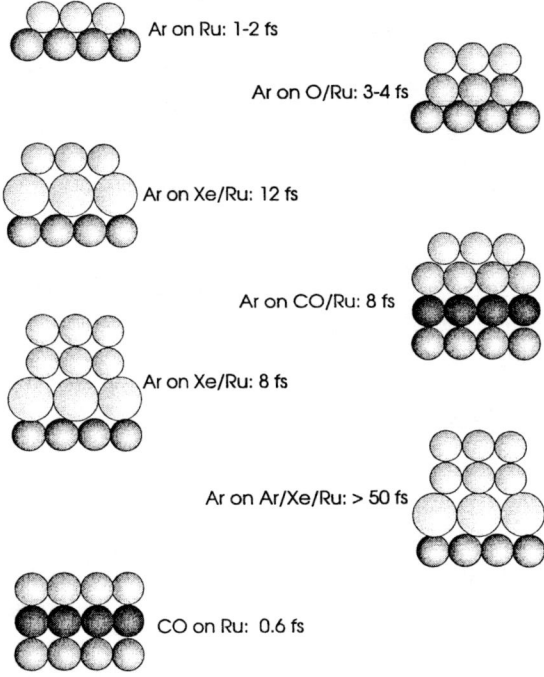

**FIGURE 5:** In the figure the results for the electron transfer times obtained for the different systems investigated in our study are summarized. (After ref. (7)).

A particularly interesting situation exists for the case of two monolayers of Ar on a single layer of Xe. Here the excitation can be placed into the top Ar or the intermediate Ar layer, because the corresponding resonances are energetically separable. The Ar in the top layer is so strongly decoupled that only Raman contributions can be measured in the decay spectra which translates into a CT time larger than 50 fs. The Ar directly on the Xe, on the other hand, has a CT time (8 fs) which is shorter than that of Ar on Xe *without* a second Ar layer above it (12 fs). This may be due to the different resonance positions, but there could also be a contribution from the squeezing of the Ar wave functions towards the substrate by the Ar "lid" on it. It is interesting to mention that CO and $N_2$ on a Xe spacer layer show pure Raman behavior (18) (and in fact retain the detuning behavior of the free molecules (19)); however, here this is not due to decoupling, but to the fact that for those layers the energetic position of the resonantly excited electron is *below* the Fermi level, so that no CT is possible.

Another comparison can be drawn to recent measurements of the CT times from image charge states in front of the surface, and the influence of decoupling spacer layers (20). While the absolute scale of the CT times is much longer here, which makes accurate laser pump-probe measurements possible, and which is understandable from the large difference of the corresponding wave functions and their penetration into the bulk from those of an Ar atom (or rather of a K 4s electron), the change introduced by a Xe spacer layer leads to a comparable elongation *factor*. This is consistent with the assumption that in both cases the attenuation of the tails of the electron wave functions extending into the substrate plays the main role.

Summarizing this section, we believe that core excitation/decay measurements under resonant Raman conditions have opened a new avenue to detailed measurements of CT times at surfaces. To be sure, laser pump-probe measurements, which in the near future are expected to be extended into the time range investigated here, are more flexible due to the tunability of the probe time. However the principle described here has the advantage that it can easily be extended into even shorter time ranges by going to shorter-living core holes, of which there are plenty. This is planned for the future.

## INFORMATION IN THE FREQUENCY DOMAIN

The clearest use of the frequency domain is constituted by the observation of broadening of spectral features well known for isolated molecules which is induced by their adsorption, i.e. by their coupling to a substrate by the surface bond. Absorption spectra for core-to-bound excitations ("NEXAFS") are the simplest example. Here a free atom has a spectral width which is determined by the lifetime of the core hole; for a free molecule vibrationally resolved spectra (determined by the corresponding Franck-Condon factors) can be observed and resolved under suitable conditions. Physisorbed atoms (e.g. of the rare gases) exhibit broadened NEXAFS features which suggest considerably shortened lifetimes of the excited electron; and for chemisorbed molecules no vibrational resolution can be achieved in NEXAFS spectra. E.g., Ar atoms on a Ru(001) surface possess $2p_{3/2} \gg 4s$ absorption features which are 300 meV broader than the gas phase line (110 meV (21)); and the $C1s \gg 2\pi$ excitation in CO, which for the isolated molecule is nicely vibrationally resolved with an individual line width of 80 meV (21) has an unstructured absorption peak 0.8 eV wide when adsorbed on Ru(001). If the additional broadening could be assigned to the limited lifetime of the excited electron (3.5 and 1.5 eV above $E_F$, respectively), then the lifetimes would be about 1.5 fs and 0.8 fs for the two cases (compare the values given above from time domain measurements), respectively. However, while in those two cases the comparison is quite good, such a conclusion is not safe at all, since it only works for absorption peaks which are dominated by a single vibronic transition. Also, inhomogeneous broadening (even the most homogeneous layers may contain species in slightly different environments, like on steps or other surface defects), and contributions of coupling to other - in particular low energy - degrees of freedom, like the frustrated translations of the adsorbate, or electron-hole excitations in the substrate, cannot be excluded. The derived values can therefore only be taken as minimum lifetimes. Comparison with the directly determined CT times (see above) does result in agreement for some cases as just mentioned before for Ar directly on Ru(001), or C1s excitation of chemisorbed CO (but not for the latter´s O1s excitation because of its broad vibrational envelope); but this cannot be taken for granted and can only be stated post factum.

The same is the case for photoelectron and Auger electron lines. Valence spectra are usually broadened considerably by adsorption, compared to the free particle, for (screened; see below) one-hole (UPS) as well as for two-hole (AES) states. For UPS lines band structure effects add width in addition to the inhomogeneous and low energy mode contributions. On the other hand, very recent results have shown that XPS lines from core ionization of strongly chemisorbed molecules (e.g., C1s of CO on Ni(100) and Ru(001), (22,23); see fig. 6) can be vibrationally resolved and contain rather small

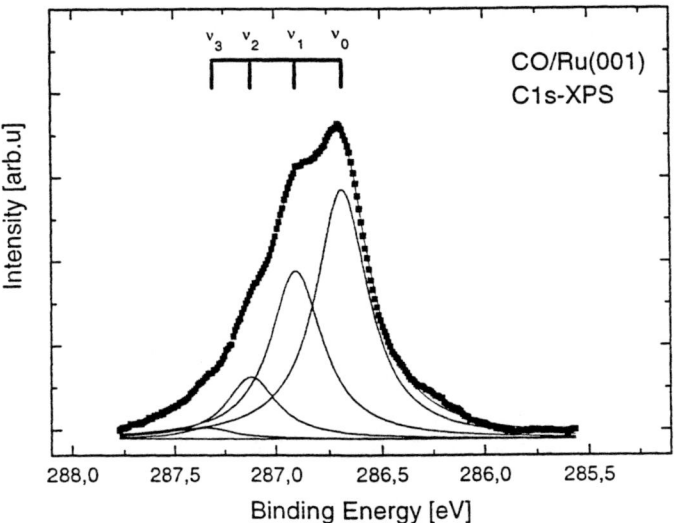

**FIGURE 6:** C1s XPS peak of adsorbed CO on Ru(001), after ref. (23). The vibrational components into which the peak can be resolved are indicated. The FWHM of the individual peaks is about 280 meV.

contributions of additional Lorentz broadening compared to the free molecule, which can in fact be explained by coupling to the low energy frustrated translation of the adsorbed molecule against the substrate. We have concluded from this finding that the overwhelmingly predominant final state of the continuum core ionisation of the adsorbate must be the fully charge transfer-screened state, i.e. even for core ionisation in the sudden limit a neutral adsorbate results. If this were not so, a contribution from electron-hole excitations by coupling of the charged molecule to the substrate would broaden the XPS line. Furthermore, the vibrational structure found for the adsorbate compares best with the NEXAFS, not the XPS, peaks of the free molecule, stressing again the neutral character of the XPS final state of the adsorbate. Thus the CT at the

Fermi level, i.e. in the bonding range of interaction, must be fast even on the time scale of sudden photoemission. While no number can be given, this makes CT at the Fermi level considerably faster than in the range of the excitation resonances discussed above, a few eV above $E_F$. Interestingly, the weighting of the various resolvable vibrational states in XPS of chemisorbed CO changes drastically with the photon energy decreasing towards threshold, and appears to approach fully vibrationally adiabatic behavior at threshold (23). This behavior does not exist in the free molecule and must therefore be caused by the coupling to the metal. We believe that it is due to predominance of multiple excitations involving metal states in the energy range in question which can be quenched into the metal bulk before their energy can be fed back into emission of a photoelectron (as would happen in an isolated species). The details are not yet really explained; the behavior poses a challenge to the understanding of dynamical screening and charge transfer at surfaces.

Another interesting threshold behavior in adsorbates that can be connected to dynamical screening is the influence of surface coupling on post-collision interaction (PCI). In free atoms and molecules, the observed shifts and broadenings close to threshold can be explained by the influence of the sudden change from a singly to a doubly charged species upon core hole decay, on the energy of the leaving photoelectron. It has recently been shown that for adsorbed Ar on Ru(001) (24) the PCI

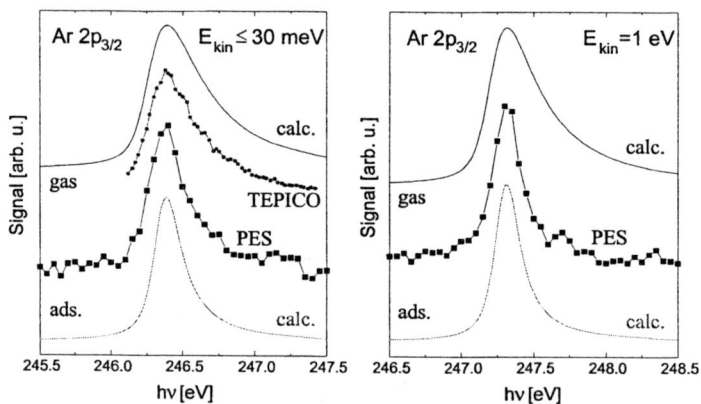

**FIGURE 7:** Comparison of calculated and experimentally obtained post-collision interaction peak profiles for isolated and physisorbed Ar atoms, for two electron kinetic energies (after ref. (24). The adsorbate peaks are seen to be much narrower than those of the atom; this can be traced back to screening of the adsorbate, as borne out by the calculational results obtained on this basis.

effect is strongly modified, although still observable. The found behavior can be explained in a classical model by partial screening of the created additional hole as seen by the leaving photoelectron (see fig. 7). For chemisorbed species such as CO on the same surface, no PCI can be found at all as one would expect from the CT times discussed so far. However it should be pointed out that it would be very difficult to observe since the photoemission cross section has been found to decrease strongly close to threshold - a behavior which appears to be closely connected to the other effects discussed above, in particular the adiabatic behavior of vibrations close to threshold. It can be seen that this field of research is in very active development.

## CONCLUSIONS

This short survey has the purpose to call attention to the new information on extremely fast charge (and energy) transfer processes between adsorbates and metallic substrates which have become accessible utilizing the new potential of high brightness SR sources and applying them to various core excitation spectroscopies. The most interesting possibility in our opinion is the application of Auger resonant Raman conditions to core excitation/decay processes of adsorbates, since this makes quantitative measurement of extremely fast (from below 1 to some 10 fs) CT processes possible. But also other, less quantitative methods which use lineshape analysis of spectra for conclusions give some - in part unexpected - information which poses challenges to the detailed understanding of dynamical screening and charge transfer at surfaces. One very interesting aspect of the utilization of the core hole clock is that it is easy to go to even shorter timescales by using shorter living core holes. We believe that there is a bright future for such measurements.

## ACKNOWLEDGEMENTS

We gratefully acknowledge numerous, very helpful discussions with Peter Feulner. He and his coworkers Bernd Kassühlke, Ralf Romberg, and Peter Averkamp provided us with the PCI results. Christian Keller, and Markus Stichler, two of our Ph.D. students, together with the help of Giovanni Comelli, Friedrich Esch and Silvano Lizzit from the Sincrotrone Trieste were essential in performing the Auger resonant Raman experiments. Alexander Föhlisch, Olov Karis, Anders Nilsson, and Nils Mårtensson are to be credited for most of the XPS results; we thank them also for very lively and helpful discussions. Very fruitful discussions with Zbigniew W. Gortel are also acknowledged.

Our part of the reported work has been supported financially by the Deutsche Forschungsgemeinschaft under project Me266/22-2 and SFB 338, and by the EC Large Scale Installation program, project ERBFMGETCT950022.

# REFERENCES

(1) See, e.g., Feulner, P., and Menzel, D., in "Laser Spectroscopy and Photochemistry on Metal Surfaces", H.-L. Dai and W. Ho, eds., World Scientific, Singapore 1995, p. 627-684; and references therein.
(2) Menzel, D., and Gomer, R., J. Chem. Phys. **40**, 1164-5 and **41**, 3311-3328 (1964); Redhead, P., Can.J. Phys. **42**, 886-94 (1964);
Menzel, D., Nucl. Instrum. Methods B **101**, 1-10 (1995); and references therein.
(3) Wolf, M., Surf. Sci. **377-379,** 343-349 (1997), and references therein.
(4) Petek, H., and Ogawa, S., Progr. Surf. Sci. **56,** 239-310 (1997).
(5) Björneholm, O., et al., Phys. Rev. Lett **68,** 1892-1895 (1992); Ohno, M., Phys. Rev. B **50,** 2566 (1994).
(6) Karis, O., et al, Phys. Rev. Lett. **76**, 1380 (1996); Wurth, W., Appl. Physics **A65**, 155 (1997); Wurth, W., Appl. Physics **A65**, 597 (1997).
(7) Wurth, W., and Menzel, D., Chem. Physics (in press), and references therein.
(8) Chen, C.T., et al., Phys. Rev. B **32**, 8434 (1985); Eberhardt, W., et al., Aust. J. Phys. **39**, 853 (1986).
(9) Wurth, W., et al., Phys. Rev. B **35**, 7741 (1987); Wurth, W., et al., Phys. Rev. B **37**, 8725-8729 (1988).
(10) Wurth, W., Feulner, P., and Menzel, D., Physica Scripta **T41**, 213-216 (1992).
(11) Björneholm, O., et al., Phys. Scr. **T41**, 217-225 (1992).
(12) Gel´mukhanov, F., and Ågren, H., Physics Reports **312**, 87-330 (1999), and references therein.
(13) Gortel, Z.W., unpublished results.
(14) Keller, C., et al., Phys. Rev. B **57**, 11951-11954 (1998).
(15) Keller, C., et al., Phys. Rev. Lett. **80**, 1774-1777 (1998).
(16) Mårtensson, N., and Nilsson, A., J. Electron Spectrosc. Rel. Phen. 75, 1 (1995).
(17) Sandell, A., et al., Phys. Rev. Lett. **78**, 4994-4997 (1997).
(18) Keller, C., et al., Phys. Rev. B (in press).
(19) Sundin, S., et al., Phys.Rev.Lett. **79**, 1451-1454 (1997).
(20) Berthold, W., et al., Chem. Physics, in press.
(21) Prince, K., et al., J. Electr. Spectr. Rel. Phen. **101-103**, 141 (1999).
(22) Föhlisch, A., et al., Phys. Rev. Lett. **81**, 1730-1733 (1998).
(23) Föhlisch, A., et al., submitted to Chem. Phys. Lett..
(24) Kassühlke, B., et al., Phys. Rev. Lett. **81**, 2771-2774 (1998).

# Probing the Nature of Hydrogen Bonds With X-Rays

P. M. Platzman

*Bell Laboratories, Lucent Technologies*
*700 Mountain Ave.*
*Murray Hill, NJ 07974*

Hydrogen bonds constitute a unique type of intermolecular interactions.[1] The smallish bond energy and the longish distance allows its formation and disruption under a wide range of physiologically interesting temperatures and conditions, i.e. it's crucial for life. In addition, hydrogen bonds are directional and therefore ideally suited to play a role in molecular recognition phenomena. Given the obvious importance and ubiquitous nature of hydrogen bonds it is remarkable that (until these experiments which I will discuss) a microscopic quantitative understanding of the hydrogen bond covalent or quantum mechanical character remained experimentally untested and controversial.

A number of groups have shown that very high momentum transfer inelastic x-ray scattering experiments unambiguously measure the Fourier transform of the ground state wave function squared.[2] This quantity, the momentum density, is most sensitive to the outer electrons. In this talk we describe our x-ray scattering studies of the hydrogen bond in ice *Ih*. The technical details of the experiment and the detailed interpretation are in Reference 3. However, we can summarize by saying that the experiments show that the measured Compton profile anisotropies are exceptionally sensitive to the phase of the electronic wave function and therefore to the covalency of the hydrogen bond. Periodic intensity variations in the anisotropy reveal two distances, one of 1.72 Å, near the hydrogen bond length of 1.75 Å, and another at 2.85 Å, close to the nearest-neighbor O-O distance of 2.75 Å. The presence of these two dominant lengths in the Compton profile anisotropy is interpreted as the first direct experimental evidence for the substantial covalent character of the hydrogen bond. Very good quantitative agreement between the data and a fully quantum mechanical bonding model for ice *Ih* and the disagreement with a purely electrostatic (classical) bonding model are strong support for this interpretation. A qualitative analysis of the quantum density of states allows us to estimate the "amount" of covalency at 10%.

We have also measured the temperature dependence of one of the profiles. It shows small ($\cong$ .2%) but rapid changes as the temperature is increased through a few hundred Kelvin. Neutron data allows us to argue that this change comes from relative rotational motion of two water molecules. This in turn suggest that bending of the hydrogen bond modifies its covalent character. In addition, Compton scattering data on urea where the bond is between O-N show quantitatively different behavior. In particular the anisotropy is much smaller, i.e. the bond is much less covalent (less than 2%).

The results presented here and in Reference 3 show the remarkable sensitivity of the Compton scattering to the phase coherence of ground state electronic wave functions. This means that modest, resolution statistically very (< 1%) accurate, Compton scattering experiments can uniquely probe a wide range of interesting hydrogen bonded systems.

# REFERENCES

1. Martin, T. W. and Derewanda, Z. S., *Nature Structural Biology*, **6**, 404 (1999).

2. "*Momentum Distributions*", ed. by Silver, R. N. and Sokol, P. E., Plenum Press, NY, 1989.

3. Isaacs, E. D., Shukla, A., Platzman, P. M., Hamann, D. R., Barbiellini, B., and Tulk, C. A., *Physical Review Letters*, **82**, 600 (1999).

# V. HIGHLY CHARGED IONS

# Inverse Photoionization Studied via Radiative Electron Capture into Highly Charged Ions

Th. Stöhlker[1,2], O. Brinzanescu[2,3], A. Krämer[1,2], T. Ludziejewski[2,4], X. Ma[2,5], P. Swiat[6], and A. Warczak[6]

[1] *Institut für Kernphysik, University of Frankfurt, 60486 Frankfurt, Germany*
[2] *Gesellschaft für Schwerionenforschung, 64291 Darmstadt, Germany*
[3] *National Institute for Laser, Plasma and Radiation Physics, 76900 Bucharest-Magurele, Romania*
[4] *Institute for Nuclear Studies, 05-400 Świerk, Poland*
[5] *Institute of Modern Physics, 730000 Landzhou, China*
[6] *Institute of Physics, Jagiellonian University, 30-059 Cracow Poland*

**Abstract.** Differential aspects of the photoelectric effect for high-Z hydrogen-like and few-electron ions are studied by the time-reversed process in ion-atom collisions, i.e. by radiative capture of a quasi-free target electron. In this time-reversed situation the capture of an electron into a bound state of the ion is accompanied by the simultaneous emission of a photon, allowing for a direct and unambiguous experimental access to this process. Therefore, radiative capture provides a unique tool to investigate the details of the photoionization process even in the case of high-Z highly charged ions where such investigations are currently not accessible in the direct channel. At high-Z relativistic effects become important for which photon angular distributions turn out to be a very sensitive probe. This was demonstrated very recently, by an angular distribution study performed for electron capture into the 1s ground state of $U^{92+}$. This particular experiment allowed us to identify spin-flip contributions to the photoionization process at large backward angles. Here, we review the shell and subshell differential photon-angular distribution studies for radiative capture into highly-charged uranium ions. The experimental data are compared with exact relativistic calculations giving detailed insight into the fundamental electron-photon interaction process involved.

## INTRODUCTION

In relativistic atomic collisions between highly-charged heavy ions and low-Z target atoms *Radiative Electron Capture* (REC) is one of the most important reaction channels. Here, the coupling between the electron and the electromagnetic field of the fast moving projectile results in an electron capture into a bound state of the ion via simultaneous emission of a photon carrying away the energy and momentum difference between the initial and final electron states [1–4]. Consequently REC can be viewed as the time-reversed photoionization process as long as the initial electron momentum distribution in the target atom can be neglected, an excellent approach when dealing with relativistic encounters between high-Z projectiles and low-Z atoms [3].

Previously we have shown that for high-Z ions detailed information about the elementary photoionization process itself can be obtained via a study of its inverse reaction occurring in energetic ion-atom collisions [3,4,6]. In this domain of high-Z ions the photon-atom interaction is strongly influenced by relativistic effects [7]. For the exploration of such details, angular distributions are a very sensitive probe. For the nonrelativistic domain of low-Z ions, this has already been stressed by Krässig and coworkers, who observed nondipolar corrections to the photoeffect [8]. In contrast, at high-Z the full retarded multipole expansion of the photon field must be considered and the strength of relativistic effects may lead to considerable corrections. Evidently, the latter effects are of particular interest when dealing with high-Z ions [9]. However, they are difficult to observe in direct photoionization experiments, since retardation leads to a strong forward peaking of the photoelectron angular distribution. Moreover, REC provides most favorable conditions because it gives experimental access to the study of photoionization even for H-like high-Z ions, i.e. for an explicitly point Coulombic system. Most remarkably, as has been first pointed out by Spindler et al. [10], a cancellation between retardation and Lorentz transformation occurs for the angular distribution of REC into the $K$ shell of bare ions. As a consequence, the $\sin^2\theta$ distribution of the simple nonrelativistic dipole-approximation is restored. Indeed a complete cancellation occurs as long as retardation is considered in all orders in a nonrelativistic treatment [10,11]. Therefore, deviations from the symmetric $\sin^2\theta$ distribution provide a direct measure of relativistic corrections. These deviations turn out to be rather unimportant at ion energies up to 300 MeV/u and nuclear charges up to $Z = 50$ or 70 [12]. However, at these energies and for $Z = 92$ a significant cross section at forward angles has been predicted for REC into the $K$ shell [3,12,13] and has been shown to be a unique signature of spin-flip transitions. Besides this unique aspect of the time-reversed situation, further peculiarities must be mentioned. By means of REC, photoionization can be studied even for excited states in high-Z H-like ions. Most important, no corrections due to electron scattering occurring in solid targets are required, leading in conventional photoionization studies for high-Z elements to a considerable broadening of the electron emission angle [14]. The absence of the these corrections in REC experiments allows one to extend photoionization studies to much lower energies than available for the direct channel. For a review of the photoionization experiments at high-Z we refer to [7,15].

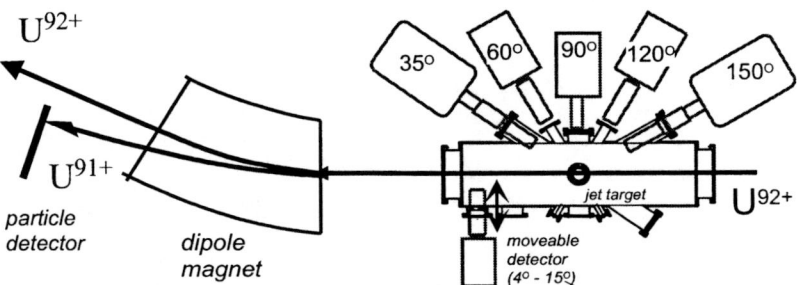

**FIGURE 1.** Layout of the experimental arrangement at the ESR-jet target. Various x-ray detectors view the target interaction zone at observation angles ranging from almost 0° to 150°. Photon emission is observed in coincidence with the down-charged ions, detected in a particle counter located behind the dipole magnet.

# EXPERIMENTAL TECHNIQUE

In the following we concentrate on the most recent experiments conducted at the relativistic heavy-ion storage ring ESR at GSI in Darmstadt [16]. This electron-cooler ring provides the basis for atomic-collision experiments dealing with beams of unprecedented quality [4,17–19]. At the ESR, the interaction of intense beams of cooled high-$Z$ ions with low-dense gaseous matter can be studied without any beam collimation, guaranteeing almost completely background-free experimental conditions. Typically, the beam diameter is close to 1 mm and the gasjet target density amounts to about $10^{12}$ particle/cm$^3$ [20], compared to $10^{21}$ particle/cm$^3$ of solid-state targets. Nevertheless, due to the high revolution frequency of the ions of about $10^6$ s$^{-1}$, even collision processes with tiny reaction cross-sections (0.1 barn or even smaller) became accessible by experiments. Until now, most atomic-collision studies at the ESR were concentrating on the study of x-ray emission occurring in relativistic collisions [4,17,21]. In Fig. 1, a scheme of the present experimental arrangement at the ESR gas-jet target is depicted. With this setup complete angular distribution studies are now accessible for the first time. Various x-ray detectors viewing the gas-jet/beam interaction zone at different observation angles whereby covering the range from almost 0° to almost 180° [22]. In addition, the dipole magnet accomplish to analyze the charge states of the emerging projectiles and the x-ray emission can be measured in coincidence with the down-charged projectiles ($U^{91+}$). A typical x-ray spectrum associated with electron capture is displayed in Fig. 2 which was obtained at an observation angle of 150° for 310 MeV/u $U^{92+} \rightarrow N_2$ collisions. Beside the Lyman ground-state transitions (Ly$\alpha_1$: $2p_{3/2} \rightarrow 1s_{1/2}$, Ly$\alpha_2$: $2p_{1/2} \rightarrow 1s_{1/2}$, M1 $2s_{1/2} \rightarrow 1s_{1/2}$) the most prominent features observed in the spectrum are due to radiative electron capture into the ground and excited projectile states. The width of these lines reflect the Compton profile of the bound target electrons (see e.g. [17]).

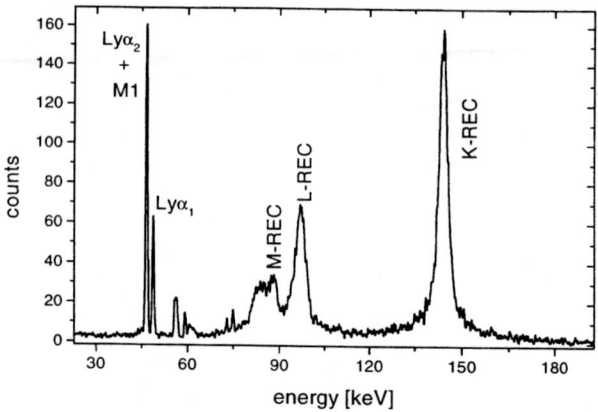

**FIGURE 2.** X-ray spectrum observed at nearly 150° for $U^{92+} \to N_2$ collisions at 310 MeV/u [22]. The data were taken in coincidence with down-charged $U^{91+}$ ions [22].

In this most recent experiment we were aiming on a complete angular distribution measurement for REC into the ground-state of bare uranium (Z=92), i.e. the heaviest stable element. Until now, similar investigations for REC were restricted to Z≤ 54. In the experiment, x-ray spectra were measured by using various Ge(i) and NaI x-ray detectors at in total six different observation angles covering the range from almost 0° to 150° [22]. Moreover, x-ray detection at almost 0° was accomplished by using a Ge(i) detector with four independent segments, each furnished with an individual readout. The horizontal and vertical size of each segment is 13 and 25 mm, respectively. The detector was mounted on a movable support, 510 mm down-stream from the projectile-target interaction region. Periodically, after the accumulation procedure was finished, the detector was positioned at a perpendicular distance of just 1 cm apart from the circulating beam. For the first stripe, this position corresponds to a mean observation angle of 4.6° with an angular acceptance of $\Delta\theta = \pm 0.7°$. Moreover, since the used Ge(i) detector possesses a good energy resolution (1.2 keV at 220 keV) but low detection efficiency for high photon energies (at almost 0° the K-REC centroid energy is close to 660 keV), we have also replaced the Ge(i) detector by an NaI counter which provides an improved detection efficiency by more than an order of magnitude. In Fig. 3, a coincident x-ray spectrum, measured with the innermost segment of the Ge(i)-detector is displayed along with the spectrum taken with the NaI counter. The mean observation angle of the latter detector corresponds to 6°. In both detectors, the K-REC line can be well identified at an x-ray energy of 660 keV. In addition, the Lyman transitions measured with the Ge(i) detector are shown separately in the inset, demonstrating the good energy resolution obtained. We have to stress that the almost background-free spectra observed close to 0° emphasize the brilliant beam conditions of the storage ring, which are in contrast to standard single-pass experi-

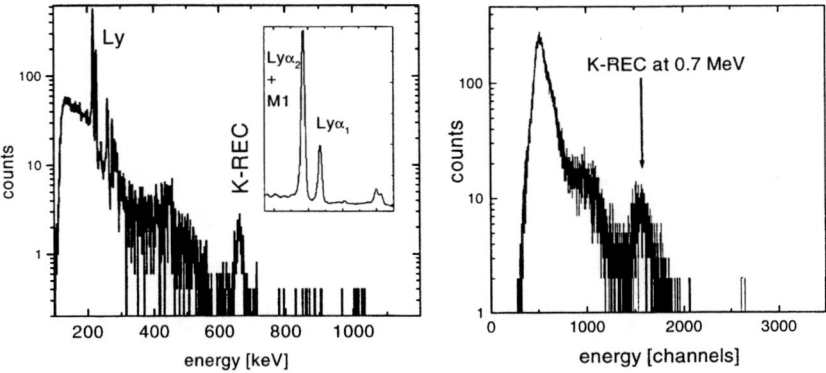

**FIGURE 3.** X-ray spectrum measured at nearly 0° for $U^{92+} \to N_2$ collisions at 310 MeV/u [22]. Spectrum at the right side: recorded with an NaI-detector; spectrum at the left side: recorded with a Ge(i)-detector.

ments with relativistic ion beams utilizing solid targets. There, the huge secondary electron production and its related bremsstrahlung prevents photon detection in the forward hemisphere.

## $K$-Shell Angular Distribution for Hydrogen-Like Uranium: The Identification of Spin-Flip Transitions

In Fig. 4, the obtained differential cross section values are given as a function of the observation angle (for details of the data analysis we refer to Ref. [17,22]). In the figure the solid circles refer to the data obtained by the Ge(i) detectors whereas the open square depicts the result from the NaI counter. In addition, the data are compared with predictions based on rigorous relativistic calculations [12,13]. These calculations use exact relativistic bound-state and continuum wave functions and take into account all multipole orders up to about $L = 20$. In order to achieve convergence, the summation was carried out over all electron partial waves with Dirac quantum numbers $|\kappa| \leq 20$. For comparison, the measured angular distribution was normalized to the theoretical prediction at 90°. As seen in the figure, a good agreement between the experimental data and the rigorous relativistic theory is obtained. In order to reveal the necessity of such a complete relativistic treatment for high-Z projectiles we also depict in the figure the $\sin^2\theta$ distribution of the nonrelativistic theory which considers the full retardation as well as Lorentz transformation to the laboratory frame. Obviously, the experimental data deviate considerably from symmetry around 90°. Most important, the large cross section observed close to 0° disproves the nonrelativistic theory which predicts that 0° photon emission for REC into states with angular momenta $\ell = 0$ is forbidden due to angular momentum conservation. In a complete relativistic treatment, however,

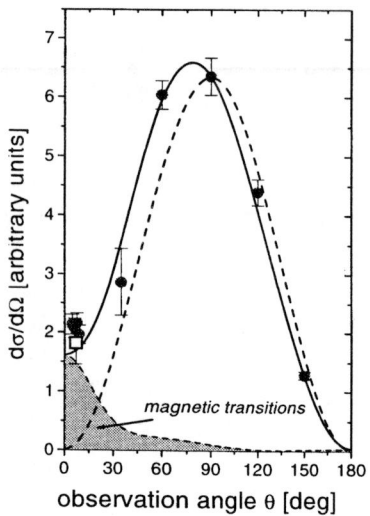

**FIGURE 4.** Angular distribution for REC into the $K$ shell of bare uranium (solid circles) as a function of the observation angle $\theta$ (310 MeV/u $U^{92+} \rightarrow N_2$). The solid line depicts the complete relativistic calculations and the shaded area refers to the spin-flip contributions [13,12]. The $\sin^2\theta$ shape of the nonrelativistic dipole-approximation is given by the dashed line.

the interaction of the electron magnetic moment with the magnetic field produced by the fast moving projectile can produce spin-flip transitions which compensate the angular momentum carried by the photon.

Consequently, our measurement close to 0° provides a clear identification of spin-flip transitions occurring in relativistic ion-atom collisions. This effect has been predicted by Eichler et al. [11,12] pointing out that its differential cross section is a precise probe for theoretical wave functions. However, until now, this effect was confirmed experimentally. Even for the case of $Xe^{54+} \rightarrow Be$ collisions at 197 MeV/u [23], no relativistic effects were observed and the measured $K$-REC angular distribution could be well reproduced by the nonrelativistic theory. We also have to emphasize that for high-Z ions our finding points to the importance of treating REC in all orders of $\alpha Z$. Considering the relativistic Sauter approximation, where the interaction between a free electron and a bare ion is treated in lowest order of $\alpha Z$, also the spin-flip contributions disappear at 0°.

In order to reveal the physics of the REC process and its relation to photoionization in more detail, we normalized the measured $K$-REC x-ray yield to the intensity of the simultaneously observed Ly-$\alpha_2$+**M1** line. Since the Ly-$\alpha_2$ and the **M1** transitions stem from the decay of the $2p_{1/2}$ and the $2s_{1/2}$ level, the corresponding line intensity follows an isotropic emission characteristic in the emitter frame. By applying this technique of normalization, the REC angular distribution in the emitter frame can be derived directly from the x-ray spectra. For this purpose only the Lorentz transformation of the observation angle is required (see e.g. Ref. [11]). The

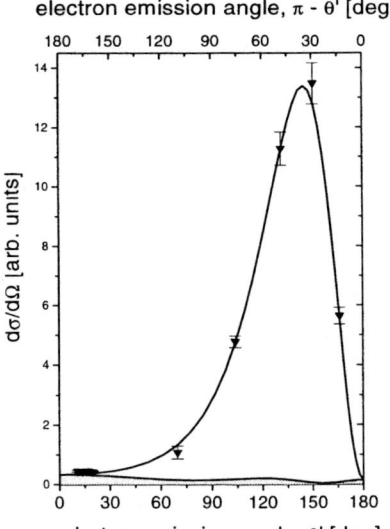

**FIGURE 5.** $K$-REC angular distribution (projectile frame) as a function of the emission angle $\theta'$ (bottom x-axis). The x-axis at the top refers to the corresponding electron angular distribution for photoionization of H-like uranium (photon energy 302 keV) obtained by the interchanging $\theta'$ by $\pi - \theta'$. The full line gives the photoionization distribution as obtained from complete relativistic calculation whereas the shaded area depicts the spin-flip contribution [22].

additional advantage of this technique is that also the uncertainty caused by the solid angle determination cancels out completely. In Fig. 5, the result for this ratio is depicted as a function of emission angle in the emitter frame (compare lower x-axis) along with the corresponding prediction by the rigorous relativistic theory. As can be observed in the figure, both the experimental and the theoretical angular distributions, exhibit a pronounced backward shift, since the strong retardation effect is no longer cancelled by the Lorentz transformation. The maximum of the distribution is now localized close to 150° and the cross section decreases drastically by more than an order of magnitude towards 0°. Here, indeed the occurrence of spin-flip transitions (compare shaded area in Fig. 5) appears to be a tiny effect with an almost isotropic emission characteristic. This is in obvious contrast to the discussed distribution in the laboratory frame where the Lorentz transformation not only compensates retardation but also amplifies the relative weight of the spin-flip transitions close to 0° with respect to the maximum of the distribution.

From the experimental REC distribution in the emitter frame the corresponding angular distribution for photoionization can also be derived. For this purpose the REC angle $\theta'$ has simply to be replaced by $\pi - \theta'$. This transformation is needed, because in relativistic ion-atom collisions, the quantization axis is determined by the direction of the projectile momentum. This is opposite to the direction of the electron (or photon) momentum as seen from the projectile. As a consequence, Fig. 5 refers also to the electron angular distribution for photoionization of H-like uranium at the corresponding photon energy of 302 keV whereby the angular transformation is given by the upper x-axis. This presentation illustrates the potential of REC experiments for precision studies of the photoionization process. In particular, it reveals the origin of the observed spin-flip transition as events related to large angle backward scattering in photoionization, i.e. where the momentum transfer to the nucleus is largest. Note, that this effect was not observed in direct photoionization experiments for high-Z ions. On the contrary, spin-flip contributions in the direct channel were observed at almost 0° (corresponding to 180° in the time reversed situation) at much larger energies.

## Radiative Capture into Excited States

### Radiative Capture into L-subshells

By means of REC, photoionization can also be studied for excited states in highly-charged heavy ions [6]. Here high-$Z$ ions are of particular interest since the large fine structure splitting provides direct experimental access even for the investigation of angular distributions associated with REC into the $j$ sublevels of the $L$ shell.

For the first experimental study of the angular distributions for REC into the $L$-shell sublevels, He-like uranium was chosen as a projectile which can be produced with sufficient intensities even at an energy of as low as 89 MeV/u (corresponding to photoionization at 79 keV). Since the width of the target Compton profile decreases with decreasing beam energy [17], the low beam energy allowed us to resolve the fine structure components of the $L$ shell of uranium. Note, that at such low photon energies, photoionization experiments for high-Z elements are presently not possible.

The results obtained within this experiment, conducted at the GSI Fragment Separator for $U^{90+} \rightarrow C$ collisions, are depicted in Fig. 6. In the figure the measured angular distributions for REC into the $j=1/2$ and $j=3/2$ levels of the $L$ shell of He-like uranium are plotted in comparison with the predictions of the exact relativistic theory. The data for capture into the $j=1/2$ levels show a considerable bending of the angular distribution into forward direction whereas the distribution for capture into the pure p-state ($j=3/2$) exhibit a slight enhancement at backward angles. An excellent agreement between measurement and relativistic theory is found for both fine-structure components [24]. This result demonstrates that for p-states the Lorentz transformation is not sufficient to cancel the backward shift caused by retardation as in the case of s-states. On the contrary, the observed slight forward shift for capture into the $j=1/2$ levels is caused by REC into the $2s_{1/2}$ state which is partially caused by spin-flip transitions. As shown in Ref. [24], the application of a non-relativistic treatment for the time reversed photo-effect yields strong deviations

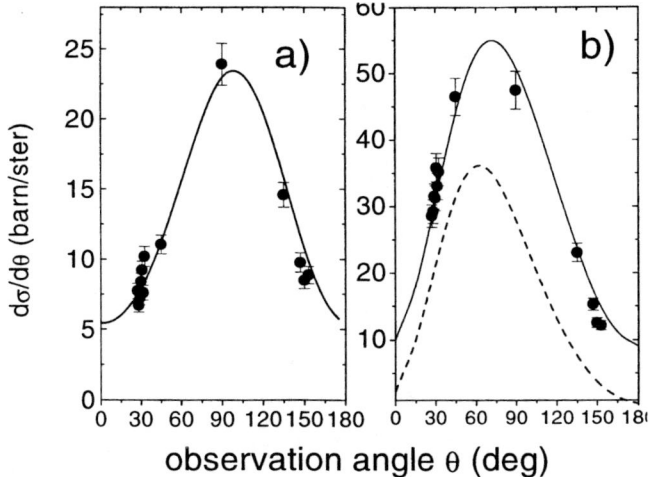

**FIGURE 6.** Experimental angular distribution of L-REC radiation for (a) REC into the $2p_{3/2}$ level and (b) for the $2s_{1/2}$ and $2p_{1/2}$ states. The solid line shows the result derived from exact relativistic calculations. The theoretical result for capture into the $2s_{1/2}$ state (dashed line in (b)) is shown separately.

especially in forward and backward directions. Again, these findings point out that angular differential REC cross-sections are extremely sensitive to details of the atomic wave functions.

## *Time-Reversed Two-Step Photoionization - Alignment of the $2p_{3/2}$ State Caused by Radiative Capture*

In the following, we consider the inverse of a two-photon/one-electron ionization process. Here, the first photon resonantly excites an electron from the $1s_{1/2}$ ground state into the $2p_{3/2}$ hydrogen-like state and, subsequently, the second photon ionizes the excited electron within the lifetime of the $2p_{3/2}$ state, i.e. for the case of H-like uranium a time of the order of $10^{-17}$ s. We are interested in the magnetic subshell population of REC into the $2p_{3/2}$ state. Usually, information about degenerated atomic states, which differ only in their magnetic quantum number, is obtained by polarization measurements. However, since for high-Z ions like uranium the Ly$\alpha_1$ transition energy appears close to 100 keV, a polarization measurement is almost impossible. The alignment of the $2p_{3/2}$ state and consequently its magnetic-substate population can also be deduced from an angular distribution measurement. Since the $2p_{3/2}$ decays by an **E1** transition to the ground state, the anisotropic emission in the emitter frame is simply given by [25]

$$I(\theta) \propto 1 + \beta_A \cdot [1 - \frac{3}{2}\sin^2\theta] \qquad (1)$$

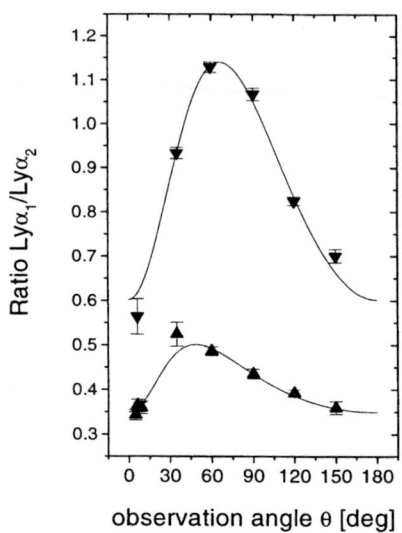

**FIGURE 7.** Experimental Ly$\alpha_1$/Ly$\alpha_2$ intensity ratio (solid triangles) measured for U$^{92+}$→N$_2$ collisions (up-triangles: 310 MeV/u; down-triangles: 79 MeV/u). The lines depict the results of a least square adjustment of Eq. (1) to the experimental data [6].

where $\theta$ denotes the emission angle in the projectile frame and $\beta_A$ gives the coefficient of the alignment which is related to the population cross-sections $\sigma(j, m_j)$ for the magnetic sublevels [25]

$$\beta_A = \frac{1}{2} \cdot \frac{\sigma_{(3/2,3/2)} - \sigma_{(3/2,1/2)}}{\sigma_{(3/2,3/2)} + \sigma_{(3/2,1/2)}}. \tag{2}$$

Consequently, by measuring the Ly$\alpha_1$ emission at only a few observation angles, the relative population of the magnetic sublevels can be experimentally determined.

Such investigations were performed recently with bare Pb$^{82+}$ and U$^{92+}$ projectiles at the gas-jet target of the ESR storage ring [6]. In order to reduce possible systematic uncertainties, we again took advantage of the close spaced, simultaneously observed Ly$\alpha_2$+**M1** transitions, which are isotropic in the emitter frame. By normalization to the latter, almost all systematic uncertainties cancel out. Moreover, the cascade population of the 2p$_{3/2}$ state was investigated experimentally (see e.g. [6]) which was found to produce only Ly$\alpha_1$ transitions. From the measured alignment of the 2p$_{3/2}$-levels follows that about 75% of the capture events populate the $|m_j|=1/2$ sublevels of the 2p$_{3/2}$ state which corresponds to a 40% linear polarization of the emitted Ly$\alpha_1$ radiation. By subtracting the cascade contribution, we were able to deduce that direct radiative capture into the 2p$_{3/2}$ state populates the $|m_j|=1/2$ magnetic sublevels by more than 90%. As a consequence, a surprisingly strong linear polarization by as much as 50% can be deduced from this finding. This result was derived from measurements at only three observation

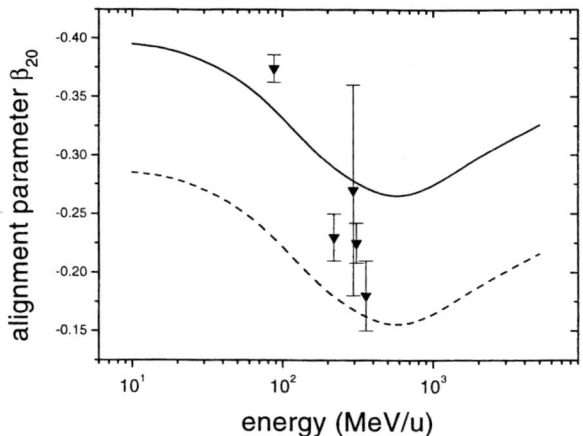

**FIGURE 8.** All alignment parameters $\beta_{20}$ (down triangles) measured for the Ly$\alpha_1$ radiation of $U^{91+}$ produced in $U^{92+} \rightarrow N_2$ collisions as a function of collision energy. The full line refers to relativistic calculations of the alignment of the $2p_{3/2}$ assuming the time-reversal of two-step photoionization. The dashed line refers to the same calculations but considers in addition feeding transitions from higher excited levels. The latter were obtained from experimental data.

angles, exploiting the fact that the shape of the angular distribution is precisely known (see Eq. 1). During the most recent beam time with the new experimental setup (see above), however, we obtained also data for the alignment of the $2p_{3/2}$ in H-like uranium by a measurement in the angular range between nearly 0° and 150° ($U^{92+} \rightarrow N_2$ collisions at 310 MeV/u). In Fig. 7, the measured intensity distribution of the Ly$\alpha_1$ radiation is depicted as function of the observation angle. In addition, the solid line depicts a least square adjustment of Eq. 2 to the experimental data points (upper triangles in Fig. 7). As a result, we obtain for the collision energy of 310 MeV/u a $\beta_A$ parameter of -0.225 ± 0.017, in accordance with the former experimental findings for energies above 220 MeV/u [6]. However, for these high energies where REC is known to be the only relevant Ly$\alpha$ production process, all experimental alignment data appear to be in variance with complete theoretical *ab initio* calculations [25,26]. This is depicted in Fig. 8 where the theoretical and the experimental data are compared as a function of beam energy. In the figure, the solid line refers to the theoretical prediction of the alignment parameter for radiative capture into the $2p_{3/2}$ of H-like uranium [26]. The dashed line includes in addition feeding transitions from higher excited levels and should be compared with the experimental observation for energies above 200 MeV/u. Although this correction was derived from the experimental data, the cascade contributions might be responsible for the found deviations.

Already at high energies (above 200 MeV/u) the observed alignment of the $2p_{3/2}$ is surprisingly strong. On the other hand, we observed an even more pronounced and almost complete alignment ($\beta_A$=-0.374 ± 0.012) of the $2p_{3/2}$ state in $U^{91+}$ for decelerated $U^{92+}$ ions in collisions with $N_2$ molecules (compare also Fig. 7). At this low-energy the competing non-radiative electron capture process can no longer be neglected. Since we expect for this process an almost statistical population of the magnetic sublevels, the reason for the almost complete alignment observed for decelerated ions appears to be unknown.

## OUTLOOK AND SUMMARY

In summary, the fundamental interaction process between an electron and a photon in the presence of a strong central atomic field is discussed from the point of view of radiative electron capture into highly-charged uranium ions. In particular the REC cases studied for bare $U^{92+}$ ions show the unique potential of these investigations in order to elucidate the subtleties of the photoionization process for bare high-Z ions where the interaction can be studied for explicitly point Coulombic systems. The angular distribution studies discussed benefit in particular from the partial cancellation of retardation which arises from relativistic transformations from the emitting fast ion system into the laboratory frame. As a consequence, deviations from the $\sin^2\theta$ distribution turn out to be a very sensitive probe for relativistic corrections to the angular distribution for photoionization. The sensitivity of this technique is illustrated by the experimental separation and identification of spin-flip contributions to the angular-differential cross-section at the correspondingly photon energy of 302 keV for the time-reversed process. At such low-energies this effect was not accessible up to now in direct photoionization studies. Moreover, the structure of the photon spectra allows us to clearly differentiate between radiative capture into the ground state and into higher shells. For the L-shell even the fine-structure components are resolved in the REC spectra, providing an additional differentiation among the $j$ values of the L-levels ($2s_{1/2}$ and $2p_{1/2}$ as well as $2p_{3/2}$). Here, the experimental results underline the importance of the angular momentum of the final state. In particular, the data confirm relativistic calculations that for capture into p-states retardation effects show up more pronounced compared to s-states, i.e. in contrast to s-states retardation is not canceled by the Lorentz transformation. Furthermore, even the magnetic substate population of the time-reversed double-photoionization could be addressed by the experiments. Here, a surprisingly strong linear polarization (almost 40%) of the emitted $Ly\alpha_1$ photons was deduced from the anisotropy of the measured angular distribution. However, in this particular case, the experimental data deviations is found to be in variance with theory. Additional studies are required in order to solve this issue.

In the near future we like to extend these studies to decelerated high-Z ions. This will allow us to perform experiments where in the time-reversed situation the corresponding photon energy is close to threshold, i.e. an energy regime where no data for photoionization of high-Z elements exists. Moreover, we like to concentrate on high-Z few electron systems where the spin coupling in multi electron systems must be considered in addition. Up to now our experimental findings confirm theoretical

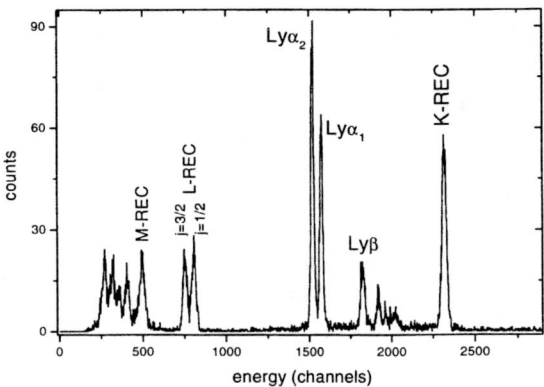

**FIGURE 9.** X-ray spectrum of $Pb^{81+}$, produced in collisions of strongly decelerated $Pb^{82+}$ projectiles with $H_2$ molecules at 25 MeV/u [27]. The spectrum was measured at an observation angle of 150°.

predictions that for $K$-shell photoionization neither screening effects nor electron correlations are of importance at high energies. However, for high-Z few electron ions at low energies the situation is not clear. As a prerequisite the deceleration mode at the ESR could already be established which allowed us very recently to decelerate bare gold ions (Z=79) to energies of as low as 25 MeV/u (corresponding to an electron moving with a kinetic energy of 13 keV in the projectile frame). In addition, a $H_2$ cluster jet-target was commissioned at the ESR [27]. As a result, x-ray spectra with an extremely narrow REC line-profile could be measured (compare Fig. 9).

This work was done in collaboration with F. Bosch, R.W. Dunford, C. Kozhuharov, P.H. Mokler, Z. Stachura, and with the members of the ESR team, K. Beckert, B. Franzke, H. Reich, F. Nolden, and M. Steck. The author would like to thank J. Eichler for the close collaboration over many years and his continuous support with new and stimulating theoretical results.

# REFERENCES

1. G. Raisbeck and F. Yiou, Phys. Rev. Lett. 4, 1858 (1971).
2. H. W. Schnopper, H. Betz, J. P. Devaille, K. Kalata, A. R. Sohval, K. W. Jones, and H. E. Wegner, Phys. Rev. Lett. 29, 898 (1972).
3. A. Ichihara, T. Shirai, and J. Eichler, Phys. Rev. A 54, 4954 (1996).
4. P. H. Mokler and Th. Stöhlker, Adv. in Atomic, Molecular, and Optical Physics, Vol.37, 297 (1996).
5. R. W. Schmieder, in *Proceedings of the NATO Workshop on the Physics of Highly Ionized Atoms, Cargese, Corsica June 1988*, edited by R. Marrus (Plenum Press: New York) 231 (1990).

6. Th. Stöhlker, F. Bosch, A. Gallus, C. Kozhuharov, G. Menzel, P.H. Mokler, H.T. Prinz, J. Eichler, A. Ichihara, T. Shirai, R.W. Dunford, T. Ludziejewski, P. Rymuza, Z. Stachura, P. Swiat, A. Warczak, Phys. Rev. Lett. 79, 3270 (1997).
7. R. H. Pratt, A. Ron, and H. K. Tseng, Rev. Mod. Phys. 45, 273 (1973).
8. B. Krässig, M. Jung, D. S. Gemmell, E. P. Kanter, T. LeBrun, S. H. Southworth, and L. Young, Phys. Rev. Lett. 75, 2736 (1995).
9. B. Craseman, X-Ray and Inner-Shell Process 1996 (AIP, Woodbury, New York, 1997), p. 3.
10. E. Spindler, H.-D. Betz and F. Bell, Phys. Rev. Lett. 42, 832 (1979).
11. J. Eichler and W. E. Meyerhof, *Relativistic Atomic Collisions* (Academic Press, San Diego, 1995).
12. J. Eichler, A. Ichihara, and T. Shirai, Phys. Rev. A 51, 3027 (1995).
13. A. Ichihara, T. Shirai, and J. Eichler, Phys. Rev. A 49, 1875 (1994).
14. S.J. Blakeway et al., J. Phys. B: At. Mol. Phys. 16, 3752 (1983).
15. H.K. Tseng, R.H. Pratt, S. Yu, A. Ron, Phys. Rev. A 17, 1061 (1978).
16. K. Blasche, B. Franzke, *Proc. of the $4^{th}$ Europ. Part. Acc. Conf.*, V. Suller and Ch. Petit-Jean-Genaz eds., World Scientific, Singapore, 1994, 133–137.
17. Th. Stöhlker, T. Ludziejewski, H. Reich, F. Bosch, R.W. Dunford, J. Eichler, B. Franzke, C. Kozhuharov, G. Menzel, P.H. Mokler, F. Nolden, P. Rymuza, Z. Stachura, M. Steck, P. Swiat, A. Warczak, Phys. Rev. A 58, 2043 (1998).
18. F. Bosch, Nucl. Instr. Meth. B 190, 23 (1987).
19. F. Bosch, Nucl. Instr. Meth. A 92, 314 (1992).
20. H. Reich, W. Bourgeois, B. Franzke, A. Kritzer, and V. Varentsov, Nucl. Phys. A 696, (1997).
21. Th. Stöhlker, C. Kozhuharov, P. H. Mokler, A. Warczak, F. Bosch, H. Geissel, R. Moshammer, C. Scheidenberger, J. Eichler, A. Ichihara, T. Shirai, Z. Stachura, P. Rymuza, Phys. Rev. A**51**, 2098 (1995).
22. Th. Stöhlker, T. Ludziejewski, F. Bosch, R. W. Dunford, C. Kozhuharov, P. H. Mokler, H. F. Beyer, O. Brinzanescu, B. Franzke, J. Eichler, A. Griegal, S. Hagmann, A. Ichihara, A. Krämer, J. Lekki, D. Liesen, F. Nolden, H. Reich, P. Rymuza, Z. Stachura, M. Steck, P. Swiat, A. Warczak, Phys. Rev. Lett. 82, 3232 (1999).
23. R. Anholt, S.A. Andriamonje, E. Morenzoni, Ch. Stoller, J.D. Molitoris, W.E. Meyerhof, H. Dowman, J.-S. Xu, Z.-Z. Xu, J.O. Rasmussen, D.H.H. Hoffmann, Phys. Rev. Lett. 53, 234 (1984).
24. Th. Stöhlker, H. Geissel, H. Irnich, T. Kandler, C. Kozhuharov, P.H. Mokler, G. Münzenberg, F. Nickel, C. Scheidenberger, T. Suzuki, M. Kucharski, A. Warczak, P. Rymuza, Z. Stachura, A. Kriessbach, D. Dauvergene, B. Dunford, J. Eichler, A. Ichihara, T. Shirai, Phys. Rev. Lett. 73, 3520 (1994).
25. J. Eichler, Nucl. Phys. A 572, 147 (1994).
26. J. Eichler, A. Ichihara, T. Shirai, Phys. Rev. 58, 2128 (1998).
27. A. Krämer et al., to be published (1999).

# Atoms in Extreme Virtual Photon Fields of Fast, Highly Charged Ions

J. Ullrich[a], B. Bapat[a], A. Dorn[a], S. Keller[b], H. Kollmus[a], R. Mann[c], R. Moshammer[a], R. E. Olson[d], W. Schmitt[a], M. Schulz[a,d]

[a] *Universität Freiburg, Hermann-Herder-Str. 3, 79104 Freiburg, Germany*
[b] *Universität Frankfurt, Institut für theoretische Physik, 60054 Frankfurt, Germany*
[c] *Gesellschaft für Schwerionenforschung, 64291 Darmstadt, Germany*
[d] *University of Missouri – Rolla, Rolla, Missouri 65401, USA*

**Abstract:** Highly-charged ions at velocities close to the speed of light generate strong (I = $10^{17}$ - $10^{20}$ W/cm$^2$), ultra-short (t = $10^{-19}$ - $10^{-17}$ s) electromagnetic pulses when passing target atoms or molecules. Single and multiple ionization occurs and can be interpreted in terms of photoionization: at small field strength, ionization is due to the absorption or the scattering of a *single* virtual photon, whereas the incoherent, "simultaneous" absorption of "*many*" field quanta gives rise to multiple ionization in strong fields. In this paper, analogies to photoionization are investigated by comparison of kinematically complete experimental data with results of various theoretical calculations for collisions of 100 MeV/u C$^{6+}$, 1000 MeV/u U$^{92+}$ and 3.6 MeV/u Au$^{53+}$ on He, Ne and Ar targets. Dynamical multiple ionization mechanisms are identified as a function of the momentum transfer. Two-electron final states are found to depend sensitively on the correlated initial state, and the many-particle Coulomb continuum is investigated for triple ionization of neon in super-strong fields.

## INTRODUCTION

### Ionization by Photons and Charged Particles

It was recognized very early by Bethe (1) that the transition matrix element for excitation or ionization of a target electron by fast charged particle impact is identical to that for photoionization in first order perturbation theory, i.e., for comparatively *weak electromagnetic fields* induced by the ion. In both cases, the transition operator is $\exp(i\vec{q} \cdot \vec{r})$, where $\vec{q}$ is the total momentum transfer $\vec{k}_i - \vec{k}_f$ either by the photon (Compton scattering) or by the charged projectile. In the limit of negligibly small $\vec{q}$, approaching the minimum momentum transfer $|\vec{q}| = E_\gamma / c$ by a photon of energy $E_\gamma$, the matrix element becomes identical to that for photoabsorption. For *strong electromagnetic fields*, i.e., at perturbations $Q_P / v_P \geq 1$ ($Q_P, v_P$: projectile charge and velocity, respectively), Fermi, Weizsäcker and Williams (2) have developed a non-perturbative approximation: the time and impact-parameter dependent field is decomposed into two pulses propagating along and transverse to the projectile velocity. Fourier transformation and subsequent quantization yield two pulses of "equivalent" photons carrying $n(\omega)$ photons incident on the atom per unit area and frequency interval $d\omega$. Multiple ionization in strong fields can then be described as the incoherent, "simultaneous" absorption of several virtual photons from the pulse. Both approximations have been widely used in atomic (see e.g. (3-6)) as well as in nuclear physics (see e.g. (7)).

Theoretical interest in the detailed understanding of the equivalence between photon and charged-particle induced ionization strongly increased after precise experimental data on total cross section ratios of helium double to single ionization became available for photoabsorption, Compton scattering and ion impact. Charged particle induced dipole transitions at a specific energy loss $\Delta E_P$ of the charged projectile have been related to the absorption of a photon with an energy of $E_\gamma = \Delta E_P$, whereas non-dipole transitions have been related to Compton scattering (see (8-10)). Thus, the ratio for charged particle impact could theoretically be derived from those for photons colliding with helium. Most recently, it has even been demonstrated that kinematically complete experiments on double ionization of helium by charged particle impact are feasible (11, 12), opening the unique possibility to explore this fundamental equivalence in unprecedented detail (13).

In this paper, kinematically complete data on single, double and multiple ionization are presented for $0.1 \le Q_P / v_P \le 4$, i.e., from the perturbative regime where one virtual photon is exchanged to strongly non-perturbative situations described by the Weizsäcker–Williams approximation. Similarities to photoionization are investigated. Evidence is provided that kinematically complete data for charged particle impact covering the entire multi-dimensional final-state momentum space might be advantageous to explore basic questions frequently posted in the past in the context of photo double ionization: dynamic ionization mechanisms become apparent by investigating differential spectra as a function of the momentum transfer by the projectile $\vec{q}$. Signatures of the initial-state wave function are observed by projecting final-state electron momentum distributions onto a subspace where $\vec{q}$ is negligibly small. Comparison of experimental two-electron momentum distributions with various theoretical model calculations illuminates that the final state sensitively depends on the correlated initial state. Finally, the three-electron – two-ion continuum is investigated for the first time using a modified Dalitz representation for the ejected electrons. On the basis of classical model calculations and auto-correlation functions, the question is addressed whether the "simultaneous" absorption of "many" virtual photons, in one collision without significant momentum transfer, might open the unique possibility to obtain information on the short time correlation of ground state wave functions.

## Kinematics

For multiple ionization by fast heavy particle impact, it can readily be shown that a negligible amount of momentum $\vec{q}$ and energy ($\Delta E_P$) are transferred compared to the initial momentum and energy of the projectile (for details see (14)). Then, momentum balances transverse ($\perp$) and longitudinal ($\parallel$) to the projectile propagation can be treated separately and one obtains for multiple ionization without electron transfer (i.e. no mass transfer):

$$q_\parallel = (p_{R\parallel} + \sum p_{e\parallel}^i); \quad q_\parallel = -\Delta E_P/v_P = -(E_R + \sum U_e^i + \sum E_e^i)/v_P \quad (1)$$

$$q_\perp = (p_{R\perp} + \sum p_{e\perp}^i) \quad (2)$$

where $E_R$ is the energy of the recoil ion, $U_e^i$ and $E_e^i$ are the binding and kinetic energy of the $i^{th}$ electron, respectively (atomic units are used: $m_e = \hbar = e = 1$; with $m_e$: electron mass, e: electron charge, $\hbar$: Planck's constant). At large velocities, close to the speed of light c, the longitudinal momentum transfer approaches the minimum momentum transfer obtained in a photoabsorption event: $q_\parallel \approx -\Delta E_P/c = -\Delta E_\gamma/c$. In general, for fast heavy ion impact, $q_\parallel$ is negligibly small, independent of $\bar{q} \approx q_\perp$ and significant momentum can only be transferred in the direction transverse to the beam propagation (see figure 2).

This leads to important conclusions. First, the signatures of different dynamic mechanisms should manifest themselves in the transverse plane, i.e. in the plane of momentum transfer. Second, the natural direction to study characteristics of the initial state wave function is along the projectile propagation since $q_\parallel \approx 0$. This provides the unique possibility to study the role of the initial state simultaneously for all momentum and energy transfers occurring in the collision. This is very different from the situation realized in photoionization experiments where the energy, momentum and angular momentum transfer is well defined. No significant signatures of the correlated initial state have been observed in five fold differential cross sections (FDCS) for photo double ionization of helium.

## EXPERIMENT

From the considerations in the previous paragraph it follows that not only kinematically complete experiments are required but, moreover, measurements are desirable that cover the entire multi–dimensional final– state momentum space. This allows one to explore the various questions in subspaces that are ideally suited as outlined above. Recently, we have developed such a device, called the REACTION MICROSCOPE, which has been described in detail (14,15). Only the salient features will be summarized here. Ion beams of 100 MeV/u $C^{6+}$, 1000 MeV/u $U^{92+}$ and 3.6 MeV/u $Au^{53+}$ were delivered by the GANIL, the SIS and UNILAC facilities at Caen and GSI in Darmstadt, respectively. As outlined in figure 1, the given beam was directed on a supersonic, internally cold (typically below 1 K) atomic gas-jet. Recoiling ions and electrons emitted in the collision are guided by homogeneous electric and magnetic fields to multi-hit, position sensitive multi-channel plate detectors, mounted in the longitudinal direction, i.e., along the axis of symmetry parallel to the ion beam propagation. From the time-of-flight (coincidence with each projectile or with the trigger for pulsed ion beams)

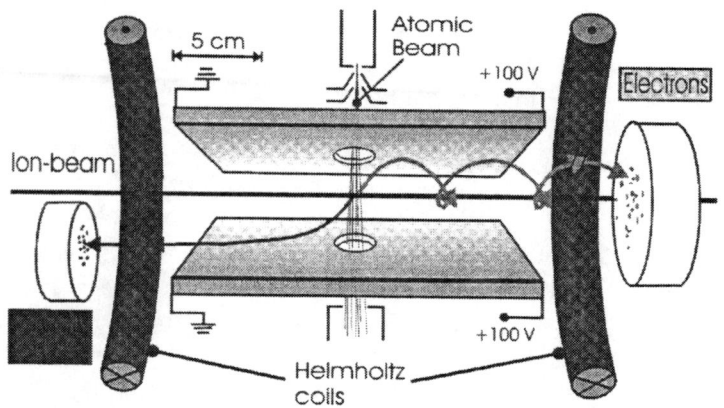

**FIGURE 1:** Outline of the REACTION MICROSCOPE (see text).

and the positions of arrival, the initial momenta of the fragments are calculated from the equations of motion for electrons and ions in the well known electric and magnetic fields.

By varying the strength of the projection fields, both the resolution as well as the fraction of the fragments in momentum space that are projected can be chosen over a wide range. Typically, all ions of interest ($|\vec{P}_R| \leq 5$ a.u.) are accepted simultaneously. At the same time all electrons with transverse energies $E_{e\perp} \leq 100$ eV, as well as with longitudinal energies $E_{e\|} \leq \infty$ eV in the forward and $E_{e\|} \leq 15$ eV in the backward direction, are detected in a typical experiment. Up to ten hits on the electron detector are accepted for a minimum time between two hits of 10 ns in case that both electrons hit the detector in a distance of less than 1 cm. For all other events, electrons can be detected on the 8 cm diameter detector even if they hit the detector at identical times.

Thus, depending on the collision dynamics and on the exact electric and magnetic fields chosen in the specific experiment, the REACTION MICROSCOPE simultaneously monitors between 60 % (100 MeV/u $C^{6+}$) and 90 % (3.6 MeV/u $U^{53+}$) of the twelve–dimensional final-state momentum space for double ionization. Care has to be taken especially if the data are compared to theoretical results. Though achieving large momentum acceptance in general, certain areas are not seen at all by the MICROSCOPE which might generate artificial structures. Usually projections are used, where the limitations are clearly visible or, for more difficult situations, the experimental acceptance is taken into account by folding it into the theoretical calculations.

Superior momentum resolution of $\Delta|\vec{P}_e| = 0.01$ a.u., corresponding to an energy resolution of $\Delta E_e = 1.4$ meV for electrons close to zero energy in the continuum, is achieved. Thus, doubly differential cross sections are obtained for the first time in a regime that is notoriously difficult to reach for conventional electron spectroscopy, i.e., using electrostatic analyzers (see, e.g., (17)). No principle limitations in resolution are at hand down to energies of a few neV.

# RESULTS

Investigations of multiple ionization by charged particle impact using REACTION MICROSCOPES offer the unique possibility to explore dynamic many particle quantum systems for "all" (see above paragraph) momentum and energy transfers in *weak* fields (exchange of one virtual photon), as well as in *super-strong* fields, where many equivalent photons are exchanged "simultaneously", within less than attoseconds ($10^{-18}$ s). Both situations have not been accessible up to now in experiments at third generation light sources or using ultra-fast (i.e., 10 femtoseconds) superintense lasers. Thus, basic questions on dynamical mechanisms, on the role of the correlated initial state as well as on the many particle Coulomb continuum, can be addressed with unprecedented completeness in domains that are entirely unexplored up to now.

## Single Ionization by Exchange of One Virtual Photon

In figure 2, final state momenta of the electron and the recoiling target ion, as well as the momentum change of the scattered projectile ($\Delta \vec{P}_P = -\vec{q}$), are shown for helium single ionization in collisions with 1000 MeV/u $U^{92+}$ projectiles. Exploiting azimuthal symmetry, all momenta are projected onto a collision plane defined by the incoming projectile momentum vector $\vec{P}_P = P_{P\parallel}$ and the momentum vector $\vec{P}_R = (-P_{Rx}, P_{R\parallel})$ of the recoiling ion. As outlined before, the momentum change of the projectile in the longitudinal direction, calculated from the collision dynamics $\Delta P_{P\parallel} = \Delta E_P / v_P$, is negligibly small ($\leq 0.06$ a.u.) for typical electron energies $E_e < 200$ eV (more than 90 % of all events). The FWHM of the $\Delta P_{P\parallel}$ – distribution in figure 2 is determined by the experimental resolution (mainly of the recoil ion) and it is about 0.2 a.u.

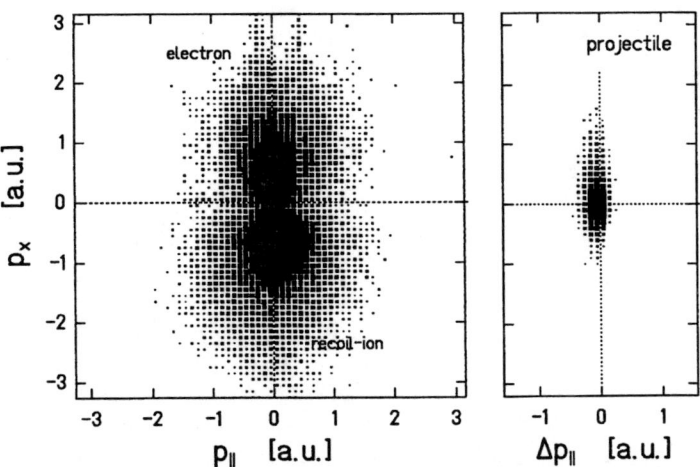

**FIGURE 2:** Distributions of the momenta for electrons and the target ions, and for the change in projectile momentum projected onto the collision plane (see text) for 1 GeV/u $U^{92+}$ on He single ionization.

Thus, essentially no momentum is transferred to the projectile in the longitudinal direction, and even the transverse momentum transfer is found to be small. Scattering angles are typically less than 20 nanorad. We therefore trivially rediscover the dynamics for the absorption of a single virtual photon of $E_\gamma = \Delta E_P$ where the ejected electron momentum is compensated by the recoiling target momentum alone (for details and the equivalent photon picture see (4,5,18)). Especially in the longitudinal direction, the measured momenta reflect the bound state properties of the helium atom, independent of the transverse momentum transfer.

This has been investigated at lower energies using Ne and Ar targets in collisions with 3.6 MeV/u $Au^{53+}$. In figure 3, experimental longitudinal electron momentum distributions are shown for different cuts in $P_{e\perp}$ for single ionization of argon (full circles) along with theoretical CDW-EIS predictions (full line, Continuum Distorted Wave – Eikonal Initial State (19,20)). Three striking features are observed in the doubly differential cross sections. First, all distributions are shifted into the forward direction (positive momenta), i.e., electron emission into the forward hemisphere is most likely. This has been observed before (21) and has been attributed to the "post collision interaction" (PCI) with the highly-charged receding projectile that pulls continuum electrons into the forward direction. In figure 2, this PCI is not visible since the perturbation is smaller and, moreover, since relativistic effects contract the longitudinal projectile potential, making it "short ranged" and, thus, less effective in the final state (see the discussion in (18)).

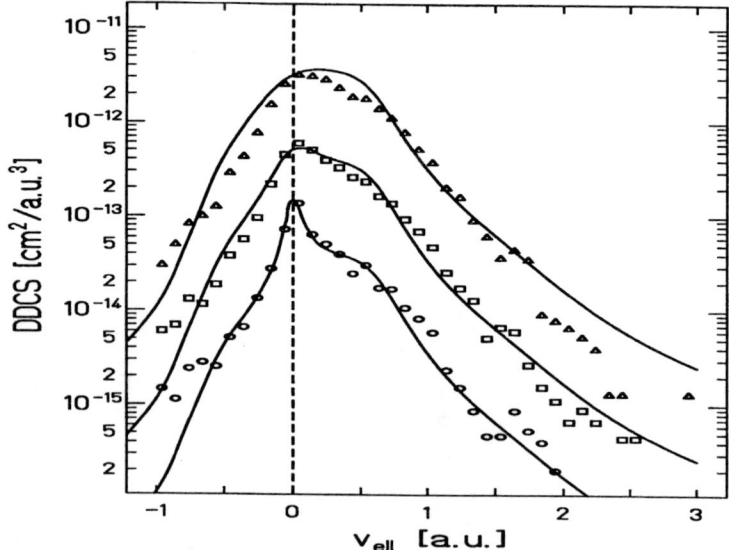

**FIGURE 3:** Doubly differential cross sections $DDCS = d^2\sigma/(dv_\parallel\, dv_\perp\, 2\pi v_\perp)$ as function of the longitudinal electron velocity for certain transverse velocity cuts in singly ionizing 3.6 MeV/u $Au^{53+}$ on Ar collisions. DDCS at different $v_\perp$ are multiplied by a factor of ten, respectively. Lines: CDW-EIS theory.

Second, for small transverse momenta, a singularity emerges at $P_{e\parallel} = 0$ which has recently been identified as a "target cusp", i.e., the expected singularity for long ranged potentials (22). Third, structures are identified near $v_{e\parallel} = \pm 0.5\,a.u.$ which are well reproduced by the theory (full line). Separate theoretical calculations for electron emission from the $3p_0$ and $3p_{\pm 1}$ magnetic subshells (Hartree–Fock–states), respectively, clearly show that the enhancements are related to the nodal structure of the $3p_0$ state (20). The $3p_{\pm 1}$ substates, which are aligned transverse to the projectile direction, have only one maximum at $p = 0$ in the momentum distribution and consequently the structure at $P_{e\parallel} = \pm 0.5\,a.u.$ does not occur.

In summary, we have provided strong evidence that the longitudinal electron momentum distributions reflect the properties of the bound state wave function independent of the momentum transfer, occurring in the transverse direction only. Moreover, since the transverse momentum transfer is small, the leading dynamic single ionization mechanism for fast heavy-ion impact can be identified, and might be described as photodissociation of the target atom in the virtual photon field of the projectile. Finally, precise quantum mechanical calculations beyond perturbation theory are at hand which reliably predict the observed features on an absolute scale.

## Double Ionization by Exchange of One Virtual Photon

Dynamical mechanisms, as well as signatures of the correlated initial state, have been explored for double ionization of He in the perturbative regime for 100 MeV/u $C^{6+}$ ($Q_P/v_P = 0.1$ ; $Q_P$: projectile charge) and 3 keV electron impact ($|Q_P|/v_P = 0.07$) in the transverse plane and along the longitudinal direction, respectively. Similarities to photo ionization are highlighted in the plane of the three emerging target fragments.

### Dynamical Mechanisms

In figure 4, azimuthal ionized electron angular distributions are shown for large ($q_\perp \geq 1.2\,a.u.$, figure 4 a) and small ($q_\perp \leq 1.2\,a.u.$, figure 4 b) momentum transfers, respectively. Arrows indicate the direction of the projectile momentum change $\Delta P_{P\perp} = -q_\perp$. Here, we distinguish the electrons by their relative energies.

For large momentum transfer, the high-energy electron (full square) is directed along the momentum transfer whereas the second is emitted isotropically. This strongly suggests a first order process for the emission of the "first" electron, followed by the shake–off for the ejection of the second one, to be the dominant dynamic mechanism. Accordingly, calculations in the Bethe-Born approximation plus shake–off have been found to be in good agreement with the measured angular distributions of the total transverse electron momentum $P_\perp = \sum p^i_{e\perp}$ with respect to the momentum transfer direction $\Phi_e = \angle(P_\perp, \Delta P_P)$ as well as with the relative emission angle spectra between the two electrons for all q (13). Agreement was achieved by using even the most simple Hartree-Fock $1s^2$ initial state wave function and a Gamov factor to account for the final state interaction between the two electrons (for details see (13)).

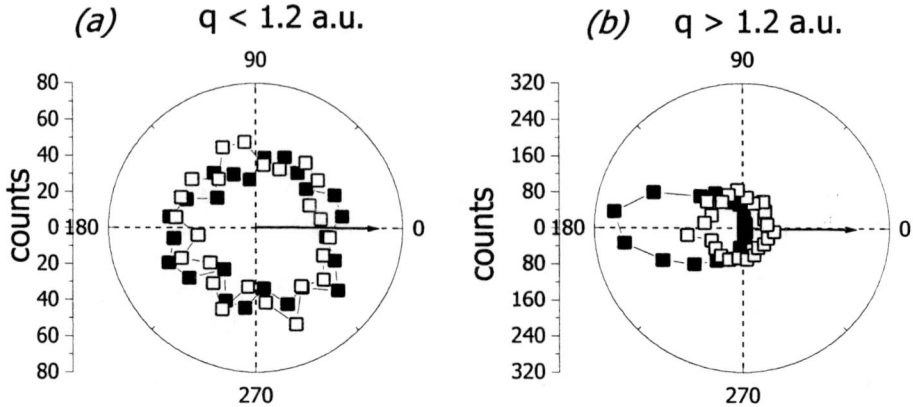

**FIGURE 4:** Azimuthal angular distributions of the fast and slow electrons at small (a) and large (b) momentum transfers for double ionization of He by 100 MeV/u $C^{6+}$ impact. Arrows: direction of projectile momentum change (negative of the momentum transfer).

For small momentum transfers, being not even sufficient to knock out one single electron by an impulsive binary projectile–electron encounter, both electrons are found to be emitted strongly isotropically (figure 4b). Here, dynamic features, being typical for double ionization after absorption of a photon, have been observed (13). Reasonable, but significantly worse, agreement with theory was reached in this domain, which is inherently included in the Bethe–Born approximation. Deviations between experiment and theory are most significant in the $\phi_{12}$-distribution (the angle between the two electrons) that is expected to be most sensitive to the electron–electron correlation.

In summary, projection of double ionization data onto the transverse plane enabled us to clearly identify the dynamical mechanisms that dominantly contribute to double ionization in the perturbative regime. When the maximum momentum transfer was limited to values below 1.2 a.u., distinct similarities to double ionization dynamics for the absorption of one real photon have been uncovered.

## Role of the Correlated Initial State

Having identified the leading dynamical mechanisms, signatures of the initial state are ideally explored along the longitudinal direction in comparison with theory. In figure 5, the electron sum-momentum in this direction is plotted versus their difference in a two dimensional spectrum integrated over all $q_\perp$. Strong correlation between the two electrons is found, being most obvious if the $P_\parallel^-$-distributions ($P_\parallel^- = p_{1\parallel} - p_{2\parallel}$)

are plotted for different sum momenta of the two electrons $P_\parallel^+ = p_{1\parallel} + p_{2\parallel}$, i.e., if doubly differential two-electron spectra are considered (figure 6). Data are shown for positive $P_\parallel^+$, where the limitations in momentum acceptance are small. The only parts of momentum space that are not accepted ($P_\parallel^- = 0 \pm 0.1$ a.u.) are indicated by the shaded areas. Whereas small momentum differences $P_\parallel^- \approx 0$ are strongly favored for small $P_\parallel^+$, experimentally one even finds a minimum at $P_\parallel^- = 0$ for large $P_\parallel^+$.

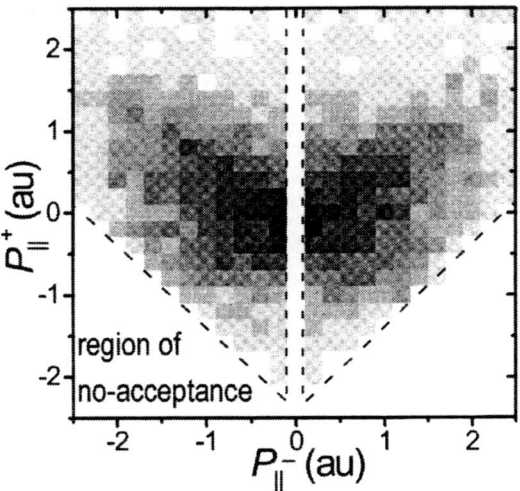

**FIGURE 5:** Double differential cross section $DDCS = d^2\sigma/(dP^- dP^+)$ for double ionization of He in collisions with 100 MeV/u $C^{6+}$. Grey scale is linear. Lines indicate regions of no momentum acceptance.

This is most surprising, since the final state repulsion between the two electrons is expected to be most effective if the relative momentum between the two electrons is small. In contrast, significant effects are found at large momentum differences, which are in qualitative agreement with the results of calculations in the Bethe–Born approximation plus shake-off (open symbols in the figure).

In figure 6a, theoretical results are plotted for two different initial state wave functions, a product of two one-electron 1s wave functions with an effective nuclear charge (full line), and a Hylleraas wave function accounting for the electron–electron interaction (dashed-dotted line). In both calculations, coming to very similar predictions, the electron–electron interaction in the final state is completely neglected. Inclusion of the repulsion by means of a Gamov–factor yields an even worse overall agreement with the experimental results (figure 6b,c).

Two features are surprising in the comparison with theory. First, by inspecting figures 6a and 6b, one finds that the effectiveness of the final state interaction sensitively depends on the details of the initial state wave function. For any assumed mechanism leading to double ionization, i.e., shake–off in our case, different parts in relative position space between the two electrons might dominantly contribute for different initial state wave functions, resulting in different effects in the final state. Thus, the final state interaction might be seen as a "magnifying glass" that enlarges tiny, subtle differences in the initial state at relative positions that are dominantly contributing to double ionization. Second, very different from the conclusions drawn for photo double ionization, different initial state wave functions clearly lead to strongly varying results in the final state for charged particle induced double ionization. Thus, as expected, such experiments provide a sensitive tool to explore correlated bound state wave functions. This is probably due to the fact that the wave function is tested for essentially "all" momentum transfers and "all" virtual photon frequencies (note that the relative difference in the prediction of the total helium binding energy between both wave functions is only one percent). For a detailed analysis using a variety of initial states see refs. (23,24).

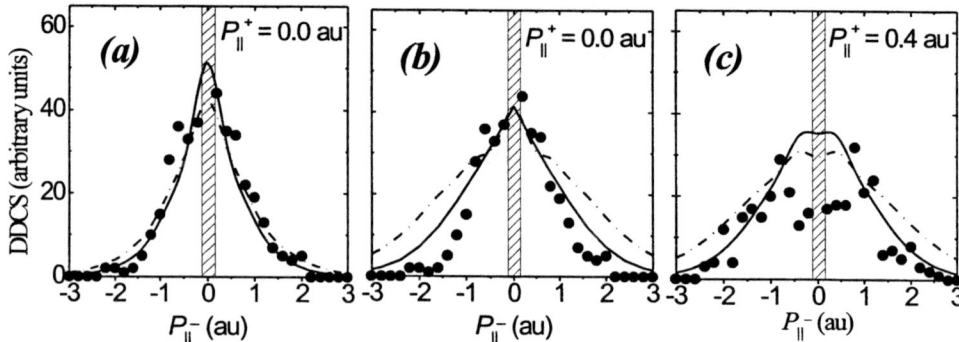

**FIGURE 6:** Doubly differential cross sections $DDCS = d^2\sigma/(dP^- dP^+)$ for double ionization of helium by 100 MeV/u $C^{6+}$ impact for various $P_\parallel^+$ cuts. Full circles: experimental data. Lines: Theory (see text).

## Similarities to Photo Double Ionization

Fivefold differential cross sections (FDCS) for photo double ionization of helium are decisively shaped by the Coulomb repulsion of the two electrons in the final state as well as by selection rules and momentum and energy conservation (25). Dörner et al (26) have explored these predictions in the plane defined by the momentum vectors of all three helium fragments. In essence, the features observed by these authors, i.e., a very small emission probability for both electrons in the same (Coulomb repulsion) and opposite directions (conservation of linear and angular momentum as well as parity) with zero probability for $\vec{p}_{e1} = -\vec{p}_{e2}$ are recovered in figure 7 for double ionization of helium by 3 keV electron impact.

Clearly, the patterns found for charged particle impact are less pronounced, since all virtual photon frequencies are allowed and, moreover, since non-dipole transitions contribute that wash out the strict minimum for photoionization at $\vec{p}_{e1} = -\vec{p}_{e2}$.

In the future, experiments for charged particle impact with significantly increased number of coincidence events (factor 50 to 100) should become feasible. They will allow for an ultimate comparison between photon and charged particle induced double ionization of helium on the level of FDCS for fixed virtual photon energy, i.e., energy transfer by the charged projectile, as well as for negligibly small higher order multipole contributions, i.e., for minimum momentum transfer realized for "zero" transverse scattering of the fast projectiles.

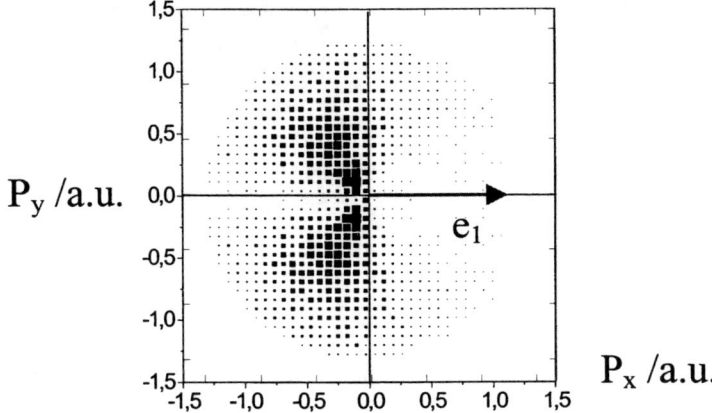

**FIGURE 7:** Momentum distribution of electrons in the fragmentation plane (see text) for double ionization of He by 3 keV electron impact. The direction of one electron is fixed along the arrow.

## Triple Ionization in Ultra-Strong Virtual Photon Fields

Relativistic charged particles represent a source of virtual photons with unique properties that are not realizable by any present source of real photons. Power densities of more than $10^{20}$ W/cm$^2$ are achievable for pulse times of a few $10^{-19}$ seconds, i.e., sub-attoseconds, which are a factor of 10,000 faster than pulses from ultra-short, super-intense femtosecond lasers. The installation of REACTION MICROSCOPES at heavy-ion storage rings will enable multiple ionization studies in completely unexplored time and intensity domains, where the removal of many electrons might be described by the "simultaneous" absorption of many virtual photons from the field (4,5).

In figure 8, the momentum vectors of triply ionized Ne recoil ions are plotted together with the vector sum momenta of all three emitted electrons for 3.6 MeV/u Au$^{53+}$ impact. The collision plane is defined as in figure 2. Obviously, even for triple ionization in a projectile velocity regime (12 a.u.), where the equivalent photon picture is not expected to be applicable, features typical for photo absorption are observed. As reported previously for double ionization (11), little momentum is transferred at considerable energy transfer, leading to an "explosion" of the atom in the field. Due to the large perturbation by the projectile in the final state, a strong effect by the PCI is observed dragging each of the electrons along but, at the same time, pushing away the Ne$^{3+}$ ions with nearly identical momenta. PCI thus can be seen as a dissociation of the target fragments in the field of the receding ion without any net-momentum transfer to the fragments.

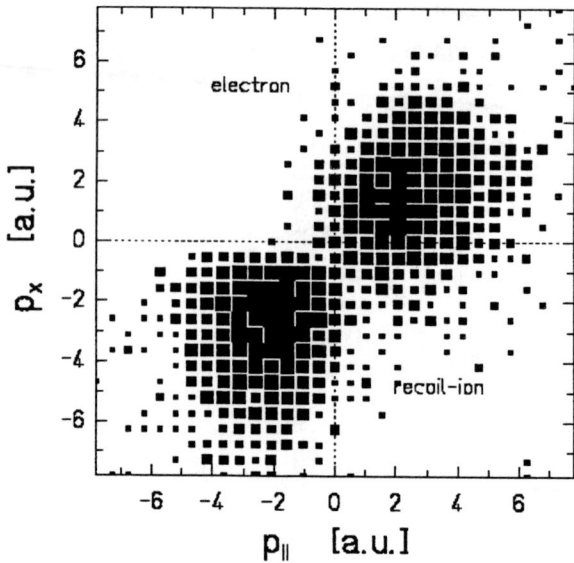

**FIGURE 8:** Momentum distributions for the Ne$^{3+}$ recoil ion and the sum momentum of three electrons in triply ionizing 3.6 MeV/u Au$^{53+}$ on Ne collisions.

On this basis, one might consider to separate the final state interaction from the signatures that depend on the initial state, i.e., the relative momenta and energies between the three electrons, by a transformation into the three-electron center-of-mass (CM) coordinate frame. Implying that all the electrons are influenced by the PCI on the same footing, independent on their momenta in the instant of ionization, such a transformation would lead to a frame where the PCI is not present at all. We have performed such a transformation and plotted the relative energies of the three electrons in the CM system $\varepsilon_{ei} = E_i^{CM} / \sum E_i^{CM}$ (with $E_i^{CM}$ being the CM energy of the i$^{th}$ electron) in a modified Dalitz–plot (27) in figure 9. This is an equilateral triangle where each triple ionization event is represented by one point inside the triangle with its distance from each individual side being proportional to the relative energy of the corresponding electron, as indicated in the figure. Only events in the inscribed circle are allowed due to momentum conservation of the three electrons in the CM frame $\sum \vec{P}_{ei}^{CM} = 0$. Numbering of the electrons is achieved exploiting information on their emission angle: Electron 1 is the one with the smallest angle relative to the projectile propagation direction in each triple ionization event, electron 3 the one with the largest angle, and electron 2 lying in between.

Obviously, the electron energies are not independent of each other, and the many electron continuum explored for the first time experimentally is found to be strongly correlated. There is an increased probability that electron one and three have large energies compared to electron two. Performing nCTMC calculations (n-body Classical Monte Carlo approach; (28)), with the electron–electron interaction not included beyond an effective potential in the initial state, these structures cannot be reproduced.

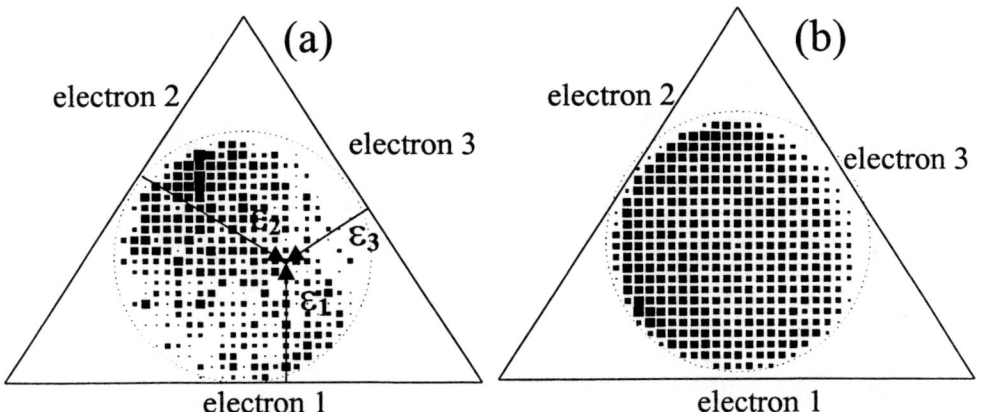

**FIGURE 9:** Dalitz-plots (see text) for the relative energies of three electrons in the CM-system emitted in triply ionizing 3.6 MeV/u $Au^{53+}$ on Ne collisions. (a): experiment; (b): theory.

One should keep in mind that such calculations have been demonstrated to accurately predict the dynamics of multiple ionization reactions. A similar result has been obtained for double ionization recently, and qualitative agreement between nCTMC and experiment was only achieved when the electron–electon interaction was explicitly implemented in the final state. Presently, such calculations are underway and will be presented in a forthcoming, detailed paper (29).

In the light of the results for single ionization by the same projectiles, where the longitudinal electron momentum spectra have been demonstrated to closely reflect the Hartree-Fock initial state wave function (see figure 3), it does not seem to be too optimistic to expect that multi-electron momentum distributions might reveal direct information about the correlated many-electron bound state wave function. Moreover, since the target disintegration occurs within attoseconds, i.e., on a time scale short compared to typical revolution times of ground state electrons, one might even hope that such experiments will provide direct information on the short-time correlation between the electrons in the initial state.

## CONCLUSION

Using our REACTION MICROSCOPE, we have performed a series of kinematically complete experiments on single, double and triple ionization of target atoms for fast highly charged ion impact, exploring in detail the relationship between photon and charged particle induced ionization processes.

Covering the whole multi-dimensional final state momentum space (within the minor limitations due to finite acceptances mentioned before), such data for the first time offer the unique possibility to project onto any suitable subspace of lower dimension, or to transform into any set of desired coordinates, where the underlying physics is appropriately elucidated. For the investigation of dynamic mechanisms that contribute to single or multiple ionization, the plane transverse to the beam propagation has been identified as an adequate subspace, and electron momenta have been projected onto this plane. Similarly, signatures of the initial state wave function are demonstrated to be sensitively reflected along the longitudinal direction, where no significant momentum can be transferred during the collision. Finally, evidence was provided that a transformation into the CM coordinate frame of the emitted electrons in multiple ionization events helps to identify the correlated electron motion even in the presence of strong, disturbing final state interaction with the receding projectile.

On this basis, it was demonstrated that new information, beyond results that have been achieved in photo ionization studies, might be obtained. For weak fields, where the projectile interacts via absorption or scattering of a single virtual photon, we have shown that longitudinal electron momentum spectra sensitively depend on the Hartree-Fock initial state for single ionization, or on the correlated bound state in case of double ionization. Our projection techniques enabled us to explore signatures of the correlated initial state at all momentum and energy transfers, i.e., at all virtual photon frequencies. For strong fields, not yet producible with real photon sources, the transformation into the CM system separates dynamical effects, due to the final state interaction of the projectile with the target fragments from the correlated motion of the electrons relative to each other. The three-electron continuum has been found to be strongly structured.

In the future, such kind of experiments will become considerably more efficient, detailed and precise using internally cold, sub-nanosecond pulsed, low-emittance and possibly polarized beams in storage rings, from EBIT ion traps or photo-emission polarized electron sources. Experiments are under way to investigate the dynamics of multiple ionization in femtosecond intense laser fields using REACTION MICROSCOPES.

## ACKNOWLEDGEMENT

We are grateful to GSI and GANIL for providing excellent beams. We acknowledge financial support from GSI, the Deutsche Forschungsgemeinschaft within the Leibniz-Programm and SFB-276 projects B7 and B8, as well as from the European Community under the program "Access to large facilities".

# REFERENCES

1. Bethe, *Ann. Phys., Lpz.* **5**, 325 (1930)
2. Fermi E., *Z. Phys.* **29**, 510 (1924); von Weizsäcker C. F., *Z. Phys.* **88**, 612 (1934); Williams E. J., *Phys. Rev.* **45**, 729 (1934);
3. Van der Wiel M. J., Wiebes G., *Physica,* **53**, 225 (1971); Wight G. R., Van der Wiel M. J., *J. Phys. B* **9**, 1319, (1976)
4. Moshammer R., Schmitt W., Ullrich J., Kollmus H., Cassimi A., Dörner r., Jagutzki O., Mann R., Olson R. E., Prinz H. T., Schmidt-Böcking H., Spielberger L., *Phys. Rev. Lett.* **79**, 3621 (1997)
5. Keller S., Lüdde H. J., Dreizler R., *Phys. Rev. A* **55**, 4215 (1997)
6. Stolterfoht N., et al., *Phys. Rev. Lett.* **80**, 4649, (1998)
7. Bertulani C., Baur G., *Phys. Rep.* **163**, 299 (1988); Baur G., *J. Phys. B.* (Topical Review, in print)
8. Wang J., McGuire J. H., Burgdörfer J., *Phys. Rev. A* **51**, 4687 (1995)
9. Manson S. T., McGuire J. H., *Phys. Rev. A* **51**, 400 (1995)
10. Wang J., et al., *Phys. Rev. Lett.* **77**, 1723, (1996)
11. Moshammer R., Ullrich J., Kollmus H., Schmitt W., Unverzagt M., Jagutzki O., Mergel V., Schmidt-Böcking H., Mann, R., Woods C. J., Olson R. E., *Phys. Rev. Lett.* **77**, 1242, (1996)
12. Dorn A., Moshammer R., Schröter C. D., Zouros T. J. M., Schmitt W., Kollmus H., Mann R., Ullrich J., *Phys. Rev. Lett.* **82**, 2496, (1999)
13. Bapat B., Moshammer R., Keller S., Schmitt W., Cassimi A., Adoui L., Kollmus H., Dörner R., Weber Th., Khayyat K., Mann R., Grandin J. P., Ullrich J., *J. Phys. B.* **32**, 1859, (1999)
14. Ullrich J., Moshammer R., Dörner R., Jagutzki O., Mergel V., Schmidt-Böcking H., Spielberger L., *J. Phys. B.* **30**, 2917, (1997), Topical Review
15. Moshammer R., Unverzagt M., Schmitt W., Ullrich J., Schmidt-Böcking H., *Nucl. Instrum. Meth.,* B **108**, 425, (1996)
16. Kollmus H., Schmitt W., Moshammer R., Unverzagt M., Ullrich J., *Nucl. Instrum. Meth.,* B **124**, 377, (1997)
17. Suarez S., et al., *Phys. Rev. A* **48**, 4339, (1993)
18. Wood C. J. Olson R. E., Schmitt W., Moshammer R., Ullrich J., *Phys. Rev. A* **56**, 3746, (1997)
19. Fainstein P. D., et al., *Phys. Rev. A* **53**, 3243, (1996)
20. Moshammer R., Fainstein P. D., Schulz M., Schmitt W., Kollmus H., Mann R., Hagmann S., Ullrich J., *Phys. Rev. Lett.* (submitted)
21. Moshammer R., Ullrich J., Unverzagt M., Schmitt W., Jardin P., Olson R. E., Mann R., Dörner R., Mergel V., Buck U., Schmidt-Böcking H., *Phys. Rev. Lett.* **73**, 3371, (1994)
22. Schmitt W., Moshammer R., O'Rourke F. S. C., Kollmus H., Sarkadi L., Mann R., Hagmann S., Olson R. E., Ullrich J., *Phys. Rev. Lett.* **81**, 4337, (1998)
23. Bapat B., Keller S., Moshammer R., Ullrich J., *J. Phys. B* (submitted)
24. Keller S., Bapat B., Moshammer R., Ullrich J., Dreizler R. M., *J. Phys. B* (accepted)
25. Huetz A., Selles P., Waymel D., Mazeau J., *J. Phys. B.* **24**, 1917, (1991)
26. Dörner R., Mergel V., Bräuning H., Achler M., Spielberger L., Jagutzki O., Weber Th., Khayyat K., Moshammer R., Ullrich J., Cocke C. L., Prior M. H., Azuma Y., Schmidt-Böcking H., Proceedings of the *International Workshop on Atomic and Molecular Physics at High Brillianz Synchrotron Radiation Facilities* (1999), accepted
27. Dalitz R. H., *Phil. Mag.,* **44**, 1068, (1953)
28. Olson R. E., Ullrich J., Schmidt-Böcking H., *Phys. Rev. A* **39**, 5572 (1989)
29. Schulz M., Moshammer R., Schmitt W., Kollmus H., Mann R., Hagmann S., Olson R. E., Ullrich J., *J. Phys. B* (accepted)

# X-rays and inner-shell processes with heavy ions channeled in thin crystals

D. Dauvergne,[1] M. Chevallier,[1] C. Cohen,[2] N. Cue,[3] J. Dural,[4]
R. Kirsch,[1] A. L'Hoir,[2] D. Lelièvre,[4] P.H. Mokler,[5] J.-C. Poizat,[1]
H.-T. Prinz,[5] J.-M. Ramillon,[4] J. Remillieux,[1] P. Roussel-Chomaz,[6]
J.-P. Rozet,[2] F. Sanuy,[1] D. Schmaus,[2] C. Stephan,[7]
M. Toulemonde,[4] D. Vernhet,[2] and A. Warczak[8]

[1] *Institut de Physique Nucléaire de Lyon and IN2P3, Université Claude Bernard Lyon-I, 43, Bd du 11 Novembre 1918, 69622 Villeurbanne Cedex, France*
[2] *Groupe de Physique des Solides, CNRS UMR 75-88, Universités Paris VII et Paris VI, 75251 Paris Cedex 05, France*
[3] *Department of Physics, The Hong Kong University of Sciences and Technology, Kowloon, Hong Kong*
[4] *CIRIL, UMR 11 CNRS-CEA, rue Claude Bloch, 14040 Caen Cedex, France*
[5] *Gesellschaft für Schwerionenforschung (GSI), D-64291 Darmstadt, Germany*
[6] *GANIL, CEA/IN2P3, BP 5027, 14076 Caen Cedex 5, France*
[7] *IPN Orsay and IN2P3, BP 1, 91406 Orsay Cedex, France*
[8] *Uniwersytet Jagielloński, Institut Fizyki, PL-30-059 Kraków, Poland*

**Abstract.** We present some of the recent developments on heavy-ion channeling experiments, in the framework of X-ray emission and inner-shell processes. We discuss the possibility to characterize the very low electron densities sampled by hyperchanneled ions, and we report on an attempt to observe exotic trielectronic recombination.

## INTRODUCTION

Positively charged particles entering a crystal at a small angle with respect to a crystallographic direction are steered away from the lattice atoms. Indeed, the so-called channeled ions are trapped in a 2D (axial) or 1D (planar) potential well, resulting from the average potential of the atomic strings or planes. Their transverse energy is given by the incidence conditions (position and angle relative to the axis or planes), and defines the area in the transverse space where they are allowed to travel inside the thin crystal. Thus close collisions with the target atomic cores are forbidden for low-transverse-energy particles. They can only interact with the quasi-free valence or conduction electron gas, the density of which is lower than the average electron density in the solid. The exclusion of close collision with core electrons implies a reduction of the energy loss of the projectile. Since no recoil of target atoms can be involved for a channeled ion, the dominant charge exchange processes are different from those occurring during classical ion-atom collisions. Direct ionization comes from Electron Impact Ionization (EII), and capture is either resonant, like Dielectronic Recombination (DR, in which the electron capture is accompanied by the electronic excitation of the ion), or requires a photon as a third body (Radiative Electron Capture, REC, that is the time-reverse of the photoelectric effect). Extensive studies of energy-loss and charge-exchange processes in channeling conditions have been performed over the last two decades. A review of such studies was made a few years ago by Krause and Datz [1]. Out of the scope of the present paper, one should also mention the possibility to study coherence effects on electronic excitations, due to the periodicity of the crystal field (Resonant Coherent Excitation or RCE), which is also reported in Ref. [1].

One essential difficulty (but certainly a very exciting aspect) is to extract quantitative results from such charge exchange experiments in channeling conditions. Indeed, one has to know the transverse energy distribution of the beam inside the crystal. It depends on the crystal quality and on the beam characteristics [2,3]. The value of the transverse energy of each projectile can be determined by measuring its energy loss [4] and/or, for thick enough targets and for incident projectiles far off the equilibrium charge state, by the charge state analysis at the exit of the target [5,6]. In the latter case, the best-channeled particles remain "frozen" in their initial charge state, whereas non-channeled ions undergo multiple charge exchanges. Thus one can relate the charge state distribution to mean sampled electron density via the transverse energy distribution. Also a detailed analysis of DR resonance profiles [1,4] and REC photon line shapes [7] allows to characterize the sampled electrons, through their longitudinal momentum distribution (Compton profile).

Two particular aspects of electron capture in channeling conditions will be stressed in the following sections. First, we discuss about the possibility to characterize the density and Compton profile of the electrons at the very center of a crystal channel, which requires to select the so-called hyperchanneled ions. Second, we use the very "thick" quasi-free electron target (typically $\sim 10^{20} e^-/cm^2$) offered by aligned thin crystals to search for very-low-cross-section processes like the trielectronic recombination.

# RADIATIVE ELECTRON CAPTURE BY CHANNELED IONS

Channeling effects on X-ray emission are illustrated in Figure 1. Details of this experiment are given in Ref. [7]. Bare $Kr^{36+}$ ions were sent onto a 37 $\mu m$ thick silicon crystal at 60.1 MeV/u. The X-ray spectra, normalized to the same number of incident particles, were recorded at 90° by a germanium detector for random and ⟨110⟩ axial crystal orientations. For a random orientation of the crystal (a), the spectrum is dominated by the projectile $K_{\alpha,\beta}$ photons, due to de-excitation after Mechani-

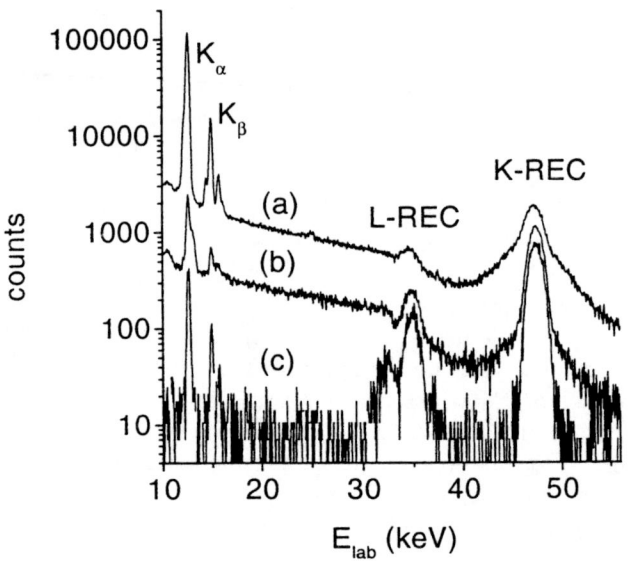

**FIGURE 1.** X-ray spectra recorded by a germanium detector at 90° from the beam direction for Kr$^{36+}$ incident ions at 60.1 MeV/u on a 37 μm thick silicon crystal (a) random crystal orientation. (b) ⟨110⟩ axial crystal orientation, single spectrum. (c) ⟨110⟩ axial crystal orientation, in coincidence with well channeled ions (low energy loss) having captured one electron. All spectra are normalized to the same number of incident ions.

cal Electron Capture (MEC) into outer-shells, and also to further excitation/de-excitation inside the rather thick target. An important bremsstrahlung continuum decreases slowly with increasing photon energy. Primary bremsstrahlung (i.e., bremsstrahlung of target electrons in the field of the projectile) extends up to about 31 keV. Photons of secondary electron bremsstrahlung (in the field of target nuclei) can have an energy up to the maximum energy transferable to a knock-on electron, i.e. $2\beta^2\gamma^2 m_e c^2 = 136$ keV for an electron initially at rest.

Photon peaks resulting from REC into the projectile K- and L-shells are also visible. The shape of these peaks is rather complex, since all electrons of the target contribute to the capture. Core electrons have a broad Compton profile, whereas the central part of the peaks is due to the capture of the quasi-free silicon valence electrons.

The situation in channeling conditions (single spectrum b) is drastically different. Mechanical Electron Capture and Nuclear Impact Excitation are out of reach for channeled ions. Thus $K_\alpha$ and $K_\beta$ peaks are reduced, by a factor 50. The remaining provides us the fraction of unchanneled ions, i.e. ions that enter the crystal close to an atomic row, and also dechanneled ions (by multiple scattering). Ion-electron interaction rates are also reduced, as one can see for the bremsstrahlung contributions. The attenuation factor for primary bremsstrahlung is close to the ratio of

4/14, which is the ratio of the number of valence electrons to the total number of electrons per silicon atom.

A more refined analysis could be performed with the REC peaks, and in particular K-REC. The valence-electron contribution is very little affected by channeling conditions, whereas the core-electron contribution is reduced both in amplitude and width. With the help of simulations based on flux distributions inside a crystal channel, we could reproduce [7] the shape of these REC peaks for various incidence angles. An important feature in this analysis was the use of Compton profiles for 2s and 2p core electrons that depend on the transverse distance to the atomic strings. The wave functions of these electrons extend up to about 0.5 Å, a distance much larger than the extension of the thermal vibrations of the lattice atoms ($\sim 0.1$ Å). Because of the non-uniform particle flux, channeling is the only tool to observe such a dependence of the electron momenta on the impact parameter.

The third spectrum of Fig. 1 (c) was recorded in coincidence with only a fraction of well channeled $Kr^{35+}$ transmitted ions. Indeed, we selected, after a magnetic analysis, about 50% of the ions that captured one electron and had the minimum energy loss inside the crystal. Such conditions enabled us to select only photons resulting from the radiative electron capture of valence electrons: the remaining $K_\alpha$ and $K_\beta$ photons are now only those consecutive to L,M,...-REC (a $n \geq 3$-REC peak is now clearly visible). The bremsstrahlung is strongly reduced because it is not correlated to an electron capture. Only the narrow valence contribution of the REC peaks is remaining. We could compare this line shape with the one obtained for a non-uniform free electron gas, the amplitude being given by the average electron density sampled by these ions, and the width by the average mean local electron density (at the capture site). However, we were not able to select hyperchanneled ions during this experiment, because multiple scattering inside the rather thick crystal led to a significant increase of the transverse energy of the ions, and the selection of the best-channeled ions by their energy loss was not severe enough. So we could observe only a very small decrease of the width of the valence electron momentum distribution between the spectra (b) and (c) of Fig. 1.

We present in Figure 2 an extension of the simulations discussed above and in Ref. [7]. On the upper left of the figure is the transverse view of the particle flux inside the $\langle 110 \rangle$ channel of silicon, once statistical equilibrium is reached. Statistical flux equilibrium means that the distribution of the projectile flux is getting constant after some oscillations close to the entrance surface. No angular divergence of the incident 60 A.MeV $Kr^{36+}$ beam and no dechanneling were taken into account in this simplified simulation. The lower left figure is the corresponding calculated distribution of K-REC photons (within the dipole approximation) emitted at 90° from the beam direction in the laboratory frame, per incident ion, per $\mu$m of crystal thickness, per unit solid angle and unit photon energy. No experimental factor is included (no shift due to energy loss, no broadening due to a detector resolution and angular aperture). A small contribution of core electron capture by poorly- or non-channeled ions is observable, and is much smaller than the valence electrons contribution. On the right side of the figure, a selection has been made of ions with

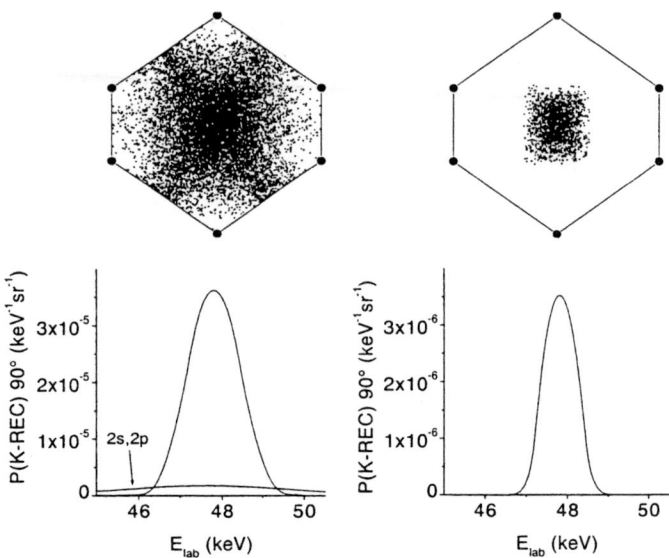

**FIGURE 2.** Simulation of particle flux distributions (upper part) and corresponding calculated K-REC photon lines (lower part), for $Kr^{36+}$ ions at 60.1 A.MeV inside the $\langle 110 \rangle$ axial channel of silicon. P(K-REC) is the number of photons emitted per incident ion (whole beam), per unit solid angle at 90°, per unit photon energy and per micron of crystal thickness. 2s, 2p corresponds to the capture of core electrons. Left: no selection in transverse energy $E_T$. Right: selection of $E_T < 1$ eV per unit charge of the ion.

transverse energy below 1 eV per unit charge (about 17% of the incident beam). The corresponding number of K-REC photons per incident ion is divided by one order of magnitude and the width of the K-REC peak is much narrower. Quantitatively, the calculated mean sampled valence electron densities are 0.12 $e^-/\text{Å}^3$ (without selection) and 0.045 $e^-/\text{Å}^3$ ($E_T <1$ eV selection). The corresponding mean local valence-electron densities are 0.39 $e^-/\text{Å}^3$ and 0.062 $e^-/\text{Å}^3$ at the place where the electron is captured, respectively.

Such K-REC lines by hyperchanneled ions could certainly be observed experimentally, if one would use a much thinner crystal than in Ref. [7], and also a very-high-resolution spectrometer. Note that Andersen et al. [4] estimated that they could select ions having sampled downto 0.1 $e^-/\text{Å}^3$ during their study of Dielectronic Recombination (DR) resonance profile. However, in contrast with K-REC photon peaks, DR resonances have a fine structure that prevents directly obtaining the Compton profile, except for very-low- or very-high-$Z$ projectiles, for which the structure is negligible or resolvable, respectively. Moreover, getting a very-high-resolution for the line-shape analysis with DR is much more time-consuming experimentally, because the beam energy has to be changed step-by-step to cover the resonance.

A very good sensitivity on the transverse energy of ions that undergo K-REC should be achievable by using slowed-down, cooled-down, very heavy, bare (or H-like) ions extracted from a storage ring. Such a charge state is very far from the equilibrium charge state, and, like during electron impact ionization experiments [5,6], only the best-channeled particles would capture only one electron inside the crystal (by REC). Higher-transverse-energy ions would capture many electrons by REC, because they sample higher electron densities, but also-and mainly-by MEC, the cross-sections of which are much higher REC ones. Under such conditions, channeling is a unique tool to study the competition between REC and MEC as a function of impact parameter. Moreover, solid-state effects (wake effects), like those predicted in Ref. [8], could be tested with high accuracy using ions with adiabaticity parameter $\eta = (E/(Z^2 Ry))(m_e/M) < 1$, where $E, Z$ and $M$ are the energy, atomic number and mass of the ion, respectively, $m_e$ the electron mass and $Ry = 13.6$ eV. Extracted beams from the GSI-ESR storage ring after deceleration are now available [9,10], and very low energies are expected soon.

## SEARCH FOR RESONANT TRIELECTRONIC RECOMBINATION

As discussed above, channeling offers the possibility to study REC, DR or EII, that are generally dominated by MEC and Nuclear Impact Ionization (NII) during ion-solid or ion-gas collisions. One can also use this very "thick" electron target to observe processes that have smaller cross-sections than REC or EII, provided one can clearly identify them.

Within this spirit, an attempt to observe the resonant trielectronic recombination in channeling conditions was performed recently at GANIL. Compared to dielectronic recombination, trielectronic recombination (also called Resonant Transfer and Double Excitation - RT2E - [11] in the case of an initially bound electron) involves a double electronic excitation of the ion during the capture of an electron. For instance KK-LLL trielectronic recombination is a capture into the L-shell of an ion, both K-electrons of which being excited also into the L-shell. This leads to a double K-hole hollow ion, with at least three electrons in the L-shell. This three-electron interaction process is expected to have much smaller cross-sections than DR [11,12]. A previous attempt by Zaharakis et al. [11] to observe KK-LLL RT2E during $Kr^{34+} - H_2$ ion-gas collisions yielded an upper limit of $\sim 10^{-25}$ cm$^2$ for the cross-section at the resonance energy. For such high Z ions, the double-K-hole state is expected to decay via two successive $K_\alpha$ (or $K_\beta$) emissions. Indeed the authors of Ref. [11] tried to observe RT2E by detecting transmitted $Kr^{33+}$ ions, and X-rays in coincidence.

Using also $Kr^{34+}$ projectiles and a 3.6 $\mu$m thick silicon crystal aligned along the $\langle 110 \rangle$ axial direction, we measured $K_{\alpha,\beta} \times K_{\alpha,\beta} \times e^-$ capture triple coincidence rates at three incidence energies: 37.1, 40.6 and 42.7 A.MeV, the central value being at the resonance, and the two other ones being out of the resonance, account being taken of its broadening by the Compton profile of valence electrons. X-rays were detected by a set of three large solid angle NaI(Tl) scintillators. The absolute efficiency for $K_{\alpha,\beta} \times K_{\alpha,\beta}$ coincidence detection was 3.7% (assuming an isotropic X-ray emission in the projectile frame). A high-resolution magnetic spectrometer enabled us to select transmitted $Kr^{33+}$ ions and to measure their energy loss. More details of this experiment will be published elsewhere [13]. Figure 3 presents the results of our measurements, with the total probability for an ion to emit two $K-$X-rays in $4\pi$, and to capture one electron in the crystal target. This result holds for ions for which the energy loss inside the crystal was between 0.45 and 0.6 times the "normal" energy loss (i.e. obtained for a random crystal orientation). The poor quality of the crystal used here prevented us from performing a more severe selection on the transverse energy of channeled ions. Due to the presence of etch pits, a small fraction of unchanneled ions could have a lower energy loss than channeled ones, because they sampled a much thinner part of the target.

No resonance is observed at 40.6 A.MeV, and from the statistical accuracy of a fit through the experimental values we can extract an upper limit for the trielectronic recombination cross-section per target electron of $1.9 \times 10^{-27}$ cm$^2$ (with 95% confidence level).

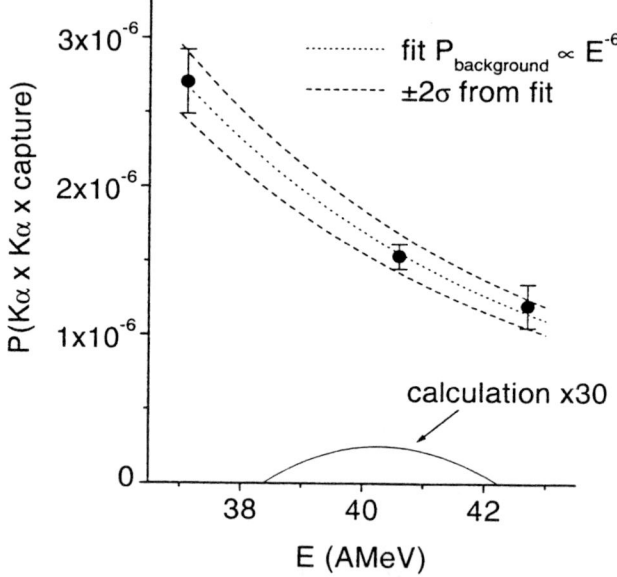

**FIGURE 3.** Probability (in the $4\pi$ solid angle), for $Kr^{34+}$ incident ions on a 3.6 $\mu$m thick silicon crystal aligned along the $\langle 110 \rangle$ axial direction, to capture one electron and to emit two $K-$X-rays. A selection of energy loss between 0.45 and 0.6 times the normal energy loss was made. A calculated energy dependence of the resonant trielectronic recombination probability is shown (based on "double" Auger rate estimates) after multiplication by a factor 30.

The requirements for obtaining the signature of trielectronic recombination were quite severe (triple coincidence and selection in energy loss). The probability for a channeled ion to undergo three uncorrelated events (one radiative electron capture and two electron impact excitations or emissions of bremsstrahlung photons in the same energy range as $K-$X-rays) increases as the cube of the target thickness. This was the reason for our choice of a thin crystal, for which this probability was very low. The dependence on the energy of the fitting function in Fig. 3 corresponds rather to the energy dependence of MEC×NIE×NIE (NIE = Nuclear Impact Excitation). Such a sequence of events is only possible for unchanneled ions that have lost little energy inside the thinner parts of the crystal target as mentioned above.

This upper limit for the trielectronic recombination cross-section is two orders of magnitude smaller than the one obtained during ion-gas experiments. However, to be observable here, this process should have a probability more than one order of magnitude higher than the one calculated following Marques et al. (calculations of the time reversed "double" Auger rates).

Using a much better quality crystal could allow one to suppress efficiently the contribution from unchanneled ions, as illustrated in the first part of this paper. The expected theoretical value of the cross-section could then be probably reached.

## SUMMARY

In summary, we illustrated here two aspects of X-rays studies with heavy-ion channeling. In the first part, we use very high statistics on K-REC to allow a very detailed study of the target electron momentum distribution through the REC photon line shape analysis. With the help of simulations based on the particle flux distribution inside a crystal, we discuss the possibility of observing the much narrower K-REC peak by hyperchanneled ions. This could be obtained with slow, very heavy, bare or H-like ions, i.e. ions with an incident charge far from the equilibrium charge state, which would ensure a very good selectivity in transverse energy. In the second part, we illustrate the use of a very "thick" electron target to try to observe very low cross-section charge exchange processes. A cross-section in the millibarn region was obtained for the resonant trielectronic recombination on krypton ions, a result out of reach of other methods.

## REFERENCES

1. H.F. Krause and S. Datz, *Advances in Atomic, Molecular and Optical Physics*, edited by B. Bederson and H. Walther, Academic, New York, **37**, 139 (1996).
2. D.S. Gemmell, *Rev. Mod. Phys.* **46**, 129 (1974).
3. J. Lindhard, *Mat. Fys. Medd. Dan. Vid. Selsk.* **34** n° 14 (1965).
4. J.U. Andersen, J. Chevallier, G.C. Ball, W.G. Davies, J.S. Forster, J.S. Geiger, J.A. Davies, H. Geissel and E.P. Kanter, *Phys. Rev. A* **54**, 624 (1996).

5. A. L'Hoir, S. Andriamonje, R. Anne, N.V. De Castro Faria, M. Chevallier, C. Cohen, J. Dural, M.J. Gaillard, R. Genre, M. Hage-Ali, R. Kirsch, B. Farizon-Mazuy, J. Mory, J. Moulin, J.C. Poizat, Y. Quéré, J. Remillieux, D. Schmaus and M. Toulemonde, *Nucl.Instr. and Meth.* **B48**, 45 (1990).
6. D. Dauvergne, C. Scheidenberger, A. L'Hoir, J.-U. Andersen, S. Andriamonje, C. Böckstiegel, M. Chevallier, C. Cohen, N. Cue, S. Czajkowski, J. S. Forster, H. Geissel, H. Irnich, T. Kandler, R. Kirsch, A. Magel, P. H. Mokler, G. Münzenberg, F. Nickel, YU. L. Pivovarov, J.-C. Poizat, M.-F. Politis, J. Remillieux, D. Schmaus, Th. Stöhlker, T. Suzuki and M. Toulemonde, *Phys. Rev. A* **59**, 2813 (1999).
7. S. Andriamonje, M. Chevallier, C. Cohen, N. Cue, D. Dauvergne, J. Dural, R. Kirsch, A. L'Hoir, J.-C. Poizat, Y. Quéré, J. Remillieux, C. Röhl, J.-P. Rozet, D. Schmaus, M. Toulemonde and D. Vernhet, *Phys. Rev. A* **54**, 1404 (1996).
8. J.M. Pitarke, R.H. Ritchie and P.M. Echenique, *Phys. Rev. B* **43**, 62 (1991).
9. H.-Th. Prinz, D. Dauvergne, S. Andriamonje, K. Beckert, M. Chevallier, C. Cohen, J. Dural, H. Eickhoff, B. Franzke, H. Geissel, R. Kirsch, A. L'Hoir, P. H. Mokler, R. Moshammer, F. Nickel, F. Nolden, J.-C. Poizat, H. Reich, J. Remillieux, F. Sanuy, C. Scheidenberger, D. Schmaus, M. Steck, Th. Stöhlker and M. Toulemonde, *Hyperfine Interactions* **108**, 325 (1997).
10. S. Andriamonje, K. Beckert, M. Chevallier, C. Cohen, D. Dauvergne, J. Dural, H. Eickhoff, B. Franzke, H. Geissel, R. Kirsch, A. L'Hoir, P. H. Mokler, R. Moshammer, F. Nickel, F. Nolden, J.-C. Poizat, H.-T. Prinz, H. Reich, J. Remillieux, F. Sanuy, C. Scheidenberger, D. Schmaus, M. Steck, Th. Stöhlker and M. Toulemonde, *J. Phys. B* **30**, 5099 (1997).
11. K.E. Zaharakis, R.R. Haar, O. Woitke, M. Zhu, J.A. Tanis and N.R. Badnell, *Phys. Rev. A* **52**, 2910 (1995).
12. J-P. Marques, F. Parente, P. Indelicato and J.-P. Desclaux, *J. Phys. B* **31**, 2897 (1998).
13. M. Chevallier et al., submitted to *Phys. Rev. A*.

# Two- And Three-Body Effects in Fast Ion-Atom Collisions: Analogies Between Photon And Charged Particle Impact

N. Stolterfoht[1], B. Sulik[2], J. A. Tanis[1,3], J.-Y. Chesnel[1,4], L. Gulyás[2], F. Frémont[4], D. Lecler[4], D. Hennecart[4], X. Husson[4], J. P. Grandin[4], M. Grether[1], Cs. Koncz[2], and B. Skogvall[5],

[1] *Hahn-Meitner-Institut GmbH, Glienicker Str. 200, D-14109 Berlin, Germany*
[2] *Institute of Nuclear Research (ATOMKI), H-4001 Debrecen, Hungary*
[3] *Western Michigan University, Kalamazoo, Michigan 49008 USA*
[4] *Centre Interdisciplinaire de Recherche Ions-Lasers, Université de Caen, F-14070 Caen, France*
[5] *Technische Universität Berlin, Hardenbergstr. 36, 10623 Berlin, Germany*

**Abstract.** Two- and three-body effects in the emission of low-energy electrons from atomic Li by fast (95 MeV/u) $Ar^{18+}$ projectiles are identified and separated experimentally. Cross sections for single electron emission have been measured for electron energies ranging from 3 to 1000 eV and angles ranging from 25° to 155°. Models based on the Born approximation are introduced to separate two- and three-body effects in the angular distributions of the ejected electrons. The two- and three-body processes are associated with Compton scattering and photoabsorption, respectively. Both the high projectile velocity and the use of the Li target are shown to be essential for performing the present analysis. The emission of the 1s electron is attributed mainly to three-body effects. The cross section for three-body processes rapidly decreases with the electronic energy transfer, in accordance with estimations based on the photoabsorption cross sections. Remarkably large contributions from two-body collisions for the low-energy emission of the 2s electrons have been found.

## INTRODUCTION

The three-body Coulomb problem is one of the fundamental topics in physics. Electron emission from a target atom by ion impact is a Coulomb three-body process [1-3], which can be systematically studied from both the experimental and the theoretical sides. Due to its importance in basic understanding and applications, electron ejection has been a subject of intense investigation since the beginning of detailed studies of ion-atom collisions [4,5].

The two-body Coulomb problem, involving an ion colliding with a free electron, may be considered as being solved within the framework of quantum mechanics, whereas three-body phenomena can only be described by means of approximate methods. Therefore, separate studies of the two- and many-body parts provide important insight into the fundamental nature of ionization mechanisms [6-13]. A large variety of three- and many-body Coulomb processes are exhibited in atomic collisions (including one center electronic transitions, electron transfer, two-center effects, post-collision effects, etc.). Experimental and theoretical studies of these phenomena are often complicated, especially in low and intermediate velocity collisions. A relatively simple three-body process is a one-center electronic transition with low-momentum-transfer, induced by the electric field of a fast charged projectile. It can be studied at high impact velocities, in the validity region of first-order perturbation theories. Moreover, as will be shown below, a good separation of two-body and the above simple three-body effects can be achieved in the collisions of very fast ions (at impact velocities $v/c \geq 0.4$) with specific atomic targets, e.g., Li.

In the present work, we provide a brief summary of our experimental and theoretical studies for separating two-body and three-body contributions in collisions of fast heavy ions with atomic Li [10-13]. In this work, more emphasis is given to the similarities with photon impact. Additionally, the role of the binding energy and the atomic structure in the relative yields of two- and three-body processes is analyzed.

## General Considerations

Electron-emission spectra, double differential in energy and emission angle, exhibit various characteristic features, which can be associated with particular ionization mechanisms. The ionization process is dominated by the emission of low-energy electrons, referred to as soft-collision electrons. They are produced mainly in large impact-parameter collisions, and they have the largest probability for ejection.

When the velocity of the projectile is much larger than the velocity of the active atomic electron, the momentum transfer in a soft collision is very small on an atomic scale. On the other hand, an ejected electron may carry away a significant amount of momentum. If the electron momentum is larger than the momentum transfer, a third body is required to balance the missing momentum. If no other particle is ejected in the collision, the target nucleus (or target core) has to take part in the ionization process. Therefore, soft collisions at high projectile energies can be attributed to three-body collisions involving the projectile, the active electron, and the residual target [14]. This particular three-body process can be considered to be the simplest case for the three-body Coulomb effect. It resembles photoionization where the incident photon is annihilated, therefore it can not ionize without interaction with the residual target ion. It is noted here that the photon has the smallest momentum belonging to a specific energy, namely $k=E/c$.

The analogy between photons and charged particles with regard to ionization was recognized in the pioneering studies by Bethe [4] and Williams [5] and have subsequently received much attention by Inokuti [8,15] and Kim and collaborators [16,17]. More recently, the construction and use of large ion accelerator facilities and advanced synchrotron light sources have led to new interest in this field [1,3,18-22].

Photoabsorption is mediated by dipole transitions with the transfer of a unit angular momentum ($\Delta l = 1$). In analogy, one may consider a soft collision process as emission of virtual photons by the fast projectile and the absorption of one of them by the *target atom as a whole* giving rise to dipole transitions [5,20]. Due to the uncertainty principle, the angular momentum transfer affects the angular distribution of the ejected electron [10]. The angular momentum $l$ and the emission angle $\theta$ are canonical quantities which are subject to the condition $\Delta l \, \Delta\theta \geq 1$ [10,11]. Hence, low-order multipole transitions produce a broad angular distribution. For instance, the ejected electrons produced by dipole transitions from an initial $s$ state exhibit a ($\sin^2\theta$)-like angular distribution symmetric around 90° [9].

Hence, ejection of slow electrons in fast ion-atom collisions at low momentum transfer has a clear three-body character, but it is a relatively simple process. It is easy to treat with the use of perturbation theories. Moreover, it provides a simple angular distribution pattern in the spectrum of the emitted electrons.

In a close encounter between the projectile and the atomic electron, significant momentum can be transferred, and two-body effects become important. The incident ion interacts with an individual electron in a binary-encounter process [23] where the target atomic core (or target nucleus) plays only a minor role. The binary-encounter approximation and, thus, the neglect of the interaction with the residual ion, is adopted in the framework of the impulse approximation [24,25]. The energy spectra of the binary-encounter electrons exhibit a pronounced peak whose properties are determined by two-body kinematics [2].

The binary-encounter process between an ion and an electron also has a counterpart in photon impact. As already pointed out by Bethe [4], the two-body process in electron emission by ion impact is analogous to the Compton scattering of photons. In the latter case a photon occurs in the final channel of the collision which may provide the missing momentum. Hence, as in the case of the binary-encounter process, the target nucleus plays a minor role in the Compton scattering of photons.

It is worth noting here, that the scattering of an ion or a photon by a free electron, initially at rest, have an important common feature. Specifically, in both cases there is a unique $E(\theta)$ connection between the energy and the emission angle of the recoil electron directly following from the two-body character of the processes. Hence, the angular "distribution" of a free-electron scattering at a specific electron-energy is infinitely sharp. The measured angular distributions originating from two-body scattering of ions or Compton scattering of photons by *atomic* electrons exhibit peaks with finite widths, reflecting the momentum distribution of the atomic electrons in their initial states. The shape of the binary-encounter peak is determined by the Compton profile of the corresponding bound orbital [26-28]. Note that the binary-encounter peak occurs in both the energy and angular distribution of the emitted electrons.

An important criterion to observe a sharp binary-encounter peak is the validity of the impulse approximation [24,25], which requires projectiles with a large velocity. Fast ions create a binary-encounter peak near 90° in the angular distribution involving high-order multipoles $\Delta l \gg 1$. A smaller binding energy corresponds to a smaller average value of the initial state orbital velocity, resulting in a narrower Compton profile with necessarily higher order multipoles. The transfer of high angular momenta and the production of a distinct peak of small angular width are consistent with the uncertainty principle mentioned above [11].

In summary, for fast projectiles two-body effects in ionization produce a sharp peak, while three-body effects lead to a broad angular distribution of the ejected electrons. These characteristic differences in the angular distributions of the binary- and soft-collision electrons provide an experimental method to separate the two- and three-body effects. In the following, we summarize the experimental method and our results in the separation of the above two effects in the angular distributions of electrons ejected in collisions of fast ions with Li [10-13]. Atomic units are used throughout, if not otherwise stated.

## EXPERIMENTAL METHODS

A detailed description of the experimental methods and the apparatus is provided in Refs. [11,13]. Here we recall only the most important information relevant to the physics studied in the present work. The measurements have been performed at the beamline of the ISL heavy ion cyclotron facility, at the Hahn-Meitner Institut, Berlin, Germany (10.7 MeV/u $N^{7+}$ ions), and at the GANIL accelerator facility in Caen, France (95 MeV/u $Ar^{18+}$ ions) using the scattering chamber shown in Fig. 1. The essential parts are the electron spectrometer [29] and the oven used to produce a lithium vapor target. The oven could be filled with a relatively large quantity of Li, providing a vapor beam for about 6 - 8 hours, with a density corresponding to several mTorr in the target region.

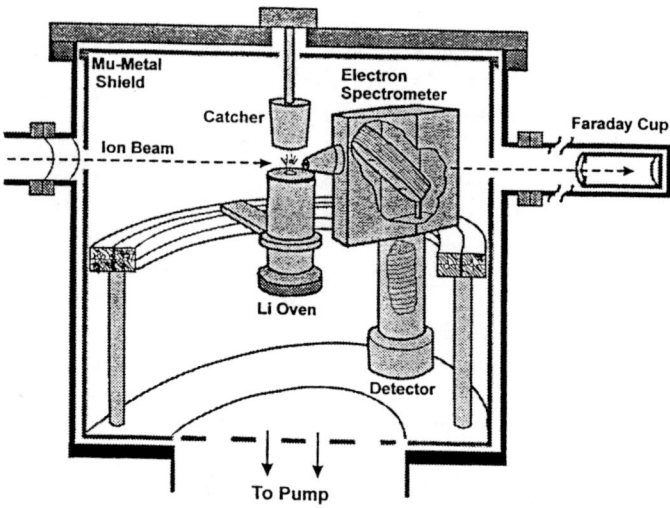

**FIGURE 1.** The experimental setup for measuring the angular distributions of the ejected electrons. The spectrometer is mounted on a movable ring. The vapor jet is directed into a cooled catcher. From Ref. [11].

The experimental difficulties when working with the lithium vapor target are treated in Ref. [11]. Here, we only mention the importance of the slow heating up of the metallic Li to drive contamination from the surface, the construction of an efficient baffle system to avoid lithium deposition on insulators and on surfaces in the spectrometer, and technical solutions to avoid disturbances from stray magnetic fields originating from the heating current of the Li oven. With the spectrometer set-up used, electron yields could be measured reliably for emission energies as low as 3 eV.

The Li-vapor target was crossed by a beam of ions whose current was typically 1-2 µA. The beam was collimated to a size of about 2 mm x 2 mm. Continuum electrons emitted from the Li target were analyzed with the spectrometer within the electron energy range 3-1000 eV. Ejected electrons were observed in an angular range 25°-155°. The maximum of the binary-encounter peak in the vicinity of 90° was measured with an angular step size of 2° and in the wings of the peak the step size was 5° to 10°.

From the measured electron intensities we determined cross sections differential in the energy and angle of the ejected electrons. To obtain absolute cross sections we integrated the measured data over the electron emission angle and normalized the results to the corresponding Rutherford cross sections [1,2]. The relative experimental uncertainties are about ±25% (±40 % below ~10 eV).

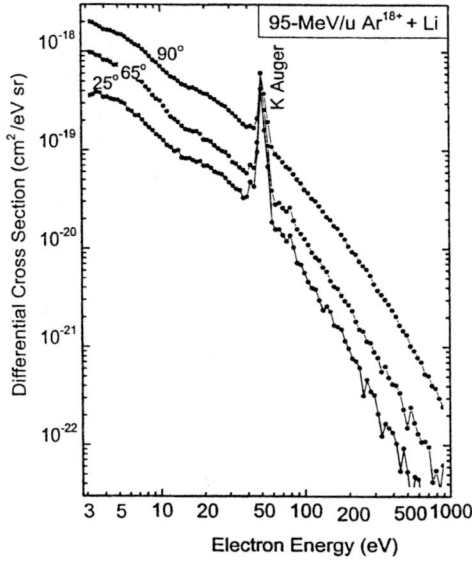

**FIGURE 2.** Energy distribution of electrons emitted in collisions of 95 MeV/u $Ar^{18+}$ on Li at three different emission angles. The continuum part of the spectra originates from single ionization of the Li target. The peak near 50-60 eV is due to KLL Auger-electrons ejected after the excitation of a $1s$ electron to higher bound states. From Ref [11].

Typical double differential cross section spectra are shown in Fig. 2. as a function of the electron energy for the electron emission angles of 25°, 65°, and 90°. The spectra show a distinct maximum near 50-60 eV which is due to KLL-Auger transitions in core excited lithium [30]. Furthermore, the cross sections for continuous electron emission increase strongly with increasing angle up to the region close to 90° where the low-energy electron emission data reach a maximum, indicating the occurrence of the binary-encounter peak. It is noted that the measured data represent the sum of the electron emission from both the $1s$ and $2s$ orbitals. The individual contributions from these shells will be separated by means of model calculations as outlined below.

## THEORETICAL CONSIDERATIONS

A detailed description of the theory is provided in Ref. [11]. In the present work we summarize the main statements and the most important formulas which define the method and the models used. The electronic structures of the initial Li states are very important for the present analysis [10]. Therefore, we first exhibit the densities and wave functions [31] of the bound orbitals 1s and 2s in Fig. 3. The 1s electron has a binding energy of about 59 eV and is localized close (< 1 a.u.) to the nucleus. The 2s electron has a much smaller binding energy of about 5.5 eV. Due to the node in the wave function, the 2s orbital has two parts, an inner part close to the nucleus (< 1 a.u.) and an outer part extending quite far from the nucleus (up to about 7 a.u.).

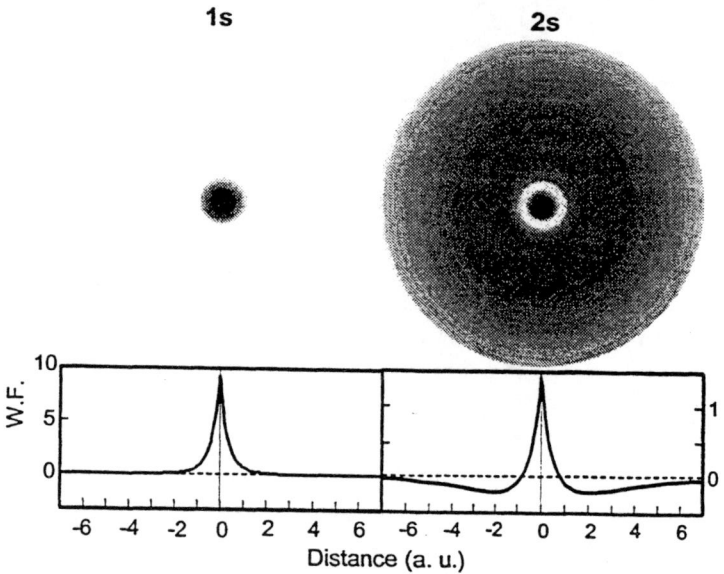

**FIGURE 3.** Electron densities (upper part) and wave functions (lower part) for the atomic orbitals 1s and 2s of Li evaluated using the Cowan code [31]. From Ref.[10].

The fast collisions in this work are treated by means of perturbation theory. For 10.7 MeV/u $N^{7+}$ the projectile velocity was 20.6 a.u., and the ratio of the projectile charge to the projectile velocity was $Z_p/v_p = 0.34$. For the faster 95 MeV/u $Ar^{18+}$ projectile the velocity is 57.5 a.u., ($v/c = 0.42$, $\gamma=1.1$), and $Z_p/v_p = 0.31$. These interaction parameters $Z_p/v_p$ are expected to be sufficiently small for the validity of the Born approximation [2]. Nevertheless, to ensure that two-center effects are negligible, we evaluated electron emission cross sections from the continuum-distorted wave (CDW-EIS) code by Gulyás et al. [32] in comparison with the Born calculations using the same code. The higher-order contributions were found to be significant for the slower (10.7 MeV/u $N^{7+}$) projectile, while their effect was found less than a few percent for the faster argon ions. Accordingly, we put the focus on the faster collision system and neglect two-center effects (which are, in turn, definitely three-body effects) in the present analysis.

Within the framework of the Born approximation, when restricting to $s$ initial states only, the double differential cross section for emission of electrons with energy $\varepsilon$ into the solid angle $\Omega$ can be expressed by expanding the final continuum state into angular-momentum partial waves $lm$ as follows [6,11]

$$\frac{d\sigma}{d\varepsilon\, d\Omega} = 8\pi \frac{Z_p^2}{v_p^2} k \left( \sum_{lm} b_{lm} |Y_{lm}(\Omega)|^2 + \sum_{l \neq l', m} c_{ll'm} Y_{lm}^*(\Omega) Y_{l'm}(\Omega) \right), \quad (1)$$

where $k$ is the momentum of the ejected electron, $Y_{lm}$ are spherical harmonics, and the coefficients

$$b_{lm} = \int_{K_{min}}^{K_{max}} \frac{|F_{lm}(\mathbf{K})|^2}{K^3} dK \quad (2a)$$

and

$$c_{ll'm} = \int_{K_{min}}^{K_{max}} \frac{F_{lm}^*(\mathbf{K})\, F_{l'm}(\mathbf{K})}{K^3} dK \quad (2b)$$

govern the contributions of the individual angular momentum partial waves and their interferences, respectively. Here $F_{lm}(\mathbf{K})$ stands for the form factor for the final continuum state partial wave with electron energy $\varepsilon = k^2/2$, and quantum numbers $lm$, as a function of the momentum transfer $\mathbf{K}$. The minimum momentum transfer is given by $K_{min} \approx \Delta E/v_p$, ($\Delta E = E_i - E_f$) while $K_{max} \approx 2 v_p$.

Due to the orthogonality of the spherical harmonics, the interference terms cancel when single differential cross sections are evaluated by integrating over the electron observation angle. This holds for one-center theories like the Born approximation [2].

We calculated individual multipole terms relevant for integrated cross sections within the framework of the Born-approximation with Hartree-Fock-Slater (HFS) wave functions for the initial and final state of the active electron. Fig. 4 shows examples for the quantity $|F_{lm}(\mathbf{K})|^2/K^3$ occurring within the integral of Eq. (2a). (after a summation over $m$, the vector $\mathbf{K}$ reduces to the modulus $K$.) Results are given for the angular momenta $l=1$ and $l \neq 1$, showing that the function $|F_{lm}(\mathbf{K})|^2/K^3$ maximizes in different ranges of the momentum transfer. The $l=1$ contribution is strongly peaked at $K \rightarrow 0$ whereas the contribution of the $l \neq 1$ multipoles exhibits a maximum near $K \approx k$ whose low-energy wing falls off rapidly at $K \rightarrow 0$. The occurrence of two distinct peaks in $K$ space supports the separation of the $l=1$ and $l \neq 1$ multipoles.

The separation of the form factor has already been proposed in the pioneering work of Bethe [4] by introducing an intermediate momentum transfer $K_0$ to split the integral over the momentum transfer into two parts as follows.

$$\frac{d\sigma}{d\varepsilon\, d\Omega} = 8\pi \frac{Z_p^2}{v_p^2} k \left( \int_{K_{min}}^{K_0} \frac{|F_{if}(\mathbf{K})|^2}{K^3} dK + \int_{K_0}^{K_{max}} \frac{|F_{if}(\mathbf{K})|^2}{K^3} dK \right). \quad (3)$$

By performing the multipole expansion in Eqs. (1-2), the meaning of the above splitting suggested by Bethe becomes obvious. In various studies [4,7,8,16], the first term in Eq. (3) has been already recognized as being due to dipole transitions. As seen from Fig. 4, the intermediate momentum transfer $K_0$ may be located at the minimum of the sum of the two contributions i.e., at the crossing point of the $l=1$ and $l\neq 1$ curves. Thus, the multipoles may be separated to good approximation and, indeed, the first term is nearly exclusively attributed to dipole transitions.

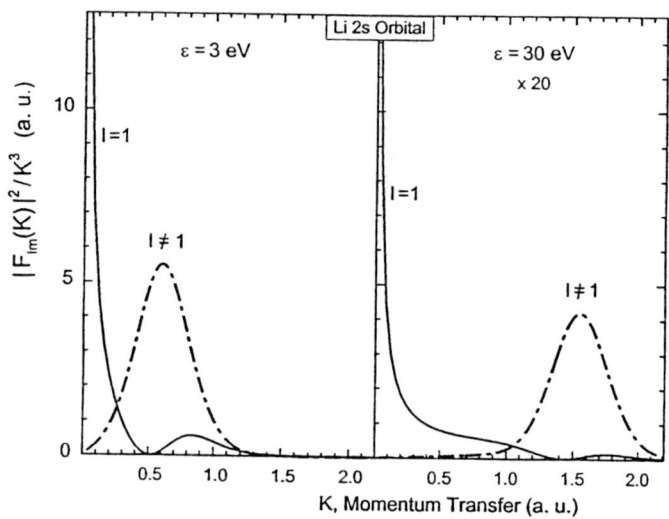

**FIGURE 4.** Expression $|F_{lm}(K)|^2/K^3$ for the Li 2s orbital summed over m for the angular momenta $l=1$ and $l\neq 1$ as a function of the momentum transfer $K$ evaluated in the Born approximation. The left figure is for 3 eV electrons, and the right-hand side for 30 eV. From Ref. [11].

Since the $l\neq 1$ peak maximizes near $k=\sqrt{2\varepsilon}$, the two peaks approach each other with decreasing energy $\varepsilon$, and their overlap may become significant. However, Fig. 4a indicates that even at $\varepsilon =3$ eV, the groups due to $l=1$ and $l\neq 1$ can be quite well separated.

Let us turn to the analogy between ionization by ions and photons. The dipole term in Eq. (3) corresponds to photoabsorption and the second term corresponds to the Compton effect. As pointed out before, photoionization cannot occur without the interaction with the target nucleus and, hence, it refers necessarily to a three-body process. On the other hand, the Compton effect may proceed without the presence of the target nucleus, so that two-body aspects become dominant. Accordingly, in ion-atom collisions the distinction of the $l\neq 1$ and $l=1$ contributions is associated with that of two- and three-body effects, respectively. We note that this separation is primarily justified by the different $K$ dependencies of the form factors demonstrated in Fig. 4.

Finally, we give approximate expressions for the two- and three-body parts which can be used to estimate the corresponding contributions [11]. The cross sections due to the two- and three-body parts will be specified by the labels 2 and 3, respectively.

First, we treat the three-body part by using the dipole term for $l=1$ from Eq. 1. The angular distribution reduces to the form $A+B\sin^2\theta$, where the constants $A$ and $B$ are given by the amplitudes $b_{10}$ and $b_{11}$. Since the smooth component of the angular distribution (three-body part) exhibits asymmetries, we also keep the monopole term $l=0$, which modifies the constant A and introduces the cross term $C\cos\theta$ with $C=c_{100}$ due to an interference effect with the dipole term. Hence one obtains

$$\frac{d\sigma}{d\varepsilon\, d\Omega} = A + B\sin^2\theta + C\cos\theta. \tag{4}$$

Since the monopole term is small due to the high projectile energy, the interference term does not play a significant role. Therefore, the sum of monopole and dipole terms is referred to as an extended dipole term. For small interference, it can be seen from Eq. (4) that the three-body part of the electron emission cross section has a broad angular distribution governed primarily by the $\sin^2\theta$ term.

The two-body part refers to the second term in Eq. (3), representing the transfer of high momenta and angular momenta characteristic of violent collisions. For such a case, it is reasonable to treat ionization as a binary collision between the incident projectile and the target electron, while neglecting the target nucleus. The *free electron peaking approximation* (FEPA) formula [2,11]

$$\frac{d\sigma}{d\varepsilon\, d\Omega} = 4\frac{Z_p^2}{v_p^2 k_c^3} J(p_z)$$

with $\quad p_z = k\cos\theta - K_{min},$ $\tag{5}$

and $\quad k_c \approx (k^2 + 2E_{binding})^{1/2}$

is based on the Compton profile of the initial state $J(p_z) = \int |\varphi_i(\mathbf{p})|^2 d^2 p_\perp$.

Eq. (5) describes a distinct peak, which is identified as the binary encounter peak. It can readily be shown that the location of the binary encounter peak is determined from the condition that the Compton profile maximizes at $p_z=0$. The formula also indicates that the electrons are ejected near 90° when the electron velocity remains much lower than the projectile velocity. The latter condition can readily be achieved when high-velocity projectiles are used.

Since angular distributions of the ejected electrons are of primary interest in this work, we estimate the angular width of the binary encounter peak from Eq. (5), forming the derivative $dp_z/d\theta = k\sin\theta$. Thus, the width of the peak is obtained as

$$\Delta\theta \approx \frac{\Delta p_z}{k\sin\theta} \tag{6}$$

It follows from Eq. (6) that the binary encounter peak is the sharpest if the maximum of the peak at low electron energies is located at $\theta\approx 90°$, where $\sin\theta$ reaches its maximum. This can be achieved again by the impact of high velocity projectiles. Moreover, it follows from Eq. (6) that the peak becomes sharper with increasing electron energy.

## RESULTS

The method of verifying the contributions from the two- and three-body processes is shown in Fig. 5 at four different electron energies. The two-body part is evaluated by means of Eq. (5), where the Compton profiles were deduced from atomic Hartree-Fock wave functions [31]. Fig. 5 shows the fractions for the $1s$ shell (label 2[$1s$]) and the total 2-body contributions (label 2) including both the $1s$ and $2s$ shells. After subtraction of the two-body part, the extended dipole term (Eq. (4)) is used to fit the experimental data taken primarily at forward and backward angles. The results of the fit (label 3) is given by the dashed curves and the sums of all the theoretical components are given by the solid curves.

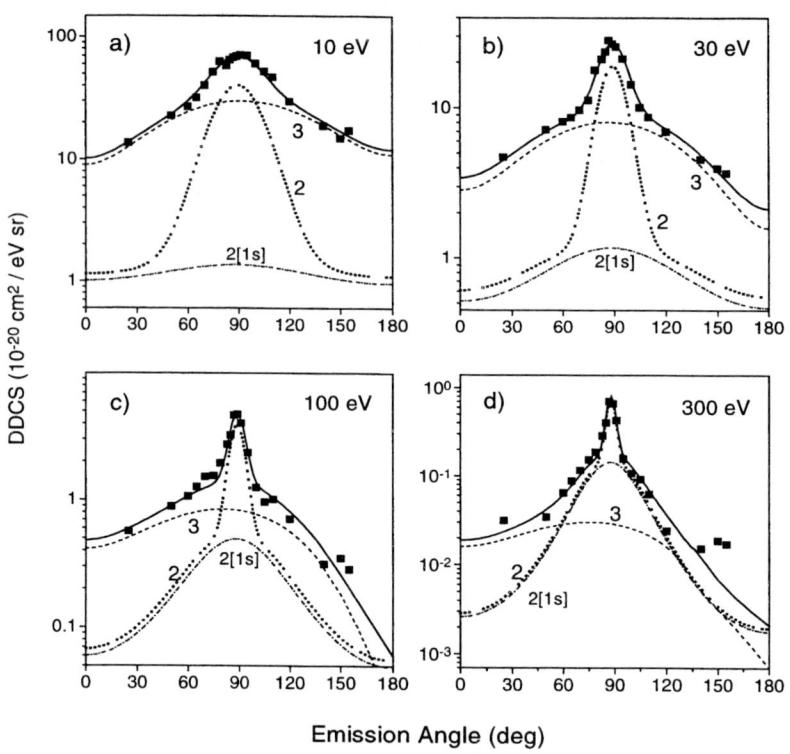

**FIGURE 5.** Angular distribution of 10, 30, 100, and 300 eV electrons emitted in 95-MeV/u $A^{18+}$ + Li collisions. Dotted curves (label 2): two-body calculations (Eq.(5)); Dot-dashed curves (label 2[$1s$]): two-body component for the $1s$ orbital only; Dashed curves (label 3): fit for the three-body part; Solid curves: sum of the contributions (see text). From Ref. [11].

In Fig. 5, the most significant feature of the spectra is the distinct binary-encounter peak near 90° which represents the two-body processes. The shape of the binary-encounter peak is determined by the Compton profile of the target electrons. The most distinct part of the binary encounter peak can be attributed to the $2s$ orbital. The $1s$

electron, since it involves higher momenta, has a broader Compton profile. Also, the inner part of the 2s wave function produces a broader component of it. In fact, it is found that the Compton profiles for the 1s wave function and the inner part of the 2s wave function are nearly proportional. Thus, the 1s wave function, as well as the node in the 2s orbital, gives rise to a "kink" in the Compton profile, which is confirmed by the experiment. It is also seen in Fig. 5, that the binary-encounter peak becomes narrower at higher electron energies, which may readily be understood from Eq. (6).

The theoretical results are seen to agree well with the experimental data. It is clearly shown by Fig. (5) that the relative importance of the two- and three-body parts changes considerably in magnitude with varying electron energy. For higher electron energies, the two-body processes dominate. At lower energies, however, the two-body part does not vanish. For 10 eV electrons, the two-body yield is still as large as the three-body part. This finding is remarkable. Since the pioneering work of Bethe [4], it has become common practice to attribute the emission of low-energy electrons by fast projectiles to dipole transitions [8,16]. The present results show that significant contributions due to two-body processes remain at the low-energy limit.

Theoretical results for individual multipoles were evaluated in the first Born approximation (B1) using a modified version of the program by Gulyás et al. [32]. The calculated double-differential cross sections were integrated over the electron emission angle. The results for the single differential cross section are shown in Fig. 6, separately for the two- and three-body components from both experiment and theory.

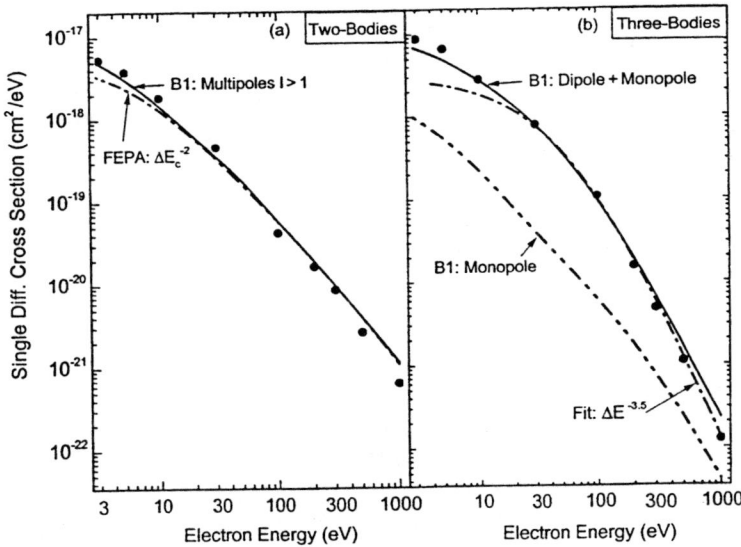

**FIGURE 6.** Angle-integrated cross sections for electron emission in 95-MeV/u $A^{18+}$ + Li collisions separated into two-body processes (a) and three-body processes (b). The curves labeled B1 are Born calculations for the sum of the $l>1$ (a) and the $l \leq 1$ components (b). The results of the FEPA model are shown for comparison in (a), and a power-function referring photoionization in (b). From Ref. [11]

The experimental two-body contributions were obtained by integration of the double differential cross sections after subtraction of the corresponding three-body parts represented by the dashed line in Fig. 6. In addition, Fig. 6 shows results from the FEPA model. Apart from discrepancies at lower energies, where the FEPA is not expected to yield accurate results, the two theoretical data sets exhibit excellent agreement. This agreement is remarkable in view of the simplicity of the FEPA model, based only on the Compton profile of the target. We emphasize here, that the two-body contribution can be described by a simple, parameter-free model with reasonable accuracy.

Similar conclusions have been drawn by Bell et al. [26] for other target species, see also the work by Zouros [28] and recently by Richard et al. [33] and Liao et al. [34]. It is noted here that accurate knowledge of the two-body part makes it possible to extract *electron-ion elastic cross sections* from ion-atom collision data with reasonable accuracy. This fact can be relevant in plasma physics, and for the physics of atomic collisions involving Rydberg atoms, as well.

A significant result of the spectral analysis is that the three-body part of the ionization process can be separated in good approximation from the two-body part (Fig. 6). At energies 100 eV and below, the spectral decomposition shows that the three-body part is well fitted by the extended dipole formula (Eq. (4)). The reliability of the separation is strongly supported by the fact that the monopole term (labeled "B1:Monopole" in Fig. 6b) is relatively small. The summed dipole and monopole terms (labeled "B1:Dipole + Monopole") is about an order of magnitude larger than the monopole term alone. The good agreement obtained between experiment and theory seen in Fig. 6 attests to the validity of attributing the fitted part of the experimental data to dipole transitions and, hence, to three-body effects.

The relatively small monopole term, however, cannot explain stronger forward-backward asymmetries by means of a monopole-dipole interference within the extended dipole term. Note that this is the only one interference term in the model used within the separation method. According to our PWBA calculations, at higher energies (above ca. 400 eV), the asymmetry becomes dominantly due to the interference between the dipole ($l=1$) and the quadrupole ($l=2$) terms. It makes the separation of two- and three-body contributions less reliable. It is noted however, that the error of angle-integrated two- and three-body contributions determined by means of the above separation method does not increase significantly compared to the experimental error in the high-energy region. This is partly due to the fact that the angular integral of the interference term is zero, partly because the roughness of the model used in the separation is covered by the increasing statistical error of the measured data at forward and backward angles. By means of our partial wave PWBA calculations, and the analysis of the statistical uncertainties, we estimated that the error of the three-body part does not exceed 40 % at the highest emitted electron energies measured.

Returning to the analogy between fast ions and photons, we note from the Einstein relation for the photoeffect, that the present electron spectroscopy measurements allow for determining the energy of the annihilated (virtual) photon, since this energy is equal to the electronic energy transfer $\omega = \Delta E = \varepsilon + E_{binding}$. At high impact energies, the photoionization cross section decreases strongly with the photon energy, following the exponential dependence $\sigma_{PI} \sim \Delta E^{-3.5}$ [18,35].

We have fitted the experimental three-body data with the function $C\,\Delta E^{-a}$ where the constants $C$ and $a$ were adjusted to achieve agreement with the experimental data between 30 to 300 eV. The binding energy of the 1$s$ shell was used to determine $\Delta E$, since the 1$s$ orbital is governed by three-body effects. For the power of the exponential function, the fit resulted in the value of $a = -3.5$. The fit results given in Fig. 6b also compare well with the curves obtained from the Born approximation.

For making an estimation of the differential cross section in fast ion-impact in the equivalent photon model [36,37,20], one should calculate first the spectrum (i.e., the number density $n(\omega)$) of the virtual photons emitted by the projectile ion. Following Ref. [36], the single differential cross section corresponding to the absorption of a virtual photon by the atom can be written as (note that $\omega = \varepsilon + E_{binding}$):

$$\frac{d\sigma}{d\varepsilon} = \frac{\sigma_{PI}(\omega)}{\omega} n(\omega). \qquad (7)$$

In the present work, we calculated $n(\omega)$ with a minimum impact parameter $R=1$ au. (see p. 304 in Ref. [36]). Since we were far from the asymptotically high photon energy region, the photoionization cross section $\sigma_{PI}$ was explicitly calculated by means of both hydrogenic and Hartree-Fock-Slater (HFS) wave functions. All the calculations have been performed for the 1$s$ shell of Li. In the 30-300 eV ejected-electron energy region Eq. (7) resulted in a functional form of the cross section which could be quite well approximated by a power function in the form $\sim \Delta E^{-3.9}$ for the hydrogenic wave function, and $\sim \Delta E^{-3.7}$ for the HFS wave function. The latter power, -3.7, is close to the value $a = -3.5$, which is the result of the fit to the measured three-body data. Note that both the measured three-body data and the HFS results are close to the asymptotic form of the total photoionization cross section, $\sigma_{PI} \sim \Delta E^{-3.5}$ [18,35].

It has become clear in the foregoing discussion that the decomposition of the experimental results provides information about two- and three-body effects. However, the experiment cannot distinguish the contributions of the 1$s$ and 2$s$ orbitals. To analyze these contributions, we use theoretical cross sections evaluated by means of the Born approximation [32]. The excellent agreement found between experiment and theory provides confidence that the theory also accurately predicts individual shell contributions. The results shown in Fig. 7 indicate that ionization of the 2$s$ electron is most important at low ejected electron energies, whereas at higher energies the 1$s$ ionization becomes dominant. As already noted from Fig. 6, at high electron energies the two-body part governs the cross sections. This is due to the fact that the three-body cross section follows an exponential law with an exponent of about -3.5, whereas the two-body part involves an exponent of -2 only.

The most significant result for the comparison of 1$s$ and 2$s$ contributions (Fig. 7) is that the two- and three-body effects are very different for the different shells. The 1$s$ ionization is dominated by three-body effects, i.e., two-body effects are found to be small in a wide range of lower electron energies. On the contrary, for the 2$s$ orbital, two-body effects remain important *even at the lowest electron energies*. The results for the Li 1$s$ orbital (Fig. 7) support the dominance of the dipole transitions, since at low energies the two-body part is found to be more than an order of magnitude lower than the three-body part. This result is in agreement with other studies [20]. However, the results for the Li 2s orbital show that the conclusion of the dominant dipole contribution for soft-collision electrons cannot be generalized.

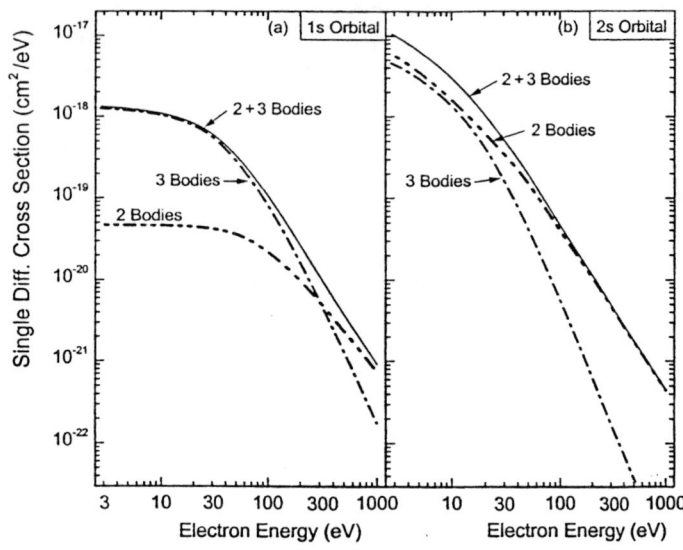

**FIGURE 7.** Angle-integrated cross sections for electron emission in 95-MeV/u Ar$^{18+}$ + Li collisions divided into contributions from the 1s and 2s orbitals, calculated in the Born approximation. Ref. [11].

Characteristic properties of the 2s orbital are its small binding energy, its node producing an inner and outer part, and its large spatial extension. The analysis of the measured angular distribution has shown that the distinct central part of the binary-encounter peak originates from the extensive outer portion of the 2s orbital. However, it is also expected that the small 2s binding energy favors the two-body processes.

To find out which property of the 2s orbital primarily causes the dominance of the two-body effects, we carried out auxiliary calculations in the Born approximation using model hydrogenic wave functions for the 1s and 2s orbitals, always with a binding energy of 5.5 eV, equal to that of the 2s electron. The results of these calculations compared with the more realistic HFS data for the Li 2s shell are shown in Fig. 8.

It is clearly seen in Fig. 8 that the hydrogenic model calculations crudely overestimate the more realistic HFS results, which are in very good agreement with experiment (see Figs. 5-7). The hydrogenic 1s and 2s models give almost identical 3-body contributions, which are about five times larger than that of the HFS calculation. The total cross sections in both (1s and 2s) hydrogenic models are also significantly larger than the HFS results. The two-body contribution is negligible for the 1s model, and relatively small for the hydrogenic 2s model. It becomes larger than the three-body components, however, when using more realistic HFS wave functions. We note here that the *absolute* values of the two-body contributions calculated with 2s wave functions and integrated over the emission angle are almost the same for the hydrogenic and the HFS case, as can be confirmed by a closer inspection of Fig. 8b.

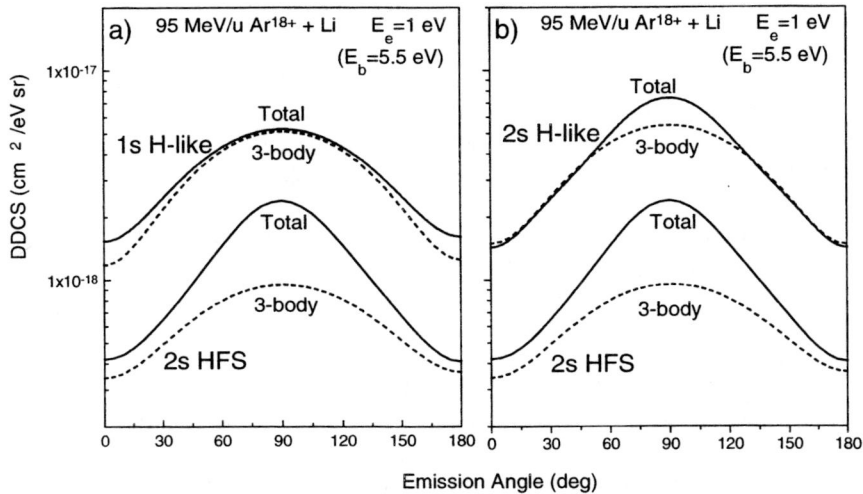

**FIGURE 8.** Calculated angular distributions of the electrons emitted with 1 eV energy from the $2s$ shell of Li in 95 MeV/u Ar$^{18+}$ + Li collisions. **a)**: comparison of the results of a hydrogenic $1s$ wave function calculation with those of a realistic HFS Li $2s$ wave function. **b)**: comparison of a hydrogenic $2s$ calculation with the same HFS Li $2s$ data. All calculations have been performed in the Born approximation by using 5.5 eV, i.e., the realistic (and HFS) value of the binding energy of the Li $2s$ shell.

From the above observations, we conclude that the small value of the binding energy is a necessary but not sufficient condition for gaining a significant two-body contribution in the present collision system. The large value of the two-body contribution can be associated with the large spatial extension of the $2s$ orbital (the two-body component is practically missing in the hydrogenic $1s$ model), but the absolute value of the two-body part is *not* sensitive to the specific form of the $2s$ wave function.

In contrast, the three-body contribution is extremely sensitive to the fact that the wave functions are H-like, or HFS-type. Rather unexpectedly, however, it appears insensitive to the structure of the wave function ($1s$ or $2s$). The absolute value of the three-body component looks primarily depending on the *shape* of the effective *atomic potential*, not the atomic wave function (at least for orbitals with the same binding energy). In the present case, the three-body component is significantly smaller for a screened atomic field than for a $Z_{eff}/r$ potential. It appears that the field of a pure Coulomb center "favors" dipole transitions over those with higher multipoles. This might be an interesting aspect in photoabsorption studies. The prominent role of the two-body effect for the Li $2s$ shell can be considered as a result of the above two findings, namely that a large two-body component appears due to the spatial extension of the $2s$ shell, *and* that the three-body component is small due to the mean-field of the other electrons. Some of the above conclusions for the system 10.7 MeV/u N$^{7+}$ + Li have been drawn by Skogvall et al. [13] in our earlier study.

# CONCLUDING REMARKS

In the present paper, we summarize our results concerning the experimental separation of two- and three-body effects in fast ion-atom collisions. We have shown for fast ion projectiles colliding with atomic Li, that such a separation can be performed to a good approximation. It has also been shown that the use of high projectile velocities is essential for the present analysis. The specific features of the Li target, namely the low binding energy and the large spatial extension of the $2s$ orbital, are rather unique for the ground state of the atoms. However, such properties are rather common for excited states of atomic species. In this respect, the present analysis may be useful e.g., in the treatment of atomic collisions involving Rydberg atoms.

The comparison of $1s$ and $2s$ contributions shows that the two- and three-body effects are very different for the different shells. The $1s$ ionization is dominated by three-body effects in a wide range of lower electron energies. On the contrary, for the $2s$ orbital, two-body effects remain important even at the lowest electron energies. Our results for the Li $2s$ orbital prove that the conclusion of the dominant dipole contribution for soft-collision electrons cannot be generalized.

We have found that two-body models can be used rather accurately in atomic collisions if the two-body effect is dominant. To this end, atomic structure data, such as initial state momentum distributions, or electron-ion scattering cross sections, can be determined from the collisions of ions with light atoms. Such data are of importance for plasma physics, and the direct measurement of them by electron-ion collisions is very difficult [33]. We conclude that accurate knowledge about two-body effects in fast collisions can be important for different aspects.

It has been shown that experimental three-body data follow a dependence which can be derived from the equivalent photon model and the photoionization cross sections. This fact shows that electron spectroscopy experiments with ion-impact provide characteristic information about the photoionization process up to rather high photon energies, thereby confirming the analogy between charged particle and photon impact. It would be desirable to study the connection between the two-body effect and the Compton scattering in more details as well.

The influence of the binding energy and the large spatial extension of the $2s$ orbital has been found to be the major reason for the enhanced two-body contributions observed for the ionization of the $2s$ electron. The yield of the three-body (dipole) component was found to be affected primarily by the shape of the atomic potential.

# ACKNOWLEDGMENTS

This work was supported by the German-French Cooperation Programme PROCOPE (Project No. 98089) and the German-Hungarian Intergovernmental Collaboration (Project No. B/129). J.-Y. C. received a grant from the Alexander von Humboldt Foundation, Germany; B. S. was supported by the Hungarian National Science Fund (Contract No. OTKA-T-020-771); J. A. T. was supported by the Division of Chemical Sciences, Office of Basic Energy Sciences, Office of Energy Research, U.S. Department of Energy.

# REFERENCES

1. M.E. Rudd, Y.K. Kim, D.H. Madison, and T.J. Gay, *Rev. Mod. Phys.* **64**, 441 (1992).
2. N. Stolterfoht, R.D. Dubois, and R.D. Rivarola, *Electron Emission in Heavy Ion-Atom Collisions* (Springer-Verlag, Berlin, (1997).
3. J.H. McGuire, *Introduction to Dynamic Correlation* (Cambridge Univ. Press, Cambridge, 1997).
4. H. A. Bethe, *Ann. Phys.* (Leipzig) **5**, 325 (1930).
5. E.J. Williams, *Phys. Rev.* **45**, 325 (1934).
6. S.T. Manson, L.H. Toburen, D.H. Madison and N. Stolterfoht, Phys. Rev. A **12**, 60 (1975).
7. U. Fano, *Ann. Rev. Nucl. Science* **13**, 1 (1963)
8. M. Inokuti, *Rev. Mod. Phys.* **43**, 297 (1971).
9. A.F. Starace, in *Atomic, Molecular, and Optical Physics Reference Book*, edited by G.W. F. Drake (American Institute of Physics, New York, 1996), Vol. 35.
10. N. Stolterfoht *et al.*, *Phys. Rev. Lett.* **80**, 4649 (1998).
11. N. Stolterfoht, *et al,. Phys. Rev. A* **59**, 1262 (1999).
12. N. Stolterfoht, *Nucl Instr. Meth.* B **145,** 1023 (1999).
13. B. Skogvall, *et al.*, *Nucl. Instrum. Methods.* B **124**, 186 (1997).
14. J.N. Madsen and K. Taulbjerg, *Phys. Scr.* Vol **T 37**, 137 (1997).
15. M. Inokuti, Y. Itakawa, and J.E. Turner, *Rev. Mod. Phys.* **50**, 23 (1978).
16. Y.-K. Kim, *Phys. Rev.* A **6**, 666 (1972).
17. Y.-K. Kim and M. Inokuti, *Phys. Rev.* A 7, 1257 (1973).
18. J. Burgdoerfer, L.R. Anderson, J.H. McGuire, and T. Ishihara, *Phys. Rev.* A **50**, 349 (1994).
19. Y.D. Wang, J. McGuire, J. Burgdoerfer, and Y. Qiu, *Phys. Rev. Lett.* **77,** 1723 (1996).
20. R. Moshammer *et al.*, *Phys. Rev. Lett.* **79**, 3621 (1997).
21. L. Spielberger *et al.*, *Phys. Rev. Lett.* 74, 4615 (1995).
22. R. Doerner *et al.*, *Phys. Rev. Lett.* **77**, 1024 (1996).
23. M. Gryzinski, *Phys. Rev.* **138**, 305, 332 and 336 (1965)
24. P. Eisenberger and P. M. Platzman, *Phys. Rev.* A **2**, 415 (1970).
25. D. Brandt, *Phys. Rev.* A **27**, 1314 (1983).
26. F. Bell, H. Boeckl, M.Z. Wu, and H. D. Betz, *J. Phys. B* **16**, 187 (1983).
27. D.H. Lee *et al.*, *Phys. Rev.* A 41, 4816 (1990).
28. T. J. M. Zouros *et al.*, *Phys. Rev.* A 49, 3155 (1994).
29. N. Stolterfoht, *Z. Physik* **248**, 81 (1971).
30. P. Ziem, R. Bruch and N. Stolterfoht, *J. Phys. B* **8**, L480 (1976). & N. Stolterfoht *et al.*, in *Abstract of Contributed Papers*, edited by I.E. McCarthy, W.R. MacGillivray, and M.C. Standage, (Griffith University Press, Brisbane, 1991) p. 393.
31. R.D. Cowan, *The Theory of Atomic Structure and Spectra* (Univ. of California Press, Berkeley, CA, 1981).
32. L. Gulyás, P.D. Fainstein, and A. Salin, *J. Phys. B* **28**, 245 (1995).
33. P. Richard, C.P. Bhalla, S. Hagmann and P. Zavodszky, *Phys. Scripta*, T **80**, 87-92, (1999).
34. C. Liao *et al.*, *Phys. Rev.* A **59**, 2773, (1999)
35. T. Ishihara, K. Hino, and J. McGuire, *Phys. Rev.* A **44** , R6980 (1991).
36. C. A. Bertulani and G. Baur, *Physics Reports* **163**, 299 (1988).
37. S. Keller, H.J. Lüdde, and R.M. Dreizler, *Phys. Rev.* A 55, 4215 (1997).

# High–Resolution Measurements of the K–Shell Spectral Lines of Hydrogenlike and Heliumlike Xenon

Klaus Widmann[1], Peter Beiersdorfer, Gregory V. Brown, José R. Crespo López–Urrutia[2], Albert L. Osterheld, Kennedy J. Reed, James H. Scofield, and Steven B. Utter

*Department of Physics, Lawrence Livermore National Laboratory, P.O.Box 808, Livermore, CA, 94551, USA*

**Abstract.** With the implementation of a transmission–type curved crystal spectrometer at the Livermore high–energy electron beam ion trap (SuperEBIT) the window on sub–eV level measurements of the ground–state quantum electrodynamics and the two–electron quantum electrodynamics of high–Z ions has been opened. High–resolution spectroscopic measurements of the $K\alpha$ spectra of hydrogenlike $Xe^{53+}$ and heliumlike $Xe^{52+}$ are presented. The electron–impact excitation cross sections have been determined relative to the radiative recombination cross sections. The electron–impact energy was 112 keV which is about 3.7 times the excitation threshold for the $n = 2 \rightarrow 1$ transitions. Although the relative uncertainties of the measured electron–impact excitation cross sections range from about 20% to 50%, significant disagreement between the measured and calculated cross section values has been found for one of the heliumlike xenon lines. Overall, the comparison between experiment and theory shows that already for xenon (Z=54) the Breit interaction plays a significant part in the collisional excitation process. The measured cross sections for the hydrogenlike transitions are in good agreement with theoretical predictions. Additionally, the $Xe^{53+}$ Ly–$\alpha_1$ transition energy has been measured utilizing the $K\alpha$ emission of neutral cesium and barium for calibration. Surprisingly, the experimental result, $(31279.2 \pm 1.5)$ eV, disagrees with the widely accepted theoretically predicted value of $(31283.77 \pm 0.09)$ eV. However, this disagreement does not (yet) call for any correction in respect to the theoretical values for the transition energies of the hydrogenlike isoelectronic sequence. It rather emphasizes the need for a re–evaluation of the commonly used x–ray wavelengths table for atomic inner–shell transitions, in particular, for the cesium $K\alpha$ lines.

---

[1] E–mail: widmann1@llnl.gov.
[2] Present address: Fakultät für Physik, Albert–Ludwigs–Universität Freiburg, D-79104 Freiburg, Germany

# INTRODUCTION

High-resolution spectroscopic measurements of the K-shell spectra of highly charged heavy ions are the key to sub-eV level measurements of the ground-state quantum electrodynamics (QED) and the two-electron QED, respectively [1]. In particular, precise measurements of the 1s-Lambshift of hydrogenlike ions are being persued by several research groups around the world [2]. The most precise tests of QED for highly charged heavy ions, however, are utilizing intra L-shell transitions of lithiumlike ions, such as $U^{89+}$, $Th^{87+}$, and $Bi^{80+}$ [3-6], where high-resolution crystal spectrometers can be employed. The experimental precision of these QED-test measurements is in the 0.1% regime [6].

The increased "visibility" of QED and relativistic effects in high-Z few-electron systems[3] is, unfortunately, also accompanied by an increased impact of the nuclear properties on the atomic structure [7]. Utilization of high-resolution spectrometers is essential. In particular, the acquisition of well resolved spectral features is necessary for the experimental determination of electron-ion collision cross sections, e.g., the electron-impact excitation cross sections, which play an important role for an accurate interpretation and, hence, for reliable predictions of the atomic processes that take place in the x-ray emitting sources. In contrast to the energy-level studies, there is only a small pool of experimental cross-section values available for highly charged ions, especially, for the electron-impact excitation [8,9], which is certainly a drawback for the development of accurate electron-ion collision models. The qualitative and quantitative analysis of the spectral information obtained from the emission of high-Z ions in very high-temperature plasmas depends on these models. Thus, an experimentally well-supported atomic database is needed [10].

High-resolution spectroscopic investigations of the K-shell emission of highly charged ions have been performed for x-ray energies up to 13 keV, i.e, the $n = 2 \rightarrow 1$ transitions in heliumlike $Kr^{34+}$ [11-13]. The implementation of a transmission-type curved crystal spectrometer at the Livermore high-energy electron beam ion trap (SuperEBIT) enables to push these high-resolution measurements into the hard x-ray regime. The observation of hydrogenlike and heliumlike xenon spectra prove the feasibility of this transmission-type crystal spectrometer. The fine-structure-resolved spectrum of the $K\alpha$ emission of the $Xe^{53+}$ and $Xe^{52+}$ ions gives access to transition-energy and relative line-intensity measurements. From the charge-specific determination of the line intensities the electron-impact excitation cross sections are inferred relative to the cross sections for the radiative recombination lines. Xenon was the element of choice due to its potential use in the next-generation fusion devices, such as the National Ignition Facility (NIF) or the International Thermonuclear Experimental Reactor (ITER), which are proposed

---

[3] The approximate scaling behavior for both, relativistic and QED effects, is $Z^4$.

to have electron temperatures in excess of 30 keV. In such a high–temperature plasma only heavy elements like xenon would still have a significant abundance in the heliumlike charge state ($Xe^{52+}$), which is the preferred charge state for plasma diagnostics based on x–ray spectroscopy. Low–Z and mid–Z elements would be completely ionized under these conditions. Another advantage of admixing a high–Z element to the fusion plasma is the strong scaling of radiative transitions and, thus, radiative cooling power with Z. The technological challenge of admixing and removing the high–Z trace element to and from the plasma chamber favors the use of a noble gas, e.g., xenon.

This paper is organized as follows. The first section, "**SuperEBIT operation**", gives a brief description about the SuperEBIT operating parameters that have been used for the production and trapping of the highly charged xenon ions. Section two, "**Transmission–type curved crystal spectrometer**", describes the design and performance of the transmission–type curved crystal spectrometer that has been used for the high–resolution measurements of the $K\alpha$ emission of hydrogenlike and heliumlike xenon. The spectra are presented in the third section, "**$K\alpha$ spectra of $Xe^{53+}$ and $Xe^{52+}$**". The contribution of the various radiative and collisional atomic processes present in the SuperEBIT trap region to the measured relative line intensities is discussed in section four, "**Determination of the electron–impact excitation cross sections**". Section five, "**Measurement of the transition energies**", is dedicated to the measurement of the Ly–$\alpha_1$ transition energy of hydrogenlike xenon utilizing the $K\alpha$ emission of neutral cesium and barium for calibration. A summary and discussion of the results is presented in the sixth and final section, "**Discussion**".

## SUPEREBIT OPERATION

The Livermore SuperEBIT facility is capable of producing, trapping, and exciting any kind of highly charged ions [14,15] up to bare uranium ($U^{92+}$) [16]. The cylindrical trap region measures about 2 cm in axial direction, i.e., along the electron beam. In radial direction, the confinement for the ions is given by the space charge of the electron beam, which is compressed by a 3–Tesla strong magnetic field to a diameter of about 100 $\mu$m [17].

The x–ray emission can be observed perpendicular to the electron beam through round apertures of 1.27 cm diameter. For the investigation of the highly charged xenon x–ray emission, neutral xenon atoms are injected into SuperEBIT by means of a pulsed gas injection system. A continuous flow of neon was also introduced into the trap region in order to supply light ions for the evaporative cooling process [18,19]. The gas injector pressure for the neon cooling was on the order of $2.6 \cdot 10^{-8}$ Torr ($\approx 3.5 \cdot 10^{-6}$ Pa). The xenon injection pressure was above $6 \cdot 10^{-8}$ Torr ($\approx 8 \cdot 10^{-6}$ Pa) during the injection pulse.

The electron beam energy was set to 112 keV in order to obtain a charge balance that is dominated by the heliumlike and hydrogenlike species[4], and the electron beam current was 250 mA. The duration of a "timing" or "trapping" cycle was 11.2 s. One cycle includes injection of xenon gas, ionization until the high charge states are obtained, a period of data acquisition, and dumping of the ions before other high–Z elements intrinsic to the SuperEBIT device, such as barium and tungsten from the electron gun assembly, accumulate in the trap.

A high–purity germanium detector is used to record the x–ray radiation emitted from the trap region. These broadband spectra — as shown in Figure 1a — include the directly excited lines $K\alpha, \beta, \gamma, \ldots$ around 30 keV, the continuous Bremsstrahlung below the electron beam energy, and the radiative recombination lines. The line character of these free–to–bound transitions is due to the narrow energy–band width of the electron beam. The radiative recombination lines are labeled according to the atomic shell into which the electron was captured. The radiative recombination spectrum gives access to the charge distribution in the trap region. Using calculated values for the ionization energy and the radiative recombination cross sections for each charge state [20], a fit of the spectral features (see Figure 1b) yields the relative abundance of the highly charged xenon ions. An absolute number density of the ions can be derived by taking into account the solid angle, the detector efficiency, and the attenuation of the x–rays along their path from the trap region to the detector. The result of this charge state distribution measurement is listed in Table 1.

**TABLE 1.** Fractional and absolute ion abundances $(f_n, N)$ of trapped highly charged xenon ions ($Xe^{54+\ldots 51+}$). The abundances were determined using the line intensities of the radiative recombination transitions onto the K–shell and L–shell (see Figure 1). The ion density values $(n)$ are based on a cylindrical trap geometry of 1.27 cm in height and 0.01 cm in diameter.

| Ion species | $f_N$ | $N$ | $n$ cm$^{-3}$ |
|---|---|---|---|
| bare $Xe^{54+}$ | 0.051 ± 0.003 | 9200 ± 1300 | (0.86 ± 0.12) ·10$^8$ |
| H–like $Xe^{53+}$ | 0.184 ± 0.010 | 33200 ± 4800 | (3.10 ± 0.44) ·10$^8$ |
| He–like $Xe^{52+}$ | 0.631 ± 0.033 | 117000 ± 19000 | (11.0 ± 1.8) ·10$^8$ |
| Li–like $Xe^{51+}$ | 0.139 ± 0.043 | 20800 ± 7100 | (1.94 ± 0.66) ·10$^8$ |

# TRANSMISSION–TYPE CURVED CRYSTAL SPECTROMETER

Besides the reduced crystal reflectivity at small Bragg angles, crystal spectrometers that depend on spatially resolved detection of the diffracted x rays also suffer from the tremendous decrease in efficiency of the position sensitive detectors at

---

[4]) Heliumlike and hydrogenlike xenon have ionization potentials of $E_{He \to H} = 40.3$ keV and $E_{H \to bare} = 41.3$ keV, respectively.

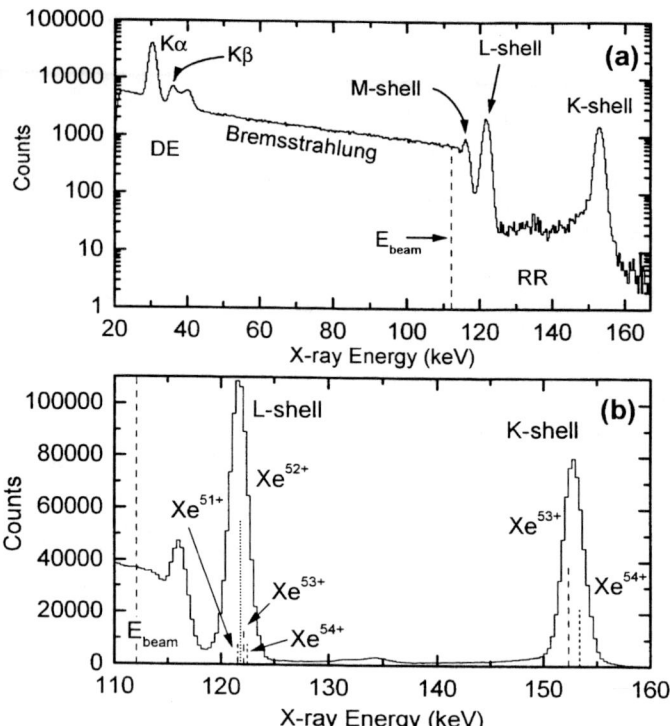

**FIGURE 1.** Direct excitation (DE) and radiative recombination (RR) spectra of highly charged xenon ions measured with a germanium detector. Spectrum (a) which was taken during an one–hour measuring period shows the ratio between the DE and RR emission. The dashed and dotted lines in the RR spectrum displayed in (b) — acquired over several days — mark the positions of the RR lines obtained by adding the ionization energy that is being released upon recombination to the beam energy, which was set to 112.1 keV.

higher x-ray energies [21]. Thus, for the hard x-ray regime, a transmission–type curved crystal spectrometer was developed based on the DuMond design [22]. The choice of a transmission–type spectrometer was favoured by the larger solid angle achievable with the same crystal geometry due to the perpendicular orientation of the lattice planes with respect to the crystal surface. The DuMond geometry, which employs a cylindrically bent crystal, requires that the radius of curvature of the crystal is the diameter of the so-called Rowland circle and that the center of the source (or spectrometer entrance slit) and the center of the crystal are part of this circle as shown in Figure 2. A crystal that is curved in such a way offers the incoming x rays a much larger area with the same Bragg angle in comparison with a flat crystal. Thus, the curved design increases the spectral throughput, i.e., the transmitted flux per energy interval, tremendously. Correct curvature and placement also strongly reduce the overall bandwidth of the diffracted x rays. The goal of the SuperEBIT transmission–type spectrometer design was to reduce this bandwidth such that the spectrometer can be used as a monochromator, where no spatially resolved detection of the diffracted x rays is necessary, and detectors with almost 100% counting efficiency can be used, like high–purity germanium detectors. The energy dispersion of the germanium detector in such an arrangement serves as a tool for the suppression of the background radiation.

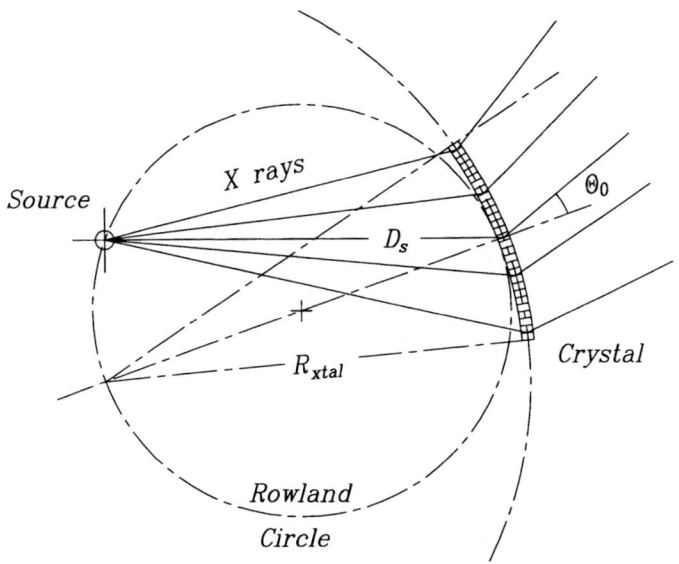

**FIGURE 2.** Geometric requirements of the DuMond–type crystal spectrometer. The curvature of the cylindrically bent crystal $R_{xtal}$ equals the diameter of the Rowland circle. Placing the center of the source and the center of the crystal on the Rowland circle defines the distance between these points as a function of the central Bragg angle $\theta_0$, i.e., $D_s = R_{xtal} \cos\theta_0$.

For the following measurements, the spectrometer utilizes a $(140 \times 40 \times 0.6)$ mm$^3$ quartz crystal with a $(13\overline{4}0)$ orientation. The 2d spacing of this crystal is 2.3604 Å [23] and, thus, the Bragg angle for the diffraction of the 30–keV x rays is around 10°. The radius of curvature of the crystal was $\approx 2.5$ m. The crystal is rotated by means of a high–precision stepper motor which is mounted on a modified rotation stage. The stepper motor has a unidirectional repeatability of $\pm 1\,\mu$m, and the uncertainty with respect to the traveled range is $\pm 5\,\mu$m. In combination with the $6\frac{3}{8}$" ($\approx 161.9$ mm) lever of the modified rotation stage, minimum angular increments of 0.00035° are achievable. To take advantage of this high unidirectional precision, all scans were performed by always rotating the crystal in the same manner, e.g., counter–clockwise.

The spectrometer performance was tested using a 1–mCi $^{133}$Ba source and a 75–$\mu$m wide and 12.7–mm tall slit made out of 2.4–mm thick tantalum. The radioisotope $^{133}$Ba decays by capturing an innershell electron, thus, creating an innershell excited cesium atom. The characteristic K-shell x rays of cesium are centered around 31.0 keV ($K\alpha_1$) and 30.6 keV ($K\alpha_2$). The spectrum was obtained by rotating the crystal in small increments and counting the number of diffracted x rays for a certain time before rotating the crystal to the next position. Figure 3 shows the result of several scans over both lines. The measured line width of 30 eV contains the instrumental width and also the natural line width of the cesium $K\alpha$ radiation, which is about 15 eV for $K\alpha_1$ [24]. Deconvolution of the measured profile leads to an instrumental line width of about 25 eV, which is equivalent to a resolving power of $\approx 1200$.

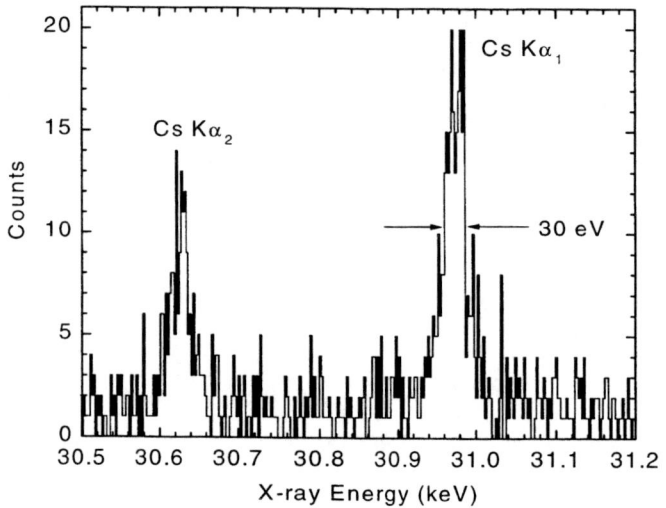

**FIGURE 3.** Highly resolved spectrum of the cesium $K\alpha$ radiation emitted by a $^{133}$Ba radioisotope, demonstrating the feasibility of the transmission-type spectrometer designed for high-resolution hard x-ray measurements at SuperEBIT.

## $K\alpha$ SPECTRA OF $XE^{53+}$ AND $XE^{52+}$

A sketch of the experimental setup, i.e., the implementation of the transmission-type curved crystal spectrometer at SuperEBIT, is shown in Figure 4. Operation of the spectrometer in monochromator mode requires a normalization of the counts, i.e., the number of diffracted x rays recorded with the detector labeled *GeX*, measured at each crystal position. The normalization which seemed best suited for these measurements is based on the number of xenon $K\alpha$ x rays emitted from the trap region. This was accomplished with the germanium detector labeled *GeSE* in Figure 4. Due to the cylindrical symmetry of the electron beam, all access ports

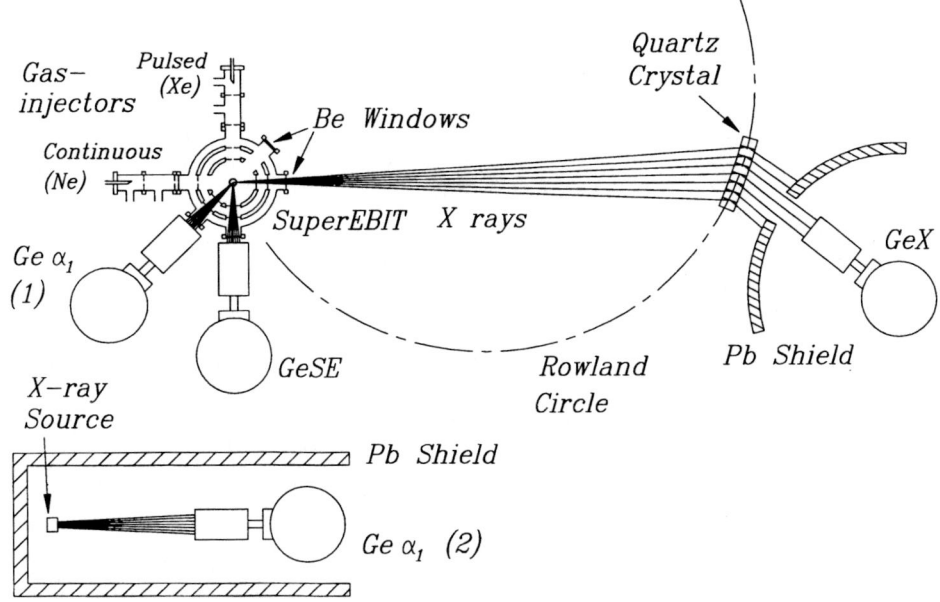

**FIGURE 4.** Sketch of the transmission–type curved crystal spectrometer setup. The electron beam, oriented perpendicular to the page, and the cylindrically bent crystal have to be part of the Rowland circle. The transmission spectrometer is operated in monochromator mode, i.e, a spectrum is obtained by rotating the crystal and recording the number of diffracted x rays with the germanium detector *GeX* as a function of the crystal position. Synchronized data acquisition between the germanium detectors *GeX* and *GeSE* is essential for the normalization of the measured high–resolution spectra. The energy of the Ly–$\alpha_1$ transition was measured with the germanium detector, labeled *Ge* $\alpha_1$. Xenon spectra *(1)* and calibration spectra *(2)* were taken successively in three–hour intervals.

receive the same type and amount of illumination over time, which makes a normalization by means of the measured flux at a different access port feasible. When the number of xenon K$\alpha$ photons measured with this germanium detector (*GeSE*) reached a preset amount, the counting process for both detectors *GeX* and *GeSE* was stopped, the crystal moved into the next position, and the counting process restarted. The spectrum shown in Figure 5 is the result of several scans over the wavelength region of interest, i.e., over the various Xe$^{52+}$ and Xe$^{53+}$ $n = 2 \rightarrow 1$ lines.

The area of the germanium detector *GeX* that was used for recording the diffracted x rays was 1018 mm$^2$, about five times larger than the detector used for the measurement of the Cs K$\alpha$ lines (Figure 3). Thus, the resolving power of the spectrometer for the highly charged xenon K$\alpha$ measurement was somewhat smaller than for the Cs K$\alpha$ measurement. In particular, $E/\Delta E \approx 900$ versus the

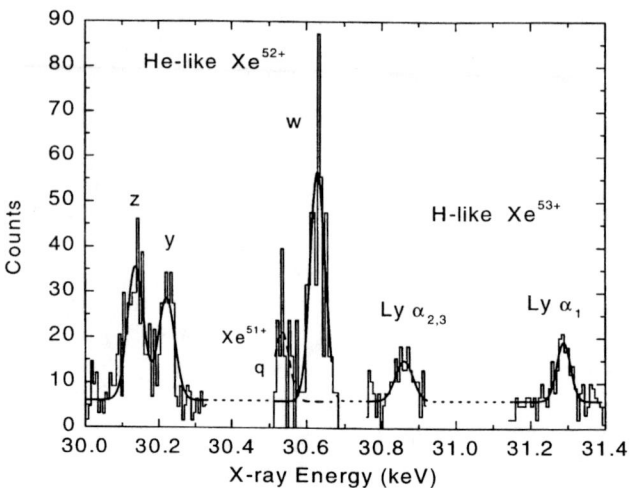

**FIGURE 5.** Spectrum of the $K\alpha$ transitions of hydrogenlike $Xe^{53+}$ and heliumlike $Xe^{52+}$. The spectrum shows the hydrogenlike $2p_{3/2} \to 1s_{1/2}$ (Ly-$\alpha_1$) transition, and a blend of the $2p_{1/2} \to 1s_{1/2}$ and $2s_{1/2} \to 1s_{1/2}$ (Ly-$\alpha_{2,3}$) transitions, and the heliumlike $1s2p\,^1P_1 \to 1s^2\,^1S_0$ (w), $1s2p\,^3P_1 \to 1s^2\,^1S_0$ (y), and $1s2s\,^3S_1 \to 1s^2\,^1S_0$ (z) transitions. The dashed line indicates the lithiumlike $1s2s2p\,^2P_{3/2} \to 1s^2\,2s\,^2S_{1/2}$ (q) innershell transition. Each channel represents the number of detected photons accumulated over three steps, i.e., three consecutive crystal positions. The dispersion is about 6.6 eV per channel. A resolving power of 930 ± 110 was inferred from the fitted line widths. The total data acquisition time was 122 hours. The measured wavelengths were converted to energies using the wavelength–voltage product of 12398.4243 eV Å.

previously measured 1200. The use of a larger detector area was necessary because of the small x–ray flux emitted from the trapped xenon ions, e.g., the count rate for the Ly-$\alpha_1$ line was only about 10 counts per hour. Overall, the xenon spectrum displayed in Figure 5 represents 122 hours of data acquisition time.

# DETERMINATION OF THE ELECTRON–IMPACT EXCITATION CROSS SECTIONS

The electron–impact excitation cross sections are determined relative to the radiative recombination cross sections. The values used for the radiative recombination cross sections are theoretical predictions by Scofield [20]. These calculations are based on a relativistic Hartree Slater model which has proven to be in excellent agreement (within 3%) with measured cross section values of the inverse process, i.e., photoionization, for photon energies in the 10–keV to 100–keV range [25]. Unfortunately, the comparison between the calculated and measured values is only available for neutral elements, e.g., neutral xenon atoms. However, one can argue that for the much simpler atomic structure of highly charged xenon ions, the accuracy of the calculated values should be at least comparable to the case of neutral xenon. In fact, measurements of the radiative electron capture of highly charged heavy ions have shown an agreement between theoretical and experimental cross section values of better than 5% [26,27].

Following steps are required for the determination of the electron–impact cross sections:

(i) Observation of the whole characteristic spectrum covering the transitions directly excited by electron impact and the radiative recombination lines. The resolution of the radiative recombination spectrum has to be sufficient for inferring the relative ion abundance.

(ii) High–resolution measurement of the direct excitation (DE) spectrum which shows the composition of the DE radiation with respect to the different ion species. Thus, the low–resolution spectrum of the DE lines (see previous step) can be divided into the various spectral features observed with the high–resolution instrument.

(iii) Conversion of the measured differential cross section values into total cross sections.

(iv) Extraction of the electron–impact excitation cross section by accounting for the various atomic processes in SuperEBIT that generate emission of the characteristic x rays.

The information obtained with the spectra displayed in Figures 1 and 5 address the points (i) and (ii) listed above. Since the high–resolution spectrum of the DE lines is taken with a crystal spectrometer, the measured line intensities have to be corrected with respect to the polarization of the line emission before the spectra are compared with the low–resolution germanium detector measurements. Because of the small Bragg angles encountered during the high–resolution measurement, i.e., from $9.67°$ to $10.04°$, the adjustment of the line intensities due to polarization is less than 3% , much smaller than the uncertainties introduced by the low number of counts in each line. The information of the polarization of the recorded spectral lines is also important for step (iii), the conversion to total cross sections. The experimental setup is designed for measuring the radiation emitted perpendicular to the electron beam. The total cross sections are inferred using the relation between the overall amount of emitted radiation of a dipole source and the intensity in perpendicular direction with respect to the dipole axis accounting for the polarization of the radiation [28].

The last step, item (iv), requires an estimate of the magnitude of contributions which populate the $n = 2$ levels due to processes other than direct electron–ion impact excitation of a ground state electron. In SuperEBIT, the important atomic processes that populate the L–shell of trapped hydrogenlike and heliumlike ions are:

- Radiative recombination onto the n = 2 levels,

- Radiative recombination onto n > 2 levels, followed by cascades that include a n′ → 2 transition,

- Charge exchange processes between neutrals and highly charged xenon ions, followed by cascades that include a n′ → 2 transition,

- Inner–shell ionization of lithiumlike xenon ions,

- Electron–impact excitation from the ground state to n > 2 levels, followed by cascades that include a n′ → 2 transition, and

- Direct electron–impact excitation, i.e., from the ground state to n = 2.

A schematic diagram regarding these processes is shown in Figure 6. The impact

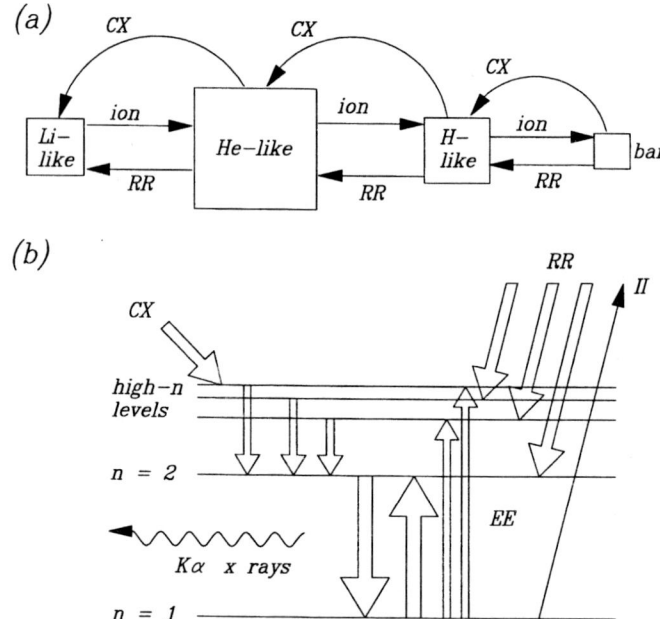

**FIGURE 6.** Schematic diagram of the main processes that effect **(a)** the charge balance and **(b)** the population of the n = 2 levels of the trapped highly charged ions. Included are charge exchange reactions between neutrals and ions ($CX$), radiative recombination ($RR$), and electron–impact ionization (*ion*). The area of each box is proportional to the abundance. The processes which populate the n = 2 levels and, thus, contribute to the measured intensity of the K$\alpha$ radiation are sketched in **(b)**. $II$ symbolizes the innershell ionization of lithiumlike xenon ions.

of the radiative recombination on the n = 2 level population was estimated using theoretical values of the radiative recombination cross sections for electron capture onto n < 5 levels. The contribution due to charge exchange processes was determined using measurements of the K–shell x–ray emission of heliumlike and hydrogenlike xenon after electron capture due to charge exchange with neutrals. For these measurements, SuperEBIT was operated in the magnetic trapping mode [29,30]. The amount of enhanced K$\alpha$ x–ray emission caused by inner–shell ionization of lithiumlike xenon has been estimated by applying electron–impact ionization cross sections derived from the Lotz formula [31,32]. The population of the L–shell in the hydrogenlike and heliumlike xenon due to electron–impact excitation of n > 2 levels followed by cascades that include a n' → 2 transition and due to direct electron–impact excitation of n = 2 levels was estimated using radiative and collisional rate calculations for all n = 2, 3, and 4 levels based on the Hebrew University / Lawrence Livermore Atomic Code (HULLAC). Table 2 gives an overview of these estimates, i.e., of the contributions due to processes other than electron–impact excitation of the n = 1 → 2 transitions, in terms of electron–ion interaction cross-section equivalent values, denoted $\tilde{\sigma}$. The extracted electron–impact excitation cross–sections are also listed in Table 2.

**TABLE 2.** Electron impact excitation ($EE$) cross sections for the n = 1 → 2 transitions in heliumlike and hydrogenlike xenon. The values are extracted from the measured total cross sections after accounting for radiative recombination ($RR$), innershell ionization ($II$), and charge exchange ($CX$) reactions with neutrals. The contributions are given in the form of an electron–ion interaction equivalent cross section (notation $\tilde{\sigma}$), e.g., the cross section values representing the contribution due to $CX$ are not the actual $CX$ cross sections. The factor $f_{pop}$ is the fraction of the n = 2 level population that is due to direct n = 1 → 2 electron–impact excitation. Thus, $1 - f_{pop}$ describes the relative population of the n = 2 levels due to electron–impact excitation into n > 2 levels followed by cascades. The uncertainties listed combine statistical uncertainties and uncertainties in the contribution from the adjustments with respect to the $RR$, $CX$, and $II$ processes. No systematic uncertainties due to the use of theoretical models have been added.

| line | $\sigma_{expt}$ barn | $\tilde{\sigma}_{RR}$ barn | $\tilde{\sigma}_{CX}$ barn | $\tilde{\sigma}_{II}$ barn | $f_{pop}$ | $\sigma_{EE}$ barn |
|---|---|---|---|---|---|---|
| Ly-$\alpha_1$ | 9.8 ± 1.6 | 0.146 ± 0.013 | 0.57 ± 0.26 | — | 0.944 | 8.6 ± 1.5 |
| Ly-$\alpha_{2,3}$ | 10.4 ± 3.7 | 0.734 ± 0.064 | 0.50 ± 0.23 | — | 0.904 | 8.2 ± 3.4 |
| w | 11.8 ± 1.8 | 0.127 ± 0.009 | 4.29 ± 0.99 | — | 0.950 | 7.0 ± 2.0 |
| y | 7.4 ± 1.4 | 0.332 ± 0.023 | 3.00 ± 0.63 | — | 0.953 | 3.9 ± 1.5 |
| z | 11.3 ± 1.8 | 0.388 ± 0.027 | 3.43 ± 0.72 | 2.71 ± 0.84 | 0.227 | 1.08 ± 0.48 |

The experimentally achieved electron–impact excitation cross sections are compared with three sets of theoretically predicted values. All three sets are based on a distorted–wave approach [33,34], two sets are relativistic calculations, and one of the two relativistic calculations also includes the Breit interaction between the free and the bound electrons [35,36]. The comparison is presented in Table 3 and in graphical form in Figure 9. The uncertainties of the measured cross sections only

**TABLE 3.** Comparison between the measured ($\sigma_{EE}$) and calculated electron–impact excitation cross sections. The calculated values are based on a distorted–wave approach [33,34]. Relativistic effects certainly play an important role in the interaction between the 112–keV beam electron and the highly charged xenon ion as seen in the large difference between the non–relativistic calculations, $\sigma_{non-rel}$, and the relativistic calculations, $\sigma_{rel}$. Additionally, the impact of the Breit interaction between the free and the bound electrons is significant in the excitation process of the heliumlike xenon. Thus, agreement between the measured and calculated cross section values can only be found when the Breit interaction, i.e., the Generalized Breit Interaction [35,36], is included in the calculations, $\sigma_{GBI}$.

| line | $\sigma_{EE}$ barn | $\sigma_{non-rel}$ barn | $\sigma_{rel}$ barn | $\sigma_{GBI}$ barn |
|---|---|---|---|---|
| Ly-$\alpha_1$ | 8.6 ± 1.5 | | 8.256 | 8.109 |
| Ly-$\alpha_{2,3}$ | 8.2 ± 3.4 | | 6.541 | 6.787 |
| w | 7.0 ± 2.0 | 21.64 | 17.45 | 8.364 |
| y | 3.9 ± 1.5 | 0.127 | 7.313 | 3.842 |
| z | 1.08 ± 0.48 | 0.123 | 0.172 | 0.152 |

reflect the uncertainties due to the low counts in the measured spectral lines. The systematic uncertainty introduced by tying the measured values to theoretical values of the radiative recombination cross sections, by using theoretical predictions for the polarization of the spectral line emission, and by applying the HULLAC code for estimating the population of the n = 2 levels due to cascades have not been included.

# MEASUREMENT OF THE TRANSITION ENERGIES

The high precision of the crystal holder rotation mechanism allows a very accurate determination of the dispersion of the recorded high–resolution spectra. However, the transmission spectrometer in its current setup is not equipped for absolute Bragg angle measurements, and the determination of the wavelengths requires the observation of at least one calibration line. Therefore, the wavelengths were measured relative to the Ly-$\alpha_1$ line, and a calibration of the Ly-$\alpha_1$ transition energy was performed in a separate experiment. This Ly-$\alpha_1$ transition energy measurement was performed parallel to the high–resolution transmission–type curved crystal spectrometer measurement discussed above. A state-of-the-art germanium detector[5] was used for successively observing the K$\alpha$ x-ray radiation emitted by the highly charged xenon ions in the trap and the spectra of x-ray sources containing the K$\alpha$ emission of cesium and barium. The detector is labeled $Ge\ \alpha_1$ in the experimental setup which is shown in Figure 4. With a resolution of about

---

[5] High–purity germanium detector with a planar crystal geometry from EG&G ORTEC, Model No. GLP 16195/10.

270 eV at photon energies of about 31 keV, this detector is suitable for resolving the xenon Ly-$\alpha_1$ transition, which is separated from the adjacent spectral features, the Ly-$\alpha_{2,3}$ blend and the $1s2p\,^1P_1 \rightarrow 1s^2\,^1S_0$ transition in heliumlike xenon by more than 400 eV.

In three hour intervals the detector was switched between the SuperEBIT observation port and a well shielded calibration stand (see Figure 4), where the detector dispersion was calibrated utilizing the K$\alpha$ fluorescence of cesium and barium. In particular, the calibration stand was equipped with the radioisotope $^{133}$Ba, which after K–shell electron capture emits the Cs K$\alpha$ x–rays, and the radioisotope $^{241}$Am, which was covered with a barium window, thus, producing the Ba K$\alpha$ fluorescence. The calibration stand was adjusted in such a way that the count rate of the Cs K$\alpha$ emission was about twice the average count rate of the Xe Ly-$\alpha_1$ transition. Therefore, the count rate for the Cs K$\alpha_1$ x rays was higher, and for the Cs K$\alpha_2$ x rays was lower than the count rate for the Xe Ly-$\alpha_1$ x rays. Matching the count rates for measurement and calibration is important for high–precision measurements, since the electrical field across the germanium crystal and, thus, the magnitude of the detected current pulse, is sensitive to the amount of electron–hole pairs created by photon impact. A spectrum combining the barium, cesium, and highly charged xenon K$\alpha$ emission is presented in Figure 7.

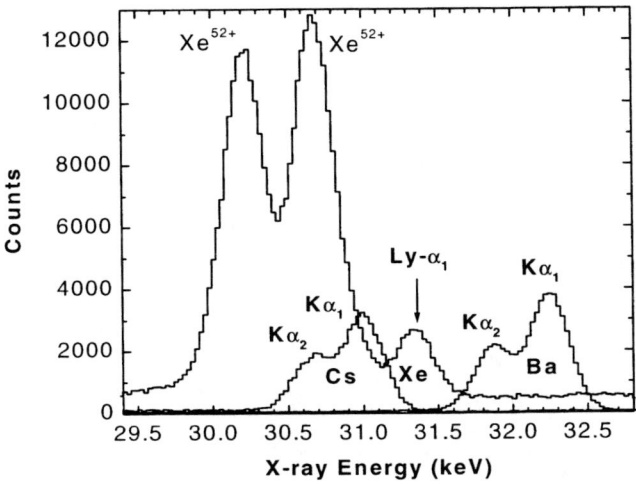

**FIGURE 7.** K$\alpha$ spectra of highly charged xenon ions, cesium, and barium. The cesium fluorescence is emitted by a $^{133}$Ba radioisotope, which after capturing a K–shell electron converts into an innershell excited $^{133}$Cs*. The barium K$\alpha$ radiation is emitted by a barium foil which is illuminated with the radioisotope $^{241}$Am causing K–shell photoionization in the barium. All three spectra together represent about 12 hours of data acquisition time.

The centroid positions of the spectral features were determined using a line–shape function that accounts for the response of the germanium detector, i.e, Compton profile and escape–peak behavior. Figure 8 shows the line centroid values as a function of SuperEBIT runtime. This plot clearly emphasizes the importance of

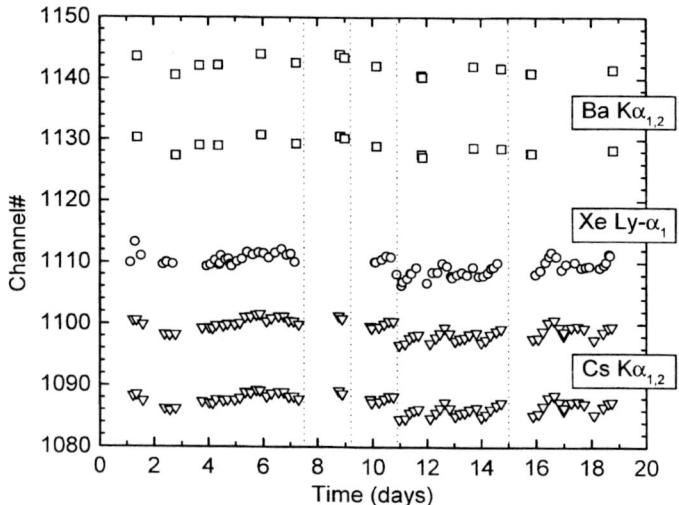

**FIGURE 8.** Centroid positions of the spectral lines given in the arbitrary quantity *Channel Number* recorded with a germanium detector as a function of the SuperEBIT runtime, i.e., the duration of the experiment. The statistical uncertainties for the centroid positions are smaller than the size of the symbols used in this plot. Thus, the shifts in the centroid positions are significant. The dashed lines indicate when the high voltage power supply for the detector was turned off.

continuous calibration. Otherwise, the long–term drift of the pulse–height signal of the detector introduces a systematic uncertainty which is more than an order of magnitude bigger than the statistical uncertainty based on the number of counts in the spectral line. The barium K$\alpha$ fluorescence, for example, was only measured twice a day and, thus, the observed shift of the line centroid in time does not contain the details necessary for an accurate tracking of the shifts in the germanium detector's pulse–height signal.

The energy values for the calibration lines are taken from the *CRC Handbook for Chemistry and Physics* [37] and are listed in Table 4. These values are taken from a paper by Bearden published in 1967 [38] and require correction due to changes in the value for the voltage–wavelength product and the value for the conversion factor of the wavelength units, $V\lambda = 12398.4243$ eV Å $\pm 0.60$ ppm, $\text{Å}^*/\text{Å} = 1.00001481 \pm 0.92$ ppm [39]. The Å$^*$/Å correction incorporates also the most recent result of the 2d spacing of the (220) oriented lattice planes in a nearly perfect silicon crystal, $d_{220} = 1.92015540$ Å $\pm 0.21$ ppm [40,41]. Additional adjustment is required for the uncertainties, which reflect the probable error (abbreviated as "p.e." in the wavelength and energy compilation by Bearden) which is defined as a 50% confidence limit, and does not include the uncertainty of the Å* unit due to the uncertainty of the W K$\alpha_1$ measurement. Even though the original paper [38] emphasizes the fact that Gaussian statistics was not applicable for most data, it seems to be the standard approach to convert the "p.e." values to a 67% confidence limit by assuming Gaussian distribution and, thus, to multiply the uncertainty values by a factor of 1.48 (e.g. [42–44]). Table 4 also lists the most recent measurement available for the barium K$\alpha$ transition energies [45] and the transition–energy values that are currently used by the Quantum Metrology Group at the National Institute of Standards and Technology (NIST) [46].

**TABLE 4.** Transition energies of the cesium and barium calibration lines. The energy values in column $E_{67}$ are taken from the compilation published by Bearden [38]. These values are also listed in the *Handbook of Chemistry and Physics* [37]. $E_{67,adj}$ are the adjusted energy values incorporating the current values for the x-ray standards [39] and the most recent measurement of the Si (220) 2d spacing [40,41]. For the K$\alpha$ emission of barium a more recently measured value is available, $E_{82,adj}$ [45]. This value, too, was adjusted to the current x-ray standards. Column $E_{99}$ lists the transition energies currently used at NIST [46].

| Line | $E_{67}$ eV | $E_{67,adj}$ eV | $E_{82,adj}$ eV | $E_{99}$ eV |
|---|---|---|---|---|
| Cs K$\alpha_2$ | 30625.1(3) | 30626.32(44) | | 30625.42(45) |
| Cs K$\alpha_1$ | 30972.8(3) | 30974.06(46) | | 30973.15(46) |
| Ba K$\alpha_2$ | 31817.1(4) | 31818.39(61) | 31816.62(6) | 31816.631(60) |
| Ba K$\alpha_1$ | 32193.6(3) | 32194.89(50) | 32193.27(7) | 32193.279(70) |

The result of the Ly–$\alpha_1$ transition energy measurement is given in Table 5 together with the theoretically predicted value from Johnson and Soff [47]. Surprisingly, there is a significant disagreement between the measured and the calculated values. The disagreement is much larger than the changes in the measured value due to using the different sets of calibration line values from Table 4. A suggested explanation for this unexpected disagreement is given in the section titled "Discussion".

**TABLE 5.** Measurement of the Xe$^{53+}$ Ly–$\alpha_1$ transition energy. The measured value $E_1$ was obtained utilizing the calibration set $E_{67,adj}$ from Table 4. Using $E_{82,adj}$ and $E_{99}$ yields the values $E_2$ and $E_3$, respectively. The measured values are compared to the theoretical prediction by Johnson and Soff, labeled $E_{J\&S}$ [47].

| $E_1$ eV | $E_2$ eV | $E_3$ eV | $E_{J\&S}$ eV |
|---|---|---|---|
| 31279.2 ± 1.5 | 31278.8 ± 1.5 | 31278.1 ± 1.5 | 31283.77 ± 0.09 |

Using both the theoretical and the measured values for the xenon Ly–$\alpha_1$ transition energy, the transition energies and wavelengths, respectively, for all spectral features observed in Figure 5 have been determined. For example, taking the theoretically predicted value [47], i.e., 31283.77(9) eV and 0.396321(1) Å, respectively, and a 2d–spacing of 2.3604 Å for the (13$\bar{4}$0)–oriented Quartz crystal yields a Bragg angle of 9.6660°. The distance to the centroid of the neighboring Ly–$\alpha_{2,3}$ blend is 380.0 steps with the stepper motor that controls the crystal position. This crystal rotation is equivalent to a Bragg angle difference of 0.136° and, thus, the wavelength of the neighboring Ly–$\alpha_{2,3}$ blend is 0.40183 Å. The wavelengths values are converted to transition energies for easier comparison with theoretical predictions, and the result is presented in Table 6 and in graphical form in Figure 10. The accu-

**TABLE 6.** Comparison of the measured and calculated energies for the $n = 2 \rightarrow 1$ transitions in hydrogenlike and heliumlike xenon. The first set of experimental results $E_{expt}$ is calibrated by utilizing the measured value of the Ly–$\alpha_1$ transition, i.e., 31279.2(27) eV. The $E_{exth}$ values have been determined relative to the theoretical value for the Ly–$\alpha_1$ line, i.e., 31283.77(9) eV [47]. The result of the hydrogenlike transitions is compared to the calculations by Johnson and Soff $E_{J\&S}$ [47]. For the heliumlike transitions the theoretical values are from Drake $E_D$ [48], Plante, Johnson and Sapirstein $E_P$ [49], and Chen, Cheng, Johnson and Sapirstein $E_C$ [50,51].

| Line | Hydrogenlike transitions | | |
|---|---|---|---|
| | $E_{expt}$ eV | $E_{exth}$ eV | $E_{J\&S}$ eV |
| Ly–$\alpha_1$ | 31279.2 ± 2.7 | 31283.8 ± 2.3 | 31283.77 |
| Ly–$\alpha_2$ | 30850.4 ± 5.7 | 30855.0 ± 5.7 | 30856.36 |
| Ly–$\alpha_3$ | | | 30863.49 |

| Line | Heliumlike transitions | | | | |
|---|---|---|---|---|---|
| | $E_{expt}$ eV | $E_{exth}$ eV | $E_D$ eV | $E_P$ eV | $E_{C\&C}$ eV |
| w | 30619.9 ± 4.0 | 30624.5 ± 4.0 | 30629.28 | 30629.68 | 30630.64 |
| x | — | — | 30593.54 | 30593.93 | 30594.96 |
| y | 30210.5 ± 4.5 | 30215.1 ± 4.5 | 30205.58 | 30205.87 | 30206.90 |
| z | 30126.7 ± 3.9 | 30131.3 ± 3.9 | 30128.40 | 30128.78 | 30129.79 |

racy is currently limited by counting statistics. In fact, an increase in the number of counts of at least a factor of 10 would be necessary to match the much lower uncertainty associated with the determination of the dispersion of the transmission–type curved crystal spectrometer.

# DISCUSSION

## Electron–impact excitation cross section measurement

The result of the electron–impact excitation cross section measurement is summarized in Figure 9. The comparison between the measured and the theoretically predicted cross sections shows that relativistic effects, in particular, the Breit interaction between the free and the bound electrons, are significant. Especially, for the

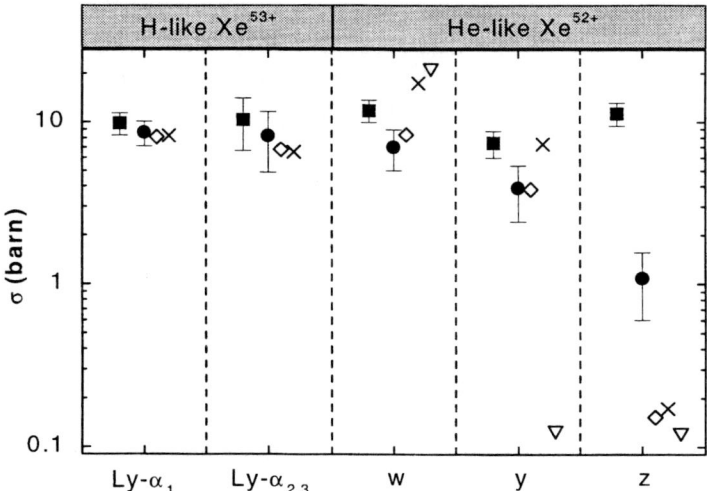

**FIGURE 9.** Comparison between the experimental and theoretical values for the electron impact excitation cross section of the n = 1 → 2 transitions in heliumlike and hydrogenlike xenon. The plot shows two sets of measured total cross section values. The first set (filled squares) represents the "overall" cross sections without accounting for the various processes that populate the L-shell of the xenon ions. The second set (filled circles) represents the extracted electron–impact excitation cross sections. The calculated values are based on a distorted–wave approach [33,34]. Neglecting relativistic effects yields the ▽ values. The symbol × denote relativistic calculations but without accounting for the Breit interaction between the free and the bound electrons. Incusion of the Breit interaction yields the values marked by the ◇ symbol [35,36]. The uncertainties shown are statistical uncertainties only.

excitation of the heliumlike $Xe^{52+}$ ground states by impact of 112–keV electrons, agreement between the theoretical and experimental values can only be found when the Breit interaction is included in the calculations, even though the uncertainties of the measured values are in the 20% to 50% range. Moreover, no agreement exists for the excitation cross section of the $1s^2\,^1S_0 \rightarrow 1s2s\,^3S_1$ transition (line z). The disagreement in case of line z reflects most likely the complexity of the population mechanism of this metastable level in the heliumlike ion. For the hydrogenlike lines we found good agreement for both calculations, with and without inclusion of the Breit interaction. Besides the differences in the calculated values depending on the inclusion of various effects, Figure 9 also shows the impact on the population of the L–shell of the xenon ions due to effects other than direct electron–impact excitation from the ground state — visualized by the difference between the two sets of experimental values. In other words, the two sets of measured cross section values in Figure 9 indicate the fraction of the n = 2 → 1 x–ray emission which is caused by direct n = 1 → 2 electron–impact excitation. While the emission in hydrogenlike $Xe^{53+}$ is clearly dominated (more than 80% ) by direct excitation, less than 60% of the emission in heliumlike $Xe^{52+}$ is due to direct excitation of the ground state. In fact, for line z, direct excitation only accounts for about 10% of the observed emission, making the cross section value for line z strongly dependent on the models utilized for simulating the population mechanism.

## Determination of the n = 2 → 1 transition energies

The predictions for the transition energies in hydrogenlike systems have shown agreement with most experimentally obtained values and, thus, the theoretical values are widely accepted as being accurate, e.g. [2]. Therefore, it is common practice to use spectral lines of hydrogenlike ions to calibrate the dispersion of the spectroscopic instrumentation, e.g. [52]. In fact, a calibration with lines emitted by hydrogenlike ions is more precise than utilizing the $K\alpha$ emission of neutral elements because the hydrogenlike lines are not affected by unresolved satellites due to shake processes. Additionally, the natural line width is much smaller for lines emitted by hydrogenlike ions than by inner–shell excited neutrals. Nevertheless, the characteristic emission of hydrogenlike ions has not been accepted as an x–ray standard, yet. Therefore, a calibration with respect to the current x–ray standard, i.e., the tungsten $K\alpha_1$ line, and to secondary standards, respectively, was performed. Surprisingly, the calibration measurement yielded a value for the Ly-$\alpha_1$ transition energy that disagreed with the theoretical value significantly. For the wavelength calibration of the observed n = 2 → 1 transitions, both the theoretical and the measured values for the Ly-$\alpha_1$ line were used. Thus, the comparison between theory and experiment shown in Figure 10 contains two sets of experimental data.

The 4.6–eV difference between the measured and calculated transition energy of the Ly-$\alpha_1$ line (see Table 5) is equivalent to a deviation of more than 3 $\sigma$[6]. However, since the calculations for the transition energies of the hydrogenlike isoelectronic sequence have proven to be successful in so many comparisons with measurements, it is very likely that this significant disagreement is due to a shortcoming in the experimentally obtained result.

One suggestion for the disagreement is a systematic uncertainty introduced by the calibration line values. A weakness of the wavelengths reference table compiled

---

[6] The probability of finding a data point outside the 3–$\sigma$ confidence limit is less than 0.3%.

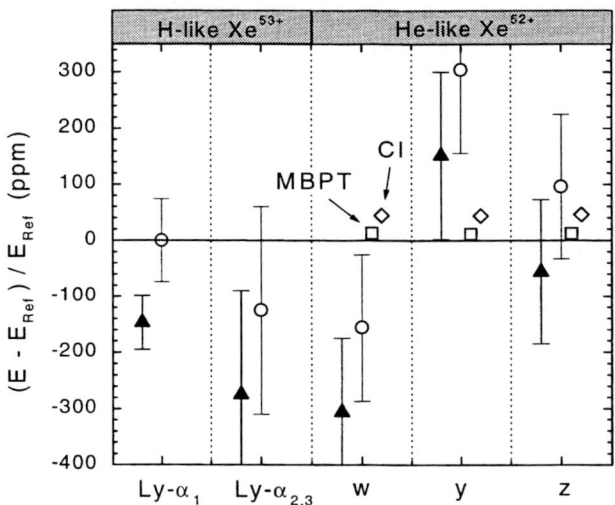

**FIGURE 10.** Relative deviation between measured and theoretically predicted transition energies for the hydrogenlike and heliumlike xenon Kα lines. The transition energies have been measured relative to the xenon Ly-α₁ line. The filled–triangle data set utilizes the measured value of the Ly-α₁ transition energy, i.e., 31279.2(27) eV, the data set marked by the open circles is obtained using the value calculated by Johnson and Soff, i.e., 31283.77(9) (see Table 5). The reference energies ($E_{Ref}$) for the hydrogenlike lines are from Johnson and Soff [47]. For the heliumlike transition energies the values calculated by Drake using the unified method are taken as a reference [48]. The □ values were calculated by Plante et al. [49] summing all–orders of many–body perturbation theory (MBPT). The ◇'s represent the values obtained by Cheng et al. and Chen et al. [50,51] applying relativistic configuration interaction (CI) calculations combined with *ab initio* QED calculations.

by Bearden [38] is the assignment of the wavelengths to the peak of the observed spectral feature. For a symmetric line profile the peak value is identical with the centroid, thus, the given value is inert with respect to applications involving spectrometers of different resolving power. For asymmetric spectral features, however, the peak value can only be reproduced utilizing the same type of spectrometer used for measuring the tabulated wavelength values. Otherwise, the peak value might not reflect the same position on the wavelength scale due to the different response function of the spectrometers. Recent high–resolution measurements of K-shell and L-shell emission of neutrals after electron impact have shown a significant contribution of satellite lines, causing an asymmetry of the observed spectral profile [53–55]. The presence of all these satellites also shows the internal complexity of the inner–shell processes, especially in high–Z elements. In contrast to the electron–impact based x-ray sources, the emission of the radioisotopes used for the calibration are much less affected by M-satellite structure. Unfortunately, the counting statistics in the spectrum shown in Figure 3 is not sufficient for a conclusive analysis regarding the contribution to the measured line emission due to satellites. In conclusion, the disagreement between the measured and the calculated transition energy of the xenon Ly–$\alpha_1$ line emphasizes the need for a re–evaluation of the commonly used x–ray wavelengths table for atomic inner–shell transitions, in particular, for the cesium K$\alpha$ lines.

## ACKNOWLEDGMENTS

We thank Ed Magee and Dan Nelson for technical support and Honglin Zhang for providing us with the theoretical cross section values that include the Breit interaction. This work was performed under the auspices of U.S.DOE by Lawrence Livermore National Laboratory under Contract No. W–7405–ENG–48.

## REFERENCES

1. H. Persson et al., Hyperfine Inter. **108**, 3 (1997).
2. H. F. Beyer, H.-J. Kluge, and V. P. Shevelko, in *X-Ray Radiation of Highly Charged Ions, Atoms and Plasmas* (Springer, Heidelberg, 1997), Chap. 3 "Atomic Structure and Spectra", pp. 61–84, comparison of Lamb–shift measurements in Table 3.16, pg. 82.
3. J. Schweppe et al., Phys. Rev. Lett. **66**, 1434 (1991).
4. P. Beiersdorfer et al., Phys. Rev. Lett. **71**, 3939 (1993).
5. P. Beiersdorfer et al., Phys. Rev. A **52**, 2693 (1995).
6. P. Beiersdorfer et al., Phys. Rev. Lett. **80**, 3022 (1998).
7. T. Beier, G. Plunien, and G. Soff, Hyperfine Inter. **108**, 19 (1997).
8. S. Chantrenne, P. Beiersdorfer, R. Cauble, and M. Schneider, Phys. Rev. Lett. **69**, 265 (1992).
9. K. Wong, P. Beiersdorfer, K. Reed, and D. Vogel, Phys. Rev. A **51**, 1214 (1995).
10. L. Horton, Phys. Scr. **T65**, 175 (1996).
11. J. Briand et al., Z. Phys. A **318**, 1 (1984).
12. P. Indelicato, J. Briand, M. Tavernier, and D. Liesen, Z. Phys. D **2**, 249 (1986).
13. K. Widmann, P. Beiersdorfer, V. Decaux, and M. Bitter, Phys. Rev. A **53**, 2200 (1996).
14. M. Levine, R. Marrs, D. Knapp, and M. Schneider, Phys. Scr. **T22**, 157 (1988).
15. M. Levine et al., in *International Symposium on Electron Beam Ion Sources and Their Applications, AIP Conference Proceedings No. 188*, edited by A. Hershcovitch (AIP Press, Woodbury, New York, 1989), pp. 82–101.
16. D. Knapp et al., Nucl. Instr. and Meth. in Phys. Res. A **334**, 305 (1993).
17. S. Utter, P. Beiersdorfer, J. Crespo López-Urrutia, and K. Widmann, Nucl. Instr. and Meth. in Phys. Res. A **428**, 276 (1999).
18. M. B. Schneider et al., in *International Symposium on Electron Beam Ion Sources and Their Applications, AIP Conference Proceedings No. 188*, edited by A. Hershcovitch (AIP Press, Woodbury, New York, 1989), pp. 158–165.

19. B. Penetrante et al., Phys. Rev. A **43**, 4873 (1991).
20. J. H. Scofield, Phys. Rev. A **40**, 3054 (1989).
21. D. Vogel, P. Beiersdorfer, V. Decaux, and K. Widmann, Rev. Sci. Instrum. **66**, 776 (1995).
22. J. W. DuMond, Rev. Sci. Instrum. **18**, 626 (1947).
23. J. Kirz et al., X-ray data booklet, Center for X-ray Optics, Lawrence Berkeley National Laboratory, University of California, Berkeley, California 94720 (1986).
24. G. Zschornack, *Atomdaten für die Röntgenspektralanalyse* (Springer-Verlag, Berlin, 1989).
25. E. Saloman, J. Hubell, and J. Scofield, At. Data Nucl. Data Tables **38**, 1 (1988).
26. T. Stöhlker, P. Mokler, C. Kozhuharov, and A. Warczak, Comments on Atomic and Molecular Phys. **33**, 271 (1997).
27. T. Stöhlker et al., Phys. Rev. Lett. **82**, 3232 (1999).
28. I. Percival and M. Seaton, Phil. Trans. Royal Soc. London **251**, 113 (1958).
29. P. Beiersdorfer, L. Schweikhard, J. Crespo López-Urrutia, and K. Widmann, Rev. Sci. Instrum. **67**, 3818 (1996).
30. P. Beiersdorfer et al., Phys. Scr. **T80**, 121 (1999).
31. W. Lotz, Astrophys. J. **XVI**, 207 (1967), supplement Series.
32. W. Lotz, Z. Phys. **216**, 241 (1968).
33. H. Zhang, D. Sampson, and R. Clark, Phys. Rev. A **41**, 198 (1990).
34. H. Zhang and D. Sampson, Phys. Rev. A **47**, 208 (1993).
35. C. J. Fontes, D. H. Sampson, and H. L. Zhang, Phys. Rev. A **47**, 1009 (1993).
36. C. J. Fontes, H. L. Zhang, and D. H. Sampson, Phys. Rev. A **59**, 295 (1999).
37. D. R. Lide and H. Frederikse, in *CRC Handbook of Chemistry and Physics*, 70 ed. (CRC Press Inc., Boca Raton, FL, 1989-1990), Chap. "X-ray Wavelengths", pp. E152-E190.
38. J. Bearden, Rev. Mod. Phys. **39**, 78 (1967).
39. E. R. Cohen and B. N. Taylor, Rev. Mod. Phys. **59**, 1121 (1987).
40. E. Kessler Jr. et al., J. Res. Nat. Inst. Stand. Technol. **99**, 1 (1994), pp. 1-18 article; p. 285 Errata to this article.
41. P. Becker et al., PTB-Mitteilungen **105**, 95 (1995).
42. E. Kessler Jr., R. Deslattes, A. Henins, and W. Sauder, Phys. Rev. Lett. **40**, 171 (1978).
43. E. Kessler Jr., R. Deslattes, and A. Henins, Phys. Rev. A **19**, 215 (1979).
44. T. Mooney et al., Phys. Rev. A **45**, 1531 (1992).
45. E. Kessler Jr. et al., Phys. Rev. A **26**, 2696 (1982).
46. Dr. Richard D. Deslattes, Quantum Metrology Group, NIST, Gaithersburg, MD 20899, private communication.
47. W. Johnson and G. Soff, At. Data Nucl. Data Tables **33**, 405 (1985).
48. G. Drake, Can. J. Phys. **66**, 586 (1988).
49. D. Plante, W. Johnson, and J. Sapirstein, Phys. Rev. A **49**, 3519 (1994).
50. M. Chen, K. Cheng, and W. Johnson, Phys. Rev. A **47**, 3692 (1993).
51. K. Cheng, M. Chen, W. Johnson, and J. Sapirstein, Phys. Rev. A **50**, 247 (1994).
52. P. Beiersdorfer, A. Osterheld, and S. Elliott, Phys. Rev. A **58**, 1944 (1998).
53. T. Ludziejewski et al., Phys. Rev. A **54**, 232 (1996).

54. J.-C. Dousse and J. Hoszowska, Phys. Rev. A **56**, 4517 (1997).
55. J.-C. Dousse, J. Hoszowska, and O. Mauron, *Poster presentation*, 18th International Conference on X-ray and Inner-Shell Processes, Chicago, 23-27 August, 1999.

# First Experimental Results and New Theoretical Calculations on Photoionization Processes in Multiply-Charged $Xe^{3+}$ to $Xe^{7+}$ Ions

J.-M. Bizau,* C. Blancard,[+] R. Marmoret,[#] D. Hitz,° J.-M. Esteva,*
D. Cubaynes,* C. Couillaud,[+] P. Ludwig,° C. Rémond,[#] A. Compant La Fontaine,[+] J. Delaunay,° J. Bruneau,[#] J. Lachkar[+] and F.J. Wuilleumier*

*LSAI, UMR-CNRS n° 8624, Bât. 350, Université Paris-Sud, 91405 Orsay cedex, France
[+]DPTA and [#]DCRE, Centre CEA-DAM/DIF, BP12, 91680 Bruyères-le-Châtel, France
°DRFMC/SI2A, CEA Grenoble, 17 rue des Martyrs, 38054 Grenoble cedex9, France

**Abstract.** We present new results obtained on photoionization processes in $Xe^{3+}$ to $Xe^{7+}$ ions, using ion spectrometry. The experiment combines for the first time a beam of multiply-charged ions produced in an electron cyclotron resonance (ECR) ion source and a beam of synchrotron radiation emitted from an undulator. These results provide a complete picture of the behavior of resonant and continuum photoionization processes along the Xe isonuclear sequence, following the complete stripping, electron by electron, of the 5p subshell.

With the progress made in the production of high flux photon-beams from insertion devices (undulators) installed on third generation synchrotron radiation (SR) sources, photoionization of ions is rapidly expanding. Since the pioneering work achieved in the beginning of the 70's, most of the experimental data available today on photoionization of ions have been obtained from photoabsorption studies in laser-produced plasma (1). To go further, and to measure absolute photoionization cross sections, ion beam-photon beam techniques have to be implemented. The first determination of some absolute cross sections was achieved in the 80's for singly-charged ions in ion spectrometry experiments using SR emitted from a bending magnet of the Daresbury Laboratory storage ring (2). More recently, renewed efforts to exploit storage rings have succeeded in providing detailed results on some singly charged ions (3,4), doubly-charged $Xe^{2+}$ ion (5), and preliminary results for the $Xe^{3+}$ ion (6).

In this paper, we present a new ion spectrometry experiment developed on the Super ACO storage ring. The experimental set-up couples, for the first time, a beam of multiply-charged ions produced in an electron cyclotron resonance ion source (ECRIS) and a monochromatized beam of SR emitted from an undulator. In order to test this set-up, we chose to study first the behavior of photoionization processes involving 4d electrons along the Xe isonuclear series. In particular, we concentrate on the effect of removing outer electrons on the collapse of the 4f orbital in the multiply-charged Xe ions. We present here the results we have obtained for $Xe^{3+}$ to $Xe^{7+}$ ions. To interpret the data, we have used a multiconfiguration Dirac-Fock (MCDF) code (7), accounting for Breit interaction and for radiative and finite nuclear mass corrections. This program has been used to calculate energy levels, photoexcitation and photoionization cross sections, as well as radiative and Auger rates.

A schematic of the experimental set-up we have built is shown in Figure 1. The ions are produced in a compact permanent magnet ECRIS. Heating of the plasma is achieved by a 10 GHz micro-wave generator, delivering a maximum power of about 30 W in the present work. The source is biased to 7 kV to extract the multiply-charged ions. One single ionic charge stage is then selected from the beam using a commercial Wien filter (Colutron model 600B). To increase the density of the ions which will interact with the photons and to reduce the interfering ion signal produced by collisional processes, the kinetic energy of the selected ions is then lowered to 4 keV per charge. The trajectory of the ion beam is further bent by 90° in an electrostatic quadrupolar deflector which makes the ion beam colinear with the photon beam. Both beams are smoothly focused over the 20 cm long interaction volume. A correct matching of the two beams is performed using two apertures (2 mm in diameter and 70 mm apart) placed before the interaction zone. Typical ion currents of 100 nA are available in the interaction volume. The charge of the ions after interaction with the photons is analyzed using a 30° toroïdal electrostatic deflector. The ions

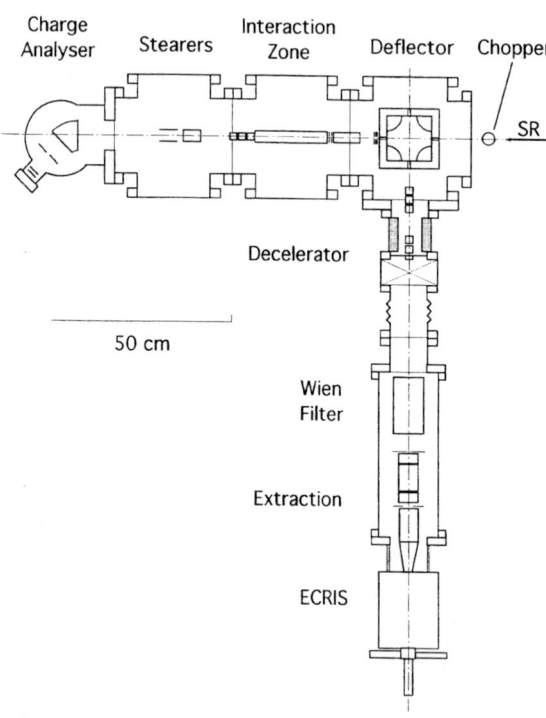

**FIGURE 1.** Experimental set-up

are then counted with a multichannel plate array.

The photons are produced in the SU6 undulator of the Super ACO storage ring in Orsay. A 1m toroïdal grating monochromator allows to select the energy of the photons. The flux of incoming photons is monitored with a Si photo-diode. A typical photon flux of a few $10^{12}$ photons/s/0.1%BP is available at the exit slit of the monochromator in the 30-180 eV photon energy range. The energy scale of the monochromator is calibrated by measuring the $L_{II,III}$ absorption edges of Al and Si thin foils (8). For the results presented here we have used a spectral resolution between 0.3 and 0.6 % of the photon energy.

A chopper placed at the exit of the monochromator allows for a sequential recording of the ions signal obtained with and without the photons, and to substract a possible contribution of collisional ionization processes. As an example, a signal to noise ratio of 1700 counts/s over 40 counts/s has been measured in the continuum with 103 eV photons in the case of $Xe^{3+}$ ions.

With this new experimental set-up, single and double ionization signals (and even triple in the case of $Xe^+$ ions) have been measured for $Xe^+$ to $Xe^{4+}$ ions, and single ionization signal for $Xe^{5+}$ to $Xe^{7+}$ ions. As an example, we show in the left part of Figure 2 the variation, as a function of photon energy, of the sum of these different contributions (ion yield) measured for $Xe^{3+}$ to $Xe^{7+}$ ions, and in the right part the results of our calculations made in the velocity form for photoexcitation as described below. The calculated positions of the 4d ionization thresholds (noted $4d^{-1}$) are indicated by vertical bars in the experimental spectra. Below these thresholds, all spectra are dominated by intense lines corresponding to discrete excitation of a 4d electron. The excited ion decays then via autoionization according to the scheme given below as an example for $Xe^{3+}$ ions:

$$Xe^{3+} [Kr] 4d^{10} 5s^2 5p^3 + h\nu \rightarrow Xe^{3+*} [Kr] 4d^9 5s^2 5p^3 n'l$$
$$\rightarrow Xe^{4+} [Kr] 4d^{10} 5s^2 5p^2 + e^-$$

with l = 1 (n' ≥ 5) or l = 3 (n' > 5). The 4d → 5p transitions, which lie around 60 eV photon energy, have been observed and calculated, but are not shown in the figures.

In the case of $Xe^{3+}$ and $Xe^{4+}$ ions, the experimental spectra display above the 4d thresholds a smooth decrease of the ions signal, corresponding mainly to direct photoionization in 4d subshell, and to a less extent in 5s and in 5p subshells. When energetically allowed, i.e. for $Xe^{3+}$ and partly for $Xe^{4+}$ ions, the 4d hole will further decay via Auger emission, contributing to the double ionization signal.

The identification of the discrete lines has been performed using our MCDF code. $Xe^{n+*}$ [Kr] $4d^9$ n'l final configurations with n' ≤ 7 have been included in the

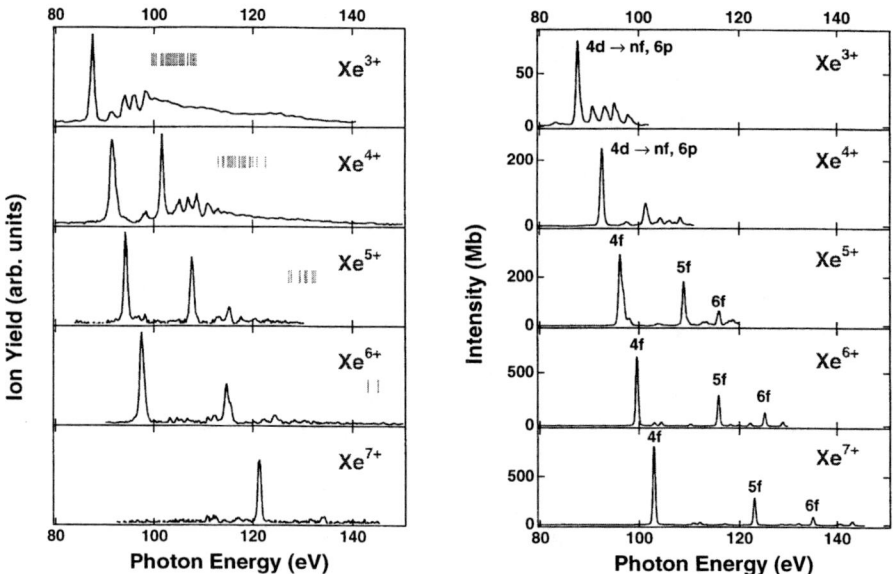

**FIGURE 2.** Left: Experimental spectra for single and double ionization of $Xe^{3+}$ and $Xe^{4+}$ ions, and single ionization for $Xe^{5+}$ to $Xe^{7+}$ ions. The vertical bars indicate the position of the calculated 4d-ionization thresholds. Right: Results of our MCDF calculations for the 4d-excitations in $Xe^{3+}$ to $Xe^{7+}$ ions.

calculation. The result is shown in the right part of Figure 2. We have assumed a Voigt profile for the shape of each excitation line with a constant full width at half maximum set to 100 meV for the lorentzian contribution, a value which is close to the natural width of the 4d hole measured for Xe I, and to 500 meV for the gaussian contribution, representing the experimental band pass. For a given ionic stage, the occupation probabilities of the initial levels were assumed to be statistical. Within our experimental resolution, each discrete line observed in the spectrum results from the sum of many transitions, especially for $Xe^{3+}$ to $Xe^{5+}$ ions. We indicate above the most intense lines the assigment of the dominating 4d → n'l transitions in the calculated spectra. Taking into account the finite excitation band pass used in the experiment, the agreement between measured and calculated spectra is quite satisfactory, except for two anomalies: the surprising intensity of the 4d → 5f transition we measured in $Xe^{4+}$, and the absence of the line at the energy of the 4d → 4f transition in the experimental spectrum of $Xe^{7+}$. For the $Xe^{4+}$ ion, we have no satisfactory explanation. If we take into account the broadening effects due to the actual values of the natural width, which is likely smaller for the $Xe^{4+*}$ [Kr] $4d^9 5s^2 5p^2$ 5f excited states than for the $Xe^{4+*}$ [Kr] $4d^9 5s^2 5p^2$ 4f states, the relative intensity of the 5f line compared to the 4f line certainly would be increased, but probably not enough to agree with the measured value. For the $Xe^{7+}$ ion, the disagreement is only apparent. The excitation energies of all $Xe^{7+*}$ [Kr] $4d^9$ 5s 4f excited states are lower than the energy of the $Xe^{8+}$ [Kr] $4d^{10}$ ground state, and then these excited states cannot autoionize.

In order to estimate the number of ions produced in a metastable state in the ECRIS, we have calculated the excitation spectrum starting from the [Kr] $4d^{10}$ 5s 4f metastable configuration in $Xe^{6+}$, i.e. in the ion for which we need one of the highest micro-wave power. This calculation predicts very intense 4d → 4f transitions, with an oscillator strength 10 times higher and with excitation energies 5 to 6 eV lower than for the same transitions in the $Xe^{6+}$ ions in the ground state. There is no evidence for such lines in the experimental spectrum. We can then conclude that the population of ions present in metastable states is very weak in our experiment.

Figure 2 provides a complete picture of the behavior of resonant and continuum photoionization processes along the Xe isonuclear sequence, when the 5p subshell is completely stripped electron by electron. We observe that the intensity of the continuum ionization is almost negligible for ions with charge stages higher than 3, indicating that the collapse of the nf orbitals is complete for the $Xe^{4+}$ ions. For higher charge stages, we observe a simplification of the spectra, showing almost pure 4d → np, nf Rydberg series. In the $Xe^{6+}$ spectrum, the contribution of continuum ionization is too weak to be measurable within our experimental sensitivity.

The present experimental set-up is still in a developing phase. Improvements are in progress to allow for measurements of absolute values of the photoionization cross sections. An oven is also being tested for the production of metallic ions.

## REFERENCES

1. F. Wuilleumier, "Many-body Theory of Atomic Structure and Photoionization" ed T.N. Chang, World Scientific, Singapore, 1993, p 349 and references therein.
2. I. Lyon et al, J. Phys. B **20**, 1471 (1987) and references therein.
3. T. Koizumi et al, J. Phys. B **28**, 609 (1995).
4. M. Sano et al, J. Phys. B **29**, 5305 (1996).
5. N. Watanabe et al, J. Phys. B **31**, 4137 (1998).
6. T. Koizumi et al, Physica Scripta **T73**, 131 (1997).
7. J. Bruneau, J. Phys. B **16**, 4135 (1983).
8. E.M. Gullikson et al, Phys. Rev. B **49**, 16283 (1994).

# Charge exchange induced X-ray transitions of hollow ions in laser field ionized plasmas

F.B. Rosmej[1], D.H.H. Hoffmann[1],
A.Ya. Faenov[2], T.A. Pikuz[2], A.I. Magunov[2], I.Yu. Skobelev[2],
T. Auguste[3], P.D'Oliveira[3], S. Hulin[3], P. Monot[3]

[1]TU-Darmstadt, Institut für Kernphysik, Schloßgartenstr. 9, D-64289 Darmstadt, Germany,
[2]Multicharged Ions Spectra Data Center of VNIIFTRI, Mendeleevo, 141570 Russia,
[3]Commissariat à l'Energie Atomique DSM/DRECAM/SPAM, Bât. 522, C.E. Saclay, 91191 Gif-Sur-Yvette Cédex, France

**Abstract** Double electron charge exchange is proposed for the formation of hollow He-like ions when laser field ionized nuclei penetrate into the residual gas. Using transitions from different configurations in hollow ions a method for the determination of the electron temperature in the long lasting recombination phase is developed.

## I. INTRODUCTION

Space and spectrally resolved registration of soft X-ray emission of laser produced plasmas by means of high quality spherical mica crystals [1] have revealed rather broad unusual structure near the usual positions of resonance lines. Stark and opacity broadening calculations could not explain the experimental observation and it was proposed [2] that not resonance line emissions have been observed but accumulated Rydberg dielectronic satellites. Particular in dense cold plasmas near the target surface, satellite transitions can exceed the intensity of usual resonance line emission [3, 4]. These phenomena are of great importance for the interpretation of the He$_\alpha$ (Li-like 1s2lnl'-satellites) and He$_\beta$ (Li-like 1s2lnl'- and 1s3lnl'-satellites) resonance emission in laser produced plasmas. Usual excitation channels failed to match with the data and new excitation channels have been developed [3-6] being in good agreement with the data.

## II. Soft X-ray radiation from hollow ions

Figure 1 shows the experimental spectra from nitrogen: focussing a CPA Ti-sapphire laser onto a laminar pulsed nitrogen gas jet. The density of atoms was $1.5\cdot10^{19}$ cm$^{-3}$ for a maximum pressure of 20 bar. The laser intensity was $10^{19}$ W/cm$^2$, energy 750 mJ, pulse width of 65 fs. Fig. 1 shows that numerous line positions coincide for quite different experiments resulting in broad emission structures near the usual position of the H-like resonance lines.

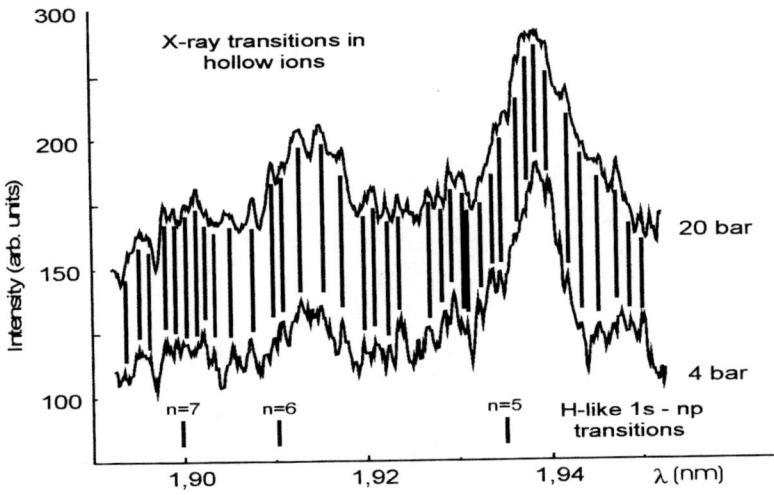

**Figure 1.** Soft X-ray transitions in hollow nitrogen ions. A clear correlation of several transitions for different gas pressures is indicated by vertical bars.

## III. Interpretation

Stark broadening calculations show [5] that the density is much too low to account for the broad emission structures near the H-like resonance lines. We propose He-like transitions in hollow ions to account for the observation: nln'l' → (1snl, 1sn'l') + hν. The population of hollow ions may occur, when field ionized nitrogen nuclei penetrate into the residual gas:

nuc + N → nl n'l' + $N^{2+}$, nuc + $N^+$ → nl n'l' + $N^{3+}$, nuc + $N^{2+}$ → nl n'l' + $N^{4+}$.

The cross section for the double electron capture is of the same order of magnitude as for single electron capture [7] (below v = 1 a.u. theoretical cross sections agree roughly within a factor of 3 with measurements), therefore the hollow ion configurations nln'l' can be populated in one step (the velocity of the ions was assumed to be given from Coulomb explosion). This is distinct to mechanisms like radiative recombination, three-body recombination, single electron charge exchange and collisional excitation requiring two step processes. Figure 2 shows an example of the various line transitions originating from the 3l6l', 4l6l' and 6l6l' configurations in the spectral window from 1.88 to 2.00 nm. Figure 3 shows the spectra simulations carried out with the MARIA-code [8] including all the configurations nln'l' with n=3-8 and n' = 5-8 (Hartree-Fock calculations including intermediate coupling, configuration interactions and relativistic corrections up to the second order [9]). The spectral distribution I(ω) is approximated according

$$I(\omega) = \frac{n_0}{g_0} \sum_k \sum_{i,j,i \neq j} g_j^k A_{ji}^k \exp\left(-\frac{\Delta E_{0k}}{kT_e}\right) \Phi_{ji}^k(\omega)$$

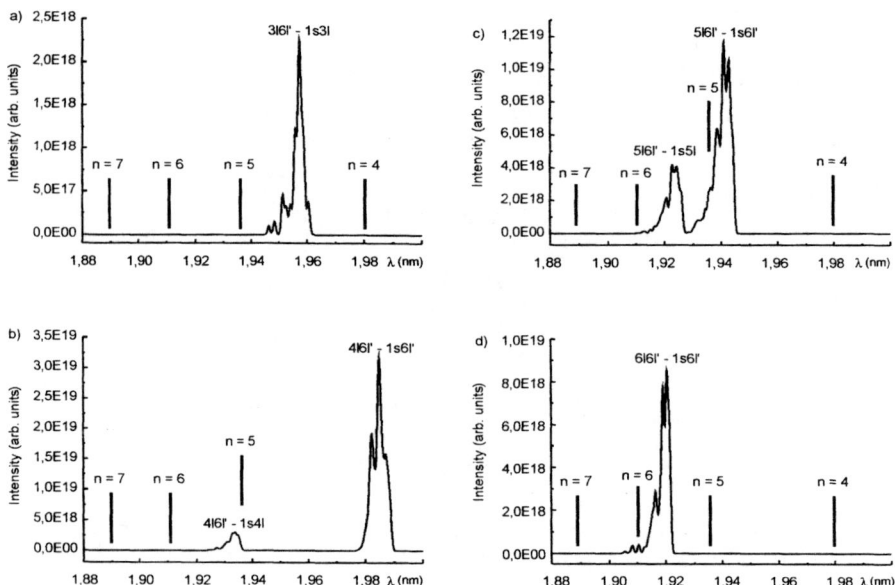

**Figure 2.** Theoretical X-ray spectra from line transitions in hollow nitrogen He-like ions.

"k" designates different configurations, e.g. k = 3l5l', the index "0" designates the lowest reference level (note, that in the limit of high densities the above spectral distribution is exact). As the energy levels of the hollow ion configurations differ as much as about 75 eV, a strong temperature dependence is obtained, see Fig. 3. As the formation of the hollow ions takes place mainly after the main laser pulse, this temperature diagnostic potentially diagnoses the long lasting recombination phase.

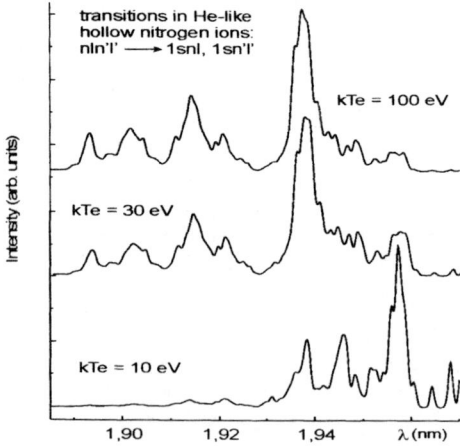

**Figure 3.** Temperature dependence of the spectral distribution originating from He-like hollow ions.

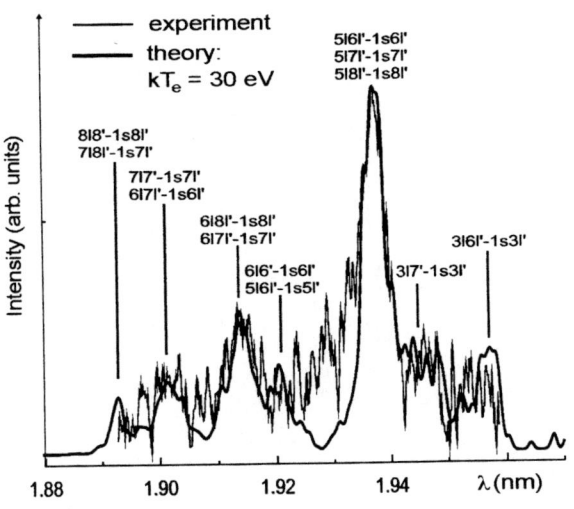

**Figure 4.** Spectrum fitting based on hollow-atom dielectronic satellite emission.

Figure 4 shows the best fit of the experimental spectrum based on He-like transitions in hollow nitrogen ions. A few configurations are indicated. It can be seen clearly from the modeling that the transitions in hollow ions reflect the broad emission structure near the usual position of the H-like resonance lines (see Fig. 1) and also the onset of numerous "fine-structures". The low temperature determined stimulated a new approach for the tunneling ionization in relativistic laser fields [10]: it is due to different e-e relaxation times caused by successive L-, K-shell field ionization.

## REFERENCES

1. Faenov, A.Ya. et al., Phys. Scripta **50**, 333 (1994)
2. Rosmej, F.B. and Faenov, A.Ya., Physica Scripta **T73**, 106 (1997)
3. Rosmej, F.B. et al., JQSRT **58**, 859 (1997)
4. Rosmej, F.B. et al., J. Phys. B Lett. **31**, L921 (1998)
5. Rosmej, F.B. et al., J. Phys. B Lett. **32**, L107 (1999)
6. Rosmej, F.B. et al., JQSRT, "X-ray radiation from ions with K-shell vacancies", in print (1999)
7. Janev, R.K., Presnyakov, L.P., Shevelko, V.P., *Physics of Highly Charged Ions*, Springer, Berlin, 1985
8. Rosmej, F.B., J. Phys. B. Lett. **30**, L819 (1997)
9. Cowan, R.D., *Theory of Atomic Structure and Spectra*, Berkeley, CA: University of California Press, 1981
10. Andreev, N.E. et al. 1998, JETP Lett. **68**, 592

# VI. NUCLEAR EFFECTS

# Inelastic Scattering of Synchrotron Radiation from Electrons and Nuclei for Lattice Dynamics Studies

E. E. Alp, W. Sturhahn, H. Sinn, T. Toellner, M. Hu, J. Sutter, and A. Alatas

*Advanced Photon Source, Argonne National Laboratory, Argonne, Illinois 60540*

**Abstract.** The inelastic scattering of x-rays, one of the the first applications of x-rays to the field of condensed matter physics, has been rejuvenated in the last decade. The availability of synchrotron radiation from wiggler and undulator sources combined with advances in monochromatization of the incident beam and analysis of the scattered beam with meV resolution led to the measurement of phonon dispersion relations. In addition, the use of Mossbauer nuclei as scatterer and analyzers has led to the discovery of the inelastic nuclear resonant scattering technique. This new method allows extraction of partial phonon density of states from amorphous materials, thin films, multilayers and interfaces, and liquids.

## INTRODUCTION

The early attempts of Debye [1] and the first determination of phonon dispersion relations by thermal diffuse scattering in Al [2] demonstrated the potential of x-rays for vibrational studies. However, there was a period of low activity in the 1960s and 70s due to lack of efficient analyzers with energy resolution commensurate with excitation energies, and not surprisingly, coincident with the development of inelastic neutron scattering [3].

The discovery of the Mössbauer effect [4] and the ensuing excitement over the possibility to use extremely monochromatic radioactive sources for lattice dynamics studies [5,6] were dampened with the realization that both the range of energy tunability, and the bandwidth of the radiation were too small. However, the Lamb-Mössbauer factor, or recoiless fraction of absorption and emission of x-rays, and the second order Doppler shift obtained from Mössbauer spectra proved to be useful for lattice dynamics studies. Similarly, element-specific Debye-Waller factors obtained from EXAFS data were used to understand lattice dynamics in addition to the more traditional techniques of infrared and Raman spectroscopy.

The use of synchrotron radiation and the subsequent improvements in the source brilliance via the introduction of wigglers, coupled with clever use of crystal optics [7], led the way for a revival of inelastic x-ray scattering, which resulted in the first successful demonstration of phonon dispersion relation in Be [8]. Similarly, synchrotron radiation was used to observe coherent nuclear Bragg diffraction in $Y_3{}^{57}Fe_5O_{12}$ [9], and a decade later inelastic incoherent nuclear resonant scattering was realized [10] to extract phonon density of states [11].

## METHODOLOGY

X-rays are a suitable probe for a variety of excitations in condensed matter because of their ability to penetrate, their wavelength being comparable to interatomic distances, and the possibility to determine the exact amount of energy and momentum transfer during the scattering process. In a typical x-ray scattering experiment, photons with energy $E_i$ with a wavevector $\mathbf{k}_i$ and polarization $\mathbf{e}_i$ are incident onto a sample. The measurement then involves determination of the final energy $E_f$, scattered into a new direction, $\mathbf{k}_f$, and polarization state, $\mathbf{e}_f$. Since the energy analysis requires a certain solid angle to be intercepted, the scattering cross section is defined as functions of both frequency, $\omega$, and solid angle, $\Omega$:

$$\frac{d^2\sigma}{d\Omega d\omega} = \frac{d\sigma}{d\Omega} S(\mathbf{Q},\omega),$$

where $\mathbf{Q} = \mathbf{k}_f - \mathbf{k}_i$ is the momentum, and $\hbar\omega = E_f - E_i$ is the energy transfer.

The first term, describing the coherent coupling of electrons, is the Thomson cross section, $d\sigma/d\Omega = r_0^2 (\mathbf{e}_i \cdot \mathbf{e}_f)^2 \omega_f/\omega_i$, where $r_0$ is the classical electron radius. The second term, which describes the overall scattering strength of the probed volume, will depend on the dynamics of the particles in this volume. It is expressed in terms of the correlation of the phase of the scattering amplitude:

$$S(\mathbf{Q},\omega) = \sum_{i,f} g_i \left| \langle i | \sum_j e^{i\mathbf{Q}\cdot\mathbf{r}_j} | f \rangle \right|^2 \delta(\omega - \omega_f - \omega_i),$$

where summation over $j$ is performed over all electrons, and $g_i$ is a weight factor for the initial states. This can also be written as a time Fourier transform of the density correlation function:

$$S(\mathbf{Q},\omega) = \frac{1}{2\pi} \int dt \, (e^{-i\omega t} \sum_i g_i \langle i | \sum_{j,l} e^{-i\mathbf{Q}\cdot\mathbf{r}_j(0)} e^{i\mathbf{Q}\cdot\mathbf{r}_l(t)} | i \rangle).$$

For *coherent elastic scattering*, where there is no energy exchange, the cross section is reduced to $S(\mathbf{Q}) = \left| \sum_j e^{i\mathbf{Q}\cdot\mathbf{r}_j} \right|^2$. The sum will vanish, unless $\mathbf{Q}\cdot\mathbf{r}_j$ is an integer multiple of $2\pi$ for all sites. It can be shown that this condition is realized under Bragg diffraction, when the scattering vector **Q** is equal to the reciprocal lattice vector **G**. Thus S(**Q**) is a measure of the static structure factor.

*Coherent inelastic scattering,* on the other hand, will involve exchange of energy, in addition to momentum. A coherent scattering experiment in which the removal rate of photon (or neutron) is measured as a function of energy and momentum transfer yields information about the frequency and spatial extent of the phonons. We can describe phonons as quantized lattice vibrations of ion cores around their equilibrium positions. Coherent elastic or inelastic scattering of x-rays (neutrons) measure static or dynamic structure factors by scattering from core and outer electrons (nuclei). In the adiabatic approximation, the electrons and the nuclei move together, and neutron scattering experiments should yield the same result as x-ray scattering experiments. However, for metals, the conduction electron gas need not behave according to the adiabatic approximation [12]. The only experimental evidence so far for the breakdown of the Born-Oppenheimer approximation is obtained from x-ray diffraction measurements [13]. The difference in charge distribution in LiH and its LiD derivative is attributed to a coupling between vibrations and electronic states. Inelastic x-ray scattering may provide more direct evidence and thus presents a unique opportunity to study the behavior of classical and quantum liquids [14].

Inelastic x-ray scattering has been used to determine the temperature dependence of a transition from normal to fast sound in liquid water and to establish a coupling between propagating density fluctuations and dynamical structural rearrangements in the liquid [15]. Similarly, positive dispersion of sound velocity is observed in liquid lithium, which can be associated with viscoelastic shear relaxation [16].

*Coherent inelastic x-ray (neutron) scattering* yields information about the dynamic structure factor. When the inverse of the transferred momentum is comparable to the interatomic distance and the transferred energy can be analyzed with meV resolution, collective excitations of ion cores, i.e., phonons can be measured. By varying the momentum transfer, the dependence of phonon energy on momentum, $\omega(k)$, can be determined. The dispersion relations can then be analyzed to develop a force model, which ultimately leads to a more accurate description of the interatomic potential. When complete dispersion relations are available, a method known as Born-von Kármán analysis provides a connection between force constants and the dispersion relations. Phonon density of states, which is a key ingredient in making the connection to the classical thermodynamic quantities like specific heat and vibrational entropy, can be derived by sampling the dispersion curves [17].

When there is no crystalline order, as in the case of liquids or amorphous solids, incoherent inelastic neutron scattering technique may be used to measure phonon density of states directly. However, when the material is in the form of a thin film or a monolayer at an interface, the traditional techniques fail to yield information about the vibrational properties. The recent discovery of *inelastic nuclear resonant scattering* of synchrotron radiation addresses some of these aspects of lattice dynamics. This method directly measures the partial phonon density of states.

The *inelastic nuclear resonant interaction* cross section is related to the nuclear resonance cross-section and phonon excitation probability as follows:

$$\sigma(E,\mathbf{k}) \approx \sigma_0 \Gamma S^*(\omega,\mathbf{k}) .$$

$$\sigma_0 = \frac{2\pi}{1+\alpha}\left(\frac{\hbar c}{E_0}\right)^2 \frac{2I_e+1}{2I_g+1}$$

Here $\sigma_0$ is the nuclear resonance cross section, $\alpha$ is the internal conversion coefficient, $E_0$ is the nuclear resonance energy, $I_g$ and $I_e$ are the total angular momentum quantum number for the ground and the excited state of the nucleus, and $\Gamma$ is the linewidth of the excited state. The function $S^*(\omega,\mathbf{k})$ is not the same as the dynamic structure factor given earlier. It can be considered as a probability density for phonon excitation. It is a function of incident energy and momentum. It can be shown that $S^*(\omega,\mathbf{k})$ is related to the pair auto-correlation function $G(r,t)$ via a space and time Fourier transform. A set of specific sum rules put forward by Lipkin provides a procedure to extract lattice dynamics information such as specific heat and average force constant [18]. The extraction of phonon density of states from the experimental data is described elsewhere [11,19].

## EXPERIMENTAL IMPLEMENTATION

The two methods mentioned above, *coherent inelastic scattering, and incoherent nuclear resonant scattering of synchrotron radiation* can be implemented in various ways. The set up used at the Advanced Photon Source 3-ID beamline combines the resources for both experiments, and they are schematically shown in Figure 1.

The *coherent inelastic x-ray spectrometer* described in Figure 2(a) utilizes an "in-line" tunable high energy monochromator as a source, and a curved crystal back-scattering Si crystal as an analyzer. The monochromator is a combination of Si (4 4 0)-(15 11 3) nested channel-cut crystals. The analyzer is a Si (18 6 0) diced crystal with a 6 m bending radius. The overall energy resolution is 2 meV, with a momentum transfer range up to 5 Å$^{-1}$. The energy of the incident photons is 21.657 keV. A typical phonon measurement obtained from single crystalline Be is shown in Figure 2.

The *inelastic nuclear resonant spectrometer* uses a similar "in-line" monochromator, consisting of either a "nested" channel-cut geometry or two-flat crystal geometry [20] to achieve meV resolution at the 14.4125 keV nuclear resonance of [57]Fe or 23.880 keV nuclear resonance of [119]Sn. The choice of these isotopes was mainly motivated by the long lifetime of intermediate states that are comparable to the duration between the electron bunches in the storage ring. This separation is typically of the order of 100 nsec. When the incident radiation is tuned around the nuclear resonance energy, which has an energy width less than 0.1 µeV, the phonon creation or annihilation may subtract or supplement the incident photon energy to excite the nuclear resonance. This can be unambiguously detected by the delayed fluorescence signal after the main pulse. A typical data set is given in Figure 3 a for amorphous and crystalline $Fe_2Tb$. The central peak corresponds to the nuclear resonance, and side peaks are due to phonon annihilation(left) and creation (right) processes. The logarithmic scale used helps to see a few unique features of this method: i) high signal-to-noise ratio allows the observation of two phonon processes. The peak at + 44 meV for crystalline $Fe_2Tb$ is a two-phonon contribution associated with the +22 meV peak. ii) The asymmetry between phonon creation (right) and phonon annihilation (left) is related to phonon occupation probability, $S(-\omega,\mathbf{k}) = e^{-\hbar\omega/kT} S(\omega,\mathbf{k})$, and provides a detailed balance for a given temperature. iii) Signal-to-background ratio over 5 orders of magnitude is due to the way measurements are done at a pulsed synchrotron source with time discrimination. In Figure 4 b the derived Fe partial phonon density of states are shown for the amorphous and the corresponding crystalline phase. These type of studies have been extended to thin films [21], multilayers, and biological systems. The amount of material required is of the order of micrograms, and therefore, in combination with microfocusing techniques, these type of measurements have been extended to materials under high pressure in diamond-anvil cells [22].

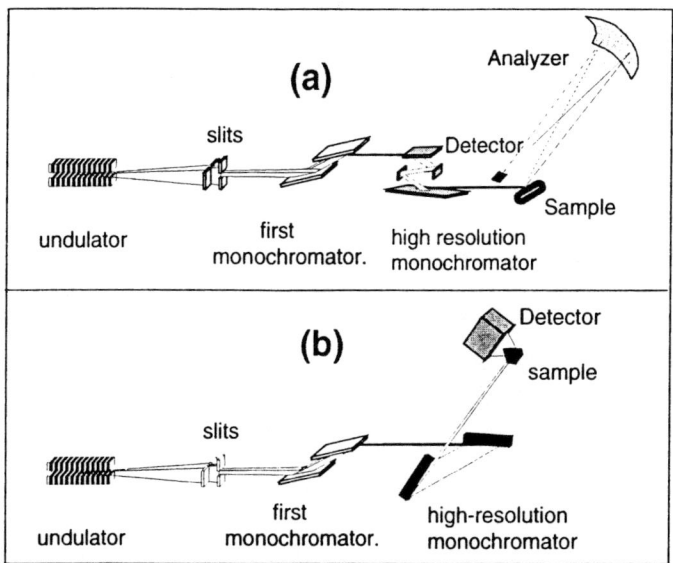

Figure 1. The schematic realization of inelastic x-ray scattering experiments at the 3-ID beamline at the APS. (a) The inelastic x-ray spectrometer with an "in-line" monochromator, and a curved crystal analyzer with a bending radius of 6 m. (b) The inelastic nuclear resonant spectrometer, with 2-flat crystal monochromator. The energy resolution is 0.6 meV at 14.4 keV $^{57}$Fe resonance, and 0.5 meV at 23.880 keV $^{119}$Sn resonance.

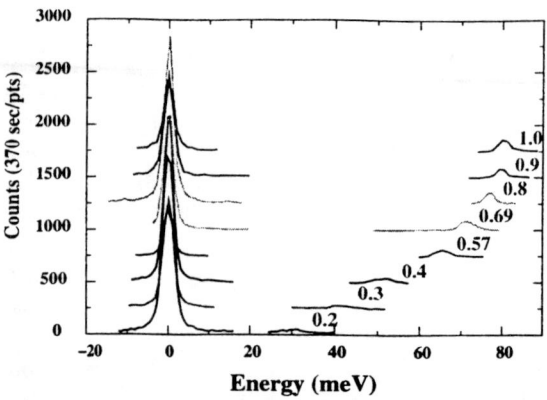

Figure 2. Phonon dispersion measurements in Be along (0 0 ξ) direction for the longitudinal phonons for ξ values indicated for each spectrum. The elastic peak is both due to the window material on the sample chamber, and impurities in beryllium.

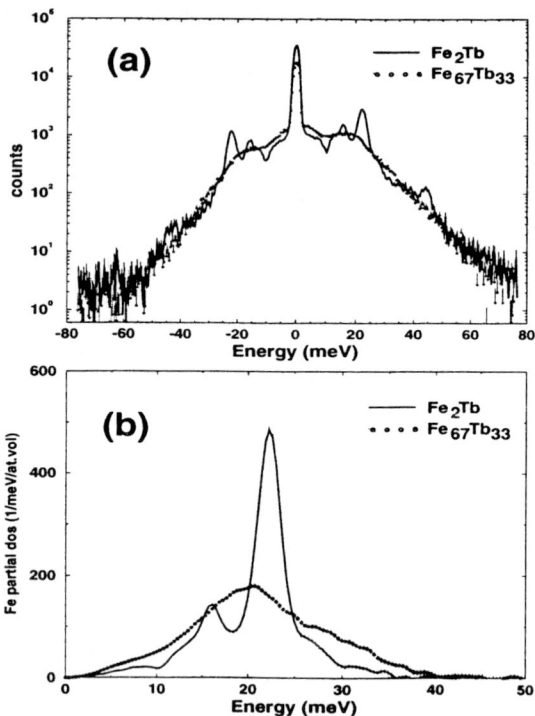

Figure 3. (a) The raw data measured around the $^{57}$Fe resonance in $Fe_2Tb$ alloy in crystalline and amorphous state, at room temperature. The samples were prepared in thin film form. The zero energy corresponds to the nuclear resonance energy of $^{57}$Fe at 14.4125 keV. (b) The Fe partial density of states extracted from the data.

## ACKNOWLEDGEMENTS

This work is supported by US-DOE BES, under contract number W-31-109-ENG-38. We would like to thank Prof. W. Keune and T. Ruckert for providing the $Fe_2Tb$ samples, and Prof. E. Burkel for the Be single crystal.

## REFERENCES

1. P. Debye, Ann. Phys. 43 (1914) 49.
2. C. B. Walker, Phys. Rev. 103 (1956) 547.
3. B. N. Brockhouse, *Inelastic Scattering of Neutrons in Solids and Liquids*, IAEA, Vienna, (1961)
4. R. L. Mossbauer, Z. Physik, 151 (1958) 124.
5. V.M. Vischer, Annals of Physics, 9 (1960) 194.
6. K. S. Singwi, A. Sjolander, Phys. Rev. **120** (1960) 1093.
7. W. Graeff, G. Materlik, Nucl. Instrum. Meth., **195** (1982) 97.
8. B. Dorner, E. Burkel, Th. Illini, and J. Peisl, Z. Phys. B, Condensed Matter **69** (1987) 179.
9. E. Gerdau, R. Rüffer, H. Winkler, W. Tolksdorf, C. P. Kalges, J. P. Hannon, Phys. Rev. Lett. **54** (1985) 835.
10. M. Seto, Y. Yoda, S. Kikuta, X. W. Zhang, M. Ando, Phys. Rev. Lett. **74** (1995) 3828.
11. W. Sturhahn, T. S. Toellner, E. E. Alp, X. Zhang, M. Ando, S. Kikuta, M. Seto, C.W. Kimball, B. Dabrowski, Phys. Rev. Lett. **74** (1995) 3832.
12. M. Rasolt, Phys. Rev. B 31 (1984) 1615.
13. G. Vidal-Valat, J. -P. Vidal, Acta Cryst. **A48** (1992) 46.
14. A.A. Louis, N. W. Aschcroft, J. Non. Cryst. Solids, **250-252** (1999) 9.
15. A. Cunsolo, G. Ruocco, F. Sette, C. Masciovecchio, A. Mermet, G. Monaco, M. Sampoli, R. Vernebi, Phys. Rev. Lett. **82** (1999) 775.
16. H. Sinn, F. Sette, U. Bergmann, Ch. Halcoissis, M. Krisch, R. Verbeni, and E. Burkel, Phys. Rev. Lett. **78** (1997) 1715.
17. G. Gilat, L, J, Raubenheimer, Phys. Rev. **144** (1965) 390.
18. H. J. Lipkin, Phys. Rev. B **52** (1995) 10073.
19. M. Hu, W. Sturhahn, T.S. Toellner, P. M. Hession, J. P. Sutter, E. E. Alp, Nucl. Instrum. Meth. **A 428** (1999) 551.
20. T. S. Toellner, M. Y. Hu, W. Sturhahn, K. Quast, and E. E. Alp, Appl. Phys. Lett., **71** (1997) 2112.
21. W. Sturhahn, R. Röhlsberger, E. E. Alp, T. Ruckert, H. Schör, and W. Keune, J. Magnetism And Magnetic Materials, **198-199** (1999) 590.
22. H. K. Mao, W. Sturhahn, et al, (to be published, Argonne National Laboratory)

# Nuclear Excitation by Electronic Transition Between Atomic Shells

Eugene V. Tkalya

*Institute of Nuclear Physics of the Moscow State University, Moscow, 119899, Russia*

**Abstract.** The current status of the process of Nuclear Excitation by Electronic Transition between atomic shells (NEET) is analyzed.

## INTRODUCTION

Emission of x-rays and Auger electrons are the usual channels of atomic deexcitation. The process of nonradiative nuclear excitation by means of direct energy transfer from the excited atomic shell to the nucleus via a virtual photon is also possible, if within the atomic shell there exists an electronic transition close in energy and coinciding in type with the nuclear one (see Figure 1).

This possibility was first pointed out in the case of muonic atoms by Wheeler [1]. The widely used label NEET was suggested by Morita [2], who first considered this process for ordinary atoms. After that, several theoretical and experimental studies were devoted to the NEET process (see Refs. [3–29]).

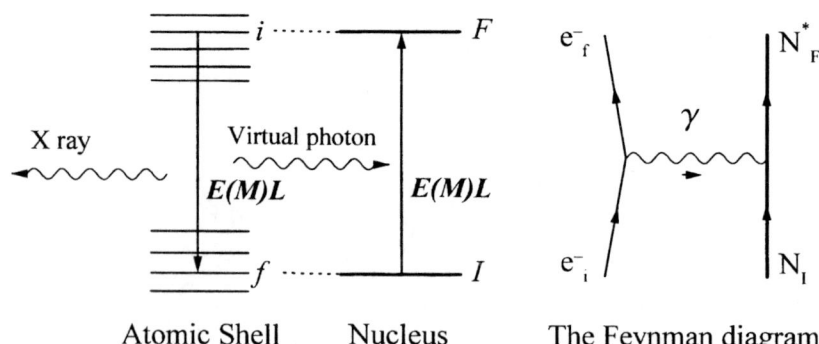

**FIGURE 1.** Diagram of the NEET process.

Some of the theoretical studies [2,4,5,9,13,14] have predicted that NEET has a relatively high probability. In several experimental studies, excitation of the low-energy levels of the nuclei $^{189}$Os, $^{197}$Au, and $^{237}$Np accompanying irradiation of targets by electrons and photons was explained by NEET [3,6,8,10–12,25]. The result of the experiment [7] on the excitation of the nucleus $^{235}$U to the isomeric level 76.8 eV in a hot surface laser plasma was also interpreted using NEET.

At the same time, there are some studies where a relatively small probability is predicted for NEET [15,17,18,29]. Recent experiments [24,29] showed that the theoretical estimates of Refs. [17,18] seem reasonable.

There are many reasons why the NEET process is interesting for atomic and nuclear physics. First, NEET is one of the fundamental processes of nuclear excitation, along with photoexcitation and inelastic electron scattering. Second, the Inverse NEET process (INEET) is a new channel of deexcitation of isomeric nuclear levels, which can strongly accelerate the decay of the isomers [23]. Third, NEET probes the atomic shell – nuclear interaction. Last, a very interesting research direction is to study the dynamical effect of the nuclear size in NEET (the so-called anomalous NEET) [21,22], because it provides a key to the physics of toroidal moments.

## THE NEET PROBABILITY

The NEET process can be described by a Feynman diagram of second order (see Figure 1). We consider here only that stage of the process when the nuclear excitation takes place during the electronic transition from the initial state $i$ to the final state $f$. The prehistory — the process of vacancy formation in the shell $i$ (let $i$ be the K-shell, for example) — is not considered, because the lifetime of the K-shell vacancy greatly exceeds the time required for the ejected K-shell electron to leave the interaction region [17,18].

The element of the $S$-matrix corresponding to the diagram in Figure 1 has the form

$$S^{(2)}_{fi} = \frac{i}{\omega_N - \omega_A - i(\Gamma_i + \Gamma_f + \Gamma_N)/2} \int d^3r d^3R (-e) \bar{\psi}_f(\mathbf{r}) \gamma^\mu \psi_i(\mathbf{r})$$
$$\times D_{\mu\nu}(\omega_N, \mathbf{r} - \mathbf{R}) \, e\Psi^+_F(\mathbf{R}) \hat{J}_\nu \Psi_I(\mathbf{R}). \quad (1)$$

Here $\omega_N$ and $\omega_A$ are the energies of the nuclear and atomic transitions (the adopted system of units is $\hbar = c = 1$), $e$ is the proton charge, $\gamma^\mu$ are the Dirac matrices, and $\hat{J}_\nu$ is the operator of the nuclear electromagnetic current. $\psi$ and $\Psi$ are the wave functions of the electronic and nuclear states, respectively. The widths of these states are denoted as $\Gamma_{i,f}$ for the atom and $\Gamma_N$ for the nucleus.

The expression for the NEET probability may be put in the form

$$W_{NEET} = \Gamma_f P_{NEET}, \quad (2)$$

where the relative probability $P_{NEET}$ for a nuclear transition into an excited state during electronic hole decay to the atomic level $f$ is

$$P_{NEET} = \left(1 + \frac{\Gamma_i}{\Gamma_f}\right) \frac{E_{int}^2}{(\omega_N - \omega_A)^2 + (\Gamma_i + \Gamma_f + \Gamma_N)^2/4}. \qquad (3)$$

The interaction energy of the electronic and nuclear currents $E_{int}$ is a key parameter in Eq. (3). $E_{int}$ is expressed in the form

$$E_{int}^2 = 4\pi e^2 \omega_N^{2(L+1)} \frac{(j_i 1/2 L0 | j_f 1/2)^2}{[(2L+1)!!]^2} \mathcal{R}_L^{E(M)L}(\omega_N) B(E(M)L; J_i \to J_f), \qquad (4)$$

where $B(E(M)L; J_i \to J_f)$ is the reduced transition probability, and $\mathcal{R}_L^{E(M)L}$ are the atomic radial matrix elements of electric ($EL$) and magnetic ($ML$) multipolarity:

$$\mathcal{R}_L^E(\omega) = \int_0^\infty dr r^2 \{h_L^{(1)}(\omega r)[g_i(r)g_f(r) + f_i(r)f_f(r)] -$$

$$\frac{h_{L-1}^{(1)}(\omega r)}{L}[(\kappa_i - \kappa_f - L)g_i(r)f_f(r) + (\kappa_i - \kappa_f + L)f_i(r)g_f(r)]\}, \qquad (5)$$

$$\mathcal{R}_L^M(\omega) = \frac{\kappa_i + \kappa_f}{L} \int_0^\infty dr r^2 h_L^{(1)}(\omega r)[g_i(r)f_f(r) + f_i(r)g_f(r)]. \qquad (6)$$

Here $\kappa = (l-j)(2j+1)$, and $g(r)$ and $f(r)$ are, respectively, the large and small components of the electronic wave functions, with the normalization condition $\int_0^\infty dr r^2 (g^2(r) + f^2(r)) = 1$.

**TABLE 1.** The NEET probability $P_{NEET}$.

| Nuclei | Experiment | Reference | Theory | Reference |
|---|---|---|---|---|
| $^{189}$Os | $1 \cdot 10^{-6}$ | [3] | $1.5 \cdot 10^{-7}$ | [4] |
| ($M1$, | $(1.7 \pm 0.2) \cdot 10^{-7}$ | [6] | $2.5 \cdot 10^{-7}$ | [13] |
| 69.5 keV) | $(4.3 \pm 0.2) \cdot 10^{-8}$ | [10] | $2.31 \cdot 10^{-7}$ | [14] |
| | $(5.7 \pm 1.7) \cdot 10^{-9}$ | [12] | $1.2 \cdot 10^{-9}$ | [15] |
| | $(2.6 \pm 1.6) \cdot 10^{-11}/b_0$[a] | [24] | $1.1 \cdot 10^{-7}$ | [16] |
| | $< 9 \cdot 10^{-10}$ | [29] | $1.2 \cdot 10^{-10}$ | [17] |
| | | | $1.3 \cdot 10^{-10}$ | [29] |
| $^{197}$Au | $(2.2 \pm 1.8) \cdot 10^{-4}$ | [11] | $3.5 \cdot 10^{-5}$ | [13] |
| ($M1$, | $(5.1 \pm 3.6) \cdot 10^{-5}$ | [25] | $2.2 \cdot 10^{-5}$ | [14] |
| 77.3 keV) | | | $4.2 \cdot 10^{-7}$ | [15] |
| | | | $1.4 \cdot 10^{-7}$ | [17] |
| $^{237}$Np | $(2.1 \pm 0.6) \cdot 10^{-4}$ | [8] | $1.5 \cdot 10^{-7}$ | [13] |
| ($E1$, | | | $2.6 \cdot 10^{-4}$ | [14] |
| 103 keV) | | | $8.5 \cdot 10^{-9}$ | [15] |
| | | | $3.2 \cdot 10^{-12}$ | [17] |

[a] $b_0$ is the unknown branching ratio for the $5/2^-$ (69.5 keV) $\to 9/2^-$ (30.8 keV) transition.

The NEET probability $P_{NEET}$ from Eq. (3) is measured experimentally. The values of $P_{NEET}$ obtained in different theoretical and experimental work are listed in Table 1.

The NEET process has most often been investigated for the nucleus $^{189}$Os, for which the level diagram is shown in Fig. 2. The value of $P_{NEET}$ determined for $^{189}$Os is not the result of direct measurements of the ratio of the number of nuclei excited to the $5/2^-$ (69.5 keV) level to the number of $K$ vacancies in the Os atoms created by external irradiation. In all cases, the efficiency of excitation of the $5/2^-$(69.5 keV) level was determined from the activity of the decay of the 5.8 h isomeric level $9/2^-$(30.8 keV) (the long-lived isomeric nuclei are accumulated during a period of irradiation). The number of these nuclei ($N_{is}$) is related to the number excited to the 69.5 keV level nuclei ($N_{69.5}$) and the number of $K$ vacancies ($N_K$) by the relations

$$N_{is} = b_0(N_{69.5}^{e,\gamma} + N_{69.5}^{NEET}) + \sum_{i \geq 1} b_i N_i^{e,\gamma}, \qquad (7)$$

$$N_{69.5}^{NEET} = P_{NEET} N_K, \qquad (8)$$

where $i$ in the sum indicates all other nuclear levels (except the 69.5 keV level) excited by photons and electrons, with branching ratios $b_i$ for population of the 30.8 keV long-lived isomeric level.

The first-order-forbidden $K$ $E2$ transition $5/2^-$ (69.5 keV)$\rightarrow$ $9/2^-$(30.8 keV) represented by the dashed line in Fig. 2 has not yet been observed experimentally. The coefficient $b_0$ is not known, and hence it must be a free parameter in Eqs. (7)–(8). Analysis of the two experiments of Refs. [6,12] shows that a more plausible value of the branching ratio is $b_0 \simeq 3.7 \cdot 10^{-3}$–$4.4 \cdot 10^{-3}$ for the NEET probability

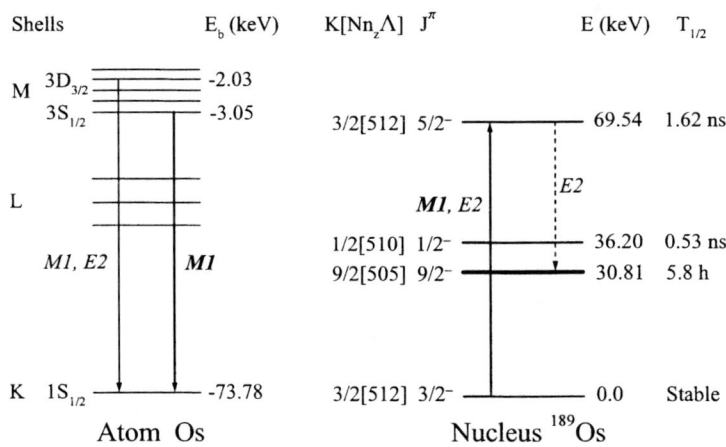

**FIGURE 2.** The NEET process in $^{189}$Os.

$P_{NEET} \simeq 1.2 \cdot 10^{-10}$ [19]. These values of $b_0$ and $P_{NEET}$ are consistent with the results of both experiments Ref. [6,12] simultaneously.

There is no agreement between experimental and theoretical results for other nuclei listed in Table 1. An exception is the model from Ref. [14], where the NEET probability does not include the nuclear matrix element. However, the experimental results [24,29] for $^{189}$Os are inconsistent with this model.

## ACCELERATED DECAY OF NUCLEAR ISOMERS VIA THE INVERSE NEET PROCESS

The dominant process for the decay of most low-energy nuclear isomeric levels is internal conversion. The internal conversion probability depends directly on the presence of electrons in the corresponding atomic states. Ionization of one of these shells leads to an increase of the nuclear state half-life. But, for some cases, the ionization is accompanied by a sharp increase in the total probability for the nuclear isomer decay. The reason for this acceleration is a new decay process - Inverse NEET (INEET).

The INEET process in $^{197}$Au is shown in Fig. 3. The main condition for accelerated decay of the $1/2^+(77.35$ keV) level is a hole in the $M_I$ $(3s_{1/2})$ shell of the Au atom. The probability of INEET in $^{197}$Au is connected with the probability of NEET by the relations

$$W_{INEET} = \Gamma_{M_I} P_{INEET}, \qquad (9)$$

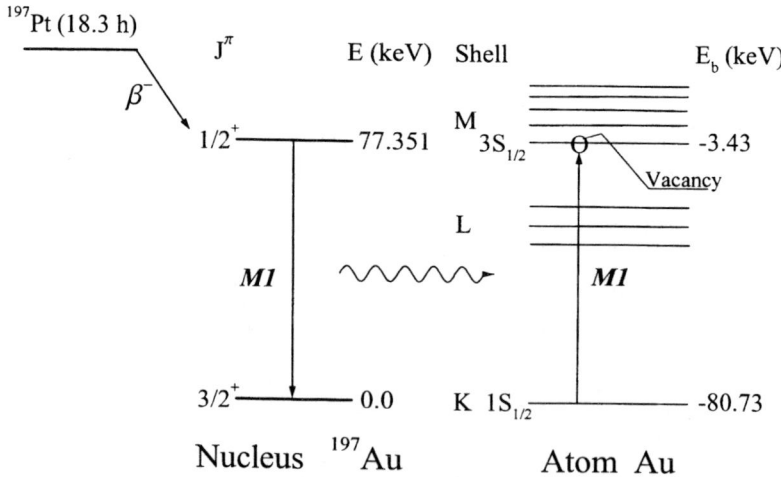

**FIGURE 3.** The inverse NEET process in $^{197}$Au.

$$P_{INEET} = \frac{\Gamma_K}{\Gamma_{M_I}} \frac{2J_{gr}+1}{2J_{is}+1} P_{NEET}, \tag{10}$$

where $J_{gr,is}$ are the nuclear spins in the ground and isomeric states, respectively. Using the values of vacancy widths of atomic shells from Ref. [30] and $P_{NEET}$ from Ref. [17], one finds that $W_{INEET} \simeq 1.4 \cdot 10^{-5}$ eV [23]. This value of $W_{INEET}$ corresponds to a half-life of $3.3 \cdot 10^{-11}$ s for the $1/2^+(77.35$ keV) isomeric level. Under ordinary conditions, the half-life of this level is $1.91 \cdot 10^{-9}$ s. When there is a vacancy in the $M_I$ shell, the probability for the decay of the $1/2^+(77.35$ keV) isomeric level in $^{197}$Au increases by a factor of more than 50 [23].

The same considerations apply to the nucleus $^{193}$Ir. A similar calculation shows that INEET increases the decay probability of the $1/2^+(73.04$ keV) level by a factor of several units [23].

A vacancy in a $M_I$ shell tends to decay by ordinary atomic processes, such as X-ray emission, for example. The relative probability $P_{INEET}$ for excitation of the atom in a nuclear transition, accompanied by the formation of a hole in the $M_I$ shell, is therefore small. Nevertheless, the value $W_{INEET}/\Gamma_{M_I} \simeq 0.6 \cdot 10^{-6}$ for $^{197}$Au appears to be measurable.

## ANOMALIES AND DYNAMICAL EFFECT OF NUCLEAR SIZE IN NEET

The NEET process occurs between $ns_{1/2} \to ms_{1/2}$ and $np_{1/2} \to ms_{1/2}$ ($n > m = 1, 2, \ldots$) atomic states for all of the nuclei in Table 1. Electronic wave functions $\psi(r)$ for such states have large amplitudes at the origin. As a consequence, the electronic current $j_{fi}^\mu = -e\bar{\psi}_f(\mathbf{r})\gamma^\mu\psi_i(\mathbf{r})$ effectively penetrates the nucleus, and an "intranuclear" NEET becomes possible.

A new phenomenon arises if the coordinates of the electronic current $j_{fi}^\mu(\mathbf{r})$ and the nuclear current $\hat{J}_{fi}^\nu(\mathbf{R})$ satisfy the condition $r < R$. The region $r < R \leq R_0$ ($R_0$ is a nuclear radius) can give a substantial contribution to the NEET probability for the $l$-forbidden $M1$ nuclear transitions and for the $E1$ transitions which are forbidden by the asymptotic quantum numbers of the Nilsson model [21,22].

This effect is analogous to the dynamical nuclear volume effect in internal conversion [31]. However, NEET is not just one more process in which it is possible to measure the same nuclear parameters as in anomalous internal conversion. In some cases, the dynamical effect will be seen much more vividly in NEET than in internal conversion. The reason is as follows. In the case of internal conversion from the $L$ shell, for example, the electron goes "up" into continuum states. In NEET, the electron goes "down" to a deep shell $K$, in which the amplitude of the electronic wave function in the nucleus may be considerably larger than those for the conversion states of the continuum. As a result, the relative contribution of the nuclear region to the radial integral of the electronic matrix element, and to the probability of the process, is greater for NEET.

The energy of the interaction between electronic and nuclear currents in NEET has the form

$$H_{int} = \int_0^\infty d^3R \int_0^\infty d^3r j^\mu(\mathbf{r}) D_{\mu\nu}(\omega_N; \mathbf{r} - \mathbf{R}) J^\nu(\mathbf{R}), \quad (11)$$

where $D_{\mu\nu}(\omega_N; \mathbf{r} - \mathbf{R}) = -g_{\mu\nu} \exp(i\omega_N |\mathbf{r} - \mathbf{R}|)/|\mathbf{r} - \mathbf{R}|$ is the photon propagator. Hence, the multipole expansion of the photon propagator should be made in both of the regions $r > R$ and $r < R$.

The region $r > R$ is usually considered for integration in Eq. (11). The formula (4) was obtained in that manner. If one includes the region $r < R$ in the integrals of Eq. (11), the interaction energy will have the form

$$H_{int} = H_{int}^R + \Delta H_{int}. \quad (12)$$

Here $H_{int}^R$ is the usual interaction energy. The square of the absolute value of $H_{int}^R$ gives formula (4) after summing over the final states and averaging over the initial states of the nucleus and electron (the superscript "$R$" in the $H_{int}^R$ indicates use of the Rose model or the penetration-free model [32]). $\Delta H_{int}$ is an "additional" interaction energy which arises due to the dynamical penetration of the electronic current into the nucleus.

There are two cases where this additional interaction energy gives a substantial contribution to the NEET probability. The first case is for $l$-forbidden $M1$ nuclear transitions. Such an anomaly should take place in the nuclei $^{197}$Au and $^{193}$Ir. The second case is for $E1$ nuclear transitions forbidden in the asymptotic quantum numbers of the Nilsson model. Such an anomaly should change the NEET probability in the nuclei $^{237}$Np and $^{181}$Ta [21,22].

Let us consider the $M1$ transitions, for example. The multipole expansion of the contraction of the currents with the photon propagator in Eq. (11) in the regions $r > R$ and $r < R$ is

$$\mathbf{J}(\mathbf{R}) \frac{exp(i\omega_N |\mathbf{r} - \mathbf{R}|)}{|\mathbf{r} - \mathbf{R}|} \mathbf{j}(\mathbf{r}) =$$

$$= 4\pi i \omega_N \sum_{L,M,a} \begin{cases} \mathbf{J}(\mathbf{R}) \cdot \mathbf{A}_{LM}^{a*}(\mathbf{R}; \omega_N) \mathbf{B}_{LM}^{a}(\mathbf{r}; \omega_N) \cdot \mathbf{j}(\mathbf{r}), & r > R \\ \mathbf{J}(\mathbf{R}) \cdot \mathbf{B}_{LM}^{a*}(\mathbf{R}; \omega_N) \mathbf{A}_{LM}^{a}(\mathbf{r}; \omega_N) \cdot \mathbf{j}(\mathbf{r}), & r < R \end{cases}, \quad (13)$$

where $A_{LM}(\mathbf{r}; \omega)$ and $\mathbf{A}_{LM}^{a}(\mathbf{r}; \omega)$ are the electromagnetic potentials ($a = E, M, Y$, $E$ — electric, $M$ — magnetic, $Y$ — longitudinal). We need a magnetic potential

$$\mathbf{A}_{Lm}^M(\mathbf{r}; \omega) = j_L(\omega r) \mathbf{Y}_{LL;m}(\mathbf{n}), \quad (14)$$

where $j_L(x)$ are Bessel functions and $\mathbf{Y}_{LJM}(\mathbf{n})$ are the vector spherical functions. The potentials $\mathbf{B}_{LM}^M(\mathbf{r}; \omega)$ are obtained from $\mathbf{A}_{LM}^M(\mathbf{r}; \omega)$ by substituting the Hankel functions $h_L^{(1)}(\omega r)$ for the Bessel functions in Eq. (14).

The regions $r > R$ and $r < R$ in the multipole expansion Eq. (13) give different $M1$ nuclear matrix elements:

$$N_{1m}^M = \int_0^\infty d^3R\, \mathbf{J}(\mathbf{R}) \cdot \mathbf{A}_{1m}^{M\,*}(\mathbf{R}; \omega_N)\,, \tag{15}$$

$$N_{1m}^{M(an)} = \int_0^\infty d^3R\, \left(\frac{R}{R_0}\right)^2 \mathbf{J}(\mathbf{R}) \cdot \mathbf{A}_{1m}^{M\,*}(\mathbf{R}; \omega_N)\,. \tag{16}$$

The selection rules for the "anomalous" nuclear matrix element (16) differ from the selection rules for the normal nuclear matrix element (15). If the normal $M1$ nuclear transition is $l$-forbidden (the $2d_{3/2} \leftrightarrow 3s_{1/2}$ transition in $^{197}$Au and $^{193}$Ir, for example), then the additional factor $(R/R_0)^2$ in Eq. (16) allows a nuclear transition described by the anomalous matrix element $N_{1m}^{M(an)}$ [33].

The final expression for the interaction energy for the $M1$ nuclear transition in NEET is

$$H_{int}(M1) = e\sqrt{2\pi\omega_N} \sum_{m=0,\pm 1} (1/2\, m_i\, 1m|1/2\, m_f)\, N_{1m}^M \Big\{ \mathcal{R}_1^M(\omega_N) +$$
$$i\frac{\lambda^{(0)}}{4\pi^2}\left(\frac{R_0}{a_B}\right)^2\left(\frac{\lambda_N}{a_B}\right)^2 (\kappa_i + \kappa_f) \frac{3c_i c_f}{10} \left( \begin{array}{c} (f_i^0 + f_f^0)\delta_{M1}(nS_{1/2} \leftrightarrow mS_{1/2}) \\ (g_i^0 + g_f^0)\delta_{M1}(nP_{1/2} \leftrightarrow mP_{1/2}) \end{array} \right) \Big\}, \tag{17}$$

where $\lambda_N = 2\pi/\omega_N$ and $a_B$ is the Bohr radius. The penetration parameter $\lambda^0 = N_{1m}^{M(an)}/N_{1m}^M$. The following expansion of the electronic wave functions of the $ns_{1/2}$ and $mp_{1/2}$ shells in the nucleus was used in Eq. 17:

$$g(r) = \frac{g(x)}{a_B^{3/2}}\,, \quad f(r) = \frac{f(x)}{a_B^{3/2}}\,, \quad x \equiv \frac{r}{a_B}\,,$$
$$S_{1/2}: g(x) = c\,, \qquad f(x) = cf^0 x\,, \tag{18}$$
$$P_{1/2}: g(x) = cg^0 x\,, \quad f(x) = c\,,$$

where $c$, $g^0$ and $f^0$ are constants to be calculated or taken from Ref. [34].

One can estimate the value of the dynamical effect of nuclear size in NEET by comparing the two terms in braces in Eq. 17. The penetration parameter is approximately known for $^{197}$Au: $\lambda^{(0)} \approx 3.4$ [35]. The new value for the NEET probability is $P_{NEET} = 1.2 \cdot 10^{-7}$ instead of $1.4 \cdot 10^{-7}$. The penetration parameter is experimentally unknown for $^{193}$Ir. Theoretical estimation gives the value $\lambda^{(0)} \approx 9.8$ [36]. The new value for the NEET probability in $^{193}$Ir (the 73 keV transition) is $P_{NEET} = 5.4 \cdot 10^{-9}$, instead of the value $7.4 \cdot 10^{-9}$ obtained without regard to the anomalies.

Concerning the anomalous $E1$ transitions in NEET, first consider the $^{237}$Np nucleus. On the one hand, the anomalies are well known in the internal conversion coefficients for the $E1$ transition from the $5/2^-, 5/2[523](59.5$ keV) level to the levels $7/2^+, 5/2[642](33.2$ keV) and $5/2^+, 5/2[642](0.0$ keV). On the other hand, the presence of these anomalies does not agree with the known selection rules in the asymptotic quantum numbers of the Nilsson model [37]. It is very interesting that anomalies in internal conversion are not observed for the $7/2^-, 5/2[523](103$ keV)$\rightarrow 5/2^+, 5/2[642](0.0$ keV) transition. That is why an experimental study of NEET might give an independent answer to the problem of the existence of anomalies in the $E1$ 103 keV transition in the nucleus $^{237}$Np [22].

There is a very large anomaly in the internal conversion coefficients for the $E1$ 6.24 keV transition in $^{181}$Ta (the decay of the first excited level). The excitation of this level in NEET is possible in the $3s_{1/2} \to 2p_{3/2}$ and $3s_{1/2} \to 2p_{1/2}$ electronic transitions. A strong anomaly appears in the second transition. The partial NEET probability for the $3s_{1/2} \to 2p_{1/2}$ electronic transition increases by a factor of approximately 170 when the anomaly is taken into account [21,22].

NEET investigations are apparently feasible only for a limited number of nuclei. However, the number of cases can be appreciably expanded by ionization of atomic shells. The binding energies of atomic levels and the energies of atomic transitions depend on the atomic charge state. That is, one can better "match" the atomic transition energy by changing the atomic charge state. It is also important that, in the ionized atom, the amplitudes of the electronic wave functions inside the nucleus exceed the corresponding amplitudes of the neutral atom. Consequently, the dynamical effect of nuclear size in NEET is more strongly manifested in the ionized atom. That is why the NEET process can be studied on nuclei with either neutral or ionized atomic shells, and can be considered as part of the research program "Atomic Shell — Nuclear Interaction".

## ACKNOWLEDGEMENTS

This work was supported partly by the Russian Foundation for Basic Research on Grants No 98-02-16070a, and Grant No 96-15-96481 in Support of the Leading Scientific Schools.

## REFERENCES

1. Wheeler, J.A., *Rev. Mod. Phys.* **21**, 133 (1949).
2. Morita, M., *Progr. Theor. Phys.* **49**, 1574 (1973).
3. Otozai, K., Arakava, R., and Morita, M. *Progr. Theor. Phys.* **50**, 1771 (1973).
4. Okamoto, K. *Laser Interaction and Related Plasma Phenomena.* **4A**, 284 (1977).
5. Okamoto, K. *J. Nucl. Sci. Tech.* **14**, 762 (1977).
6. Otozai, K., Arakava, R., and Saito, T. *Nucl. Phys.* **A297**, 97 (1978).
7. Izawa, Y., and Yamanaka, C. *Phys. Lett* **B88**, 59 (1979).
8. Saito, T., Shinohara, A., and Otozai, K. *Phys. Lett* **B92**, 293 (1980).
9. Okamoto, K. *Nucl. Phys.* **A341**, 75 (1980).
10. Saito, T., Shinohara, A., Miura, T., and Otozai, K. *J. Inorg. Nucl. Chem.* **43**, 1963 (1981).
11. Fujioka H., Ura, K., Shinohara, A., et al. *Z. Phys.* **A315**, 121 (1984).
12. Shinohara, A., Saito, T., Shoji, M., et al. *Nucl. Phys.* **A472**, 151 (1987).
13. Pisk, K., Kaliman, Z., and Logan, B.A. *Nucl. Phys.* **A504**, 103 (1989).
14. Ljubicic, A., Kekez, D., and Logan, B.A. *Phys. Lett.* **B272**, 1 (1991).

15. Ho, Y.-K., Zhang, B.-H., and Yuan, Z.-S. *Phys. Rev.* **C44**, 1910 (1991).
16. Bondarkov, M.D., and Kolomiets, V.M. *Izv. Akad. Nauk SSSR* ser. fiz. **55**, 983 (1991) [in Russian].
17. Tkalya, E.V. *Nucl. Phys.* **A539**, 209 (1992).
18. Tkalya, E.V. *Sov. Phys. JETP* **75**, 200 (1992) [*Zh. Eksp. Teor. Fiz.* **102**, 379 (1992)].
19. Tkalya, E.V. *JETP Lett.* **56**, 131 (1992) [*Pis'ma Zh. Eksp. Teor. Fiz.* **56**, 137 (1992)].
20. Ho, Y.-K., Yuan, Z.-S., Zhang, B.-H., et al. *Phys. Rev.* **C48**, 2277 (1993).
21. Tkalya, E.V. *JETP Lett.* **59**, 13 (1994) [*Pis'ma Zh. Eksp. Teor. Fiz.* **59**, 15 (1994)].
22. Tkalya, E.V. *Sov. Phys. JETP* **78**, 239 (1994) [*Zh. Eksp. Teor. Fiz.* **105**, 449 (1994)].
23. Tkalya, E.V. *JETP Lett.* **60**, 627 (1994) [*Pis'ma Zh. Eksp. Teor. Fiz.* **60**, 619 (1994)].
24. Lakosi, L., Tam, N.C., and Pavlicsek, I. *Phys. Rev.* **C52**, 1510 (1995).
25. Shinohara, A., Saito, T., Otozai, K., et al. *Bull. Chem. Soc. Jpn.* **68**, 566 (1995).
26. Shinohara, A., Saito, T., Taniguchi, K., et al. *Chem. Lett. Jpn.*, 19 (1995).
27. Typel, S., and Leclercq-Willain, C. *Phys. Rev.* **A53**, 2547 (1996).
28. Tkalya, E.V., Eremin, N.V., and Giardina, G. In: *Proc. Int. Conf. Large-Scale Collective Motion of Atomic Nuclei*, Italy, 1996. Singapore: World Sci. Publ. Co. Pte. Ltd., 1997, p..633-636.
29. Ahmad, I., Dunford, R.W., Esbensen, H., et al. In: *Proc. 1998 Int. Workshop on Atomic and Molecular Phys. at High Brilliance Synchrotron Radiation Facilities*, Japan, 1998.
30. Bambynek, W., et al. *Rev. Mod. Phys.* **4**, 716 (1972).
31. Church, E.L., and Weneser, J. *Phys. Rev.* **104**, 1382 (1956).
32. Green, T.A., and Rose, M.E. *Phys. Rev.* **110**, 105 (1958).
33. Rose, M.E. In *Alpha-, Beta- and Gamma-Ray Spectroscopy*. Vol. 3. Ed. K. Siegbahn, North-Holland Publish. Comp., Amsterdam, 1965.
34. Band, I.M., and Fomichev, V.I. *At. Data Nucl. Data Tabl.* **23**, 295 (1979).
35. Krpic, D., et al. *Z. Phys.* **243**, 452 (1971).
36. Band, I.M., Listengarten, M.A., and Feresin, A.P. *Anomalies in the Internal Conversion Coefficients for Gamma Rays*. Nauka, Leningrad, 1976. [In Russian].
37. Voikhanskii, M.E., and Listengarten, M.A. *Izv. Akad. Nauk SSSR, Ser. Fiz.* **23**, 238 (1959). [In Russian].

# Non-Resonant Excitation of Nuclear Levels by Photons

## E. G. Drukarev

*Petersburg Nuclear Physics Institute*
*Gatchina, St. Petersburg 188350, Russia*

**Abstract.** Possible mechanisms of the non-resonant excitation process of nuclear levels are analyzed. It is shown that the process can be viewed mainly as Compton scattering on the electrons of the atomic shell, followed by resonant excitation.

We analyze excitation of nuclear levels with excitation energy $\omega_R$ during the interaction between an incoming photon with energy $\omega > \omega_R$ and the atom. Interest in this process increased some time ago. Both theoretical and experimental aspects of the problem were reviewed recently in (1,2). Here I present a simple investigation of the theory of the process. At some points, the conclusions differ from those obtained by others.

In the process considered here, the energy $\omega - \omega_R$ should be transferred to the atom. One can single out two main mechanisms. The first one is Raman scattering of the photon on the nuclei. Here the energy $\omega - \omega_R$ is carried by the outgoing photon. The energy $\omega - \omega_R$ can be transferred also to the atomic shell. In the lowest order of expansion in powers of the fine structure constant $\alpha = 1/137$, all the energy $\omega - \omega_R$ is carried by the outgoing electron. In the Feynman graph technique, the two mechanisms are described by the diagrams of Fig. 1.

It is very helpful that the inverse process, in which electromagnetic decay of the nucleus is followed by irradiation of the photon with the energy $\omega < \omega_R$, was investigated earlier. In the first case, the inverse process is the two-quantum transition of the nucleus — Fig. 1a,b. The results of measurements of the double-to-single $\gamma$ radiation ratio $\tau = W_{\gamma\gamma}/W_\gamma$, supported by numerical estimations for a number of nuclei (see, e.g., (3) for the references), provide the upper limit $\tau \lesssim 10^{-6}$. On the other hand, the parametrical analysis gives a much weaker limitation $\tau \lesssim \alpha$. Here we do not consider the case of dipole excitations, when there is additional suppression of the value of $\tau$ caused by the selection rules for angular momentum. Thus, there is some additional quenching of nuclear matrix elements.

In the second case, it is the internal Compton effect which plays the role of the

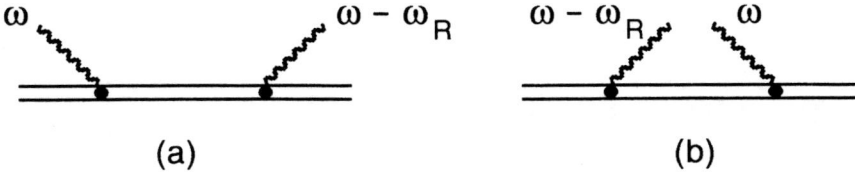
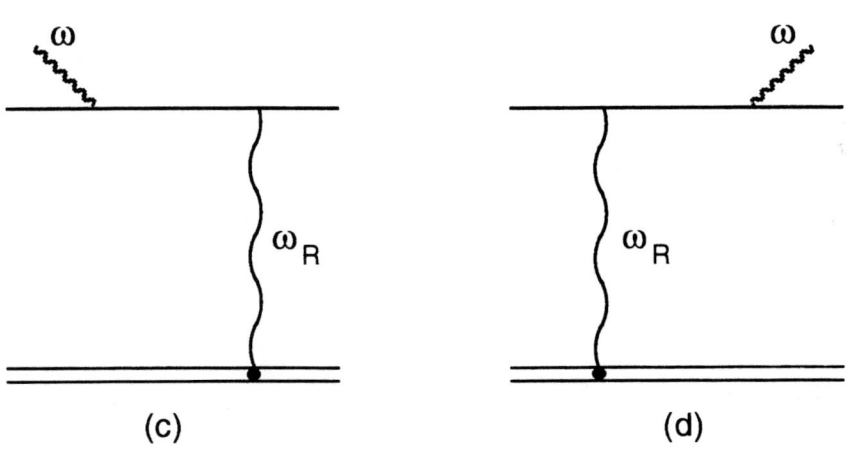

FIGURE 1. Feynman diagrams of non-resonant excitation in the lowest order of the fine structure constant. Figs 1a,b show the nuclear Raman scattering. Figs 1c,d show the contribution of atomic electrons. Double lines show the nucleus. Solid lines show the electrons, wavy lines denote the photons.

inverse process — Fig. 1c,d. Here the energy of the electromagnetic transition of the nucleus is shared between the knocked out electron and the irradiated photon. Now we limit ourselves to the case, when the energy $\omega_R$ is large enough, greatly exceeding the binding energies of the internal bound electrons $I$, i.e.

$$\zeta^2 = \frac{I}{\omega_R} \ll 1. \tag{1}$$

This condition is true for all the cases studied in the experiments. For example, for excitation of $^{115}$In with $\omega_R = 1078$ keV, we find $\zeta^2 \approx 0.03$. Thus, there are two scales of momenta: $p \sim \omega_R$ and $p \sim (2mI)^{1/2} \ll \omega_R$ with $m$ standing for the electron mass. For the $K$-shell electrons, $I \approx m(\alpha Z)^2/2$ with $Z$ being the charge of the nucleus and $(2mI)^{1/2} \approx \eta = m\alpha Z$. For the external bound electrons, $(2mI)^{1/2} \ll m\alpha Z$.

Our approach to investigation of the process is based on the observation that each act of transferring of large momenta $Q \gg \eta$ to the nucleus leads to a small factor of the order $\zeta^2$ in the amplitude. These ideas, developed first by Bethe (4), enable one to single out the region of kinematics which dominates the cross section of the process. Also, the amplitude of the process is enhanced if one of the intermediate particles is close to the mass-shell. A general approach to this type of process was developed in (5).

In our case, we cannot avoid transferring of momentum $q$ to the nucleus during the process of excitation of the latter. However, momentum, exchanged between the nucleus, treated as the static source of the external field and the electrons, can be made small. This means that $Q = k - q - p$, with $k(p)$ being the momentum of the incoming photon (outgoing electron), can be of the order of $\eta$. Both the intermediate electron and the photon can be close to the mass-shell. In the former case, a large momentum $Q \gg \eta$ should be transferred to the static nucleus, since there is no photoeffect on a free electron. In the latter case, the momentum $Q$ can be made small if

$$(\bar{q} - \bar{p})^2 = \omega_R^2 \tag{2}$$

The kinematical region, determined by Eq. (2), provides the leading contribution to the cross section.

Thus we came to the two-step picture of the process, which is very much like that of internal Compton effect (6). In the first step, Compton scattering on the bound electron takes place with the photon, carrying momentum $q$ and energy $\omega_R$ and the outgoing electron in the final state. Only small momentum $Q \sim \eta$ is transferred to the nucleus. The process takes place at distances $r \sim Q^{-1} \sim \eta^{-1}$, i.e. those of the order of the size of the atom. Being approximately on the mass-shell — see Eq.(2) — the photon passes the distances of the order of the size of the atom, approaching the nucleus at the distance of the order $q^{-1} \ll \eta^{-1}$. Here the excitation takes place as the second step. Both Compton scattering and the excitation take place at small time intervals $t \sim \omega^{-1}, \omega_R^{-1}$. The two steps are separated by a time interval $t_1 \sim \eta^{-1}$, and $t_1 \gg t$ if $\omega_R \gg \eta$.

Using the technique developed in (5), we find for the cross section of non-resonant excitation

$$\sigma_{NR}(\omega, \omega_R) = \kappa \frac{d\sigma_C(\omega, \omega_R)}{d\omega_R} \cdot \sigma_R(\omega_R) . \tag{3}$$

Here $\sigma_R$ is the cross section of resonant excitation, and $\sigma_C$ stands for the Compton scattering on the free electron. All of the dependence on atomic parameters is contained in the factor

$$\kappa = \frac{\langle \psi | \Sigma r_i^{-2} | \psi \rangle}{4\pi} \tag{4}$$

with $\psi$ being the total wave function of the atom.

Note that other two-step channels, i.e., photoionization, followed by excitation by electron impact or creation of an electron-positron pair by the incoming photon with further annihilation of the positron with the bound electron and excitation of the nucleus by a virtual photon (1), provide smaller contributions. This is because they require transferring of additional large momentum to the nucleus. As to virtual creation of $e^+e^-$ pairs, they are contained in Eq. (3), since the relativistic Green function describes propagation of both electron and positron.

Using Eq. (3), we can estimate

$$\sigma_{NR}(\omega, \omega_R) \sim \alpha^2 (\alpha Z)^2 \frac{m}{\omega^2} \sigma_R(\omega_R) . \tag{5}$$

If the excitation energy $\omega_R$ is large enough, i.e., $\omega_R > m/2 \approx 255$ keV, the momentum $Q$ can be small for any value of $\omega$, while the intermediate photon is on the mass-shell. However, for smaller values of $\omega_R$, the intermediate photon is far from the mass-shell if $\omega$ is not too large

$$\omega < \frac{\omega_R}{1 - 2(\omega_R/m)} . \tag{6}$$

In this case, the ratio $\sigma_{NR}/\sigma_R$ obtains an additional small factor $\alpha Z$. On the other hand, there is an enhancement in the non-relativistic limit $\omega_R \ll m$. This is because the amplitude of the excitation is proportional to the factor $q^L$ for a $2^L$ multipole nuclear transition. For excitation by real photons $q = \omega_R$, while in the graphs of Fig. 1b,c $q \approx (2m\omega_R)^{1/2}$, if inequality (6) is true. Thus, in the latter case,

$$\sigma_{NR}(\omega, \omega_R) \sim \alpha^2 (\alpha Z)^3 \left(\frac{m}{\omega}\right)^{L-1} \cdot \frac{m}{\omega^2} \sigma_R(\omega_R) . \tag{7}$$

Comparing Eqs. (5) and (7) to estimate the Raman scattering contribution $\sigma_{NR}(\omega, \omega_R) \sim \tau \frac{1}{\omega} \sigma_R(\omega_R)$, one can expect the electron mechanism to dominate.

Most of the experimental data are obtained for photon energies of the order of 1 MeV. In this case, intermediate electrons in the diagrams of Figs. 1c,d can be described by free relativistic propagators. If the difference $\omega - \omega_R$ is large enough, the outgoing electrons can be described by plane waves. In this case, all of the calculations can be carried out analytically. Assuming also $(\omega - \omega_R)/\omega_R \ll 1$, we find

$$\sigma_{NR} = 2\pi\alpha^2 \frac{m}{\omega^2} \kappa \sigma_R . \qquad (8)$$

In the real experimental situation, the energy $\omega - \omega_R$ is not large enough for using plane waves for description of the outgoing electron.

As it was noted in (1), several closely laying nuclear states can contribute to the process. For example, in the case of photoexcitation of the 1078 keV nuclear level in $^{115}$In by 1330 keV gamma rays, one should include the possible E2 and M7 transitions (1). Taking this into account, we come to reasonable agreement between the existing experimental data and the results of the analysis, developed above. Thus, we may hope that understanding of the main mechanism of the process is achieved. Future calculations with more precise treatment of the dynamics of the outgoing electron would provide the possibility for more detailed comparison with experimental data.

## ACKNOWLEDGEMENTS

I am indebted to Dr. D. A. Bradley for stimulating discussions. I also thank the Organizing Committee of the X-99 Conference, whose financial support made my participation in the Conference possible.

## REFERENCES

1. A. Ljubičič, *Radiat. Phys. Chem.*, **51**, 341 (1998).
2. D. A. Bradley, Ithnin Abdul Jalil, M. Krcmar and A. Ljubičič, to be published.
3. E. G. Drukarev, *Zeit. f. Phys.*, **359**, 133 (1997).
4. H. Bethe, in *"Handbuch der Physik"*, Springer, Berlin, 1933, 24/1, 273.
5. E. G. Drukarev, V. G. Gorshkov, A. I. Mikhailov, S. G. Sherman,
   *Phys. Lett.*, **46A**, 467 (1974);
   E. G. Drukarev, *Nucl. Phys. A*, **541**, 131 (1992).
6. E. G. Drukarev, *Sov. J. Nucl. Phys.*, **17**, 174 (1973).

# VII. FUNDAMENTAL PHYSICS

# Testing cosmological variability of fundamental constants

D. A. Varshalovich, A. Y. Potekhin, and A. V. Ivanchik

*Ioffe Physico-Technical Institute, 194021 St. Petersburg, Russia*

**Abstract.** One of the topical problems of contemporary physics is a possible variability of the fundamental constants. Here we consider possible variability of two dimensionless constants which are most important for calculation of atomic and molecular spectra (in particular, the X-ray ones): the fine-structure constant $\alpha = e^2/\hbar c$ and the proton-to-electron mass ratio $\mu = m_p/m_e$. Values of the physical constants in the early epochs are estimated directly from observations of quasars – the most powerful sources of radiation, whose spectra were formed when the Universe was several times younger than now. A critical analysis of the available results leads to the conclusion that the present-day data do not reveal any statistically significant evidence for variations of the fundamental constants under study. The most reliable upper limits to possible variation rates at the 95% confidence level, obtained in our work, read:

$$|\dot\alpha/\alpha| < 1.4 \times 10^{-14} \text{ yr}^{-1}, \quad |\dot\mu/\mu| < 1.5 \times 10^{-14} \text{ yr}^{-1}$$

on the average over the last $10^{10}$ yr.

## INTRODUCTION

Contemporary theories (SUSY GUT, Superstring and others) not only predict the dependence of fundamental physical constants on energy[1], but also have cosmological solutions in which low-energy values of these constants vary with the cosmological time. The predicted variation at the present epoch is small but non-zero, and it depends on theoretical model. In particular, Damour and Polyakov [1] have developed a modern version of the string theory, whose parameters could be determined from cosmological variations of the coupling constants and hadron-to-electron mass ratios. Clearly, a discovery of these variations would be a great step in our understanding of Nature. Even a reliable upper bound on a possible variation rate of a fundamental constant presents a valuable tool for selecting viable theoretical models.

---

[1] The prediction of the theory that the fundamental constants depend on the energy of interaction (the "running" constants) has been confirmed in experiments. In this paper, we consider only the space-time variability of their low-energy limits.

Historically, a hypothesis that the fundamental constants may depend on the *cosmological time t* (that is the age of the Universe) was first discussed by Milne [2] and Dirac [3]. The latter author proposed his famous "large-number hypothesis" and suggested that the gravitational constant was directly proportional to $t$. Later the variability of fundamental constants was analyzed, using different arguments, by Gamow [4], Dyson [5], and others. The interest in the problem has been revived due to recent major achievements in GUT and Superstring models (e.g., [1]).

Presently, the fundamental constants are being measured with a relative error of $\sim 10^{-8}$. These measurements obviously rule out considerable variations of the constants on a short time scale, but do not exclude their changes over the lifetime of the Universe, $\sim 1.5 \times 10^{10}$ years. Moreover, one cannot rule out the possibility that the constants differ in widely separated regions of the Universe; this could be disproved only by astrophysical observations and different kinds of experiments.

Laboratory experiments cannot trace possible variation of a fundamental constant during the entire history of the Universe. Fortunately, Nature has provided us with a tool for direct measuring the physical constants in the early epochs. This tool is based on observations of quasars, the most powerful sources of radiation. Many quasars belong to most distant objects we can observe. Light from the distant quasars travels to us about $10^{10}$ years. This means that the quasar spectra registered now were formed $\sim 10^{10}$ years ago. The wavelengths of the lines observed in these spectra ($\lambda_{\rm obs}$) increase compared to their laboratory values ($\lambda_{\rm lab}$) in proportion $\lambda_{\rm obs} = \lambda_{\rm lab}(1+z)$, where the *cosmological redshift* $z$ can be used to determine the age of the Universe at the line-formation epoch. In some cases, the redshift is as high as $z \sim 3-5$, so that the intrinsically far-ultraviolet lines are registered in the visible range. The examples are demonstrated in Fig. 1. Analysing these spectra we may study the epoch when the Universe was several times younger than now.

Here we review briefly the studies of the space-time variability of the fine-structure constant $\alpha$ and the proton-to-electron mass ratio $\mu$.

## FINE-STRUCTURE CONSTANT

Various tests of the fundamental constant variability differ in space-time regions of the Universe which they cover. *Local tests* relate to the values of constants on the Earth and in the Solar system. In particular, *laboratory tests* infer possible variation of certain combinations of constants "here and now" from comparison of different frequency standards. *Geophysical tests* impose constraints on combinations of fundamental constants over the past history of the Solar system, although most of these constraints are very indirect. In contrast, *astrophysical tests* allows one to "measure" the values of fundamental constants in distant areas of the early Universe.

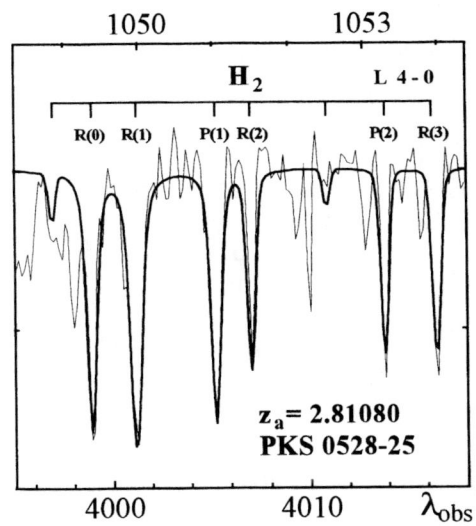 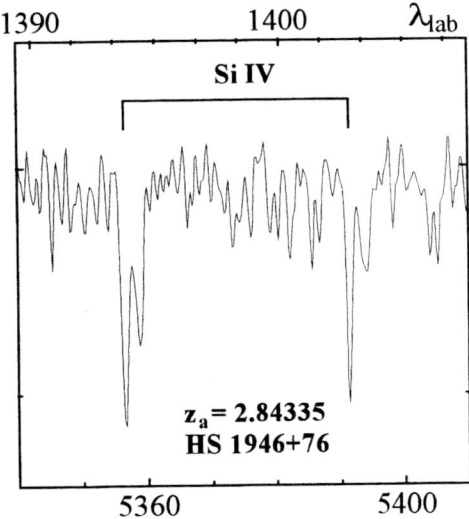

**FIGURE 1.** Portions of quasar spectra which show the absorption lines of $H_2$ and Si IV with large redshifts, $z = (\lambda_{obs} - \lambda_{lab})/\lambda_{lab} \approx 2.8$. The lower horizontal axis gives the wavelengths in the observer's frame ($\lambda_{obs}$) and the upper axis gives the wavelengths in the quasar's frame $\lambda_{lab}$ (in Å). (a) The spectrum (thin line) of the quasar PKS 0528–250, obtained with the 4-meter CTIO telescope (Chile), containing $H_2$ lines which belong to the L 4–0 branch of the spectrum; thick line plots the spectral fit. (b) The spectrum of the quasar HS 1946+76, obtained with the 6-meter SAO telescope (Russia), containing Si IV doublet lines which correspond to the $^2S_{1/2} \rightarrow\, ^2P_{3/2}$ and $^2S_{1/2} \rightarrow\, ^2P_{1/2}$ transitions.

## Local tests

### Laboratory experiments

There were a number of laboratory experimets aimed at detection of trends of the fundamental constants with time by comparison of frequency standards which have different dependences on the constants. We mention only two of the published experiments.

Comparison of H-masers with Cs-clocks during 427 days revealed a relative (H–Cs) frequency drift with a rate $1.5 \times 10^{-16}$ per day, while the rates of (H–H) and (Cs–Cs) drifts (i.e., the drifts between identical standards, used to control their stability) were less than $1 \times 10^{-16}$ per day [6]. A similar result was found in comparison of a $Hg^+$-clock with a H-maser during 140 days [7]: the rate of the relative frequency drift was less than $(2 \pm 1) \times 10^{-16}$ per day.

Such a drift is treated as a consequence of a difference in the long-term stability of different atomic clocks. In principle, however, it may be caused by variation of $\alpha$. That is why it gives an upper limit to the $\alpha$ variation [7]: $|\dot\alpha/\alpha| \leq 3.7 \times 10^{-14}$ yr$^{-1}$.

## *Geophysical tests*

The strongest bound on the possible time-variation rate of $\alpha$ was derived in 1976 by Shlyakhter [8], and recently, from a more detailed analysis, by Damour and Dyson [9], who obtained $|\dot\alpha/\alpha| < 0.7 \times 10^{-16}$ yr$^{-1}$, The analysis was based on measurements of isotope ratios in the Oklo site in Gabon, where a unique natural uranium nuclear fission reactor had operated 1.8 billion years ago. The isotope ratios of samarium produced in this reactor by the neutron capture reaction $^{149}$Sm+$n$ $\to$ $^{150}$Sm+$\gamma$ would be completely different, if the energy of the nuclear resonance responsible for this capture were shifted at least by 0.1 eV.

Another strong bound, $|\dot\alpha/\alpha| < 5 \times 10^{-15}$ yr$^{-1}$, was obtained by Dyson [5] from an isotopic analysis of natural radioactive decay products in meteorites.

A weak point of these tests is their dependence on the model of the phenomenon, fairly complex, involving many physical effects. For instance, Damour and Dyson [9] estimated possible shift of the above-mentioned resonance due to the $\alpha$ variation, assuming that the Coulomb energy of the excited state of $^{150}$Sm*, responsible for the resonance, is not less than the Coulomb energy of the *ground* state of $^{150}$Sm. In absence of experimental data on the nuclear state in question, this assumption is not justified, since heavy excited nuclei often have Coulomb energies smaller than those for their ground states [10]. Furthermore, a correlation between the constants of strong and electroweak interactions (which is likely in the frame of modern theory) might lead to further softening of the mentioned bounds by 100-fold, to $|\dot\alpha/\alpha| < 5 \times 10^{-15}$ yr$^{-1}$, as noted by Sisterna and Vucetich [11].

In addition, the local tests cannot be extended to distant space regions and to the early Universe, since the law of possible space-time variation of $\alpha$ is unknown *a priory*. It is the extragalactic astronomy that allows us to study these remote regions of space-time, in particular the regions which were causally disconnected at the epoch of formation of the observed absorption spectra.

## Astrophysical tests

To find out whether $\alpha$ changed over the cosmological time, we have studied the fine splitting of the doublet lines of Si IV, C IV, Mg II and other ions, observed in the spectra of distant quasars. According to quantum electrodynamics, the relative splitting of these lines $\delta\lambda/\lambda$ is proportional to $\alpha^2$ (neglecting very small corrections). Consequently, if $\alpha$ changed with time, then $\delta\lambda/\lambda$ would depend on the cosmological redshift $z$. This method of measuring $\alpha$ in distant regions of the Universe had been first suggested by Savedoff [12] and was used later by other authors. For instance, Wolfe et al. [13] derived an estimate $|\dot\alpha/\alpha| < 4 \times 10^{-12}$ yr$^{-1}$ from an observation of the Mg II absorption doublet at $z = 0.524$.

**TABLE 1.** Variation of $\alpha$ value estimated from redshifted Si IV fine-splitting doublets.

| Quasar | $z$ | $\Delta\alpha/\alpha$ | Ref. |
|---|---|---|---|
| HS 1946+76 | 3.050079 | 1.58 | [16] |
| HS 1946+76 | 3.049312 | 0.34 | [16] |
| HS 1946+76 | 2.843357 | 0.59 | [16] |
| S4 0636+76 | 2.904528 | 1.37 | [16] |
| S5 0014+81 | 2.801356 | -1.80 | [16] |
| S5 0014+81 | 2.800840 | -1.70 | [16] |
| S5 0014+81 | 2.800030 | 1.11 | [16] |
| PKS 0424−13 | 2.100027 | -4.51 | [15] |
| Q 0450−13 | 2.230199 | -1.48 | [15] |
| Q 0450−13 | 2.104986 | 0.02 | [15] |
| Q 0450−13 | 2.066646 | 1.03 | [15] |
| J 2233−60 | 1.867484 | -1.92 | [17,18] |
| J 2233−60 | 1.869756 | -2.21 | [17,18] |
| J 2233−60 | 1.871074 | -1.41 | [17,18] |
| J 2233−60 | 1.925971 | 1.11 | [17,18] |
| J 2233−60 | 1.941979 | 0.48 | [17,18] |

An approximate formula which relates a deviation of $\alpha$ at redshift $z$ from its current value, $\Delta\alpha_z$, with measured $\delta\lambda/\lambda$ in the extragalactic spectra and in laboratory reads

$$\Delta\alpha_z \approx \frac{c_r}{2}\left[\frac{(\delta\lambda/\lambda)_z}{(\delta\lambda/\lambda)_0} - 1\right], \tag{1}$$

where $c_r \sim 1$ takes into account radiation corrections [14]: for instance, for Si IV $c_r \approx 0.9$.

Many high-quality quasar spectra measured in the last decade have enabled us to significantly increase the accuracy of determination of $\delta\lambda/\lambda$ at large $z$. An example of the spectra observed is shown in Fig. 1. For the present report, we have selected the results of high-resolution observations [15,16,18], most suitable for an analysis of $\alpha$ variation. The values of $\Delta\alpha/\alpha$ calculated from these data according to Eq. (1) are given in Table 1.

As a result, we obtain a new estimate of the possible deviation of the fine-structure constant at $z = 2$–$4$ from its present ($z = 0$) value:

$$\Delta\alpha/\alpha = (-4.6 \pm 4.3\,[\text{stat}] \pm 1.4\,[\text{syst}]) \times 10^{-5}, \tag{2}$$

where the statistical error is obtained from the scatter of astronomical data (at large $z$) and the systematic one is estimated from the uncertainty of the fine splitting measurement in the laboratory [19,20] (at $z = 0$, which serves as the reference point for the estimation of $\Delta\alpha$). The corresponding upper limit of the $\alpha$ variation rate averaged over $\sim 10^{10}$ yr is

$$|\dot\alpha/\alpha| < 1.4 \times 10^{-14} \text{ yr}^{-1} \qquad (3)$$

(at the 95% confidence level). This constraint is much more stringent than those obtained from all but one previous astronomical observations. The notable exception is presented by Webb et al. [21], who have analysed spectroscopic data of similar quality, but estimated $\alpha$ from comparison of Fe II and Mg II fine-splitted walelengths in extragalactic spectra and in the laboratory. Their result indicates a tentative time-variation of $\alpha$: $\Delta\alpha/\alpha = (-1.9 \pm 0.5) \times 10^{-5}$ at $z = 1.0$–$1.6$. Note, however, two important sources of a possible systematic error which could mimic the effect: (a) Fe II and Mg II lines used are situated in different orders of the echelle-spectra, so relative shifts in calibration of the different orders can simulate the effect of $\alpha$-variation, and (b) were the relative abundances of Mg isotopes changing during the cosmological evolution, the Mg II lines would be subjected to an additional $z$-dependent shift relative to the Fe II lines, quite sufficient to simulate the variation of $\alpha$ (this shift can be easily estimated from recent laboratory measurements [22]). In contrast, the method based on the fine splitting of a line of the same ion species (Si IV in the above example) is not affected by these two uncertainty sources. Thus we believe that the restriction (3) is the most reliable at present for the long-term history of the Universe.

According to our analysis, some theoretical models are inconsistent with observations. For example, power laws $\alpha \propto t^n$ with $n = 1$, $-1/4$, and $-4/3$, published by various authors in 1980s, are excluded. Moreover, the Teller–Dyson's hypothesis on the logarithmic dependence of $\alpha$ on $t$ [23,5] has also been shown to be inconsistent with observations.

Many regions of formation of the spectral lines, observed at large redshifts in different directions in the sky, had been causally disconnected at the epochs of line formation. Thus, no information could have been exchanged between these regions of the Universe and, in principle, the fundamental constants could be different there. However, a separate analysis [24] has shown that $\alpha$ value is the same in different directions in the sky within the $3\sigma$ relative error $|\Delta\alpha/\alpha| < 3 \times 10^{-4}$.

## PROTON-TO-ELECTRON MASS RATIO

The dimensionless constant $\mu = m_p/m_e$ approximately equals the ratio of the constants of strong interaction $g^2/(\hbar c) \sim 14$ and electromagnetic interaction $\alpha \approx 1/137.036$, where $g$ is the effective coupling constant calculated from the amplitude of nucleon–$\pi$-meson scattering at low energy.

In order to check the cosmological variability of $\mu$ we have used high-redshift absorption lines of molecular hydrogen $H_2$ in the spectrum of the quasar PKS 0528–250. This is the first (and, in a sense, unique) high-redshift system of $H_2$ absorption lines discovered in 1985 [25].

A possibility of distinguishing between the cosmological redshift of spectral wavelengths and shifts due to a variation of $\mu$ arises from the fact that the electronic,

**TABLE 2.** Comparison of wavelengths of electron-vibro-rotational lines for $H_2$, $D_2$, and $T_2$.

| $i$ | $\lambda_i(H_2)$ | $\lambda_i(D_2)$ | $\lambda_i(T_2)$ | $K_i$ |
|---|---|---|---|---|
| L 0–0 R(1) | 1108.633 | 1103.351 | 1101.021 | $-8.18 \times 10^{-3}$ |
| L 2–0 R(1) | 1077.697 | 1081.153 | 1082.760 | $+5.35 \times 10^{-3}$ |
| L 9–0 R(1) | 992.013  | 1015.610 | 1027.218 | $+3.80 \times 10^{-2}$ |

vibrational, and rotational energies of $H_2$ each undergo a different dependence on the reduced mass of the molecule. Hence comparing ratios of wavelengths $\lambda_i$ of various $H_2$ electron-vibration-rotational lines in a quasar spectrum at some redshift $z$ and in laboratory (at $z = 0$), we can trace variation of $\mu$. The method had been used previously by Foltz et al. [26], whose analysis was corrected later in our papers [27,24,28]. In the latter papers, we calculated the sensitivity coefficients $K_i$ of the wavelengths $\lambda_i$ with respect to possible variation of $\mu$ and applied a linear regression analysis to the measured redshifts of individual lines $z_i$ as function of $K_i$. An illustration of the wavelength dependences on the mass of the nucleus is given in Table 2, where a few resonance wavelengths of hydrogen, deuterium, and tritium molecules are listed. One can see that, as the nuclear mass increases, different wavelengths shift in different directions. More complete tables, as well as two algorithms of $K_i$ calculation, are given in Refs. [24,29].

Thus, if the proton mass in the epoch of line formation were different from the present value, the measured $z_i$ and $K_i$ values would correlate:

$$\frac{z_i}{z_k} = \frac{(\lambda_i/\lambda_k)_z}{(\lambda_i/\lambda_k)_0} \simeq 1 + (K_i - K_k)\left(\frac{\Delta\mu}{\mu}\right). \qquad (4)$$

We have performed a $z$-to-$K$ regression analysis using a high-resolution spectrum of PKS 0528−250 [29]. A set of the $H_2$ lines (82) have been identified; a portion of the spectrum which reveales some of the lines is shown in Fig. 1. The redshift estimates for individual absorption lines with their individual errorbars are plotted in Fig. 2 against their sensitivity coefficients. The resulting parameter estimate and $1\sigma$ uncertainty is

$$\Delta\mu/\mu = (-11.5 \pm 7.6\,[\text{stat}] \pm 1.9\,[\text{syst}]) \times 10^{-5}. \qquad (5)$$

The $2\sigma$ confidence interval to $\Delta\mu/\mu$ is

$$|\Delta\mu/\mu| < 2.0 \times 10^{-4}. \qquad (6)$$

Assuming that the age of the Universe is $\sim 15$ Gyr the redshift of the $H_2$ absorption system $z = 2.81080$ corresponds to the elapsed time of 13 Gyr (in the standard cosmological model). Therefore we arrive at the restriction

$$|\dot\mu/\mu| < 1.5 \times 10^{-14} \text{ yr}^{-1} \qquad (7)$$

on the variation rate of $\mu$, averaged over 90% of the lifetime of the Universe.

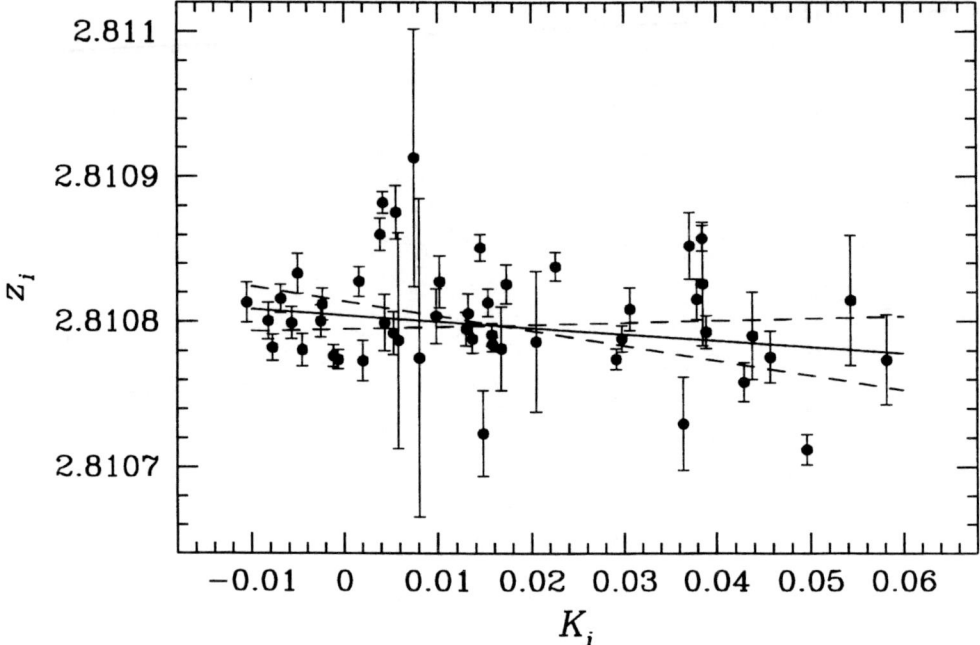

**FIGURE 2.** Redshift values inferred from an analysis of separate spectral features in the $H_2$ absorption system in the spectrum of the quasar PKS 0528−250, plotted vs. $\lambda_i(\mu)$-sensitivity coefficients $K_i$. The slanted solid line shows the most probable regression and the dashed ones corespond to $\pm 1\sigma$ deviations of the slope.

## CONCLUSIONS

Despite the theoretical prediction of the time-dependences of fundamental constants, a statistically significant variation of any of the constants have not been reliably detected up to date, according to our point of view substantiated above. The upper limits obtained indicate that the constants of electroweak and strong interactions did not significantly change over the last 90% of the history of the Universe. This shows that more precise measurements and observations and their accurate statistical analyses are required in order to detect the expected variations of the fundamental constants.

**Acknowledgements.** This work was performed in the frames of Program "Fundamental Metrology" and supported by the grant RFBR 99-02-18232.

# REFERENCES

1. Damour, T., and Polyakov, A.M., *Nucl. Phys.* **B 423**, 532 (1994).
2. Milne, E., *Proc. R. Soc.* **A 158**, 324 (1937).
3. Dirac, P.A.M., *Nature* **139**, 323 (1937).
4. Gamow, G., *Phys. Rev. Lett.* **19**, 759 (1967).
5. Dyson, F.J., in *Aspects of Quantum Theory*, edited by A. Salam and E. P. Wigner, , p. 213 (Cambridge Univ., Cambridge, 1972).
6. Demidov, N.A., Ezhov, E.M., Sakharov, B.A., et al., in *Proc. of 6th European Frequency and Time Forum*, p. 409 (1992).
7. Prestage, J.D., Tjoelker, R.L., and Maleki, L., *Phys. Rev. Lett.* **14**, 3511 (1995).
8. Shlyakhter, A.I., *Nature* **25**, 340 (1976).
9. Damour, T., and Dyson, F.J., *Nucl. Phys.* **B 480**, 37 (1996).
10. Kalvius, G.M., and Shenoy, G.K., *Atomic and Nuclear Data Tables* **14**, 639 (1974).
11. Sisterna, P.D., and Vucetich, H., *Phys. Rev. D* **41**, 1034 (1990).
12. Savedoff, M.P., *Nature* **264**, 340 (1956).
13. Wolfe, A.M., Brown, R.L., and Roberts, M.S., *Phys. Rev. Lett* **37**, 179 (1976).
14. Dzuba, V.A., Flambaum, V.V., and Webb, J.K., *Phys. Rev. A*, submitted (e-print: physics/9808021).
15. Petitjean, P., Rauch, M., and Carswell, R.F., *Astron. Astrohys.* **91**, 29 (1994).
16. Varshalovich, D.A., Panchuk, V.E., and Ivanchik, A.V., *Pis'ma Astron. Zh. (Engl. transl.: Astron. Lett.)* **22**, 8 (1996).
17. Ivanchik, A.V., *PhD thesis, Ioffe Phys.-Tech. Institute*, (1998).
18. Outram, P.J., Boyle, B.J., Carswell, R.F., Hewett, P.C., and Williams, R.E., *Mon. Not. Roy. Astron. Soc.* **305**, 685 (1999). (http://www.ast.cam.ac.uk/AAO/hdfs/)
19. Morton, D.C., *Astrophys. J. Suppl.* **77**, 119 (1991).; (E) **81**, 883 (1992)
20. Kelly, R.L., *J. Phys. Chem. Ref. Data NBS* **16**, Suppl. 1 (1987).
21. Webb, J.K., Flambaum, V.V., Churchill, C.W., Drinkwater, M.J., and Barrow, J.D., *Phys. Rev. Lett.* **82**, 884 (1999).
22. Pickering, J.C., Thorne, A.P., and Webb, J.K, *Monthly Not. R. Astron. Soc.* **300**, 131 (1998).
23. Teller, E., *Phys. Rev.* **73**, 801 (1948).
24. Varshalovich, D.A. and Potekhin, A.Y., *Space Sci. Rev.* **74**, 259 (1995).
25. Levshakov, S.A., and Varshalovich, D.A., *Monthly Not. R. Astron. Soc.* **212**, 517 (1985).
26. Foltz, C.B., Chaffee, F.H., and Black, J.H., *Astrophys. J.* **324**, 267 (1988).
27. Varshalovich, D.A. and Levshakov, S.A., *JETP Lett.* **58**, 237 (1993).
28. Varshalovich, D.A., and Potekhin, A.Y., *Pis'ma Astron. Zh. (Engl. transl.: Astron. Lett.)* **22**, 3 (1996).
29. Potekhin, A.Y., Ivanchik, A.V., Varshalovich, D.A., Lanzetta, K.M., et al., *Astrophys. J.* **505**, 523 (1998).

# Energies and QED Effects in Few-Electron Atoms and Ions

## G. W. F. Drake* and Z.-C. Yan[†]

*Department of Physics, University of Windsor, Windsor, Ontario N9B 3P4, Canada [1]
†Department of Physics, University of New Brunswick Fredericton, New Brunswick E3B 5A3

**Abstract.** QED corrections in few-electron atomic systems can be extracted from measured transition frequencies if the lower order nonrelativistic energies and relativistic corrections can be calculated to sufficient accuracy. This paper will particularly emphasize the region of low nuclear charge $Z$, where essentially exact results are available for the nonrelativistic energy and lowest-order relativistic corrections of order $\alpha^2 Z^4$. Helium and He-like ions have been particularly well studied. Spin-dependent corrections of $O(\alpha^4)$ and $O(\alpha^5 \ln \alpha)$ have been calculated, and work on terms of pure $O(\alpha^5)$ is in progress. A comparison with experiment at the 1 kHz level of accuracy for the fine structure splittings of the $1s2p\,^3P_J$ states will yield a significant new measurement of the fine structure constant $\alpha$. Other comparisons with experiment are presented for the QED shifts in total energies. Recently, important progress has been achieved for lithium and Li-like ions, including the use of isotope shifts to measure the nuclear charge radius of the exotic neutron-rich "halo" nucleus $^{11}$Li.

## I INTRODUCTION

There has been much recent interest in the study of quantum electrodynamic (QED) effects in few-electron atoms. For example, helium is the simplest atomic system more complicated than hydrogen. Just as for hydrogen, there are electron self-energy and vacuum polarization corrections to the atomic energy levels, but unlike hydrogen, they are obscured by the large electrostatic splittings between energy levels. If these large non-QED contributions can be accurately calculated and subtracted, then the residual QED shift is revealed.

Recent advances in the nonrelativistic theory of atomic helium, and other related three-body systems, allow essentially exact calculations of the nonrelativistic energy and lowest-order relativistic corrections for a wide range of atomic states. The purpose of this paper is to give a brief survey of the variational and asymptotic expansion methods used to solve the nonrelativistic problem, and the principal

---

[1] Research supported by the Natural Sciences and Engineering Research Council of Canada

effects that must be taken into account in order to estimate the higher-order QED corrections.

There are two basic approaches to carrying out this program, depending on whether the nuclear charge $Z$ is small or large. For low-$Z$ atoms and ions, the principal challenge is the accurate calculation of nonrelativistic electron correlation effects. Relativistic corrections can then be included by perturbation theory. For high-$Z$ ions, relativistic effects become of dominant importance and must be taken into account to all orders via the one-electron Dirac equation. Corrections due to the electron-electron interaction can then be included by perturbation theory. The cross-over point between the two regimes is approximately $Z = 27$ where correlation effects (proportional to $Z$) are about the same size as relativistic effects (proportional to $\alpha^2 Z^4$). Both methods yield useful results over a substantial range of $Z$, leading to interesting comparisons between them. The main emphasis in this paper is on the region of low $Z$ where an expansion of relativistic and QED effects in powers of $\alpha Z$ as well as $\alpha$ is useful.

The paper is organized as follows. Section II outlines the principal effects that must be taken into account, and summarizes the principal high precision measurements in helium and He-like ions. Then Sects. III and V describe the main ideas concerning the calculation of high precision nonrelativistic wave functions and energies, and the lowest-order relativistic corrections. These results are extended to Rydberg states with high angular momentum $L$ by means of asymptotic expansion methods in Sect. IV. Since all of these contributions can be calculated to high precision, the central issue is the calculation of quantum electrodynamic (QED) effects as discussed in Sect. VI. The results are then compared with experiment in Sect. VII, and other applications such as the determination of nuclear radii and measurement of the fine structure constant are discussed in Sect. VIII. The latter section also briefly reviews recent progress for the case of lithium, and especially the use of isotope shifts to measure the nuclear charge radius of the exotic neutron-rich "halo" nucleus $^{11}$Li. The aim throughout is to give the main ideas and results, together with appropriate references.

## II  PRINCIPAL EFFECTS

The principal effects that must be taken into account, and their relative orders of magnitude, are as listed in Table 1. In the table, $\mu/M$ is the ratio of the reduced electron mass to the nuclear mass for $^4$He, and $\alpha^2$ is the square of the fine structure constant. Since these basic expansion parmeters are about the same size for helium, the corresponding contributions to the energy are comparable in magnitude. The nonrelativistic energy refers to the energy for a hypothetical atom with infinite nuclear mass, and the mass polarization corrections (specific mass shift) arise from the fact that the dynamics of the actual nucleus in the center-of-mass frame must also be taken into account. The lowest-order relativistic corrections come from the Breit-Pauli interaction, and the relativistic recoil terms are finite nuclear mass

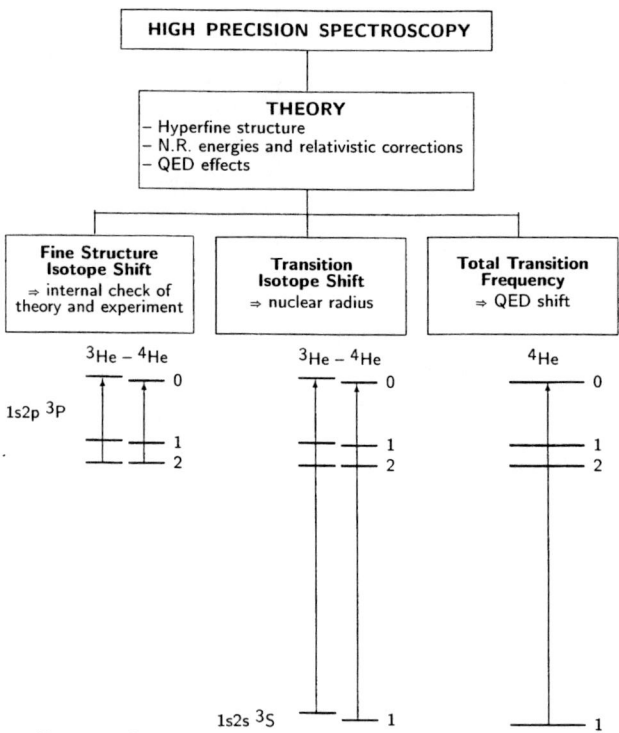

**FIGURE 1.** Flow diagram showing types of high precision measurements and their interpretation.

corrections to these. The anomalous magnetic moment terms are simply taken into account as corrections to the basic Breit-Pauli interaction. All of these terms can be calculated to very high precision and subtracted from the observations, leaving the Lamb shift (QED) terms as the principal additional effect to be taken into account in comparing theory with experiment. This term gives by far the largest contribution to the theoretical uncertainty.

The types of information that can be extracted are illustrated in Fig. 1 in the form of a 'flow diagram', for a typical experiment involving transitions between the $1s2s\,^3S$ state and the fine structure levels of the $1s2p\,^3P_{0,1,2}$ manifold of states. Beginning at the right-hand-side, the total transition frequency gives the QED shift, for which theory is not yet fully developed and experimental checks are very valuable. However, if one measures the $^3$He – $^4$He isotope shift for the same transition (middle panel), then the QED uncertainty largely cancels, allowing the differential nuclear radius to be accurately determined. Finally, if one measures the isotope shift in the fine structure intervals themselves, then the nuclear size correction also becomes negligible, providing an internal consistency check on both theory

**TABLE 1.** Contributions to the energy and their orders of magnitude. The expansion parameters are $Z$, $\mu/M = 1.370\,745\,620 \times 10^{-4}$, and $\alpha^2 = 0.532\,513\,6197 \times 10^{-4}$.

| Contribution | Magnitude |
|---|---|
| Nonrelativistic energy | $Z^2$ |
| Mass polarization | $Z^2 \mu/M$ |
| Second-order mass polarization | $Z^2 (\mu/M)^2$ |
| Relativistic corrections | $Z^4 \alpha^2$ |
| Relativistic recoil | $Z^4 \alpha^2 \mu/M$ |
| Anomalous magnetic moment | $Z^4 \alpha^3$ |
| Lamb shift | $Z^4 \alpha^3 \ln \alpha + \cdots$ |
| Finite nuclear size | $Z^4 \langle R_N/a_0 \rangle^2$ |

(especially for hyperfine structure) and experiment.

The availability of high precision theory has stimulated a number of recent experiments of the above types, as summarized in Table 2. Most of these will not be discussed in detail here, but a few of them will be quoted as examples.

## III  NONRELATIVISTIC WAVE FUNCTIONS

The basic two-electron problem to be solved is illustrated in Fig. 2. A nucleus of charge $Z$ is located at the origin, and the two electrons have position vectors $\mathbf{r}_1$ and $\mathbf{r}_2$ with an angle $\theta$ between them. The distance between the two electrons is $r_{12} = |\mathbf{r}_1 - \mathbf{r}_2|$. Assuming (for the moment) infinite nuclear mass, the Hamiltonian is (in atomic units)

$$H = -\frac{1}{2}\nabla_1^2 - \frac{1}{2}\nabla_1^2 - \frac{Z}{r_1} - \frac{Z}{r_2} + \frac{1}{r_{12}}, \tag{1}$$

and the Schrödinger equation to be solved is

$$H\Psi(\mathbf{r}_1, \mathbf{r}_2) = E\Psi(\mathbf{r}_1, \mathbf{r}_2). \tag{2}$$

Without the $1/r_{12}$ Coulomb repulsion term in Eq. (1), the Schrödinger equation would be separable, and the solution could be trivially written as a simple (anti)symmetrized product of hydrogenic orbitals. In fact, the Hartree-Fock (HF) approximation can be thought of as the best possible wave function for the full Schrödinger equation that can be written in separable product form. The correlation energy is then defined as the difference between the HF energy and the exact energy. For helium, the correlation energy of about 0.8 eV (0.03 a.u.) is huge in comparison with spectroscopic accuracies of $\pm 100$ kHz ($1.5 \times 10^{-11}$ a.u.), rendering

**TABLE 2.** Summary of high precision measurements for helium and He-like ions.

| Group | Measurements |
|---|---|
| Amsterdam[a] | He $1s^2\ ^1S - 1s2p\ ^1P$ |
| NIST[b] | He $1s^2\ ^1S - 1s2p\ ^1P$ |
| Harvard[c] | He $1s2s\ ^3S - 1s2p\ ^3P$ |
| North Texas[d] | He $1s2s\ ^3S - 1s2p\ ^3P$ |
| Florence[e] | He $1s2s\ ^3S - 1s3p\ ^3P$ |
| Paris[f] | He $1s2s\ ^3S - 1s3d\ ^3D_1$ |
| York[g] | He $1s2p\ ^3P_1 - 1s2p\ ^3P_0$ |
| NIST[h] | He $1s2s\ ^1S - 1snp\ ^1P$ |
| Yale[i] | He $1s2s\ ^1S - 1snd\ ^1D$ |
| Colorado State[j] | He $10\ ^{1,3}L - 10\ ^{1,3}(L+1)$ |
| York[k] | He $10\ ^{1,3}L - 10\ ^{1,3}(L+1)$ |
| Strathclyde[l] | Li$^+$ $1s2s\ ^3S - 1s2p\ ^3P$ |
| U. of Western Ontario[m] | Be$^{++}$ $1s2s\ ^3S - 1s2p\ ^3P$ |
| Argonne[n] | B$^{3+}$ $1s2s\ ^3S - 1s2p\ ^3P$ |
| Florida State[o] | N$^{5+}$ $1s2s\ ^3S - 1s2p\ ^3P$ |
| Oxford[p] | F$^{7+}$ $1s2p\ ^3P$ fine struct. |

[a]Eikema et al. [1]
[b]Bergeson et al. [2]
[c]Wen and Gabrielse [3]
[d]Shiner et al. [4]
[e]Marin et al. [5]
[f]Dorrer et al. [6]
[g]Storry and Hessels [7]
[h]Sansonetti and Gillaspy [8]
[i]Lichten et al. [9]
[j]Claytor, Hessels, and Lundeen [10]
[k]Storry, Rothery, and Hessels [11]
[l]Riis et al. [12]
[m]Scholl et al. [13]
[n]Dinneen et al. [14]
[o]Thompson et al. [15]
[p]Myers et al. [16]

the HF approximation of little use for high precision applications. For more complicated atoms, configuration interaction (CI) methods have been introduced, but even here the accuracy seldom exceeds $\pm 10^{-6}$ a.u.

Early in the history of quantum mechanics, Hylleraas [17] suggested expanding the wave function in the form

$$\Psi(\mathbf{r}_1, \mathbf{r}_2) = \sum_{i,j,k}^{i+j+k \leq \Omega} a_{ijk} r_1^i r_2^j r_{12}^k e^{-\alpha r_1 - \beta r_2} \mathcal{Y}_{l_1,l_2,L}^M(\hat{\mathbf{r}}_1, \hat{\mathbf{r}}_2)$$

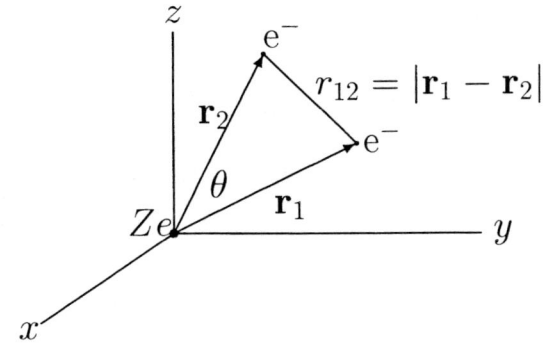

**FIGURE 2.** Geometry of the two-electron helium problem.

$$\pm \text{ exchange term,} \tag{3}$$

where the $a_{ijk}$ are linear variational coefficients, and $\mathcal{Y}_{l_1,l_2,L}^M(\hat{\mathbf{r}}_1,\hat{\mathbf{r}}_2)$ represents a vector-coupled product of spherical harmonics with angular momenta $l_1$ and $l_2$ to form a state with total angular momentum $L$. $\alpha$ and $\beta$ are additional nonlinear scale factors that can be separately adjusted to optimize the energy. The usual procedure is to include all powers $i, j, k$ such that $i + j + k \leq \Omega$, and to study the convergence as the integer $\Omega$ is progressively increased. If all powers are included, the number of terms in the basis set is

$$N = \frac{1}{6}(\Omega + 1)(\Omega + 2)(\Omega + 3) \tag{4}$$

and so grows rapidly with $\Omega$.

The principal computational step is to diagonalize the $\mathbf{H}$ matrix in the nonorthogonal basis set defined by

$$\varphi_{ijk}(\alpha,\beta) = r_1^i r_2^j r_{12}^k e^{-\alpha r_1 - \beta r_2} \mathcal{Y}_{l_1,l_2,L}^M(\hat{\mathbf{r}}_1,\hat{\mathbf{r}}_2)$$
$$\pm \text{ exchange term.} \tag{5}$$

This is equivalent to satisfying the variational condition

$$\delta \int \Psi^*(H - E)\Psi \, d\tau = 0. \tag{6}$$

The first several variational eigenvalues are then upper bounds to the true eigenvalues, provided only that the correct *number* of variational eigenvalues lies below (Hylleraas-Undheim-MacDonald Theorem [18]), and the eigenvector coefficients are the optimum values of the $a_{ijk}$ coefficients in Eq. (3). For fixed $\alpha$ and $\beta$, all the eigenvalues move inexorably downward toward the exact energies as $\Omega$ is progressively increased.

Early calculations with small basis sets containing just a few powers of $r_{12}$ easily recovered nearly all the correlation energy (see Bethe and Salpeter [19] for a review). These results demonstrated the great efficiency of Hylleraas-type basis sets in describing electron correlation; but they also showed that the *odd* powers of $r_{12}$ are much more effective than the *even* powers. This fact provides a simple argument to explain why CI calculations are much more slowly convergent as follows. From the geometry of the triangle in Fig. 2, one has

$$r_{12}^2 = r_1^2 + r_2^2 - 2r_1 r_2 \cos\theta, \qquad (7)$$

and from the spherical harmonic addition theorem

$$\cos\theta = \frac{4\pi}{3} \sum_{m=-1}^{1} Y_1^{m*}(\hat{\mathbf{r}}_1) Y_1^m(\hat{\mathbf{r}}_2). \qquad (8)$$

The angular coupling here is the same as for two $p$-electrons coupled to form an S-state in a CI calculation. Similarly, $r_{12}^4$ would contain $s$-waves, $p$-waves and $d$-waves coupled to form an S-state. In general, a CI calculation containing terms up to $\ell$-wave coupling for an S-state is equivalent to a Hylleraas calculation containing the powers $r_{12}^0, r_{12}^2, \ldots, r_{12}^{2\ell}$. However, the odd powers, which are much more efficient then the even powers, are never included by the CI method.

Further Hylleraas-type calculations with basis sets of increasing size and sophistication, culminating with the work of Pekeris and coworkers in the 1960's (see Accad, Pekeris, and Schiff [20]) showed that nonrelativistic energies accurate to a few parts in $10^9$ could be obtained by this method, at least for the low-lying states of helium and He-like ions. However, these calculations also revealed two serious numerical problems. First, it is difficult to improve upon this accuracy of a few parts in $10^9$ without using extremely large basis sets where roundoff error and numerical linear dependence become a problem. Second, as is typical of variational calculations, the accuracy is best for the lowest state of each symmetry, but rapidly deteriorates with increasing $n$.

## A  Recent Advances

Over the past 15 years, both of the above limitations on accuracy have been resolved by "doubling" the basis set so that each combination of powers $i$, $j$, $k$ is included twice with different exponential scale factors [21–23]. Explicitly, each basis function $\varphi_{ijk}(\alpha,\beta)$ defined by Eq. (5) is replaced by

$$\tilde{\varphi}_{ijk} = a_A \varphi_{ijk}(\alpha_A, \beta_A) + a_B \varphi_{ijk}(\alpha_B, \beta_B) \qquad (9)$$

where $a_A$ and $a_B$ are independent variational parameters, and $(\alpha_A, \beta_A)$, $(\alpha_B, \beta_B)$ are two sets of exponential scale factors that are common to all the basis set members. A complete optimization with respect to all the exponential scale factors leads to a

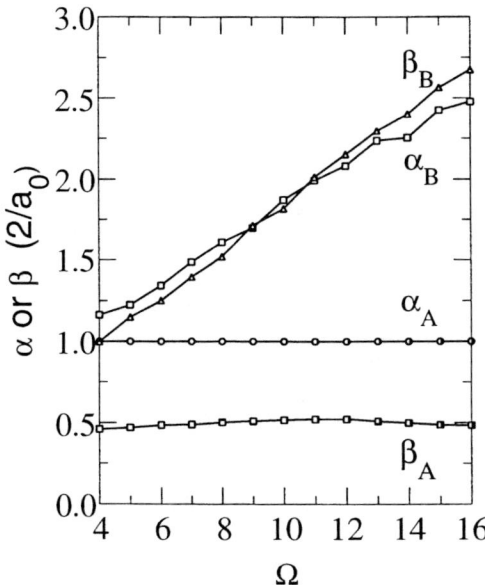

**FIGURE 3.** Variation of the exponential scale factors with basis set size for the helium $1s2p\,^1P$ state.

natural partition of the basis set into two distinct distance scales—one appropriate to the long-range asymptotic behavior of the wave function, and one appropriate to the complex correlated motion near the nucleus. The greater flexibility in the available distance scales allows a much better physical description of the atomic wave function, especially for the higher-lying Rydberg states where two sets of distance scales are clearly important. However, the multiple distance scales also greatly improve the accuracy for the low-lying states. With care, the basis set size can be reduced by omitting some of the powers $i$, $j$, $k$ from one of the two sectors (see Ref. [22] for further details).

As an example, Fig. 3 shows the systematic way that the $(\alpha_A, \beta_A)$, $(\alpha_B, \beta_B)$ vary as a function of $\Omega$ for the $1s2p\,^1P$ state. The $\alpha_A$, $\beta_A$ are nearly independent of $\Omega$ and describe the asymptotic (screened hydrogenic) behavior of the wave function, while the $\alpha_B$, $\beta_B$ increase linearly with $\Omega$. These terms are appropriate to a description of complex inner correlation effects.

As a final subtlety, the screened hydrogenic wave function $\psi_{1s}^Z(\mathbf{r}_1)\psi_{nL}^{Z-1}(\mathbf{r}_2) \pm$ exchange is included as an additional independent member of the basis set. Without this term, rather large basis sets are required just to recover the screened hydrogenic energy $-2 - (Z-1)^2/(2n^2)$ for Rydberg states.

As a typical example, Table 3 shows the convergence pattern for the nonrelativistic energy of the $1s^2\,^1S$ state. The quantity $R(\Omega)$ in the last column is the ratio of successive differences defined by

**TABLE 3.** Convergence study for the ground state energy of helium (in atomic units).

| $\Omega$ | $N_{\text{tot}}(\Omega)$ | $E(\Omega)$ | $R(\Omega)$ |
|---|---|---|---|
| 4 | 44 | −2.903 724 131 001 531 809 52 | |
| 5 | 67 | −2.903 724 351 566 477 006 22 | |
| 6 | 98 | −2.903 724 373 891 109 909 02 | 9.88 |
| 7 | 135 | −2.903 724 376 548 959 509 68 | 8.40 |
| 8 | 182 | −2.903 724 376 960 412 587 10 | 6.46 |
| 9 | 236 | −2.903 724 377 018 168 461 60 | 7.12 |
| 10 | 302 | −2.903 724 377 030 786 217 38 | 4.58 |
| 11 | 376 | −2.903 724 377 033 426 036 92 | 4.78 |
| 12 | 464 | −2.903 724 377 033 966 492 28 | 4.88 |
| 13 | 561 | −2.903 724 377 034 076 499 72 | 4.91 |
| 14 | 674 | −2.903 724 377 034 107 875 42 | 3.51 |
| 15 | 797 | −2.903 724 377 034 116 018 94 | 3.85 |
| 16 | 938 | −2.903 724 377 034 118 518 36 | 3.26 |
| 17 | 1090 | −2.903 724 377 034 119 239 36 | 3.47 |
| 18 | 1262 | −2.903 724 377 034 119 478 92 | 3.01 |
| 19 | 1446 | −2.903 724 377 034 119 553 56 | 3.21 |
| 20 | 1652 | −2.903 724 377 034 119 582 74 | 2.56 |
| 21 | 1871 | −2.903 724 377 034 119 592 06 | 3.13 |
| 22 | 2114 | −2.903 724 377 034 119 595 82 | 2.48 |
| Extrapolation | $\infty$ | −2.903 724 377 034 119 598 13(23) | |
| Goldman [24] | 8066 | −2.903 724 377 034 119 593 82 | |
| Bürgers et al. [25] | 24 497 | −2.903 724 377 034 119 589(5) | |
| Baker et al. [26] | 476 | −2.903 724 377 034 118 4 | |

**TABLE 4.** Variational energies for the $n = 10$ singlet and triplet states of helium.

| State | Singlet | Triplet |
|---|---|---|
| 10 S | −2.005 142 991 747 919(79) | −2.005 310 794 915 611 3(11) |
| 10 P | −2.004 987 983 802 217 9(26) | −2.005 068 805 497 706 7(30) |
| 10 D | −2.005 002 071 654 256 81(75) | −2.005 002 818 080 228 84(53) |
| 10 F | −2.005 000 417 564 668 80(11) | −2.005 000 421 686 604 88(26) |
| 10 G | −2.005 000 112 764 318 746(22) | −2.005 000 112 777 003 317(21) |
| 10 H | −2.005 000 039 214 394 532(17) | −2.005 000 039 214 417 416(17) |
| 10 I | −2.005 000 016 086 516 1947(3) | −2.005 000 016 086 516 2194(3) |
| 10 K | −2.005 000 007 388 375 8769(0) | −2.005 000 007 388 375 8769(0) |

$$R(\Omega) = \frac{E(\Omega - 1) - E(\Omega - 2)}{E(\Omega) - E(\Omega - 1)}. \tag{10}$$

If $R(\Omega)$ were a constant, then the series could be extrapolated to $\Omega = \infty$ as a geometric series. However, the tabulated values show a relatively smooth decrease,

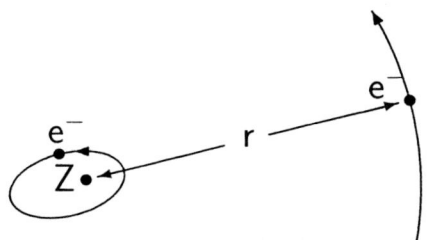

**FIGURE 4.** Illustration of the physical basis for the asymptotic expansion method in which the Rydberg electron moves in the field generated by the polarized core.

and a systematic even-odd alternation that becomes much more pronounced for Rydberg states. The extrapolated energy is obtained by separately extrapolating the even and odd values of $R(\Omega)$ to sufficiently high $\Omega$. The results are in good agreement with other recent calculations using entirely different basis set methods, suggesting that the indicated accuracy of 8 parts in $10^{20}$ is meaningful. The results in Table 3 are an update of those presented in Ref. [27].

## IV ASYMPTOTIC EXPANSIONS

Results of similar accuracy are now available for all the higher-lying $1snl\ ^{1,3}L$ Rydberg states of helium up to $n = 10$ and $L = 7$ (see Drake [28] and earlier references therein). One might object that these long strings of figures are just numerology with little physical content. However, with increasing $L$, one can give a full physical account of the variational results by means of a simple (in concept) core polarization model largely developed by Drachman [29] (see also Drake [23]). An examination of the eigenvalues for the $n = 10$ Rydberg states listed in Table 4 reveals two significant features. First, with increasing $L$, the first several figures are accounted for by the screened hydrogenic energy

$$E_{\rm SH} = -\frac{Z^2}{2} - \frac{(Z-1)^2}{2n^2} \qquad (11)$$
$$= -2.005 \quad \text{for } Z = 2,\ n = 10$$

corresponding to the energy of the inner $1s$ electron with the full nuclear charge $Z$, and the outer $nl$ electron with the screened nuclear charge $Z - 1$. Second, the singlet-triplet splitting goes rapidly to zero with increasing $L$. This suggests that for sufficiently high $L$, one can treat the Rydberg electron as a distinguishable particle moving in the field of the polarizable core consisting of the nucleus and the tightly bound $1s$ electron. As illustrated in Fig. 4, the various multipole moments of the core then give rise to an asymptotic potential of the form

$$\Delta V(r) = \frac{c_4}{r^4} + \frac{c_6}{r^6} + \frac{c_7}{r^7} + \cdots \qquad (12)$$

where $r$ is the coordinate of the Rydberg electron. In first order, the correction to the energy is then $\langle \Delta V(r) \rangle$, where the expectation value is with respect to the Rydberg electron. Since the core is a hydrogenic system, all the $c_i$ coefficients and expectation values can be calculated analytically. For example, $c_4$ is related to the core polarizability $\alpha_1 = (9/32)a_0^3$ by $c_4 = -\alpha_1/2$ ($a_0$ is the Bohr radius), and $c_6$ is related to the quadrupole polarizability $\alpha_2 = (15/64)a_0^5$ and a nonadiabatic correction to the dipole polarizability $\beta_1 = (43/512)a_0^5$ by $c_6 = -\alpha_2/2 + 3\beta_1$. Detailed expressions for the higher order terms up to $c_{10}$ have been derived (see Drachman [29] for further discussion). Each term can be calculated analytically by repeated use of the perturbation methods of Dalgarno and Lewis [30]. However, the expansion must be terminated at $i = 2(L+1)$ because the expectation values $\langle r^{-i} \rangle$ diverge beyond this point. In this sense, the series must be regarded as an asymptotic expansion.

As an example, Table 5 shows that the terms up to $c_{10}$, together with a second-order perturbation correction [31], account for the variationally calculated energy of the $1s10k$ state to within an accuracy of only a few Hz. All the entries can be expressed analytically as rational fractions. For example, the $c_4 \langle r^{-4} \rangle$ contribution is exactly (in atomic units)

$$c_4 \langle r^{-4} \rangle = -\frac{3 \times 61}{2^{10} \times 5^6 \times 7 \times 13 \times 17}$$
$$= -7.393\,341\,95\cdots \times 10^{-9}. \qquad (13)$$

Since the accuracy of the asymptotic expansion rapidly gets even better with increasing $L$, there is clearly no need to perform numerical solutions to the Schrödinger equation for $L > 7$. The entire singly excited spectrum of helium is covered by a combination of high precision variational solutions for small $n$ and $L$, quantum defect extrapolations for high $n$, and asymptotic expansions based on the core polarization model for high $L$. The complete asymptotic expansion for helium up to $\langle r^{-10} \rangle$ is [29,23]

$$E_{nL} = -2 - \frac{1}{2n^2} + \frac{1}{2}\left\{-\frac{9}{32}\langle r^{-4}\rangle + \frac{69}{256}\langle r^{-6}\rangle + \frac{3833}{7680}\langle r^{-7}\rangle\right.$$
$$- \left[\frac{55923}{32768} + \frac{957}{5120}L(L+1)\right]\langle r^{-8}\rangle - \frac{908185}{344064}\langle r^{-9}\rangle$$
$$+ \left[\frac{3824925}{524288} + \frac{33275}{14336}L(L+1)\right]\langle r^{-10}\rangle\right\}$$
$$+ e_{2,0}^{1,1} - \frac{23}{20}e_{2,0}^{1,2} \qquad (14)$$

**TABLE 5.** Asymptotic expansion for the energy of the 1s10k state of helium.

| Quantity | Value |
|---|---|
| $-Z^2/2$ | $-2.000\,000\,000\,000\,000\,00$ |
| $-1/(2n^2)$ | $-0.005\,000\,000\,000\,000\,00$ |
| $c_4 \langle r^{-4} \rangle$ | $-0.000\,000\,007\,393\,341\,95$ |
| $c_6 \langle r^{-6} \rangle$ | $0.000\,000\,000\,004\,980\,47$ |
| $c_7 \langle r^{-7} \rangle$ | $0.000\,000\,000\,000\,278\,95$ |
| $c_8 \langle r^{-8} \rangle$ | $-0.000\,000\,000\,000\,224\,33$ |
| $c_9 \langle r^{-9} \rangle$ | $-0.000\,000\,000\,000\,002\,25$ |
| $c_{10} \langle r^{-10} \rangle$ | $0.000\,000\,000\,000\,003\,73$ |
| Second order | $-0.000\,000\,000\,000\,070\,91$ |
| Total | $-2.005\,000\,007\,388\,376\,30(74)$ |
| Variational | $-2.005\,000\,007\,388\,375\,8769(0)$ |
| Difference | $-0.000\,000\,000\,000\,000\,42(74)$ |

The last two terms are small second-order dipole-dipole and dipole-quadrupole perturbation corrections. The numerical values of all these terms for the example of the 1s10k state are as listed in Table 5.

## V   RELATIVISTIC CORRECTIONS

The lowest order relativistic corrections are given by expectation values of the well-known Breit interaction [19]

$$H_{\rm rel} = B_1 + B_2 + B_{3Z} + B_{3e} + B_5 \\ + \frac{Z\pi\alpha^2}{2} \sum_{i=1}^{2} \delta(\mathbf{r}_i) - \pi\alpha^2 \left(1 + \frac{8}{3} \mathbf{s}_1 \cdot \mathbf{s}_2 \right) \delta(\mathbf{r}_{12}) \quad (15)$$

where $B_1 = -(\alpha^2/8) \sum_{i=1}^{2} \nabla_i^4$, $B_2$ is the orbit–orbit interaction, $B_{3Z}$ is the spin–orbit interaction, $B_{3e}$ is the spin-other-orbit interaction proportional to the spin sum $s_1 + 2s_2$, and $B_5$ is the spin–spin interaction. In addition, finite-nuclear-mass corrections of $O(\alpha^2 \mu/M)$ au come from the mass scaling of these terms, cross terms with the mass polarization operator, and the relativistic recoil terms $\tilde{\Delta}_2$ and $\tilde{\Delta}_3$ first derived by Stone [32]. All of these effects can be calculated to high precision and included in the final results as described in Refs. [21–23,28]. Asymptotic expansions analagous to those discussed in Sect. IV are also known for all these terms [21–23]. As with the nonrelativistic energy, they provide accurate numerical values for $L \geq 7$.

## VI   QUANTUM ELECTRODYNAMIC CORRECTIONS

All the terms up to this point can be calculated to high precision, leaving a finite residual piece due to higher order relativistic and quantum electrodynamic effects which lie at the frontier of current theory.

## A  Electron-Nucleus Terms

The leading QED term of $O(\alpha^3)$ is the first term to present new computational challenges. It contains contributions coming from both the electron-nucleus interactions of leading order $\alpha^3 Z^4$, and the electron-electron interaction of leading order $\alpha^3 Z^3$. As derived by Kabir and Salpeter [33], the general form of the electron-nucleus part $\Delta E_{L,1}$ for helium is simply obtained from the corresponding hydrogenic case by inserting the correct electron density at the nucleus in place of the hydrogenic quantity $\langle \delta(\mathbf{r}) \rangle = Z^3/(\pi n^3)$. The lowest-order QED shift is then

$$E^0_{3,Z} = \frac{4\alpha^3 Z}{3} \langle \delta(\mathbf{r}_1) + \delta(\mathbf{r}_2) \rangle \left[ \ln(Z\alpha)^{-2} + \frac{19}{30} - \beta(nLS) \right] \quad (16)$$

This part is easily done, but the Bethe logarithm $\beta(nLS)$, representing the emission and absorption of virtual photons, is much more difficult to calculate. It is defined in terms of a sum over virtual two-electron intermediate states by

$$\beta(nLS) = \frac{\sum_m |\langle 0|\mathbf{p}_1 + \mathbf{p}_2|m\rangle|^2 (E_m - E_0) \ln[2Z^{-2}(E_m - E_0)]}{\sum_m |\langle 0|\mathbf{p}_1 + \mathbf{p}_2|m\rangle|^2 (E_m - E_0)} \quad (17)$$

The accurate calculation of $\beta(nLS)$ is one of the most challenging problems in atomic structure theory. The problem is that the sum in the numerator is very nearly divergent, and so the dominant contribution comes from states lying high in the scattering continuum (both one- and two-electron). In monumental calculations based on earlier work by Schwartz [34], Baker et al. [35] and Korobov and Korobov [36] have obtained accurate values of $\beta(nLS)$ for the low-lying S-states of helium ($1\,^1S$, $2\,^1S$, and $2\,^3S$). These results have an important impact on bringing theory and experiment into agreement.

Although no other direct calculations of similar accuracy are available for other states, for sufficiently high $L$, one can use instead the core polarization model. This picture shows that the dominant contribution to the change in $\beta(nLS)$ from the hydrogenic $\beta(1s)$ comes from the perturbing effect of the Rydberg electron on the $1s$ electron, rather than from the Rydberg electron itself. The dipole polarization result allows $\beta(nLS)$ to be expressed in terms of the known hydrogenic Bethe logarithms, plus a correction term proportional to $\langle r^{-4} \rangle_{nl}$ calculated with respect to the screened hydrogenic wave function of the Rydberg electron. The final result is [37,38]

$$\beta(nLS) = \beta(1s) + \frac{(Z-1)^4}{Z^4 n^3} \beta nL + \frac{0.316\,205(6)}{Z^6} \langle r^{-4} \rangle_{nl}, \quad (18)$$

where $\beta(1s)$ and $\beta(nL)$ are the accurately known hydrogenic Bethe logarithms [39]. This result is of pivotal importance because it allows the QED part of the

D-state energies to be calculated to sufficient accuracy that these states can be taken as absolute points of reference in the interpretation of measured transition frequencies. In particular, the much larger S-state QED shift can then be extracted from measured $n\,\text{S} - n'\,\text{D}$ transition frequencies by subtraction of the other known terms.

## B  Electron-Electron Terms

The corresponding QED shift coming from the electron-electron interaction is [40,41]

$$\Delta E_{\text{L},2} = \alpha^3 \left( \frac{14}{3} \ln \alpha + \frac{164}{15} \right) - \frac{14}{3} \alpha^3 Q \tag{19}$$

where

$$Q = \left( \frac{1}{4\pi} \right) \lim_{\epsilon \to 0} \langle r_{ij}^{-3}(\epsilon) + 4\pi(\gamma + \ln \epsilon)\delta(\boldsymbol{r}_{12}) \rangle \tag{20}$$

$\gamma$ is Euler's constant, and $\epsilon$ is the radius of a sphere centered about $r_{12} = 0$ that is excluded from the range of integration. The above is the sum of several contributions coming from one- and two-photon exchange, vertex terms, vacuum polarization terms, Coulomb corrections, and the anomalous magnetic moment of the electron. Since all the terms in Eq. (19) can be accurately calculated, they do not introduce additional sources of uncertainty.

## C  Higher Order Terms

Relativistic and QED terms of order $\alpha^4$ a.u. and $\alpha^5$ a.u. are also important in the comparison with experiment. The theory of these terms is incomplete. A complete treatment requires a systematic reduction of the Bethe-Salpeter equation in order to find equivalent nonrelativistic operators whose expectation values in terms of Schrödinger wave functions yield the correct coefficients for a given order of $\alpha$. The result then represents an extension of the Breit interaction to higher order. To date, this ambitious program has been carried to completion only for the spin-dependent parts. However, a comparison with experiment indicates that for S-states, these higher order terms are dominated by large QED contributions analagous to the corresponding terms in the one-electron Lamb shift. For example, the hydrogenic term of order $\alpha^4 Z^5$ (corrected for the electron density at the nucleus) contributes $-771.1$ MHz, $-51.995$ MHz, and $-67.634$ MHz respectively, to the (positive) ionization energies of the helium $1s^2\,^1\text{S}$, $1s2s\,^1\text{S}$, and $1s2s\,^3\text{S}$ states, while the experimental uncertainties are more than an order of magnitude smaller. There would be large disrepancies between theory and experiment without this QED term. Approximation methods for the calculation of these higher order terms are discussed in detail and tabulated by Drake and Martin [42].

Ionization energies for all the states of helium up to $n = 10$ and $L = 7$ have been tabulated by Drake and Martin [42]. These results can be extended to states of higher $n$ by the use of quantum defect methods, as discussed by Drake [28,43].

## VII  COMPARISON WITH EXPERIMENT

The comparison between theory and experiment (see Drake and Martin [42], and Baker et al. [35]) shows that, for the ionization energy of the $1s^2\,^1S$ ground state of helium, the two agree at the $\pm 100$ MHz level (1.7 parts in $10^8$) out of a total ionization energy of 5 945 204 226(100) MHz. The total QED contribution is $-41\,233(100)$ MHz. For the $1s2s\,^1S$ state, the agreement is spectacularly good. The difference between theory and the experimental average [8,9] is only 0.1 MHz (1.2 parts in $10^{10}$) out of a total theoretical ionization energy of 960 332 040.9(25) MHz. Here, the total QED contribution is $-2807(25)$ MHz. Both of these results rely on the calculated ionization energies of the higher-lying P- and D-state energies as absolute points of reference. In view of the large $\pm 25$ MHz uncertainty assigned to unclaculated QED terms, the agreement for the $1s2s\,^1S$ state is much better than what one might expect. However, the experimental uncertainty of $\pm 0.15$ MHz for the $1s2s\,^1S$ state corresponds to an accuracy of $\pm 53$ parts per million (ppm) in the total QED shift. This exceeds the accuracy of the best microwave resonance measurement of the Lamb shift in $He^+$ hydrogenic ion ($\pm 86$ ppm) [44].

## VIII  OTHER APPLICATIONS

### A  Nuclear Radius Determinations

For low-$Z$ atoms and ions, the correction to the energy due to the finite nuclear radius is given by

$$\Delta E_{\text{nuc}} = \frac{2\pi Z (R/a_0)^2}{3} \left\langle \sum_i \delta(\mathbf{r}_i) \right\rangle \tag{21}$$

where $R$ is the root-mean-square radius of the nuclear charge distribution and $a_0$ is the Bohr radius. For $^4$He, the value from measurements of the muonic Lamb shift is $R = 1.673(1)$ fm [45].

Normally, a value for $R^2$ cannot be extracted from a comparison of measured transition frequencies with theory because the energy shift $\Delta E_{\text{nuc}}$ is about a factor of 10 smaller than the QED uncertainty. This problem would be cured by improved QED calculations. In the mean time, recall from Fig. 1 that the QED uncertainty largely cancels when one calculates the isotope shift in transition frequencies. The residual theoretical uncertainty is then sufficiently small ($< 10$ kHz for $n = 2$) that the *difference* in $R^2$ values for difference isotopes can be determined from the measured isotope shift.

This method, first discussed in Ref. [22], has now been applied to determine the nuclear radius of $^3$He relative to $^4$He, and of $^6$Li relative to $^7$Li [12]. The results for the fundamental three-nucleon system $^3$He are compared in Table 6 with values obtained from high-energy electron-nuclear scattering, and with nuclear structure theory. The interpretation of course depends on the comparison value for $^4$He being correct, but the spectroscopic results as they stand are in good agreement with nuclear theory, and disagree with the older nuclear scattering measurements. As discussed by Shiner [46], the theoretical nuclear structure models contain the proton radius as a parameter [47], and so sufficiently accurate results would provide an indirect measure of the proton radius. There is considerable potential for further applications of this method to studies of nuclear structure, for example to the exotic "halo nuclei" of heavy isotopes of lithium [48].

Recent work by Pachucki and Sapirstein [49] shows that nuclear recoil corrections of order $\alpha^3 m/M$ to the QED shift are significant in the interpretation of high precision isotope shift measurements. For example, the contribution to the $^3$He$-^4$He isotope shift is 3327.1 kHz, 200.2 kHz and 261.6 kHz respectively for the $1\,^1S_0$, $2\,^1S_0$, and $1\,^3S_1$ states respectively. Part of this comes from the mass polarization correction to the Bethe logarithm, and part from an electron-nucleus "$Q$"-term similar to Eq. 20 but with $1/r_{12}^3$ replaced by $1/r_1^3$. Further work is in progress to evaluate these contributions for other states so that a comparison can be made with experiment.

## B  Measurement of the Fine Structure Constant

The helium $1s2p\,^3$P manifold of states has three fine-structure levels levels labeled by the total angular momentum $J = 0$, 1, and 2. If the largest $J = 0 \to 1$ interval of about 29 617 MHz could be measured to an accuracy of $\pm 1$ kHz, this would determine the fine structure constant $\alpha$ to an accuracy of $\pm 1.7$ parts in $10^8$, provided that the interval could be calculated to a similar degree of accuracy. This degree of accuracy would provide a significant test of other methods of measuring $\alpha$, such as the ac Josephson effect and quantum Hall effect, where the resulting values of $\alpha$ differ by 15 parts in $10^8$ (see Kinoshita and Yennie [65]). Groups at Harvard, North Texas, Florence and York Universities are now working toward the achievement of a $\pm 1$ kHz measurement of the fine structure interval.

Theory is also close to achieving the necessary accuracy. In lowest order, the dominant contribution of order $\alpha^2$ a.u. comes from the spin-dependent part of the Breit interaction. This part is known to an accuracy of better than 1 part in $10^9$, and the corrections of order $\alpha^3$ a.u. and $\alpha^4$ a.u. have similarly been calculated to the necessary accuracy. At each stage, the principal challenge is to find the equivalent nonrelativistic operators whose expectation value in terms of Schrödinger wave functions gives the correct coefficient of the corresponding power of $\alpha$. This analysis has been completed for the next higher-order $\alpha^5 \ln\alpha$ and $\alpha^5$ terms, and numerical results obtained for the former. A full evaluation of the remaining $\alpha^5$ terms should

**TABLE 6.** Comparison of values for the rms nuclear charge radius $R$ of $^3$He obtained by various methods. (IS: isotope shift)

| Method | $R$ (fm) | Year | Author |
|---|---|---|---|
| e$^-$ scattering | 1.87(5) | 1965 | Collard et al. [50] |
| e$^-$ scattering | 1.88(5) | 1970 | McCarthy et al. [51] |
| e$^-$ scattering | 1.844(45) | 1977 | McCarthy et al. [52] |
| e$^-$ scattering | 1.89(5) | 1977 | Szalata et al. [53] |
| e$^-$ scattering | 1.935(30) | 1983 | Dunn et al. [54] |
| e$^-$ scattering | 1.877(30) | 1984 | Retzlaff et al. [55] |
| e$^-$ scattering | 1.976(15) | 1985 | Ottermann et al. [56] |
| e$^-$ scattering | 1.959(30)[a] | 1994 | Amroun et al. [57] |
| Theory | 1.92 | 1983 | Hadjimichael et al. [58] |
| Theory | 1.92 | 1986 | Schiavilla et al. [59] |
| Theory | 1.93 | 1986 | Chen et al. [60] |
| Theory | 1.95 | 1987 | Strueve et al. [61] |
| Theory | 1.92 | 1988 | Kim et al. [62] |
| Theory | 1.958(6) | 1993 | Wu et al. [63] |
| Theory | 1.954(7) | 1993 | Friar et al. [47] |
| Atomic IS | 1.938(10)[b] | 1993 | Drake [22] |
| Atomic IS | 1.9506(14) | 1994 | Shiner et al. [46] |
| Atomic IS | 1.956(42) | 1994 | Marin et al. [5] |

[a] Analysis and fit to all previous electron scattering data.
[b] From the measurements of Zhao et al. [64], and corrected for hyperfine mixing of the $2\,^3S_1$ state with the $2\,^1S_0$ state.

be sufficient to reduce the theoretical uncertainty from the present $\pm 20$ kHz to less than 1 kHz. Recent work by Pachucki [66] represents a major advance in the technology for identifying the appropriate equivalent nonrelativistic operators for these terms. Once both theory and experiment are in place to the necessary accuracy, a new value for $\alpha$ can be derived. At present, theory and experiment agree at the $\pm 20$ kHz level (see Storry and Hessels [7], and earlier references therein). This already represents a substantial advance in the accuracy that can be achieved for spin-dependent effects in helium.

## C  Applications to Lithium

The methods described in Sect. III for the calculation of accurate nonrelativistic wave functions and energies can in principle be applied to more complex atoms and molecules. The principal difficulties are that the number of terms required in the basis set to reach a given level of accuracy grows extremely rapidly with the number of particles, and the correlated integrals become much more difficult to evaluate. Only in the case of lithium (and Li-like ions) have results of spectroscopic accuracy been obtained (see Ref. [67] for a review). However, the demand on computer resources increases by about a factor of 6000 to reach the same level of accuracy.

The evaluation of matrix elements of the Breit interaction requires the calculation of even more difficult singular integrals, and this remained an unsolved problem until the recent development of new algorithms [68,69]. With these results in hand, it is now possible to include all the relativistic and QED terms as in the helium case. The resulting theoretical ionization energy for the $1s^2 2s\,^2S$ ground state of $0.198\,142\,09(2)$ a.u. is larger than the experimental value by only $0.000\,000\,06(3)$ a.u. ($0.013 \pm 0.07$ cm$^{-1}$). The fine structure splitting for the $1s^2 p\,^2P$ also agrees with experiment at the $\pm 0.000\,05$ cm$^{-1}$ level of accuracy [70].

Also as in the case of helium, asymptotic expansion methods can be applied to the Rydberg states of lithium and compared with high precision measurements [71,72]. This case is more difficult because the Li$^+$ core is a nonhydrogenic two-electron ion for which the multipole moments cannot be calculated analytically, and variational basis set methods must be used instead. However, the method is in principle capable of the same high accuracy as for helium.

Finally, for all of these cases, once accurate wave functions are available, they can be used to calculate a wide variety of atomic properties, such as oscillator strengths, multipole moments, long range interactions, etc. A great deal of work has been done in this area, some of which is reviewed in various chapters throughout the *Atomic, Molecular, and Optical Physics Handbook* [73].

## IX  CONCLUDING REMARKS

The results described here indicate the high degree of understanding that has been achieved for few-electron systems. For the heliumlike systems, the Schrödinger equation has been solved and lowest order relativistic corrections calculated to much better than spectroscopic accuracy. To a somewhat lesser extent, accurate solutions also exist for lithiumlike systems, but here theory is much less well developed. The residual discrepancies between theory and experiment determine the higher order relativistic and QED (Lamb shift) contributions to nearly the same accuracy as in the corresponding hydrogenic systems. Interest therefore shifts to the calculation of these contributions, for which theory is far from complete for atoms more complicated than hydrogen. New theoretical formulations are needed, such as the simplifications recently discussed by Pachucki [66]. Each theoretical advance provides a motivation for parallel advances in the state-of-the-art for high precision measurement. The results obtained to date provide unique tests of both theory and experiment at the highest attainable levels of accuracy. In addition to tests of fundamental theory, the application of isotope shift measurements to the determination of relative nuclear radii promises to become an important new technique.

# REFERENCES

1. Eikema, K. S. E., Ubachs, W., Vassen, W., and Horgorvorst, H., *Phys. Rev. A* **55**, 1866 (1997).
2. Bergeson, S. D., Balakrishnan, A., Baldwin, K. G. H., Lucatorto, T. B., Marangos, J. P., McIlrath, T. J., O'Brian, T. R., Rolston, S. L., Sansonetti, C. J., Wen, J., and Westbrook, N., *Phys. Rev. Lett.* **80**, 3475 (1998).
3. Wen, J., and Gabrielse, G., Ph.D. Thesis, Harvard University, 1996 (unpublished).
4. Shiner, D., and Dixson, R, **72**, 1802 (1994); Shiner, D. L/l., and Dixson, R., *IEEE Trans. Instrum. Meas.* **44**, 518 (1995).
5. Marin, F., Minardi, F., Pavone, F. S., Inguscio, M., and Drake, G. W. F., *Z. Phys. D* **32**, 285 (1995).
6. Dorrer, C., Nez, F.,de Beauvoir, B., Julien, L., Biraben, and F., *Phys. Rev. Lett.* **78**, 3658 (1997).
7. Storry, C. H., and Hessels, E. A., *Phys. Rev. A* **58**, R8 (1998).
8. Sansonetti, C. J., and Gillaspy, J. D., *Phys. Rev. A* **45**, R1 (1992).
9. Lichten, W., Shiner, D., and Zhou, Z. -X., *Phys. Rev. A* **43**, 1663 (1991).
10. Claytor, N. E., Hessels, E. A., and Lundeen, S. R., *Phys. Rev. A* **52**, 165 (1995), and earlier references therein.
11. Storry, C. H., Rothery, N. E., and Hessels, E. A., *Phys. Rev. A* **55**, 967 (1997).
12. Riis, E., Sinclair, A. G., Poulsen, O., Drake, G. W. F., Rowley, W. R. C., and Levick, A. P., *Phys. Rev. A* **49**, 207 (1994).
13. Scholl, T. J., Cameron, R., Rosner, S. D., Zhang, L., and Holt, R. A., *Phys. Rev. Lett.* **71**, 2188 (1993).
14. Dinneen, T. P., Berrah-Mansour, N., Berry, H. G., Young, L., and Pardo, R. C., *Phys. Rev. Lett.* **66**, 2859 (1991).
15. Thompson, J. K., Howie, D. J. H., and Myers E. G., *Phys. Rev. A* **57**, 180 (1998).
16. Myers, E. G., Kuske, P., Andrä, H. J., Armour, I. A., Jelley, N. A., Klein, H. A., Silver, J. D., and Träbert, E., *Phys. Rev. Lett.* **47**, 87 (1981); E. G. Myers, *Nucl. Instrum. Methods Phys. Res.* Sect. B **9**, 662 (1985).
17. Hylleraas, E. A., *Z. Phys.* **48**, 469 (1928); **54**, 347 (1929). See also *ibid.*, *Rev. Mod. Phys.* **35**, 421 (1963).
18. Hylleraas, E. A., and Undheim, B., *Z. Phys.* **65**, 759 (1930); J. K. L. MacDonald, *Phys. Rev.* **43**, 830 (1933).
19. Bethe, H. A., and Salpeter, E. E., *Quantum Mechanics of One- and Two-Electron Atoms* Springer-Verlag, New York. 1957, Sect. 32.
20. Accad, Y., Pekeris, C. L., and Schiff, B., *Phys. Rev. A* **4**, 516 (1971).
21. Drake, G. W. F., and Yan, Z.-C., *Phys. Rev. A* **46**, 2378 (1992).
22. Drake, G. W. F., in *Long Range Casimir Forces: Theory and Recent Experiments in Atomic Systems*, Edited by F. S. Levin and D. A. Micha, Plenum Press, New York, 1993, pp. 107–217.
23. Drake, G. W. F., *Adv. At. Mol. Opt. Phys.* **31**, 1 (1993).
24. Goldman, S. P., *Phys. Rev. A* **57**, R677 (1998).
25. Bürgers, A., Wintgen, D., and Rost, J.-M., *J. Phys. B: At. Mol. Opt. Phys.* **28**, 3163 (1995).

26. Baker, J. D., Freund, D. E., Hill, R. N., and and Morgan, J. D., III, *Phys. Rev. A* **41**, 1247 (1990).
27. Drake, G. W. F., and Yan, Z.-C., *Chem. Phys. Lett.* **229**, 486 (1994).
28. Drake, G. W. F., in *Atomic, Molecular, and Optical Physics Handbook*, Edited by G. W. F. Drake, AIP press, New York, 1996, pp. 154– 171.
29. Drachman, R. J., in *Long-Range Casimir Forces: Theory and Recent Experiments on Atomic Systems*, Edited by F. S. Levin and David Micha, Plenum Press, New York, 1993, pp. 219–272, and earlier references therein.
30. Dalgarno, A., and Lewis, J. T., *Proc. Roy. Soc. (London) Ser. A* **233**, 70 (1956), and Dalgarno, A., and Stewart, A. L., *ibid.*, **238**, 269 (1956).
31. Swainson, R. A., and Drake, G. W. F., *Can. J. Phys.* **70**, 187 (1992).
32. Stone, A. P., *Proc. Phys. Soc.* (London) **77**, 786 (1961); **81**, 868 (1963).
33. Kabir, P. K., and Salpeter, E. E., *Phys. Rev.* **108**, 1256 (1957).
34. Schwartz, C., *Phys. Rev.* **123**, 1700 (1961).
35. Baker, J. D., Forrey, R. C., Jerziorska, M., and Morgan, J. D., III, private communication. See also Baker, J. D.,Forrey, R. C., Morgan, J. D., Hill, R. N., Jeziorska, M., and Schertzer, J., *Bull. Am. Phys. Soc.* **38**, 1127 (1993).
36. Korobov, V. I., and Korobov, S. V., *Phys. Rev. A* **59**, 3394 (1999).
37. Goldman, S. P., and Drake, G. W. F., *Phys. Rev. Lett.* **68**, 1683 (1992).
38. Goldman, S. P., *Phys. Rev. A* **50**, 3039 (1994).
39. Drake, G. W. F., and Swainson, R. A., *Phys. Rev. A* **41**, 1243 (1990).
40. Araki, H., *Prog. Theor. Phys.* **17**, 619 (1957).
41. Sucher, J., *Phys. Rev.* **109**, 1010 (1958).
42. Drake, G. W. F., and Martin, W. C., *Can. J. Phys.* **76**, 597 (1998).
43. Drake, G. W. F., *Adv. At. Mol. Opt. Phys.* **32**, 93 (1994).
44. Dewey, M. S., and Dunford, R. W., *Phys. Rev. Lett.* 60, 2014 (1988).
45. Borie, E., and Rinker, G., *Phys. Rev. A* **18**, 324 (1978). See van Wijngaarden, A., Kwela, J., and Drake, G. W. F., *Phys. Rev. A* **43**, 3325 (1991) for a discussion of nuclear size measurements.
46. Shiner, D., Dixson, R., and Vedantham, V., *Phys. Rev. Lett.* **74**, 3553 (1994).
47. Friar, J. L., *Czech. J. Phys.* 43, 259 (1993).
48. Yan, Z.-C., and Drake, G. W. F., *Phys. Rev. A* in press (1999).
49. Pachucki, K., and Sapirstein, J., *J. Phys. B*, submitted (1999).
50. Collard, H., Hofstadter, R., Hughes, E. B., Johansson, A., and Yearian, M. R., *Phys. Rev.* **138**, B57 (1965).
51. McCarthy, J. S., Sick, I., Whitney, R. R., and Yearian, M. R., *Phys. Rev. Lett.* **25**, 884 (1970).
52. McCarthy, J. S., Sick, I., and Whitney, R. R., *Phys. Rev. C* **15**, 1396 (1977).
53. Szalata, Z. M., Finn, J. M., Flanz, J., Kline, F. J., Peterson, G. A., Lightbody, J. W., Jr., Maruyama, X. K., and Penner, S., *Phys. Rev. C* **15**, 1200 (1977).
54. Kowalski, P. C. Dunn S. B., Rad, F. N., Sargent, C. P., Turchinetz, W. E., Goloskie, R., and Saylor, D. P., *Phys. Rev. C* **27**, 71 (1983).
55. Retzlaff, G. A., and Skopik, D. M., *Phys. Rev. C* **29**, 1194 (1984).
56. Ottermann, C. R., Köbschall, G., Maurer, K., Röhrichm, K., Schmitt, K., Röhrichm Ch., and Walther, V. H., *Nucl. Phys. A* **436**, 1200 (1985).

57. Amroun, A., et al., *Phys. Rev. Lett.* **69**, 253 (1992); *Nucl. Phys. A* **579**, 596 (1994).
58. Hadjmichael, H., Goulard, B., and Bornias, R., *Phys. Rev. C* **27**, 831 (1983).
59. Schiavilla, R., et al., *Nucl. Phys.* A449, 219 (1986).
60. Chen, C. R., Payne, G. L., Friar, J. L., and Gibson, B. F., *Phys. Rev. C* **33**, 1740 (1986).
61. Strueve, W., Hajduk, Ch., Sauer, P. U., and Theis, W., *Nucl. Phys.* A436, 651 (1987).
62. Kim, Kr. T., et al., *Phys. Rev. C* **38**, 2366 (1988).
63. Wu, Y., Ishikawa, S., and Sasakawa, T., *Few-Body Systems* **15**, 145 (1993).
64. Zhao, P., Lawall, J. R., and Pipkin, F. M., *Phys. Rev. Lett.* **66**, 592 (1991).
65. Kinoshita, T., and Yennie, D. R., in *Quantum Electrodynamics*, Edited by T. Kinoshita, World Scientific, Singapore, 1990, pp. 1–14.
66. Pachucki, K., and 1998, *J. Phys. B, At. Mol. Opt. Phys.* **31**, 2489, 3547 (1998).
67. Yan, Z.-C., Tambasco, M., and Drake, G. W. F., *Phys. Rev. A* **57**, 1652 (1998).
68. Yan, Z.-C., and Drake, G. W. F., *Phys. Rev. Lett.* **81**, 774 (1998).
69. King, F. W., Ballegeer, D. G., Larson, D. J., Pelzl, P. J., Nelson, S. A., Prosa, T. J., and Hinaus, B. M., *Phys. Rev. A* **58**, 3597 (1998).
70. Yan, Z.-C., and Drake, G. W. F., *Phys. Rev. Lett.* **79**, 1646 (1997).
71. Storry, C. H., Rothery, N. E., and Hessels, E. A., *Phys. Rev. A* **55**, 128 (1997).
72. Bhatia, A. K., and Drachman, R. J., *Phys. Rev. A* **55**, 1842 (1997).
73. *Atomic, Molecular, and Optical Physics Handbook*, Edited by G. W. F. Drake, AIP press, New York, 1996.

# Atomic Collisions with Ultra Slow Antiprotons

Yasunori Yamazaki

Institute of Physics, University of Tokyo, Komaba, Meguro, Tokyo, Japan 153-8902
&
Atomic Physics Laboratory, RIKEN, Wako, Saitama, Japan 351-0198

**Abstract.** Recently, a new project, ASACUSA (Atomic Spectroscopy And Collisions Using Slow Antiprotons), has started, which aims at studying collision dynamics with slow antiprotons and high precision spectroscopy of antiprotonic atoms. It is expected that the ultra slow monoenergetic antiproton beams will open a new physics regime relating to antimatter science under very well controlled conditions. In the present report, several research plans, which includes ionization, excitation, stopping power, channeling, antiprotonic atom formation, high resolution antiprotonic atom spectroscopy, etc. are discussed. The way to prepare ultra slow antiprotons is also presented.

## INTRODUCTION

When a slow heavy particle with a negative charge such as an antiproton($\bar{p}$), a negative kaon ($K^-$), a negative pion ($\pi^-$), or a negative muon ($\mu^-$) approaches an atom, an outermost electron is repelled by the Coulomb force of the incoming particle and the binding energy gets shallower and shallower. In such a case, even a very tiny "kick" is strong enough to liberate the electron from the atom, which is expected to be a universal scenario of ionization at its initial stage induced by slow heavy particles with negative charge. This behavior is essentially different from that induced by positively charged particles. In this respect, slow heavy particles with negative charge could open a new branch of atomic collision processes. However, such a scenario has never been experimentally confirmed simply because no such beam has been available. Until now, ionization processes with antiprotons of energies from several MeV down to a few tens keV have been studied(1). In this energy range, both the experimental findings and the theoretical predictions are consistent with each other. On the other hand, there is still a lot to be studied below that energy range as is discussed below.

When the kinetic energy of the incoming particle is less than the binding energy of the outermost electron, the particle is captured by the atom after the release of the electron, which is a naïve picture of an "exotic atom" formation process. The binding energy of the particle just after the capture is comparable to or even less than the binding energy of the released electron, *i.e.*, the particle is in a high Rydberg state and possibly in a high angular momentum state, which is expected to be metastable with cascading lifetimes of the order of µsec or even longer. However, in the real world, it has been well-known that energetic hadrons with negative charge annihilate immediately after their injection into dense media. Inter- and intra-Auger transitions and Stark mixing induced by neighboring atoms have been considered to be responsible to accelerate the cascading processes. Because of this, it was a big surprise when metastable particle-He complexes were found (2)(3). Particularly, in the case of antiprotons, a considerable fraction of $\bar{p}He^+$ has been found to survive more than ~µsec even in liquid He. This extreme metastability allows to determine the binding energies of $\bar{p}He^+$ with the precision of ppm or even better, which allows to compare the Rydberg constant (proportional to $mq^2$) of $\bar{p}$ with that of p with the ppm precision. The charge ratio and the mass ratio between $\bar{p}$ and p have been evaluated to be unity with the precision of sub ppm (4) employing the fact that the charge to mass ratio (q/m) of $\bar{p}$ with respect to that of p is known to be the same with the accuracy of $9 \times 10^{-11}$ (5).

There have been vast progresses in catching and cooling antiprotons in an electromagnetic trap, which opens a way to develop ultra slow monoenergetic antiproton beams (6). Once developed, various antiprotonic atoms can be produced in vacuum keeping their intrinsic metastability, which allows to make, e.g., a high precision laser spectroscopy of various antiprotonic atoms such as protonium ($\bar{p}p$), the simplest pure hadronic atom, $\bar{p}He^{++}$, $\bar{p}Li^+$, etc. (2) (7). The ultra slow antiproton beams will be prepared by the combination of AD (antiproton decelerator), RFQD (radio frequency quadrupole decelerator), and a Multi-Ring electrode trap (MRT) (8). These findings and developments are the major motive forces to start the new project, Atomic Spectroscopy And Collisions Using Slow Antiprotons (ASACUSA) (7).

In the present report, several research plans are discussed together with a procedure to prepare ultra slow antiprotons, which includes in itself various interesting subjects relating with accelerator physics and non-neutral antimatter plasma physics.

# ATOMIC COLLISIONS WITH ANTIPROTONS

## Ionization

One of the most fundamental process in atomic collisions is ionization. In particular, ionizing processes in $\bar{p}$-H collisions provide the simplest and accordingly the ideal case to study collision dynamics, which could be a benchmark test of our understanding of collision dynamics (1)(9). Single ionization cross sections of D have been studied intensively, which are summarized in the upper part of fig.1 together with several theoretical predictions (10)(11)(12). It is seen that the theoretical predictions agree more or less with one another for energies higher than ~ 50 keV as well as with the experimental results. At lower impact energies, although the scatter among different theoretical predictions gets larger, many theoretical results predict that the cross section gets almost energy independent. Already in the late 40', Fermi and Teller (13) discussed that the binding energy of an outermost electron of the atom gets smaller when an antiproton approaches an atom, and at a certain distance, $d_{cr}$, the binding energy vanishes ($d_{cr}$ is called the critical distance), i.e., the atom is ionized even when the collision evolves adiabatically as far as the distance of closest approach is smaller than $d_{cr}$. It is noted that such a behavior is quite different from that of the ordinary ion-atom collisions, where the ionization cross section decreases as the projectile velocity decreases unless a resonant charge transfer process comes into play. The expected ionization cross section is then ~1.3 a.u. (Fermi-Teller limit) considering that $d_{cr}$ for H is 0.63 a.u. Figure 1 shows that the experimental and theoretical results are about three times bigger than the Fermi-Teller limit. It is predicted that "non-adibatic effects in adiabatic collisions" play important roles because the binding energy gets very shallow during the collision, i.e., the Massey criteria, $2\pi a \Delta E/h\upsilon \gg 1$ (a is of the order of the impact parameter, $\Delta E$ the energy difference, h the Planck constant, and $\upsilon$ the projectile velocity), is not necessarily satisfied for negatively charged particles even at low velocities.

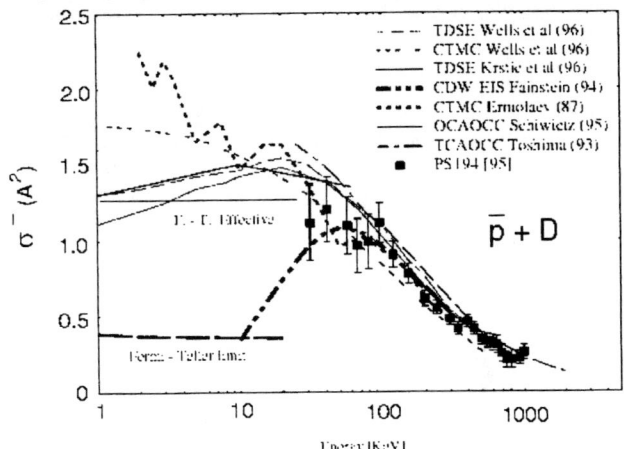

The single ionization cross section of He is drawn in the lower part of fig.1, which shows a clear peak at ~70 keV in contrast to the D target case (14). It is noted that the critical distance for He is "negative", i.e., the electronic binding energy stays negative finite even under their unified atom limit, which is known as H⁻. Ionizations of atoms and molecules with different critical distances are expected to be very important for the comprehensive understanding of collision dynamics.

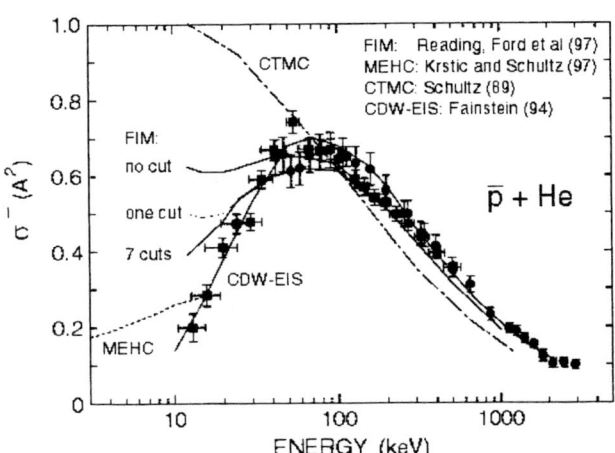

**FIGURE 1.** Single ionization cross section of D(upper) and He(lower) by antiprotons (see the text for details)

Kinematically complete ionization experiments of hydrogen and helium targets by antiprotons are under discussion employing a technique of Recoil-Ion Momentum Spectroscopy (15), which will be combined with a table top electrostatic storage ring (16). The double ionization process of helium by antiproton has been intensively studied, which revealed that $\sigma^{++}(\bar{p})$ is about two times larger than $\sigma^{++}(p)$ of the same velocity even for projectile energies as high as 10 MeV/u (1). It has been shown that the electron-electron correlation plays an essential role to reproduce the observations. In this direction, the study of double excitation process combined with the kinematically complete experiment described above should be very interesting, because a collision system with (quasi-)bound states can be handled more accurately as compared with those involving two continua like in the case of double ionization, and could be more sensitive to details of the collision dynamics (17).

The stopping power of ~ keV antiprotons (7), the channeling of ~100 keV antiprotons through a single crystal target, etc. will be studied with slow antiproton beams directly from the RFQD and from the MRT. It is predicted that ~100keV antiprotons show a characteristic channeling pattern (18), which will be experimentally studied for the first time with antiproton beam from the RFQD (see below).

## Antiprotonic Atom Formation

When an antiproton with very low kinetic energy ionizes an atom, it can be trapped into an atomic orbital with a large principal quantum number $n$. The energy conservation before and after the collision in the center of mass system requires,

$$K_{\bar{p}A} - \varepsilon_{eA} = K_{e(\bar{p}A)} - \varepsilon_{\bar{p}A}, \tag{1}$$

where $K_{AB} = (1/2)\mu_{AB} \upsilon_{AB}^2$, $\mu_{AB}$ is the reduced mass of particles A and B, $\upsilon_{AB}$ is the relative velocity between A and B, $\varepsilon_{AB}$ is the binding energy between A and B. Because the ionization takes place more or less adiabatically, the electron is oozing out with $K_{e(\bar{p}A)} \sim 0$ (19), i.e.,

$$\varepsilon_{\bar{p}A} (\sim (\mu_{\bar{p}A}/m_e)(\varepsilon_R/n^2)) \sim \varepsilon_{eA} - K_{\bar{p}A}, \tag{2}$$

where $\varepsilon_R$ is the Rydberg constant (~13.6 eV). For an antiprotonic atom to be formed, $\varepsilon_{\bar{p}A}$ should be positive, i.e., the trapping cross section is expected to be finite when $K_{\bar{p}A}$ satisfies

$$0 < K_{\bar{p}A} < \varepsilon_{eA}. \tag{3}$$

Equation (2) tells that $n$ is a function of $K_{\bar{p}A}$, which varies from $n_{min}$ ($\sim((\mu_{\bar{p}A}/m_e)(\varepsilon_R/\varepsilon_{\bar{p}A}))^{1/2}$) to infinity as $K_{\bar{p}A}$ increases from 0 to $\varepsilon_{eA}$. Because the momentum taken away by the released electron is fairly small, the momentum of $\bar{p}A^+$ is practically that of the incident $\bar{p}$. In other words, the $\bar{p}A^+$ so prepared could be used as a high quality beam maintaining the quality of the incident $\bar{p}$ beam. The basic mechanism to form $\bar{p}A^+$ is relatively simple and is expected to be general for atomic collisions involving "heavy electrons" such as $\mu^-$, $\pi^-$, $K^-$, and $\bar{p}$.

Cohen has employed a CTMC (Classical Trajectory Monte Carlo) method and an FMD (Fermion Molecular Dynamics) method to treat $\bar{p}$-H and -$H_2$ collisions in a low energy region (20)(21), the prediction of which is shown in fig. 2. The solid line and the dotted line shows protonium ($\bar{p}p$) formation cross section, $\sigma_{\bar{p}p}$, and total (formation and ionization) cross section, $\sigma_t$, in $\bar{p}$-H collisions, respectively. A clear threshold is seen at around 30eV for $\sigma_{\bar{p}p}$, although $\sigma_t$ varies smoothly. The same calculation predicts that the $n$ distribution peaks at around 30, 38, and 60 for the antiproton energy of 2.7, 10.8, and 21.8eV, respectively, which is consistent with eq.2. The $l$ distribution peaks at around 25, 35, and 30 for the same antiproton energies (21). At very low energies, a long range potential induced by a target polarization causes the antiproton orbiting around the atom, which makes the antiprotonic atom formation cross section to be inversely proportional to the antiproton velocity, which is actually observed in fig.2.

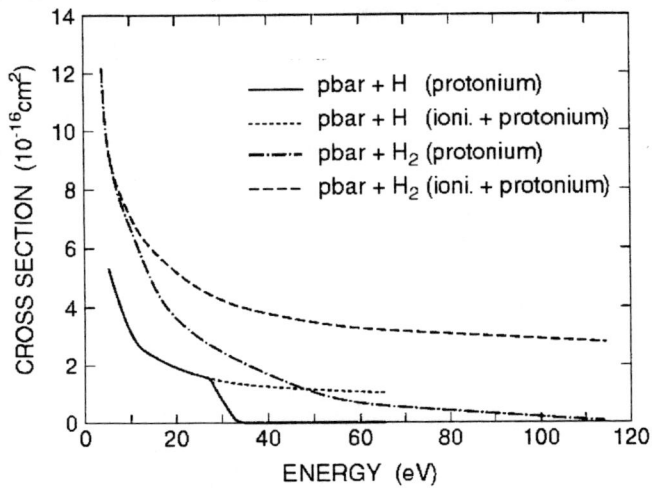

**FIGURE 2.** Protonium formation cross section for H and H2 (see text for details).

In the case of molecular targets, the electron release process leads to a temporary formation of an antiprotonic molecule, which then dissociates into an antiprotonic atom and a residual atom (ion) (20). The initial internal motion of the molecule plays an important role in the antiprotonic atom formation process, and the residual atom (ion) takes care not only of the energy balance but also of the momentum balance. Because of this, the antiprotonic atom formation cross section stays finite even beyond the threshold energy given in eq.3. The dash-dotted line and the dashed line in fig.2 show $\sigma_{\bar{p}p}$, and $\sigma_t$, respectively for $\bar{p}$-$H_2$ collisions. The role of the third body in enhancing the $\bar{p}A^+$ formation cross section above the threshold has also been predicted for multi-electron system like Ne (22), where multiple Auger electron emission is predicted to play a role. Further, molecular targets could provide an interesting chance to study a "dynamic Stark effect", because the temporary antiprotonic molecule described above will dissociate into an antiprotonic atom and a spectator atom (ion), *i.e.*, the antiprotonic atom evolves in the electric field of the spectator atom (ion) for a finite time. The electric field will increase the fraction of s-state components due to Stark mixing, and accordingly increase the annihilation rate, which may provide a new and sensitive measure of collision dynamics.

Various multielectron antiprotonic atoms are also expected to have intrinsic metastability, which can be exclusively realized only when they are produced under single collision conditions isolated in vacuum except for $\overline{p}He^+$. Like in the case of H, Li has a positive critical distance for antiproton ($d_{cr}$=0.79 a.u) (23). As the antiproton primarily replaces the 2s electron of Li, it is far outside of the residual two 1s electrons, *i.e.*, the Auger transition rates will be fairly small because the transition energies are large and the spatial overlap between the initial and the final orbits is small. It is further noted that the antiprotonic states with the same principal quantum number but with different orbital angular quantum numbers do not degenerate, *i.e.*, $\overline{p}Li^+$ is strong against annihilation induced by Stark mixing. From eq.3, a possible smallest principal quantum number, $n_{min}$, of the formed $\overline{p}Li^+$ is estimated to be ~62-63 (The binding energy of Li 2s electron is ~5.4 eV.). The right half of fig.3 shows the correlation diagrams of $\overline{p}Li^+$ for $l$=60-64 and $\overline{p}Li^{++}$ for $l$=37 and 38 (23). As is seen, there is practically no spatial overlap between $\overline{p}Li^+$ states and $\overline{p}Li^{++}$ states, *i.e.*, the formed $\overline{p}Li^+$ can de-excite only radiatively. A naive consideration indicates that such a metastability could be a general aspect for alkali metals. As a reference, the correlation diagrams of $\overline{p}He^+$ for $l$=35-39 and $\overline{p}He^{++}$ for $l$=30-34 are given in the left half of fig.3, which shows a considerable overlap between antiproton wavefunctions before and after the Auger decay, indicating that the Auger rates could be very sensitive on $n$ and $l$ as they actually are (2).

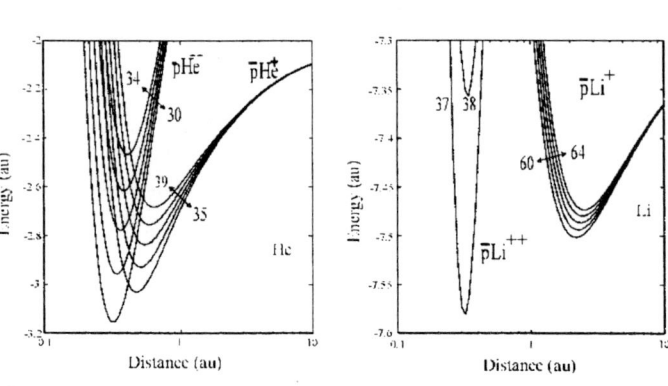

**FIGURE 3.** Correlation diagrams of $\overline{p}He^+$ and $\overline{p}He^{++}$ (left) and $\overline{p}Li^+$ and $\overline{p}Li^{++}$ (right) (23)

The annihilation of $\overline{p}He^{++}$ was used as an important monitor of the laser spectroscopy of $\overline{p}He^+$ (2), because $\overline{p}He^{++}$ has no electron shield and is very weak against external electric fields, *i.e.*, it annihilates immediately after it is produced in dense media. On the other hand, under a single collision condition, not only $\overline{p}He^+$ but also $\overline{p}He^{++}$ can keep its intrinsic metastability. A similar process will be effective to produce a meta-stable antiprotonic ion ($\overline{p}A^{++}$) using alkali earth metal. Such an antiprotonic ion (*e.g.* $\overline{p}Be^{++}$) has a closed electron shell, and accordingly the spatial overlap of the electrons with the antiproton is small, which again decays primarily via radiation and is expected to be strong against external electric field. It might be stored even in a trap.

# High Precision Spectroscopy of Antiprotonic Atoms

The discovery of meta-stable $\bar{p}He^+$ has made it possible to study the nature of antiprotonic atoms with high precision laser spectroscopy (2)(24). This field has developed rapidly from the level of identification of the principal and angular momentum quantum numbers (*i.e.*, $n$ and $l$) to the level of determination of the transition energies with a fraction of ppm (4), which agrees with theoretical predictions, where relativistic and QED effects on the bound electron are taken into account (25)(26)(27). The above finding tells that if the theoretical treatment of the Coulomb three-body system is correct, the mass and the charge difference between proton and antiproton has been determined with the accuracy better than ppm, which is an order of magnitude more accurate than before (4).

In the ASACUSA project, studies in this direction will further be pursued employing *e.g.*, laser and microwave double resonance, which enables to determine the magnetic moment of antiprotons with much higher accuracy than before (7).

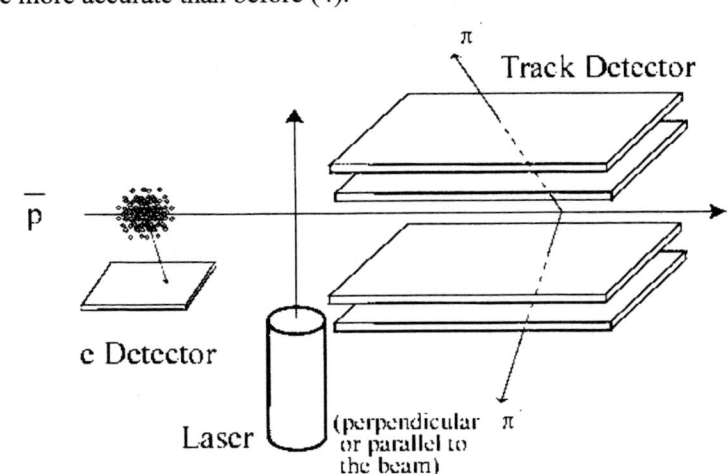

**FIGURE 4.** A possible experimental setup of $\bar{p}p$ lifetime measurements and Laser spectroscopy of $\bar{p}p$, $\bar{p}He^+$, $\bar{p}He^{++}$, etc.

Among various antiprotonic atoms to be available with the ultra slow $\bar{p}$ beams, $\bar{p}p$ is particularly interesting because it is the simplest two body system consisting of a particle and an antiparticle with strong interaction. It is noted again that a monoenergetic $\bar{p}p$ is available only with an atomic hydrogen target, the density of which cannot be very high. In this respect, a possible alternative of a pure two body system is $\bar{p}He^{++}$, the principal quantum number of which could be tunable not only with the incident energy of $\bar{p}$ but also with a He target in metastable excited states (*e.g.*, 1s2s$^{1,3}$S, 1s2p$^3$P) (The binding energy of the resultant $\bar{p}He^{++}$ is comparable to that of the He(*) target, the transition energies of which are again in the accessible range of lasers). Such a two body system in a Yrast state can decay only via slow radiative transitions when the external electric field is negligible, and its lifetime can be much longer than 1μsec, *i.e.*, a high resolution laser spectroscopy becomes applicable to $\bar{p}p$ and/or $\bar{p}He^{++}$ for the first time. A sketch of a possible experimental setup to measure the formation cross section and to make high precision laser spectroscopy is drawn in fig.4. The formation cross section and the $(n,l)$ distribution of $\overline{p}p$ will be determined by measuring the time difference between the electron signal and the annihilation signal together with the position of annihilation. In the case of Laser spectroscopy, again the emitted electron triggers a Laser, which excites one of the formed states into a high $n$ state with a much longer lifetime, *i.e.*, a high precision spectroscopy of $\overline{p}p$ can be made via the lifetime measurements.

One may compare $\overline{p}p$ with positronium ($e^+e^-$), which are similar with each other in the sense that (a)they are two body systems consisting of the same mass with opposite charge, (b) they have no magnetic moment accompanied by orbital angular momentum because the charge and the mass center coinsides with each other. On the other hand, the interaction of positronium is pure electromagnetic although strong interaction plays a vital role in the case of protonium. For example, the annihilation widths of the ns, np, and nd states of $\overline{p}p$ are $\sim n^{-3}$ keV, $\sim 0.3 n^{-3}$eV, and $\sim 10 n^{-3}$μeV, respectively (28).

## PRODUCTION OF ULTRA SLOW ANTIPROTONS

In order to realize various experiments described in the previous section, the production of a high quality antiproton beam of ~10 eV is essential, which is going to be realized via the following 5 steps (see fig.5).

In the first step, antiprotons are produced with 26 GeV/c protons of $1.5 \times 10^{13}$/pulse supplied from the CERN PS (proton synchrotron) colliding with a antiproton production target, which gives $\sim 5 \times 10^7$ $\overline{p}$. The momentum of the antiprotons at its maximum yield is ~4 GeV/c, *i.e.*, the antiproton energy should be reduced by 9 digits to reach our "goal" of the eV antiproton beam.

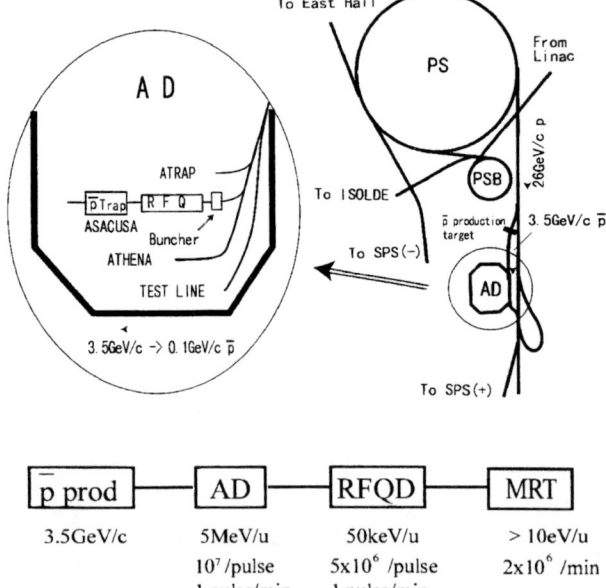

FIGURE 5. A schematic drawing of the proton synchrotron complex at CERN.

In the second step, the 3.5 GeV/c antiprotons from the production target are accumulated by the AD (Antiproton Decelerator), cooled and decelerated down to

0.1 GeV/c (5.3 MeV/u), gaining the first 3 digits. It takes ~1min for stochastic and electron coolings.

In the third step, the 5.3 MeV antiprotons are extracted from the AD as a pulse of ~1x10$^7$ antiprotons once a minute with the pulse width of ~250 ns, transported to the ASACUSA beamline, and then injected in the RFQD (radio frequency quadrupole decelerator). The RFQD can accept ~50% of the bunched beam and decelerate it down to 50 keV, gaining another 2 digits. As the electrodes of the RFQD can be floated by 50kV, the energy of antiprotons from the RFQD is in principle continuously tunable from 0 to 100 keV.

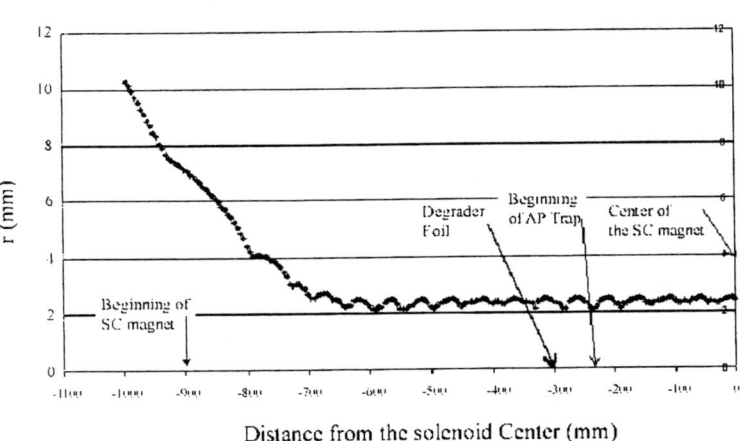

FIGURE 6. An envelope of simulated $\bar{p}$ beam trajectories from the RFQD to the center of the 5T superconducting solenoid (25).

In the fourth step, pulsed antiprotons of several tens keV from the RFQD are transported to the MRT, guided by two normal conducting solenoid and a superconducting solenoid (5T) where the antiproton trap is installed. Transport features of the antiprotons from the RFQD to the MRT have been simulated intensively by the RFQD group at CERN. An example of the optimized result is shown in fig.6, which tells that the envelope (at 5 standard deviations) of the antiproton beam trajectories is compressed down to ~2 mm in radius at the center of the solenoid (29).

In the fifth step, antiprotons are captured, electron-cooled, and then radially compressed in the MRT, and finally extracted. A schematic diagram of the procedure is given in fig.7. In order to capture antiprotons with rather low trapping potential, a thin degrader foil will be inserted to reduce the antiproton energy below 10 keV. The foil can be positively floated which effectively reduces the energy straggling of the degraded beam and eventually increases the trapping efficiency. The right end electrode of the trap (fig.9) is negatively biased so that it reflects antiprotons. Before the reflected antiprotons reach the left end of the trap, the left end electrode is biased from 0 to <-10 kV, which results in trapping the antiprotons. In the trap, electrons are pre-loaded, and are cooled via synchrotron radiation with a time constant $\gamma_{rad}$ till the electron temperature reaches the environmental temperature, which is in the present case several K. At 5T, $\gamma_{rad}$ is about 0.1sec. The cooled electrons then cool the injected antiprotons via the Coulomb interaction in the time range of 10 sec (6). In this way, the remaining 4 digits are gained. The MRT has been designed so that it can stably store as many as ~5x10$^6$ antiprotons and about 100 times more electrons in a prolate spheroid with a radius ~1 mm and its axial length ~50 mm. The overall trapping efficiency of the present setup is expected to be about two orders of magnitudes higher than that obtained by a conventional trapping scheme, *i.e.*, the combination of a degrader foil and an electromagnetic trap. The solenoid is designed so that the bore is bakable while keeping the superconducting solenoid at liquid helium temperature. Recently, slow positrons of 0.1eV to several tens eV with an energy width of about 18meV have been successfully prepared (30) employing a method similar to what is discussed here, which supports the scheme described here.

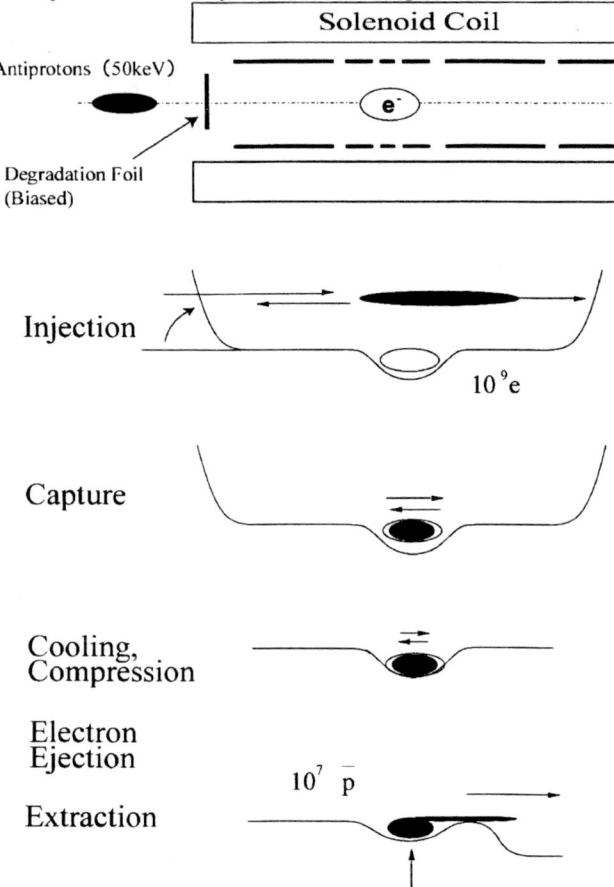

**FIGURE 7.** A schematic procedure to trap and to cool antiprotons, and then to extract as an ultra slow monoenergetic beam.

For charged clouds with a temperature below about an eV, the Debye length, $(\varepsilon_0 k_B T/\rho e^2)^{1/2}$ ($\varepsilon_0$ is the vacuum dielectric constant, $k_B$ is the Boltzmann constant, and $\rho$ is the electron density), is shorter than the size of the cloud, *i.e.*, it behaves like a plasma. The radial electric field produced by the trap and the plasma itself together with the magnetic field induces an ExB drift rotation (31) as shown in fig.8. The rotation is governed by the balance of three forces,

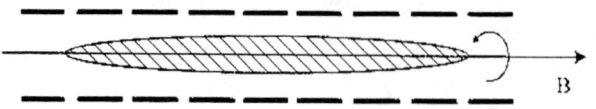

**FIGURE 8.** ExB drift rotation of a prolate spheroid plasma in magnetic field B.

i.e., the electric field, the ExB force, and the centrifugal force. When the plasma is axially symmetric, the angular frequencies of the rotation for electrons and antiprotons, $\omega_e$ and $\omega_{\bar{p}}$, are given by

$$\omega_e = (1/2)(\omega_{ce} +- (\omega_{ce}^2 - 2\omega_{pe}^2)^{1/2}), \qquad (4)$$

$$\omega_{\bar{p}} = (m_e/2m_p)(\omega_{ce} +- (\omega_{ce}^2 - 2(m_e/2m_p)\omega_{pe}^2)^{1/2}), \qquad (5)$$

respectively, where $\omega_{ce}$ is the cyclotron angular frequency of electron ($=eB/m_e$), $\omega_{pe}$ is a plasma angular frequency of the electron, which is defined as $((\rho_e+\rho_{\bar{p}})e^2/m_e\varepsilon_0)^{1/2}$ ~ $5\times10^4 (\rho_e(cm^{-3}))^{1/2}$ (sec$^{-1}$). Equations (4) and (5) tell that there exists a maximum density that can be confined at a given magnetic field B, which is called the Brillouin limit. Furthermore, in the present condition (i.e.., $\rho_e \gg \rho_{\bar{p}}$), $\rho_e$ governs the Brillouin limit for antiprotons as well as that for electrons, which are $3\times10^9 B(T)^2/cm^3$ and $5\times10^{12}B(T)^2/cm^3$, respectively. The solenoid is designed to yield 5 T, which keeps the expected plasma density well-below the Brillouin limit even for antiprotons. In this case, $\omega_e^+$ and $\omega_{\bar{p}}^+$ are approximately given by $\omega_{ce}$ and $(m_e/m_{\bar{p}})\omega_{ce}$ ($= \omega_{c\bar{p}}$), respectively, and $\omega_e^-$ and $\omega_{\bar{p}}^-$ are

$$\omega_e^- \sim (\omega_{pe}^2/2\omega_{ce})(1+\omega_{pe}^2/2\omega_{ce}^2), \qquad (6)$$

$$\omega_{\bar{p}}^- \sim (\omega_{pe}^2/2\omega_{ce})(1+m_{\bar{p}}\omega_{pe}^2/2m_e\omega_{ce}^2), \qquad (7)$$

respectively. This rotation could cause a serious problem if one wants to make very cold antiprotons in the laboratory frame. For example, at $\rho_e \sim 5\times10^9/cm^3$ and B=5T, $\omega_{e(\bar{p})}^-$ is ~$10^7$/sec, which corresponds to ~ 0.5 eV for antiprotons at the periphery of the plasma in the present conditions. In other words, the effective temperature of the plasma in the laboratory frame is unfortunately far above the possible temperature in the rotating frame which could be as low as the envirionmental temperature. Equations 6 and 7 tell that (a) $\omega_{e(\bar{p})}^-$ is proportional to $\rho_e/B$, i.e., the kinetic energies of antiprotons and electrons due to the rotation are higher for lower B, and (b) $\omega_{\bar{p}}^-$ is ~2% larger than $\omega_e^-$, which causes the antiproton cloud to be extruded out from the electron cloud (32).

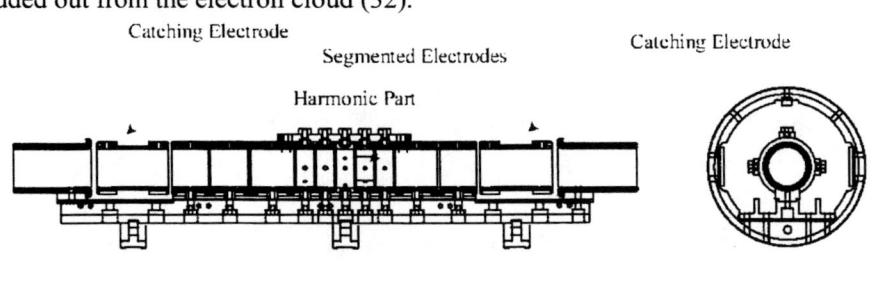

**FIGURE 9.** A drawing of the multi ring antiproton trap.

A drawing of the ASACUSA MRT is shown in fig.9, which consists of 14 cylindrical electrodes of 40 mm in inner diameter and total length of ~500 mm. Seven electrodes near the center are to form a harmonic potential to stably store and to cool the antiprotons (8). Actually, one of them is segmented into four so that a rotating field can be applied on the plasma to increase or decrease its rotation frequency. As is seen from eqs. (6) and (7), the higher the frequency, the higher the density and compressed more (33).

To test the radial compression of the plasma, a coaxial double Faraday cup with 35mm$\phi$ and 3mm$\phi$ diameter are installed at the end of the MRT. The radial distribution of electrons stored in the trap were estimated by the ratio of the charge received by this double faraday cup, which is plotted in fig.10 as a function of the rotating field frequency (34). The rotation field is seen to be very effective to radially compress the electron cloud.

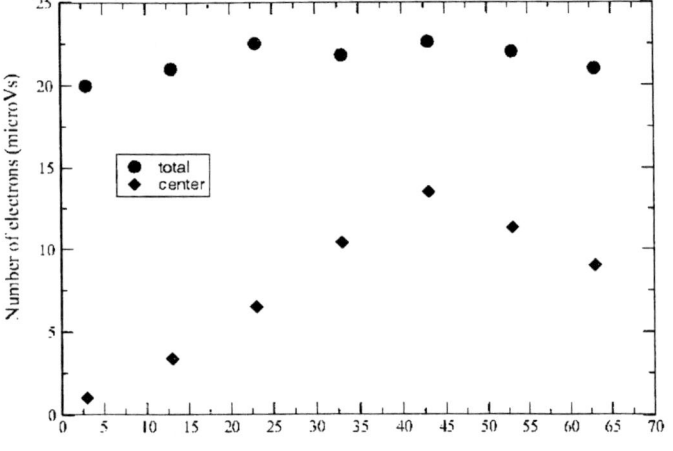

**FIGURE 10.** Charges accumulated by the center faraday cup (3mm$\phi$) (diamond) and by the total (solid circle) as a function of RF frequency.

Cooled antiprotons will be extracted at ~keV as a continuous or a pulsed beam from the trap and transported to the target area, where the antiprotons are decelerated down to 10 eV range. To keep the trap area at UHV and at the same time to use a gas cell of $10^{-3}$Torr in the target chamber, several differential pumping stages separated by small holes (~3 mm$\phi$) are necessary on the way from the trap to the target chamber. Simulations of extracted antiproton trajectories have shown that the radial distributions of the antiprotons in the trap should be equal to or smaller than ~1 mm for a reasonable transportation (35).

It is noted that unstable particles such as $\mu^-$ or $\pi^-$ are not available as high quality slow beams, *i.e.*, only antiprotons provide the chance to pursue the subjects described above.

## ACKNOWLEDGEMENTS

The author is deeply indebted to K. Ohtsuki, A. Lombardi, H. Totsuji, J. Cohen, G.Korenman, and colleagues of the ASACUSA project, particularly the trap group members, T.Ichioka, H. Higaki, M.Hori, N.Ohshima, A. Mohri, and K. for their fruitful and vivid discussions. I am grateful to J. Cohen for his critical reading of the manuscript and helpful comments. The work is supported by the Grant-in-Aid for Creative Basic Research (10NP0101), Ministry of Education, Science, and Culture.

# REFERENCES

1. Knudsen,H., and Reading,J., Phys. Rep.**212**, 107(1992).
2. Yamazaki,T., et al., Nature **361**,238(1993), Morita, N., et al., Phys. Rev. Lett. **72**,1180(1994), Hayano, R.S.,et al., Phys. Rev. A **55**,1(1997).
3. A possible existence of metastable $\overline{p}He^+$ was discussed by Condo (Phys. Lett. **9**, 65 (1964)).
4. Torii,H.A., et al., Phys. Rev. A **59**, 223 (1999).
5. Gabrielse, G., Phys.Rev.Lett. **82**, 3198(1999).
6. Feng,X., Holzscheiter,M.H., Charlton,M., Hangst,J., King,N.S.P., Lewis,R.A., Rochet,J., and Yamazaki, Y., Hyperfine Interactions **109**, 145(1997).
7. ASACUSA collaboration proposal, 1997 CERN/SPSC 97-19, SPSC P-307
8. Mohri,A., et al., Jpn. J. Appl. Phys. **37**, 664(1998).
9. Knudsen,H., et al., Phys. Rev. Lett. **74**, 4627(1995).
10. Toshima,N., Phys. Lett. A **175**, 133(1993).
11. Krstic,P.S., Schultz,D.R.,and Janev, R.K., J. Phys. B **29**, 1941(1996).
12. Wells,J.C., Schultz,D.R., Gravras,P., and Pindzola,M.S., Phys. Rev. A **54**, 593(1996).
13. Fermi, E. and Teller, E., Phys. Rev. **72**, 399(1947).
14. Hvelplund,P., et al., J. Phys. B **27**, 925 (1994), Andersen,L.H., et al., Phys. Rev. A **41**, 6536(1990).
15. Ullrich,J., et al., Topical Review,J. Phys. B **30**, 2917(1997).
16. Schumidt-Boecking, H., et al., Abstract of International Workshop on Atomic Collisions an Atomic Spectroscopy with Slow Antiprotons, (Tsurumi, Japan, 1999), and Moller, S.P., Nucl.Instrum.Methods A**394**, 281(1997).
17. Morishita,T., et al., J. Phys. B **30**, 2187(1997).
18. Kabachnik, N.M., Balashova, L.L., and Trikalinos, Ch., Abstract of International Workshop on Atomic Collisions an Atomic Spectroscopy with Slow Antiprotons, (Tsurumi, Japan, 1999), and private communications, E.Uggerhoj, Nucl.Instrum.Methods B135(1998).
19. Dolinov,V.K., et al., Muon Catalyzed Fusion **4**, 169(1989).
20. Cohen, J.S., Phys. Rev. A **56**, 3583(1997), and private communication.
21. Cohen, J.S. and Padial, N.T., Phys. Rev. A **41**, 3460(1990), Cohen, J.S., Phys. Rev. A **36**, 2024 (1987).
22. Cohen, J.S., to be published in the Proceedings of the XXIth ICEAC (Sendai, 1999)
23. Ohtsuki, K., Abstract of International Workshop on Atomic Collisions an Atomic Spectroscopy with Slow Antiprotons, (Tsurumi, Japan, 1999), and private communication
24. Widmann, E.,et al., Phys. Lett. B **404**, 15(1997).
25. Korobov,V.I., Phys. Rev. Lett.to be published.
26. Elander, N., and Yarevsky, E., Phys. Rev. A **56**, 1855(1997).
27. Kino,Y., Kamimura, M.,and Kudo,H., Proc.XV. Int. Conf. Few-Body Problems in Physics, Groningen, 1997.
28. West, D., Rep.Prog.Phys. 21, 271(1958).
29. Lombardi,A., Private communication.
30. Gilbert, S.J., Greaves, R.G, and Surko C.M., Phys.Rev.Lett. 84(1999)5032.
31. O'Neil, T.M., Non-Neutral Plasma Physics, p1, ed. C. W. Roberson & C. F. Driscoll (American Institute of Physics, New York, 1988)
32. Larson, D.J., et al., Phys. Rev. Lett. **57**, 70(1986).
33. Huang,X.-P., et al., Phys. Rev. Lett. **78**, 875(1997).
34. Ichioka, T., et al., to be published in AIP conference proceedings on 1999 Workshop on Nonneutral Plasmas. (July, 1999, Princeton, ed. Bollinger,J.)
35. Ichioka, T., Master Thesis, University of Tokyo (1996).

# VIII. IMAGING AND MEDICAL APPLICATIONS

# Atom-Resolving X-Ray Holography

B. Adams, T. Hiort, G. Materlik, Y. Nishino [1], and D. V. Novikov

*Hamburger Synchrotronstrahlungslabor HASYLAB at Deutsches Elektronen-Synchrotron DESY, Notkestrasse 85, D-22603 Hamburg, Germany*

**Abstract.** The current state of atomic resolution x-ray holography is discussed on the basis of theory and experimental results. X-ray holography is theoretically described in quantum theory. Presently-used experimental implementations are shown together with the data analysis used. Reconstructions of experimental and simulated holograms are compared for a $Cu_3Au$ crystal structure. Rigorous experimental realizations of pure direct and reciprocal x-ray holography methods are demonstrated, and future developments and applications of the method are suggested.

## I INTRODUCTION

Atom-resolving holography was first proposed by Szöke in 1986 [1]. It is based on the principle of lensless imaging as suggested by Gabor back in 1948 [2]. In Szöke's original form, atom-resolving holography employs characteristic fluorescence x-rays, photoelectrons, or Auger electrons from atoms inside the sample to produce the reference wave. Its coherence volume is large enough to image the neighbors of the reference-wave emitter atom. The reference wave is scattered from neighboring atoms producing the object waves. These interfere in the far field with the reference wave and form a holographic interference pattern. The reconstruction of the image is done numerically, using a Fourier-based transform which was introduced by Barton [3]. Experimentally the method was first realized with electrons in 1990 [4–6].

The implementation with x-rays followed in 1996 [7], when x-ray fluorescence holography (XFH) was demonstrated on a strontium titanate single crystal providing images of the heavy strontium atoms. In XFH the radiation source is a fluorescing atom inside the sample and the detector registers the holographic intensity pattern that is formed in the far field outside the sample, see Fig. 1. X-ray holography is also possible with the reciprocal version of XFH, multiple energy x-ray holography (MEXH) [8]. In MEXH, the positions of source and detector are interchanged in comparison to XFH. A radiation source outside the sample emits

---

[1] on leave from Japan Synchrotron Radiation Research Institute (JASRI), 1-1-1 Kouto, Mikazuki-cho, Sayo-gun, Hyogo, 679-5198, Japan.

a monochromatic plane wave that serves as the holographic reference wave. It is scattered from the atoms in the sample producing the holographic object waves. The incident and the scattered waves interfere inside the sample. The intensity of the resulting x-ray standing wave (XSW) is registered through the fluorescence yield of an atom incorporated in the structure [9]. The holographic interference pattern is obtained by rotating the sample relative to the incoming beam (Fig. 1).

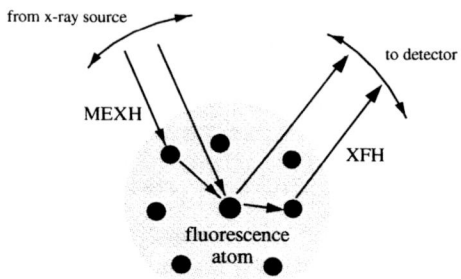

**FIGURE 1.** Principles of XFH and MEXH. In MEXH, the holographic reference wave is the wave from an external x-ray source. It is scattered from the atoms of the sample, thus forming the holographic object waves. A fluorescing atom inside the sample is a detector of the local interference field intensity. In XFH, the fluorescence wave from this atom is the reference wave. It is scattered from neighboring atoms, thus forming the holographic object waves. An energy dispersive detector outside the sample records the interference field intensity. Both processes occur together in the sample. The experimental conditions have to ensure that only one of the field variations is measured. In MEXH, the incident beam is rotated relative to the sample whereas in XFH, the outside detector is rotated with respect to the sample.

If a hologram is measured only at one energy, the reconstructed image suffers from twin images, which are inherent to the in-line Gabor holography schemes [10]. If a sample is centro-symmetric about the reference-wave emitter atom, real and twin images appear at the same position, and interference of these images produces an oscillation of atomic image intensity with energy [11–13]. When destructive interference occurs, the atom is not imaged. In order to obtain an improved reconstruction, Barton developed a multiple-energy algorithm which is a phased sum of single energy reconstructions and is supposed to suppress twin images [10]. In XFH, it is only possible to sum over reconstructions at the energies of the characteristic fluorescence lines available. In the multiple-energy method, the energy of the reference wave can be freely chosen above an absorption edge of the detecting atom, and a number of MEXH holograms can be obtained at several wavelengths.

In this paper, the conventional terminology is used in the same way as in the first experimental papers [7,8]. The internal source holography is called "direct" holography or XFH and the internal detector holography is referred to as "reciprocal" holography or MEXH. Note however that it is possible to measure in XFH at multiple energies if different characteristic lines are available.

The paper is organized as follows. In section II, a formulation of atomic resolution x-ray holography is presented together with a new formulation from quantum theory. Section III describes experimental arrangements and the analysis of holograms. As an example, the experimental data is shown for a $Cu_3Au$ single crystal. In section IV new improved arrangements for XFH and MEXH experiments are presented. A three dimensional reconstruction of MEXH and XFH holograms is shown.

## II  FORMULATION

### A  Overview

A hologram is formed by interference of the reference wave $R(\mathbf{k})$ with the object waves $O(\mathbf{k})$, $\mathbf{k}$ being the wave vector. The observed intensity is given by

$$I(\mathbf{k}) = |R(\mathbf{k}) + O(\mathbf{k})|^2 \simeq |R(\mathbf{k})|^2 + 2\mathbf{Re}R^*(\mathbf{k})O(\mathbf{k}), \qquad (1)$$

neglecting terms of order $|O(\mathbf{k})|^2$. This is justified by the small scattering cross sections of atoms at x-ray wavelengths. The approximation is valid for clusters and for crystals far away from Bragg angles. The holographic interference term is given by the second term on the right side of eq. (1). It can be obtained in a normalized form via

$$\chi(\mathbf{k}) \simeq \frac{I(\mathbf{k}) - |R(\mathbf{k})|^2}{|R(\mathbf{k})|^2}. \qquad (2)$$

For scalar waves at energies far from any absorption edges, $\chi(\mathbf{k})$ in the Born approximation is given by [14]

$$\chi(\mathbf{k}) \simeq -2r_e \mathbf{Re} \int d\mathbf{r}\, \frac{\rho(\mathbf{r})}{r} e^{i(kr - \mathbf{k} \cdot \mathbf{r})}, \qquad (3)$$

where $r_e$ is the classical electron radius and $\rho(\mathbf{r})$ is the electron charge density. Multiple scattering is present but very weak for x-rays [15,16]. For vector waves, one should take into account polarization effects [17], and also near-field effects [14]. A detailed formulation in the frame of classical electrodynamics is described in [14]. The quantum calculations [18–20] are consistent with classical electrodynamics according to [20].

The reconstruction algorithm imaging the atom positions was suggested by Barton [3,10]. The atomic image intensity $|\psi(\mathbf{r})|^2$ is obtained via the Barton algorithm

$$\psi(\mathbf{r}) = \sum_k \int d\Omega_k\, \chi(\mathbf{k}) e^{-i(kr - \mathbf{k} \cdot \mathbf{r})} \qquad (4)$$

The atomic image intensity gives structural information on neighbors of a specific atomic element due to characteristic fluorescence detection. If more than one fluorescing atom of the same kind is present in a sample, the images originating from these atoms are incoherently superposed.

## B  Quantum Theory of X-Ray Holography

Recently x-ray holography was formulated using quantum electrodynamics (QED) in [20], where MEXH is explained in terms of its underlying photoionization process. In the photoionization process, a pair of photoelectron and core-hole state is created by the incident photon (Fig. 2). In [20] the photoionization cross section is calculated including the initial-state photon interaction (Fig. 2 b) and the final-state photoelectron interaction (Fig. 2 c) with neighboring atoms. It explains not only MEXH, but also photoelectron holography (PEH), and extended x-ray absorption fine structure (EXAFS). The photoionization process is described in a single-step picture including the initial-state photon and the final-state photoelectron. In conventional treatments of MEXH in classical theory, however, it is implicitly assumed that the initial state of the photoionization process is a single-step process, and the final-state photoelectrons are disregarded. On the other hand, in EXAFS and PEH, it is usual to disregard the effect of the photon interference in the initial state. In fact, there are contributions both of the initial-state photon interaction and of the final-state photoelectron interaction in all of these experimental methods.

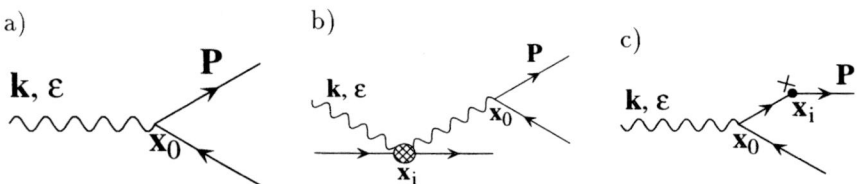

**FIGURE 2.** Feynmann diagrams of photoionization process [20]. The wavy lines are photons. The incident photon is indicated with its momentum $k$ and polarization $\epsilon$. The straight lines are electrons. The photoelectron is shown with its momentum $p$. The electron line with the arrow to the left is the core-hole state. a) The leading order contribution. The incident photon scatters an electron bound in the atom at $X_0$ and creates a pair of photoelectron and core-hole state. b) The scattering of the photon in the initial state. The photon is scattered by an electron of the atom at $X_i$. In QED, the virtual photon travels in both directions: from the atom at $X_i$ to the one at $X_0$ and from the atom at $X_0$ to the one at $X_i$. The circle with hatch contains all photon-electron interaction diagrams. For the non-relativistic electron, they are the uncrossed, the crossed and the seagull diagrams. Among them, the Thomson scattering corresponds to the seagull diagram. c) The scattering of the photoelectron in the final state. The photoelectron is scattered by the Coulomb field of the atom at $X_i$.

In MEXH, the contribution of the final-state photoelectron interaction was found to be not negligible compared to the MEXH signal. For energy 5 keV above the absorption edge, it is typically in the order of 10% of the MEXH signal. It, however, has a simple functional form, appearing only through the factor $[\epsilon \cdot (\boldsymbol{X}_i - \boldsymbol{X}_0)]^2$ where $\epsilon$ is the polarization vector of the incident photon, $\boldsymbol{X}_0$ is the position vector of the fluorescing atom, and $\boldsymbol{X}_i$ is the position vector of a neighboring atom. If the polar coordinate is taken for incident photon wave vector $\boldsymbol{k}$, $\boldsymbol{k} = (k \sin\theta \cos\varphi_s, k \sin\theta \sin\varphi_s, k \cos\theta)$, the fluorescence signal modulation due to the final-state photoelectron interaction is expressed as sinusoidal functions of $\varphi_s$ with periodicity one and two for given energy and $\theta$. In order to obtain the MEXH signal, it should be eliminated from the observed signal during the data analysis.

The XFH transition probability is also calculated in [20]. The obtained hologram functions of MEXH and XFH are found to be consistent with the calculations in classical electrodynamics.

## III  EXPERIMENTAL HOLOGRAMS OF A CU$_3$AU SINGLE CRYSTAL

### A  Experimental Set-up

Experimental holograms were measured for several energies at the bending magnet beamline CEMO at the Hamburger Synchrotronstrahlungslabor HASYLAB [21]. A Cu$_3$Au single crystal with a flat polished surface served as a sample. Cu$_3$Au has a fcc crystal lattice with a lattice constant of 3.75Å. We used two fast energy dispersive silicon drift detectors [22–26]. Each detector recorded separately the intensities of the Cu K$_\alpha$ and Au L fluorescence lines. At count rates up to $10^5$/s the energy resolution was $\sim 0.3\%$ at room temperature. The schematic set-up is shown in Fig. 3. The beam from the DORIS III storage ring is monochromatized by a Si (220) double crystal monochromator and passes a slit of size 1.2mm vertical by 2.0mm horizontal. The incident intensity as monitored by an ionization chamber was kept constant within $\sim 0.2\%$ at $\sim 10^9$photons/s/mm$^2$. It was stabilized using the ionization chamber signal in a feedback loop together with a piezo crystal acting on the angular alignment of the first monochromator crystal [27]. Sample and detectors were mounted on a multi circle diffractometer. The holograms were measured in incremental scans of step size 1°, $\theta$ ranging from 19° to 90° and $\varphi_s$ ranging from 0° to 359°. The range was limited by geometrical restrictions. Holograms were measured for the energies from 15 keV to 30 keV in a step size of 3 keV.

For a pure MEXH method it is necessary to rotate the detectors together with the sample in both angular coordinates. For technical reasons, this requirement was fulfilled only in one angular coordinate. In section IV, we show results of pure MEXH and XFH experiments. In this experiment, the detectors were rotated

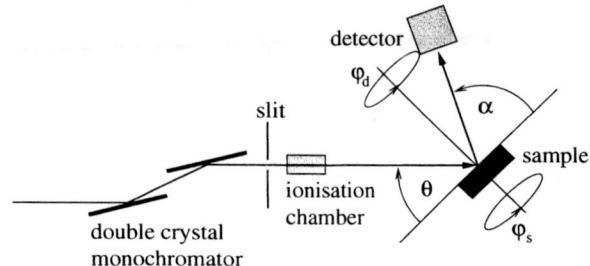

**FIGURE 3.** Experimental set-up (for details, see text).

together with the sample in the elevation angle $\theta$, leaving $\alpha$ constant but only the sample was rotated in the azimuthal angle $\varphi_s$. This means that the detectors moved relative to the sample in $\varphi_s$. On the side of the incident radiation, MEXH scans were recorded at different $\theta$ values. Simultaneously, an XFH scan corresponding to the constant exit angle $\alpha$ was recorded on the side of the fluorescence radiation. The holograms obtained in this geometry are a product of a MEXH hologram and one XFH scan. This kind of hologram can be called mixed MEXH [28–30].

A pure XFH method requires the sample to be fixed with respect to the incident beam while the detectors are moved around the sample so that only the interference pattern of the fluorescence radiation is measured. If this requirement is not fulfilled then a mixed XFH hologram is obtained.

## B  Data Analysis

The unscattered reference wave $|R(\boldsymbol{k})|^2$ is removed from the measured signal $I(\boldsymbol{k})$ according to eq. (2) in order to obtain the holographic signal. A fit to the experimental data that takes only a constant background and angular dependent absorption in the crystal into account was used to determine $|R(\boldsymbol{k})|^2$ [14,28].

Since the measurements were performed in a mixed MEXH method, each single $\varphi_s$ scan of the MEXH hologram is multiplied by the same XFH scan. This XFH scan is obtained due to the relative movement of detector and sample. Its shape depends on the structure of the sample and the exit geometry. The effect of the XFH scan can be seen in the experimental hologram as a modulation of the intensity with $\varphi_s$ which is the same for all values of $\theta$ in the whole mixed MEXH hologram [21,28].

The resulting holograms obtained for the Cu $K_\alpha$ and Au L fluorescence lines at 24 keV are shown in Fig. 4. A line pattern can be seen in the holograms. These lines are the reciprocal equivalent of Kossel lines [31,9]. They are caused by Bragg diffraction on the side of the incoming beam and are normally used in the XSW method to obtain long-range structural information [32]. They can be seen in holograms of crystals that are measured with a resolution of $\leq 1°$. MEXH and XFH are reciprocal to each other in the same way as XSW and Kossel lines are [8]. In XFH holograms of crystals, Kossel lines can be seen just as XSW patterns can be seen in MEXH holograms of crystals. The holograms obtained after removal of the XFH scan are MEXH holograms. At the angular step size chosen for the experiment and the present state of data evaluation, it was not necessary to remove the x-ray standing wave pattern from the holograms. The information about the structure symmetry obtained from XSW patterns can be used to extend experimental holograms to a full $4\pi$ sphere, making isotropic spatial resolution in the reconstructed images possible [33–35].

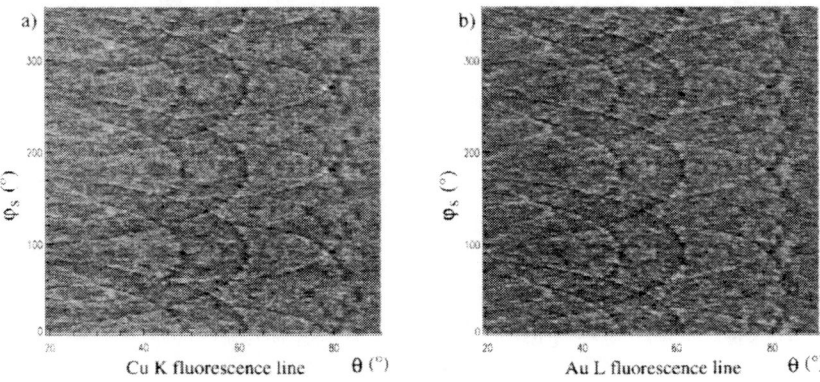

**FIGURE 4.** Reciprocal experimental holograms after subtracting background and XFH trace, measured at E=24 keV [21]. Comparing the holograms taken at the Cu $K_\alpha$ and Au L fluorescence lines shows that the XSW line pattern does not depend upon the fluorescence energy.

The reconstructions were performed without using the symmetry information available from Kossel lines or Fourier filtering the experimental holograms in any way. The experimental multiple energy reconstruction obtained from adding the single energy reconstructions according to the Barton algorithm is shown in Fig. 5 and compared to the multiple energy reconstruction obtained from the calculated holograms. The holograms at incident energies from 15 keV to 30 keV with a step size of 3 keV were used in both cases. All maxima that are strong in the image reconstructed from the calculated holograms are present in the experimental reconstruction as well. These are exclusively maxima that occur at Au atom positions. For the energies chosen, the peaks at Cu atom positions are very weak in the reconstruction of the experimental and the calculated holograms. The reconstruction of the experimental holograms was obviously successful and shows that the removal of the contributions from reference wave and XFH scan was appropriate.

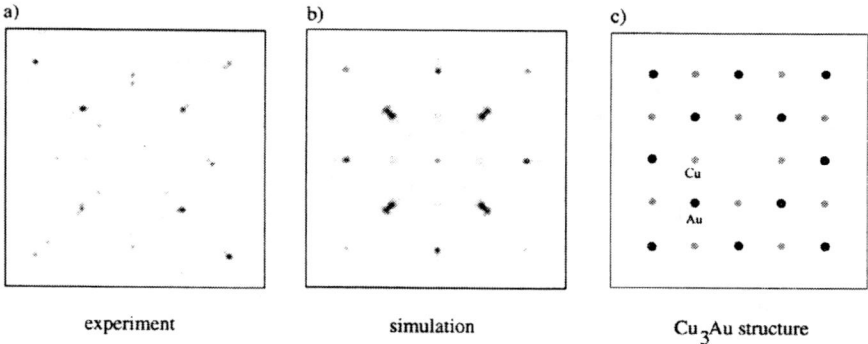

**FIGURE 5.** Reconstructed images obtained in the plane passing through the fluorescence emitter atom and parallel to the (001) lattice planes from a) the experimental holograms, b) the simulated holograms, image size = 15 × 15 Å$^2$, background removed. The reconstructed images were added over the energies 15, 18, ... 30 keV and taken at the Au L fluorescence line. c) Positions of neighbors of the central Au atom in the imaging plane [21].

## IV  EXPERIMENTAL STATUS

When the atoms within a sample are excited to fluorescence by incident monochromatic x-rays, two standing wave fields are formed: one at the energy of the incident wave and another at the energy of the fluorescence wave. For MEXH, being based on the interference at the energy of the incident radiation, the interference field contribution of the fluorescence waves should be kept constant in the detected signal. An experimental solution for pure MEXH would be to keep the detector system stationary in space with respect to the sample [36]. XFH requires, consequently, that the interference pattern due to the incident wave is held constant, for example by keeping the sample stationary with respect to the incident beam. In this case, the interference pattern has to be recorded by a detector which moves around the sample in the far field.

In all previous x-ray holography measurements, these requirements for the experimental setup were not strictly fulfilled [7,8,14,28–30,37–40], because pure experiments are much more complicated to implement. This resulted in so called mixed holograms which contain information from XFH and from MEXH [28,21,29]. In mixed techniques, the danger of introducing systematic artifacts during background subtraction in the holograms and successively in their reconstructions is always present. Subtraction of the XFH scan is especially difficult when the holography method is applied to crystals since Kossel or XSW lines with fast intensity fluctuations are present in every $\varphi$ scan. It is thus highly desirable to carry out pure MEXH or pure XFH measurements.

# A Pure MEXH Experiment with High Angular Resolution

An example for a pure MEXH implementation is given in [36]. The sample and the experimental scheme were the same as in section III A. A detector was positioned right on the $\varphi_s$ axis at $\alpha = 90°$. The unwanted XFH scan at $\alpha = 90°$ is reduced to a constant so that this detector recorded a pure MEXH hologram. This is a new experimental technique for recording a pure MEXH hologram. Another method to get rid of the influence of the mixed method will be integrating over the full XFH scan. This can be achieved using highly oriented pyrolytic graphite (HOPG) cylindrical monochromators [41,42], compare also [8,29].

The first pure reciprocal measurement was performed in [43,44] with $\gamma$-ray Mössbauer holography, where the detector moved together with the sample in two angular coordinates.

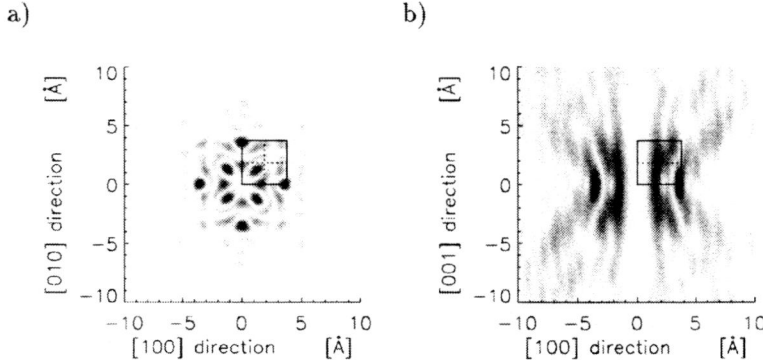

**FIGURE 6.** Holographic reconstruction of $Cu_3Au$ crystal with fcc structure at incident x-ray energy 10.5 keV [36]: a) in the plane parallel to the (001) lattice plane and cutting through a fluorescing Cu atom; b) in the plane parallel to the (010) lattice plane and cutting through a fluorescing Cu atom. The reconstructed atomic intensity is shown in a linear gray scale (darker for higher intensity) without background cutoff. The coordinate origin is the position of a fluorescing Cu atom. The square with solid lines has a side length of the actual lattice constant 3.75 Å, and the dashed lines are crossed at the center of the square.

Fig 6 shows reconstructed images of $Cu_3Au$ in the planes parallel to (001) and to (010) lattice planes. For $Cu_3Au$, the $\varphi$ rotation axis was chosen parallel to the [001] fourfold rotational symmetry axis. It is apparent that the resolution of reconstructed images is better in the plane parallel to (001). In addition, the obtained atomic image in the plane parallel to (001) possesses a fourfold rotational symmetry in a good approximation, although the data were not symmetrized. The dependence of the resolution on reconstruction planes simply stems from the data region of the measurement [34,35]. The projection of the hologram on the (010) lattice plane covers a small area and is extremely asymmetric (Fig. 7).

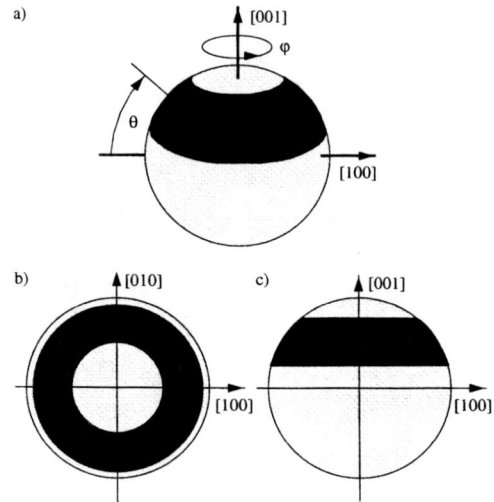

**FIGURE 7.** a) Schematic illustration of the angular range in an x-ray holography experiment. In this case, the azimuthal $\varphi$ rotation axis and the crystal [001] axis coincide. b) The projection of the hologram on the (001) lattice planes keeps the symmetry around the [001] axis and covers wide area. c) The projection of the hologram on the (010) lattice planes is not symmetric with respect to the [010] axis.

In order to obtain isotropic spatial resolution in the reconstructed images, we utilize the fourfold rotational symmetry and the mirror symmetry of the sample. In principle, this symmetry information can be obtained from the XSW peak pattern. After averaging $\psi(\mathbf{r})$ over symmetrically equivalent points in real space, we obtain the three dimensional image of fig. 8.

In the reciprocal method, the angular resolution depends on the divergence of the incoming beam and can be readily reduced by monochromators to few mrads so that a pattern at very high angular resolution can be recorded. In the direct method, the angular resolution depends on the size of the beam footprint on the sample surface and the detector acceptance angle. A very high resolution on the exit side will decrease the count rate such that it becomes really tedious to record a hologram with present radiation sources and detectors.

When a hologram of a crystal is recorded at an angular resolution of $\leq 1°$ Kossel lines or XSW patterns become visible (see Fig. 4). We performed a fast MEXH measurement with high angular resolution in $\varphi_s$ [36] in order to reveal additionally the corresponding line shapes in intensity vs. angle plots. Fig. 9 shows a detailed profile of a reciprocal Kossel line measured in this way. The sample was rotated

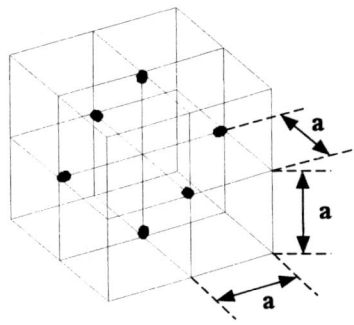

**FIGURE 8.** Three-dimensional holographic reconstruction of $Cu_3Au$ obtained after symmetrization (fourfold rotational symmetry and the mirror symmetry) [36]. Black dots are plotted at the positions where the reconstructed atomic image intensity $|\psi(r)|^2$ is larger than 50% of the maximum intensity. The cubes have a side length of the actual lattice constant $a = 3.75$ Å. They are drawn in order to indicate the actual positions of atoms. The center of the figure is the position of a fluorescing Cu atom. The images of the next-nearest neighbors of a fluorescing Cu atom peak at 3.4 Å from the center. The nearest neighbors are weaker in intensity and not visible.

continuously in $\varphi_s$ while the fluorescence counts were added in a histogramming memory at angle-proportional addresses. This way, the data were integrated over a small angular range instead of sampling at discrete points only, as is the case for point to point scans. If the integration is performed for an angular range that is comparable to the divergence of the incident beam no angular resolution is lost in the integration. If the integration over a larger angular range is desired it can be easily implemented.

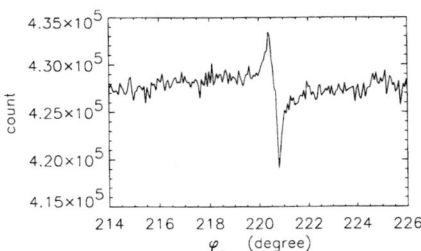

**FIGURE 9.** A cut through a XSW line [36]. The scan was measured on a $Cu_3Au$ single crystal at an incident energy of 10.5 keV and at $\theta = 53°$.

An additional advantage of the fast rotation technique is that the effect of source instabilities on the hologram can be reduced by adding the fluorescence counts in each of the angular intervals in $\varphi$ over many full revolutions. Single scans that show strong instabilities can be discarded without much loss of statistics in the resulting hologram.

## B  Pure XFH Experiment

The first pure direct x-ray fluorescence holography measurement was performed on a bcc Fe single crystal [45]. The sample stayed fixed with respect to the incident beam while the detectors were moved around it (angles $\varphi_d$ and $\alpha$ in Fig. 3) at a constant distance. The Fe $K_\alpha$ and $K_\beta$ fluorescence intensities were recorded by two energy dispersive silicon drift detectors. The symmetry information from the Kossel line pattern was used to extend the holograms to $4\pi$. The three dimensional reconstruction obtained is reproduced in Fig. 10 and shows a remarkable image of the unit cell together with the neighbors of the fluorescing atom which are the center atoms of the neighboring cells. This gives a glimpse into what atomic holography is able to provide in the future.

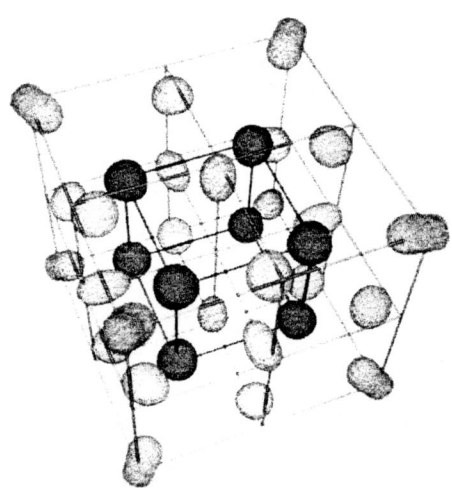

**FIGURE 10.** Three-dimensional reconstruction of the extended hologram in a volume of $(7 \text{ Å})^3$ [45]. The nearest neighbors of the emitter atom are displayed in a darker grey than its more distant neighbors. For this plot, the experimental reconstruction was multiplied by the distance from the origin to make atoms at larger distances visible and is displayed with 50% background cutoff. Atom positions in the Fe lattice are indicated by the cross points of coordinate lines.

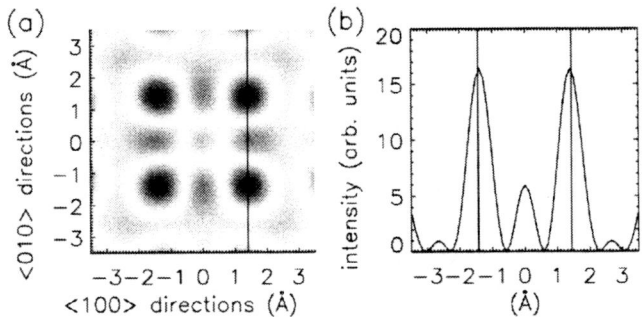

**FIGURE 11.** (a) Reconstructed image obtained from the fully extended XFH holograms. The image plane is parallel to the {100} lattice planes and taken at a distance of 1.4 Å from the fluorescence emitter atom. The image is displayed without background cutoff in a linear grey-scale where dark color corresponds to high peak intensity. (b) One-dimensional cut parallel to the < 100 > axes as indicated in (a). Vertical lines indicate the atom positions in the Fe structure [45].

Fig. 11 (a) shows a plane parallel to the {100} lattice planes taken at a distance of 1.4 Å ≈ $a/2$ from the plane containing the emitter atom. It clearly displays the expected four nearest neighbors. For a quantitative analysis, one-dimensional cuts through the three-dimensional reconstruction were evaluated. An example is given in fig. 11 (b). The one-dimensional cut passes through fig. 11 (a) as indicated.

In order to further check the accuracy of the measurement, holograms of a Fe bcc cluster of radius $2a$ were also calculated [14,21] for the Fe $K_\alpha$ and $K_\beta$ line energies. Their three-dimensional reconstruction was compared to that of the extended experimental holograms. The atom positions obtained from calculated and measured reconstructions agree with each other within the errors of ±0.2 Å. The obtained values for the FWHM vary between $(0.7 \pm 0.2)$ Å and $(1.2 \pm 0.2)$ Å for different maxima in both reconstructions. These values are close to the theoretical estimate.

As an example, the nearest neighbors of the emitter have a distance of $(2.4 \pm 0.2)$ Å from the emitter in the reconstruction of the experimental holograms and a distance $(2.2 \pm 0.2)$ Å in the reconstruction of the calculated holograms. Both values are slightly smaller than the tabulated value $a/\sqrt{3} = 2.49$ Å. As another example, the third-nearest neighbors of the emitter have a distance of $(3.4 \pm 0.2)$ Å from the emitter in the reconstruction of the experimental holograms and a distance $(3.2 \pm 0.2)$ Å in the reconstruction of the calculated holograms. Both values agree within the errors with each other, but they are distinctly smaller than the tabulated $\sqrt{2}a = 4.06$ Å. The calculation thus shows that the shift is mostly due to twin images. Note, however, that this also proves that model calculations based on the results obtained from few energies can be used for a detailed refined structural analysis. In addition, a larger number of counts reduces the experimental errors and makes a more detailed analysis of the peak shapes reasonable. All this will in the future increase the precision of the measurements considerably.

## C  Future Developments

A rigorous theory of atom resolving x-ray holography is still under development. Special reconstruction algorithms need to be found to remove or use the influence of many-particle effects in crystalline objects and to obtain additional information on the long range order from the Kossel or XSW lines.

From the experimental point of view, faster energy resolving detectors will help to improve the experimental statistics. This is a crucial problem for the holographic method, as the signal to background ratio stays at a level of $10^{-4} - 10^{-5}$ due to the low scattering cross section of x-rays. Current developments of multi-element silicon drift detectors [46] can improve the statistics and reduce the data acquisition time at least by a factor of 100. Another promising way is the application of wavelength dispersive optics based on curved analyzer crystals.

Although present-day third-generation synchrotron radiation sources already provide for crystalline samples, scattered fluorescence intensities which can hardly be handled with state-of-the-art detectors, problems of reduced dimension, e.g. the study of surfaces, interface strain and impurities or defects, will require higher source brilliances, which can in the future be supplied by x-ray free-electron-lasers using the excitation principle of self-amplified spontaneous emission [47–50]. These sources can emit in one 100 fs long x-ray flash a similar number of photons as are presently provided by the best x-ray undulator sources in one second. The source diameter is only 20 $\mu$m and the photon beam divergence 1 $\mu$rad. It will therefore become possible to generate a full XFH hologram with one pulse only. Of course, the registration of such interference patterns needs new detector systems which are not yet available today.

Since standard monochromatic parallel beam diffraction methods cannot be applied with such an extreme time resolution - the sample needs to be rotated while the beam is on - the principle of XFH can overcome this complication.

## V  CONCLUSION

X-ray holography with atomic resolution is a new method at an early stage of development. It has the potential to become an effective tool for direct investigations of atom arrangements in clusters and solid matter. The holographic approach does not depend on long-range translational order, which provides a unique opportunity to study new classes of objects. It is highly tolerant to misorientation and sensitive to the near environment of definite sorts of atoms independent of their concentration. This makes the method applicable to atomic structures which can hardly be studied by traditional methods, for example active sites in biological materials of low crystalline quality or crystals with very low dopant concentrations.

With the arrival of new 4th generation x-ray sources the holographic approach to structure determination can also provide down to fs time resolution.

## REFERENCES

1. A. Szöke, in *Short Wavelength Coherent Radiation: Generation and Applications*, edited by D. T. Attwood and J. Boker, AIP Conference Proceedings No. 147 , page 361 (American Institute of Physics, New York, 1986).
2. D. Gabor, Nature **161**, 777 (1948).
3. J. J. Barton, Phys. Rev. Lett. **61**, 1356 (1988).
4. D. K. Saldin, P. L. de Andres, Phys. Rev. Lett. **64**, 1270 (1990).
5. G. R. Harp, D. K. Saldin, and B. P. Tonner, Phys. Rev. Lett. **65**, 1012 (1990).
6. G. R. Harp, D. K. Saldin, and B. P. Tonner, Phys. Rev. B **42**, 9199 (1990).
7. M. Tegze, G. Faigel, Nature **380**, 49 (1996).
8. T. Gog, P. M. Len, G. Materlik, D. Bahr, C. S. Fadley, and C. Sanchez-Hanke, Phys. Rev. Lett. **76**, 3132 (1996).
9. T. Gog, D. Bahr, and G. Materlik, Phys. Rev. B **51**, 6761 (1995).
10. J. J. Barton, Phys. Rev. Lett. **67**, 3106 (1991).
11. D. K. Saldin, Surface Review and Letters **4**, 441 (1997).
12. P. M. Len, T. Gog, C. S. Fadley, G. Materlik, Phys. Rev. B Rapid Comm. **55**, 3323 (1997).
13. P. M. Len, F. Zhang, S. Thevuthasan, A. P. Kaduwela, C. S. Fadley, M. A. Van Hove, J. Electron Spectrosc. Relat. Phenom. **85**, 145 (1997).
14. B. Adams, D. V. Novikov, T. Hiort, and G. Materlik, E. Kossel, Phys. Rev. B **57**, 7526 (1998).
15. P. M. Len, S. Thevuthasan, and C. S. Fadley, A. P. Kaduwela and M. A. Van Hove, Phys. Rev. B **50**, 275 (1994).
16. P. M. Len, C. S. Fadley, G. Materlik, in *X-Ray and Inner Shell Processes*, edited by R. L. Johnson, H. Schmidt-Böcking, B. F. Sonntag, AIP Conference Proceedings No. 389, page 295 (Woodbury, New York, 1997).
17. P. M. Len, T. Gog, D. Novikov, R. A. Eisenhower, G. Materlik, and C. S. Fadley, Phys. Rev. B **56**, 1529 (1997) .
18. G. A. Miller and L. B. Sorensen, Phys. Rev. B **56**, 2399 (1997).
19. L. Fonda, Phys. Stat. Sol. (b) **201**, 329 (1997).
20. Y. Nishino and G. Materlik, Phys. Rev. B, in print.
21. B. Adams, T. Hiort, E. Kossel, G. Materlik, Y. Nishino, and D. V. Novikov , Phys. stat. sol. (b) **215**, 757 (1999).
22. E. Gatti and P. Rehak, Nucl. Instrum. Meth. **225**, 608, (1984).
23. P. Rehak, E. Gatti, A. Longoni, J. Kemmer, P. Holl, R. Klanner, G. Lutz and A. Wylie, Nucl. Instrum. Meth. A **235**, 224, (1985).
24. L. Strüder and H. Soltau, Radiation Protection Dosimetry **61**, 39 (1995).
25. P. Lechner, L. Strüder, Nucl. Instrum. Meth. A **354**, 464 (1995).
26. P. Lechner et al, Nucl. Instrum. Meth. A **377**, 346 (1996).

27. A. Krolzig, G. Materlik, M. Swars and J. Zegenhagen, Nucl. Instrum. Meth. A **219**, 430 (1984).
28. D. V. Novikov, B. Adams, T. Hiort, E. Kossel, G. Materlik, R. Menk and A. Walenta, J. Synchrotron Rad. **5**, 315 (1998).
29. G. Faigel and M. Tegze, Rep. Prog. Phys. **62**, 355 (1999).
30. M. Tegze, G. Faigel, S. Marchesini, M. Belakhovsky, A. I. Chumakov, Phys. Rev. Lett. **82**, 4847 (1999).
31. W. Kossel, H. Loeck, H. Voges, Z. Physik **94**, 139 (1935).
32. M. J. Bedzyk and G. Materlik, Phys. Rev. B **32**, 6456 (1985).
33. D. V. Novikov, B. Adams, T. Hiort, E. Kossel, and G. Materlik, HASYLAB Annual Report I, 899 (1997).
34. B. P. Tonner, Zhi-Lan Han, G. R. Harp, and D. K. Saldin, Phys. Rev. B **43**, 14423 (1991).
35. D. K. Saldin, G. R. Harp, B. L. Chen, & B. P. Tonner Phys. Rev. B **44**, 2480 (1991).
36. B. Adams, Y. Nishino, and G. Materlik, submitted to J. Synchrotron Rad.
37. T. Gog, R. H. Menk, F. Arfelli, P. M. Len, C. S. Fadley, G. Materlik, Synchrotron Radiat. News **9**, 30 (1996).
38. T. Gog, R. A. Eisenhower, R. H. Menk, M. Tegze, G. LeDuc, J. Electron Spectrosc. Rel. Phenom. **92**, 123 (1998).
39. J. Kawai, K. Hayashi, T. Yamamoto, S. Hayakawa, and Y. Gohshi, Analytical Sciences **14**, 903 (1998).
40. K. Hayashi, T. Yamamoto, J. Kawai, M. Suzuki, S. Goto, S. Hayakawa, K. Sakurai, and Y. Gohshi, Analytical Sciences **14**, 987 (1998).
41. I. G. Grigoryeva, A. A. Antonov, V. B. Baryshev, Synchrotron Radiation News **3**, 15 (1990).
42. B. Beckhoff, B. Kanngießer, J. Scheer, W. Swoboda, Advances in X-Ray Analysis, **37**, 523 (1994).
43. P. Korecki, J. Korecki, and T. Slezak, Phys. Rev. Lett. **79**, 3518 (1997).
44. P. Korecki, J. Korecki, and W. Karas, Phys. Rev. B **59**, 6139 (1999).
45. T. Hiort, D. V. Novikov, E. Kossel, and G. Materlik, Phys. Rev. B, accepted for publication.
46. Ch. Gauthier et al, Nucl. Instrum. Meth. A **382**, 524 (1996).
47. A. M. Kondratenko and E. L. Saldin, Part. Accel. **10**, 207 (1980).
48. R. Bonifacio, C. Pellegrini and I. M. Narducci, Opt. Commun. **50**, 373 (1984).
49. J. Arthur, G. Materlik, R. Tatchyn, and H. Winick, Rev. Sci. Instrum. **66**, 1987 (1995).
50. R. Brinkmann, G. Materlik, J. Rossbach, J. R. Schneider, B.-H. Wiik, Nucl. Instrum. Meth. A **393**, 86 (1997).

# Live X-Ray Refraction Imaging Using Vertically and Horizontally Wide X-Rays

Junji Matsui*, Yasushi Kagoshima*, Yoshiyuki Tsusaka*,
Kazushi Yokoyama*[1], Kengo Takai*, Shingo Takeda*
and Katsuhito Yamasaki†

*Faculty of Science, Himeji Institutu of Technology
2-1, Kouto 3-chome, Kamigori-cho, Ako-gun, Hyogo, 678-1297 Japan
†Department of Radiography, Kobe University Hospital,
5-2, Kusunoki-cho 7-chome, Chuo-ku, Kobe City, Hyogo, 650-0017 Japan

**Abstract.** Since the synchrotron X-rays are intense enough even after the sequential expansion of the X-ray beam by successive asymmetric Bragg reflections, live refraction images of internal structure for biological materials can be seen on an X-ray image sensor. Video images of some living insects or a frog or a mouse show clearly their internal structures of the body, for instance, cellular structures in a lung.

## INTRODUCTION

The internal structure of relatively thick materials can be imaged by using hard X-rays because of low absorption coefficient for many materials. Since Röntgen's discovery of X-rays more than 100 years ago [1], X-rays have provided the excellent imaging means for inspection of the human bodies. The contrast obtained has been interpreted on the basis of absorption contrast.

The absorption contrast is related to the imaginary part $\beta$ of the X-ray refractive index $n = 1 - \delta - i\beta$. The image contrast, however, is sometimes poor for the case that only small difference in the absorption coefficients is concerned in the material. Especially, low absorption objects such as carbon-based light element compounds do not appear as clear image contrast on a film. In addition, a finite size of the X-ray light source also leads to poor image contrast due to the blurring. The closer the distance between the sample and an X-ray detector, e.g. an X-ray film, is, the sharper the spatial resolution becomes.

Recently, phase contrast imaging techniques using hard X-rays have been of much interest world-wide, since the phase of the X-ray waves transmitting the materials

---

[1] University Placement from the New Indutory Reserch Organization (NIRO).

is very sensitive to the density difference or the thickness variation even if the total absorption by the object is weak. This is due to the phase change of X-rays during penetration into the object, involving the real part $\delta$ of the refractive index $n$. The phase contrast images thus obtained are sometimes simply called "refraction images" regardless of a critical definition of the refraction (gradient) contrast. Investigation of the internal structure by the refraction imaging has been first carried out using a laboratory X-ray generator of mostly small focus size [2-5].

In order for the contrast to be clear at a boundary of density variation in the object, incoming X-ray waves should be as parallel as possible to minimize shadowing by the object. Using synchrotron radiation from a bending magnet, brighter hard X-rays make the recording time short. Also phase gradients are highly resolvable by putting the detector at a large distance from the sample because of high parallelism of the X-ray beam [6,7].

After the quality of semiconductor silicon crystals has been much improved, tomographic method combined with a Bonse-Hart type interferometer [8] to reconstruct the three-dimensional distribution of $\delta$ has been successful [9,10]. Nearly perfect silicon crystals (float-zone silicon with only small quantities of oxygen) can be used as components of the X-ray optics such as monochromators or collimators and can provide highly coherent and parallel incident X-rays for the object.

In this article, an advanced optical system composed of sequential double-crystal expanders which employ successive asymmetric reflections from silicon crystals is proposed to obtain wide-area refraction images with a high spatial resolution [11], in combination with brilliant hard X-rays from a new undulator installed at SPring-8. Using an X-ray video camera, live images of some biological objects are presented at the conference.

# EXPERIMENTAL

## X-ray Light Source

The present experiments have been carried out at the Hyogo Beamline (BL24XU) at SPring-8, which is the first contract beamline installed in 1996 [12]. The beamline has employed an *in-vacuum* type 'figure-8' undulator [13] for the hard X-ray region, which can provide both vertically and horizontally polarized X-rays as integer and half-odd integer harmonics, respectively. The undulator has been tuned to a horizontally polarized X-ray beam of 15 keV photon energy. The effective horizontal and vertical beam sizes, $\Sigma_x$ and $\Sigma_y$, and the effective horizontal and vertical angular divergences of the light source, $\Sigma_{x'}$ and $\Sigma_{y'}$, (all those values are given as standard deviations for Gaussian distribution) are 76.7 $\mu$m, 24.3 $\mu$m, 76.7 $\mu$rad and 6.5 $\mu$rad, respectively. FWHM's (2.335$\Sigma$) are thus 179 $\mu$m, 57 $\mu$m, 179 $\mu$rad, and 15.2 $\mu$rad, respectively.

A four-quadrant slit with an aperture size of $1mm \times 1mm$ has been installed in the front end at 31 m away from the light source. Therefore, taking the beam sizes and the angular divergences of the light source into account, only the horizontal divergence of a single axis system is narrowed by the slit, down to about 38 $\mu$rad.

## X-ray Optics

A silicon double-crystal monochromator of horizontal dispersion with 111 symmetric reflection has been placed at a distance of 61 m from the light source. In the experimental hutch 'C' of BL24XU, two double-crystal beam expanders have been installed consecutively at around 66 m, as schematically shown in Fig. 1. The expanders are expected not only to expand the X-ray beam but also to achieve much higher parallelism both in the vertical and horizontal directions [14].

According to the two-wave X-ray dynamical theory [15], a full width $\omega_s$ of reflectivity profile for the symmetric reflection is proportional to the real part $\chi'_h$ of the relevant Fourier coefficient and is expressed as

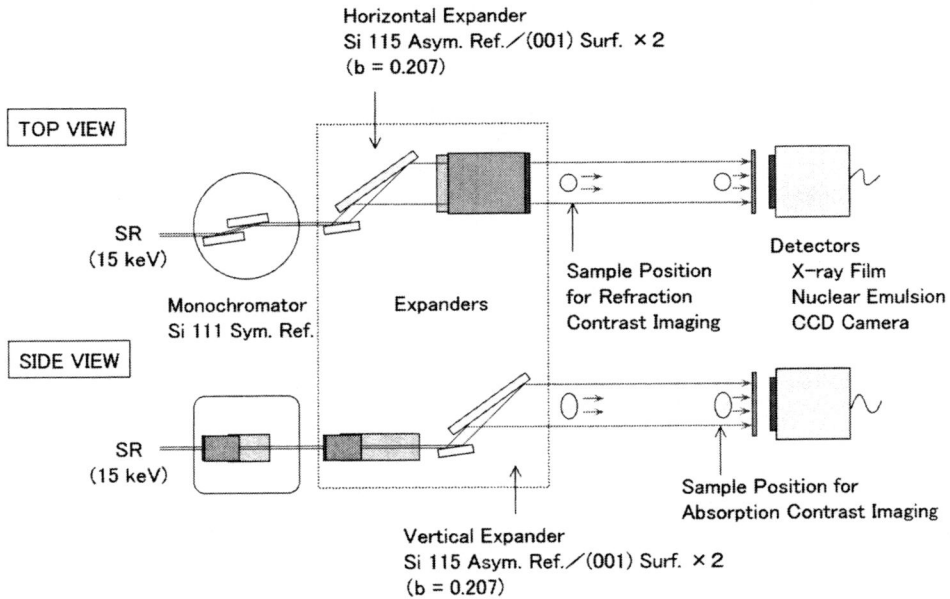

**FIGURE 1.** Schematic view of X-ray optics used for refraction images. Double-crystal expanders with successive (+, -) arrangement of asymmetric reflections are placed behind a double-crystal monochromator. Asymmetric factor $b$ is 0.207 for each reflection. Absorption contrast images are also taken by putting the object just in front of the image detector.

$$\omega_s = \frac{2|\chi'_h|P}{\sin 2\theta_B}, \qquad (1)$$

where $P$ is a polarization factor ($P = 1$ for $\sigma$ polarization and $P = \cos 2\theta_B$ for $\pi$ polarization) and $\theta_B$ is a Bragg angle of the reflection. $\chi'_h$ can be replaced by

$$\chi'_h = \frac{r_e \lambda^2}{V_c} F'_h, \qquad (2)$$

where $r_e = e^2/mc^2$ is the classical electron radius, $\lambda$ is a wavelength of X-rays, $V_c$ is a volume of unit cell and, $F'_h$ is the real part of the structure factor.

When the reflection is asymmetric, the full widths $\omega_0$ and $\omega_h$ of reflectivity profiles for the incident and diffracted X-rays are, repectively,

$$\omega_0 = \frac{\omega_s}{\sqrt{b}} \quad \text{and} \quad \omega_h = \sqrt{b}\,\omega_s, \qquad (3)$$

where $b$ is so called 'asymmetric factor' and is given by

$$b = \frac{\sin(\theta_B - \alpha)}{\sin(\theta_B + \alpha)}, \qquad (4)$$

the value varying between -1 for the symmetric Laue case and +1 for the symmetric Bragg case, being dependent upon the inclination angle $\alpha$ of the reflecting lattice plane from the crystal surface taking note of the incident X-rays.

For 115 asymmetric reflection from a (001) surface of the cubic crystal (hereafter, this will be denoted simply as 115/(001)), $b$ is calculated to be 0.207 for $\theta_B = 23.3°$ at 15 keV and $\alpha = 15.8°$. Since $\chi'_h$ is $-1.42 \times 10^{-6}$ for 115 reflection and then $\omega_{s,115} = 3.8$ $\mu$rad for $\sigma$ polarization, the full widths $\omega_{0,115}$ and $\omega_{h,115}$ for 115/(001) asymmetric reflection are $\omega_{0,115} = 8.4$ $\mu$rad and $\omega_{h,115} = 1.7$ $\mu$rad, respectively. For 111/(111) symmetric reflection at the monochromator, $\omega_{h,111}$ ($= \omega_{0,111} = \omega_{s,111}$) is calculated to be 18.0 $\mu$rad.

## X-ray Beam Size and Angular Divergence for the Sample

Figure 2(a) shows a duMond diagram [16] explaining intersection of the X-ray band diffracted from the 111/(111) monochromator with the angular width of $\omega_{h,111}$ and the X-ray band acceptable to the 115/(001) horizontal expander with the angular width of $\omega_{0,115}$. A gray zone surrounded by a parallelogram ABCD gives a convolution of the divergence of the incident beam and the dynamical width of the reflection at the first crystal of the horizontal expander. The total angular divergence $\Omega_{0,\parallel}$ is the angular width between the center of AB and the center of CD ($\simeq 25$ $\mu$rad). After the 115 asymmetric reflection, the parallelogram ABCD is transposed to the parallelogram A'B'C'D'. It should be noticed that the total horizontal angular divergence $\Omega_{h,\parallel}$ between the center of A'B' and the center of C'D' is almost the same as $\Omega_{0,\parallel}$, even if $\omega_{h,115}$ becomes much smaller than $\omega_{0,115}$.

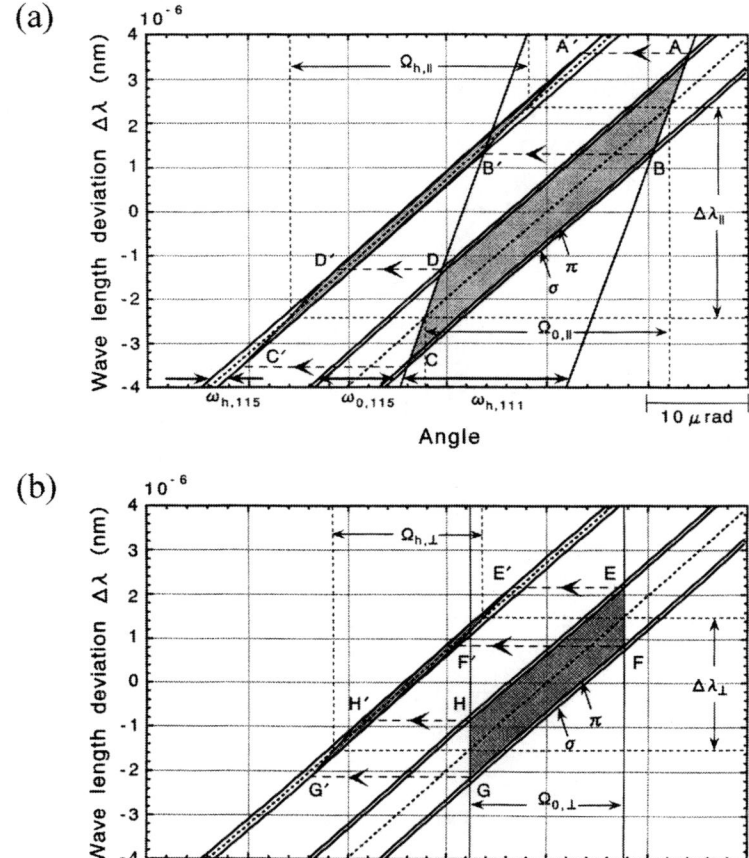

**FIGURE 2.** (a) duMond diagram for outgoing X-rays from the 111/(111) monochromator and acceptable incident X-rays by the first crystal of 115/(001) horizontal expander. Both angular widths of intersecting bands are given by $\omega_{h,111}$ ($= \omega_{s,111}$) and $\omega_{0,115}$ for $\sigma$ polarization. $\Delta\lambda_{\parallel}$ denotes a wavelength allowance corresponding to the total angular width $\Omega_{h,\parallel}$ ($\approx \Omega_{0,\perp}$) of 115 reflection. (b) duMond diagram for acceptable incident X-rays by the first crystal of 115/(001) vertical expander. $\Delta\lambda_{\perp}$ denotes a wavelength allowance corresponding to the total angular width $\Omega_{h,\perp}$ ($\approx \Omega_{0,\perp}$) which is limited by $\Sigma_{y'}$.

As already stated, an angular divergence of the X-rays incoming to the first crystal of the horizontal expander behind the four-quadrant slit is about 38 $\mu$rad in the horizontal direction. As $\Omega_{0,\parallel}$ is less than 38 $\mu$rad, the incident X-rays with a full angular width $\Omega_{0,\parallel}$ can be diffracted by the first crystal of the horizontal expander. By repeating the same reflection from the second crystal, the angular width of X-rays of a single wavelength becomes $b \cdot \omega_{h,115} = b\sqrt{b}\,\omega_{s,115} = 0.36$ $\mu$rad.

We next consider the case of the vertical expander placed behind the horizontal one. Figure 2(b) shows a duMond diagram which gives an acceptable region for the incident X-rays by the first crystal of 115/(001) vertical expander. The X-rays of $\sigma$ polarization for the horizontal expander turn to be those of $\pi$ polarization for the vertical expander. A gray zone surrounded by a parallelogram EFGH in Fig. 2(b) is limited by the angular divergence of the light source, e.g. $\Omega_{0,\perp} = 2.335\Sigma_{y'} = 15.2$ $\mu$rad.

Because the X-rays from the horizontal expander of a wavelength within $\Delta\lambda_\parallel$, but out of $\Delta\lambda_\perp$, cannot reflect at the first crystal of the vertical expander and also the dynamical width of X-rays after the 115 successive asymmetric reflections is very small, the total angular divergence in both the horizontal and the vertical directions is almost equal to $\Omega_{h,\perp}$ ($\approx \Omega_{0,\perp}$). The similar situation also holds true for the X-rays of $\pi$ polarization coming to the horizontal expander. Thus, the X-ray beam size is totally determined to be $1.0 \times 1.0$ mm$^2$ at a distance of 66 m from the light source.

As far as X-rays of a single wavelength are concerned, the X-ray beam size is expanded by a factor of $1/b$ after 115/(001) asymmetric reflection. Therefore, the double asymmetric reflections make the beam size larger by a factor of $1/b^2 = 23.3$. Since the angular width of X-rays of each wavelength is as narrow as 0.36 $\mu$rad, the total beam size behind the two expanders is expected to be about $23.3 \times 23.3$ mm$^2$. The local angular divergence of X-rays for the object, however, is 0.36 $\mu$rad independent of the wavelength, as given by the dynamical width of each expander crystal.

## Imaging Procedure

After aligning whole axes of the goniometers, the crystals were carefully held on the goniometers. Rotational accuracy of each goniometer on the expanders is 0.01 arcsec ($= 5 \times 10^{-2}$ $\mu$rad). All operations were controled by step motors driven by a programmed computer. The object like an ant or a moth or a frog was placed behind the second expander. When a nude mouse was treated, it was bound by strings to a metal frame under an anesthetic.

Refraction images were taken by an X-ray camera or an X-ray film and, if necessary, a nuclear emulsion plate positioned on the beam, at about $4 \sim 4.5$ m apart from the vertical expander. When taking an absorption image, the image sensor was placed closely behind the object.

## RESULTS AND DISCUSSION

To adjust the Bragg reflection of glancing angle incidence, each goniometer stage on the expanders was carefully rotated by monitoring the intensity measured in the ion chamber or by watching a diffracted pattern on a TV display. Thus, four crystals put on the both expanders were subsequently adjusted in an asymmetric reflection arrangement.

## Image on Nuclear Emulsion Plate

In order to secure a spatial resolution of the images taken by the present X-ray optics system, some insects were tried to take refraction images on a nuclear emulsion plate of Ilford L4 type. Figure 3 reproduces an X-ray refraction image of an ant taken in 30 seconds. As seen in an enlarged picture, a branching tip of the legs is clearly observed. The refraction gives rise to a clearer black-white contrast at any parts in the body bounded by air or gas, or probably, at boundaries where a more abrupt change of density occurs.

Nonuniformity of the background in Fig. 3 has been found to be not due to the remaining damages or morphological fluctuation at the silicon surface or the poor bulk crystallinity, but due to other imperfection of optical element in the front end of the beamline. Further details are now under investigation. Exposed area of $6 \times 5$ mm$^2$, which is smaller than the expected one, may be caused by a slight curvature of the silicon crystals used for the expanders and also by incomplete optical alignment.

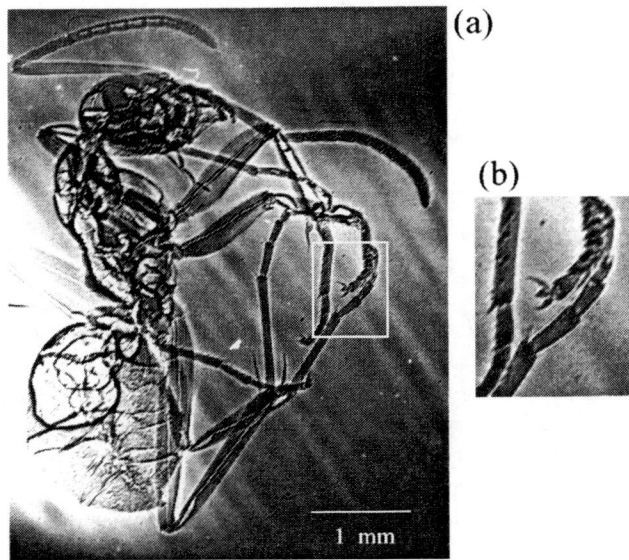

**FIGURE 3.** Refraction image of an ant and enlarged picture of its legs taken on a nuclear emulsion plate, Ilford L4 type.

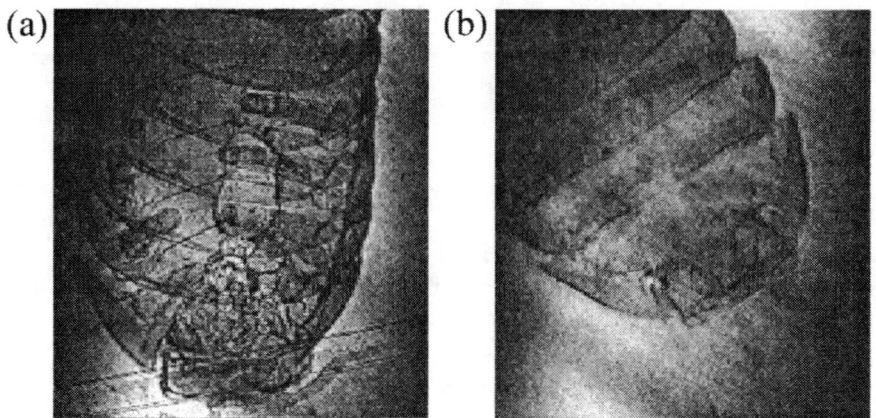

**FIGURE 4.** Live X-ray images of a pillbug ('*armadillidium vulgare*') captured from a videotape. (a) is a refraction contrast image and (b) is an absorption contrast image of its body.

## Live Images Using X-ray TV Camera

So far, the merit of using hard X-rays has been to image the internal structure of living biological objects on an X-ray detector. Refraction imaging with a high spatial resolution for the living objects will be possible, as far as the incident X-ray beam keeps high intensity and parallelism. The use of an X-ray camera combined with synchrotron radiation enables one to take an image of the internal structure in real time and to record their movement on a videotape.

Figure 4(a) shows one of the video frames reproduced from a videotape recording the refraction images of a pillbug ('*armadillidium vulgare*'). Figure 4(b) shows an absorption contrast image of the same object for comparison. It is obvious that, in Fig. 4(a), the boundary structures in its body are clearly observed with higher black-white contrast than those in Fig. 4(b). The fact means that the materials composed of light elements are appropriate samples for the refraction contrast imaging. The black-white contrast is again much emphasized at the regions bounded by air or gas.

Figure 5 shows successive images captured from a videotape used for a kind of moth, appearing its behavior to gulp an air bubble down into its body. A contrast of thin wings with some network patterns has also been observed on a TV display. It is said that the air is essentially important for insects, especially when they spawn or cast off the skin besides breathing. But, no effective tools to visualize their behavior to catch the air has been reported.

Some animals having lungs were next investigated to observe detail of the internal structures. Figure 6 shows a refraction image taken in real time of a living frog of 2 cm body length. Nostrils and an eyeball are recognized in Fig. 6(a), and some

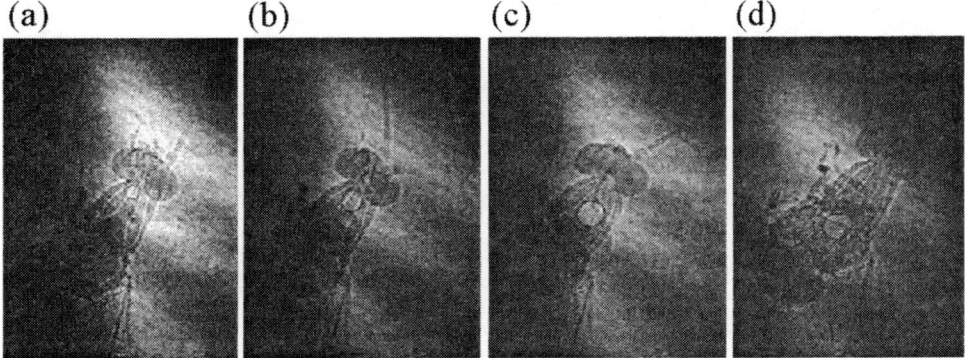

**FIGURE 5.** Live X-ray refraction contrast images of a kind of moth captured from a videotape. The pitures show how the moth gulps an air bubble down into its body in Figs. (a) to (d).

cellular structure inside the lung are imaged in good contrast in Fig. 6(b). It is surprising that even the lung cells behind the spine also make contrasts. This means that the refraction does not matter the existence of absorbing media or organs which affect the absorption contrast. As it has been well known, the refraction contrast is formed when X-rays impinge the boundary at which density variation exists and

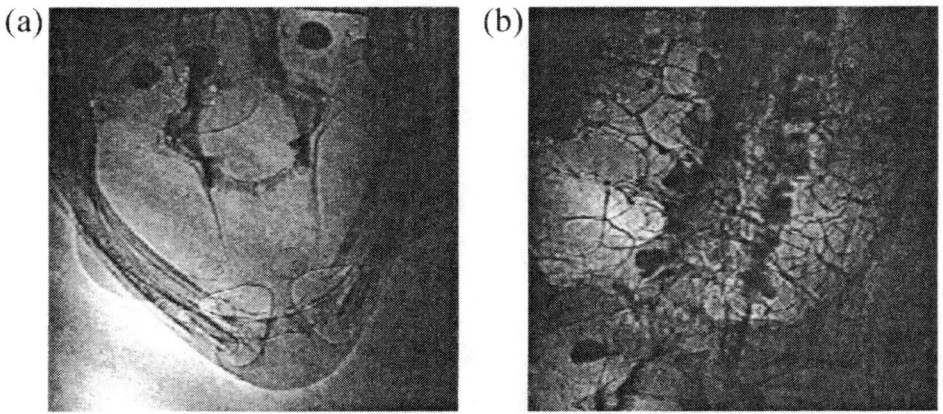

**FIGURE 6.** Live X-ray refraction contrast images of a frog, captured from a videotape. (a) A part of head showing clear images of nostrils in particular, and (b) a part of chest showing images of cellular structures inside the lung. Note that cell walls even behind the spine are still observable.

phase of the X-ray wave changes there. The density variation is mostly abrupt at the boundary between the organ and the air, enhancing the refraction contrast there more than other parts of a smaller density variation.

In terms of medical applicability of the refraction contrast imaging, it will be an important goal to investigate a mammal, aiming to apply the present method to human bodies finally. Thomlinson et al. reported refraction images of a mouse with an implanted tumor by using an analyser crystal tuned to various positions on the rocking curve [17].

We have also tried to apply this method (but the energy is 25 keV and 117/(001) expanders are used) to a nude mouse having several tumor regions originated intentionally by the intravenous injection of a certain kind of the cancer cells [18]. In order to obtain a wider area of exposure than the previous cases, the sample and the film were scanned through the X-ray beam [17] instead of using the vertical expander. Figure 7 shows an enlarged image of the chest in a nude mouse. Boundaries of the lungs (indicated by L) with inside cellular structures and a branching bronchus (indicated by B) are clearly observed.

As far as the spatial resolution is concerned, the blurring by the light source, in this case, can be negligible since the angular divergence of illuminating X-ray beam is very small. The Fresnel diffraction at an edge of the object, however, is dominant and causes the blurring on the detector. When the object edge stands on the beam

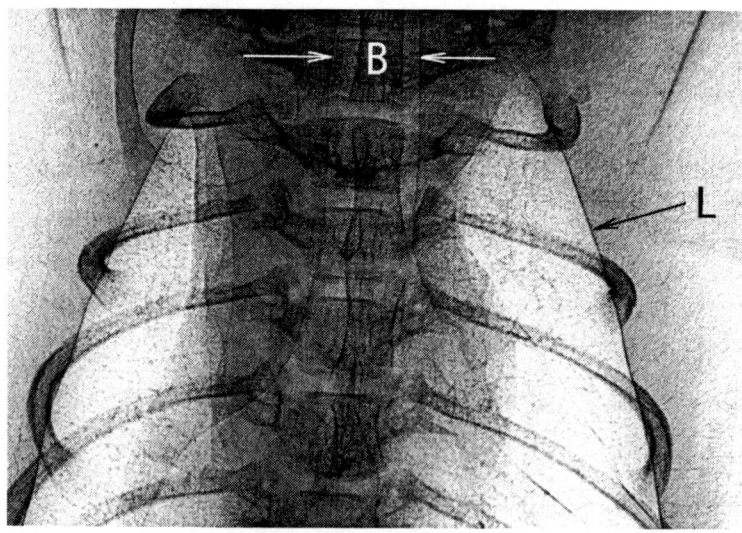

**FIGURE 7.** Live X-ray refraction contrast image, captured from a videotape, of chest in a nude mouse, showing a contrast of bronchus (B) as well as lungs (L).

axis, the first intensity maximum appears at a distance of $p \simeq \sqrt{\lambda(l_2 - l_1) \cdot l_2/l_1} \simeq$ 18 μm from the beam axis [19], where $l_1$ is a distance between the source and the detector and $l_2$ is a distance between the source and the object edge. It was found that the diffraction fringes appear almost in this distance in actual refraction images, which means that the blurring of the object edge gives a spatial resolution limit in the present case.

The present refraction imaging method is likely to be an inspection technology for an initial stage of cancer or other respiratory deseases in future. There still remain, however, various issues to overcome, for example, a radiation dosage problem, before the method will be actually applied to human bodies as a real diagnosis.

The technique might be a powerful tool for various application to not only biological materials but also inorganic materials, such as metals or plastics including minute voids or cracks.

## SUMMARY

X-ray optics consisting of sequentially located expanders have been installed to perform refraction contrast imaging using a highly brilliant synchrotron radiation from the 'figure-8' undulator at SPring-8. The X-ray optics system using expanders makes it possible to achieve a wide-area exposure with vertically and horizontally expanded X-rays of high parallelism. Some living insects or a frog or a mouse have been tried to take their refraction images in real time and recorded on the videotapes. Those live images give interesting information about the internal structures or their movement in the body. For example, lung cells or bronchi can be seen, being undisturbed by the presence of other organs.

## ACKNOWLEDGEMENTS

The authors would like to thank Dr. J. Chikawa for his continuous encouragement and Mr. M. Ochiai for his financial arrangement from a very initial stage of planning and construction of BL24XU, the first contract beamline at SPring-8. Many persons in SPring-8 should be appreciated for their collaboration in the construction and the operation of the BL24XU.

## REFERENCES

1. Röntgen, W. C., *Nature* **53**, 274 (1896).
2. Davis, T. J., Gao, D., Gureyev, T. E., Stevenson, A. W., and Wilkins, S. W., *Nature* **373**, 595 (1995).
3. Davis, T. J., Gureyev. T. E., Gao, D., Stevenson, A. W., and Wilkins, S. W., *Phys. Rev. Lett.* **74**, 3133 (1995).
4. Ingal, V. N., and Beliaevskaya, E. A., *J. Phys. D* **28**, 2314 (1995).

5. Wilkins, S. W., Gureyev, T. E., Gao, D., Pogany, A., and Stevenson, A. W., *Nature* **384**, 335 (1996).
6. Snigirev, A., Snigireva, I., Kohn, V., Kuznetsov, S., and Schelokov, I., *Rev. Sci. Instr.* **66**, 5486 (1995).
7. Di Michiel, M., Olivo, A., Tromba, G., Arfelli, F., Bonvicini, V., Bravin, A., Cantatore, G., Castelli, E., Dalta Palma, L., Longo, R., Pani, S., Pontoni, D., Poropat, P., Prest, M., Rashevsky, A., Vacchi, A., and Vallazza, E., *Medical applications of Synchrotron Radiation*, eds. Ando, M., and Uyama, C. Tokyo: Springer-Verlag, 1998, pp. 78.
8. Bonse, U., and Hart, M., *Appl. Phys. Lett.* **6**, 155 (1965).
9. Ando, M., and Hosoya, S., *Proc. 6th Int. Conf. On X-ray Optics and Microanalysis*, eds. Shinoda, G., Kohra, K., and Ichinokawa, T. Tokyo: University of Tokyo, 1972, pp. 63.
10. Momose, A, Takeda, T., Itai, Y., and Hirano, K., *Nature Medicine* **2**, 473 (1996).
11. Kagoshima, Y., Tsusaka, Y., Yokoyama, K., Takai, K., Takeda, S., and Matsui, J., *Jpn. J. Appl. Phys.* **38**, L470 (1999).
12. Matsui, J., Kagoshima, Y., Tsusaka, Y., Katsuya, Y., Motoyama, M., Watanabe, Y., Yokoyama, K., Takai, K., Takeda, S., and Chikawa, J., *SPring-8 Annual Report 1997* Hyogo: Japan Synchrotron Radiation Research Institute, 1997, pp. 125.
13. Tanaka, T., and Kitamura, H., *Nucl. Instr. & Methods* A **364**, 368 (1995).
14. Kohra, K., and Ando, M, *Nucl. Instr. & Methods* **177**, 117 (1980).
15. Zachariasen, W. H., *Theory of X-ray Diffraction*, New York: John Wiley & Sons, 1945.
16. duMond, J. W., *Phys. Rev.* **52**, 872 (1937).
17. Thomlinson, W., Chapman, D., Zhong, Z., Jhonston, R. E., *Medical applications of Synchrotron Radiation*, eds. Ando, M., and Uyama, C. Tokyo: Springer-Verlag, 1998, pp. 72.
18. Yamasaki, K., et al. (in preparation).
19. Cosslett, V. E., and Nixon, W. C., *J. Appl. Phys.* **24**, 616 (1953).

# Frontiers of X-Ray Spectromicroscopy in Biology and Medicine: Gadolinium in Brain Cancer

Gelsomina De Stasio [a,b], B. Gilbert [c], P. Perfetti [b], G. Margaritondo [c], D. Mercanti [d], M. T. Ciotti [d], P. Casalbore [d], L. M. Larocca [e], A. Rinelli [e], and R. Pallini [e]

[a] *University of Wisconsin-Madison, Department of Physics, Madison WI 53706, USA,*
*e-mail: pupa@src.wisc.edu*
[b] *Istituto di Struttura della Materia, CNR, Rome, Italy.*
[c] *IPA-EPFL, CH-1015 Switzerland.*
[d] *Istituto di Neurobiologia, CNR, Rome, Italy.*
[e] *Universita' Cattolica del Sacro Cuore, Rome, Italy.*

**Abstract.** We present the first feasibility test of spectromicroscopy on the microlocalization of gadolinium in brain cancer tissue. A gadolinium compound was injected to the patients before the brain tumor was extracted with surgery, and we looked for Gd in the tumor tissue. The goal of the experiment was to understand if Gd Neutron Capture Therapy (GdNCT) is viable for clinical tests, i.e. if there is enough Gd, and it is localized near the nuclei of tumor cells. The experiments were performed using the MEPHISTO X-ray PhotoElectron Emission Microscope (X-PEEM) at the Wisconsin Synchrotron Radiation Center. The present results demonstrate the feasibility of the experiment, and suggest how to improve the sample preparation and data acquisition to achieve the goal.

## INTRODUCTION

### Gadolinium Neutron Capture Therapy (GdNCT)

Boron Neutron Capture Therapy (BNCT) is a non-invasive experimental therapy for glioma and glioblastoma. It is based on two steps: first the patient is intravenously injected with a tumor-seeking $^{10}$B-enriched compound; second, the patient's skull is exposed to thermal neutrons, which induce a short range, biologically destructive nuclear reaction $^{10}B(n,\alpha)^7Li$. $^{10}$B has a capture cross section for thermal neutrons many times greater than other elements present in tissue, so if $^{10}$B is present in tissue irradiated by a neutron flux, almost all of the radiation dose results from the boron neutron capture reaction. If compounds containing $^{10}$B can be delivered only to regions of tumor tissue, the neutron capture reaction will result in the selective destruction of tumor, leaving neighboring healthy tissue unharmed. Clinical experiments on BNCT are currently underway.[1-4]

Gadolinium neutron capture therapy (GdNCT) is an alternative therapy to BNCT, which has never been clinically tested. In this case the locally destructive nuclear reaction is ($^{157}$Gd(n,$\gamma$)$^{158}$Gd). [5,6] The n-$\gamma$ reaction is accompanied by internal conversion and Auger electron emission. The Auger electrons may induce double strand damage and hence cell death if Gd is in the proximity of DNA.[7]

GdNCT appears to be a good potential therapy for several reasons:

1. $^{157}$Gd (15.7% of natural abundance) is the most effective element in terms of neutron capture, having the largest thermal neutron cross section of all the stable isotopes (254,000 barn, $^{10}$B has 2,840 barn).
2. Gd compounds are not toxic.
3. Gadolinium compounds are known to accumulate selectively in brain tumor tissues. They are in fact used as tumor contrast-enhancing agents for magnetic resonance imaging (MRI), given the large magnetic moment of the Gd$^{3+}$ ion. [8]
4. The pharmacokinetics, biodistribution and tolerance of Gd compounds used for MRI are well documented, and such compounds can easily be injected to patients.

For the success of GdNCT, Gd must penetrate into tumor cell nuclei. If the Gd complexes stay in the extracellular matrix, in fact, only the $\gamma$ component and conversion electrons contribute to the radiotoxic effects, while the short-range Auger electrons that would induce non reparable DNA damage, cannot reach the DNA.

In this feasibility test we address this issue with spectromicroscopy, a synchrotron technique applied to the microchemical analysis of physiological and trace elements in biological specimens. The results presented here are the first experimental evidence that x-ray spectromicroscopy can detect gadolinium in brain tissue sections, although the Gd concentration was so low that its subcellular localization could not be addressed.

## EXPERIMENTAL METHODS

### X-Ray PhotoElectron Emission Spectromicroscopy (X-PEEM)

Microchemical analysis of brain tissue was performed with the MEPHISTO synchrotron spectromicroscope [9,10], which uses an electron optics system (SpectroMicroTech, Orlando, FL, USA) to form a magnified image of the photoelectrons emitted by a specimen under soft x-ray illumination. The electron image intensity is amplified by a series of two microchannel plates, and converted into a visible image by a phosphor screen (Chevron mounting by Galileo, CA, USA). This image is captured by a video camera (Dage, MTI, USA) linked to a PC for display and data acquisition. The image magnification is continuously variable up to 8,000 times, and the optimum lateral resolution has been measured to be 20 nm [10]. The photoelectrons are not energy filtered, so the total photoelectron yield, per unit area per unit time, is recorded as a function of photon energy. Such spectra reflect the x-ray absorption coefficient of the specimen surface and are therefore referred to as x-ray absorption spectra [11]. The position and lineshape of spectral features provide element identification and chemical state information. Spectra can be acquired simultaneously from regions selected on the real time image of the sample surface (the probed depth is on the order of 100 Å). For this work, MEPHISTO was mounted on the HERMON beamline of the University of Wisconsin-Madison Synchrotron Radiation Center.

Direct MEPHISTO micrographs were manipulated in Adobe Photoshop® 5.0 for Macintosh to enhance the contrast and add a scale bar to photoelectron micrographs acquired at a specific photon energy. MEPHISTO spectra were saved as text files and plotted in Kaleidagraph® 3.0.4 for Macintosh. Spectra taken from cell structures and substrate areas were normalized by dividing by a third-order polynomial fit to the raw data.

Although in general the x-ray absorption spectra acquired in MEPHISTO may contain complicated dependencies on sample, monochromator and beamline characteristics, the Gd 3d peaks are far from absorption structures of other elements, and the output of the HERMON monochromator is extremely smooth at these photon energies. Therefore spectra normalized as described, produce a plot dependent only on the Gd lineshape and local concentration. This allows us to make comparisons of relative Gd concentrations between different cells or sub cellular structures.

We acquired more than 800 MEPHISTO spectra from tissue regions of various sizes, ranging between 5-300 μm, and from substrate areas, on a total of 12 tissue sections from 12 different patients.

## Sample Preparation

The brain tissue sections were obtained from 12 meningioma patients that had to undergo surgery independent of our experiments. Nine of the patients were injected a gadolinium compound (Gd(III), in gadopentetic acid also known as diethylene triamine pentetic acid (DTPA), the same compound used for tumor contrast enhancement in MRI) before surgery. Each patient was injected at a different time before tumor removal (1.5-72 hours). Three control patients were not injected Gd before extraction of meningioma. Before surgery, MRI images of the tumors were taken from all patients. Immediately after surgical extraction, the meningioma bulk tissue was fixed in paraformaldehyde, embedded in paraffin and stored.

Subsequently the tumor tissues were microtomed to 2.5 µm thick, 10-20 mm wide sections and deposited on gold coated silicon wafers for MEPHISTO analysis. An immediately subsequent tissue section was deposited on a microscope glass slide, and stained with ematoxilin (blue) and eosin (red), to stain the cell nuclei and cytoplasm respectively. This stained section corresponds perfectly to the MEPHISTO section and is used to identify cell structures at the visible light microscope. The next adjacent section was digested in 1 ml $HNO_3$ for Inductively Coupled Plasma Atomic Emission Spectroscopy (ICP-AES) quantitative analysis of Gd.

The brain tissue sections for MEPHISTO analysis were ashed by exposure to UV light from a low-pressure mercury lamp in the presence of ozone. $UV/O_3$ ashing selectively removes carbon and nitrogen from the tissue, while preserving the microlocalization of all other elements (including Gd). [12] In the present case it was employed to enhance the local concentration of Gd which would otherwise not be detectable with spectromicroscopy. The tissue was ashed in air for 119 hours, at a distance of 5 mm from the UV lamp.

## ICP-AES

We performed ICP-AES analysis of the tissue sections immediately adjacent to those used for MEPHISTO analysis. ICP-AES is a quantitative analysis of aqueous samples that can reach a sensitivity of a few ppb for some elements. The analyte concentration is proportional to the intensity of a specific atomic emission line of the vaporized element at 8000 C. The analysis of tissue sections requires digestion of the biological material in nitric acid with sonication.

ICP-AES analysis was unable to detect Gd in any of the tissue sections from the 12 patients. The detection limit of the technique approaches 2 ppb for Gd, but the small tissue samples were digested in 1 ml nitric acid, the minimum sample volume for ICP-AES analysis. In each case, the negative results indicate a Gd concentration of less than 2ppm. This, however, assumes a uniform distribution of Gd in the tissue, whereas the local concentrations may be higher on a microscopic scale.

## RESULTS AND DISCUSSION

We analyzed all of the 12 tissue samples with MEPHISTO, systematically acquiring spectra in the Gd3d energy region. We could detect Gd spectral signal from only one sample, the one taken from the patient injected with gadolinium the shortest time (1.5 hours) before surgery to extract the meningioma.

The region in which Gd was observed is shown in Fig. 1, and the corresponding Gd3d spectra are reported in Fig. 2. Note that there was no gadolinium signal detectable from any substrate region, nor from any of the control patient tissue sections.

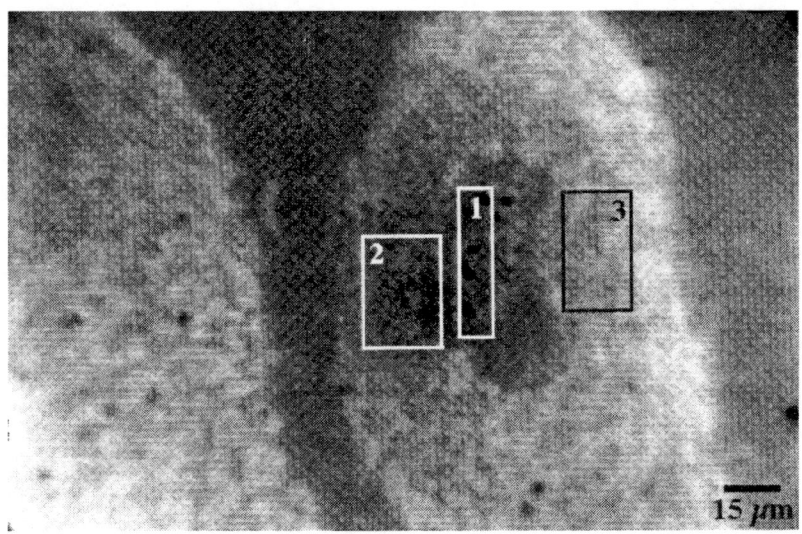

**FIGURE 1.** MEPHISTO micrograph of a region of meningioma brain tissue section, acquired at 1175 eV photon energy. The boxes indicate the microscopic regions in which the spectra of Fig. 2 were acquired. The vertical oval structure visible on the right hand side of the image is a psammomatous body, an area in which meningioma cells align to each other in a vortex. Such formations are typical of meningioma tissues.

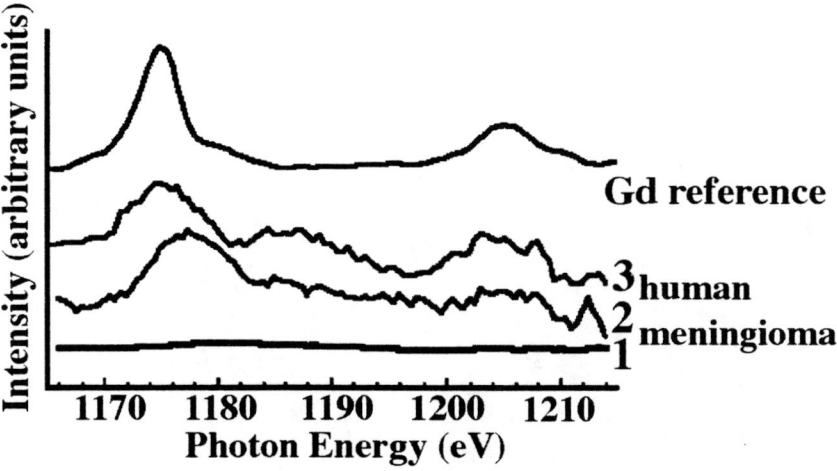

**FIGURE 2.** MEPHISTO absorption spectra of gadolinium acquired in the box regions of Fig. 1. The top spectrum was taken on a gadopentetic acid dried droplet, and is reported here for reference. Note the presence of the Gd3d peak at 1175 eV on regions 2 and 3, and the absence of this peak from region 1. Spectra 1, 2 and 3 were acquired simultaneously.

Also on the sample on which the data of Figs. 1 and 2 were acquired, we only found Gd spectral signal in the region shown here.

Nevertheless, we were able to observe Gd3d peaks in this sample, despite the negative ICP-AES.

The discovery of Gd in a single location of a single sample is difficult to interpret, given the lack of ICP-AES data. Extensive MEPHISTO analysis of the samples covered their surfaces almost completely. Before we can interpret this result as evidence of non-uniform Gd localization, however, we must discover if Gd is absent from other tissue regions or merely present at a concentration below the detection limit of MEPHISTO and ICP-AES. The goal of this work remains to discover if Gd may enter the nuclei of tumor cells, which also requires that we maximize the Gd signal in the samples we analyze.

It is certain that gadopentetic acid was accumulated in the tumor volume at some time point after being administered to the patients since MRI images were taken of each brain.

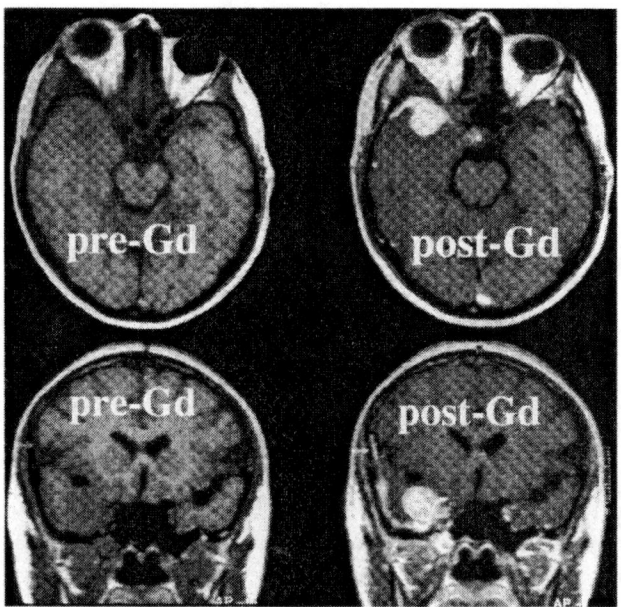

**FIGURE 3.** MRI images of the same patient from whom the tissue section of Fig. 1 was extracted. Note that before injection of gadolinium compound (images on the left-hand side), the brain tumor is not clearly visible, whereas after Gd injection, Gd is accumulated preferentially in the tumor tissue, and acts as a contrast enhancing medium. On the right hand side images the tumor appears brighter than the rest of the brain. Such contrast is due to the large magnetic moment of Gd. MRI shows association of gadopentetic acid with tumor, but does not allow high magnification to determine if Gd penetrates tumor cell nuclei.

The MRI images of Fig. 3 were acquired on the same patient from whom the tissue of the above results was extracted. The patient was injected gadopentetic acid during the MRI analysis, to enhance the image contrast in the tumor region. These images demonstrate that there was Gd in the tumor, approximately 30 minutes after Gd injection.

MRI is a very sensitive, but non-quantitative imaging technique which does not provide any information on the concentration of Gd present in the tumor tissue. In addition, drugs present in a human body are removed with time, following specific pharmacokinetic routes in each organ. There is currently no model for the retention or expulsion of Gd from the body, but the net concentration in the tumor tissue will decrease in the time interval between injection and surgery. A short interval minimized Gd loss, as we were able to observe Gd in MEPHISTO only in the patient for whom this interval was the shortest (1.5 hours). It is possible that Gd concentration was a lot higher after 30 minutes (at which time the images on the right hand side of Fig. 3 were taken) than after 90 minutes, when the tumor was extracted. Note that we cannot improve on this delay, since Gd can be injected to the patient immediately before the induction of narcosis, but it will always take time to extract the tumor with careful brain surgery. The 90 minutes delay of the case reported here is as short as could practically be achieved.

The other 8 patients were injected Gd 3 or more hours before tumor extraction, and in those Gd was undetectable.

Another possible effect which could reduce the Gd signal is the extraction of gadopentetic acid during fixation in paraformaldehyde solution. Experiments to rule out this possibility are currently underway. Quick freezing of the bulk tissue immediately after extraction, as well as fixation in paraformaldehyde vapors are currently being attempted.

## CONCLUSIONS

We performed synchrotron spectromicroscopy experiments to detect the microscopic localization of Gd in tissue sections, for the preliminary optimization of GdNCT, before this novel therapy can be applied to clinical trials.

In conclusion, we proved that at least in one case MEPHISTO could detect Gd in human meningioma brain tissue sections. The failure to detect Gd in all other areas and cases may depend on insufficient sensitivity of the technique, on the unavoidable delay between Gd injection and tumor extraction, or on the fixation technique. A more specialized sample preparation is currently being studied. Furthermore, we are considering the possibility of performing the same experiments on a brighter synchrotron source, such as the Advanced Photon Source of the Argonne National Laboratories, to achieve a higher signal to noise ratio and consequently higher sensitivity to low concentrations of Gd.

## ACKNOWLEDGMENTS

The experiments were performed at the Wisconsin Synchrotron Radiation Center, a facility supported by NSF. We are indebted to Didier Perret for allowing us to use the ICP-AES instrument at the University of Lausanne, and to Mark Bissen for his expert help during the experiments on the SRC-HERMON beamline.

## REFERENCES

1. Yang, W., Barth, R. F., Adams, D. M., Soloway, A. H., *Cancer Res.* **57**, 4333-4339 (1997).

2. Barth, R. F., Yang, W., Rotaru, J. H., Moeschberger, M. L., Joel, D. D., Nawrocky, M. M., Goodman, J. H., Soloway, A. H., *Cancer Res.* **57**, 1129-1136 (1997).

3. Takagaki, M., Ono, K., Oda, Y., Kikuchi, H., Nemoto, H., Iwamoto, S., Cai, J., Yamamoto, Y., *Cancer Res.* **56**, 2017-2020 (1996).

4. Smith, D. R., Chandra, S., Coderee, J. A., Morrison, G. H., *Cancer Res.* **56**, 4302-4306 (1996).

5. Shih, J. A., Brugger, R. M., "Gadolinium as a Neutron Capture Therapy Agent" in *Progress in Neutron Capture Therapy*, edited by B. J. Allen, D. E. Moore and B. V. Harrington, New York: Plenum Press, 1992, pp. 183-186.

6. Rivard, M. J., Waid, D. S., Wierzbicki, J. G., Yudelev, M. "Mearured gadolinium neutron capture dose enhancement using $^{252}$Cf brachytherapy sources" in *Progress in Neutron Capture Therapy*, edited by B. Larsson, J. Crawford and R. Weinrich, Amsterdam: Elsevier, 1997, pp. 430-435.

7. Martin, R. F., D'Chuna, G., Pardee, M., Allen, B. J., *Pigment Cell Res.* **2**, 330-332 (1989).

8. Niendorf, H. P., Felix, R., Laniado, M., Schorner, W., Kornmesser, W., Claussen, C., *Acta Radiol. Suppl.* **369**, 561-563 (1986).

9. De Stasio, G., Capozi, M., Lorusso, G.F., Baudat, P.A., Droubay, T.C., Perfetti, P., Margaritondo, G., and Tonner, B.P., *Rev. Sci. Instrum.* **69**, 2062-2067 (1998).

10. De Stasio, G., Perfetti, L., Gilbert, B., Fauchoux, O., Capozi, M., Perfetti, P., Margaritondo, G., Tonner, B. P., *Rev. Sci. Instrum.* **70**, 1740-1742 (1999).

11. Gudat, W., and Kunz, C., *Phys. Rev. Lett.* **29**, 169-173 (1972).

12. De Stasio, G., Gilbert, B., Perfetti, L., Hansen, R., Mercanti, D., Ciotti, M. T., Andres, R., White, V. E., Perfetti, P., and Margaritondo, G., *Anal. Biochem.* **266(2)**, 174-180 (1999).

# High-Resolution X-Ray Imaging for Microbiology at the Advanced Photon Source

B. Lai[1], K. M. Kemner[1], J. Maser[1], M. A. Schneegurt[2], Z. Cai[1],
P. P. Ilinski[1], C. F. Kulpa[2], D. G. Legnini[1], K. H. Nealson[3], S. T. Pratt[1],
W. Rodrigues[1], M. Lee Tischler[4], W. Yun[5]

1. Argonne National Laboratory, Argonne, IL 60439, USA
2. University of Notre Dame, Notre Dame, IN 46556, USA
3. Jet Propulsion Laboratory, Pasadena, CA 91109, USA
4. Benedictine University, Lisle, IL 60532, USA
5. Lawrence Berkeley National Laboratory, Berkeley, CA 94720, USA

**Abstract.** Exciting new applications of high-resolution x-ray imaging have emerged recently due to major advances in high-brilliance synchrotron sources and high-performance zone plate optics. Imaging with submicron resolution is now routine with hard x-rays: we have demonstrated 150 nm in the 6-10 keV range with x-ray microscopes at the Advanced Photon Source (APS), a third-generation synchrotron radiation facility. This has fueled interest in using x-ray imaging in applications ranging from the biomedical, environmental, and materials science fields to the microelectronics industry.

One important application we have pursued at the APS is a study of the microbiology of bacteria and their associated extracellular material (biofilms) using fluorescence microanalysis. No microscopy techniques were previously available with sufficient resolution to study live bacteria ($\approx$ 1 µm x 4 µm in size) and biofilms in their natural hydrated state with better than part-per-million elemental sensitivity and the capability of determining chemical speciation. *In vivo* x-ray imaging minimizes artifacts due to sample fixation, drying, and staining. This provides key insights into the transport of metal contaminants by bacteria in the environment and potential new designs for remediation and sequestration strategies.

# INTRODUCTION

X-ray microscopy and imaging in the hard x-ray regime has emerged as one of the most important applications of high-brilliance third-generation synchrotron sources such as the Advanced Photon Source (APS). With the advent of high-resolution microfocusing optics, such as Kirkpatrick-Baez mirrors (1) and Fresnel zone plates (2), it is now possible to focus 8-keV x-rays to a spot size of only 150 nanometers with a gain of > 30,000 in the flux density (3). This means $10^9$-$10^{10}$ photon/sec can be delivered into a tight submicron spot. With the improved performance in both x-ray sources and optics, it is now possible to build practical x-ray microprobes, which open up many new applications never previously considered. One such applications is the environmental study of the interaction between bacteria and heavy-metal contaminants described here.

Understanding the fate of heavy-metal contaminants in the environment (4) is of fundamental importance in the development and evaluation of effective remediation and sequestration strategies. Among the factors influencing the transport of these contaminants are their chemical speciation and the chemical and physical attributes of the surrounding medium. Bacteria and the extracellular material associated with them are thought to play a key role in determining a contaminant's speciation and thus its mobility in the environment. In addition, the microenvironment at and adjacent to actively metabolizing cell surfaces can be significantly different from the bulk environment. Thus, the spatial distribution and chemical speciation of contaminants and elements that are key to biological processes must be characterized at micron and submicron resolution in order to understand the microscopic physical, geological, chemical, and biological interfaces that determine a contaminant's macroscopic fate. Hard x-ray microimaging is a powerful technique for the element-specific investigation of complex environmental samples at the needed micron and submicron resolution. This paper presents results of studies of the spatial distribution of naturally occurring metals and a heavy-metal contaminant (Cr) in and near hydrated bacteria (*Pseudomonas fluorescens*) in the early stages of biofilm development. The experiments were performed at the Advanced Photon Source 2-ID-D x-ray microscopy beamline.

# X-RAY MICROPROBE STUDY

The objectives of our studies are 1) to determine the spatial distribution and chemical speciation of metals near bacterial-geosurface interfaces and 2) to use this information to identify the interactions occurring near these interfaces among the metals, mineral surfaces, and bacterially produced extracellular materials under a variety of conditions. The microprobe used an APS undulator source, which supplies a very high-brilliance x-ray beam above 3 keV. The broad spectrum x-rays were then monochromatized by a pair of Si(111) crystals. A phase zone plate, located at 71 m from the source, was used to focus the x-rays to a cross section of 0.15 µm. The zone plate used in these microscopy experiments had an effective focal length of 12.5 cm at 10.0 keV (3).

We used hard x-ray phase zone plates to investigate the spatial distribution of $3d$ elements in single hydrated *Pseudomonas fluorescens* bacteria adhered to a Kapton film. Another layer of Kapton film was used to enclose the bacteria and maintain them in a hydrated condition. The samples were mounted on a computer-controlled XY piezo-stage at 10 degrees to the incident beam, thus negligibly affecting the x-ray footprint on the sample in the horizontal dimension. The intensity of the fluorescence radiation from the sample was monitored by a single-element solid-state detector that enables efficient detection of fluorescent x-rays with energies greater than 1.5 keV. To exploit the horizontal polarization of the synchrotron x-ray beam, the fluorescence detector was placed in the horizontal plane at 90 degrees to the incident beam in order to reduce scattered radiation from the sample. To reduce scattering from other parts of the station, a conical aperture was installed on the detector to limit its angular acceptance. Spatial maps of several elements were then obtained by scanning the sample in 0.15-µm steps through the focused monochromatic x-ray beam and integrating the selected K$\alpha$ fluorescence for 5 sec/pt. The total data collection time was approximately 6 hours.

Figure 1 shows results of the x-ray microprobe measurements, qualitatively indicating the spatial distributions of Cr, K, and Ca in and near a hydrated *Pseudomonas fluorescens* bacterium adhered to a Kapton film at ambient temperature that was exposed to 1000 ppm Cr in solution for 6 hours. Observation of these images indicates that monitoring the spatial distribution of the K and Ca K$\alpha$ fluorescent radiation coming from the sample enables identification of the rod-shaped *Pseudomonas fluorescens* as well as the extracellular exudes associated with it. Additionally, comparison of the distribution of Cr, relative to that of K or Ca, indicate that the majority of the Cr in this sample is associated extracellularly. These results indicate that the majority of the Cr(VI) that was introduced to the sample was probably not actively metabolized. Finally, although these results demonstrate the utility of imaging hydrated bacteria at ambient temperature, in the future, a cryostat may be required to quick-freeze the samples in order to reduce the effects of radiation damage when performing spectromicroscopy studies.

This study illustrates several unique capabilities of a x-ray microprobe compared to conventional charged particle microprobes. For instance, fluorescence cross sections

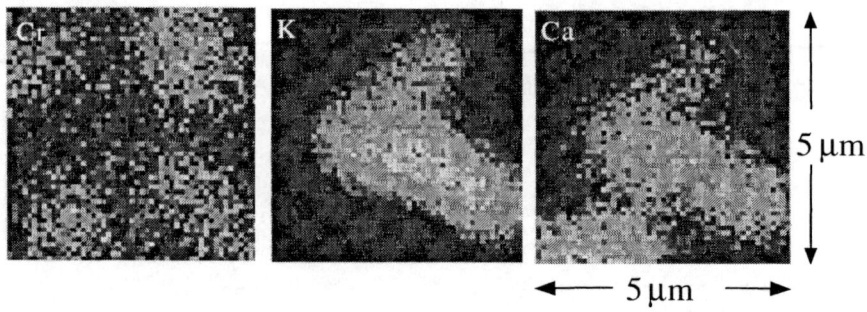

**FIGURE 1.** Grey-scale elemental maps of hydrated *Pseudomonas fluorescens* bacterium treated with Cr(VI) solution. See text for further details.

for excitation by x-rays are typically 10 to $10^3$ times larger than those by charged particles. Thus, the elemental sensitivity of the x-ray microprobe can reach the tens of part-per-billion (ppb) level with substantially reduced energy deposition on the sample. In particular, the fluorescence yield is very high for third-row and heavier elements, many of which are important nutrients, micronutrients, and environmental contaminants. For this reason, we are pursuing x-ray microprobe studies of the subcellular distribution of drugs and trace elements in different biological cells. Another advantage of x-rays is the minimal sample preparation needed, with no special staining nor fixing procedures required. In fact, biological samples can be examined in their natural hydrated state because there is no vacuum requirement and because of the high penetrating power of x-rays in water. For instance, hydrated plant roots and fungi have also been studied with the x-ray microprobe (5), in a way similar to the bacteria study presented here.

## SUMMARY

We have demonstrated the utility of x-ray microbeams, particularly those produced by hard x-ray phase zone plates, for investigating biological and environmental systems. Specifically, we have illustrated the use of submicron hard x-ray beams (0.15 µm) for determining the spatial distribution of metals in a hydrated bacterium that was exposed to 1000 ppm Cr for six hours. The further development of these techniques for such applications promises to provide unique opportunities in the fields of microbiology and environmental research.

## ACKNOWLEDGMENTS

This work was supported by the U. S. Department of Energy, Office of Energy Research, Office of Basic Energy Sciences and Office of Biological and Environmental Research (NABIR Program), as well as internal Argonne National Laboratory (LDRD) funds. Additional support was from the Center for Environmental Studies and Technology, University of Notre Dame. Use of the Advanced Photon Source was supported by the U.S. Department of Energy, Basic Energy Sciences, Office of Science, under Contract No. W-31-109-Eng-38.

## REFERENCES

1. P. Eng, M. Newville, M. Rivers, S. Sutton, SPIE Proc. **3449**, 145 (1998); A. MacDowell, C. –H. Chang, G. Lamble, R. Celectre, J. Patel, H. Padmore, SPIE Proc. **3449**, 137 (1998).
2. B. Lai, W. B. Yun, D. Legnini, Y. Xiao, J. Chrzas, P. J. Viccaro, V. White, D. Denton, F. Cerrina, E. Di Fabrizio, L. Grella, M. Baciocchi, Appl. Phys. Lett. **61**, 1877 (1992).
3. W. Yun, B. Lai, Z. Cai, J. Maser, D. Legnini, E. Gluskin, Z. Chen, A. Krasnoperova, Y. Valdimirsky, F. Cerrina, E. Di Fabrizio, M. Gentili, Rev. Sci. Instrum. **70**, 2238 (1999).
4. *Molecular Environmental Science: Speciation, Reactivity, and Mobility of Environmental Contaminants*, Report of DOE Molecular Environmental Science Workshop, Airlie Center, Virginia, July 5-8, 1995.
5. W. Yun, S. Pratt, R. Miller, Z. Cai, D. Hunter, A. Jarstfer, K. Kemner, B. Lai, H-R. Lee, D. Legnini, W. Rodrigues, C. Smith, J. Synchrotron Rad. **5**, 1390 (1998).

# The Possibility of Using X-Ray Diffraction With Hair to Screen for Pathologic Conditions Such as Breast Cancer

Veronica James* and David Cookson†

*Research School of Chemistry, The Australian National University, ACT 0200, Australia
†Australian Nuclear Science and Technology Organisation, Private Mail Bag 1, Menai, NSW 2234, Australia

**Abstract.** Mammalian hair exhibits a complex structure on length scales ranging from a few to hundreds of Angstroms. High-quality synchrotron x-ray images have yielded new insight about the structure and packing of the intermediate keratinous filaments that represent the bulk of a hair's volume. When comparing human hair diffraction patterns from healthy individuals and breast cancer patients significant differences have been seen, raising the possibility that fiber diffraction may be useful as a screening technique for certain pathologic conditions.

## INTERMEDIATE FILAMENTS IN HAIR

On a scale resolvable with visible microscopy, mammalian hair can exhibit obvious differences between species, and between individuals within a species [1]. Such differences include pigmentation and the thickness of the various layers that comprise a single hair, namely the medulla (core), the cortex (middle layer) and the cuticle (outer layer). On an atomic and molecular level, however, there is a remarkable consistency in the way that 'hard' alpha-keratin forms intermediate filaments (IFs) which then pack together to form the hair cortex. The high degree of order exhibited by the IF structure has made x-rays diffraction a useful tool in its study, although the relatively low scattering power of hair can make obtaining good signal-to-noise data challenging.

Using high and low-angle scattering data obtained from conventional x-ray sources, Fraser *et al.* were the first to postulate an IF structure for the hard α-keratin micro-fibrils set in the non-filamentous matrix of the hair[2,3]. Their model proposed that the α-keratin formed dislocated helices with finite spacing of 19.8nm and an infinite lattice of 47nm. This model, however, was based on very limited data and could not account for a number of observed reflections, which led them to suggest that the true lattice spacing was some multiple of 47nm.

Using synchrotron radiation, James et al. were able to record many more reflections than had previously been seen from a range of keratinous fibers including baboon hair, human hair, sheep's wool and echidna quill[4-6] (Fig. 1).

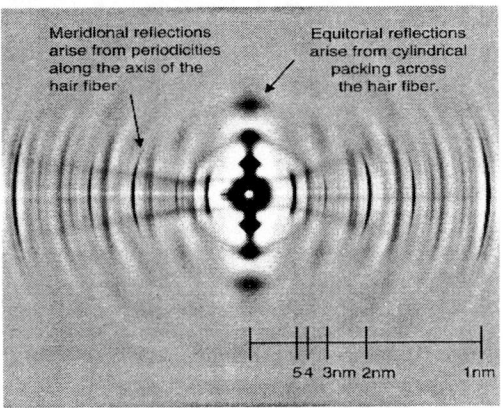

**FIGURE 1.** Synchrotron x-ray scattering pattern (background subtracted) from a single baboon hair ($\lambda = 1.5$Å) showing many meridional reflections arising from periodicity along the hair's axis. The hair was oriented horizontally for this image.

These data, when combined with electron microscopy and fiber swelling data, allowed Feughelman et al. to propose a far more detailed and consistent model of the basic IF structure[6]. In this model the helical portions of two α-keratin molecules twist around each other to form a double helix which in turn, pairs with another double helix to form a single tetramer of α-keratin. The tetramers in turn form the IF; a large, slowly twisting 8 stranded helix with the non-helical tetramer tails pointing outwards in a staggered hexagonal arrangement (Fig. 2). These tails cross-link with tails from adjacent IFs resulting in hexagonal packing of the filaments.

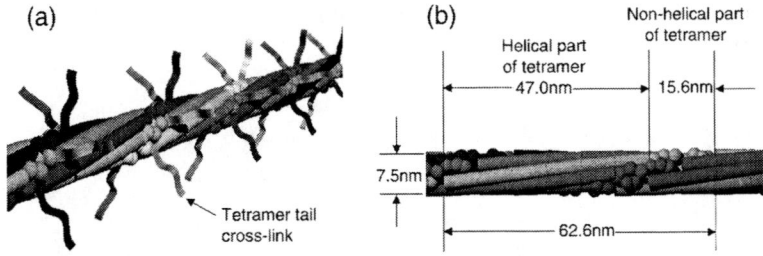

**FIGURE 2.** Eight-strand helix formed by tetramers of α-keratin. (a) shows tetramer tails which form staggered hexagonal cross links with adjacent IFs and (b) shows the helical and non-helical tetramer tails that form the repeated unit in the IF (the part of the α-keratin tetramer tail that extends away from the IF is not shown)

The cross linking between adjacent IFs allow them to maintain their overall orientation even when a hair is swelled in formic acid solution. The fact that a 30% increase in the diameter of a hair during swelling results in a 2.3% contraction in its length is a consequence of the staggered hexagonal cross linking that results in a gentle helical deformation of each IF [6]. In a subsequent paper[7] James et al. demonstrated that the 62.6nm reflections seen in diffraction data could be explained by a set of six infinite lattices with normals at 7° to the IF orientation. These infinite lattices arise as a consequence of the nearest neighbor distance between IFs being exactly three times the IF radius.

## PATHOLOGIC CONDITIONS AND HUMAN HAIR

While the state of an individual's hair is often associated with their overall health, most evidence to this effect is anecdotal at best. In 1997 James et al. described a clear difference seen in the x-ray diffraction patterns of hair taken from healthy and insulin-dependent diabetic humans and baboons. The differences were ascribed to a disruption in the normal process of IF cross-linking during the formation of the hair in its follicle[8]. The nature of this extra-cellular bonding suggested that the change in the hair is endogenous, occurring via the blood during the aggregation of IFs in the follicle. After the hair cells keratinize and die this change is 'locked in' to the hair shaft, providing a long-term record of changes in the blood's biochemistry in a tissue that can be easily obtained.

The possibility that other pathologic conditions may also be reflected in the structure of dead hair was tested in four studies on hair from controls and patients with breast carcinoma[9]. The first double blind study (samples provided by A. Howel, Christie CRC Research Center, Manchester) was carried out on beamline 15A at the Photon Factory (Japan). After correlating the diffraction patterns with patient data, the results were encouraging and hinted that changes in the hair diffraction pattern (Fig. 3) indicated not just a propensity to malignancy, but also the actual presence of breast cancer. To test this hypothesis, another double-blind study was performed on a larger sample of hair from the same source. This time, the x-ray measurements were performed on Australian National Beamline Facility and later confirmed on the Biophysics Collaborative Access Team at the Advanced Photon Source (Argonne National Laboratory).

The initial results showed some inconsistencies, which were eliminated when hair samples that had recently (within three months of collection) been "permed" were excluded from the sample set. To avoid further problems caused by hair treatments, pubic hair collected by J. Kearsley at the Oncology Unit, St. George Hospital (Sydney, Australia) was used in subsequent tests.

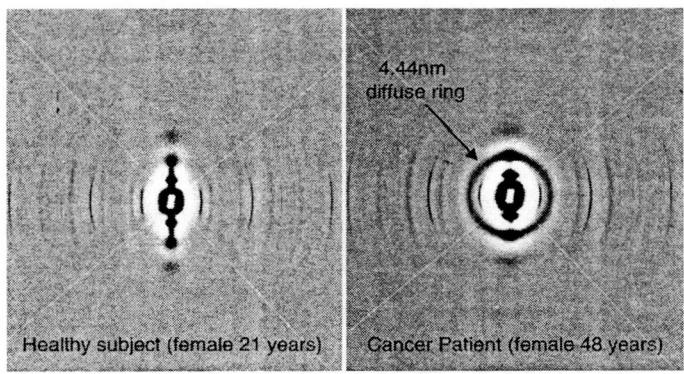

**FIGURE 3.** Comparison of background-corrected diffraction images from single strands of hair. The reflections are less intense than those for baboon hairs, which are usually thicker and therefore have a greater scattering power.

TABLE 1. Correlation of breast cancer with hair structure (reproduced with permission from *Nature*)

| Breast cancer diagnosed? | Samples | Samples with changed structure | Familial History? (Pathologic BRCA1 mutation) | Hair origin |
|---|---|---|---|---|
| Yes | 15 | 15 | Yes (Positive) | Scalp |
| Yes | 8 | 8 | None | Pubic |
| No | 3 | 3 | Yes (Positive) | Scalp |
| No | 2 | Partial 2 | Yes (Positive) | Scalp |
| No | 8 | 1 | Yes (Negative) | Scalp |
| No | 16 | 3 | None | Scalp |
| No | 4 | 0 | None | Pubic |

Table 1 shows that all samples (23 out of 23) taken from breast cancer patients exhibited a change in the form of the appearance of one or more diffuse rings. Of the samples taken from patients not suspected of having breast cancer, the scattering pattern of 86% (24 out of 28) were normal. In addition, of those not yet diagnosed with breast cancer, but suspected of being at risk (because of familial history or the presence of a pathologic mutation of the BRCA1 gene), 60% (3 out of 5) showed the full change while the remainder showed a partial change.

# CONCLUSION

The diffuse ring shown in Fig. 3b represents a molecular spacing of 4.44±0.06nm, placing it directly on the equitorial arc arising from the plasma membrane in normal hair. This suggests that the ring may arise from some randomly oriented lipid bilayers in either the plasma membrane or in membranous inclusions in the hair cells.

A great deal of additional research will be required before any conclusions can be drawn on the nature of this reported change in hair diffraction patterns, and how it may reflect subtle changes in the biochemistry of the blood nourishing the hair follicle. The possibility that such work could lead to a relatively simple screening tool for breast cancer and other pathological conditions, makes further study imperative.

## ACKNOWLEDGMENTS

We wish to acknowledge the generous support of the Photon Factory and Professor Y. Amemiya for the use of their facility BL15A. Financial support from the Australian Synchrotron Research Program (funded by the Commonwealth of Australia via the Major National Research Facilities Program) and the use of the Australian National Beamline Facility is acknowledged. We would also like to acknowledge the assistance of T. Irving and the use of the BioCAT beamline at the Advanced Photon Source which is supported by the U.S. Department of Energy, Basic Energy Sciences, Office of Energy Research under Contract No. W-31-109-ENG-38. BioCAT is a NIH supported Research Center under grant # RR08630.

Finally, we would like to express our thanks to M. Feughelman for valuable discussions, to R. Derrick, M. Read, D. Miller, G. James and F. Rawan for assistance in data collection, and to S. James and E. P. Baranov for data handling and processing.

## REFERENCES

1. Feughelman, M., *Mechanical Properties and Structure of Alpha-Keratin Fibres*, UNSW Press, Sydney, 1997.

2. Fraser, R.D.B., and MacRae, T.P., *Int. J. Biol. Macromol.* **10**, 178-184 (1988)

3. Fraser, R.D.B., MacRae, T.P., and Parry, D.A.D., "The Three-Dimensional Structure of IF" in *Cellular and Molecular Biology of Intermediate Filaments* edited by R.D. Goldman and P.M.Steinert, Plenum Press, NY. 1990.

4. James, V.J., Wilk, K.E., and McConnell, J.F., Int. J. Biol. Macromol. **17**, 99-104 (1995)

5. Wilk, K.E., James, V.J., and Amemiya, Y., *Biochim. Biophys. Acta* 1245, 392-396 (1995).

6. Feughelman, M. and James, V.J., *Textile Res. J.* **68**, 110-114 (1998)

7. James, V.J. and Amemiya, Y., *Textile Res. J.* **68**, 167-170 (1998)

8. James, V.J., Yue, D.K., and Lennan, M., *Biophys. Biochim. Res. Com.* **233**, 76-81 (1997)

9. James, V.J., Kearsley, J., Irving, T., Amemiya, Y., and Cookson, D., *Nature* **398**, 33-34 (1999)

# IX. NEW SOURCES AND TECHNIQUES

# Prospects for an X-Ray FEL Light Source and some Possible Scientific Applications

John Arthur

*Stanford Synchrotron Radiation Laboratory*
*Stanford Linear Accelerator Center, Stanford, CA 94309*

**Abstract.** Free electron lasers are now being designed which will operate at wavelengths down to about 1 Å. Due to the physics of the high-gain, single pass FEL process that these sources will exploit, the radiation produced will have unique properties. In particular:
• The FEL peak intensity and peak brightness will be many orders of magnitude higher than can be produced by any other source.
• The pulse length will be less than 1 picosecond, orders of magnitude shorter than can be achieved with any other bright source such as a synchrotron.
• The FEL radiation will have full transverse coherence and a degeneracy parameter (photons/coherence volume) equal to $10^9$ or more. No other source can produce hard x-radiation with a degeneracy parameter significantly greater than 1.
These properties offer the chance to study chemical, biological, and condensed matter dynamical processes with sub-picosecond time resolution and angstrom spatial resolution. The high peak power of the FEL radiation (greater than $10^{14}$ W/cm$^2$) could be used to create precisely-controlled chemical and structural modifications inside samples. There is also the possibility that nonlinear x-ray interactions could be used to give increased resolution for spectroscopic studies, to greatly expand the parameter space for atomic physics studies, and to permit new fundamental tests of quantum mechanics. The exploration of these new x-ray techniques will require considerable development, not only in technical areas such as optics and detectors, but also in understanding the basic physics of the interaction of very intense x-radiation with matter. A large collaboration of US institutions is now conducting preliminary research and development in these areas, with the intention of creating an FEL operating at 1.5 Å in about the year 2006. Germany also has a strong short-wavelength FEL research program, with a soft x-ray FEL under construction and a proposal for a future large facility which would produce a variety of hard and soft x-ray laser beams.

# INTRODUCTION

There have been tremendous changes over the last 30 years in all areas of science which depend on radiation with wavelength shorter than about 10 Å (throughout this paper the x-rays being discussed are those in this short-wavelength region of the spectrum). These scientific areas have benefited tremendously from the development of x-ray sources based on high-energy electron accelerators. The most advanced of these, the so-called third generation synchrotron light sources, are about 11 orders of magnitude brighter than the x-ray sources of the 1960's. Many of the papers included in this volume present results which depend on the capabilities of modern synchrotron sources.

The continued increase in synchrotron brightness cannot continue indefinitely, however. Though improvements to existing sources will probably yield a few orders of magnitude more, it is most likely that the next large jump in x-ray source capability will come from a different type of high-energy electron machine, the linear-accelerator-based free-electron laser (FEL). This new technology offers an increase in brightness over today's synchrotron source comparable to the synchrotron's increase over the lab sources of 30 years ago.

As bright as they are, synchrotron light sources are not lasers. The radiation that is produced comes from a spontaneous emission process, and is not amplified. The coherence properties of the radiation reflect the incoherence of the source. Because the wavelength is much smaller than the source size, the coherent fraction of the radiation is quite small. Also, the number of x-ray photons emitted in a single pulse is relatively small, so that nearly all time-resolved studies must be done in a stroboscopic way, requiring samples with repeatable behavior.

On the other hand, the coherence characteristics of FEL radiation derive from the phase matching criteria for the amplification process, which leads to a diffraction-limited, highly degenerate pulse of photons. In addition, the pulse length is orders of magnitude shorter than that produced by a synchrotron source, and the number of x-ray photons per pulse is high enough to do complete measurements in single flash mode. The characteristics of the radiation are in fact so unusual that, rather than replacing other x-ray sources such as synchrotrons, FELs will most likely be used to address completely new scientific questions, using completely new experimental methods.

This paper consists of a brief review of physics and technology issues related to the development of an x-ray FEL, and a more detailed discussion of the important aspects of FEL radiation. This is followed by a speculative look at possible applications in various scientific areas. Finally, the US and German plans for developing x-ray FELs are discussed.

# THE FEL PROCESS

A free-electron laser relies on the interaction between a high-energy electron beam and a co-propagating electromagnetic field to produce coherent amplification of the field [1]. This interaction requires the presence of a periodic magnetic field (such as in an undulator insertion device), which causes the electron beam to follow an oscillatory path, and produce synchrotron radiation. A strong electromagnetic field moving with the electron beam slightly deflects the oscillatory motion of the electrons. With proper phase matching (a condition satisfied at the odd harmonics of the undulator fundamental emission frequency), this deflection leads the electrons to become longitudinally bunched, with the period of the bunching matching the wavelength of the electromagnetic field. The bunched electrons then tend to produce synchrotron radiation which is coherently phased with the electromagnetic field. Thus the field produces some bunching which produces a stronger field, which in turn leads to more bunching, and so on. Eventually the process saturates when enough energy is removed from the electron beam to cause it to lose phase matching with the electromagnetic field.

Since the 1970's more than 50 FELs have been constructed, with operating wavelengths ranging from the infrared to the near ultraviolet. About 40 are still in operation [2]. All of these devices include resonant optical cavities to contain the electromagnetic field, surrounding an undulator insertion device. They require multiple passes of electron pulses through the undulator/optical cavity before the FEL radiation field reaches full intensity. Some of these machines are operated as user facilities, for experiments needing very intense, short pulses of infrared radiation.

# AN X-RAY FEL BASED ON SELF-AMPLIFIED SPONTANEOUS EMISSION

In order to extend the FEL concept into the x-ray region of the spectrum, it will be necessary to do without a resonant optical cavity. A scheme to achieve strong FEL action in a single pass of a high-energy electron pulse through a long undulator is known as self-amplified spontaneous emission (SASE) [3]. This scheme relies on spontaneous undulator radiation to form the electromagnetic field that initiates electron bunching. It also requires the FEL gain to be large enough to lead to an exponentially-growing radiation intensity along the length of the undulator. To achieve the high-gain FEL regime (also called the FEL instability), four conditions must be met:

*1.* The electron beam transverse emittance must not be much larger than the wavelength of the FEL radiation. In other words, the phase space of the electron beam must be matched to that of the radiation field.

*2.* The electron beam energy spread must be small, to give a sufficiently narrow line width to the spontaneous undulator radiation peaks.

*3.* The FEL power gain length must be short enough so that the increase in radiation intensity along the undulator is not lost due to diffraction spreading.

*4.* The electron pulse length must be longer than the FEL radiation wavetrain that is produced from it.

Of these conditions, the most difficult to achieve for an x-ray FEL is the first. Even the "low emittance" storage rings used in third-generation synchrotron sources have emittance values that are measured in tens of angstroms, much too large to support an angstrom-wavelength SASE FEL. The emittance limits for these machines are ultimately due to recoil effects on the circulating electrons from the spontaneous emission of synchrotron radiation.

The most promising way to achieve the required beam emittance is by using a linear electron accelerator. Linacs do not produce synchrotron radiation, and in principle the emittance can be reduced without limit, scaling inversely with the energy of the electrons.

In practice, it is still not easy to achieve an electron emittance of an angstrom and create an FEL operating in the angstrom range. However, three recent technical advances give confidence that such an FEL will soon be built.

*RF Photocathode Gun.* In a perfect linac the output emittance is simply proportional to the source emittance divided by the output energy, so a smaller source emittance yields a smaller output emittance. A new type of electron source has recently been developed, using a photocathode built into an RF-accelerator section. A laser pulse synchronized to the accelerator radio-frequency wave creates photoelectrons which are nearly instantly accelerated to relativistic energy. This suppresses the photoelectrons' tendency to disperse due to electrostatic repulsion, providing a very bright electron source which can be easily coupled to a linac.

*Control of Emittance in the Linac.* To achieve angstrom emittance, the electron pulse from even an RF photocathode gun must be accelerated to more than 10 GeV. To achieve this acceleration without having the emittance spoiled requires a detailed understanding of the interactions between the electron pulse and the accelerator. This level of sophistication in accelerator physics has come from recent studies related to future high-energy electron-positron colliders based on linear accelerators.

*Undulators.* A SASE x-ray FEL will require an electron beam to be guided through tens of meters of undulator, while maintaining alignment with the radiation field that it is creating. The ability to assemble and adjust thousands of blocks of permanent magnet material so as to create an undulator of the required precision is a product of research associated with the third-generation synchrotron sources.

## A TYPICAL SASE X-RAY FEL

Meeting the conditions for SASE FEL operation in the angstrom wavelength region will require some advancement of the state of the art in all three areas mentioned above: electron gun technology, emittance preservation in the linac, and

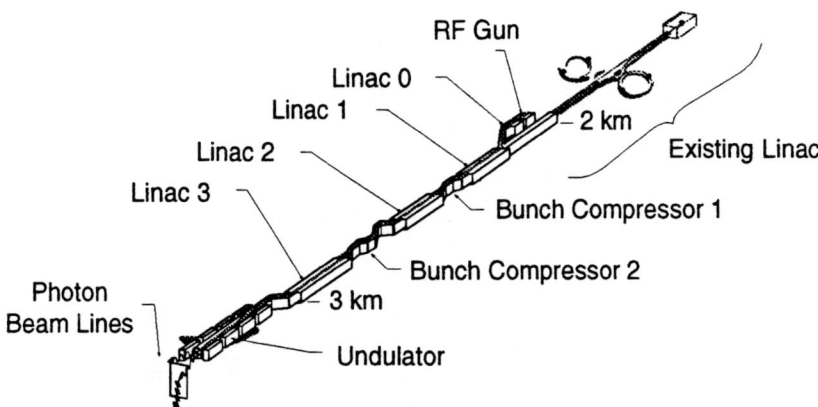

**FIGURE 1.** The Linac Coherent Light Source (LCLS), a proposed SASE FEL that would utilize one third of the 3-km linac at the Stanford Linear Accelerator Center. An RF photocathode electron gun would be added, along with electron bunch compressors and a 100 m undulator. The design FEL radiation wavelength is 1.5 Å.

electron beam control in a long undulator. Two groups have recently prepared conservative designs minimizing innovation in all three areas [4,5], and their proposals are quite similar. These proposals give a good picture of what an x-ray FEL is likely to look like. Figure 1 shows a schematic of the design prepared by the SLAC group.

The x-ray FEL will begin with an RF-photocathode gun producing an electron pulse with a normalized transverse emittance of about 3 $\mu$m. This pulse will be accelerated to 15–20 GeV in a linear accelerator. At one or more points during the acceleration process, the pulse will be routed through dispersive magnets in order to compress it longitudinally (increasing the instantaneous electron current). At the end of the linac about $10^9$ electrons will be contained in a pulse lasting about 100 fs. This high-energy pulse will be directed through an undulator similar in design to those used at synchrotron sources, but roughly 100 m long. At the end of the undulator, the electron pulse will be deflected away from the radiation pulse.

## A SEEDED X-RAY FEL

A SASE FEL operates as an oscillator, amplifying particular spectral components of the spontaneous fluctuations in the density of the electron pulse in order to produce coherent emission of radiation. This approach is straightforward, although as described above the constraints on the beam parameters are stringent. However, the startup from noise gives a few undesirable characteristics to the FEL radiation. The random variations in noise amplitude at the FEL frequency lead to large pulse-to-pulse variations in FEL radiation intensity. Also, the temporal coherence length of the radiation is much shorter than the length of the pulse.

A more controlled output pulse could be achieved by using the FEL process to amplify a well-controlled x-ray seed pulse. This would require a seed with coherent power of 200 kW or more, sufficient to dominate the power in the spontaneous undulator radiation. It would be possible to use a SASE FEL as a source of intense radiation, which could be filtered through a monochromator to produce a strong seed. A more attractive idea is to use as a seed an x-ray pulse obtained from a less expensive source. Perhaps within a few years x-ray sources based on intense optical lasers will be able to produce seed pulses of sufficient power.

A seeded FEL could in principle produce a completely coherent pulse of radiation, with a pulse length adjustable from sub-fs to several fs, based on the bandwidth of the seed. This would make the FEL even more attractive for very fast time-resolved studies. While it is most likely that the very first x-ray FELs will rely on SASE, in the long run FEL light sources are more likely to involve seeded FELs.

## PROPERTIES OF THE RADIATION

The radiation from a SASE x-ray FEL will be unlike any seen before in this spectral region. The peak brightness of the FEL radiation will be about 10 orders of magnitude greater than the next brightest x-ray sources, which are the third-generation synchrotron sources (see Fig. 2). The radiation will also have unprecedented coherence properties. Coming from a laser which amplifies only a single spatial mode, the radiation will have complete transverse coherence, with a beam diameter of about 100 $\mu$m. This beam will be diffraction limited, with a divergence of about 1 $\mu$rad. In the direction of propagation, the roughly 1000 undulator periods comprising the final few gain lengths of the amplifier will yield a coherence length of about 0.1 $\mu$m. The coherence volume will contain nearly $10^{10}$ photons. A quantum mechanical description of the FEL wavefield would have an expectation value of nearly $10^{10}$ photons in the same quantum state. This expectation value, called the degeneracy parameter [6], plays an important role in connection with photon correlations and nonlinear optical processes. Its large value is remarkable in that no previous x-ray source has produced a degeneracy parameter significantly greater than 1.

Current designs for SASE FELs envision an electron pulse about two hundred times longer than the FEL longitudinal coherence length, producing an output that consists of two hundred independent, sequential micropulses in a macropulse lasting roughly 100 fs. Note that this macropulse length is at least two orders of magnitude shorter than the x-ray pulses created by other bright sources such as synchrotrons. The amplitudes of the micropulses will vary greatly, representing amplified stochastic variations in the electron density. A seeded FEL could provide a totally coherent output pulse, of length determined by the bandwidth of the seed.

The linac repetition rate would determine the firing rate for the FEL system. A superconducting linac could provide several thousand nanosecond-spaced pulses in a train separated by milliseconds from other pulse trains. Though this pulse spacing

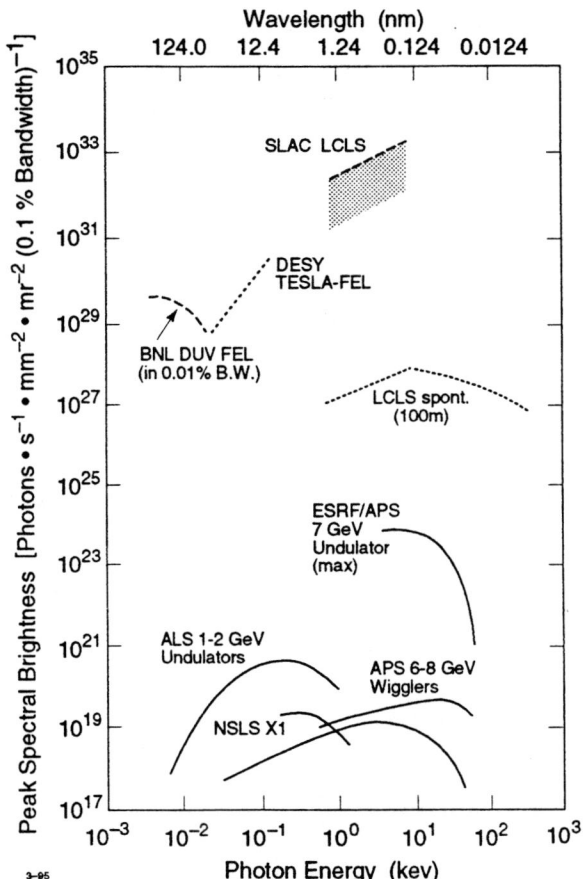

**FIGURE 2.** A comparison of the peak spectral brightness (photon density in phase space) of several hard and soft x-ray sources. The top curves all apply to FEL sources that are either proposed or under construction. These curves represent calculated values (the shading around the LCLS curve indicates an estimated range of uncertainty, allowing for possible errors in the electron optics). The curves at the bottom apply to existing synchrotron sources, but they are also calculated values incorporating expected future improvements in brightness. One curve in the middle shows the expected peak brightness of the spontaneous (non-FEL) radiation from an x-ray FEL source.

**TABLE 1.** Calculated characteristics of the LCLS radiation.

| Parameter | Value | Unit |
|---|---|---|
| FEL wavelength | 1.5 | Å |
| FEL bandwidth ($\delta E/E$) | 0.002 | |
| Pulse duration (FWHM) | 233 | fs |
| Pulse length (FWHM) | 67 | $\mu$m |
| Peak coherent power | 9 | GW |
| Peak coherent power density | $1.5 \times 10^{14}$ | W/cm$^2$ |
| FEL energy/pulse | 2.6 | mJ |
| Peak brightness | $1.2 \times 10^{33}$ | flux/(mm$^2$ mrad$^2$ 0.1% BW) |
| FEL photons/pulse | $2. \times 10^{12}$ | |
| FEL photons/second | $2.4 \times 10^{14}$ | |
| Degeneracy parameter | $3.3 \times 10^9$ | |
| Peak EM field (unfocused) | $3.4 \times 10^{10}$ | V/m |
| Average FEL power | 0.31 | W |
| Average FEL brightness | $4.2 \times 10^{22}$ | flux/(mm$^2$ mrad$^2$ 0.1% BW) |
| Transverse size of FEL beam | 78 | $\mu$m (FWHM) |
| Divergence of FEL beam | 1 | $\mu$rad (FWHM) |
| Peak power of spontaneous radiation | 81 | GW |

is much more sparse than that produced by a synchrotron source, nevertheless the average brightness of the FEL will still be several orders of magnitude higher than today's third-generation synchrotron sources.

Table 1 lists a number of calculated properties for the radiation to be produced by the LCLS, a SASE FEL proposed by the Stanford Linear Accelerator Center. The properties of most interest to an experimenter are:

- The FWHM of the radiation pulse is 230 fs.
- Each pulse contains about $10^{12}$ photons in the FEL fundamental peak.
- The degeneracy parameter (photons per coherence volume) for these photons is nearly $10^{10}$. The beam is circular in cross-section (diameter 80 $\mu$m), and has complete transverse coherence. The longitudinal coherence length is about 0.1 $\mu$m.

In addition to the FEL laser radiation, the SASE device will produce an intense, broadband pulse of spontaneous undulator radiation (see Fig. 3). Though without the coherence properties of the FEL radiation, the spontaneous radiation will be scientifically useful because of its broad spectrum and sub-picosecond pulse length. The spontaneous radiation spectrum will extend to several MeV; above about 100 keV its average brightness will exceed that of any other radiation source.

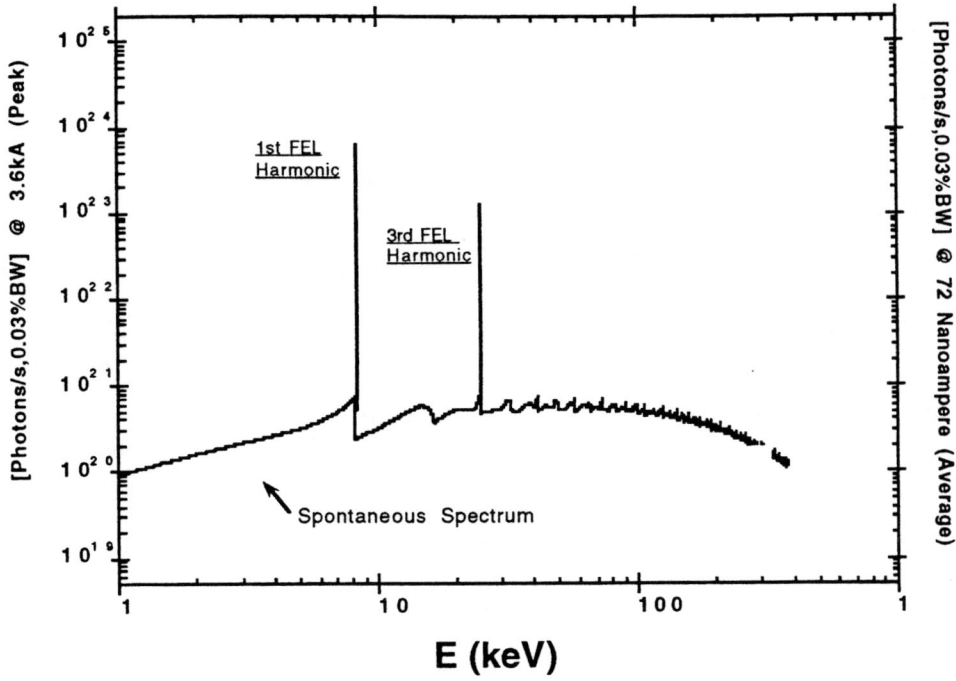

**FIGURE 3.** Calculated spectrum for the LCLS, a SASE FEL source. The FEL fundamental wavelength is 1.5 Å. FEL amplification is also possible at the third harmonic wavelength, though the gain is much less than for the fundamental, and may be suppressed by errors in the electron beam alignment.

# SCIENTIFIC APPLICATIONS

The exceptional characteristics of FEL radiation lead one to envision applications requiring very good time resolution or very high coherence. To assess the possibilities in these areas, it is useful to carefully compare the capabilities of an x-ray FEL with those of other short-pulse or highly-coherent x-ray sources.

## Short-pulse x-ray sources

The brightest x-ray sources today, the third-generation synchrotron sources, produce pulses whose duration is about 100 ps. The x-rays are highly collimated, with 10–100 $\mu$rad divergence, and the pulses are repeated at a rate measured in MHz. The time-averaged monochromatic flux (in a bandwidth of about $10^{-4}$) is about $10^{14}$ $s^{-1}$. In order to use such a source for sub-ps time-resolved experiments, schemes have been developed which utilize a sub-ps optical laser to serve as a sort of gate [7], subdividing the longer synchrotron x-ray pulse. These approaches, and any time-resolved techniques that make use of pulsed optical lasers, suffer from the incommensurability between the synchrotron pulse rate and that of the laser (which is typically in the kHz range). Thus, for an experiment requiring 100 fs time resolution, the synchrotron flux is effectively reduced by at least 6 orders of magnitude due to the need to subdivide the synchrotron pulse and to match the repetition rate of an optical laser. The net useful flux is around $10^6$–$10^8$ $s^{-1}$.

Other sources have been developed to produce sub-ps pulses of x-rays, including pulsed plasma sources [8], and inverse Compton effect sources [9]. These sources can be fast (100 fs), but are not very bright. The flux that they can send into an experiment with mrad divergence requirements is typically $10^4$–$10^6$ $s^{-1}$. There have also been extensions of traditional atomic lasers toward shorter wavelengths, utilizing inner-shell transitions or non-linear generation of very high harmonics of optical lasers. These sources can be quite fast (10 fs) and intense, but are so far confined to the very soft x-ray region, with their performance deteriorating rapidly at wavelengths shorter than 100 Å.

These fast sources produce enough useful x-ray flux to perform sub-ps time resolved experiments in which the signal intensity is a fair fraction of the incident beam intensity, such as studies of Bragg reflections of good-quality crystals. The x-ray FEL, on the other hand, can produce enough flux to study much weaker effects with sub-ps resolution. The FEL is highly collimated, and its pulse repetition rate is compatible with that of a pulsed optical laser, so that essentially all of its average flux of $10^{14}$–$10^{16}$ $s^{-1}$ can be effectively used for fast time-resolved experiments. And because of the very high flux delivered by a single pulse ($10^{12}$), the FEL allows the possibility of completing an experiment with a single pulse.

## Coherent x-ray sources

Coherent x-ray experiments typically require spatial coherence lengths of about 10 $\mu$m [10]. A synchrotron source can provide an average flux of about $10^9$ s$^{-1}$ with this level of coherence. Other existing x-ray sources cannot approach this level of coherent flux in a collimated, monochromatic beam.

An x-ray FEL, with complete transverse coherence, can provide a spatially coherent flux of $10^{14}$–$10^{16}$ s$^{-1}$. In addition, the FEL radiation exhibits coherence beyond the single-photon level. The degeneracy of the photon wavefunction produced by the FEL (the number of identical, coherent photons in a coherence volume) is greater than $10^9$. No other x-ray source produces radiation with a degeneracy significantly greater than 1. This means that multi-photon coherence experiments should be practical with an FEL, whereas they are impractical with any other x-ray source.

Since the FEL is a fast pulsed source as well as a coherent one, it offers the possibility of conducting coherent experiments with sub-ps time resolution, or of conducting single-shot coherence experiments.

With such a leap in capability over existing sources, it is nearly impossible to correctly identify the most scientifically fruitful applications that will emerge. Nevertheless, several international workshops have been convened during the past five years to explore ideas for applications of an x-ray FEL [11–16]. Below is a summary of the ideas presented at these workshops.

## Fundamental Quantum Mechanics

Nonlinear scattering of the high-brightness FEL beam could be used to produce large numbers of entangled multi-photon states, which could be used to study correlation effects such as the Bell inequality. It would be advantageous to use x-rays for experiments of this type, rather than optical photons, because the high quantum efficiency of x-ray detectors would make it easier to keep track of every photon.

The ability to create intense, short-wavelength electric field patterns using an x-ray FEL would be useful for atom-interferometry experiments. FEL standing wave interference patterns could be used as beam splitters and mirrors for the atom beams. This is achieved today with visible lasers; using the FEL would allow these optics to be created with shorter periods, allowing larger scattering angles for the atom beams, and thereby giving the interferometry experiments much higher sensitivity.

If the full peak power of the x-ray FEL were focused into a submicron spot, the peak electric field could approach $10^{14}$ V/m, and the power density could approach $10^{21}$ W/cm$^2$. Even higher power densities could be achieved by using a tapered FEL undulator. FEL amplification in a uniform undulator saturates when enough energy is removed from the electron pulse that it loses its phase matching with the electromagnetic field. A properly tapered undulator could maintain the phase match even as the electron pulse energy decreases, increasing the FEL gain by several orders of magnitude. It is possible that power densities might be reached which would be useful for tests of quantum electrodynamics.

## Atomic, Molecular, and Plasma Physics

The extremely high peak brightness of a focused FEL x-ray beam offers the chance to extend the study of photon interactions with core atomic electrons well into the nonlinear regime. The new experimental capability would provide tests of theoretical analysis that goes beyond the dipole approximation. For example, it would allow tests of nonlinear processes such as core-resonant ionization in helium, and adiabatic stabilization [17,18], in which under certain conditions transition rates are predicted to decrease with increasing intensity.

In the area of molecular physics, the high brightness and short pulses of the FEL source would allow one to better understand ionization and dissociation processes by controlling them coherently. One could also study chemical reactions in the time domain with pump/probe techniques.

The plasma physics community could use an FEL x-ray source to greatly extend their studies of the interactions between matter and extremely high-power-density electromagnetic fields. This would allow new tests of the scaling properties of nano-plasma multiple ionization, and inner-shell excitations.

## Chemical Physics

The x-ray FEL has the potential to become a powerful monitor of surface chemical reactions, using x-ray fluorescence or XPS as the system probe. This application has the ability to fully characterize the kinetics (reactants, products and rates) of surface reactions.

By using the x-ray pulse as a probe beam, a fast-pulse optical laser could be used to set up femtosecond-resolution structural dynamics studies. Local structure would be obtained by EXAFS and global structure would be obtained via diffraction measurements. The intensity of the FEL would allow both dilute and complex systems to be analyzed.

Eventually, one can expect that x-ray FEL sources will be developed which have sub-femtosecond time resolution, which would allow probing the next level of temporal dynamics in atomic systems.

## Condensed Matter Physics and Materials Science

The characteristic distances important for studies of condensed materials typically range from micrometers down to angstroms, and the typical interaction energy runs from about 1 eV down to 1 $\mu$eV (corresponding to interaction times in the femtosecond to nanosecond range). These values match very well with the characteristic length and time scales of an x-ray FEL pulse: Angstrom wavelength with coherence length of many micrometers, and pulse duration measured in femtoseconds.

In addition, the brightness is much higher than that produced by any other x-ray source. This feature will allow a large number of standard x-ray techniques to be applied to smaller samples, and to record the relevant signal in less time. Examples include surface scattering studies of very small samples, diffraction from samples in very high pulsed magnetic fields, time-resolved studies of crack propagation, diffraction from materials undergoing shock wave distortion, diffraction from single grains in complex polycrystalline materials, studies of critical phenomena, and scattering studies of laser pulse-induced charge modulations in materials [5]. One large class of experiments that can take advantage of the x-ray FEL are those involving x-ray photon correlation spectroscopy (XPCS) [10]. This technique today is limited by source brightness to studying length scales larger than 100 nm and time scales longer than 1 ms. Both the length and time limits could be reduced by orders of magnitude with an x-ray FEL. In particular, time scales shorter than 1 ps could be probed, allowing the XPCS technique to complement energy-resolved inelastic x-ray scattering techniques. Initial FEL designs, such as the LCLS, are particularly suited for XPCS measurements in the ps-ns time range. This range is complementary to that covered by the neutron spin echo technique, which is now being used to study the dynamics of the glass transition. The vastly higher brightness of the FEL source would allow experiments to be performed in much less time and with much smaller samples.

In addition to these extensions of existing techniques, the x-ray FEL should allow the development of some completely new ways to study condensed matter. Some of these new techniques will most likely involve the high degeneracy of the FEL photon state. By analogy with visible laser science, one can envision gaining higher spectroscopic resolution through multi-photon excitations, or nonlinear interactions between the FEL x-rays and synchronized visible laser pulses [19]. If suitable resonances can be identified, x-ray photon echo experiments could provide a new sensitive probe of internal fields in materials.

The very high peak power of the x-ray FEL could also be used to induce desired permanent changes in materials. For example, a spatial interference pattern created from the FEL radiation could have enough intensity to carve a high-quality Fresnel optic into a smooth block of material. The focused FEL beam could also be used to create small holes (microexplosions) deep inside a sample.

# Biology

Structural biologists wish to determine the atomic structures and to observe the dynamical interactions between large molecules (mass between 5 kDa and

$5 \times 10^6$ kDa). The dynamical time scales of interest are typically microseconds or longer, but for some extremely interesting systems (*e.g.*, photosynthetic reaction centers, light-harvesting complexes, photosystem II, and light sensors such as photoactive yellow protein and bacteriorhodopsin) the interesting time scale can be as short as a few femtoseconds.

The high brightness of the x-ray FEL would allow structures to be determined using very small samples. It should be possible to study two-dimensionally ordered crystals (*e.g.*, membrane proteins), which are notoriously hard to crystallize in three dimensions and which are both numerous and of keen biological interest. In addition to conventional crystallographic techniques, it might be possible to exploit the spatial coherence of the FEL radiation to get structural information holographically.

The short time structure of the FEL pulse could be used to probe sample dynamics on a femtosecond to nanosecond time scale. There is interest in both time correlation studies of thermal fluctuations, and pump-probe relaxation studies (using as a pump either an external synchronized laser or the FEL x-ray pulse itself).

## X-RAY FEL DEVELOPMENT PLANS

Since the realization in the early 1990's that it is probably technically feasible to construct a SASE FEL operating at angstrom wavelengths, the x-ray community has come to embrace the idea of an FEL machine as the "fourth generation" in the progression of large x-ray facilities based on high-energy electron accelerators. At the ICFA Workshop on 4th Generation Light Sources in 1996 in Grenoble, France [12], the working group on hard x-ray applications of future x-ray sources concluded that "[The] hard x-ray group [is] unanimously excited about the FEL project as [the] 4th generation". This sentiment was echoed in the report of a Dept. of Energy (DOE) working group considering the state of synchrotron radiation in the US in 1997 [20], which stated that "We have also considered 'fourth generation' x-ray sources which will in all likelihood be based on the free electron laser concept... It is our strong view that exploratory research on fourth generation x-ray sources must be carried out and we give this item very high priority". More recently, a DOE panel charged with setting priorities in the area of novel coherent light source development [21] decided that "The Panel found that the most exciting potential advance in the area of innovative science is most likely in the hard X-ray region in the range of 8–20 keV, and even higher. This is especially the case if a light source can be built with a high degree of coherence, temporal brevity, and high pulsed energy".

In response to this rising enthusiasm, the DOE has established a plan for development of FEL light sources. The ultimate goal of this plan is to build a working SASE x-ray source in the 1–10 Å range. This will be accomplished through R&D efforts toward satisfying the technical requirements of such a machine, and through studying SASE FEL physics at a sequence of FEL projects with progressively shorter wavelength. A number of DOE National Laboratories are collaborating in this effort, along with several university labs. If adequate funding is provided, an x-ray FEL based on the SLAC linear accelerator (the only existing linac capable of reaching the necessary 15 GeV electron energy) will begin producing 1.5 Å radiation in about 2006.

In parallel with the US FEL program, Germany is also building short-wavelength SASE FELs. The TESLA project at DESY in Hamburg is building an FEL facility based on a new superconducting linac. Proceeding in stages, this project is expected to become an FEL scientific user facility within a few years, with radiation wavelengths as short as 60 Å. DESY has also developed an ambitious plan for a large x-ray FEL user facility, with 50 experimental stations operating simultaneously, based on a huge superconducting linac which would also supply electrons for high-energy physics experiments [5]. This plan is not funded yet, but enjoys political support.

Though many technical and political hurdles remain, it now seems quite likely that one or more fourth-generation x-ray source, based on FELs, will be built within the next ten years. It will provide a basis for a fantastic new generation of x-ray science.

## ACKNOWLEDGMENTS

Many people have contributed substantially to the development of the ideas presented here. Among the hundreds of participants at workshops on the applications of x-ray FELs (references 10–15), I wish to recognize Bernhard Adams, Robert Birgeneau, Pisin Chen, Fred Dylla, Keith Nelson, and Gerhard Materlik, for their particularly cogent contributions. I owe my education in FEL physics to the members of the LCLS collaboration, in particular Max Cornacchia, Claudio Pellegrini, and Roman Tatchyn. And, above all, I would like to acknowledge the energy and vision of Herman Winick, who forced x-ray scientists to pay attention to what accelerator scientists were saying about FELs.

This work was supported by the US Department of Energy, Office of Basic Energy Sciences, under contract DE-AC03-76SF00515.

## REFERENCES

1. Madey, J.M.J., *J. Appl. Phys.* **42**, 1906 (1971).
2. An internet site at the University of California at Santa Barbara contains links to information about FELs around the world. The address is: http://sbfel3.ucsb.edu/www/vl_fel.html
3. Murphy, J.B. and Pellegrini, C., *J. Opt. Soc. Am.* **B2**, 259 (1985).
4. *LCLS Design Study Report*, SLAC Report 521 (1998) [available online at http://www-ssrl.slac.stanford.edu/lcls/design_report/e-toc.html]

5. *Conceptual Design Report of a 500 GeV e+e- Linear Collider with Integrated X-Ray Laser Facility.* DESY Report 1997-048 (1997) [available online at http://www.desy.de/~schreibr/cdr/cdr.html]
6. Mandel, L., *J. Opt. Soc. Am.* **51**, 797 (1961).
7. Zholents, A., et al., *Nucl. Instrum. and Methods* **A425**, 385 (1999).
8. Barty, C.P.J., et al., in *Time-Resolved Diffraction*, J. Helliwell and P.M. Rentzepis, eds., New York: Oxford Univ. Press, 1988, ch. 2, p.44.
9. Schoenlein, R.W., Leemans, W.P., et al., *Science* **274**, 236 (1996).
10. Gruebel, G., ed., *Workshop on Perspectives of X-Ray Photon Correlation Spectroscopy*, ESRF 1996.
11. Arthur, J., Materlik, G, and Winick, H., eds., *Workshop on Scientific Applications of Coherent X-Rays*, SLAC Report 437 (1994).
12. Laclare, J.-L., ed., *4th Generation Light Sources*, ESRF (1996).
13. Schneider, J., ed., *X-Ray Free Electron Laser Applications*, DESY (1996).
14. Knotek, M., Arthur, J., Dylla, F., and Johnson, E., *Workshop on Scientific Opportunities for Fourth Generation Light Sources*, Argonne (1997).
15. Lindau, I., and Arthur, J., eds., *Workshop on Scientific Applications of the LCLS*, SLAC (1999).
16. Shenoy, G., ed., *Future Light Sources*, Argonne (1999).
17. Zuo, T., and Bandrauk, A.D., *Phys. Rev.* **A51**, R26 (1995).
18. Gavrila, M., and Shertzer, J., *Phys. Rev.* **A53**, 3431 (1996).
19. Namikawa, K., *X-ray parametric scattering stimulated by an intense laser field*, contribution to Ref. 10 (1994).
20. *Report of the Basic Energy Sciences Advisory Committee, Synchrotron Radiation Light Sources Working Group* (1997).
21. *Report of the Panel on Novel Coherent Light Sources for the Basic Energy Sciences Advisory Committee* (1999).

# X-ray Lasers Driven by Optical Lasers

Yoshiaki Kato, Akira Nagashima, Keisuke Nagashima, Masataka Kado,
Tetsuya Kawachi, Noboru Hasegawa, Momoko Tanaka,
Akira Sasaki, and Kengo Moribayashi

*Advanced Photon Research Center, Kansai Research Establishment*
*Japan Atomic Energy Research Institute*
*8-1 Umemidai, Kizu, Souraku, Kyoto, 619-0215 Japan*

**Abstract.** Recent topics in the optical-laser-driven x-ray lasers are reviewed. With the collisional excitation x-ray lasers, pumping laser energy has been reduced over 100 times by reducing the pumping pulse width from ns to ps. A high gain of 30-40 cm$^{-1}$ has been achieved with the transient collisional excitation using ps pumping. A new scheme for a charge exchange x-ray laser using clusters such as $C_{60}$ irradiated with ultra short laser pulses has been introduced. In the inner shell ionization x-ray lasers, creation of a population inversion utilizing the difference in the Coster-Kronig decay rates between the upper and lower laser levels has been proposed as a scheme to enable pumping with short bursts of electrons. High gain over long duration is expected with hollow atoms created under intense broad-band x-ray irradiation emitted from oscillating electrons under relativistic intensities (Larmor radiation). The x-ray laser program at the Advanced Photon Research Center at the new site in Kyoto is presented.

## INTRODUCTION

There has been significant demand for applications of intense, short duration x-ray pulses. X-ray lasers driven by optical lasers, which can be generated with small scale facilities, have the properties of high photon flux per pulse, short temporal duration and high coherence. High brightness x-ray lasers are very useful for probing high density and high temperature plasmas: drilling of a high intensity laser pulse into a plasma (1) and growth of hydrodynamic perturbations during acceleration (2) have been measured by x-ray laser refraction and x-ray laser back-lighting methods, respectively. Using the coherent property of the x-ray laser, small surface deformation of metallic Nb due to intense electric field has been measured with 4-nm accuracy (3). Soft x-ray lasers will also be useful for studies of surface physics and photo-ionization processes, trace element analysis in semiconductor materials, inspection and writing of nm-scale structures in lithographic applications, observation of biological cells, and nonlinear optics in the x-ray region.

Recently, propagation of coherent acoustic phonons in GaAs crystals were observed by time-resolved x-ray diffraction using incoherent Cu K$\alpha$ x-rays (1.54 A) generated with 30-fs laser irradiation of Cu targets (4). This work indicates that structural dynamics of various materials (solids, liquids and molecules) with the time scale of atomic motions will be revealed if x-ray lasers of keV photon energies with ps-fs durations become available.

Some of the important issues in x-ray laser research are x-ray lasings at shorter wavelengths, improvement of the coherence, and the possibility for high repetition rate to generate high time-averaged photon flux. In this paper we review recent topics in the x-ray laser research, and present the x-ray laser program at the Advanced Photon Research Center (APRC).

## COLLISIONAL EXCITATION X-RAY LASERS

After extensive research over various possible x-ray laser schemes, it turned out that the collisional excitation (CE) method (5) results in the most robust x-ray laser. Lasing has been obtained with almost any mid- to high-Z atomic elements (from S: Z=16 to Au: Z=79), with various electronic configurations (Ne-like, Ni-like and Pd-like ions), over wide wavelength ranges (from 60.8 nm to 3.56 nm), and under various excitation conditions (ns- to fs-laser excitation, as well as capillary discharges).

Reflecting these findings, most of the recent x-ray laser research is focused on improving the performance of CE x-ray lasers. The important progress in reducing the size of the CE x-ray laser is the "double pulse pumping", where a small laser pulse is used to produce a preplasma with a controlled density profile, which is then heated with a short-duration main laser pulse to create a population inversion for x-ray lasing (6). In comparison to the previous approach, where a single ~kJ ns-pulse was used for plasma generation and heating, the required laser energy has been reduced to ~100 J for 100-ps pumping, and further to less than 10 J for 1-ps pumping.

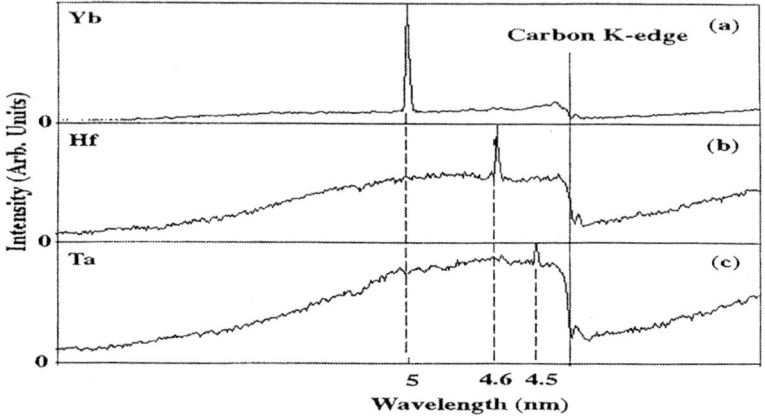

**FIGURE 1.** Spectra of the Ni-like Yb, Hf and Ta x-ray lasers (unsaturated amplification) at 5.0, 4.7, and 4.5 nm, respectively. The shorter wavelength side of the carbon K-edge at 4.38 nm is the "water window".

With the double pulse pumping method, saturated amplification has been achieved in Ni-like Sm at 7.3 nm by a 75-ps, 130-J, 1-µm laser, with an x-ray laser output energy of 313 µJ, an output intensity of $2.0 \times 10^{11}$ W/cm$^2$ and the conversion efficiency of $2 \times 10^{-6}$ (7). Extension to shorter wavelengths with this scheme is under investigation at ILE, Osaka University (8), where amplification at 5.0 nm with Yb, 4.7 nm with Hf, and 4.5 nm with Ta, although not saturated, has been observed with 100-ps, 240-J and 1-µm laser pumping (Fig. 1).

Further reduction of the pumping energy is possible by reducing the pumping pulse duration while keeping the irradiation intensity high enough for sufficient heating to create a population inversion. When the pulse duration is longer than tens of ps, the upper laser state collisionally mixes with nearby excited states, resulting in a quasi-steady state gain of typically 5-10 $cm^{-1}$. However, when the pulse duration is reduced to less than the collisional relaxation time of the upper laser state, which is typically a few ps, a very high gain coefficient of 30-40 $cm^{-1}$ is achieved, because the population inversion density during this transient state is higher than that in the quasi-steady state. Based on this transient collisional excitation (TCE) scheme (10, 11), saturated amplification in Ni-like Pd at 14.7 nm has been achieved with a 1.1-ps, 5.2-J pumping pulse which was preceeded by a 800-ps, 4.2-J prepulse (12). With TCE, lasing has been obtained in Ne-like ions of Ti (32.6 nm), V (30.5 nm) and Fe (25.5 nm), and in Ni-like ions of Y (24.0 nm), Zr (22.0 nm), Nb (20.3 nm), Mo (18.9 nm), Pd (14.7 nm) and Sn (12.0 nm).

Saturated amplification has also been achieved in discharge-pumped CE lasers in Ne-like Ar at 46.9 nm (13) and Ne-like S at 60.8 nm (14) in a capillary of 4 mm diameter and up to 16.4 cm gain length. The small capillary-discharge device has been continuously operated at a 7 Hz repetition rate resulting in the 46.9-nm Ar laser with an average power of approximately 1 mW (15). The peak coherent power of this laser has been estimated to be about 2.3 kW, more than 5 orders of magnitude higher than an Advanced Light Source (ALS) undulator radiation of 4 mW.

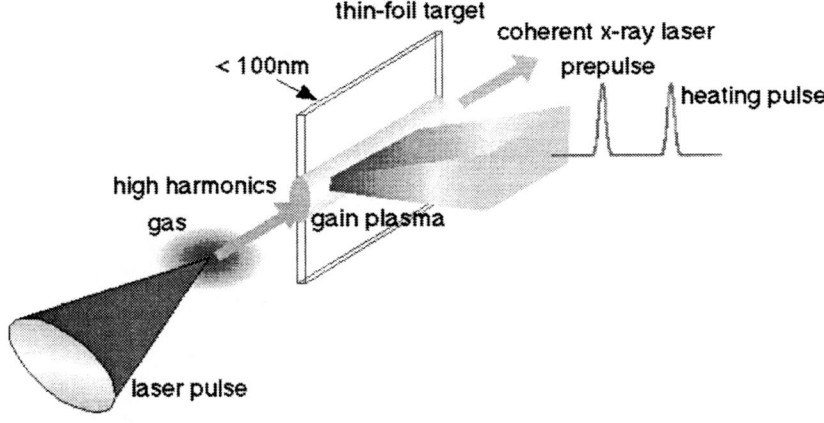

**FIGURE 2.** Schematic layout of coherent x-ray laser generation with a thin foil target plasma injected with high harmonics.

In order to develop a compact high rep-rate x-ray laser at shorter wavelengths, we are studying the possibility for a TCE-pumped x-ray laser with a thin foil target (Fig. 2). In contrast to standard slab targets where pumping laser energy is thermally conducted to cold high-density plasmas, the laser energy in this design is efficiently used to heat the plasma in the lasing region. By irradiation of a thin foil with two short-duration laser pulses of approximately 10-J energy at appropriate intensities and pulse separation, a plasma is created with an electron density of $4 \times 10^{21}$ $cm^{-3}$ (higher than the critical density for the 1-µm pumping laser) and an electron temperature of close to 1.5 keV, which is sufficient to create a population inversion for x-ray lasing in the 150-250 eV region. The gain guiding, due to the high gain coefficient of the TCE x-ray laser, will overcome the loss due to refractive defocusing of the x-ray laser beam during propagation in a symmetrically expanding plasma. In order to generate an x-ray laser beam with high coherence, injection and amplification of a high-order harmonic of a long laser pulse from a tunable Ti:sapphire laser is planned. The CPA Nd:glass laser system which we are preparing for this experiment at the Advanced Photon Research Center (APRC) is shown in Fig. 3. The output energy of each laser beam before compression is approximately 20 J. A preliminary study shows that a rep-rated slab amplifier to replace the amplifier chain is possible for continuous operation of the TCE x-ray laser.

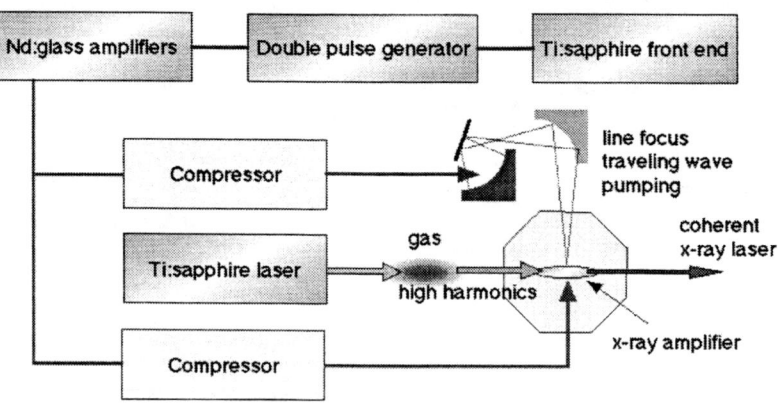

**FIGURE 3**. Schematic layout of the 2-beam CPA laser for x-ray laser experiments at the Advanced Photon Research Center.

## CHARGE EXCHANGE X-RAY LASER

With the recombination scheme to create a population inversion, ground-state lasing has been obtained in Li III at 13.5 nm in microcapillary plasmas (16). A preplasma, formed in a 0.3-mm diameter and 5-mm long LiF microcapillary by irradiation of a long laser pulse, was field-ionized with a 250-fs, 0. 25-μm, 50-mJ KrF laser pulse to create a population inversion. A gain-length product of ~6.5 has been achieved in this medium.

Another interesting recombination scheme is a charge exchange, which is an efficient process for populating particular excited states of ions. For example, the n=3 level in H-like C will be populated in the following reaction:

$$C^{6+} + C^{2+} \rightarrow C^{5+} (n=3) + C^{3+}.$$

X-ray lasing could be obtained in the n=3 →2 transition at 18.2 nm, or possibly in the 3 →1 transitions at 2.89 nm. The major difficulty in this scheme is that the high-charge and the low-charge ionic states have to coexist at the boundary where charge exchange takes place. Since the charge exchange cross section is very large, of the order of $10^{-15}$ $cm^2$, the boundary has to be very sharp (~10 μ m). In the collision of 2 plasmas, it is not easy to create such a sharp boundary over an extended length required for x-ray amplification.

An interesting proposal has been made by Chichkov and others (17), where $C_{60}$ is used as a source of $C^{6+}$. It has been demonstrated that highly charged ions are produced when a cluster is irradiated with a high intensity, short laser pulse due to collisional ionization in a cluster before the cluster explodes (18). The $C_{60}$ molecule, which is regarded to be a small size cluster, will be efficiently ionized under irradiation of a short-wavelength, short-duration laser pulse. The $C^{6+}$ ions thus generated will directly collide with the surrounding low-density weakly-ionized gas such as $C^{2+}$, or neutral He gas for the similar charge transfer to take place (see Fig. 4). This scheme could be extended to shorter wavelengths if lasing is obtained in the ground state transitions.

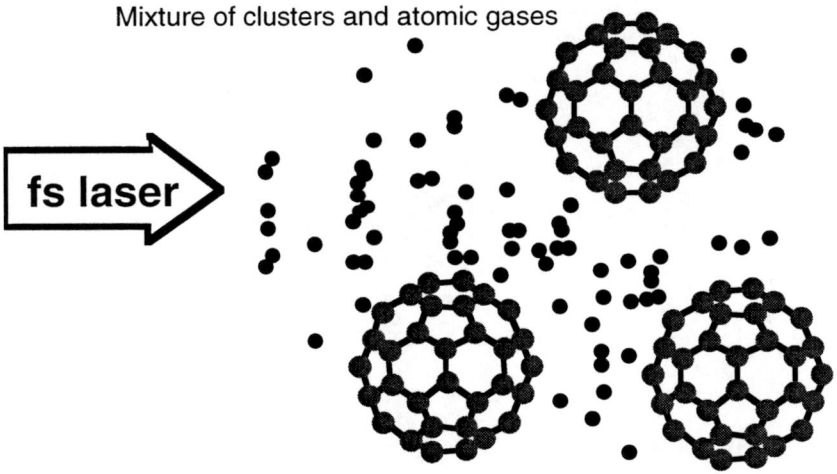

**FIGURE 4.** Charge exchange x-ray laser using $C_{60}$ irradiated with a short laser pulse.

## INNER-SHELL IONIZATION X-RAY LASERS

X-ray lasing at keV photon energies may require transitions to inner-shell ionization states. Theoretical investigation on the photo-ionization pumped carbon x-ray laser at 4.5 nm in the 2p-1s transition of neutral C has shown that a 1-J, 40-fs laser is required for pumping (19). The x-ray laser energy will be ~0.1 µJ ($2 \times 10^9$ photons) per pulse, with an average power of 1 µW at 10 Hz. The high-power, ultra-short pulse lasers required for pumping are available at several laboratories, including the 100-TW, 20-fs, 10-Hz Ti:sapphire laser at APRC (20).

Electron pumping, instead of photon pumping, for inner-shell ionization has been proposed at University of California, San Diego (UCSD). Although population inversion is not created in the electron collisional ionization, the difference in the Coster-Kronig decay rates between the upper and lower lasing levels can be utilized to create the population inversion. Many possibilities satisfying this criterion have been found, with a typical candidate being the Ti $L_2 \rightarrow M_1$ transition at 3.09 nm (21). In this case also, a short pumping duration of less than 10 fs is required to obtain significant gain. This scheme is also useful for photoionization pumping to search for a larger number of candidates for inner-shell-ionization x-ray lasers (22).

The possibility for lasing in hollow atoms has been studied at APRC, where high pumping photon flux is used to create multiple vacancies in an inner shell (23) (Fig. 5). With hollow atoms, where a particular inner shell is completely ionized, high gain is sustained over a long duration. The high photon flux with broad bandwidth required for pumping may be generated by Larmor radiation (24), which is emitted as the intense, directional radiation when the electrons oscillate along the propagation direction of the laser beam under relativistic intensities of $a_0 = eE/m_e c\omega \gg 1$, where E is the amplitude of the laser electric field, and $\omega$ is the angular frequency of the laser light. As an initial study, the high intensity soft x-ray laser beam may be used to generate hollow atoms to investigate the possibility of the hollow atom x-ray laser.

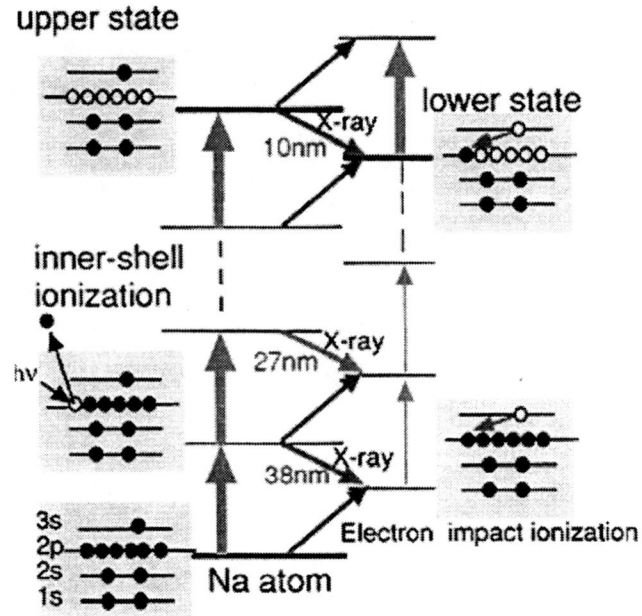

**FIGURE 5.** Multi inner-shell ionization x-ray laser with Na vapor.

## X-RAY LASER RESEARCH AT APRC

In the x-ray laser program at APRC, advanced laser facilities such as a CPA Nd:glass laser and a 100 TW, 20 fs Ti:sapphire laser are devoted to the development of high performance and compact x-ray lasers. Intense, short duration and coherent soft x-ray lasers will be developed based on TCE pumping. Emphases will be placed to start applications of the soft x-ray lasers, including interaction of the intense soft x-ray radiation with solids (crystals and polymers), which forms the basis for material science, and applications in photochemistry, lithography and biology. Extension of the

TCE laser to shorter wavelengths will be pursued using thin foil targets. New schemes which have potential for efficient and short wavelength lasings, such as the charge-exchange x-ray laser and the inner-shell ionization x-ray laser, will be investigated using the ultrashort-pulse lasers developed at APRC. It is intended that this work will lead to the development of new schemes of x-ray lasers operating in the keV region, which will have broad applications due to the high brightness, short pulse duration and high coherence (Fig. 6).

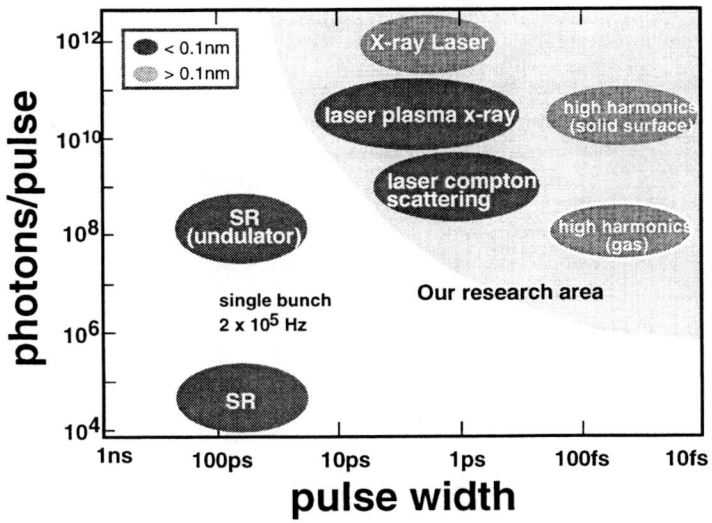

FIGURE 6. High intensity short pulse x-ray sources

## ACKNOWLEDGEMENTS

We would like to thank G. Tallents, C. Tóth and S. Suckewer for providing us with recent articles to help in writing this review.

## REFERENCES

1. Takahashi, K., et al., *Proc. SPIE* **3156**, 146 (1997).
2. Wolfrum, E., et al., *Phys. Plasmas* **5**, 227 (1998).
3. Zeitoun, Ph., et al., *Nucl. Instrum. Methods*, **A 416**, 189 (1998).
4. Rose-Petruck, Ch., et al., *Nature* **398**, 310 (1999).
5. Matthews, D. L., et al., *Phys. Rev. Lett.* **54**, 110 (1985).
6. Nilsen, J., and Moreno, J. C., *Phys. Rev. Lett.* **74**, 337 (1995).
7. Zhang, J., et al., *Science* **276**, 1097 (1997).
8. Daido, H., et al., *Proc. SPIE* **3767** (1999), to be pubilshed.
9. Tallents, G., et al., *Proc. SPIE* **3767** (1999), to be pubilshed.

10. Afanasiev, Yu. V., and Shylaptsev, V. N., *Sov. J. Quant. Electron.* **19**, 1606 (1989).
11. Nickles, P. V., et al., *Phys. Rev. Lett.* **78**, 2748 (1997).
12. Dunn, J., et al., *Phys. Rev. Lett.* **80**, 2825 (1998).
13. Rocca, J. J., et al., *Phys. Rev. Lett.* **73**, 2192 (1994).
14. Rocca, J. J., et al., *Phys. Rev. Lett.* **77**, 1476 (1996).
15. Rocca, J. J., et al., *Phys. Rev. Lett.* **81**, 5804 (1998).
16. Korobkin, D., et al., *Phys. Rev. Lett.* **77**, 5206 (1998).
17. Chichkov, B., et al., *Jpn. J. Appl. Phys.* **38**, 1975 (1999).
18. Ditmire, T., et al., *Phys. Rev. Lett.* **78**, 2732 (1997).
19. Moon, S. J., and Eder, D. C., *Phys. Rev.* **A 57**, 1391 (1998).
20. Yamakawa, K., et al., *Opt. Lett.* **23**, 1468 (1998).
21. Kim, D., Tóth, C., and Barty, C. P. A., *Phys. Rev.* **A 59**, R4129 (1999).
22. Kim, D., et al., *Proc. SPIE* **3767** (1999), to be pubilshed.
23. Moribayashi, K., et al., *Phys. Rev.* **A 58**, 2007 (1998).
24. Ueshima, Y., et al., Laser Part. Beams **17**, 45 (1999).

# X-Rays in Curved Spaces

J.A. Golovchenko* and Chien Liu[†]

*Department of Physics and *[†]Division of Applied Sciences,
Harvard University, Cambridge, MA 02138
*[†]Rowland Institute for Science, Cambridge, MA 02142

**Abstract.** In this review we discuss and demonstrate three new phenomena connected with the binding and propagation of X-rays on curved surfaces. The first involves X-ray propagation, over very large distances, of surface bound modes, made possible by the extrinsic curvature of the surface. Under proper conditions transmission of the ground state alone can be achieved. The second effect involves strong dynamical diffraction that surface bound modes can experience when the surface contains a periodic modulation. The last effect is the in-surface deflection of surface bound X-rays induced by the intrinsic, rather than extrinsic, curvature of the surface. We briefly discuss some of the implications of these discoveries.

## INTRODUCTION

The purpose of this paper is to review some recent progress made in the field of X-ray optics. Specifically we focus here on describing a curious set of phenomena associated with the propagation of X-rays bound to curved surfaces. The binding or confinement of X-rays to the near surface region of a curved dielectric mirror is closely related to the acoustic whispering gallery modes at much longer wavelengths observed and explained by Lord Rayleigh about a century ago (1). In the X-ray regime ($\lambda \approx 1$ Å), at large radii of curvature ($> \sim 10$ cm), a ray optical view of this phenomena, in which propagation occurs as a result of multiple optical reflections, can be adequate. However, a wave optical study suggests that for smaller radii of curvature ($< \sim 10$ cm), the propagation is best viewed as resulting from the interference of individual radial eigenmodes of the radiation field, each of which conforms to the curvature of the dielectric surface in a cylindrical geometry. As the radius of curvature decreases, the number of modes that can be confined decreases. Ultimately only one mode survives, and a view of ray propagation by multiple bounces is totally unwarranted.

Further study shows that these radial eigen modes are "leaky", due partly to photoelectric absorption, and partly to a "tunneling" process by which X-rays escape into the bulk of the dielectric. Calculations show that by using either a sufficiently long surface propagation distance or a sufficiently small radius of curvature, a single coherent transverse mode of the X-radiation field can be prepared which extends over several hundred Angstroms above the curved surface. We show these conditions can be achieved experimentally.

A wave optical view of the surface X-ray propagation is also necessary when the crystal planes of the dielectric from which the X-ray mirror/wave guide is made are perpendicular to the bounding surface. The surface bound X-ray eigen modes may then undergo Bragg diffraction parallel to the surface which dramatically influences the propagation along the surface. We will demonstrate that the diffraction can be so strong for a mirror fabricated from single crystal silicon that a complete dynamical theory of surface X-ray wave propagation is required.

Finally, by taking advantage of this ability to excite strong Bragg diffraction of surface bound X-rays, we have been able to reveal an additional and initially unexpected aspect of the surface propagation. This phenomena involves the gradual deflection of surface bound X-rays "within" the surface induced stresses in the dielectric medium that results in a non Euclidean geometry in the surface. The resulting effective force on the X-ray photons resembles that induced on matter (and light) by gravity, only it is the spatial curvature, rather than the space-time curvature of the propagation space, that plays a crucial role.

Various components of the above introduction were reported in two recent publications in Physical Review Letters (2) and Optics Letters (3). A more extended discussion of many of the mathematical and experimental methods can be found in the Ph.D. thesis of Dr. Chien Liu. (4) Much of our interest in this field was stimulated by synchrotron radiation experiments on flat dielectric wave guides discussed in references (5) and (6). We further acknowledge theoretical work by Howells and coworkers dealing with short wavelength X-ray propagation on curved surfaces (7) and those of Hagelstein and Vinogradov focused on cavities for X-ray lasers (8, 9).

In the following we present the highlights of a simplified theoretical discussion and a series of experimental results that demonstrate the binding, diffraction and "in-surface" deflection of X-rays discussed above.

## BINDING X-RAY PHOTONS TO SURFACES

We begin with a simple discussion of the optics of total reflection of X-rays from a flat dielectric mirror. We shall try to cast the relevant electromagnetic field equations in a form that reminds the reader of Schrodinger's equation in quantum mechanics, thereby calling to mind mechanical analogies. In particular, the notion of an "effective potential" for the light waves will be very helpful in more advanced examples.

All of the X-ray optics we will discuss has its foundations in the mean field Maxwell's equations in which linear response theory is applied to describe the induced currents and charges excited by electromagnetic waves. When all the induced effects can be hidden through the device of a (spatially dependent) dielectric function $\varepsilon$, the fields obey

$$\vec{\nabla} \times \vec{E} = -\frac{1}{c}\frac{\partial \vec{B}}{\partial t},$$

$$\vec{\nabla} \times \vec{B} = \frac{1}{c}\frac{\partial \vec{D}}{\partial t},$$

$$\vec{\nabla} \cdot \vec{B} = 0,$$

$$\vec{\nabla} \cdot \vec{D} = 0,$$

and $\vec{D}$ is related to $\vec{E}$ by the constitutive relation $\vec{D} = \varepsilon \vec{E}$. The dielectric function $\varepsilon$ is given by

$$\varepsilon(\omega) = 1 + \frac{4\pi N e^2}{m} \sum_i \frac{f_i}{\omega_i^2 - \omega^2 - i\omega \Gamma_i}$$

where $N$ is the number of atoms per unit volume in the dielectric material, $f_i$ is the effective number of electronic oscillators at frequency $\omega_i$, and $\Gamma_i$ is the oscillator damping constant. We shall write the dielectric function in the form

$$\varepsilon = 1 - \delta + i\eta.$$

For vacuum $\varepsilon = 1$ and in silicon for 17.4 keV X-rays,

$$\varepsilon = 1 - 3.17 \times 10^{-6} + i\, 1.71 \times 10^{-8}.$$

Referring to Figure 1, which depicts an X-ray approaching and being reflected by a dielectric mirror, we cast the solution in the form of a wave field whose $x$ and $t$ dependencies reflect the translational invariance of the problem in those dimensions, i.e.

$$E_y = \psi(y) e^{i k_x x - i k_0 c t}$$

where $k_0 = \omega/c$.

With this ansatz the Maxwell equations can be converted into an equation for $\psi$:

$$-\frac{\partial^2 \psi}{\partial y^2} + k_0^2\, \bar{\delta}(y)\psi = k_y^2 \psi,$$

where

$$\bar{\delta}(y) \equiv \begin{cases} 0, & y < 0 \\ \delta, & y \geq 0 \end{cases}.$$

The spatial part of the wave field is like that of a particle approaching a potential step like that shown in Figure 2 below. (Damping has been neglected in this simple discussion). As in elementary quantum mechanics, there will be a total reflection of the wave as long as its "energy" is below that of the potential step, i.e. $R = 1$ when $k_y^2 \leq k_0^2 \delta$, where $R$ is the reflection coefficient for the surface. Since $\delta$ is much less than 1, we find using the geometry indicated in Figure 1 that total reflection only occurs for angles of incidence $\theta$ such that

$$\theta \leq \sqrt{\delta}.$$

Including the effects of damping give a reflection coefficient of

$$R = \left| \frac{\sin\theta - \sqrt{\sin^2\theta - \delta + i\eta}}{\sin\theta + \sqrt{\sin^2\theta - \delta + i\eta}} \right|^2.$$

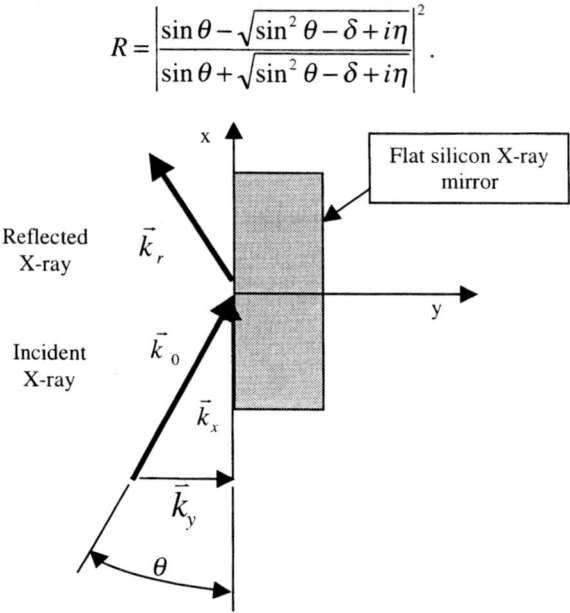

**FIGURE 1.** Specular reflection of an X-ray beam from a flat silicon mirror.

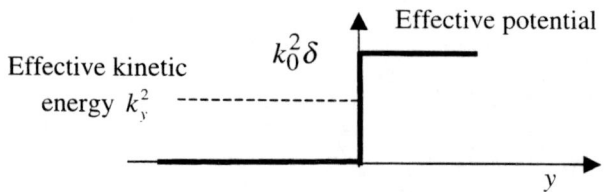

**FIGURE 2.** Effective potential for a flat silicon mirror shown in Figure 1.

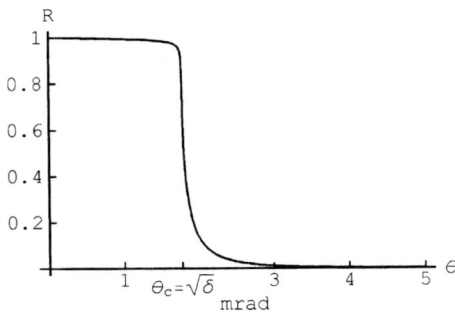

**FIGURE 3.** Reflectivity curve for a flat silicon mirror surface.

Figure 3 shows qualitatively how the reflectivity changes as a function of incidence angle. Quantitatively for silicon the critical angle $\theta_c = 1.78$ mrad for 17.4 keV X-rays and $R = 0.994$ at $\theta_c/2$.

We now attempt to predict how far along a curved mirror surface we can expect X-rays to propagate as a result of the multiple successive reflections depicted in Figure 4. If we consider propagation by multiple bounces along a 10 cm radius of curvature mirror we find after 1 cm of travel at $\theta = \theta_c/2 = 0.89$ mrad and $R = 0.994$/bounce we get 55 bounces and a total transmission of $R^{55} = 71\%$. These X-rays "hug" or are bound just above the mirror surface and never get more than about 2000 Å from it. Including all X-rays entering the mirror between angles of zero to $\theta_c$ on their first bounce, about 60% survive after 1 cm of travel.

We now show that for radii of curvature much below 10 cm this ray optical picture based on plane waves reflecting from a flat surface is unsatisfactory. To see why, let's return to the wave equation but this time deal with it directly in cylindrical coordinates. Assume the mirror has a radius of curvature $R$, and the axis direction is $z$, along which the electric field is assumed polarized for simplicity, and the azimuthal direction tangent to the mirror surface is $\hat{\phi}$ (see Figure 5). Writing for the electric field

$$E_z = \frac{1}{\sqrt{r}} \psi(y) e^{ik_\phi R\phi - ik_0 ct}$$

where $y = r - R$, we find from Maxwell's equations in a first order expansion around $r = R$ that $\psi$ again satisfies a simple one dimensional Schrodinger-like equation:

$$-\frac{\partial^2 \psi}{\partial y^2} + k_0^2 \left( \delta(y) - \frac{2r}{R} \right) \psi = \left( k_0^2 - k_\phi^2 \right) \psi.$$

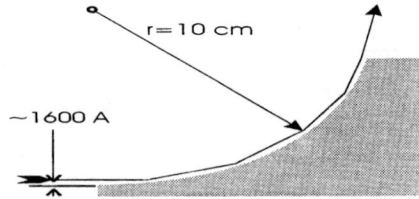

**FIGURE 4.** An X-ray beam undergoes multiple bounces on a curved mirror.

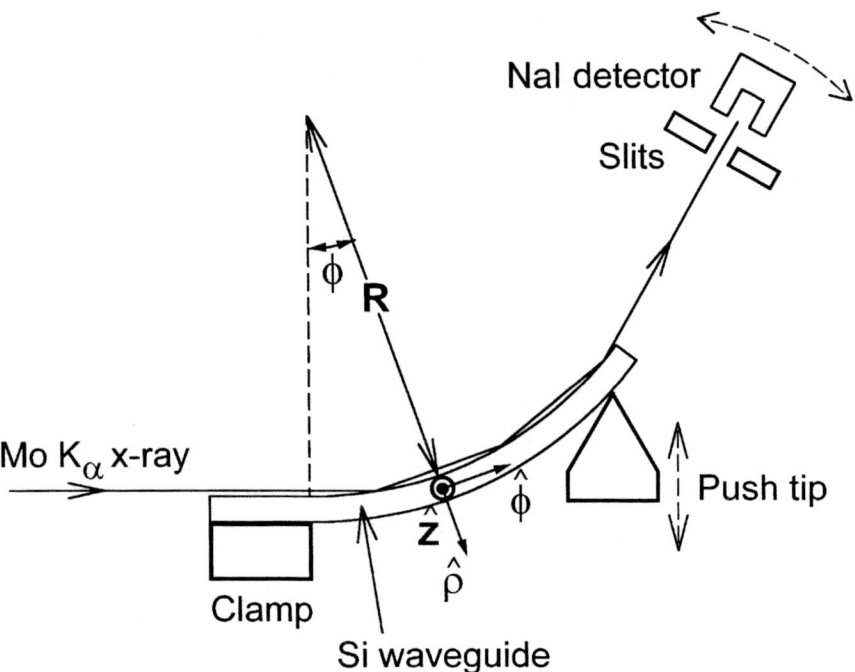

**FIGURE 5.** Implementation of a silicon wave guide with variable radius of curvature and the associated cylindrical coordinates.

Figure 6 shows the new effective potential along the radial coordinate with a few eigen function solutions. The centrifugal barrier corresponding to photons with angular momentum $\sim \hbar k_0 R$ together with the potential step caused by the index change at the mirror surface, combine to form a radially confining potential that supports quasi bound states that hug the mirror surface as they propagate. Figure 6 is calculated for a 10 cm radius of curvature silicon mirror. Note that in principle none of the bound states remain in the surface well indefinitely; they all ultimately tunnel into the mirror. In addition, when photoelectric absorption by the mirror is taken into account (through the imaginary part of the dielectric constant in the mirror), these eigen functions will decay in time at a rate proportional to the overlap with the mirror. Both of these effects are clearly minimized for the "ground state" solution which should have the longest lifetime, or equivalently the longest propagation distance along the surface.

The correspondence between the ray optical view and the wave optical view is most easily seen to occur for very large radius of curvature surfaces. At a large radius of curvature the slope of the "effective potential" in the vicinity of the bound states becomes very small and consequently the well width increases as does the number of surface bound states. Eventually one might expect there to be enough states to form a wave packet that simulates the behavior of a classical ray that that bounces back and forth radially in the well reflecting the "classical" multi bounce ray motion. Also, the distance over which states must tunnel to escape increases at large radius and this wave optical feature disappears for all states but those right near the top of the well.

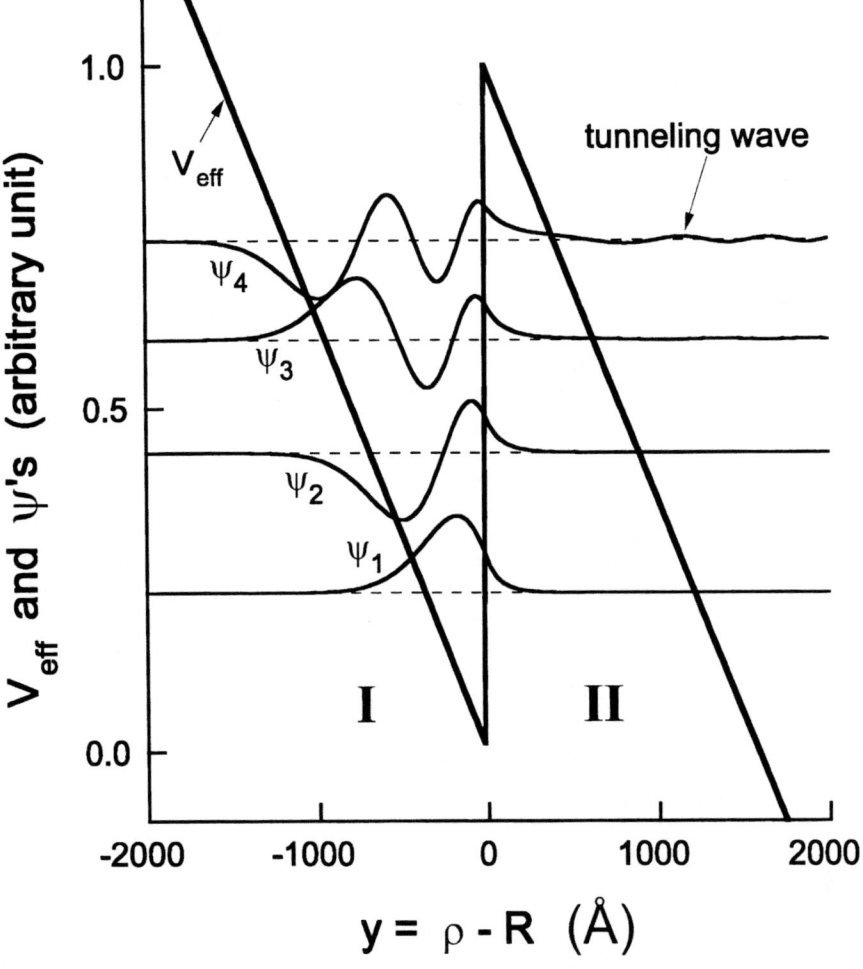

**FIGURE 6.** Effective potential and eigen functions for 17.4 keV X-rays on a 10 cm radius silicon wave guide.

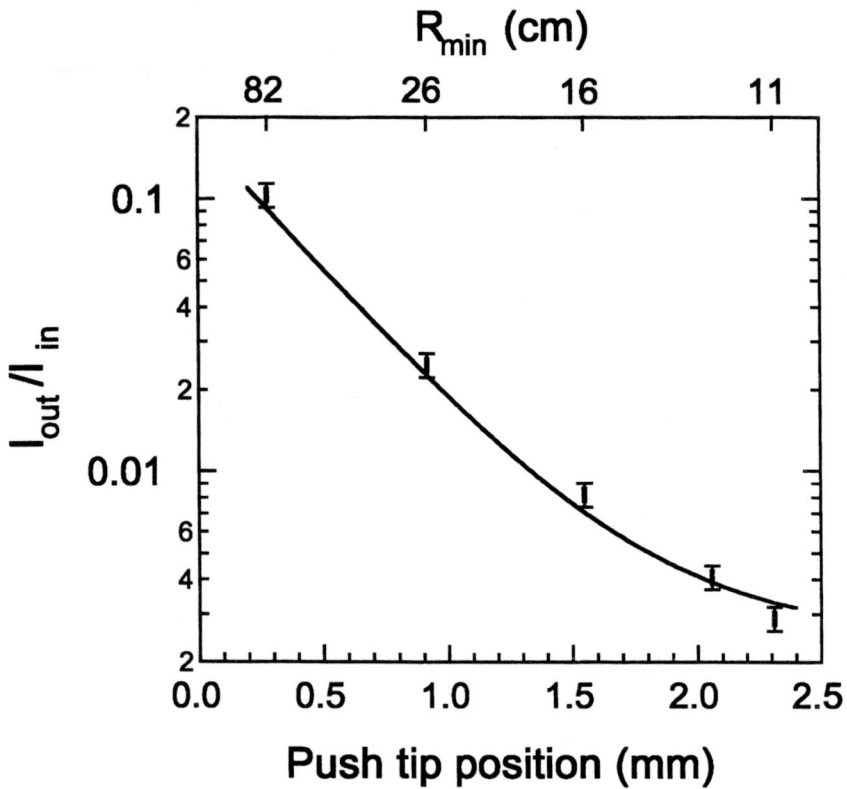

**FIGURE 7.** Transmission versus radius of curvature for an 18 mm long silicon wave guide.

Clearly the place to look for wave optical effects is at very small radius of curvature. As a first test of these ideas we performed experiments measuring the transmission of a curved X-ray mirror as a function of its radius of curvature, expecting to see significant deviations from the classical ray theory at small radii of curvature. Figure 5 shows a schematic of our experimental setup. The x-ray mirror is made of 100 micron thick silicon and mounted on a holder with a fixed clamp and a movable push tip that produces an adjustable radius of curvature in the silicon surface. A triangular shaped sample is used because a displacement of a corner relative to its clamped opposite side results in a curved mirror with nearly constant radius of curvature. A collimated photon counter is used to measure the transmitivity of the mirror at various radii of curvature. The solid curve in Figure 7 is the ray theory prediction of the surface transmitivity of the silicon mirror for radii of curvature extending down to about 11 cm, for an 18 mm long silicon mirror. The data agrees very well with the theory.

Figure 8 shows results for another mirror of length 5 mm, but whose radius of curvature could be reduced to nearly 1 cm. Here it will be seen that the data falls increasingly below the ray theory for radii below about 10 cm. Also plotted in Figure 8 is the prediction of the wave optical theory. The contributions to the transmitivity from each of the three important surface bound states are indicated. It can be seen that for decreasing radii below 10 cm only the ground state is important and the predicted reflectivities are significantly below those of the ray theory. This suppression is due to the tunneling loss of X-rays into the bulk of the mirror. The data lies significantly below the classical result, but is not quite as depressed as the wave optical theory. There are a number of possible experimental problems that can account for this, but clearly more research is needed on samples whose surface structure, oxide thickness, and macroscopic geometrical structure are more precisely characterized than has been achieved in these preliminary experiments.

Nevertheless, these result are highly encouraging and suggest that single mode x-ray operation of a curved mirror can be easily achieved by going to small radius of curvature and sufficiently long propagation distances.

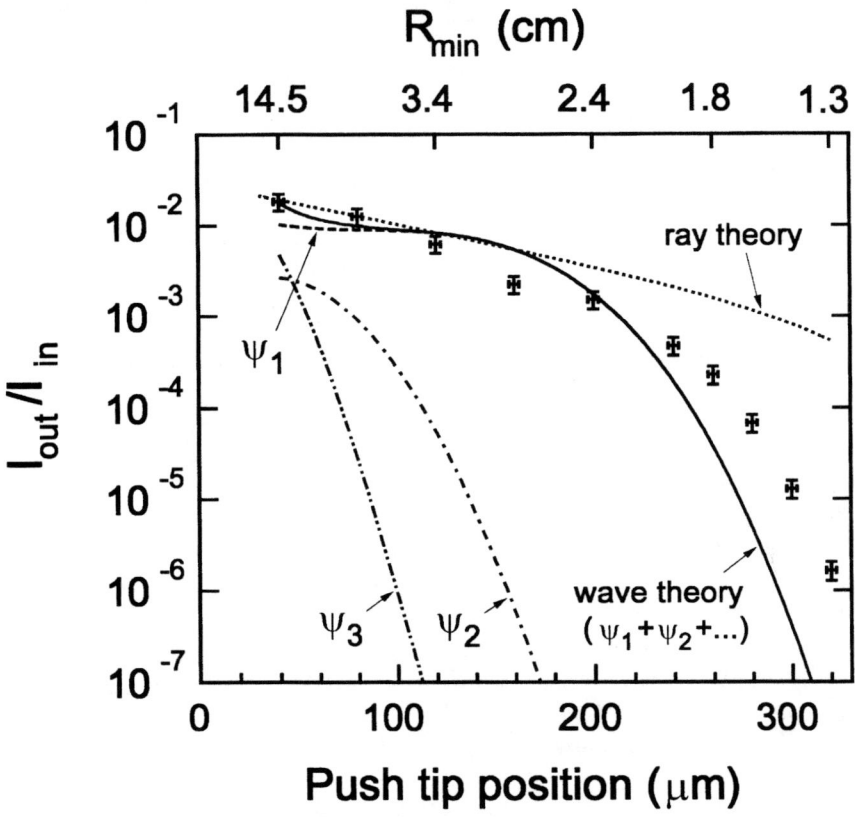

**FIGURE 8.** Transmission versus radius of curvature for a 5 mm long silicon wave guide.

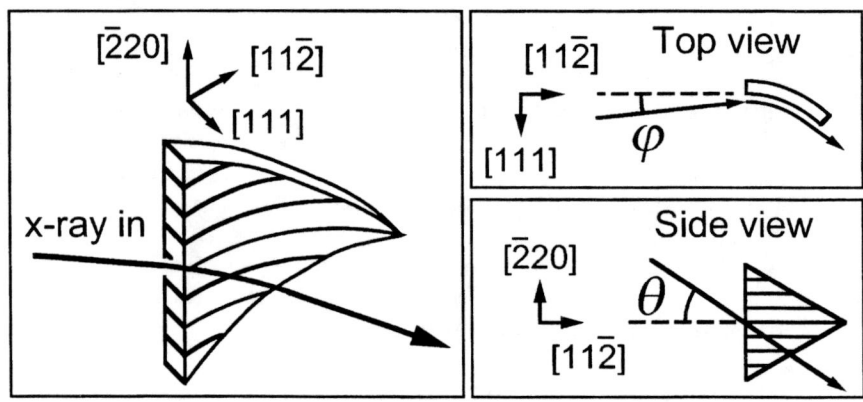

**FIGURE 9.** Geometry for studying diffraction effects of surface bound X-rays.

## DIFFRACTION OF SURFACE BOUND STATES

Another manifestation of a wave optical view of X-rays bound to curved surfaces is that they should be capable of experiencing strong diffraction effects in a periodic field that might exist at the bounding surface. Such periodicities could be artificially imposed on the surface of the mirror. In our experiments this periodicity is imposed by atomic planes that terminate at the surface of the single crystal silicon mirrors. Figure 9 shows the geometry we have used in our surface bound state diffraction studies. The lines across the mirror surface indicate the [220] atomic planes that intersect the surface perpendicularly. Thus a vector normal to the bulk atomic planes of interest here point along the z axis of our cylindrical coordinate system.

We might expect that when the incidence angle $\theta$ of a surface bound X-ray relative to the atomic planes satisfies the Bragg condition, a new surface bound diffracted wave will be excited. The coupling between these states might be expected to depend on the strength of the Fourier component of the charge density at the periodicity of the atomic planes, and the degree to which the evanescent, or "tunneling" part of the wave field in the mirror, overlaps with that charge density component. As indicated in Figure 10, Bragg's law implies that an incident wave vector $\vec{k}_1$ will excite a diffracted wave vector

$$\vec{k}_2 = \vec{k}_1 + \frac{2\pi}{d}\hat{e}_z$$

where $d$ is the planar spacing of the diffracting planes. When we are exactly at the Bragg condition we assume that for a general position along the surface, the electric field will take the form

$$E = \psi_1(y)e^{i(\vec{k}_1 \cdot \vec{x} + \vec{k}_1 \cdot \vec{z} - k_0 ct)} + \psi_2(y)e^{i(\vec{k}_2 \cdot \vec{x} + \vec{k}_2 \cdot \vec{z} - k_0 ct)},$$

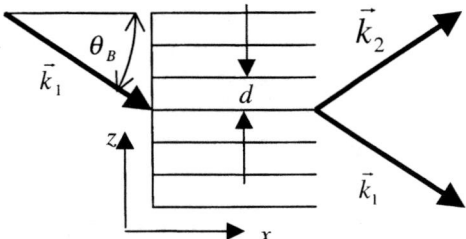

**FIGURE 10.** Lattice spacing and beam directions for Bragg condition considerations.

where for simplicity we consider a scalar field and we have also separated off the "radial" part of the wave field which has a confined eigen function shape as before. The intensity of the wave field oscillates back and forth between the incident and diffracted plane wave fields. The only influence on the equation of motion for the wave field we shall consider here is in the overlap of the evanescent wave component with the periodic charge density in the mirror. To take this into account the effective potential for the diffraction takes the form

$$V(z) = \alpha \delta_G \cos\left(\frac{2\pi}{d} z\right),$$

where $\alpha$ measures the overlap mentioned above and $\delta_G$ is the diffracting Fourier component of the dielectric function perpendicular to atomic planes. Assuming that all the fast variations in the wave function are contained in the exponentials, we can again deduce coupled Schrodinger like wave equation for the $\psi$s:

$$-\frac{d^2}{dx^2}\begin{pmatrix}\psi_1\\\psi_2\end{pmatrix} - ik_x \frac{d}{dx}\begin{pmatrix}\psi_1\\\psi_2\end{pmatrix} + \begin{pmatrix} 0 & \alpha\delta_G k_x^2 \\ \alpha\delta_G k_x^2 & 0 \end{pmatrix}\begin{pmatrix}\psi_1\\\psi_2\end{pmatrix} = (k_0^2 - \vec{k}'\cdot\vec{k})\begin{pmatrix}\psi_1\\\psi_2\end{pmatrix}.$$

Using the boundary conditions implied in Figure 10 the solutions, $\psi_1$ and $\psi_2$ satisfy

$$\psi_1 = \cos(\alpha\delta_G k_0 x / 2\cos\theta_B),$$

$$\psi_2 = i\sin(\alpha\delta_G k_0 x / 2\cos\theta_B).$$

Apparently, the two wave fields continually exchange intensity with one another with a periodicity $2\cos\theta_B / \alpha k_0$. The important point we want to emphasize here is that if one calculates the Poynting vector for the total wave field under this perfect Bragg reflection condition we expect it to oscillate back and forth between the direction of $k_1$ and $k_2$, and hence on the average, point along the atomic planes responsible for the diffraction. When the X-ray finally leaves the crystal surface it will break up into components along $k_1$ and $k_2$ with amplitudes $\psi_1$ and $\psi_2$ respectively.

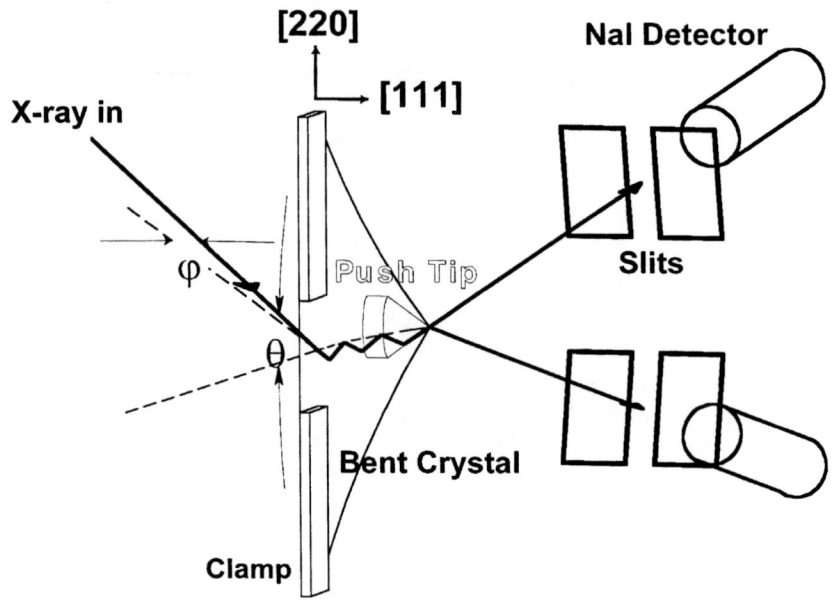

**FIGURE 11.** Experimental setup for studying diffraction effects of surface bound X-rays.

Figure 11 illustrates what we expect to see in a surface diffraction experiment under perfect Bragg reflection. Note the propagation of the "on Bragg" solution along the [220] atomic planes rather than along the incident beam direction when the wave is trapped in the surface mode. Note also how on leaving the surface, the beam breaks up into two plane waves, one in the incident beam direction and one in the diffracted direction. Of course both beams are also deflected horizontally from the incident beam direction by the extrinsic curvature of the X-ray mirror.

We have performed the experiment shown in Figure 11 with collimators and counters, but overall preliminary results showing the effects described above are most economically obtained in experiments where the counters are replaced by X-ray film which shows the many kinds of transmitted beams that can be observed as the incidence angle is changed.

The first panel of Figure 12, labeled (a), shows a film exposure when $\theta - \theta_B = -1.745$ mrad, well below the Bragg condition. The rays resulting in the numbered images on the film are illustrated in the upper part of the figure. We see a straight through beam (0) that passed just under the thin mirror and is not deflected at all, and a deflected, but undiffracted beam at (3) which follows the 15 cm radius of curvature of this particular mirror. In the next panel (b) we approach the Bragg angle and $\theta - \theta_B = -0.122$ mrad. Although we are not on the Bragg condition we observe a beam (4) that has been deflected in a direction that corresponds to a diffracted beam emanating off the mirror near its rear boundary, as illustrated in the drawing above this pane. This is a very curious result and will be dealt with in the next section. The next panel (c) is taken directly at the surface Bragg condition: $\theta = \theta_B$. The most important feature here is the presence of beam at position (2) which is a diffracted beam that emanates from the far tip of the triangular mirror. This X-ray beam has traveled along the surface parallel to the (220) planes, just as expected from the preceding discussion of the surface dynamical diffraction expected from surface X-rays. The companion ray that should also emanate from the tip of the mirror in the undiffracted direction is missing from panel (c) because of an absorber placed in its path. Panel (d) was taken at a later date without any absorbers blocking beams and both the expected surface diffraction rays are present (at positions 1 and 2), as well as a transmitted undiffracted beam (3), and a beam (6) that was diffracted near the position where X-ray beam first encounters the mirror surface. The gaps in both of these spots is due to the X-rays that are transferred to beams (1) and (2).

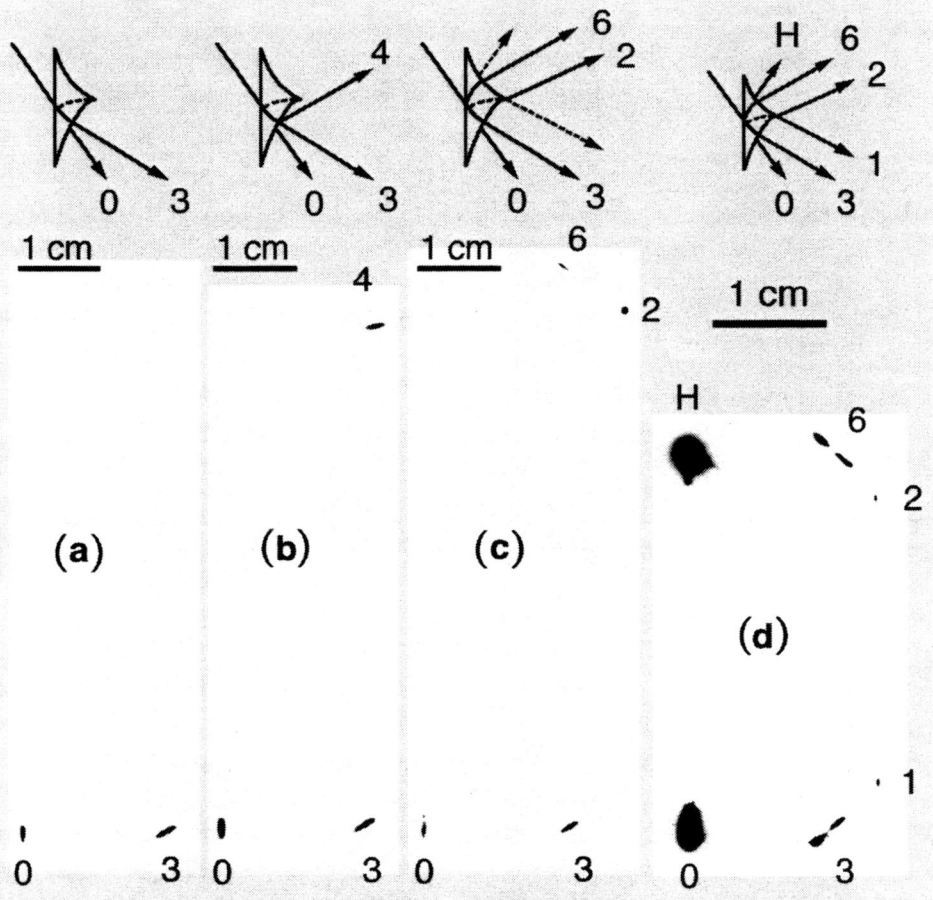

**FIGURE 12.** Schematic beam path drawings and film exposures taken at $\theta - \theta_B$ = (a) -1.745, (b) – 0.122, and (c) 0 mrad for 17.5 keV X-ray on a $R$ = 15 cm Si(111). (d): $\theta = \theta_B$, $R$ = 10 cm.

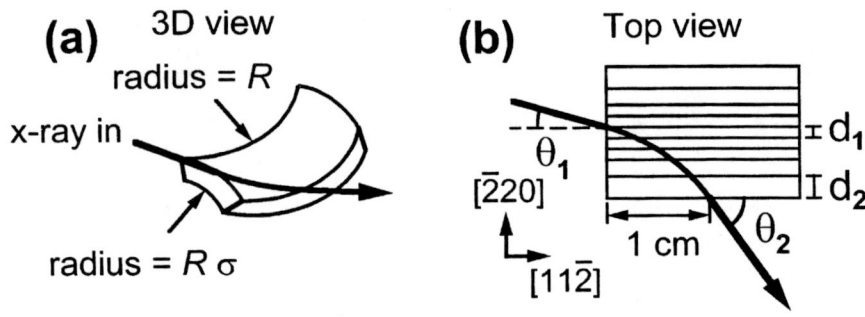

**FIGURE 13.** X-ray path on a bent silicon plate with saddle-like surface and relaxed lattice.

While much more work needs to be done to quantitatively understand the exact intensities and angular dependencies of the observed diffraction phenomena reported here, these observations are definitive examples of the strong diffraction effects which surface bound X-rays can experience.

## DEFLECTIONS DUE TO INTRINSIC SURFACE CURVATURE

The third and final phenomena covered in this review explains the "prematurely" diffracted beam observed in panel (b) at position (4). How can we explain a diffraction from the [220] planes that occurs more than a tenth of a milliradian in angle before it is expected? One can envision two possible explanations, both of which are illustrated in Figure 13. In (a) above, we have depicted the fact that in spite of our attempts to bend an X-ray mirror into a cylindrical surface of constant curvature, it will, because of internal stresses in the finite thickness plate, take on the shape of a saddle. This is due to the finite value of the Poisson ratio for silicon (and all other materials). In part (b) of the above figure we have also depicted the fact that the resulting surface stresses in the saddle shaped surface will lead to a change in the planar spacing near the surface. The spacing increases as one goes further from the horizontal symmetry plane of the saddle. We now discuss both of these effects.

Finding the new shape of the surface and the change in the planar spacing is an exercise in continuum mechanics which will not be described in detail here. A consequence of the changing lattice constant is clear. For an X-ray entering the crystal surface region where the lattice constant is slightly too small for a Bragg reflection to occur, the ray will proceed unaffected by diffraction. If this ray is generally directed towards a region where the lattice constant is increasing, it will ultimately encounter a place where the Bragg condition is satisfied and at that spatial position diffract. From the geometry indicated above this effect is of the right sign to explain our observed "premature" diffracted beam in panel (b). Our quantitative estimates of the changing lattice constant on the surface have shown that this effect can only explain a few percent of the angular deviance of our anomaly.

The remaining and largest part of the effect is more interesting. It comes from the saddle-like geometry of the surface. In comparison with a cylindrical surface (which has an extrinsic but not intrinsic surface curvature), the new feature here is that a saddle also has an intrinsic curvature. The geometry on such a surface is no longer Euclidean. The straight line trajectories we assume for plane wave solution of Maxwell's equations in free 3–space or cylindrical 2–space need to be re-examined. ("Straight lines" on the cylinder turn into rectilinear paths on the plane formed by unrolling the cylinder as if it were formed from a flexible sheet of paper.) The simplest assumption to make on a surface with intrinsic curvature is that the path will be a geodesic, which corresponds to the assumption that Fermat's principle is still valid. Under this assumption, if the surface is characterized by knowledge of the differential distance function

$$ds^2 = g_{uu}(u,v)du^2 + 2g_{uv}(u,v)dudv + g_{vv}(u,v)dv^2$$

where $u(x,y,z)$ and $v(x,y,z)$ are a parameterization of the surface, and $g(u,v)$ is the surface metric tensor, then the differential equations for the surface trajectory parameterized in terms of the arc length parameter s given above are

$$\frac{d^2u}{ds^2} + \Gamma^u_{uu}\left(\frac{du}{ds}\right)^2 + 2\Gamma^u_{uv}\frac{du}{ds}\frac{dv}{ds} + \Gamma^u_{vv}\left(\frac{dv}{ds}\right)^2 = 0,$$

$$\frac{d^2v}{ds^2} + \Gamma^v_{uu}\left(\frac{du}{ds}\right)^2 + 2\Gamma^v_{uv}\frac{du}{ds}\frac{dv}{ds} + \Gamma^v_{vv}\left(\frac{dv}{ds}\right)^2 = 0.$$

The affine connection functions $\Gamma^i_{jk}$ are determined by differentiation of the $g(u,v)$ functions:

$$\Gamma^i_{jk} = \frac{1}{2}g^{ia}\left(\partial_k g_{aj} + \partial_j g_{ak} - \partial_a g_{jk}\right),$$

where $i,j,k,a = u,v$, and summation over repeated indices is assumed. We have solved these equations for the trajectories on our surfaces using $g(u,v)$ functions of our surfaces, and discovered that the resulting gradual in surface deflection of the X-ray beam caused by the intrinsic curvature is the major cause of the "premature Bragg reflections." Figure 14 shows the results of our measurement and calculations of the $\Delta\theta$, the deviation from $\theta_B$ of the observed "premature" diffraction as a function of extrinsic radius of curvature of the mirror. The dashed lines show the theoretical contribution to the deviation from lattice constant variations on the mirror. The dotted curve shows the theoretical contribution from intrinsic curvature, and the solid line is the total calculated result as a function of radius of curvature $R$. The agreement with the data for three different radii of curvature is remarkably good.

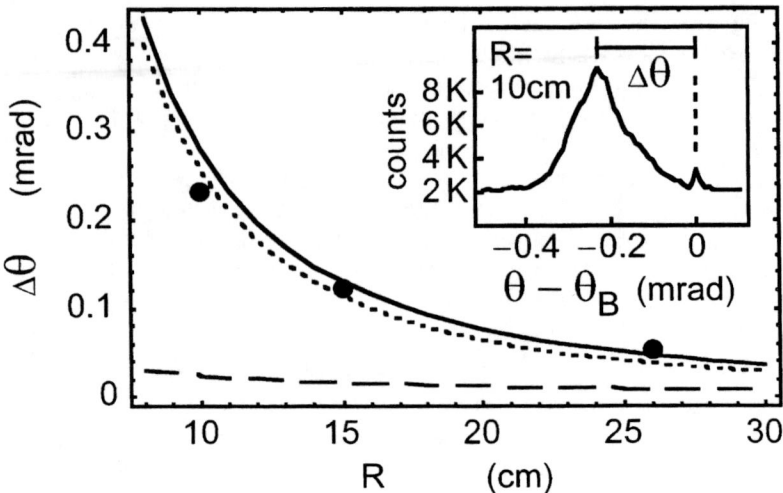

**FIGURE 14.** Inset: A $\theta$ scan shows the dynamical diffraction peak at $\theta = \theta_B$, and the edge diffraction peak at $\theta - \theta_B = -0.231$ mrad, for 17.5 keV WG X-rays on a $R = 10$ cm wave guide. Main figure: The angular separation of the two peaks $\Delta\theta$ as a function of $R$.

## CONCLUSIONS

In this review we have discussed and demonstrated three new phenomena connected with the propagation of X-rays on curved surfaces; binding, diffraction and in-surface deflection. Much work remains to be done quantifying and exploring the implications of each. Only the most preliminary consequences of the theory have been examined experimentally so far. Quantitative measurements of the detailed angular structure and coherence properties of the surface bound states after they exit the mirror would be very helpful. Likewise, studies of the angular widths of the surface Bragg diffractions and periodicities of Poynting vector directional changes during dynamical diffraction propagation should be pursued. The theory assumed perfectly smooth surfaces and the consequences of surface inhomogeneities, oxide layers, atomic steps etc. need to be understood. In addition, methods need to be developed to provide surface curvatures of X-ray mirrors that are well controlled and accurately understood if quantitative aspects of the theory are to be examined in detail. Finally, the use of Fermat's principle to deduce the in-surface deflections needs a much closer look, particularly from the theoretical side at first. A rigorous derivation of the appropriate effective two dimensional wave equations for surface propagation and the derivation of the ray equations, perhaps within the eikonal approximation, in curved spaces needs to be developed. In addition to intrinsic curvature influencing the field equations through the affine connection we expect other controllable parameters like composition gradients along the surface to play a similar role. Though perhaps a small effect for X-rays, even the changing extrinsic curvature of the surface might also play a role in the effective field equations due to the changing binding energies of the bound modes.

We anticipate that the pursuit of the X-ray physics discussed in this paper will eventually result in new optical capabilities for synchrotron and laser generated X-ray sources. In addition, because the phenomena involved are so strongly dependent on the physical state of the X-ray mirror surface, it seems inevitable that they will also contribute directly to our ability to probe and characterize the electronic and structural nature of such surfaces when a thorough understanding of the X-ray physics is in hand. Finally we anticipate that the ideas presented here for X-rays may have applications in other regions of the electromagnetic spectrum as well as for appropriate de Broglie matter waves.

## ACKNOWLEDGEMENTS

We wish to thank Mr. Carl Steinke for his kind assistance in preparing this manuscript. Experiments demonstrating the surface diffraction and in-surface deflection of x-ray surface bound states were performed at NSLS beam line 15A at Brookhaven National Laboratory. We thank Prof. M.J. Bedzyk for his cooperation and help in making these measurements possible.

## REFERENCES

1. Lord Rayleigh, *The Theory of Sound*, first published 1877; American edition: New York: Dover, 1945.
2. Liu, C., and Golovchenko, J. A., *Phys. Rev. Lett.* **79**, 788 (1997).
3. Liu, C., and Golovchenko, J. A., *Optics Lett.* **24**, 587 (1999).
4. Liu, C., Doctoral Thesis, Harvard University, 1997.
5. Wang, J., Bedzyk, M. J., and Caffrey, M., *Science* **258**, 775 (1992).
6. Feng, Y. P., Sinha, S. K., Deckman, H. W., Hastings, J. B., and Siddons, D. P., *Phys. Rev. Lett.* **71**, 537 (1993).
7. Smith, N. V., and Howells, M. R., *Nucl. Instr. Meth. Phys. Res.* **A347**, 115 (1994).
8. Braud, J. P., and Hagelstein, P. L., *IEEE J. Quan. Elec.* **27**, 1069 (1991).
9. Bukreva, I. N., Kozhevnikov, I. V., and Vinogradov, A. V., *SPIE* **2453**, 80 (1995).

# Thermal Calorimeters for High Resolution X-ray Spectroscopy

M. Galeazzi[a,b], D. McCammon[a] and W. T. Sanders[a]

[a]*University of Wisconsin, Department of Physics, 1150 University Ave., Madison, WI 53706 USA*
[b]*University of Genoa, Department of Physics, via Dodecaneso 33, I 16146 Genova, Italy*

**Abstract.** In spectroscopic experiments where high energy resolution and high efficiency are required the characteristics of conventional detectors may not be satisfactory. A new generation of detectors for X-ray and particle spectroscopy, thermal microcalorimeters, has been developed in the last 15 years in order to satisfy these requirements. The operating principle of microcalorimeters and their performances in terms of energy resolution, counting rate and detection efficiency are described in this paper. Some examples of applications are given.

## WHY CRYOGENIC MICROCALORIMETERS?

Detectors commonly used in X-ray spectroscopy are of four kinds, solid state detectors, proportional counters and Bragg crystals or grazing-incidence diffraction gratings with position-sensitive detectors. Solid state detectors are the most used because they are easy to use and inexpensive to operate. However their energy resolution is limited by the statistical fluctuations in the number of electron-hole (e-h) pairs that are created. In a silicon detector for example the energy necessary to create an e-h pair is equal to 3.66 eV at room temperature, while the energy gap is only 1.2 eV. This means that only about the 32% of the total energy is converted into e-h pairs, while the rest of the energy is converted into phonons. The energy resolution of such a detector is given by $\Delta E_{RMS} = \sqrt{FEw}$ where $w$ is the energy necessary to create one e-h pair, $F$ is the Fano factor and $E$ is the energy. In practice the energy resolution of a solid state detector is not better than 100-120 eV FWHM at 6 keV. In a proportional counter the energy necessary to create an electron-ion pair is of the order of 30 eV, and the energy resolution is even worse.

Grazing incident gratings can provide good resolution for soft X-rays, on the order of 1 eV, but their low dispersion means throughputs are very small at high resolution. Grating efficiency also falls off rapidly at higher energies. The case of Bragg crystals is different. The energy resolution is again very good and the dispersion and efficiency are higher. However, only one resolution element is reflected, so all multiplex advantage is lost and the throughput is very small, unless single lines are being observed. Moreover the energy interval covered by a single crystal is limited, in practice, to a factor of two or so.

About 15 years ago the idea of detecting the increase in temperature produced by incident photons instead of the charged pairs has been proposed (1,2). If all the energy that is deposited is thermalized (converted in thermal phonons), there is no branching and no inherent statistical limitation to the resolution. Moreover the detection of phonons means that the choice is no longer restricted to materials with good electron transport properties, such as germanium or silicon, but different materials desirable for the particular experiment can be used.

## OPERATING PRINCIPLES

Instead of reporting here the theory of cryogenic microcalorimeters that has been reported in detail by many authors (1-4), in this paragraph we want to give you a qualitative idea of the working principles. We will report some of the main quantitative guidelines in the next paragraphs, when the description of the detector characteristics requires it. A cryogenic microcalorimeter is composed of three parts, an absorber, a sensor that detects the temperature variations of the absorber and a weak thermal link between the detector and a heat sink. The operating principle of a microcalorimeter is simple. When an X-ray hits the absorber its energy is eventually converted in thermal phonons and the temperature of the detector first rises and then returns to its original value due to the weak thermal link with the heat sink. The temperature change is proportional to the energy of the incident X-ray and is detected by the sensor. The sensor is generally a resistor whose resistance has a strong dependence on the temperature at the working point.

Although the operating principle is simple, the construction of a cryogenic microcalorimeter is not as simple. The temperature variation of the detector is inversely proportional to its heat capacity, which must then be as small as possible in order to have a detectable temperature variation. This can be achieved in two different ways, reducing the size of the detector and/or reducing its working temperature. Typical working temperatures in the range of tens of mK and volumes of the order of $10^{-3}$ mm$^3$ are used for best energy resolution, but there are some applications with much higher temperature and/or much larger size. Working with cryogenic microcalorimeters usually means dealing with small objects and very low temperatures. The temperature variations to be detected may be of the order of μK, so good sensors are necessary to detect them. Despite of this unpromising preface, many groups are now working in this field and the results obtained up to now are absolutely astonishing.

Requirements for a good microcalorimeter are an absorber with small heat capacity able to convert the incident radiation into thermal phonons quickly and with high efficiency, and a sensor with low heat capacity and a high sensitivity to temperature variations. We discuss these characteristics, together with the performances of cryogenic microcalorimeters in the next paragraphs.

# DETECTOR CHARACTERISTICS

## The absorber

The choice of the absorber is one of the more delicate parts of an experiment with cryogenic detectors. Constructing a detector, the main characteristics that must be taken into account are the stopping power at the energy of interest, the collecting area and the efficiency and speed of thermalization of the incident energy. This must be combined with a sufficiently small heat capacity. Several different kinds of materials such as metals, semiconductors, superconductors and insulators, have been tested and used with relatively good results. We report here the main characteristics of these materials as absorbers.

Metals have very good thermalization; the conversion of the incident energy into thermal phonons is fast and very efficient. The stopping power is also generally very good, if a high Z material is chosen. Despite these advantages they are often undesirable as absorbers because of their very high electronic contribution to the heat capacity. Some metals with very small electronic density of states and consequently small electronic heat capacity, such as bismuth, can be used with very good results (5).

Semiconductors and insulators have no electronic contribution to the heat capacity, which can then be very small. On the other hand, part of the incident energy is converted into electron-hole pairs that can be very slow to recombine. It is then difficult to have good energy resolution since the thermalization is limited and it is affected by the statistical and positional variation in the electron-hole pair creation and trapping. Some small or zero gap semiconductors such as HgTe can have very good performance due to the negligible energy tied up in charge carrier production (6).

A promising alternative to metals and semiconductors are the superconductors. Their heat capacity can be small due to the absence of an electronic contribution at temperature below $0.1\ T_C$, while the stopping power is very good for high Z materials like lead or rhenium. In superconductors the incident energy is first converted into quasi-particle that then recombine releasing phonons. This process is complicated and depends strongly on the characteristics of the specific material (7). Superconductors as absorbers are often characterized by incomplete thermalization and a very long tail in the pulses (~100 ms) which strongly affects the detector speed and energy resolution. The best results up to now have been obtained using tin (8), but interesting results have also been obtained using other materials like rhenium (9).

In conclusion, a main advantage of cryogenic microcalorimeters is the possibility of using a large variety of materials with good results. The choice of the absorber can be optimized depending on the requirements of the experiment.

# The sensor

We already pointed out that the main characteristic of a sensor is the sensitivity to temperature variations. This characteristic is described by the quantity α, defined as:

$$\alpha = \frac{T}{R}\frac{dR}{dT} \qquad (1)$$

T and R are respectively temperature and resistance of the sensor. The parameter α is dimensionless and describes the fractional resistance variation versus the temperature variation. Higher α means higher sensitivity to temperature variations, corresponding to higher output signals; we will discuss in the next paragraph how this affects the performance of the detectors.

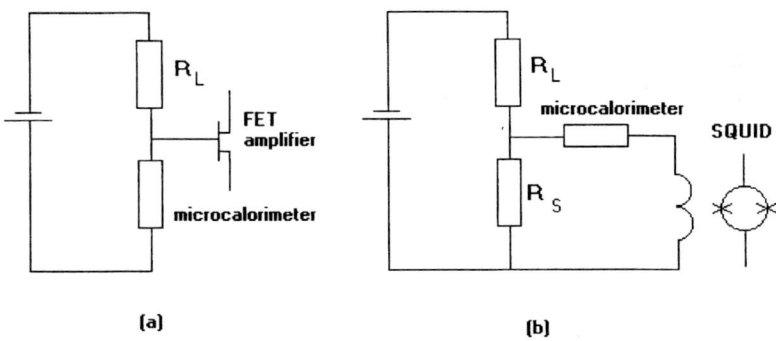

**FIGURE 1.** Schematic of a current biased microcalorimeter (a) and of a voltage biased microcalorimeter. In (a) $R_{bolometer} \ll R_L$, in (b) $R_S \ll R_{bolometer}$.

**TABLE 1.** Principal characteristics of commonly used sensors.

| Sensor | α | Temperature range | Resistance | Usual bias | Read-out electronics |
|---|---|---|---|---|---|
| Thermistors | Negative/ Small | Wide | Large | Constant current | Cold FET |
| TES | Positive / Large | Narrow | Small | Constant voltage | SQUID |

We said before that the sensors generally are resistors whose value has a strong dependence on the temperature. Such resistors can be biased either at constant current

641

or at constant voltage, the temperature variation is thus read out respectively as voltage variation or current variation. The schematics of typical circuits for the current and the voltage bias are reported in Fig. 1. We can distinguish two main categories of such sensors in current use, solid state thermistors and transition edge sensors (TES), the characteristics of both kinds of detectors are reported in Table 1.

Two kinds of semiconductor thermistors are used: neutron transmutation doped (NTD) germanium thermistors and ion implanted silicon thermistors. Their $R$ versus $T$ behavior is approximately $R = R_0 \exp(\sqrt{T_0/T})$, where $R_0$ and $T_0$ are constant parameters characteristic of the sensor. As reported in Table 1 they have negative $\alpha$ and they can be used over a broad range of temperature. They are generally biased near constant current and characterized by a very high resistance (tens of 1 MΩ at the working point) where they are well matched to junction field effect transistors (JFET) operated near 100 mK.

Transition edge sensors are superconducting thin films working at the transition temperature, where $\alpha$ can be very high. The transition is generally very narrow, so the working temperature for a given TES is fixed, but it is possible to tune the transition temperature of a TES during its fabrication using the proximity effect in superconductor-normal metal bilayers (10). The most commonly used TES are Al-Ag, Mo-Au, Mo-Cu and Ir-Au bilayers or W thin films. Their $\alpha$ is generally very high (at least a factor of 10 higher than in thermistors) and positive, while their resistance is very low. For these reasons they are generally biased at constant voltage and their current is read out using SQUID electronics at liquid helium temperature.

## PERFORMANCES

### Energy Resolution

Four different noise contributions can affect the energy resolution of a microcalorimeter: the Johnson noise of the sensor and any excess noise it has, the phonon shot noise produced by the random flow of energy carriers through the weak thermal link, and the electrical noise of the read out electronics. Current technology allows the construction of read out electronics whose noise is in general negligible with respect to the other contributions. This is done using low temperature (around 100 K) JFET as first stage plus room temperature amplifiers in the read out of the signal from a semiconductor thermistor, and liquid helium temperature D.C. SQUIDs to read out the signal of a TES. In the case of JFETs, the total voltage noise is typically a few $nV/\sqrt{Hz}$. With SQUID electronics it is possible to obtain current noise of about $10^{-12} \, A/\sqrt{Hz}$ using commercial D.C. SQUIDs.

The combination of Johnson noise plus phonon noise limits the intrinsic energy resolution of the detector. The Johnson noise is a voltage noise across the detector with voltage spectral density $\sqrt{4Rk_BT}$ where $k_B$ is the Boltzman constant. The phonon noise essentially consists of statistical fluctuations in the temperature of the detector due to the link between the detector and the heat sink. This shows as a temperature noise that is then converted into a voltage or current noise by the sensor. Skipping here the mathematical passages that can be found in (1), the intrinsic energy resolution of an ideal cryogenic microcalorimeter operating in the linear, small signal regime, can be written as:

$$\Delta E_{RMS} = \xi \sqrt{\frac{k_B T^2 C}{\alpha}} \qquad (2)$$

where $T$ and $C$ are respectively the temperature and the heat capacity of the detector and $\xi$ is dimensionless and depends on the detector characteristics and bias power and has an optimized value of 1-2.

Looking at the formula (2) we want to point out a few considerations. First, as intuitively pointed out before, the intrinsic energy resolution depends on the heat capacity of the detector. What wasn't intuitive before is the fact that the energy resolution also depends directly on the temperature of the detector; it is then immediately clear how important it is to work at very low temperature. Second, the energy resolution also depends on the detector sensitivity $\alpha$, which points out the importance of working with a good sensor. Two other important results are the fact that the intrinsic energy resolution does not depend on the energy of the incident radiation, nor on the thermal conductivity $G$ between the detector and the heat sink. The consequences of this last conclusion will be pointed out in the next subparagraph.

## Count Rate

Before discussing the performances of a microcalorimeter in terms of count rate we must introduce the effect of the detector non-linearity. Up to now we discussed the linear, small-signal regime, which assumes that whenever an incoming particle warms the detector the temperature change is so small that the characteristics of the detector don't change. That is not true in the case of relatively large temperature changes, since the thermometer sensitivity $\alpha$, the detector heat capacity and the thermal conductivity to the heat sink are temperature dependent. This shows up as a detector non-linearity that can be relatively easily taken into account in the analysis, but that greatly affects the detector count rate. In fact if the count rate is to high it is highly possible to have a pulse on the tail of another pulse. The starting temperature of this pulse is then higher than the working temperature of the detector; thus the pulse amplitude is different (smaller) than the amplitude of a normal pulse generated by the same amount of energy, worsening the performances of the detector. This means that the detector count rate is limited by the time constant of the detector.

A microcalorimeter with heat capacity $C$ and thermal conductivity to the heat sink $G$ is characterized by an intrinsic time constant $\tau = C/G$. This is the time necessary to the microcalorimeter to return to the working temperature after an X-ray warms it. The temperature rise, determined by the thermalization time, is in general much faster and it is a characteristic of the absorber. Because, as just pointed out, the energy resolution of a microcalorimeter does not depend on the thermal conductivity $G$, it is in principle possible to build a detector with very high G and consequently very small time constant. In practice, because of technical limitations, typical time constants of cryogenic microcalorimeters are in the range 0.1-10 ms or even longer in the case of very large detectors.

Great improvements can be obtained using the effect known as electro-thermal feedback. When there are no events in the detector, the temperature of the microcalorimeter is higher with respect to the one of the heat sink due to the power dissipated into the sensor by the bias. This power is simply equal to $I^2 R$ in the case of constant current bias and to $V^2/R$ in the case of constant voltage bias. When an X-ray is absorbed into the detector the resistance of the sensor changes, then also the power dissipated into it by the bias changes. This effect can be quantified introducing a parameter with the dimension of a thermal conductivity $G_P$. The real time constant of the microcalorimeter is then given by $\tau_{eq} = C/G_{eq}$, where $G_{eq} = G + G_P$. $G_P$ can be either positive or negative, depending on the sign of the parameter $\alpha$ and on the kind of polarization used, then the time constant of the detector can be either longer or shorter than the intrinsic time constant $\tau$. In general if the detector is current biased $G_P$ is positive if $\alpha$ is negative and negative if $\alpha$ is positive. The opposite is true if the microcalorimeter is voltage biased. In particular in the case of voltage biased TES, the parameter $G_P$ is not only positive, thus reducing the time constant, but, choosing the appropriate working point, it can be much larger than $G$ (11). In this way it is possible to obtain relatively fast detectors. Using the strong electro-thermal feedback effect cryogenic microcalorimeters with count rate of about 500 Hz have already been built (12).

This is the status of the art in terms of count rate for a single detector. With the use of arrays of microcalorimeters it is possible to extend this limit. In particular, arrays of 1000 elements are under construction. This would allow a count rate of the final detector of about 500 kHz.

## Detector efficiency

As pointed out previously, the detection efficiency of a single micro-calorimeter is generally very good, more that 99%, due to the possibility of using high Z materials and to the fact that the thickness of the absorber is not an intrinsic limit to the detector performance. Nevertheless, we want to point out some considerations that must be taken into account when defining the characteristics of an experiment with cryogenic microcalorimeters. First consideration is the fact that, even if it is possible to make the detector thick enough to have high stopping power, in order to have a small heat capacity the area of the detector must be reduced. A solution to have large areas and good stopping power is the use of an array of microcalorimeters. Up to now arrays of tens of microcalorimeters are already used (13) and, as already mentioned much larger arrays are under development (14).

A second important factor that may affect the efficiency of a cryogenic microcalorimeter is that, if the source is at a temperature that is higher than that of the detector, the detector must be shielded to reduce the shot noise due to the impinging infrared photons and the heat load due to the black body emission of the source. A series of filters is installed between the detector and the source to reflect the infrared radiation that would warm up the detector and the heat sink. These filters are generally thin organic films covered by aluminum. Well designed filters have transmission efficiency above 95% for energies above 1 keV, while the efficiency tends to drop below 400-500 eV (13).

## APPLICATIONS

We describe now, divided into categories, some of the experiments that use cryogenic microcalorimeters as X-ray and particle detectors. The number of such experiments is large and is increasing at a high rate, so we can give only a few examples. For further descriptions we suggest the Proceedings of the International Workshop on Low Temperature Detectors (15,16).

## Astrophysics

X-ray astrophysics is one of the fields where most of the efforts in the development of cryogenic microcalorimeters are spent. The first microcalorimeters for x-ray astronomy were developed by the University of Wisconsin / NASA Goddard Space Flight Center collaboration (13). They were first employed on a sounding-rocket experiment (the x-ray quantum calorimeter – XQC) that had an array of microcalorimeters for the study of X-rays from the interstellar medium in the energy range 30-1000 eV. We skip here the description of the scientific goals of the experiment (for those who are interested we suggest (17)), while we want to spend a few words on the detector characteristics. The array is composed of 36 elements, each with a collecting area of 1 mm$^2$, for a total sensitive area of 0.36 cm$^2$. Each pixel is composed of a silicon thermistor implanted in a micro-machined silicon chip, thermally connected to a HgTe absorber (a schematic of a detector pixel is shown in Fig. 2). The absorber is 0.75 µm and has more than 99% quantum efficiency of the detector below 1 keV. The detector is installed in an adiabatic demagnetization refrigerator at a base temperature of 60 mK; the refrigerator is capable of maintaining this temperature for a period of more than 12 hours. The detector efficiency is limited by the five infrared blocking filters mounted on the different stages of the refrigerator in order to limit the heat load on the cold stages and shot noise on the detector. The energy resolution achieved by a prototype detector with a thicker absorber is 7.3 eV FWHM for the 5.9 keV $^{55}$Fe line (see Fig. 3).

A similar array of detectors (18) is used in the x-ray spectrometer instrument (XRS) which will be launched at the beginning of the year 2000 on the Japanese satellite Astro-E. The geometry and characteristics of the detector are very similar to those of the XQC detector, but it is optimized for energies up to 10 keV (the absorbers are thicker). The energy resolution of the array ranges from 8 to 12 eV FWHM in the 0.2-10 keV energy range.

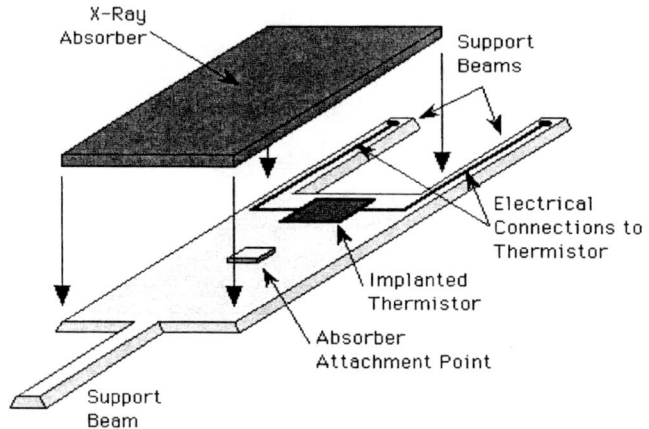

**FIGURE 2.** Schematic of a detector element of the XQC sounding rocket detector array.

Much effort now goes into creating second generation detectors capable of improving this performance. A major project is aimed at the development of large detector arrays (more than 1000 elements) with TES thermometers. At the National Institute of Standards and Technology (NIST) in Boulder Colorado, they have made TES detectors using Al-Ag and Mo-Au bilayers that achieve energy resolution of 4.5 eV FWHM at 5.9 keV and 2 eV FWHM at 1.5 keV (5). Other groups are using NTD germanium thermistors, and have achieved energy resolutions around 5 eV FWHM (8).

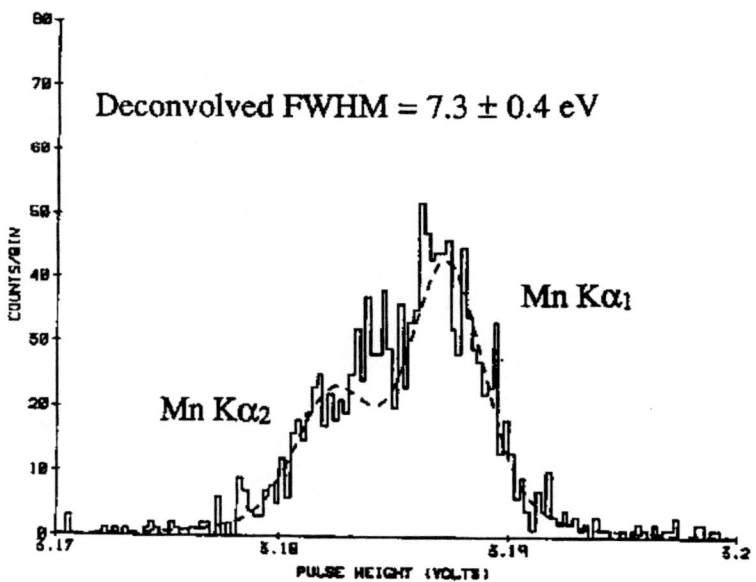

**FIGURE 3.** Mn $K_{\alpha 1}$ and $K_{\alpha 2}$ lines obtained with a prototype of the XQC/XRS detectors.

## Microanalysis

The NIST group is also using detectors based on TES thermometers and bismuth absorbers for high-resolution x-ray microanalysis (12). X-ray microanalysis is a powerful and widely used technique for providing spatially resolved chemical analysis. A scanning electron microscope (SEM) creates a finely focused electron beam that excites X rays with characteristic energy from the sample under analysis. The X rays are then analyzed with a spectrometer to have information about the chemical composition of the sample. Both solid state detectors and Bragg crystals with position sensitive detectors are commercially used, neither ones fully satisfying the request of the market. While the energy resolution of conventional solid state detectors does not allow to always resolve the X-ray lines due to different materials, qualitative analysis with Bragg crystals is severely limited by the low efficiency and the long time needed to serially scan over the entire energy range using multiple diffraction crystals. Using a cryogenic microcalorimeter coupled with X-ray optics it is possible to have at the same time the good energy resolution of a Bragg crystal and the efficiency of a semiconductor detector.

## The Lamb Shift in Heavy H-like Ions

The precise determination of the Lyman-α transitions in hydrogen-like heavy ions, such as $U^{91+}$, can provide a sensitive test of quantum electrodynamics in very strong Coulomb fields ($Z\alpha \rightarrow 1$) and especially of contributions to the self energy of higher order in $Z\alpha$ (19). Such Lamb shift measurements became possible recently by x-ray spectroscopy using highly charged ions, stored and cooled in the heavy ion storage ring ESR of GSI in Darmstadt. In order to improve the experimental precision, presently limited by the poor energy resolution of conventional germanium detectors, a high resolution experiment using cryogenic microcalorimeters for hard X rays (E ~ 100 keV) is under development. The detector employs ion-implanted silicon thermistors as sensors and superconducting tin as the absorber. Other superconductors with higher Z are also under investigation. Preliminary results using an $^{241}$Am calibration source have shown an energy resolution of 75 eV FWHM for the 60 keV Am photo-peak, about a factor of 10 better than the energy resolution of 800 eV FWHM obtained in previous experiments with conventional germanium detectors.

## Beta Environmental Fine Structure

We briefly report here an example where cryogenic microcalorimeters are used to detect electrons. In beta decays, a nucleus with Z protons and A nucleons decays to a nucleus with Z+1 protons, emitting an electron and an antineutrino. Experiments on beta decay detect the energy of the emitted electrons that have a continuous energy distribution from zero to the end-point energy. If the emission occurs in a crystalline structure the emitted electron is influenced by the structure of the surrounding atoms and the beta spectrum is modulated by an oscillatory structure whose characteristics depend on the crystal parameters.

This effect is known as Beta Environmental Fine Structure (BEFS) and it was predicted in 1991 (20). The oscillation has a very small energy period, and it has not been detected using conventional detectors. The beta decay of $^{187}$Re has been studied with a cryogenic microcalorimeter composed of a NTD germanium thermistor and a superconducting rhenium crystal as the source and absorber (the natural abundance of $^{187}$Re is about 63%). It has been possible to detect the BEFS for the first time in 1998 (21) (see Fig. 4). This effect can, in principle, be used in the analysis of crystalline and molecular structures as an alternative to EXAFS.

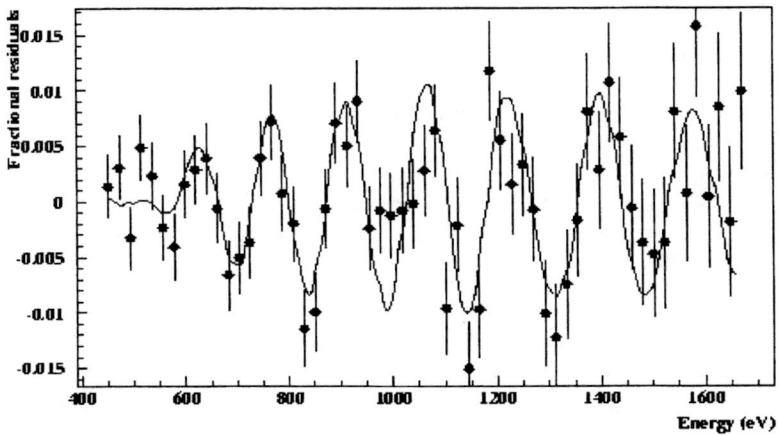

**FIGURE 4.** Beta environmental fine structure in a rhenium crystal with a cryogenic microcalorimeter.

## ACKNOWLEDGMENTS

We want to thank all the people of the University of Wisconsin / NASA GSFC XQC collaboration and all the people present at the 8$^{th}$ International Workshop on Low Temperature Detectors (LTD8) for helpful discussions and suggestions.

## REFERENCES

1. Moseley, S. H., *et al.*, *J. Appl. Phys.* **56**, 1257-1262 (1984).

2. Fiorini, E., and Niinikoski, T. O.,. *Nucl.Instr. Meth. Phys. Res.* **224**, 83 (1984).

3. Mather, J. C., *Appl. Optics* **21**, 1125-1129 (1982).

4. Galeazzi, M., *Rev. Sci. Instr.* **69**, 2017-2023 (1998).

5. Irwin, K. D., *et al.*, in *Proceedings of the 8$^{th}$ International Workshop on Low Temperature Detectors LTD8*, edited by P. de Korte et al., Nucl. Instr. Meth. Phys. Reas. A (special issue), in press, 2000.

6. McCammon, D., *et al.*, Nucl. Instr. Meth. Phys. Res. **A326**, 157-165 (1993).

7. Cosulich, E., *et al.*, *J. Low Temp. Phys.* **93**, 263 (1993).

8. Alessandrello, A., *et al.*, *Phys. Rev. Lett.* **82**, 513 (1999).

9. Galeazzi, M., *Nucl. Phys. B (Proc. Suppl.)* **66**, 203-206 (1998).

10. Irwin, K. D., *et al.*, *Nucl. Instr. Meth. Phys. Res.* **A370**, 177-179 (1995)

11. Irwin, K. D., *Appl. Phys. Lett.* 66, 1998-2000 (1995)

12. Wolman, D. A., *et al.*, "Cryogenic microcalorimeters for x-ray microcanalysis" in *Proceedings of the $8^{th}$ International Workshop on Low Temperature Detectors LTD8*, edited by P. de Korte et al., Nucl. Instr. Meth. Phys. Reas. A (special issue), in press, 2000.

13. Porter, F. S., *et al.*, "Observations of the soft x-ray background with the XQC microcalorimeter sounding rocket" in *Proceedings of the $8^{th}$ International Workshop on Low Temperature Detectors LTD8*, edited by P. de Korte et al., Nucl. Instr. Meth. Phys. Reas. A (special issue), in press, 2000.

14. Chervenak, J. A., *et al.*, "Superconducting multiplexer for arrays of transition edge sensors" in *Proceedings of the $8^{th}$ International Workshop on Low Temperature Detectors LTD8*, edited by P. de Korte et al., Nucl. Instr. Meth. Phys. Reas. A (special issue), in press, 2000.

15. *Proceedings of the $7^{th}$ International Workshop on Low Temperature Detectors LTD8*, edited S. Cooper, Munich: Max Plank Institute of Physics, 1997.

16. *Proceedings of the $8^{th}$ International Workshop on Low Temperature Detectors LTD8*, edited by P. de Korte et al., Nucl. Instr. Meth. Phys. Res. A (special issue), in press, 2000.

17. Sanders, W. T., *et al.*, *SPIE* **3114**, 636-647 (1997).

18. Kelley, R. L., *et al.*, "The Astro-E/XRS high resolution spectrometer", to appear in *Proceedings of SPIE*, vol. 3765, 1999.

19. Egelhof, P., "High resolution calorimetric low temperature detectors for applications in atomic and nuclear physics" in *Proceedings of the $8^{th}$ International Workshop on Low Temperature Detectors LTD8*, edited by P. de Korte et al., Nucl. Instr. Meth. Phys. Reas. A (special issue), in press, 2000.

20. Koonin, S. E., *Nature* **354**, 468-469 (1991).

21. Gatti, F., *et al.*, *Nature* **397**, 137-139 (1999).

# Coherent X-Ray Generation via Ultrafast Coster-Kronig Decay in Solid Targets Excited by Table-Top Lasers

Cs. Tóth[a], S.H. Son[b], D. Kim[b], and C. P. J. Barty[a]

[a]*University of California, San Diego, 9500 Gilman Dr., La Jolla, CA 92093-0339, USA*
[b]*Pohang University of Science and Technology, Pohang, 790-784, Republic of Korea*

**Abstract.** New keV X-ray laser schemes based on fast energy deposition (1-20 fs) and inner-shell atomic processes are analyzed. It is shown that population inversion between inner-shell vacancy levels of medium-Z elements can be created using electron collisional ionization only, or also by using fast, incoherent X-ray pumping if decay of lower level states is mediated by Coster-Kronig or super-Coster-Kronig Auger decay. Because of such fast decay of the lower state, the $L_2M_1$ transitions of elements Z=22 to 32 are the most stable against collisional processes. We also describe here a state-of-the-art chirped pulse amplification (CPA) laser system capable to deliver >1 J optical energy in less than 25 fs with 10 Hz repetition rate. The design of sandwiched multilayer target structure and possible experimental configurations are suggested. Preliminary experimental results of ultrafast (~25 fs), high intensity (>$10^{19}$ W/cm$^2$) excitation of layered metal targets show anomalous enhancement of specific Ti lines in the 2-14 nm wavelength range.

## INTRODUCTION

New territories in the study, understanding and exploitation of high-field light-matter interactions have been opened up due to the recent advances in 10-fs-range, high-peak-power lasers (e.g. 1-4). One of them is the generation of femtosecond coherent X-ray sources using high harmonics (5,6) and another is the development of X-ray lasers based on inner-shell atomic transitions (7-12). While there has been significant progress in the past 15 years in X-ray laser research and development, saturation in the water window spectral region has not yet been achieved. Most X-ray lasers developed up to now have pulse durations on the picosecond time scale and are low repetition rate devices. A high-repetition-rate, femtosecond X-ray laser will be very useful to study ultrafast dynamical phenomena in nature.

The first inner-shell X-ray laser scheme was proposed by Duguay and Rentzepis (13), in which the lower state of the lasing transition was the ground state of the first ion and could not decay. The idea of utilizing a decaying lower state to insure inversion was suggested by Stankevich (14), elaborated by Arecchi et al. (15) and Elton (16), and calculated in detail by Axelrod for K-shell transitions (17). The technical barrier to be overcome for the successful realization of these schemes is the development of a sufficiently fast and energetic X-ray pump source whose time-scale is of the order of the lifetime of the keV lasing transitions, i.e., in the 10-fs-range (7,8,12,17). Such pumping sources may become available soon due to the advent of femtoscond high power (a few-ten's of TW) lasers. The key benefit of these sources is the potential for ultrafast energy deposition into the target on a timescale comparable to the typical lifetimes of the inner-shell states (0.1-10 fsec). In addition to this, the intrinsic problem with any K-shell transition scheme is that the electrons produced via photo-ionization and subsequent Auger-decay are both energetic enough to collisionally ionize neutral atoms and thereby produce the lower state of the lasing transition thereby destroying the inversion. If it were possible to create an inner-shell population inversion via atomic processes involving electrons only, then photo-ionization pumped X-ray laser schemes based on the same transitions would be far less sensitive to secondary electron collisional filling of the lower state, require less demanding X-ray pumping conditions, and result in inversions which could be maintained longer, compared to equivalent photo-inner-shell ionization pumping schemes based on $KL_3$ transitions.

In this paper we discuss first our survey of a wide range of inner-shell transitions which are insensitive to secondary collisional processes and are thus possible candidates for X-ray laser emission. Then we summarize the results of computer simulations carried out in order to understand the comparative advantages/disadvantages of electron- or photo-pumping schemes in particular atomic systems. We then describe a state-of-the art, table-top CPA laser system constructed in a university laboratory environment capable of providing sufficiently energetic and fast enough laser pulses to excite inner-shell transitions effectively. And finally, we present preliminary experimental results achieved by using this particular laser system and specifically designed multilayer targets.

## MODELLING OF INNER-SHELL X-RAY LASER DYNAMICS

### Population Inversion by Fast Electron Pumping

In a recent study (7) we have inspected pairs of inner-shell vacancy levels for X-ray lasers in which the lower level rapidly decays through Coster-Kronig and super-Coster-Kronig processes. These processes are extremely fast, so that the life time of the lower level is much shorter than that of the upper level. We have surveyed possible transition pairs, and pointed out that the $L_{23}M_1$ transition in the elements of Z=20 to 30 are the best candidates for further X-ray laser studies (Table 1).

We have also shown, that the population inversion between the $L_{23}$ and $M_1$ inner-shell vacancy levels can be created even by electron collisional ionization processes only. Due to the larger electron collisional ionization cross section for all outer shells (corresponding to the lower level of a laser transition, see Fig. 1), seemingly it is not

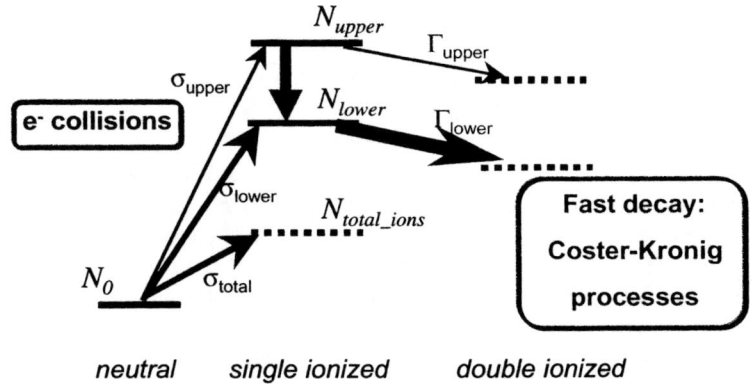

**FIGURE 1.** Energy level scheme for generic electron pumping of inner-shell vacancy levels. The initially unfavorable population can be transiently inverted via fast Coster-Kronig decay of the lower levels.

**TABLE 1.** Inner-Shell Transitions With Fast Lower State Decay (see text for explanation).

| Transition | Z range | Example Atom (Z) | λ (nm) | Decay rate (fs$^{-1}$) | D ratio | P ratio | $N_{inv}$ ($10^{19}$ cm$^{-3}$) |
|---|---|---|---|---|---|---|---|
| L$_1$→M$_3$ | 52-75 | Cs(55) | 2.63 | 6.68 | 1.2 | 46 | <10$^{-10}$ |
| L$_2$→M$_1$ | 20-90 | Ti(22) | 3.09 | 0.37 | 13.5 | 11.8 | 1.1 |
| L$_3$→M$_1$ | 20-90 | Ti(22) | 3.15 | 0.37 | 13.5 | 11.8 | 0.5 |
| L$_3$→N$_5$ | 65-95 | Er(68) | 0.15 | 7.13 | 1.3 | 288 | <10$^{-10}$ |
| M$_4$→N$_2$ | 44-90 | Sn(50) | 3.10 | 0.79 | 28.5 | 26 | 0.012 |

possible to create an inversion between inner-shell vacancy levels by electron collisional ionization processes only. On the other hand, the decay rates of the vacancy states which may undergo Coster-Kronig or super-Coster-Kronig decay can be significantly larger than those of the next deeper vacancy states, i.e. the potential upper state of a lasing transition. For selected states in numerous atomic systems in which the lower state decay is mediated by Coster-Kronig or super-Coster-Kronig decay while the upper state is not, a transient, femtosecond-timescale inversion can be achieved. The expected inversion densities according to a rate-equation-based computer simulation are listed in the last column of Table 1. Other parameters in Table 1.: 'λ' is the wavelength of the transition; 'Decay rate' is for the upper state; 'D-ratio' is the ratio of decay rates of the lower state to that of the upper state; 'P-ratio' is the ratio of collisional ionization rates of the lower state to that of the upper state. The inversion densities were calculated for the electron pulse with duration of 10 fs FWHM and the maximum electron density of $10^{21}$cm$^{-3}$.

# Inversion and Gain by Photopumping

In the case of photo-ionization pumping (Fig 2.), the ionization cross section is larger for a deeper lying vacancy ('upper state') than for an outer-shell vacancy ('lower state') if a photon energy is larger than the ionization threshold of the deeper shell. Hence the degree of initial inversion between the $L_{23}$ and $M_1$ inner-shell vacancy levels can be very large in the case of photo-ionization pumping, leading to a very high gain, as discussed below. The fast decay rate of the $M_1$ inner-shell vacancy level leads to a higher operational neutral density than in the case of the $KL_3$ transition for similar transition wavelengths. To avoid the significant pumping by collisional ionization to the lower level (the $M_1$ inner-shell vacancy level), the collisional ionization rate, $N_0 \sigma \upsilon$, where $N_0$ is the neutral density, $\sigma$ the collisional ionization cross section to the $M_1$ inner-shell vacancy level, $\upsilon$ the velocity of an electron, should be equal to or less than the decay rate of the $M_1$ inner-shell vacancy level, which is 5 $fs^{-1}$ for Ti. In the case of Ti, $N_0$ can be as large as $7 \times 10^{23}$ $cm^{-3}$, which is 10 times higher than the solid density of Ti. For the $L_{23}M_1$ transition, a target with a solid density can be used. On the other hand, as Moon et al. estimated in the case of the $KL_3$ transition in carbon (8), $N_0$ is limited to about $10^{20}$ $cm^{-3}$ due to the long lifetime of the $L_3$ level and other physical time scale. In this case a diluted special target has to be prepared as discussed in Ref. (8). This high operational density limit in the case of the $L_{23}M_1$ transition leads to a high gain (a few tens or larger $cm^{-1}$) for a pumping photon flux above a threshold. A few mm or less length of a gain medium can lead to saturation. The traveling-wave pumping over this distance is easily realizable in experiments. In the case of the $KL_3$ transition in carbon, a long gain medium in the order of 1 cm is required for saturation. The realization of the traveling-wave pumping over 1cm with femtosecond accuracy has turned out to be extremely difficult and is a practical limitation (18).

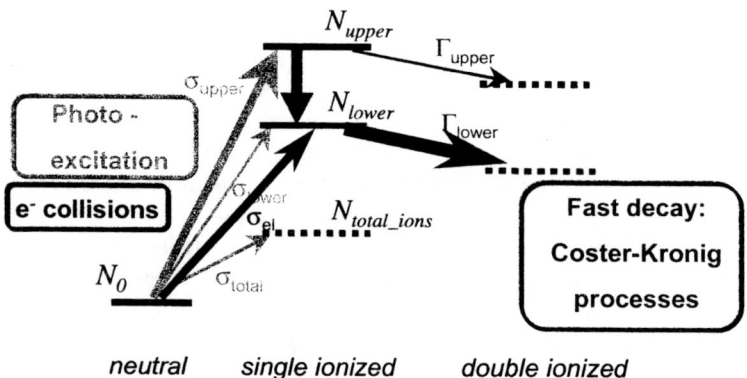

**FIGURE 2.** Energy level scheme for generic photo pumping. The subscript 'upper' designate deeper lying, high energy inner-shell hole state ($L_2$ in our simulation), while the 'lower' corresponds to an inner-shell hole state further outward from the core, with relatively lower energy ($M_1$ in the simulation).

In contrast to the pure electron pumping case, described in the previous section, we present here the gain characteristics of the $L_{23}M_1$ inner-shell transition of Ti in the case of incoherent X-ray photo-pumping. The energy levels considered in the simulations are one neutral level, 6 single vacancy levels and 21 double vacancy levels (when the energy differences for some levels are quite small, they are treated as one level). Figure 2 obviously does not show all the levels included in the simulation. Populations are calculated only for the single vacancy states. The double vacancy states are considered to include the decay channels from the single vacancy states such as photo-ionization, electron collisional ionization, Auger and Coster-Kronig decay and to calculate the energy distribution of electrons generated. Radiative decay channels are included among the single vacancy states. All the Auger and Coster-Kronig decay, photo-ionization, and electron-collisional ionization channels are included from the single vacancy levels to the possible double vacancy levels. The electron-collisional ionization rate is evaluated using an electron energy distribution calculated during the simulation. As radiation sources, both monochromatic (slightly above the $L_{23}$ absorption edge) and black-body radiation sources with Gaussian pulse shapes are considered.

Figure 3 shows one of the simulation results performed in the case of Ti for a monochromatic radiation source of 470 eV with a flux of $1.4 \times 10^{18}$ photons/cm$^2$ and a pulse duration of 2 fs FWHM. The neutral density of Ti used is $6 \times 10^{22}$ cm$^{-3}$. The time t=0 is the time when the radiation flux reaches its maximum. $G_{inv}$ is the gain due to population inversion and $G_{eff}$ the difference between $G_{inv}$ and the absorption of the lasing line by remaining neutrals: $G_{eff}=G_{inv}-G_{abs}$. A large $G_{inv}$ is obtained before the radiation flux reaches its maximum. However, there is still a significant absorption, preventing the amplification of the light. Later in time when the neutrals are sufficiently depleted, $G_{eff}$ becomes positive and amplification becomes possible. The magnitude of $G_{eff}$ is much smaller than $G_{inv}$, but still large. Table 2 shows the summary of simulation performed with monochromatic radiation source.

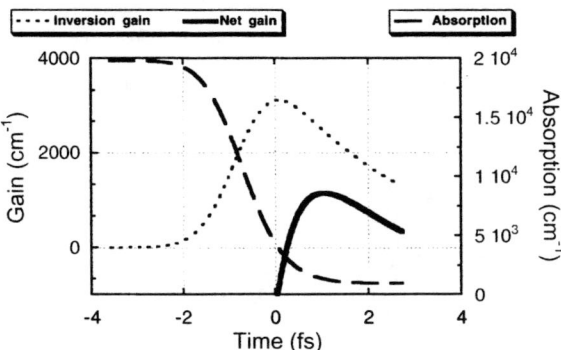

**FIGURE 3.** Time dependence of the inversion related and net (corrected by the absorption) gain for the simulation parameters of 2 fs excitation, $1.4 \times 10^{18}$ cm$^{-2}$ photon flux, $N_0 = 6 \times 10^{22}$ cm$^{-3}$ initial neutral density for the $L_2M_1$ transition in titanium.

Numerically we can also neglect or include secondary electron collisional ionization in order to evaluate its role in the inversion process. Since the electron collisional ionization preferentially populates the lower level relative to the upper level, the neglect of the collisional ionization will increase the population difference and hence the gain due to inversion. One should note, however, that in this latter case no effective gain is achieved. This is because without collisional ionization, the depletion of the neutrals is so slow that absorption is still larger than $G_{inv}$ during the entire period of time during which inversion is created. If the photon flux is increased, the depletion of the neutrals by photons only is sufficiently fast that a significant effective gain is obtained.

Since the lifetime of the lower level (e.g., 0.2 fs for the $M_1$ inner-shell vacancy level in Ti) is much shorter than that of the upper level (e.g., 2.7 fs for the $L_{23}$ level in Ti), the population inversion can be created with the radiation source of a long pulse width, even, in principle, as great as 200 fs in duration. This is one of the advantages of the $L_{23}M_1$ transition compared to the $KL_3$ transition. In the case of the $KL_3$ transition, the lifetime of the lower level (the $L_{23}$ inner-shell vacancy level) is much longer than that of the upper level (the K inner-shell vacancy level), and thus a much shorter pulse duration of a radiation source is demanded. According to Ref. (8), no gain on the $KL_3$ transition in carbon can be created for a radiation pulse longer than 80 fs at any power.

In Table 2. we summarized the main results of the simulations. The first three columns show the input parameters, the 4$^{th}$ and 5$^{th}$ refer to the achievable maximum net ($G_{eff}$) and inversion related ($G_{inv}$) gain values for the full model, when all the electron collisions are included. The last two columns list the same gain value pairs, neglecting the collisional effects in the calculations.

**TABLE 2. Summary of Simulation Results for Photopumping.**

| $N_0$ (cm$^{-3}$) | Photon flux @470 eV (#/cm$^2$) | Pulse width (fs) | $G_{eff,max}$ (cm$^{-1}$) w/collisions | $G_{inv,max}$ (cm$^{-1}$) w/collisions | $G_{eff,max}$ (cm$^{-1}$) w/o collisions | $G_{inv,max}$ (cm$^{-1}$) w/o collisions |
|---|---|---|---|---|---|---|
| 6x10$^{22}$ | 1.4x10$^{18}$ | 5 | 180 | 340 | <0 | 2100 |
| 6x10$^{22}$ | 1.4x10$^{18}$ | 10 | 30 | 125 | <0 | 1300 |
| 6x10$^{22}$ | 1.4x10$^{18}$ | 20 | 10 | 42 | <0 | 800 |
| 6x10$^{22}$ | 1.4x10$^{18}$ | 50 | ~0 | 8 | <0 | 400 |
| 6x10$^{22}$ | 1.4x10$^{18}$ | 100 | ~0 | 2 | <0 | 200 |
| 6x10$^{22}$ | 1.4x10$^{18}$ | 10 | 30 | 125 | <0 | 1300 |
| 3x10$^{22}$ | 1.4x10$^{18}$ | 10 | 25 | 90 | <0 | 700 |
| 6x10$^{21}$ | 1.4x10$^{18}$ | 10 | 10 | 50 | <0 | 200 |
| 6x10$^{22}$ | 1.4x10$^{17}$ | 10 | 5 | 70 | <0 | 300 |
| 6x10$^{22}$ | 1.4x10$^{18}$ | 10 | 30 | 125 | <0 | 1300 |
| 6x10$^{22}$ | 1.4x10$^{19}$ | 10 | 80 | 180 | 900 | 2100 |
| 6x10$^{22}$ | 1.4x10$^{20}$ | 10 | 140 | 280 | 1350 | 2700 |

The required flux and rise time of X-rays can be provided by a high-temperature, high-density plasma produced by a high-intensity femtosecond lasers. To estimate the input energy of such a laser to achieve the above output of an X-ray laser, the width of the line focus is assumed to be 10 μm. The length of the gain medium is then 2.5 mm, which leads to the total energy of 1.44 mJ of pumping X-rays at 470 eV. Assuming the conversion efficiency of 0.1 % from a visible photon to an X-ray photon under interest (8), the required energy of an optical laser becomes 1.44 J. This level of lasers is already available in several laboratories (e.g. 1-3, 19, 20), making this scheme realizable in the near future.

## EXPERIMENTAL

In this section we describe a) our chirped pulse amplification (CPA) laser system as an example of those table-top high intensity sources currently coming on line and capable of delivering ultrafast energy deposition on target at high repetition rate, (>1 Joule of optical energy in less than 25 fs with 10 Hz repetition rate); b) suggestion for experimental setup and design of X-ray laser targets; and c) preliminary results achieved on Au/Cu/Ti multilayer targets.

## Table-Top Multi-TW Femtosecond Laser System for Fast Energy Deposition

To obtain ultrashort, high-energy laser pulses in practice, a pulse amplification scheme has been developed: the chirped-pulse amplification (CPA) technique (21). This revolutionary method was applied first for picosecond Nd:glass lasers (21-23), later on for a variety of amplifier materials (24) and most successfully for femtosecond Ti:sapphire lasers (25, 26). Recently, various laboratories put together laser amplifier systems capable of producing pulses of 10-100 TW peak-power in the range of 20-30 fs ('full width of half maximum' — FWHM) pulse length (1-4, 19, 20, 27).

The 10 Hz repetition rate, 60-TW CPA laser system at UCSD (19) is specifically designed for reliable operation at high repetition rates in high-field interaction studies and for generation of incoherent and coherent X-ray radiation (7, 28-30). The essence of CPA is to avoid nonlinear propagation effects and to amplify ultrashort pulses to the TW peak power level by stretching initially ultrashort seed pulses to a nanosecond level, then amplifying them in a series of solid state amplifiers, and finally recompressing them to the original ultrafast pulse duration. Our Ti:sapphire amplifier chain consists of a 2 nJ, 15 fs Ti:sapphire oscillator, cylindrical mirror expander, and three (regenerative/4-pass/2-pass) Ti:sapphire amplification stages followed by a hybrid vacuum-atmosphere compressor (see Figure 4.).

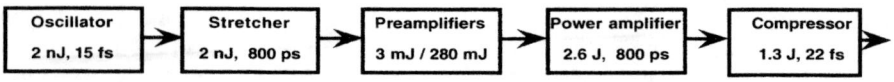

**FIGURE 4.** Block diagram of the laser amplifier chain with the main pulse parameters at the output of each stage.

The 15-fs seed pulses of the system are generated by a prism compensated Ti:sapphire oscillator (31, 32) with ~2 nJ/pulse energy. The seed pulses are stretched in time by a factor of ~50,000 in a pulse stretcher, which controls up to fourth-order dispersion using off-axes cylindrical mirrors and 1200 l/mm gratings (33). The evolution of the dispersion in the form of higher order Taylor-series dispersion terms and with a time-of-flight method has been followed and optimized by computer simulation to match the predicted distortions in the subsequent stages of the system.

Three Ti:sapphire amplifiers – each pumped by second harmonic (SH) radiation of Nd:YAG lasers – amplify the energy of every 40,000th of the series of stretched pulses from the nanojoule to the joule level. The first amplifier at 20 Hz repetition rate uses regenerative amplification in a linear resonator, which involves pulse insertion and dumping by a Pockels cell switch and thin film polarizers. An intracavity thin film etalon (34) helps to form the required broad spectral shape by pre-compensating for the spectral narrowing that occurs in the amplification process. The output energy of this stage is ~ 3 mJ.

The second amplifier is arranged in a 4-pass, 'bow-tie' configuration (1), where a 12-mm diameter Ti:sapphire crystal is pumped on both faces by spatially symmetrized beams of the green pump laser at 20 Hz repetition rate again, reaching ~280 mJ output energy at the Ti:sapphire wavelengths centered at $\lambda$=795 nm.

A 2-pass amplifier further amplifies every other pulse from the 4-pass amplifier to give a 10 Hz pulse repetition rate. This final amplifier consists of a 30-mm diameter, 30-mm long Ti:sapphire crystal pumped by the SH converted output of a custom designed 10 Hz Nd:YAG laser system. The output energy of the Ti:sapphire laser beam after two passes is 2.6 J/pulse. At this level the shot-to-shot amplitude fluctuation are ±7% and the 12 mm diameter beam has a 'top-hat' near-field intensity distribution.

The diameter of the laser beam has to be expanded again between the final amplifier and the pulse compressor by a factor of 4 to avoid any damage to the surfaces of the pulse compressor gratings. The last grating and all subsequent beam steering elements are placed in vacuum. We note here, that in contrast to other ultra-high power, femtosecond CPA systems, we put only the last grating into an evacuated chamber. Previous measurements show (35) that the performance of the system is the same as that of the 'all-vacuum' compressors within 1 fs precision, while the complexity of the setup is greatly reduced and the adjustability improved relative to fully enclosed grating systems. The final compressed energy is 1.3 (±7%) J/pulse in a 50 mm diameter beam.

To characterize the amplitude and phase properties of the final compressed pulses, we used a modified version of the commonly used frequency resolved optical gating (FROG) technique (36) in a polarization-gating arrangement (PG-FROG) based on the electronic Kerr effect. In our PG-FROG arrangement (37), we used a pair of broadband low-dispersion thin-film polarizers (TFPs) to eliminate the dispersion problems associated with the rather unusual higher-order phase content of the conventionally used calcite polarizers (38). This produces a high-contrast, low-dispersion PG-FROG for use on high peak-power ultrafast pulses. We have used this TFP based PG-FROG device to first optimize the best (shortest) pulse by visually inspecting the real-time images of the pulses while changing the separation of the compressor gratings. Then we were able to determine the full time and wavelength dependence of the laser amplitude and phase by running a retrieval algorithm. The measured average FWHM of the pulse duration is 22 ±1 fs, and the weighted RMS spectral phase error (39) is 0.5 radian.

The focusability of the 5 cm diameter beam was tested by focusing it with a 10-cm focal length off-axis parabola and imaging the focal volume onto a CCD camera. The measurements imply a best focal spot with 3±0.4 µm FWHM diameter.

Combining the results of the energy, pulse duration and focal spot measurements, one predicts that a peak laser intensity in excess of $10^{20}$ W/cm$^2$ is possible. In addition to this conventional estimation, a more direct measurement has also been performed in order to better asses the actual peak laser intensity. Photo-ionization experiments in the optical tunneling regime (19) suggest, that peak intensity of $>5 \times 10^{19}$ W/cm$^2$ has actually been reached in the focal region of the f=10 cm off-axis parabola mirror.

## Layout of the experiments

Figure 5(a) shows the schematic of the solid-target experiments carried out at the UCSD laser facility. The 5 cm diameter beam is focused by an f=10 cm off-axis parabola onto the surface of the target at close to perpendicular angle of incidence. The target has been shifted after each laser shot parallel with its surface by motorized, vacuum compatible controllers. The surface of the focusing mirror has to be protected from the debris of the laser plasma by thin (3 µm) mylar foil. The emitted X ray radiation is observed tangentially with the target surface from a distance of ~40 cm by a custom-made, grazing incidence, flat-field, soft-X-ray spectrometer (40). The spectrometer is situated in a differentially pumped vacuum chamber, separated from the main interaction chamber by a 25 µm slit.

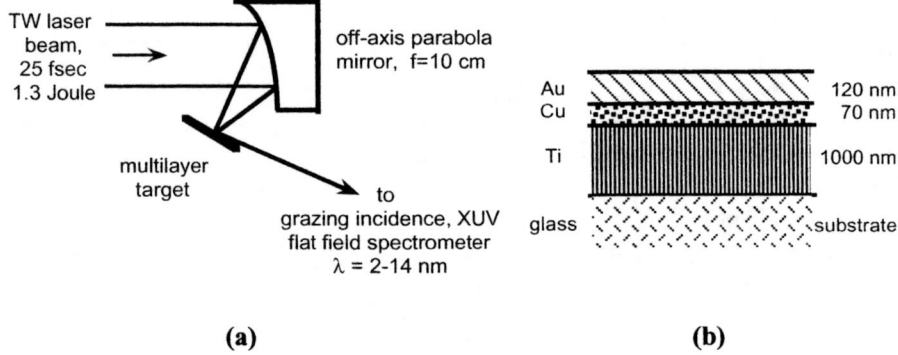

**FIGURE 5.** Optical layout (a) and target structure (b) of experiments designed for studies of inner-shell inversion in solid targets.

One example of the multilayer targets is shown on Figure 5(b). The purpose of the top gold layer is to operate as an X-ray radiation emitter excited directly by the high power, ultrashort Ti:sapphire laser pulses The thin copper layer filters the low energy part of the spectrum to avoid direct photopumping of the lower states of the Ti. The Ti layer (evaporated on the glass substrate) is the active material. The thickness of each layer had been designed taking into account the optimal ratios of the absorption values of the actual materials at the wavelengths of the optical excitation, the desired x-ray pumping and the final X-ray emission.

## Preliminary results

Typical spectra of the emission observed by single shot excitation of the multilayer targets is shown in Figure 6. The three spectra on the top (a, b, c) correspond to three different excitation energy of the multilayers. On the bottom of the composite image one can see the spectrum of bulk targets for calibration and comparison purposes. Carbon and its 'CVI' and 'CV' spectral lines (f) were used to calibrate the flat-field spectrometer. The spectrum of bulk gold (e) and bulk Ti (d) and their dependence on the pump-intensity will serve as reference spectra in a detailed comparative study of the multi-layer structures.

The spectra in Figure 6 were taken in a slightly elongated focal arrangement, in the direction of the longer axis of an elliptical focal spot. At several wavelengths of the Ti spectra anomalous line enhancement were observed by increasing pumping laser intensities (arrows on the image). Further measurements using line focus arrangement are in progress in our laboratory in order to reach gain along an elongated focal area.

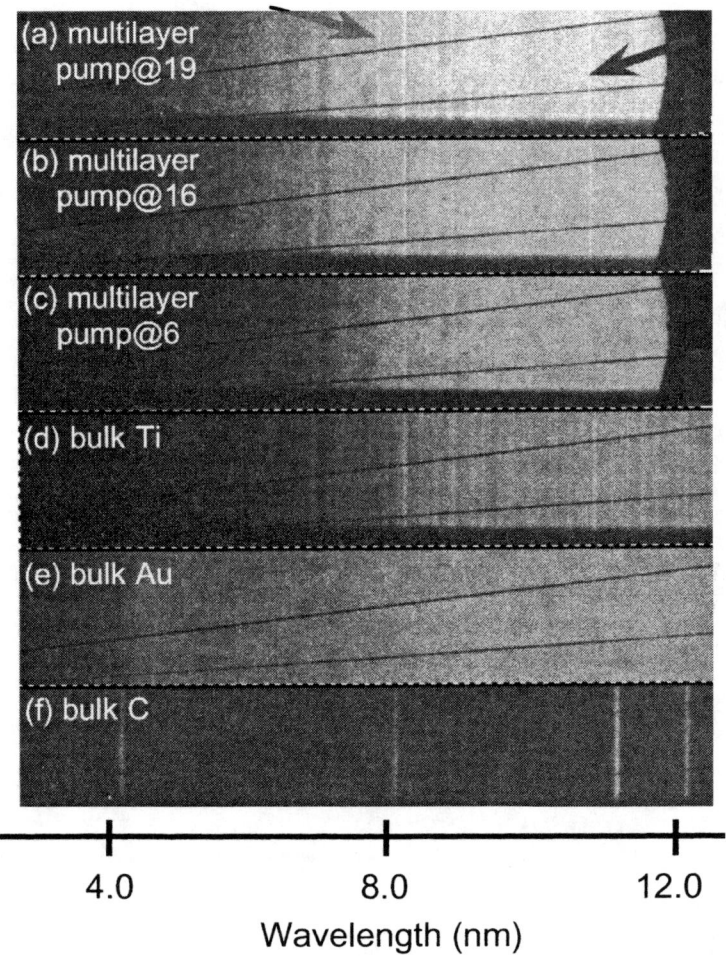

**FIGURE 6.** Emission spectra of Au/Cu/Ti multilayer target excited by 25 fs, 1.3 J energy laser pulses. (Note: the diagonal black lines are the shadows of the fiducial wires used to optimize to focusing the imaging system of the spectrometer.)

## CONCLUSIONS

In summary, we have shown that super-Coster-Kronig and Coster-Kronig mediated lower state decay can be effectively used for production of inner-shell population inversion both via electron collisional ionization and photo-ionization. In the latter case net gain at water window wavelengths is predicted to be large. Such schemes have less sensitivity to secondary electron collisional ionization and significantly less stringent pumping requirements than K-shell schemes. Lasers which are capable of pumping these schemes have been constructed and preliminary experiments, while not demonstrating explicit gain, do show anomalous line emission on the transitions of interest. Further improvement in pumping geometry and pump beam characteristics may lead to convenient, extremely bright, high repetition rate source of coherent radiation in the x-ray water window.

## REFERENCES

1. Barty, C.P.J., Guo, T., Blanc, C.L., Ráksi, F., Rose-Petruck, C., Squier, J., Wilson, K.R., Yakovlev, V.V., and Yamakawa, K. *Optics Letters* **21**, 668-670 (1996).
2. Antonetti, A., Blasco, F., Chambaret, J.P., Cheriaux, G., Darpentigny, G., Le Blanc, C., Rousseau, P., Ranc, S., Rey, G., and Salin, F. *Applied Physics B (Lasers and Optics)* **B65**, 197-204 (1997).
3. Yamakawa, K., Aoyama, M., Matsuoka, S., Takuma, H., Barty, C.P.J., and Fittinghoff, D. *Optics Letters* **23**, 525-527 (1998).
4. Backus, S., Durfee, C.G., III, Murnane, M.M., and Kapteyn, H.C. *Review of Scientific Instruments* **69**, 1207-23 (1998).
5. Rundquist, A., Durfee, C.G., III, Zenghu, C., Herne, C., Backus, S., Murnane, M.M., and Kapteyn, H.C. *Science* **280**, 1412-15 (1998).
6. Schnürer, M., Spielmann, C., Wobrauschek, P., Streli, C., Burnett, N.H., Kan, C., Ferencz, K., Koppitsch, R., Cheng, Z., Brabec, T., and Krausz, F. *Physical Review Letters* **80**, 3236-3239 (1998).
7. Kim, D., Tóth, C., and Barty, C.P.J. *Physical Review A* **59**, R4129-4132 (1999).
8. Moon, S.J. and Eder, D.C. *Physical Review A* **57**, 1391-1394 (1998).
9. Moribayashi, K., Sasaki, A., and Tajima, T. *Physical Review A* **58**, 2007-2015 (1998).
10. Li, Y., Schillinger, H., Ziener, C., and Sauerbrey, R. *Optics Communications* **144**, 118-24 (1997).
11. Strobel, G.L., London, R.A., and Eder, D.C. *Applied Physics B (Lasers and Optics)* **B60**, 513-18 (1995).
12. Kapteyn, H.C. *Applied Optics* **31**, 4931-4939 (1992).
13. Duguay, M.A. and Rentzepis, P.M. *Applied Physics Letters* **10**, 350-352 (1967).
14. Stankevich, Y.L. *Soviet Physics Doklady* **15**, 356-357 (1970).
15. Arecchi, F.T., Banfi, G.P., and Malvezzi, A.M. *Optics Communications* **10**, 214-218 (1974).
16. Elton, R.C. *Applied Optics* **14**, 2243-2249 (1975).
17. Axelrod, T.S. *Physical Review A* **13**, 376-382 (1976).
18. Snavely, R.A., DaSilva, L.B., Eder, D.C., Matthews, D.L., and Moon, S.N. *SPIE Proceedings* **3156**, 109-113 (1997).
19. Walker, B.C., Tóth, C., Squier, J.A., Fittinghoff, D.N., Guo, T., Kim, D.-E., Wilson, K.R., and Barty, C.P.J. in *Conference on Lasers and Electro-Optics, CLEO '99*, Paper CTuD2 (1999).
20. Yamakawa, K., Aoyama, M., Matsuoka, S., Kase, T., Akahane, Y., and Takuma, H. *Optics Letters* **23**, 1468-70 (1998).
21. Strickland, D. and Mourou, G. *Optics Communications* **56**, 219-221 (1985).
22. Maine, P., Strickland, D., Bado, P., Pessot, M., and Mourou, G. *IEEE Journal of Quantum Electronics* **24**, 398-403 (1988).
23. Ferray, M., Lompré, L.A., Gobert, O., L'Huillier, A., Mainfray, G., Manus, C., Sanchez, A., and Gomes, A.S. *Optics Communications* **75**, 278-281 (1990).
24. Perry, M.D. and Mourou, G. *Science* **264**, 917-923 (1994).
25. Kmetec, J.D., Macklin, J.J., and Young, J.F. *Optics Letters* **16**, 1001-1003 (1991).
26. Sullivan, A., Hamster, H., Kapteyn, H.C., Gordon, S., White, W., Nathel, H., Blair, R.J., and Falcone, R.W. *Optics Letters* **16**, 1406-1408 (1991).
27. Chambaret, J.P., Blanc, C.L., Chériaux, G., Curley, P., Darpentigny, G., Rousseau, P., Hamoniaux, G., Antonetti, A., and Salin, F. *Optics Letters* **21**, 1921-1923 (1996).

28. Barty, C.P.J., Guo, T., Blanc, C.L., Ráksi, F., Rose-Petruck, C., Squier, J., Walker, B.C., Wilson, K.R., Yakovlev, V.V., and Yamakawa, K. in *X-ray Lasers* (eds. Svanberg, S. & Wahlström, C.-G.) 282-288 (IOP, Lund, Sweden, 1996).
29. Rose-Petruck, C., Jimenez, R., Guo, T., Cavalleri, A., Siders, C.W., Ráksi, F., Squier, J.A., Walker, B.C., Wilson, K.R., and Barty, C.P.J. *Nature* **398**, 310-312 (1999).
30. Kim, D.-E., Tóth, C., and Barty, C.P.J. in *X-ray Lasers* (eds. Kato, Y. & Takuma, H.) 309-312 (IOP, Kyoto, Japan, 1998).
31. Spielmann, C., Curley, P.F., Brabec, T., and Krausz, F. *IEEE Journal of Quantum Electronics* **30**, 1100-1114 (1994).
32. Zhou, J., Taft, G., Huang, C.-P., Murnane, M.M.. and Kapteyn, H.C. *Optics Letters* **19**, 1149-1467 (1994).
33. Lemoff, B.E. and Barty, C.P.J. *Optics Letters* **18**, 1651-1653 (1993).
34. Barty, C.P.J., Korn, G., Raksi, F., Rose-Petruck, C., Squier, J., Tien, A.C., Wilson, K.R., Yakovlev, V.V., and Yamakawa, K. *Optics Letters* **21**, 219-221 (1996).
35. Walker, B.C., Squier, J.A., Fittinghoff, D.N., RosePetruck, C., and Barty, C.P.J. *IEEE Journal of Selected Topics in Quantum Electronics* **4**, 441-444 (1998).
36. Kane, D.J. and Trebino, R. *Optics Letters* **18**, 823-825 (1993).
37. Tóth, C., Fittinghoff, D.N., Walker, B.C., Squier, J.A., and Barty, C.P.J. *Ultrafast Phenomena XI* **9**, 109-111 (1998).
38. Fittinghoff, D.N., Walker, B.C., Squier, J.A., Tóth, C.S., Rose-Petruck, C., and Barty, C.P.J. *IEEE Journal of Selected Topics in Quantum Electronics* **4**, 430-40 (1998).
39. Walker, B.C., Tóth, C., Fittinghoff, D.N., and Guo, T. *Journal of Optical Society of America B* **16**, 1292-1299 (1999).
40. Nakano, N., Kuroda, H., Kita, T., and Harada, T. *Applied Optics* **23**, 2386-2392 (1984).

# Time-resolved x-ray photoabsorption and diffraction on timescales from ns to fs

P.A. Heimann[1], T. Missalla[2], A. Lindenberg[3], I. Kang[3], S. Johnson[3], Z. Chang[4], H.C. Kapteyn[4], R.W. Lee[2], R.W. Falcone[3], R.W. Schoenlein[5], T.E. Glover[1], A.A. Zholents[6], M.S. Zolotorev[6] and H.A. Padmore[1]

[1] Advanced Light Source, Lawrence Berkeley National Laboratory, Berkeley, CA 94720
[2] Lawrence Livermore National Laboratory, Livermore, CA 94551
[3] Physics Department, University of California at Berkeley, Berkeley, CA 94720
[4] Center for Ultrafast Optical Science, University of Michigan, Ann Arbor, MI 48109
[5] Materials Sciences Division, Lawrence Berkeley National Laboratory, Berkeley, CA 94720
[6] Center for Beam Physics, Accelerator and Fusion Research Division, Lawrence Berkeley National Laboratory, Berkeley, CA 94720

Time-resolved x-ray diffraction with picosecond time resolution is used to observe scattering from coherent acoustic phonons in laser-excited InSb crystals. The observed oscillations in the crystal reflectivity are in agreement with a model based on dynamical diffraction theory. Synchrotron radiation pulses of ~ 300 fs in duration have been generated by femtosecond laser pulses modulating the electron beam in the Advanced Light Source.

## 1. INTRODUCTION

Time-resolved x-ray photoabsorption and diffraction enable one to probe the electronic and structural changes associated with phase transitions, solid state dynamics and chemical reactions. Important structural dynamics occur on the timescales from nanoseconds to femtoseconds with one fundamental limit set by vibrational periods, ~100 fs. Recent experiments, using both synchrotron and laser-plasma based sources, have observed phase transitions and chemical reactions on picosecond time-scales. In biology, real-time studies of photo-initiated reactions in complex molecules such as photoactive yellow protein (PYP) have been performed (1). Diffraction experiments utilizing laser-plasma x-ray sources have observed laser-induced disorder in Langmuir-Blodgett films (2) and a coherent acoustic pulse in GaAs (3).

Two approaches are being pursued at the Advanced Light Source (ALS): the first utilizing ultrafast detectors and the second developing a femtosecond x-ray source. The long range order of the semiconductor InSb has been probed by the diffraction of x-rays into a streak camera. Oscillations in the InSb crystal reflectivity are observed because of scattering from coherent acoustic phonons. A femtosecond x-ray source has been produced by co-propagating a femtosecond laser pulse with an electron bunch in a wiggler. In recent proof-of-principle experiments, we have successfully generated ~ 300 fs synchrotron pulses for the first time.

## 2. TIME-RESOLVED X-RAY DIFFRACTION OF InSb

An ALS bending magnet beamline (7.3.3) and a Si (111) monochromator crystal provide x-rays at a wavelength of 2.4 Å (4). The diffracted beam is then directed onto an InSb crystal oriented near the Bragg angle for the (111) reflection. To better match the penetration depths of the laser and x-rays, the crystal is asymmetrically-cut so that the diffracted beam leaves the crystal at a grazing angle of about 3 degrees. We use a Ti:Al$_2$O$_3$-based 150 fs, 1 kHz, 800 nm laser, synchronized to the synchrotron radiation time structure with jitter less than 5 ps. The time-resolved x-ray diffracted intensity following laser excitation is then measured using a streak camera triggered by a GaAs photoconductive switch (5). The time resolution of the camera is 3 ps.

We now discuss the observation of laser-induced coherent phonons at a fluence 20 % below the damage threshold of 15 mJ/cm$^2$. Figure 1 shows the time-dependent diffracted intensity measured at 0, +20, and +40 arcseconds from the Bragg peak.

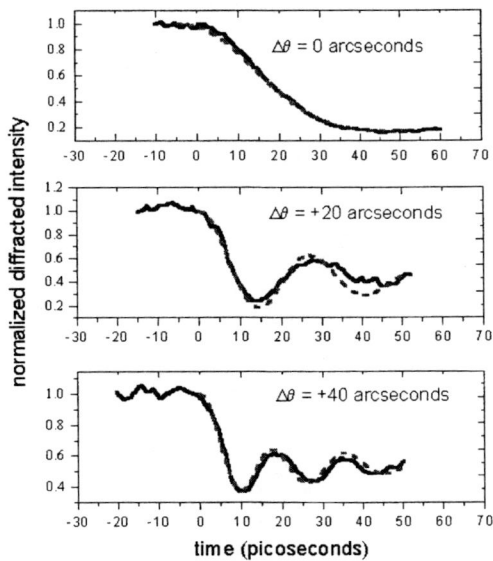

**FIGURE 1.** Experimentally measured (solid line) and simulated (dashed line) time-resolved diffracted intensity at crystal angles of 0, +20, and +40 arcseconds from the Bragg peak.

Impulsive excitation of a solid on a timescale shorter than the material's acoustoelastic response time generates coherent acoustic phonons across a range of wavevectors peaked at a wave-number of order one inverse laser penetration depth, 100 nm. As the crystal angle is varied, different phonon modes are probed out of the spectrum of excited modes. By wavevector matching considerations, the phonon frequency ω observed at a deviation $\Delta\theta$ from the Bragg angle θ of a symmetric crystal reflection is given by

$$\omega = v \, |\mathbf{G}| \, \Delta\theta \cot\theta, \qquad (1)$$

where v is the speed of sound within the crystal and **G** the reciprocal lattice vector.

Figure 1 includes calculated (normalized) diffracted intensities based on dynamical diffraction theory coupled to analytic solutions for the laser-induced strain profile. For an instantaneously heated crystal with assumed exponential temperature profile near the surface, the time-dependent strain profiles have been derived by Thomsen et al (6). There are three adjustable parameters in the model: the time for thermal transfer of energy to the lattice and the amplitudes of the thermal and deformation potential generated strain. The best fits correspond to a coupling time of 12 ps, a thermal strain of 0.17 % (just below that of InSb at its melting temperature) and a non- thermal contribution a factor of two smaller. We extract a sound velocity of ~ 4000 m/s, in agreement with the known value for InSb 3900 m/s (7).

At a slightly higher laser fluence, 10 % below the damage threshold, no

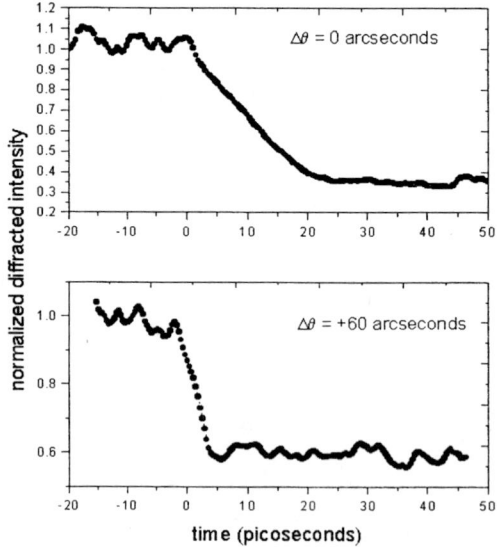

**FIGURE 2.** Time-resolved diffracted intensity at 0 and +60 arcseconds from the Bragg peak near the laser damage threshold. At 60 arcseconds, the diffracted intensity falls within 3 ps.

temporal oscillations occur, as shown in Fig. 2. At +60 arcseconds from the Bragg maximum, a 40 % reduction in intensity occurs on a time-scale limited by the streak camera time resolution (3 ps). In effect, only the first half-period of the oscillations induced at lower fluences are observed. We interpret these results as follows. At a critical fluence, the lattice is driven into a disordered state on a timescale set by one-half a phonon period. Since the observed 3 ps drop in the diffracted intensity occurs faster than the time for thermal transfer of energy to the lattice, we conclude that the first step in the observed disordering transition is essentially non-thermal in nature.

## 3. A FEMTOSECOND X-RAY SOURCE

The technique for generating femtosecond x-rays from the ALS is based on extracting a ~100 fs slice of the long (30 ps) electron bunch using a femtosecond laser pulse. (8) Co-propagation of a femtosecond laser pulse with the electron bunch through an appropriately tuned wiggler results in acceleration (and deceleration) of an ultrashort slice of electrons. These electrons will separate spatially from the main electron bunch in a dispersive section of the storage ring. Radiation from the modulated electrons at a bend magnet source can be imaged onto a slit isolating the ultrafast x-ray pulse.

The laser and wiggler emission in the near and far fields are overlapped using diagnostics at the wiggler front end. The spectrum of the laser is also matched to the fundamental of the wiggler. The efficiency of the interaction between the laser and electron beam is tested by measuring the gain in the laser pulse energy.

Femtosecond duration synchrotron pulses are directly measured by cross-correlating the visible light from bend-magnet beamline (6.3.2) at the ALS with the synchronized laser pulses in a non-linear crystal. Figure 3(a) shows a laser synchrotron cross-correlation measurement on a long time scale. The measured pulse duration, 39 ps FWHM, corresponds to the overall electron bunch duration. Measurement with higher time resolution (Fig. 3b) shows the femtosecond "dark"

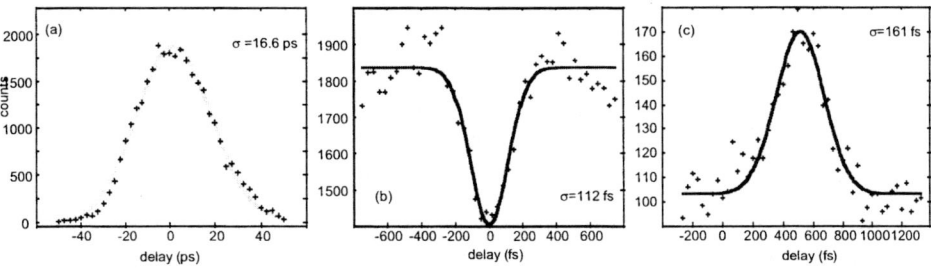

**FIGURE 3.** Cross correlation of the visible synchrotron pulse with a femtosecond laser pulse: (a) overall synchrotron pulse, (b) femtosecond dark pulse from on-axis radiation, (c) femtosecond pulse from off-axis radiation.

pulse (260 fs FWHM) originating from the central core of the electron bunch. Figure 3(c) shows a measurement with a knife edge at the $3\sigma_x$ position, where the femtosecond "bright" pulse is observed (380 fs FWHM). The femtosecond time structure will be invariant over the entire spectral range of bend-magnet emission from the near infrared to x-ray wavelengths. The observed pulse duration is limited by the position of the bend magnet beamline used in these characterization measurements. For a bend magnet immediately following the wiggler, an x-ray pulse duration of ~ 100 fs is calculated.

## ACKNOWLEDGMENTS

This work was supported by the U.S. Department of Energy through Lawrence Berkeley National Laboratory and the Institute for Laser Science and Applications at Lawrence Livermore National Laboratory. The ALS is supported by the Director, Office of Science, Office of Basic Energy Sciences, Materials Sciences Division of the U.S. Department of Energy under Contract No. DE-AC03-76SF00098.

## REFERENCES

1. Perman, B., Srajer, V., Ren, Z., Teng, T., Pradervand, C., Ursby, T., Bourgeois, D., Schotte, F., Wulff, M., Kort, R., Hellingwerf, K., Moffat, K., *Science* **279**, 1946-50 (1998).
2. Rischel, C., Rousse, A., Uschmann, I., Albouy, P.-A., Geindre, J.-P., Audebert, P, Gauthier, J.-C., Forster, E., Martin, J.L.,Antonetti, A., *Nature* **390**, 490-492 (1997).
3. Rose-Petruck, C., Jimenez, R., Guo, T., Cavalleri, A., Siders, C.W., Raksi, F., Squiers, J.A., Walker, B.C., Wilson, K.R., Barty, C.P.J., *Nature* **398**, 310-312 (1999).
4. Larsson, J., Heimann, P.A., Lindenberg, A.M., Schuck, P.J., Bucksbaum, P.H., Lee, R.W., Padmore, H.A., Wark, J.S., and Falcone, R.W., *Appl. Phys. A* **66**, 587-591 (1998).
5. Chang, Z., Rundquist, A., Zhou, J., Murnane, M.M., Kapteyn, H.C., Liu, X., Shan, B., Liu, J., Niu, L., Gong, M., Zhang, X., Appl. *Phys. Lett.* **69**, 133-135 (1996).
6. Thomsen, C., Grahn, H.T., Maris, H.J., and Tauc, J., *Phys. Rev. B* 34, 4129-4138 (1986).
7. Madelung, O., (Ed.), *Semiconductors - Basic Data,* Berlin Heidelberg: Springer-Verlag, 1996, p. 147.
8. Zholents, A.A., Zolotorev, M.S., *Phys. Rev. Lett.* **76**, 912-915 (1996).

# A Novel Type of X-Ray Interferometer

O. Kettig[1], H. Backe[1], N. Clawiter[1], S. Dambach[1], Th. Doerk[1],
N. Elbai[1], H. Euteneuer[1], F. Hagenbuck[1], P. Holl[3], H. Jacobs[1],
K.-H. Kaiser[1], J. Kemmer[3], Th. Kerschner[4], G. Kube[1], H. Koch[4],
W. Lauth[1], H. Mannweiler[1], H. Matthäy[4], H. Schöpe[1], D. Schroff[1],
M. Schüttrumpf[4], R. Stötter[3], L. Strüder[2], Th. Walcher[1],
A. Wilms[4], C. v. Zanthier[3] and M. Zemter[4]

[1] *Institut für Kernphysik, Johannes Gutenberg–Universität D-55099 Mainz, Germany*
[2] *MPI–Halbleiterlabor, D-81245 München, Germany*
[3] *KETEK GmbH, D-85764 Oberschleißheim, Germany*
[4] *Institut für Experimentalphysik, Ruhr–Universität D-44780 Bochum, Germany*

**Abstract.** Novel types of interferometers have been developed at the Mainz Microtron MAMI with which the complex index of refraction of thin self–supporting foils can be measured. For the vacuum ultraviolet and soft x–ray region the interferometer consists of two collinear undulators, between which a foil can be placed, and a grating spectrometer. For the hard x-ray region, up to an energy of about 40 keV, it consists of two foils in which the electron beam produces transition radiation, and a single crystal spectrometer in Bragg geometry. Taking advantage of the low emittance 855 MeV electron beam distinct intensity oscillations have been observed as a function of the distance between the undulators and foils, respectively. The complex index of refraction has been investigated at the K– and L–absorption edges of carbon and nickel.

## INTRODUCTION

Resonant anomalous x-ray scattering plays an increasingly important role in many disciplines of physics, biology, and material sciences. Using the brilliant and tuneable x-ray beams from modern synchrotron radiation sources it is now possible to fully exploit the information in the strong energy and polarisation dependence of the atomic scattering amplitude $f(\omega, \vec{q}) = f_0(\vec{q}) + f'(\omega) + if''(\omega)$ near absorption edges [1]. This is the ratio of the scattering amplitude of an atom to that of a free (Thomson) electron. With these quantities the complex index of refraction $n(\omega) = 1 - \delta(\omega) - i\beta(\omega)$ can be determined using the relations for the dispersion $\delta(\omega) = (1/2)(\omega_p/\omega)^2(f_0(0) + f'(\omega))/Z$ and the absorption $\beta(\omega) = (1/2)(\omega_p/\omega)^2 f''(\omega)/Z$. In these expressions $Z$ is the atomic number, $\omega_p$ the plasma frequency with $\omega_p^2 =$

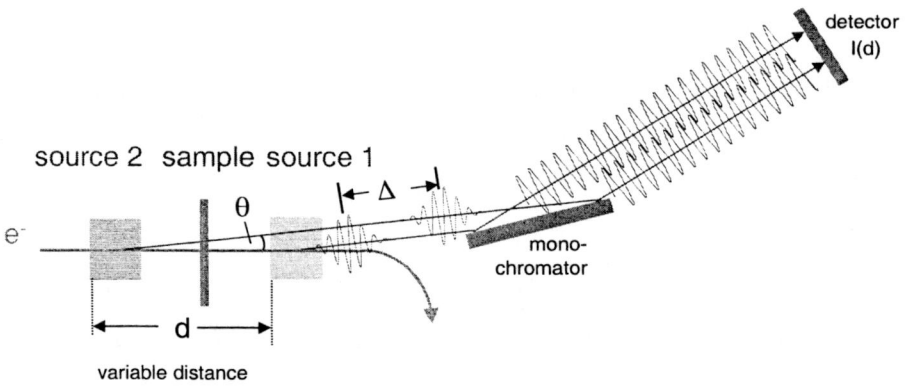

**FIGURE 1.** Interferometry with spatially separated coherent x-ray beams.

$4\pi r_0 c^2 n_a Z$, $r_0$ the classical electron radius, $n_a$ the number of atoms per volume, and $f_0(0) = Z$ neglecting relativistic corrections.

The imaginary scattering factor $f''$ can be directly determined from the total photon cross section $\sigma(\omega)$ by employing the optical theorem: $f''(\omega) = \omega\,\sigma(\omega)/(4\pi r_0 c)$. The total cross section is well approximated by the absorption cross section which can be measured by a transmission experiment. The real part $f'$ can be calculated from $f''$ by means of Kramers–Kronig dispersion relations. However, this method is suited for a relative comparison only, since it requires precise absorption data for all frequencies from zero to infinity [2]. If precise absolute values are needed, a direct measurement of $f'(\omega)$ is required. Direct measurements are based on x-ray interferometry [3], refraction through a prism [4,5], diffraction from perfect crystals and pendellösung fringes [6,7], determination of the angle of total reflection [2,8], and Fresnel bi–mirror interferometry [9]. Whereas most of these methods are based on splitting of either wave amplitudes or wave fronts the novel type of interferometer which is described here uses two spatially separated coherent x-ray emitters.

The basic idea of the interferometer will be explained by means of the schematic experimental setup shown in fig. 1. Relativistic electrons create two wave trains in source 1 and source 2, the relative distance $\Delta$ of which is given in leading order by $\Delta(\theta,d) = \frac{1}{2}(\gamma^{-2} + \theta^2)d$. Here $d$ is the distance of the sources, $\gamma$ the Lorentz factor of the electron, and $\theta$ the observation angle with respect to the electron beam direction. The monochromator serves as a Fourier analyser of the wave trains. The two resulting plane waves with a phase difference $\Phi = \frac{\omega}{v}\Delta(\theta,d)$, $v$ is the velocity of the electron, interfere in the detector, resulting in oscillations of the intensity $I(d)$, if the distance $d$ is varied. A sample foil placed between the two sources produces an additional phase shift and attenuation of wave 2. Consequently, both quantities, i.e. dispersion and absorption, can be extracted from the measured interference oscillations $I(d)$ with and without the foil between the sources. This holds independently of the nature of the emission process provided that the produced x-rays remain coherent.

# THE SOFT X-RAY INTERFEROMETER

For photon energies in the range of about 100 eV and 2 keV we use two identical undulators with period length $L_U=12$ mm, number of periods 10, undulator parameter $K=1.1$ and a variable line spacing grating spectrometer. The recorded intensity with a foil between the undulators is given by

$$I(d) = |A_1|^2 + |A_2|^2 e^{-2\frac{\omega}{c}\beta(\omega)t_0} + 2|A_1||A_2|e^{-\frac{\omega}{c}\beta(\omega)t_0} \times \cos\left\{\frac{\omega}{c}\left[\Delta(\theta,d) + \delta(\omega)t_0 + \frac{K^2}{4\gamma^2}L_U\right]\right\} \quad , \quad (1)$$

with $A_2$ being the amplitude of the upstream undulator, $A_1$ that of the downstream undulator and $t_0$ the thickness of the foil. All other quantities have been defined above.

The interferometer has been developed at the Mainz Microtron MAMI and its performance was demonstrated with measurements at the K-absorption edge of carbon at 284 eV, see Fig. 2. The visibility (coherence), defined by $C = (I_{max} - I_{min})/(I_{max} + I_{min})$ without sample foil, is close to its maximum value $C = 1$. No loss of coherence was observed over the scanning distance of 15 cm. Therefore, the optical constants $\delta$ and $\beta$ could be extracted by a fit with simple cosine functions. Details of this experiment can be found elsewhere [10].

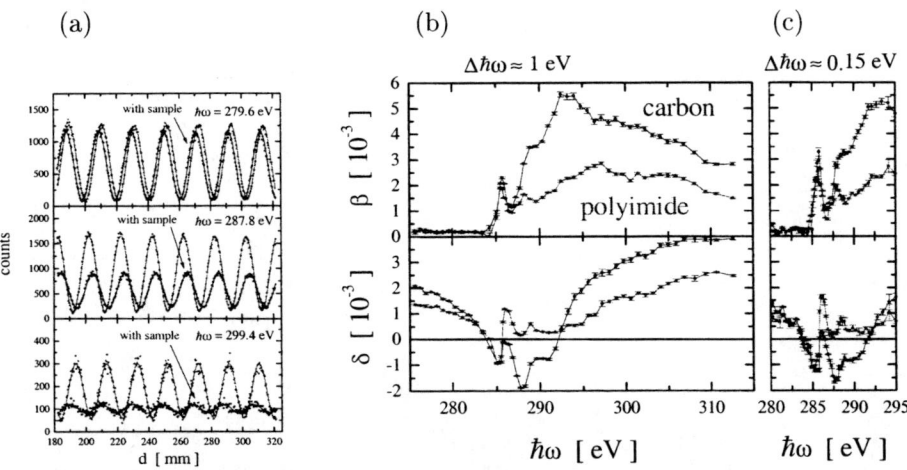

**FIGURE 2.** (a) Intensity oscillations with and without a carbon sample at three different energies. Note the change of sign of the phase shift. (b) Extracted optical constants of a carbon sample (65 $\mu$g/cm$^2$) and a polyimide sample (35 $\mu$g/cm$^2$). (c) Same as (b) with improved energy resolution of 0.15 eV.

# THE HARD X–RAY INTERFEROMETER

High photon energies can be produced at MAMI, if the 855 MeV electron beam traverses the interface between two media of different polarisabilities. The emitted transition radiation (TR) is sharply peaked into the forward direction with a characteristic opening angle of $2/\gamma$ and features broadband characteristics with a cut–off energy at about 40 keV.

The interferometer consists of two foils only. The sample foil serves simultaneously as the downstream emitter. The coherently added amplitudes of the four interfaces result in an intensity

$$I(\theta,d) = |A_1|^2 + |A_2|^2 e^{-\sigma_1} + 2|A_1||A_2|e^{-\sigma_1/2}$$
$$\times \cos\left\{\frac{\omega}{v}\Delta'(\theta,d) + \phi_1 + \arctan\left(\frac{e^{-\sigma_1/2}\sin(\phi_1)}{1 - e^{-\sigma_1/2}\cos(\phi_1)}\right)\right.$$
$$\left. - \arctan\left(\frac{e^{-\sigma_2/2}\sin(\phi_2)}{1 - e^{-\sigma_2/2}\cos(\phi_2)}\right)\right\} , \quad (2)$$

$$\text{with} \quad A_i = \frac{\sqrt{\alpha}}{\pi}\theta\left[\frac{1}{\gamma^{-2}+\theta^2+2\delta_i} - \frac{1}{\gamma^{-2}+\theta^2}\right]\left(1 + e^{-\sigma_i} - 2e^{-\sigma_i/2}\cos\phi_i\right) \quad (3)$$

and $\Delta'(\theta,d) = 1/2(\gamma^{-2}+\theta^2)(d-t_2)$, $\phi_i = \frac{\omega t_i}{2v}(\gamma^{-2}+\theta^2+2\delta_i)$, and $\sigma_i = 2\frac{\omega}{c}\beta_i t_i$. The quantities $A_i$, $t_i$, $\delta_i$ and $\beta_i$ are the TR amplitude [11], thickness, dispersion and absorption, respectively, of foils $i=1,2$. The parameters of the upstream foil $i=2$ have to be measured independently or can be taken out of the literature [12]. The dispersion $\delta_1$ of the downstream sample foil has to be determined assuming that the absorption $\beta_1$ and foil thickness $t_1$ are known from independent measurements.

The monochromator is a flat silicon single crystal in [111] orientation at a distance of 5.5 m from the foils. As detector we use a silicon $1 \times 3$ cm$^2$ pn–CCD with active thickness of 300 $\mu$m and a pixel size of $150 \times 150$ $\mu$m$^2$ [13]. The CCD is located at a distance of 5.5 m from the monochromator crystal.

Test measurements with a beryllium foil ($t_2$=10 $\mu$m) and a nickel sample ($t_1$=2 $\mu$m) have been performed at an energy of 10 keV, well above the K–absorption edge of nickel at 8340 eV. The foil distance was varied between 50 $\mu$m and 5 mm in 160 steps. Fig. 3 shows recorded interference patterns at three different distances

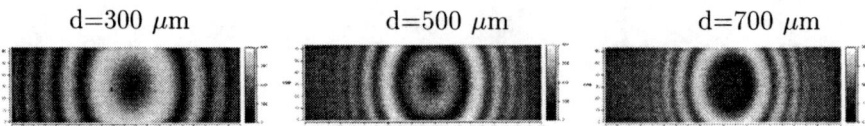

**FIGURE 3.** Interference pattern as measured with a $1 \times 3$ cm$^2$ CCD detector for a beryllium and a nickel foil with thicknesses of 10 $\mu$m and 2 $\mu$m, respectively. Distances between the foils are indicated. The photon energy amounts to 10 keV.

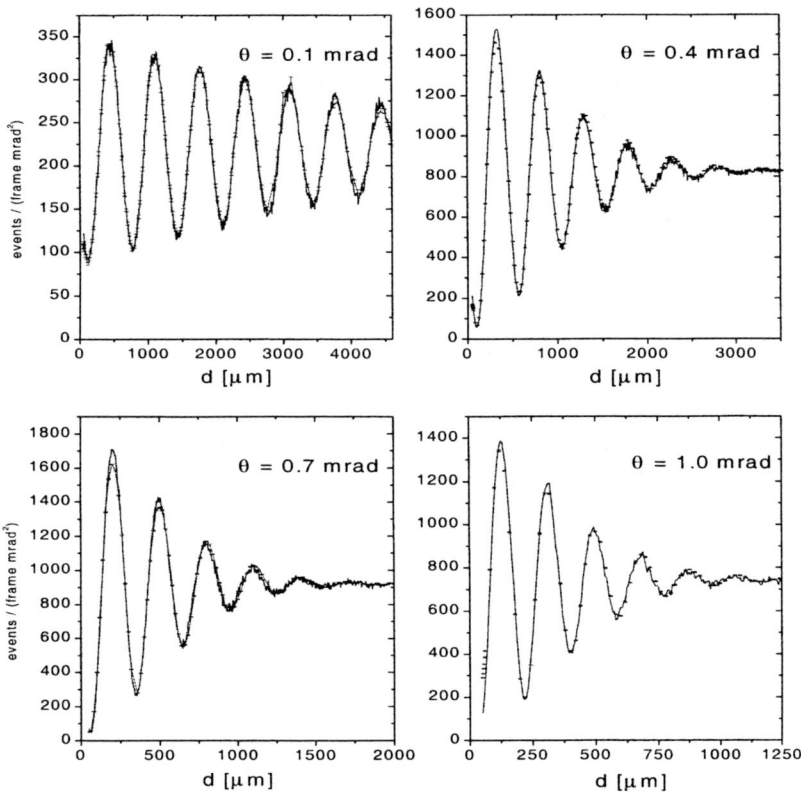

**FIGURE 4.** Measured intensity oscillations at fixed $\hbar\omega$=9975 eV and four different observation angles. Note the different scalings of the abscissa. The fits show excellent agreement with the measured data.

between the foils. Intensity oscillations as a function of the distance between the foils are plotted in fig. 4 for four different emission angles. The damping of the oscillations with increasing distance is mainly caused by small angle scattering of the electrons in the foils and the electron beam emittance.

Direct extraction of the dispersion $\delta_1$ of the downstream nickel foil from such oscillation pattern is hampered by the fact that a goniometric stage allows a measurement of the distance $d$ between the foils only on a relative scale, i.e. with an unknown offset $d_0$. However, this offset and the dispersion $\delta_1$ can be determined, in principle, from the angular oscillation pattern as shown in fig. 3, provided the parameters $t_2$, $\beta_2$, $\delta_2$ of the beryllium foil and $t_1$, $\beta_1$ of the nickel foil are known. This can be concluded from eqn.(2). The offset $d_0$ causes phase changes proportional to $\theta^2 d_0$ and is, therefore, very sensitive to variation of the observation angle $\theta$, while $\delta_1$ is not; but this analysis has not yet been performed. Instead, we took

$\delta_1$ from literature [12] and determined the offset $d_0$ by a best fit to the oscillation spectrum at $\theta=1.0$ mrad, see fig. 4. The fit is a Monte Carlo simulation based on eqn.(2) taking into account the electron beam emittance and small angle scattering of the electrons in the foil materials. For the other three spectra shown in fig. 4 this offset parameter $d_0$ was kept constant and the dispersion $\delta_1$ was set free in the fit. The best fit values of $\delta_1$ agree within a few percent with the literature value for all observation angles. This result demonstrates that the dispersion $\delta_1$ of the nickel foil can be determined with high precision also at the K-absorption edge, at least relative to the reference value at 10 keV.

## ACKNOWLEDGEMENTS

This work has been supported by the Deutsche Forschungsgemeinschaft DFG, the Bundesministerium für Bildung, Wissenschaft, Forschung und Technologie under contract 06 MZ 863 I/TP: 2, and the Ministerium für Wissenschaft und Forschung des Landes Nordrhein-Westfalen, NRW-Verbund Mikrosensorik.

## REFERENCES

1. *Resonant Anomalous X-ray Scattering*, eds. G. Materlik, C.J. Sparks, and K. Fischer (North Holland, Amsterdam, London, New York, Tokyo 1994).
2. B. Lengeler, in [1], p. 35.
3. U. Bonse and M. Hart, Appl. Phys. Lett. **7**, 238 (1965).
4. W.K. Warburton and K.F. Ludwig, Phys. Rev. **B 33**, 8424 (1986).
5. M. Deutsch and M. Hart, Phys. Rev. **B 30**, 643 (1984).
6. A. Freund, in *Anomalous Scattering*, eds. R. Ramaseshan, S.C. Abrahams, Munksgaard Copenhagen p. 69 (1975).
7. N. Kato and S. Tanemura, Phys. Rev. Lett. **19**, 22 (1967).
8. R.L. Blake *et al.*, in [1], p. 79 (1994).
9. F. Polack *et al.*, Rev. Sci. Instrum. **66** (2), 2180 (1995).
10. S. Dambach *et al.*, Phys. Rev. Lett. **80**, 5473 (1998).
11. M.L. Cherry *et al.*, Phys. Rev. **D 10**, 3594 (1974).
12. B.L. Henke, E.M. Gullikson and J.C. Davis, Atomic Data and Nuclear Data Tables **54**, 181 (1993).
13. H. Soltau *et al.*, Nucl. Instr. Meth. **A 377**, 340 (1996).

# X. CONFERENCE SUMMARY

# Closing Remarks for X99

### Richard D. Deslattes

*National Institute of Standards and Technology, Gaithersburg, MD 20899 USA*

**Abstract.** The following brief essay summarizes some of my remarks at the closing session of X-99. The three sections below include a well-deserved appreciation for the effective work of the organizers and the conference staff. A second somewhat longer section indicates a few of the themes that in this author's view epitomized the conference content. Finally there is a modest speculation on themes suggested for consideration beyond X-2002.

## INTRODUCTION

The closing remarks on which this short contribution is based began with an expression of heartfelt thanks to the organizers of X-99 and to the Conference staff. All arrangements were exceptionally well prepared so as to provide the environment needed for a productive and enjoyable X-99.

Section 2 contains a few of my overall observations and lists certain themes that seemed to me to reflect outstanding new developments in both the theoretical and experimental aspects of x-ray and inner-shell physics. In Section 3, I have very briefly summarized and updated the more detailed history of the "X-" conferences that was originally presented at X-90 (R. D. Deslattes in "X-Ray and Inner-Shell Processes", T. A. Carlson, M. O. Krause, S. T. Manson, Ed., AIP Conference Proceedings 215, 1990, 3-12). Finally, in Section 4, I offer some reflections suggested by my attendance at the XVIII General Congress of the International Union of Crystallography held in Glasgow, Scotland, earlier in the month.

## IMPRESSIONS OF THE CONFERENCE AND SOME OF ITS THEMES

Before coming to specific themes, I would like to indicate a few more general impressions. One thing that emerged from a preview of the Abstracts kindly furnished by the organizers, and was amply confirmed at the conference itself, was an impression of widespread diversity. Among the contributions in these proceedings are a modest number having only single authors, while most have multiple authorship, and a few have author complements that verge on the familiar situation of particle physics. The x-rays themselves were produced in a diverse array of sources ranging from Coolidge tubes (happily mentioned in the first of the two Compton lectures) and EBITs, to third-generation synchrotron radiation facilities, plasmas, particle accelerators, ion sources and more.

At its outset, the conference was graced by two Argonne-sponsored Compton Lectures focussing on the area of magnetic x-ray scattering and dichroism. The first of these (Blume)[1] considered the long and distinguished career of A. H. Compton and his many roles in the formation of the US science enterprise as we see it today. The second lecture (Cooper) emphasized Compton's very fruitful period in Rutherford's laboratory in Cambridge. Both lectures were of exceptional scope and offered excellent perspectives of the highly active areas in which x-rays probe the magnetic structure of materials.

## A Sampling of Some of the New Developments

Considering Physics as a spectator sport, one of its real pleasures for me is to watch the evolution of a new phenomenology. In the most characteristic of these passages, what begins as unusual, and even exotic, is gradually seen to be ubiquitous, and then, in the best of these stories, the exotic (step-) child becomes a useful tool for the investigation of other problems. To the store of such tales I think we should now add the story of hollow atoms, such strange oddities only a few years ago, which have now been convincingly elucidated and already applied to new problems, particularly those associated with surface interactions (Briand).

Another example of the evolution from initial observation to effective understanding and utilization is to be seen in the case of resonant Raman scattering. Here we have not only new and more complete experimental studies but also numerically detailed theoretical efforts (Shirley) and an overall perspective that includes the vexing problem of line narrowing within a larger comprehensive framework (Carra).

The notion of approaching "complete" experiments through recoil momentum spectroscopy and rather general coincidence methods has also matured as is seen in several contributions derived from the COLTRIMS approach (Cocke). Enhanced completeness was also approached by means of spectroscopy of oriented molecules (Becker).

While the practical realization of x-ray microscopy with sub-atomic scale resolution is yet to be realized, the results of highly developed fluorescence holography appear to point to one of the ways in which this goal may be approached (Materlik).

Current efforts to realize the exceptional potential of cryogenic micro-calorimetry for high resolution energy-dispersive spectroscopy have reached resolutions near 5 eV, and impressively high data rates with somewhat degraded resolution (Galeazzi).

---

[1] Names appearing in parentheses refer to contributions by the indicated authors found elsewhere in these proceedings.

The conference also was the occasion for the introduction of an entirely new type of x-ray interferometry based on coherent generation of x-ray pulses by either microundulators or foils producing transition radiation (Kettig). Its initial application to optical constants (particularly the index of refraction decrement) was demonstrated over a range from 0.25 keV to 28 keV.

## New Teaching and New Learning

Although each of the recent "X-" conferences have contained reports of theoretical progress and improved understanding, X-99 was particularly notable in this regard. One of the presentations examined the asymptotic behavior of photo-ionization in the high- and low-energy regimes (Pratt). The main messages are that correlation, although emphasized in the low-energy range, has effects that persist to very high energies. At the same time, non-dipole effects that are most prominent at high energy, nevertheless persist in the low-energy region.

Another theoretical overview concentrating on the case of He and He-like systems brought together in a unified way the hierarchical representations of relativistic, nuclear and QED effects on the one hand, and progressively higher-order correlation effects on the other (Drake). Not only was this viewpoint of exceptional pedagogical value, but it also showed that calculations have reached a level of completeness and accuracy that allows using the experimental data to extract a value for the fine structure constant with useful accuracy and for the determination of nuclear radii.

Although not primarily theoretical, the current status of experimental studies of the photo-ionization of highly charged ions is one of notable theoretical interest (Bizau). Similarly, the approach to strong-field limits offered by the use of virtual photon fields in collisions of fast, highly charged ions continues to develop (Dorn).

## Geometrical and Electronic Structure

In this sub-section, I note a certain duality between geometrical and electronic structure that is perhaps less emphasized in X-99 than it might be in the future. Naturally, the main body of work on spectroscopy focuses on the electronic structure of the system under study while scattering (diffraction, etc.) methods tend to focus on the geometrical structure of the target system. These are, of course, overly simplistic statements, but they do reflect the way in which most work in spectroscopy and diffraction proceeds at the present time.

On the other hand, we are well aware that certain spectroscopic methods do yield information on structural properties, albeit primarily local geometry. The main case in point is EXAFS, a spectroscopic method that gives a fairly good picture of the distribution of electron scattering power in the neighborhood of an atom undergoing photo-ionization. Over the past twenty years this has become a widely used approach that is by now so effectively routine that it was not emphasized at this conference. On the other hand, a closely related technique, namely DAFS (Diffractive Anomalous Fine Structure), allows separation of the local (EXAFS) environment in the neighborhood of two (or more) instances of the same element occupying non-equivalent sites within the same unit cell (Felici).

The mirror images of these processes, namely the application of structure-dependent scattering to modify spectra, has not received as much attention. Two obvious examples are the use of x-ray standing waves to enhance the visibility of non-dipole effects and the use of such standing waves to enhance "atomic" resonances. While the persistence of atomic effects in condensed matter spectra was specifically addressed during the conference (Sonntag), the potential modification of atomic resonances by simultaneously tuning the crystalline environment to the Bragg-Laue condition was not.

## SOME HISTORICAL REMARKS CONCERNING THE "X-" CONFERENCES

Particularly when considering plans for the future of this conference series, it seems important to have in mind how they have developed up to X-99 and its already planned successor, X-2002, scheduled to be held in Rome, Italy. The general history of joint X-ray and Inner-Shell Physics began with the merger in 1978 of two separate predecessor series. The first of these, focusing on X-ray spectroscopy, began in 1965 with a meeting at Cornell University in Ithaca, NY, USA chaired by Lymann Parratt, Leonard Jossem, and Joyce A. Bearden. The Inner Shell Physics conferences began in 1972 in Atlanta, GA, USA in 1972 and had a pre-merger successor in Freiburg, Germany in 1976. This early history is indicated in Fig. 1.

Our present conference series, beginning with the first merged meeting in Sendai, Japan in 1978, is indicated in Fig. 2. My sense is that the quality of the conference series has steadily improved, with each successive host organization giving its task the degree of attention and support needed to secure an effective outcome and provide a very positive experience for the participants.

The future of this or any other conference series depends on the extent to which it serves its constituent community, as well as the health and vitality of the community itself. Considering the rise of alternative forums, particularly driven by new sources and application areas, the participation levels indicated by the numbers shown in Fig. 3 may be considered as encouraging its continuation. On the other hand, continuation should not preclude thoughtfully considered change.

### X-ray Spectroscopy Conferences
Ithaca, NY, USA, 1965, Leipzig, DDR, 1965

Kiev, USSR, 1968

Paris, FR, 1970

Munich, FRG, 1972

Otaniemi, Finland, 1974

Gaithersburg, MD, USA, 1976

### Inner Shell Physics Conferences
Atlanta, GA, USA, 1972

Freiburg, FRG, 1976

**Figure 1. Early X-ray and Inner-shell conferences**

### Merged conferences
Sendai, Japan, 1978

Stirling, Scotland, UK, 1980

Eugene, OR, USA, 1982

Leipzig, DDR, 1984

Paris, FR, 1987

Knoxville, TN, USA, 1990

Debrecen, Hungary, 1993

Hamburg, Germany, 1996

Chicago, IL, USA, 1999

Rome, Italy, 2002

**Figure 2. List of the combined conferences in the X-series.**

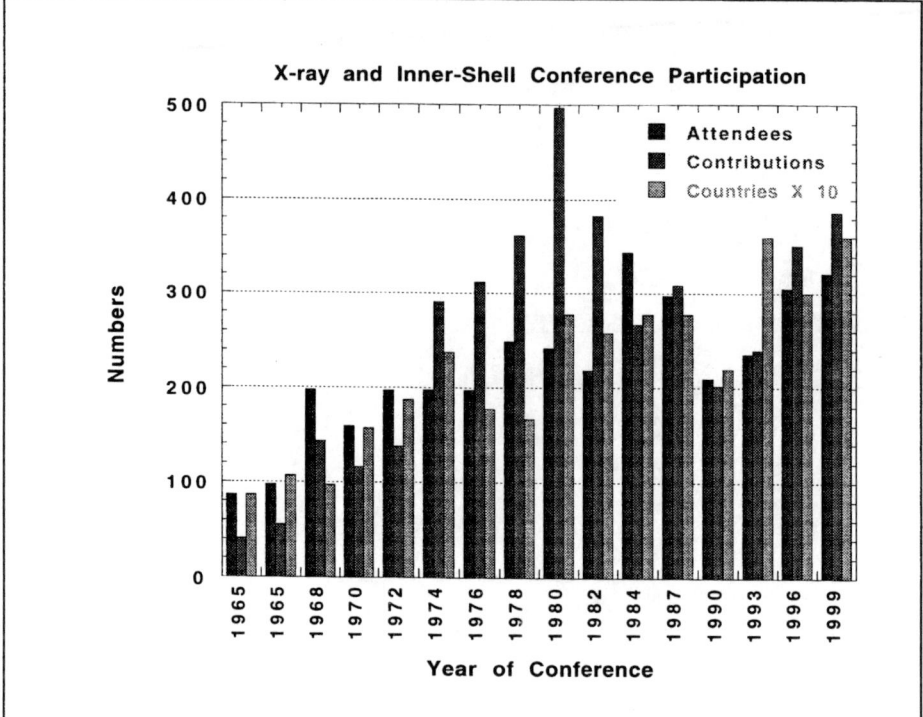

Figure 3. Participation in the "X-" versus year.

(Early Inner-Shell meetings are not shown.)

## A FEW THOUGHTS ABOUT THE FUTURE

X-2002 is scheduled to be held in Rome with its organization in very capable hands. These few thoughts are less concerned with specifics than with some very general thoughts. Some of these arise from my attendance earlier in August 1999 at the XVIII International Union of Crystallography Congress and General Assembly held in Glasgow, Scotland. There are evident differences in scale and scope between IUCr XVIII and X-99. The IUCr involved more than 2500 participants, lasted nearly two weeks, and had a large commercial exhibition. The IUCr publishes several journals as well as the International Tables for Crystallography and a series of authoritative monographs. Given the growing activity in biomolecular structures, it is not surprising that there was a large representation of such work at this meeting. What I did find very interesting was the scope, variety and sophistication of work in the inorganic domain, including a particular emphasis on self-organization and polymorphism.

As is well known, after a brief period in which they were unified, spectroscopic and structural research took rather independent paths starting as early as 1915. For the early part of this story, particularly through the development of quantum mechanics, spectroscopy was at the forefront. Today, I think the leadership role has passed to structural research both in terms of intellectual ferment and practical impact. On the other hand, in looking over the entire program of the IUCr meeting, I can find almost no instances in which the spectroscopic aspect of structural study has achieved real prominence. In looking at X-99, I see only limited coupling to structural research and concepts.

My very simple, possibly simplistic, thought is that it may be timely to re-examine this functional separation. This is not to suggest a merger of meetings, since our small community and modest numbers are very attractive aspects of the X- conferences. Perhaps what I see rather dimly is that there are problems in which spectra and structure are conjoined and suggest that it may be fruitful to look at these with a conscious effort at establishing productive linkages with the larger structural community. A concrete example of such a structure-spectra conjunction, namely Multi-Atom Resonant Photoemission (MARPE), is described in these proceedings (Fadley). The earliest stages of such an exploration may require no more than inviting some of the more physically oriented structural workers to consider contributing to a joint effort to define a conference session or "mini-symposium" at X-2002 that might begin to focus on such linkages explicitly.

# Scientific Program Monday, August 23

## Opening Session
Grand Ballroom
Chairman: **B. Crasemann** (University of Oregon, USA)

09:00 Welcome to the X-99 Conference

09:30 **J. P. Briand** (Université Pierre et Marie Curie, Paris, France)
*"X-Ray and Inner-Shell Processes: Their Impact on our Understanding of Atomic Physics and of Atoms Interacting with Solids"*

10:30 *Coffee Break*

## Parallel Sessions

Grand Ballroom
Chairman: **Y. Awaya** (Musashino Art University, Japan)

Walton Room
Chairman: **P. M. Platzman** (Lucent Technologies, USA)

11:00 **C. L. Cocke** (Kansas State University, USA)
*"Dynamics of Atomic and Molecular Collisions with Ions and Photons"*

**E. Shirley** (National Institute of Standards and Technology, USA)
*"X-Ray Absorption and Electronic Raman Scattering for Simple Solids: Modeling"*

11:30 **D. Dauvergne** (Institut de Physique Nucléaire de Lyon, France)
*"X-Ray Emission from Fast Channeled Heavy Ions"*

**J. P. Hill** (Brookhaven National Laboratory, USA)
*"Resonant Inelastic X-Ray Scattering Studies of Strongly Correlated Electron Systems"*

12:00 **M. Galeazzi** (University of Wisconsin, USA)
*"Thermal Calorimeters for High Resolution X-Ray Spectroscopy"*

**A. Kaprolat** (European Synchrotron Radiation Facility, France)
*"Electronic States of Metals and Alloys Investigated by High-Resolution Bloch k-Selective X-Ray Raman Scattering"*

12:30 *Lunch Break*

## Parallel Sessions

Grand Ballroom
Chairman: **A. Warczak** (Jagiellonian University, Poland)

Walton Room
Chairman: **S. Aksela** (University of Oulu, Finland)

14:00 **A. R. Dorn** (Universität Freiburg, Germany)
*"Atoms in Extreme Virtual Photon Fields of Fast Highly Charged Ions"*

**M. Murnane** (University of Michigan, USA)
*"Extreme Non-Linear Optics: Phase Matching in the UV and X-Ray Regions"*

14:30 **B. Sulik** (Hungarian Academy of Sciences, Debrecen, Hungary)
*"Two- and Three-Body Effects in FastIon-Atom Collisions: Analogies Between the Impact of Photons and Fast Charged Particles"*

**J. A. Golovchenko** (Harvard University, USA)
*"X-rays in Curved Spaces"*

15:00 Poster Session A

# Scientific Program Tuesday, August 24

**Plenary Session**
Grand Ballroom
Chairman: **F. P. Larkins** (University of Melbourne, Australia)

09:00 **R. H. Pratt** (University of Pittsburgh, USA) *"Frontiers of X-Ray-Atom Interactions"*

09:45 **W. Mehlhorn** (Universität Freiburg, Germany) *"Auger Spectroscopy: Highlights and Historical Perspective"*

10:30 *Coffee Break*

## Parallel Sessions

| Grand Ballroom<br>Chairman: **N. Berrah** (Western Michigan University, USA) | Walton Room<br>Chairman: **A. Marcelli** (INFN, Frascati, Italy) |
|---|---|
| 11:00 **V. Schmidt** (Universität Freiburg, Germany) *"Unique Features of Photo Electron/Auger-Electron Coincidence Experiments"* | **P. Carra** (European Synchrotron Radiation Facility, France) *"Theory of X-Ray Resonant Raman Scattering"* |
| 11:30 **Y. Azuma** (Photon Factory, KEK, Japan) *"Three-Electron Photo-Processes in Lithium"* | **A. Rogalev** (European Synchrotron Radiation Facility, France) *"Hard X-Ray Circular Dichroism"* |
| 12:00 **U. Becker** (Fritz-Haber Institute, Berlin, Germany) *"Photoelectron Emission from Oriented Molecules"* | **A. P. Moewes** (Center for Advanced Microstructures and Devices, USA) *"Resonant Inelastic Scattering at the 3d and 4d Thresholds of Lanthanides"* |

12:30 *Lunch Break*

## Hot Topics Sessions

| Grand Ballroom<br>Chairman: **N. M. Kabachnik** (Moscow State University, Russia) | Walton Room<br>Chairman: **R. Szargan** (University of Leipzig, Germany) |
|---|---|
| 14:00 **O. A. Hemmers** (University of Nevada, Las Vegas, USA) *"Beyond the Dipole Approximation: Angular-Distribution Effects in the 1s Photoemission from Small Molecules"* | **R. Felici** (European Synchrotron Radiation Facility, France) *"Diffracted Anomalous Fine Structure (DAFS) to Study Surfaces and Interfaces"* |
| 14:20 **K. Ueda** (Tohoku University, Japan) *"Angular Distributions and Correlations in Auger Cascades of Atomic Argon Following 2p to 4s Excitation"* | **D. Menzel** (Technische Universität München, Germany) *"Charge Transfer at Surfaces on Femtosecond Timescales: New Information from Electron Spectroscopies"* |

14:40 **J. -M. Bizau** (Université Paris-Sud, France)
*"First Experimental Results and New Theoretical Calculations on Photoionization Processes in Multiply-Charged $Xe^{4+}$ to $Xe^{7+}$ Ions"*

**A. Bianconi** (Universita di Roma La Sapienza, Rome, Italy)
*"Charge Ordering in Doped Perovskites by X-Ray Scattering and Spectroscopy: New Physics at Third Generation Synchrotron Sources"*

15:00 **O. Kettig** (University of Mainz, Germany)
*"A Novel Type of X-Ray Interferometer"*

**P. M. Platzman** (Lucent Technologies, USA)
*"Probing the Nature of Hydrogen Bonds with X-Rays"*

15:20 Poster Session B
French Room

International Scientific Committee Meeting
Venetian Room

18:00 *ISC Dinner*

19:30 *ISC Business Meeting*

## Scientific Program Wednesday, August 25

### A. H. Compton Plenary Session
Grand Ballroom
Chairman: **S. T. Manson** (Georgia State University, USA)

09:00 **M. Blume** (Brookhaven National Laboratory, USA)
*"X-Rays and A. H. Compton: From Fundamental Physics to Experimental Techniques"*

09:30 **M. J. Cooper** (University of Warwick, United Kingdom)
*"Exploiting Compton Scattering: Studies of Spin Density Distributions"*

10:30 *Coffee Break*

### Parallel Sessions

Grand Ballroom
Chairman: **T. Mukoyama** (Kyoto University, Japan)

Walton Room
Chairman: **P. Beiersdorfer** (Lawrence Livermore National Laboratory, USA)

11:00 **E. E. Alp** (Argonne National Laboratory, USA)
*"Nuclear Resonant Inelastic Scattering"*

**G. Drake** (University of Windsor, Canada)
*"QED Effects in Few-Electron Atoms and Ions"*

11:30 **E. V. Tkalya** (Moscow State University, Russia)
*"Nuclear Excitation by Electronic Transition Between Atomic Shells"*

**K. Widmann** (Lawrence Livermore National Laboratory, USA)
*"High-Resolution Measurements of the K-Shell Spectral Lines of Hydrogen-Like and Helium-Like Xe"*

12:00 **Y. Yamazaki** (Tokyo University, Japan)
*"Atomic Spectroscopy and Collisions Using Slow Antiprotons"*

**T. Stöhlker** (GSI, Darmstadt, Germany)
*"Inverse Photoionization Studied via Radiative Electron Capture into Heavy Highly Charged Ions"*

| | | |
|---|---|---|
| 12:30 | *Lunch Break* | |
| 14:00 | Poster Session C<br>French Room | |
| | Architecture River Cruise | |
| 16:45 | Board Boats at Michigan Avenue Bridge<br>at the Chicago River | |
| 17:00 | Boats Depart | |
| 19:00 | Boats Return | |

## Scientific Program Thursday, August 26

**Plenary Session**
Grand Ballroom
Chairman: **T. Suric** (Ruder Boskovic Institute, Zagreb, Croatia)

08:30  **D. A. Varshalovich** (A. F. Ioffe Physical-Technical Institute, St. Petersburg, Russia)
*"Testing the Cosmological Variability of Fundamendal Constants"*

09:15  **G. Materlik** (HASYLAB, Hamburg, Germany) *"Atomic Resolution X-Ray Fluorescence Holography"*

10:00  Poster Session D
French Room

12:30  *Lunch Break*

### Parallel Sessions

| | Grand Ballroom<br>Chairman: **P. A. Montano**<br>(Argonne National Laboratory, USA) | Walton Room<br>Chairman: **M. N. Piancastelli** (Uppsala University, Sweden) |
|---|---|---|
| 14:00 | **C. S. Fadley** (UC Davis and Lawrence Berkeley National Laboratory, USA)<br>*"Multi-Atom Resonant Photo-Emission (MARPE)"* | **O. Björneholm** (Uppsala University, Sweden)<br>*"Dynamics of Competing Ultra-Fast Fragmentation and Resonant Auger Processes"* |
| 14:30 | **J. Matsui** (Himeji Institute of Technology, Japan)<br>*"Live X-Ray Refraction Imaging Using Vertically and Horizontally Wide X-Rays"* | **M. Simon** (LURE, Orsay, France)<br>*"Fragmentation Processes Following Core Ionization of Free Molecules"* |
| 15:00 | **G. De Stasio** (Istituto di Struttura della Materia, CNR, Rome, Italy)<br>*"Frontiers of X-Ray Spectromicroscopy in Biology and Medicine"* | **J. D. Bozek** (Lawrence Berkeley National Laboratory, USA)<br>*"High-Resolution Inner-Shell Electron Spectroscopies of Simple Molecules"* |
| 15:30 | *Coffee Break* | |

# Scientific Program Thursday, August 26

## Hot Topics Sessions

Grand Ballroom
Chairman: **D. L. Ederer** (Tulane University, USA)

Walton Room
Chairman: **D. W. Lindle** (University of Nevada, Las Vegas, USA)

16:00 **F. B. Rosmej** (Technische Universität Darmstadt, Germany)
"Charge Exchange Induced Formation of Hollow Atoms in High Intensity Laser Produced Plasmas"

**D. Cookson** (Australian Nuclear Science and Technology Organization, Australia)
"The Possibility of Using X-Ray Diffraction With Hair to Screen for Pathologic Conditions Such as Breast Cancer"

16:20 **Cs. Toth** (University of California, San Diego, USA)
"Coherent X-Ray Generation Via Ultrafast Coster-Kronig Decay in Solid Targets Excited by Table-Top Lasers"

**E. G. Drukarev** (St. Petersburg Nuclear Physics Institute, Russia)
"Non-Resonant Excitation of Nuclear Levels by Photons"

16:40 **P. A. Heimann** (Lawrence Berkeley National Laboratory, USA)
"Time-Resolved X-Ray Photoabsorption and Diffraction on Timescales from ns to fs"

**B. Lai** (Argonne National Laboratory, USA)
"High-Resolution X-Ray Imaging for Microbiology and Microelectronic Devices at the Advanced Photon Source"

17:00 **A. Belkacem** (Lawrence Berkeley National Laboratory, USA)
"K-shell Ionization and Double-ionization of Au Atoms with 1.33 MeV Photons"

**A. Rüdel** (Fritz-Haber Institute, Berlin, Germany)
"Oscillating Partial Cross Sections in $C_{60}$: Evidence for a Beating Frequency"

18:00 *Cocktail Reception*
French Room

19:30 *Banquet*
Gold Coast Room

# Scientific Program Friday, August 27

**Plenary Session**
Grand Ballroom
Chairman: **M. Ya. Amusia** (Hebrew University, Israel)

08:30　**B. F. Sonntag** (Universität Hamburg, Germany)
　　　　*"Atomic Effects Seen in Solid Phases"*

09:15　**E. Weigold** (Australian National University, Australia)
　　　　*"Current Status of (e,2e) Measurements of Energy and Momentum Densities"*

10:00　*Coffee Break*

**Plenary Session**
Grand Ballroom
Chairman: **F. J. Wuilleumier** (Université Paris-Sud, France)

10:30　**Y. Kato** (JAERI Kansai Research Establishment, Japan)
　　　　*"X-Ray Lasers Driven by Optical Lasers"*

11:15　**J. R. Arthur** (Stanford University, USA)
　　　　*"Prospects for an X-Ray FEL Light Source and Some Possible Scientific Applications"*

12:00　**R. D. Deslattes** (National Institute of Standards and Technology, USA)
　　　　Conference Summary

12:25　Remarks by Chairman of Next Conference

12:30　*Lunch Break*

Advanced Photon Source Tour

13:30　Buses leave Drake Hotel for Argonne

18:00　Buses from Argonne arrive at Drake Hotel

# Author Index

## A

Abdallah, M. A., 101
Achler, M., 101
Adams, B., 549
Aksela, H., 148
Alatas, A., 479
Alp, E. E., 479
Arenholz, E., 251
Arthur, J., 597
Auguste, T., 472
Azuma, Y., 116

## B

Backe, H., 669
Bapat, B., 403
Barty, C. P. J., 651
Becker, U., 205, 217
Beiersdorfer, P., 444
Belkacem, A., 153
Benfatto, M., 351
Berrah, N., 188
Bianconi, A., 358
Bianconi, G., 358
Bizau, J.-M., 467
Björneholm, O., 161
Blancard, C., 467
Blankenship, S. R., 283
Bozek, J. D., 188
Braeuning, H., 101
Braeuning-Deminian, A., 101
Briand, J.-P., 5
Brinzanescu, O., 389
Brown, G. V., 444
Bruneau, J., 467

## C

Cai, Z., 585
Callcott, T. A., 283
Carlisle, J. A., 283
Carra, P., 273
Carroll, T. X., 188
Casalbore, P., 577

Ceolin, D., 177
Chang, Z., 664
Chesnel, J.-Y., 427
Chevallier, M., 418
Chiba, H., 148
Ciotti, M. T., 577
Clawiter, N., 669
Cocke, C. L., 101
Cohen, C., 418
Colapietro, M., 358
Comin, F., 351
Compant La Fontaine, A., 467
Cookson, D., 590
Cooper, M. J., 18
Couillaud, C., 467
Crasemann, B., 3
Crespo López-Urrutia, J. R., 444
Cubaynes, D., 467
Cue, N., 418
Czasch, A., 101

## D

Dambach, S., 669
Dauvergne, D., 153, 418
Delaunay, J., 467
Deslattes, R. D., 677
De Stasio, G., 577
Di Castro, D., 358
Doerk, T., 669
Doerner, R., 101
D'Oliveira, P., 472
Dorn, A., 403
Drake, G. W. F., 512
Drukarev, E. G., 496
Dural, J., 418

## E

Ederer, D. L., 283
Elbai, N., 669
Enkisch, H., 327
Esteva, J.-M., 467
Euteneuer, H., 669

## F

Fadley, C. S., 251
Faenov, A. Y., 472
Falcone, R. W., 664
Frémont, F., 427
Feinberg, B., 153
Felici, R., 351
Focke, P., 222

## G

Galeazzi, M., 638
Garcia de Abajo, J., 251
Gilbert, B., 577
Glover, T. E., 664
Golovchenko, J. A., 621
Goulon, J., 336
Grandin, J. P., 427
Grether, M., 427
Guillemin, R., 177
Gulyás, L., 427

## H

Hagenbuck, F., 669
Hasegawa, N., 613
Heimann, P. A., 664
Hemmers, O. A., 222
Hennecart, D., 427
Hentges, R., 217
Hill, J. P., 312
Hiort, T., 549
Hitz, D., 467
Hoffmann, D. H. H., 472
Holl, P., 669
Hu, M., 479
Hulin, S., 472
Hussain, Z., 251
Husson, X., 427
Huttula, S.-M., 148

## I

Ilinski, P. P., 585
Ionescu, D., 153
Ivanchik, A. V., 503

## J

Jacobs, H., 669
James, V., 590
Jia, J. J., 283
Johnson, S., 664

## K

Kabachnik, N. M., 148
Kado, M., 613
Kagoshima, Y., 565
Kaiser, K.-H., 669
Kang, I., 664
Kaprolat, A., 327
Kapteyn, H. C., 664
Kato, Y., 613
Kawachi, T., 613
Kay, A. W., 251
Keller, S., 403
Kemmer, J., 669
Kemner, K. M., 585
Kerschner, T., 669
Kettig, O., 669
Kim, D., 651
Kirsch, R., 418
Kitajima, M., 148
Koch, H., 669
Kollmus, H., 403
Koncz, C., 427
Krämer, A., 389
Krisch, M. H., 327
Kube, G., 669
Kukk, E., 188
Kulpa, C. F., 585

## L

Lachkar, J., 467
Lai, B., 585
Landers, A., 101
Langhoff, P. W., 188, 222
Larocca, L. M., 577
Lauth, W., 669
Lecler, D., 427
Leclercq, N., 177
Lee, R. W., 664
Legnini, D. G., 585

Le Guen, K., 177
Lelièvre, D., 418
L'Hoir, A., 418
Lindenberg, A., 664
Lindle, D. W., 222
Liu, C., 621
Ludwig, P., 467
Ludziejewski, T., 389

## M

Ma, X., 389
Maddi, J., 153
Magunov, A. I., 472
Mann, R., 403
Mannweiler, H., 669
Margaritondo, G., 577
Marmoret, R., 467
Maser, J., 585
Materlik, G., 549
Matsui, J., 565
Matthäy, H., 669
McCammon, D., 638
Mehlhorn, W., 33
Menzel, D., 372
Mercanti, D., 577
Mergel, V., 101
Mills, J. D., 222
Miron, C., 177
Missalla, T., 664
Moewes, A., 304
Mokler, P. H., 418
Monot, P., 472
Moribayashi, K., 613
Morin, P., 177
Moshammer, R., 403
Mun, B. S., 251

## N

Nagashima, A., 613
Nagashima, K., 613
Nealson, K. H., 585
Nishino, Y., 549
Novikov, D. V., 549

## O

Olson, R. E., 403
Osipov, T., 101
Osterheld, A. L., 444

## P

Padmore, H. A., 664
Pallini, R., 577
Perfetti, P., 577
Pifferi, A., 358
Pikuz, T. A., 472
Platzman, P. M., 385
Poizat, J.-C., 418
Potekhin, A. Y., 503
Pratt, R. H., 59
Pratt, S. T., 585
Prinz, H.-T., 418
Prior, M., 101

## R

Ramillon, J.-M., 418
Reed, K. J., 444
Remillieux, J., 418
Rémond, C., 467
Rinelli, A., 577
Rodrigues, W., 585
Rogalev, A., 336
Rosmej, F. B., 472
Roussel-Chomaz, P., 418
Rozet, J.-P., 418
Rüdel, A., 217

## S

Saethre, L. J., 188
Saini, N. L., 358
Sanders, W. T., 638
Sanuy, F., 418
Sasaki, A., 613
Sato, Y., 148
Sazhina, I. P., 148
Schmaus, D., 418
Schmidt, V., 132
Schmidt-Boecking, H., 101

Schmitt, W., 403
Schneegurt, M. A., 585
Schoenlein, R. W., 664
Schöpe, H., 669
Schroff, D., 669
Schülke, W., 327
Schulz, M., 403
Schüttrumpf, M., 669
Scofield, J. H., 444
Sellin, I. A., 222
Sheehy, J. A., 188, 222
Shigemasa, E., 177
Shimizu, Y., 148
Shirley, E. L., 283
Simon, M., 177
Singh, M., 101
Sinn, H., 479
Skobelev, I. Y., 472
Skogvall, B., 427
Smith, R. N., 283
Son, S. H., 651
Sonntag, B., 231
Sorensen, A. H., 153
Stephan, C., 418
Stöhlker, T., 389
Stolterfoht, N., 427
Stötter, R., 669
Strüder, L., 669
Sturhahn, W., 479
Sulik, B., 427
Sutter, J., 479
Swiat, P., 389

## T

Takai, K., 565
Takeda, S., 565
Tanaka, H., 148
Tanaka, M., 613
Tanis, J. A., 427
Terminello, L. J., 283
Thomas, T. D., 188
Tischler, M. L., 585
Tkalya, E. V., 486
Toellner, T., 479
Tóth, C., 651
Toulemonde, M., 418
Tsusaka, Y., 565

## U

Ueda, K., 148
Ullrich, J., 403
Utter, S. B., 444

## V

Van Hove, M., 251
Varshalovich, D. A., 503
Vernhet, D., 418

## W

Walcher, T., 669
Wang, H., 222
Warczak, A., 389, 418
Weber, T., 101
Weigold, E., 81
Widmann, K., 444
Wilms, A., 669
Wolf, H. E., 101
Wolff, W., 101
Wuilleumier, F. J., 467
Wurth, W., 372

## Y

Yamasaki, K., 565
Yamazaki, Y., 533
Yan, Z.-C., 512
Yang, S.-H., 251
Yokoyama, K., 565
Yun, W., 585

## Z

Zanthier, C. v., 669
Zemter, M., 669
Zholents, A. A., 664
Zolotorev, M. S., 664